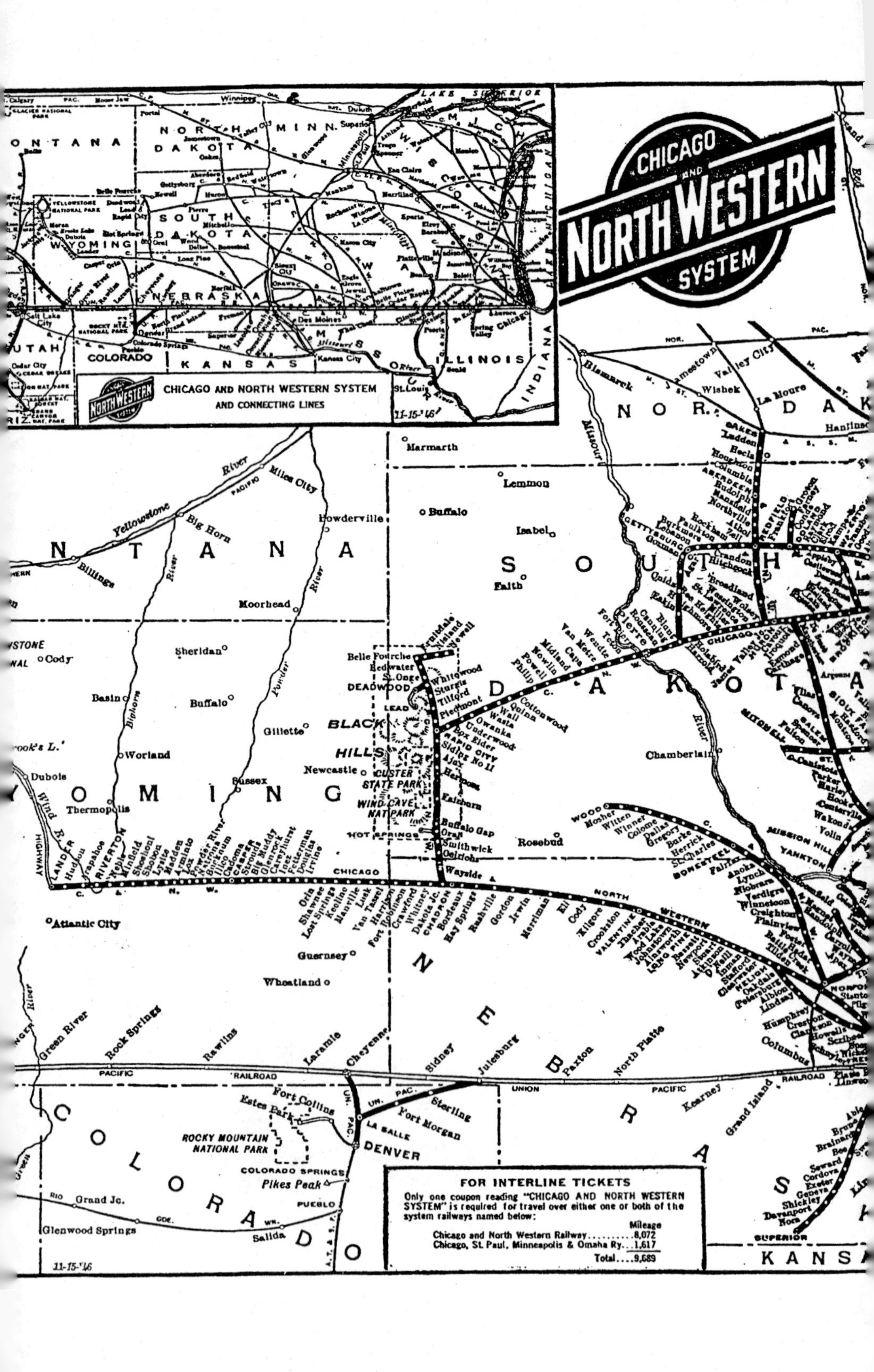
CHICAGO AND NORTH WESTERN SYSTEM
CHICAGO AND NORTH WESTERN SYSTEM
AND CONNECTING LINES
11-15-'16
FOR INTERLINE TICKETS
Only one coupon reading "CHICAGO AND NORTH WESTERN SYSTEM" is required for travel over either one or both of the system railways named below:
Mileage
Chicago and North Western Railway..........8,072
Chicago, St. Paul, Minneapolis & Omaha Ry...1,617
Total....9,689
NORTH DAKOTA
SOUTH DAKOTA
WYOMING
NEBRASKA
MINN.
IOWA
ILLINOIS
KANSAS
COLORADO
MONTANA
UTAH
INDIANA
MISSOURI
WISCONSIN
MICHIGAN
LAKE SUPERIOR
Chicago
Des Moines
Kansas City
St. Louis
Minneapolis
St. Paul
Duluth
Superior
Omaha
Denver
Salt Lake City
Winnipeg
Bismarck
Jamestown
Valley City
La Moure
Wishek
Marmarth
Lemmon
Buffalo
Isabel
Faith
Pierre
Fort Pierre
Huron
Redfield
Aberdeen
Gettysburg
Chamberlain
Mitchell
Yankton
Rosebud
Belle Fourche
Deadwood
Lead
Whitewood
Sturgis
Piedmont
Rapid City
Hermosa
Fairburn
Buffalo Gap
Oral
Smithwick
Oelrichs
Hot Springs
BLACK HILLS
CUSTER STATE PARK
WIND CAVE NAT. PARK
Newcastle
Miles City
Billings
Big Horn
Powderville
Moorhead
Sheridan
Buffalo
Gillette
Basin
Worland
Thermopolis
Sussex
Cody
Dubois
Lander
Hudson
Arapahoe
Riverton
Shoshoni
Lysite
Casper
Glenrock
Douglas
Lusk
Chadron
Valentine
Atlantic City
Guernsey
Wheatland
Green River
Rock Springs
Rawlins
Laramie
Cheyenne
Sidney
Julesburg
Paxton
North Platte
Kearney
Grand Island
Columbus
Norfolk
Superior
UNION PACIFIC RAILROAD
Fort Collins
Estes Park
Sterling
Fort Morgan
La Salle
DENVER
ROCKY MOUNTAIN NATIONAL PARK
COLORADO SPRINGS
Pikes Peak
PUEBLO
Grand Jc.
Glenwood Springs
Salida
11-15-'16

12,000 DAYS

ON THE NORTH WESTERN LINE

THE LIFE AND TIMES
OF A RAILROAD CIVIL ENGINEER
1947–1980

EUGENE M. LEWIS

DRAWN FROM DAILY JOURNALS,
INTERVIEWS, AND COLLECTED MATERIALS.

CHICAGO & NORTH WESTERN HISTORICAL SOCIETY
2005
SECOND PRINTING 2007

ISBN 0-9776088-0-8

Library of Congress Control Number: 2005937729

Designed by Studio Blue, www.studioblue.us

Printed in Canada

Cover photograph: *This is MY Railroad,* Gene Lewis at Glen Ellyn, Illinois, 1978. Author's collection.

Inside front cover: 1948 Chicago & North Western system map

Inside back cover: 1968 Chicago & North Western system map

CONTENTS

INTRODUCTION: BEFORE RAILROADING, 1936–1947

This is a personal account in more or less chronological order of my 33 year railroad career. The center of my life then was the Chicago and North Western Railway Company, an old and venerable name in this country's railroad history. After the second World War the railroad operated two main traffic arteries north and west of Chicago. Interlaced between these two strong arms was a tangle of secondary lines and branches—a legacy of the company's agrarian past. The railroad's pedigree was old by American railroad standards dating to 1836. Chicago, then a rude frontier village, had just become an incorporated city two years before and elected William Butler Ogden as its first mayor. He believed that this cluster of cabins along the swampy banks of the Chicago River would become an important transportation center. And he was to be proven correct. If only Chicago could be linked with the booming town of Galena located in the farthest northwest corner of the state. The 150 miles of roadless prairie had to be spanned by an all-weather route and what better method available was a railroad? Ogden spent the next twelve years traveling the trails between Chicago and Galena preaching railroad to the new farmers. He watched Galena's lead mines dwindle causing a rapid decline of that city. But his creation, the Galena and Chicago Union Rail Road, persevered and in 1848 opened its first ten miles west to the Des Plaines River. The train was powered by a diminutive steam locomotive, the *Pioneer,* purchased as a third hand machine in Michigan and brought to Chicago by sailing schooner. It would be another eight years before an eastern railroad would connect Chicago with the rest of the country.

The G&CU pushed its railroad to Freeport, Illinois, 45 miles short of Galena, on September 1, 1853. Talk of a Pacific railroad drew the G&CU's eyes to the west. At Turner's Junction (later West Chicago) some thirty miles west of its Chicago terminal, the G&CU turned its construction crews to building straight across Illinois on the so-called Dixon Air Line in late 1853. The track reached the Mississippi River December 16, 1855. The next year Iowa interests organized a railroad between Clinton and Cedar Rapids which opened in June 1859. The G&CU leased the line in 1863. Reaching ever westward the G&CU also leased another Iowa road planned to span the country between Cedar Rapids and the Missouri River. At this point the G&CU was merged in 1864 with the bankrupt Chicago, St. Paul and Fond du Lac to become the Chicago and North Western Railway. Two years earlier in 1862 Congress passed the Pacific Railroad Act which represented the national desire for a transcontinental railroad. Council Bluffs, Iowa, was designated as the starting point for the line westward. The Cedar Rapids and Missouri River Railroad, leased to the C&NW, was the first railroad to reach Council Bluffs from the east. The recorded date of January 17, 1867, was just in time to benefit mightily from the flow of men and materials for the great enterprise of the Union Pacific. The C&NW's line across Illinois was a part of this first western transcontinental railroad and that portion is where I learned my trade as a railroad civil engineer.

Construction of the original railroad was a major reason the country swiftly filled-up with settlers. The railroad on its part was quick to establish profitable town lot companies and make great efforts to bring in a population for them. The company was conservative, business-like, and dedicated. Its officials included many former Union Army officers whose organizational skills were useful in guiding the growth of the company's expanding system. These managers ran the company like a military organization, echoes of which still lingered seventy years later when I began my life with the company. In the post Civil War days the old description of wooden cars and iron men was more truth than poetry.

The C&NW briefly considered joining the Pacific coast marathon in the first decade of the 20th century, but the bankruptcy of its archrival, Chicago, Milwaukee and St. Paul, which had beggared itself to thrust a railroad across the Dakota and Montana plains and the tangled mountains of Montana, Idaho and Washington, sobered the C&NW's ambitions. Construction west stopped at Lander, Wyoming, despite the surveys that had been made on the Snake River to the Idaho line. When the Great War came, the C&NW was strong and established; when the Armistice came, it was wornout and leaderless. The times changed but the North Western in its complacent certitude did not. The railroad fell from its leadership role to being a mere player in the railroad business. Its underpowered steam locomotives, aging rolling stock, deteriorated railroad plant and lack of understanding of the growing importance of customer service, the growth of the nation's highway system—underscored the difficulties facing the railroad, difficulties that translated to bankruptcy and court-ordered operations during the Depression. The surge of traffic experienced in World War II only accentuated the overall economic problems facing this midwestern system. At the end of the war the company emerged from bankruptcy lead by officers who had no idea of how to get this property to function profitably. The bankruptcy court did the railroad no favor when it appointed the former trustee, Roland L. Williams, as president of the reorganized company. Bud Williams began his railroad career as a messenger boy and graduated to telegraph operator at Salem, Illinois, on the old Chicago and Eastern Illinois Railroad. His knowledge of railroad operations seemed to be frozen to that time and place. His lack of modern railroad management skills did not help the C&NW develop a recovery plan nor solve the problem of finding the capital funds so urgently needed to get the superstructure of the company replaced. Not knowing anything about these built-in problems, it was about this time that I thought I wanted to be a railroad man and the C&NW was close at hand.

On the basis of annual revenues the C&NW qualified as Class I railroad. It was not a leader; it was a faded rose living on past glory. Freight operations were characterized by long drags running between small yards powered by an overaged steam locomotive fleet. Passenger trains ran everywhere. A bright spot, perhaps like the ruby red cheeks of a tubercular patient, were the jointly-owned proud Streamliners to the West Coast and the sleek yellow and green 400 trains to the Twin Cities. Out on the rural branchlines, ancient coaches rattled over light iron branch lines. In Chicago, an aging steam powered commuter fleet, served a big commuter base. Communications were still reliant on telegraph mixed with a company telephone system. There were too many people on the payroll. The corporate attitude toward the customers who paid the bills was confrontational. Yet, management felt that somehow they would once again achieve their rightful place among the nation's top Class I's.

Who were these managers? The president was a country boy of limited vision. His vice presidents were mostly promoted from the ranks as rewards for being long time tame and faithful servants who certainly would not rock the boat. The few newcomers were viewed with suspicion until they proved that they could be like the incumbents. Woe betide any new idea. Department managers were old hands that had grown-up with the company. Yet, the postwar expansion of business forced a need for a lot of new people. The new blood and old order soon collided with frustration to the first and alarm to the second. Why would I want to become involved in such stormy seas? The answer was the sheer ignorance of youth and the need for a good job. This then is an account of the battle of the new people with the old railroad institutions and how a new and hungry competitive railroad was born and flourished—for a time.

At this point my personal pre-railroad history may provide clues as to why I devoted 33 years to the C&NW. I did not come from a railroad family. Other than an uncle by marriage, Hinkle Wilson, who was the Missouri Pacific's cashier at the end of a 16 mile branchline at Jackson, Missouri, I had no one to blame for any latent railroad gene. Family legend says I was conceived in the upper berth of a Pullman car on the Wabash's night train from St. Louis to Chicago. My parents were on their honeymoon. Perhaps, that is where the railroad fascination began. I was born June 5, 1926 to Denver and Wilma (Martin) Lewis in Chicago, Illinois. My father, a statistician for the Quaker Oats Company, supervised the installation and management of unit record accounting machines in their offices in the 1930s. Until I was five we lived in Chicago and then moved to suburban Downers Grove, Illinois, not far from the CB&Q's main line west. I did get to see the Royal Scot locomotive and train whiz through town in 1933. In 1938 we moved north to another suburb, Lombard, Illinois, immediately adjacent to the C&NW's triple main lines. Three years later the family moved east a couple of miles to Villa Park again very close to the C&NW's tracks. At this time we often rode the more frequently scheduled Chicago, Aurora and Elgin electric interurban to Chicago as its terminal was quite close to Dad's business office. The C&NW was not considered as our family's transportation to the city despite our nearness to its tracks.

The "Third Rail", as the CA&E was styled, was a fascinating operation during my youth. I converted my tinplate O-27 railroad to two rail with a powered third rail using old newspaper baling wire soldered to brass screws. After graduation from high school in 1943, I enlisted in the United States Army Air Corps (August 17, 1943) as an aviation cadet. As the air force training command schools were full, I had to wait for call to active duty. To fill in the time I worked for the Quaker Oats Company as a freight accounting clerk and managed to get in one semester of college at Bradley Polytechnic Institute in Peoria, Illinois before reporting for active duty November 30, 1944. From Fort Sheridan, Illinois, I rode a troop train to St. Louis via the C&NW and the Alton RR; thence to Biloxi, Mississippi, on the L&N. After basic training we were shipped to Childress, Texas, via troop train routed: L&N New Orleans; St. Louis, Brownsville & Mexico (MP) to Houston, TX; MKT to Fort Worth; Fort Worth and Denver to Childress. Childress was a bombardier training station. While waiting for my turn to attend flight school (there were 35,000 ahead of me), I served as a statistical clerk, got married, saw the end of the war, and went home with my bride in November 1945, never having boarded an aircraft much less experiencing the thrill of that first flight.

At home the draft board sniffed at my 11 months and 15 days of active service and made unpleasant sounds about drafting me for occupation service in Japan or Germany. I went to Chicago immediately on the CA&E and signed up in the Army Air Force reserve for three years. My old job at Quaker Oats was available, but soon my wife and I were off to college in Peoria thanks to the G.I.bill. My old school had changed its name to Bradley University and I pursued an engineering course while doing part-time work in the drafting room of the Caterpillar Tractor Company in East Peoria. At the end of the school year, my wife was pregnant and I needed to get a real job that payed real money. I calculated I needed about $200.00 per month to successfully provide for my coming family. Barbara, my gravid wife, pointed out that I was always interested in railroads so why didn't I try to find a job with one? The Pennsylvania Railroad was willing to hire me as a ticket clerk in the Chicago Union Station but not at the magic level of 200 bucks. I tried the Illinois Central (which had a centralized hiring office). They were hoping to have a training class for car draftsmen but why didn't I try the C&NW as two guys had just been in from that company looking for IC jobs?

The general offices of the C&NW were located in the Daily News Building at 400 W. Madison Street in Chicago. Totally ignorant of railroad hiring practices, I picked the first engineering department listed in the lobby display and was interviewed by Leonard R. Lamport, Engineer of Maintenance, for all of five minutes. "Don't call us, we'll call you"—and they did, two days later. There was a job available as a tapeman on the Galena Division. Not knowing what a tapeman was, I asked what the duties included. Mr. Lamport, a bit taken aback, said that a tapeman was the man on the survey crew that held the end of the chain or tape. I thought I could do that and agreed to take it. Only then did I ask what the pay would be—$207.88 per month. I was hooked for the next 12,023 days of my life.

Most of the material for the following account was drawn from my daily journals and time books. Many events, both personal and family, are not recounted unless they were deemed to influence the railroad theme. My comments are strictly personal and were not intended to embarrass or discomfort anyone. I have used real names throughout as I view my role as a historian rather than as a teller of tales. The distance of years may cloud some impressions for which I apologize in advance. I feel honored to have participated in the saga of the C&NW and salute all who suffered through its troubles and rebirth in the last half of the twentieth century.

1

THE NEW BOY, 1947 . . . At eight o'clock on a warm July morning, July 31, 1947, my railroad career began when I reported for duty to the division engineer's office. The Galena Division and the Wisconsin Division engineering departments shared a common reception and clerical room, #300, on the third floor of the old Chicago Passenger Terminal at 500 W. Madison Street in Chicago. About twenty feet square, a high ceiling and a crowding of wooden desks were my first impressions. The Galena side of the room occupied the left side of the space; the Wisconsin had windows overlooking Canal Street. Two pairs of windows extended almost to the high ceiling. Lighting was furnished by incandescent lamps hanging on long chains. The visitor was restrained from entering the room by a low wooden railing and a double-hinged swinging gate. Behind the fence was the chief clerk's desk. The next desk, piled high with manila file folders, belonged to the file clerk. Behind her was another desk for the clerk-typist. On the Wisconsin side three desks were arrayed parallel to the street wall looking for all the world like a defense position protecting the solid row of steel file cabinets immediately under the windows. The Galena's chief clerk then was Tony Daleiden; the file clerk, Catherine Daley; and the stenographer's name I do not recall. On the Wisconsin side, the chief clerk was Edward Marquardt; his file clerk was Vera Scharenberg, and the steno was a young lady of 17 named Shirley May (later Vavra). There were no electric machines in sight; everything was manual: typewriters, adding machines and steno pads. This office in shades of brown and tan with narrow board tongue and groove maple flooring, wooden desks, yellowish illumination and yellow copy paper was a scene from 1925 frozen in time.

I announced my reason for being there and was escorted through the swinging gate, past the warren of desks and into the drafting room. There I met the assistant engineer in charge, William "Bill" Wilbur. Bill was laconic to say the least. Years later I was to learn that he had been the oldest second lieutenant in the United States Army in World War II. He had served in the Persian Gulf command with the Iranian State Railways moving war materials to the USSR from the port of Bandar Shahpur in 1943–1944.

The drafting room, which was to be the focus of my life for the next ten years, was another high ceiling wooden floored space. Just as the clerk's area was a shared space for the two divisions, so was the drafting room. The Galena occupied the northern leg of an "L" shaped room along the windows, about thirty feet, while the Wisconsin area was at right angles toward the south. Two pairs of windows ran from the top of the iron radiators almost to the ceiling. The view was over the trainshed roof—a black expanse punctuated by rows of sooty skylights extending over the passenger platforms underneath. Opening the windows invited locomotive gas and soot, especially from inbound trains arriving on tracks 14, 15 and 16. Another pair of windows on the west side of the space overlooked the passenger concourse. The great ornate clock and the newspaper stand under it were plainly visible from our office.

In July 1947 the Galena's engineering staff consisted of a division engineer and twelve other ranks made up of an assistant engineer, 5 instrumentmen, 4 rodmen and 2 tapemen. The Wisconsin's staff was smaller. Their assistant engineer was Mort Smith whom I recognized from my youth as a Boy Scout counselor from Lombard. Draftsman, Max Bear, two instrumentmen, Gene Bangs and Alec Gilman, and rodmen, Jeff Edee and Bob Witt, made up their crew. At the south end of the Wisconsin's area was the entrance to the map vault. Inside this room, steel map drawers were stacked ceiling high. The first four feet were all flat drawers while the upper ones were 6 inch × 6 inch pullout drawers designed for rolled maps and blueprints. To reach the highest drawers required use of tracked library ladders. In the middle of the space was a double row of steel lockers where the men kept their outdoor rough clothes and surveying equipment. The entire west wall consisted of steel pocket shelves about 5 inches deep containing old field books—hundreds if not thousands of them. Considered legal documents, these books represented a lot of engineering effort even though reading the ancient pencilled notes usually did not prove to be of much value after the passage of so much time. Pencil? Indeed, notes had to be kept in pencil because ink in addition to being unhandy in the field tended to run when the paper became wet.

Returning to the drafting room, I was struck by how crowded it was. Two genuine drafting tables with tilting boards were supplemented by other tables propped up on trestle legs and cabinets. Most of the furniture seemed to have come over with Noah during the flood, especially the huge double desk shared by Bill Wilbur on one side and an instrumentman on the other. There were a couple of smaller wooden desks jammed into odd corners. Everyone had a desk lamp as the ceiling fixtures were useless for drafting work. In one corner of the room was a tiny lavatory cubicle containing a wash basin and a roller towel. At the side of each desk was a real spittoon. These nasty things were always being accidentally bumped or kicked, slopping the contents on the wooden floor. The only calculator was an ancient hand operated Monroe. The hand-powered adding machine was also a museum piece that sat in solitary splendor on its own wheeled cart. There was one telephone in the office—at the assistant engineer's desk (extension 582). On each desk and drafting table were stacks of working files. The rule of files said that no file could be placed in a drawer—it must be left on top of the desk in case it was wanted and the possessor was away. Add to the piles of files all the tools of the draftsman—vellum, India ink in their cast iron anti-spill rings, pens, pencils, triangles, straight edges, ink rags, soapstone, brushes, erasers, ash trays and bits of railroad iron used as paperweights, you get some idea that this was not a neat and tidy place.

Directly behind Bill Wilbur's seat was a pair of swinging doors just like the ones on the saloons so often shown in western movies. These breast height louvered gates led to the division engineer's inner sanctum. His office was about 20 feet square—an enormous space for one man compared to the crowded drafting room. A pair of the high windows faced north and another pair overlooked Canal Street. HIS desk sat right in the middle of the room. There was a leather couch along one wall and several leather covered chairs for guests. Since there was a rug on the floor, one could really be called on the carpet. HE had his own complete bathroom. A couch and a bathroom? There were times when this office became his hotel room. Snow, tornados, floods, strikes were often reasons that he needed this place. The Wisconsin Division engineer had a similar office just across the clerk's room along Canal Street.

This tour of the department was daunting. I began to think that I could never become part of this chaotic and awesome business. Overwhelmed by my ignorance, I was somewhat relieved to be assigned a small desk jammed into a corner and given instructions by rodman Edward J. Malo on the proper coloring of blueprints. Ed explained that I was to make six exact copies of the print he had already prepared. The seven prints were to accompany his handwritten estimate for some track changes somewhere out on the division. There were strict rules about the use of color on blueprints issued by the chief engineer. All new work was to be shown in yellow, removals were to be green, right-of-way and main tracks were colored red. Explanatory notes were to be done in yellow on the blueprint. The labels identifying location, job, scale and date were lettered in black India ink The red was easy—red pencils and a straightedge. Curves were done using celluloid templates from a special wooden box holding a family of different arcs. Green also was not a problem when using green pencils or green colored ink with a small paintbrush. Yellow was the challenge. Bottles of yellow pigment dissolved in water had to be shaken often to keep the material in suspension. As both rails in new construction were shown, the yellow "ink" had to be handled delicately to avoid running the closely-spaced wet lines together. At 1 inch equals 100 foot scale (our usual working scale) the rails in a track are about $\frac{1}{8}$ inch apart. Of course the water base color was slow to dry which compounded the problem by smearing at the slightest opportunity. I soon learned to do one small bit on each copy which gave the first copy time to dry. Lettering a print using a steel pen nib and yellow "ink" was a skill that had to be learned. All of the proposition prints had a white area about 3 × 4 inches in the lower right-hand corner. A rubber stamp applied here gave the division name and the type of print being made. We added location, name of project, scale and date in black ink. I began to feel better as the completed prints began to pile up. Now came lesson number 2 from Ed Malo—how blueprints must be folded.

All files in the engineering department were $8\frac{1}{2}$ × 11 inches. The oldest dated item was on the bottom—the newest, on top. It was all fastened together with a brass paper fastener in the upper left corner. Some of those files got to be like blocks of wood using every bit of a 4 inch fastener. Blueprints had to be folded to fit into this format and to be fastened just like a piece of ordinary letter paper. The secret was an accordion-like fold from right to left and from top to bottom. Properly done a large blueprint could be opened by pulling on the lower right hand corner without removing the paper fastener. The system worked well even though it seemed like a Japanese origami trick.

Now that I was more confidant, Ed introduced me to the art of making blueprints. We went down the hall to room #309 where the blueprint machine lived. All my newly gained confidence drained into my shoes when I first saw IT. A thick glass cylinder about 24 inches in diameter and 6 feet high crouched in the center of the room. Around the glass was a heavy canvas cover on rollers. Inside was an artificial sun—a carbon arc lamp. This refugee from an 1890s street light was suspended on a steel cable attached to an inertial gearing device that allowed the lamp to descend steadily at a pre-determined rate. Somewhat dismayed at the sight of this contraption, I made some remark about its antiquity. Alec Gilman, an older instrumentman from the Wisconsin Division said this was a modern machine compared to the sun frames they once had to use on the roof on sunny days. To make a blueprint on this machine the tan canvas glass cover was rolled back exposing the surface of the curved glass cylinder. The tracing to be copied was placed against the glass, a piece of unexposed blueprint paper taken from a closed bureau drawer was placed over the tracing and the canvas cover was then rolled over the stack to hold everything firmly in place. Sometimes a cardboard template was inserted between the tracing and the unexposed blueprint paper to isolate the area to be copied and make a white area for the description label. The lamp was cranked up to the top, the high voltage switch was thrown making an authoritative crash, the carbon electrodes flashed and glowed as the lamp slowly descended down the center of the glass cylinder. Timing of the exposure was learned by experience. Too little yielded a light, indistinct print; too much and the white lines were "burned out" in a dark blue background. The condition of the tracing was also a factor. An old dirty drawing took a lot more exposure to get a decent print. Over time, I learned to precut my paper, use the cardboard masks and get the right exposures. I also learned to clean dirty tracings by using gasoline, but that trick came later. Cleaning that arc lamp and reloading the electrodes was another task that had to be learned and avoided whenever possible.

The exposed papers now had to be developed. Just like photography, the image had to be fixed. This was accomplished by dipping the exposed paper into a tray of potassium permanganate, a nasty orange colored brew. The chemical reacted with the unexposed blueprint leaving the white lines in a blue background. After a few moments, the prints were tossed into a wide and shallow tub of running water to rinse the permanganate out of the paper. What with the constantly rushing water competing with hiss and sizzle of arc light and the dripping of drying blueprints, it was a good thing that the men's room was next door at #307. Drying prints meant hanging the wet paper on wooden sticks stuck into a rack over the washing tub and letting them drain and dry. Sometimes to speed up the process, we would put the wet prints on the glass of the arc light machine and let the heat generated dry the prints. A few months later the chief engineer's office sent a cast-off electric dryer to us. It was new to them in 1929 and new to us in 1947. That machine with its endless canvas belt revolving around an electrically heated drum was a great improvement to the drip-and-dry method. The last operation in the blueprint room was trimming the prints to size using a long blade paper cutter. Even this operation had its hazards; there was one new recruit who ended up with a shorter necktie.

My first day ended at 5 p.m. It was a short walk to my 5:17 train to Elmhurst. For this exhausting day I received, two weeks later, a check for $6.73 less 39¢ for Railroad Retirement. Forty four years later in 1991 I began collecting benefits from the retirement fund.

The engineering staff was not meant to sit in the office when there was work to be done out on the railroad. Being OUT IN THE FIELD was a desired activity for these men. During bad weather the men grew restive and cranky when penned up in that drafting room. Rodman Ed Malo, who had undertaken my instruction, told me to be ready for a field trip the next day. On August 2, 1947 we boarded a morning westbound commuter train and rode to Lombard. We were met at the station by a section man carrying a ballast fork and a shovel. The three of us tramped west down the triple main track passing the house where I had lived from 1937 to 1941 at the highway underpass of Illinois highway 53. We came to Deadman's Curve where the railroad swings gently to the left as it enters the sandy hills west of the east branch of the DuPage River. Deadman's Curve was named for a long ago fatal accident at a grade crossing that is no longer in existence. Ed was very safety conscious and made sure that I understood about watching for trains and that I should never step on the ball (head) of the rail. This trip, I thought, was a lovely way to spend

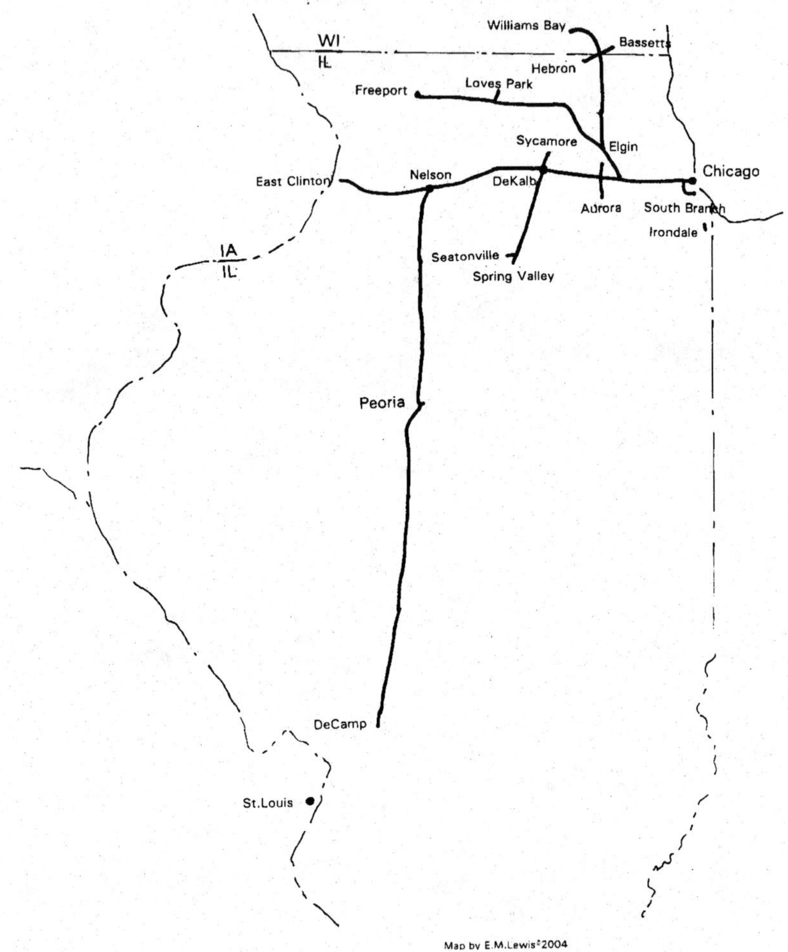

The old Galena Division's Chicago-Clinton, Iowa main traffic line with its sprinkle of light-density branch lines was augmented by the addition of the Southern Illinois Division in 1938.

the morning out on the railroad with trains coming past every so often. Arriving at milepost 22, Ed instructed the section man to dig into the ballast of the track 1, the south main. The limestone rock had to be pried out of the tie crib with the ballast fork before he could shovel a hole about two feet deep. With a six foot folding rule, Ed measured the depth of the limestone rock below the base of rail. He then measured the thickness of a layer of sand and gravel which, in turn was on top of a cinder base, again using the base of the rail as a datum. Standing up, Ed told the section man to fill up the hole and moved over to track 2, the center main to do the same thing again. By now the August sun was beginning to be felt and the digging slowed down. Another hole was needed in track 3, the northernmost track. This was getting to be a drag, but I did not volunteer to do any digging. Finished at this site we tramped west another quarter mile and repeated the process. By now it was noon and we found a shady bank where we ate our sandwiches. Ed had enough information by mid afternoon and the section man had enough digging. I was sunburned and somewhat mystified at what this ritual was all about. We hiked west to the Glen Ellyn station and caught an eastbound to Chicago.

Life settled into a routine of constant learning the varied and intricate rituals of railroad engineering. I was the blueprint donkey because I was the new boy. Making prints for the men, caring for the priceless tracings, learning the requirements for forms 1404 and 647, reading license forms and contract files, unraveling the arcane filing system of mousy Miss Daley and discovering the quirks and idiosyncracies of my associates, filled my days. Edward J. Malo was my first mentor. A quiet-spoken graduate of Notre Dame and a U.S.Navy veteran, Ed later left the railroad to become an architect. My second mentor was Magnus C. Christensen, instrumentman. Chris grew-up in Iowa Falls, Iowa, graduated from Iowa State at Ames, worked for the C&NW briefly before joining the U. S. Army Air Corps and serving as a B-24 navigator in the Eighth Air Force in England. Another instrumentman at that time was John Shanklin. I never went on a field trip with John and did not get to know him very well before he transferred to the bridge department as an inspector.

John Perry, another instrumentman, was with us such a short time that I can barely recall him now. He left the railroad to go into his family's printing business. Arthur P. Olson, the oldest man in our office, was also an instrumentman. Rumpled is the best description for this untidy and somewhat scatterbrained old codger. Arthur E. Humburg, rodman, was an older man with dark intelligence and a bitter sarcastic manner. Marion Boyd "Bob" Lithgow, instrumentman, had an interesting background. A native of Boone, Iowa, Bob graduated at Ames when the Great Depression was in full sway. He had an Army commission and went on active duty to direct activities at CCC camps in the Upper Peninsula of Michigan until such time as a railroad job would open. When laid off by the railroad, he went back to the Army. Consequently, when World War II broke out, he quickly rose to Captain and Major with assignment to an Air Corps base in Malden, Missouri. Bob was a rather stocky individual who much preferred office work to tramping about in the weeds doing a survey. So there he sat, doing paperwork amid an ever present cloud of cigarette smoke. We will meet him again at several points in this account. And then there was Harold W. Jensen, the Galena Division engineer.

Harold was a high-energy, forceful man full of invective and good intentions reminding one very much of President Theodore Roosevelt. The first time I met him, he scared the hell out of me. One summer day the quiet of the drafting room was shattered by an increasingly loud series of "god damns" emanating from somewhere behind those swinging doors. The doors flew aside as a burly figure wearing rimless glasses hurtled through them accompanied by that continuous god damn chant. He marched through our section, wheeled left across the Wisconsin division to the map vault, disappeared briefly before coming out with a station map tracing which he unrolled as he strode back to his office. Quiet again descended on the room until the buzzer at Wilbur's desk sounded. Bill grabbed a yellow pad and went in to see what was wanted. I was thunderstruck at this performance, but was reassured that it was quite normal. Harold Jensen was revered by all who knew him. He had served as an officer in both World Wars and was a longtime member of the C&NW organization. Tact was not his strength, but he was a leader who was not afraid to speak out at absurdity. Later, when I got to know him, I joined his fan club. He had some simple advice for us young fellows that served us well. He told us to never hesitate to take action when action was needed—he would back us to the hilt whether we were right or wrong. Do something was his motto. He may have lacked polish, but he was as honest and forthright a man as I was to ever meet.

We worked Saturdays in 1947. Just before I started, that day was reduced to a half day. Field trips were rarely taken on the sixth day.

It was a time for catching up on the routine paperwork such as time sheets, indexing field books, tending to reporting duties and the like. Railroad management felt that its employees should work like the railroad- all the time. Salaried people were paid for 24 hours a day and seven days a week. Many operating officials found that their duties required them to put in many hours a week in excess of their union hourly staff people. As salaried employees we were subject to the 24 hour daily requirement and had to be happy with our half day Saturday and Sunday time off. The end of my first week found me Elmhurst bound on the 12:27 p.m. train.

Having actually participated in a real field trip, I did not feel so green. Getting out on the real railroad and officially doing the company's business somehow raised my confidence level. My solo flight, however, came immediately. Now it was my turn to secure ballast sections at River Forest, about 10 miles west of the Chicago Passenger Terminal. There was a young Mexican section hand armed with a pick and shovel waiting at the station when I got off the train on a warm and bright August morning. I indicated that I was the one he was waiting for and we tramped off west along track 1 to Vale where the four suburban mains changed into a double track line. This was the spot where the very first Galena and Chicago Union train had ended its first run October 25, 1848, going no further because the Des Plaines River had not been bridged. I pointed out where the first hole was to be dug and my sturdy companion lit into it with vigor. The top layer of limestone was not as deep here as it had been further west at Deadman's Curve. He soon got into a deep strata of sand and gravel. My digger was rather surprised when I had measured the depth of the materials, straightened up and told him to fill the hole, but quickly began digging the second one without comment in either English or Spanish. I kept a sharp lookout for train movements here because there were a lot of transfer freight runs to and from the Proviso Yards to the west and the many interchanges in Chicago proper. The digging went well with practically the same results from each test. The two of us trudged across the river bridge to the Maywood station between 4th and 5th Avenues to catch a westbound commuter train to Elmhurst for some more sampling. I was pleased with this first independent foray: the weather had been nice, the job was within my limited capabilities and the assignment was successfully completed. What a good feeling to be working on the railroad.

Years later I discovered why there was so much sand and gravel in the ballast soundings at Vale. About 1895 the city councilmen of Chicago in their infinite wisdom ordered all the railroads within the city limits to separate their tracks from the public way either by building bridges, elevating their grades, or a combination of both. At the turn of the century most of the big railroads were quite wealthy and began to elevate their Chicago right-of-ways after the usual period of legal grumbling and attempts to buy their ways out of the council's decree. Raising the main line tracks was a fairly simple operation in concept, but onerous in execution. Under the special engineering organization called the Track Elevation Department, concrete retaining walls were constructed on either side of the railroad tracks and pitrun sandy gravel was dumped in vast quantities from many work trains. The tracks were jacked up through the newly dumped material until the desired elevation was reached. Street under crossings were built with steel viaducts to carry the newly raised tracks. It took a long time to raise the tracks in this fashion and required a lot of fill material. Remarkably, this work was carried out under traffic, albeit very slow traffic. The C&NW elevated its mainlines west as far as River Forest in 1908 and then descended to the Des Plaines River bridge. At some point in my Galena Division career, I found the plans for the track elevation through Maywood and Melrose Park west of the river. Because of disagreements with the municipalities about how many streets would be bridged and how many were to be closed and the lack of a legal reason to force the railroad to elevate its tracks through these towns, the elevation work never occurred. The sand and gravel for the west line elevation came from pits west of Lombard at Deadman's Curve where Ed and I first dug holes in the ballast.

On August 13, 1947 I received my first full C&NW Railway pay check. The gross was $104.39 less $6.00 for railroad retirement and $10.00 for Federal income tax, leaving a net of $88.39.

My job description was tapeman, and I had not yet learned anything about that function. On August 15th that deficiency was remedied. Instrumentman Chris Christensen, rodman Art Humburg and I were sent to stake an industrial track change in Dixon, Illinois, some

90 miles west of Chicago. We rode train #13 from the Chicago Passenger Terminal at 9:15 a.m. The train was a motley mix of mail, baggage and coaches hauled by an E-4 class 4-6-4. Arriving in Dixon about noon with all our surveying equipment, I soon learned that tapeman also meant being the donkey in the crew. In addition to the wooden boxed transit instrument, we had a transit tripod, a folding 12 foot level rod, a six foot long red and white steel sighting picket, a coiled 100 foot steel tape, field book, markers, an 8 pound maul (sledge hammer), a set of chaining pins, and a heavy bundle of 18 inch wooden "toothpick" stakes made of ¾ inch lumber bound up in coarse brown cord. In 1947 the C&NW expected its survey crews to ride the ubiquitous passenger trains and manhandle all the equipment. We even had permits to carry track motor cars (speeders) in baggage cars, but I never had to do that. On arrival at the station nearest the job site, it was up to the head of party to arrange, finagle, cajole or otherwise to get the men and equipment to the proper location. Luckily, at Dixon the job was just down the street behind the depot. The track change in question required moving an old siding that served Purity Mills which was in business in an ancient brick structure on the east side of the street. It was not much of a surveying job and could have been done just as well by the practiced eye of Tony Karras, the section foreman, who stood there with an amused look on his face watching these earnest guys from the Chicago office. Chris took this opportunity to teach me my tapeman duties. The first and most fundamental lesson involved undoing the 100 foot steel tape without having a massive uncontrolled tangle which would be hard to make right. Under Art Humburg's critical eye, I carefully unsnapped the coil and payed out the quarter inch wide steel band until the chain lay straight and unkinked. After we did our little survey, set a couple of stakes, I carefully drew the tape into exact five foot loops, bound them with the leather pulling thong and, after a demonstration by Art, managed to throw the loops into a compact coil about 15 inches across. Once accomplished this is a skill that is never forgotten—just like riding a bicycle. Chris made me undo and redo the chain four or five times to make sure I knew how to work it. After all, I was a tapeman.

In this session I also learned to set exact points on the heads of the stakes with a yellow pencil. Only a high visibility yellow pencil will do as that is the object being sighted by the transitman. Woe to the tapeman who does not carry a long yellow pencil. The exact point is then established with a survey tack which has an indented top. How does one carry those sharp tacks without getting stuck? Easy. Stick them in a soft rubber ball on a leather thong which is then fastened to the belt like a watch fob. Meanwhile Karras just stood there in the shade of his straw hat watching our layout work, his generous belly overhanging his belt. We had to wait for eastbound train #14 which arrived shortly after the westbound *City of Denver*, train #111, made its station stop. Train #14 sometimes made a station stop at Wheaton, 25 miles west of Chicago, if a revenue passenger wished to get off there. This would have been quite convenient for me as I wanted to go to Elmhurst, just 16 miles from downtown. Sometimes, #14 would make a stop at Oak Park, 8 miles west, which also could work well for me to catch a westbound commuter train. Since we were deadheads (non-revenue passengers), I could only hope there was someone that wanted either Wheaton or Oak Park. Tonight, however, we had all that surveying equipment to take back to the office, and I went all the way into town to help handle the bulky objects. The track changes at the Purity Mills in Dixon were never made.

I was to be indebted to Chris Christensen for his thoughtfulness in teaching me the rudiments of railroad civil engineering. Over the next few months he instructed me in the care and use of the optical surveying equipment, how to lay out track curves and tangents, establish profiles, perform derailment and accident surveys, how to make good field notes, and to be clear about what the processes were. Over the next thirty years Chris and I were to be involved in the same projects from time to time but from different jurisdictions. Art Humburg, the rodman on the Dixon job, was a star crossed individual sometimes given to drink and always to sarcasm which did not sit well with his superiors. Art was 36 when I first met him (born August 13, 1911). Intelligent and good natured when sober, Art was burdened by a spinal deformity that gave him a bent and crooked posture which seemed to cause him to adopt a rather truculent attitude when dealing with people. Perhaps, the death of his only son at the age of 14, contributed to this restless man's unrewarding career with the railroad. Art realized that he would never prosper in this job and eventually moved to Southern California where he worked for the state highway department. (Social security records indicated that he died July 7, 1995 at the age of 84).

That summer of 1947 I seem to have been chosen as the digger of holes. Off I went on train #13 to Sterling, Illinois, 109 miles west. I met a section hand at roadmaster Arthur E. Benson's office in the freight house. We hiked down the double main to milepost 112 where he dug the holes in the ballast section. It was a hot day and the work went slowly. There was plenty of time to catch #14 eastbound. Sterling's biggest industry was the Northwestern Steel and Wire Company whose sprawling works occupied the land south of the C&NW to the banks of the Rock River. As we dug and measured the ballast depth the steel plant and its rail yard were immediately adjacent. The railroad and mill yard tracks were full of gondola cars filled with parts of retired steam locomotives waiting to be fed into the electric furnaces for reduction to fence posts and steel wire. The steel company was using decrepit steam locomotives for switching the cars of scrap. These tired old engines would one day end up in the same furnaces.

To help my on-the-job engineering education, I purchased my own copy of the railroad track engineer's bible: C.Frank Allen's *Railroad Curves and Earthwork*, 7th edition, 1931. Commonly known as Allen's, this $6\frac{1}{2} \times 4\frac{1}{2}$ inch book was a compendium of curve, earthwork and track formulae enhanced by many tables of functions. This valued companion was just coat pocket sized and traveled with me on most field trips. Later, I purchased William H. Searles and Howard Chapin Ives's handbook *Field Engineering, a Handbook of the Theory and Practice of Railway Surveying, Location and Construction*, 21st edition, 1946. Originally published in 1880 and repeatedly updated, this book was a better office reference than Allen's. These two books are still in my library. When I see them, I am reminded of the many times that field calculations of complicated curve relationships, sometimes made under very adverse weather conditions, were done using only logarithmic tables and slide rules. We were an old fashioned operation little different from the methods used in 1880 when that Searles and Ives book was first published. Our maps and drawings were prepared with pencil and India ink on tracing paper or vellum using straightedges, triangles, protractors and a variety of measuring scales.

George Washington, Meriwether Lewis, Charles Mason and Jeremiah Dixon would have been right at home using our tools and equipment.

Late summer and I was recruited by another instrumentman for one of his surveys. Arthur P. Olson, John Van Horne, another tapeman, and I rode train #703, the Freeport Express, to Rockford, Illinois, 90 miles northwest of Chicago.

Express was an inappropriate name for this local. It ran as a commuter train for the first 30 miles making all stops to West Chicago. East of the West Chicago depot at tower NI, #703 left the Galena mains and entered the Freeport line. After a stop at the West Chicago station, the four car train headed by an elderly class D Atlantic (4-4-2) headed up the branch line. The steel coaches were of 1910 vintage complete with Daniel Boone air conditioning that provided plenty of cinders and smoke when in operation (when the windows were opened), green plush seats, a stove at the end of each coach, and very few passengers. The train stopped at every station. Most of the larger ones still had agents, telegraph communication, train order signals, and brick platforms complete with steel-wheeled express wagons. A considerable amount of express and freight was handled at the larger stations like Huntley, Marengo, Belvidere, Rockford and Freeport. The train ambled along the light rail (72 and 80 pound) in its cinder ballast. This was the original main line of the Galena and Chicago Union Railroad constructed in 1850. The original track had been replaced over the years, but surely did not look much different from the first railroad constructed over these prairies. Many of the old wooden depots were the original structures. Our conductor on #703 this day was a veteran, Bill Harvey. After leaving West Chicago, Bill settled himself down with the few passengers and told stories of the old days on the railroad. One of the stories was that old saw about how to determine what was company money in a cash fare: throw it up against the car ceiling. What sticks there belongs to the company. That story would surface again and again. Old Bill was sober westbound, but eastbound was another story.

Our job this day was to stake out a new side track to serve the Clair Barber Lumber Company at Loves Park, Illinois. This suburb of Rockford, newly incorporated in 1947, was to become noted as one of the few municipalities in Illinois to have no local taxes because of the industrial base located there. The lumber yard was located about half the way up this 5 mile fragment of the old Kenosha Division now known as the KD line. We taxied out to the site where we set line stakes with our "toothpicks" and grade stakes with 1½ square 30 inch oak staves. The latter stakes were called blue tops because we used a liberal amount of blue crayon (keel) to mark them. The grading contractor used them to guide his work when cutting and filling the subgrade for the new track. Packing up our gear, we caught a ride back to the old brick Rockford station before #706 came in from Freeport. The train consist was the same one we had ridden in the morning, but the conductor did not seem to notice us as he was beyond noticing much of anything. The engineer must have had a hot date that evening as that Atlantic bounded down the branch at a rollicking pace.

The next day, September 3rd, Art, John and I rode train #703 again with Atlantic #397 and the same old consist along with the same old conductor. Art did not seem too anxious to tell us what we were to do this day at Huntley which was the second station beyond Elgin. At his direction John and I measured buildings and various features around the Anderson and Fencil Company south of the village while Art wrote in his field book. It was a short job on a hot day and we had a lot of time to wait for #706 back to Chicago. A pint of whiskey appeared from some pocket in Art's baggy clothes. He and John polished it off quickly. Perhaps, they wanted to be in conductor Harvey's condition on the homeward trip. I left them on the train and got off at Elmhurst. Showing no effects from yesterday's tippling, Art and John were on #703 the next morning when I boarded at Elmhurst. This time we went to Rockford again to measure vertical clearances at the Kishwaukee Street overpass east of the Rock River. We set up the level and took inverted rod readings from the lowest bridge members over the track. To this reading we added the distance from top of rail and thus derived the total vertical distance. What a waste of time. A simple job like this should not have taken three men on an all day excursion. We trudged over the Rock River bridge to the station and waited a long time for #706.

Instrumentman Arthur P. Olson could be described as "rumpled". He was an untidy person, careless in dress with clothes flecked with bits of burned and unburned tobacco. Squinty eyes peered through smudged glasses, his muscle reactions were abrupt, instructions to his crews were vague and his working habits were like his personality. He seemed to be afraid to share information even though the jobs would have been accomplished more easily. An inveterate pipe smoker and gambler, his idea of a great time was visiting a race track with his wife, have plenty of booze and tobacco available, and bet on every race. His engineering skills were subject to question as he had a propensity to adjust field data to fit the design. As the new generation of college educated engineers came on the division, Art's instrumentman job became precarious and the assignments became more and more trivial. When he screwed up a simple ditch profile at Proviso by altering his field notes, he was urged to leave the railroad.

Good news, the tapeman's monthly rate went to $246.48—an increase of $37.71. Encouraged by the rise in fortune, I went with Ed Malo to Halsted Street on the CNW Rockwell Street line which branched south from the Galena main line at Kedzie Avenue. Our task was a track inventory at an old warehouse siding. We had learned that there was a change of ownership of the property. The track serving the site had been installed before World War I and was entirely owned by the C&NW. Under the rules governing track agreements between the railroad and industry, the railroad could own only so much of the siding that served two or more customers, or in the case of a single customer, only the turnout and track as far as the point the centerlines of the diverging tracks were 12 feet apart. Ed and I had to determine how much of this siding belonged to the railroad and what track material was contained in each section. Accounting would then price the material for the sale to the industry of their portion of the siding. In this case the material was so old, had not been used in many years and was almost buried in cinders. In older days cinders were used extravagantly as ballast in track because there was a lot of this waste material from the many steam locomotives and weeds did not grow in track ballasted in cinders. The cinders, however, had a tendency to corrode iron rails and fittings. There was another problem with cinder fills that I would learn about in Peoria, Illinois, and Savannah, Missouri, some years later

The Galena Division acquired a new instrumentman, Richard W. Bailey, who replaced John Perry. Dick was a practical and energetic young man who hailed from Peoria, Illinois, where his father had been an engineer with Keystone Steel and Wire Company in nearby

Notes for Smith Road Accident Survey

Survey: Made in afternoon, November 23, 1949 by J.S.Bach, A.M.Godfrey, R.L.Stone

Weather: Overcast, 38°F, light wind from west

Location CNW Ry. River Division, milepost 87.6

Incident Motor vehicle northbound collided with westbound CNW train #437
November 22, 1949.

Railroad single track tangent on 100' right-of-way
45 mph passenger and freight speed, no slow orders at Smith Road
grade 0.3% descending to west, jointed rail 112#, rock ballast
line and surface, average
whistle post 1320' east of road crossing
12 wire communication pole line along south side of R.O.W.
48" woven wire fence on both sides of R.O.W.

Road Smith Road, 18' asphalt pavement, 4' gravel shoulders, in center of 66' R.O.W.
Pavement in average condition and surafce.
No indicated speed limit on this rural road
Road makes a 105 degree angle with the railroad in survey quadrant
profile level to crossing and rising at 0.25% north beyond crossing.
Crossing protection, advance warning sign 423 feet south of railroad, standard crossbuck signs at railroad. Signs in good repair.
Roadway dry and clear at time of survey.

Crossing Standard 24' timber planking, in good repair,

Vision Lines Sightings made from a point 4' above the roadway surface at 50' intervals to the west.
Target, a 12' pole topped by a white flag.

Features Farmhouse and garage located in survey quadrant.
Land in quadrant used as open pasture

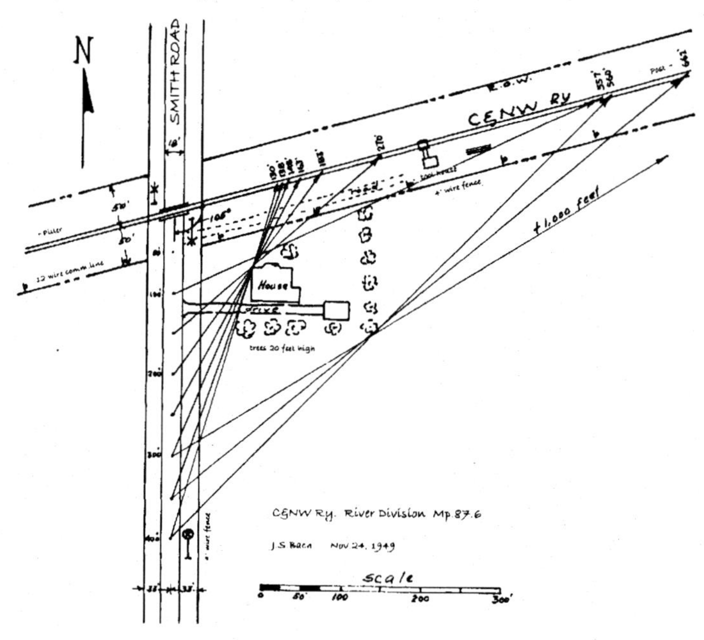

The accident survey was an important service performed by the Division Engineers for the Law Department of the railroad. Obstructions to vision were systematically identified at grade crossings.

Bartonville. We hit it off right away not only because of our common Peoria links, but because we both had enthusiasm for our work. Our first job together was an accident survey at the First Avenue grade crossing in Maywood, Illinois. An accident survey was conducted according to written instruction issued by the chief engineer and authored by the claims department. The object was production of an exact physical description of the physical site where injury or loss had occurred to company employees or outsiders. Most commonly, the accident was a grade crossing incident involving a motor vehicle and a train although pedestrians and non-motored vehicles sometimes were involved. All railroads are involved with moving big machinery at speed over great distances making them targets for legal action by affected parties or the public as a class. The resulting product from an accident survey was a scale drawing, or drawings, depicting the physical site and surroundings. Exact measurements were made to locate all features, describe the grade and elevation of tracks and roadways, detail the location and describe any obstructions to vision based on a line of sight about four feet above a roadway to a point twelve feet high on the railroad. All signs, poles, heaps of material, bushes, structures that might obstruct vision had to be described and located because the engineer often was called into court to testify as to the accuracy of his work. The fan-like sight lines plotted on the drawing at a grade crossing accident quickly identified vision obstructions. Usually only one quadrant of a railroad-highway intersection was surveyed in an accident incident. One member of the survey crew was positioned sequentially at fifty foot intervals along the centerline of the highway while the second member of the party carried a twelve foot long pole with a white rag tied to its end along the railroad. The man on the track marked each point where vision was interrupted so that measurement along the track could be recorded. On the Maywood survey Dick stood in the middle of First Avenue dodging the heavy vehicular traffic while I paraded up and down the main line with my long flagged pole. First Avenue, 11 miles out, was the first public road crossing encountered west of the Chicago Passenger Station. A heavily used four lane road, the crossing was protected by mechanical gates operated by a gateman located nearby. In retrospect Dick and I were idiots to take measurements on and along this busy highway; we could have been killed. To do such a survey in the present day would require a crew of flagmen, orange pylons, two-way radios for traffic control and lots of permits from authorities. Dick and I, however, managed to avoid being struck. When a train came along and the gates went down stopping the cars, vans, trucks and buses, we sprang into action to get our measurements on the road. The main obstruction to vision was an old brick power station in the southeast quadrant. The drawing came out beautifully; we never heard what happened in any court action.

Rodman Ed Malo and I went out on train #13 headed by a class E-4 Hudson on September 22nd. Our destination was DeKalb, 57 miles west. We walked down to the California Packing Company's plant to measure clearances along their siding for a license agreement Ed was preparing. We then measured a new machinery loading platform and ramp across the main line from the distinctive old brick depot building. Ed had another small job to do in Sycamore which was five miles north on the remnant of the original Northern Illinois Railway, long a C&NW property. That company's lines once extended from Spring Valley on the south to Belvidere on the north, but during World War II the line was abandoned and dismantled beyond the Chicago Great Western crossing west of Sycamore. Not finding anyone at the roadmaster's office in the depot that could take us north, we happened to spot an old class R-1, 4-6-0, with waycar 11588 about to head up the branch to Sycamore to do some local switching. The DeKalb switch crew was glad to accommodate us when they learned we were company men. They were en route to collect some cars at the big Anaconda Wire and Cable plant located north of the CGW main. That explained why the C&NW left the diamond crossing and a piece of track in place north of the CGW. We stayed with the crew as they waited for a westbound CGW 2-8-2 #708 to rumble out of town with its short consist. While they did their switching, Ed and I checked the side clearances at the Sycamore Preserve Works and found them to be legal. Reboarding the local, we arrived in DeKalb in time to catch train #14 eastbound. All in all, that was a great way to spend a lovely autumn day—riding trains and being paid for it.

At the end of September instrumentman Christensen, rodman Humburg and I were sent to the westernmost end of the Galena Division at East Clinton. We loaded our gear on train #13 in the morning and arrived in Clinton, Iowa, in the early afternoon. The section foreman whose territory included the East Clinton yard met us with a company truck for the trip across the Mississippi River to the job site. The American Agricultural Chemical Company (AGRICO) had purchased a big tract of the railroad's surplus land on the east side of the river for a new fertilizer distribution facility. This area had an interesting history. The AGRICO property lay just south of the city of Fulton, Illinois, which was the western terminal of the old Galena and Chicago Union Railroad in 1855. Here a ferry was operated across the Mississippi River to provide access to Lyons, Iowa, a village that was merged with Clinton in 1895. In 1859 the G&CU built a short line south from Fulton to a point opposite Clinton where the first river bridge was under construction. Until 1909 the C&NW operated a double track main line through Fulton, around a sweeping curve and south to the river bridge. In 1909 a cutoff line was built from a point called Bluffs straight across the lowlands directly to the river bridge leaving Fulton off the main line. With the main flow of traffic on the new railroad, the old route gradually dried up. The original main line was dismantled leaving just a stub line into Fulton for the few industries located there. By 1947 the railroad land south of Fulton included the remains of a major locomotive servicing facility dating from 1910 and a substantial yard. Soon after completion, the railroad realized that the roundhouse and support facilities were in an awkward location for efficient operations and would be better sited across the river in the Clinton area. The whole complex was replaced and razed by 1941. The only vestige remaining was the adjacent freight yard lying to the east of the engine house area. The old East Clinton yard remained as a storage area for derelict freight cars. AGRICO needed trackage to serve their new fertilizer warehouse, and we were here to lay them out.

Unfortunately, AGRICO chose to place their new building just where the old turntable and engine house foundations lurked under the layers of cinders and dirt. The old concrete had grown harder as the years slipped by making removal a difficult job. The new lead track from the existing yard crossed these hidden ruins so that more concrete had to be removed. As our first day on the job was well advanced, we only had time to make a careful reconnaissance for tomorrow. The section foreman drove us to a modest hotel in Fulton for the night where after supper we roughed out a general plan for the next day. In the morning the accommodating foreman collected us with his truck for the short trip down to the yard. Chris set up the transit on the westernmost yard track and sighted in a line toward the AGRICO building. As the tapeman in the party it was my job to drag the chain (the 100 foot tape) through the weeds and rubble while carrying the striped lining pole along the projected alignment following the hand signals of the transit operator who waved his hands left and right to keep me on line. The bundle of toothpick stakes steadily diminished as I pounded them into the soft earth. I broke a few of them on undiscovered pieces of concrete rubble hidden under the surface. Dusty and sweaty we set the grade stacks with their blue tops carefully marked, before walking the AGRICO people over the setup to instruct them about the grading work their contractor would perform. He was to have a hard job of preparing the track bed through all that hidden debris. Hot and dusty, we were hauled over the river to the Clinton depot with all our equipment minus those heavy bundles of wooden stakes.

AGRICO had an idea that building this distribution warehouse on the bank of a navigable river would give them the opportunity to use barge service should railroad rates prove to be too high. Later, as I became more knowledgeable about the river and its potential, the likelihood of water competition forcing reductions in rail rates was understood to be unlikely at best. The AGRICO site was located on the east bank where the Mississippi's channels were shallow and filled with bars and wooded islands. Over the succeeding years Corps of Engineers river management policies became ever more stringent as environmental rules were established about disturbances to river systems especially with regard to dredging barge channels through areas away from the cleansing current of the main navigation channel. Barge service to the east bank was never established.

Saturday's half days were usually devoted to catching up on office work. October 4, 1947 was an exception. An eastbound drag freight on the north track (#2) experienced an unintended emergency brake application. The longitudinal stresses transmitted through the train broke the center sill of an old CMO (Chicago, St. Paul, Minneapolis and Omaha Ry.) gondola causing it to hump up on the Illinois highway #47 grade crossing in the village of Elburn, 47 miles west of Chicago. Curiously, the car did not derail despite its broken center sill, but the train could not proceed. The blocked crossing created an enormous traffic backup. Exactly what had caused the brakes to go into emergency was the hot potato tossed about from department to department. The car department maintained that the brakes were in perfect operating condition. The operating department denied train handling was the cause. Engineering was accused of having rough track that had caused the brake application. To prove that track was not the cause of this incident division engineer Jensen sent Chris and me out on the next commuter train to Geneva with our leveling gear. The roadmaster met us there and drove us twelve miles to Elburn. Jensen had given Chris explicit instructions on what he wanted measured. We first chained west for a mile on track #2. Then we ran levels on the top of the 112 pound rail with special care at each joint which meant that we made readings every 19½ feet on both rails. Going over that mile again, we measured the gap, if any, between the base of the rail and the tieplate at every joint. That is, I measured these gaps with a folding rule while kneeling on the limestone ballast rock while Chris took the notes. My thin trousers were no match for the sharp faced rocks. After 270 readings, my pants were worn through and my knees were getting pretty raw. As a further indignity we were working alongside a county forest preserve where a loud party was in progress. The sound of German drinking songs and jollity did not lessen the discomfort of being very thirsty, hot, gritty and aware of spending a perfectly lovely Saturday afternoon on my knees. The first thing Monday morning Chris and I began to plot the results of our Saturday work on a long roll of profile paper. About an hour after we started, Harold Jensen came through those swinging doors and said never mind as the car department had admitted that examination of the car sill had revealed an old and rusty crack. We discarded the incomplete drawing and turned to other work.

On October 7th instrumentman Art Olson and I went to Maywood to set grade stakes for the devil strip in the Fifth Avenue crossing. The devil strip was that piece of concrete pavement between the two main tracks in a highway grade crossing. We rode train #703 headed by class D Atlantic #1037. It was quick work to set up the level and pound in a couple of oak hubs and color their tops with blue keel. We packed up and rode the next eastbound back to the office. Signs of the times: an eastbound commuter sailed past lead by Pacific 4-6-2 #1547 while train #13 went west behind a couple of E-7 diesels headed by #5010-A.

Instrumentman Bob Lithgow roused himself the next day to take me on a short trip to the California Avenue coachyard. It seems that the new General Motors 1,000 horsepower diesel switch engines which were to replace the venerable M class 0-6-0 steam kettles were reducing some of the wooden walkways to splinters and creating water geysers by knocking off many of the coach watering hydrants around the yard. Motive power claimed the yard tracks were at fault. Bob and I rode a commuter train to the first stop which was Kedzie Avenue and hiked east to the coachyard. We soon determined that the leading footboards on the new diesels were causing the problem as they were set lower than on the old steam hogs. John Wilkinson, the local roadmaster, arranged to raise a few tracks while the B&B department rebuilt the broken walkways and relocated a couple of the hydrants to provide more clearance. There was some comment that the steam men feeling that the new diesels were job threats were taking careful note of every deficiency to make complaints which were blown all out of proportion. The California Avenue coachyard was like an operating museum complete with smoky little switch engines, ancient heavy weight steel vestibule coaches (1910–1914 era), and a fleet of grimy olive green lightweight cars from 1927. This latter bunch of cars featured painted aluminum sides, a very early use of that metal in passenger car construction. Basically, California Avenue was two yards separated by double ladder tracks. All of the Galena (west line) suburban trains were yarded here between the commuter rush hours. Most of the Wisconsin and Milwaukee Division train sets were yarded at Erie Street along the north line. The coachyard was a self contained community of trades whose only job was caring for the commuter and official business car fleet. Laced throughout the yard were air, water and Pintsch gas lines supplemented by a drainage system. There was so much maintenance work here that a separate B&B crew at the yard was required to make quick repairs on this complex system. The Pintsch gas lines were still in use in 1947 as there were a substantial number of older coaches in the fleet that still required the gas for illumination. These old beauties usually were on the late night trains which made reading by the passengers a real hazard to their eyes. The gas was locally produced at the coachyard from coal. These old gaslit cars usually had a coal stove in one end. Not used anymore in Chicago service because the locomotive steam lines provided heat, the cold stoves reminded us of the cars' former use on the prairie branch lines at the end of mixed trains. Years ago before the commuter fleet was electrified, this illuminating gas was piped to the Chicago Passenger Terminal; the remains of the pipeline were clearly visible along the pipe railing on the north side of the elevated tracks. I expect there are still stretches of it hung on the pipe railing east of Noble Street.

Operation of the suburban fleet was a carefully choreographed and controlled movement of empty car sets and locomotives. After the first inbound trains had unloaded their passengers at the CPT, most of the equipment backed the three miles west to California Avenue, or north to Erie Street, under the control of a tailhose man. The tailhose was a six foot long air hose with a lever action control valve on one end. The other was attached to the train air line when in use. After parking the coaches on a yard track at California Avenue, the locomotive power was cut off and the engine crew backed down the second mains to the engine servicing facilities at Fortieth Street. The tailhose men were then finished with their work until afternoon when they would take the empty trains back downtown. Many men in commuter service held second jobs in the city as financial messengers or in other part time occupations to fill-in the hours between morning and evening assignments. At the coachyard cleaners went through the cars collecting the discarded newspapers, forgotten lunches and lost umbrellas. Sometimes more extraordinary things were found—wallets, valises, false teeth, eyeglasses, and things hardly imaginable. Minor repairs were made, gas lines filled, seats turned, and windows washed on the outside (rarely and only in good weather). While most train sets stayed together, some were broken up to provide coaches for middle of the day trains. The little M-1, M-2 and M-3 class steam engines shuffled the cars around and occasionally took some to the Chicago Shops for more serious repairs such as installing new plush seats or even for a new paint job. Around

3 p.m. the road power began to appear at the coachyard to couple onto their cars and the backup men hooked up their hoses. In a predetermined order the trainsets began backing toward the CPT on secondary mains 3, 4 and 5 where they awaited their turns to enter the terminal to load their passengers and become real trains. In 1947 the principal suburban power was the class E Pacific, 4-6-2, with a few class E-1s which were a lighter Pacific type. The trainsets were often a mixture of the old 1910–1914 cars and the 1927 lightweights. On the Milwaukee line there was the so-called bankers' special car included in a popular late afternoon train that served the wealthy northern suburbs. A parlor car also was operated on the Northwest line to Lake Geneva. Patrons paid the regular fare plus a special supplement which included an assigned bar steward.

This coachyard operation was only one part of the suburban service that had to change or perish. As suburban traffic grew the old equipment and yarding techniques began to resemble a complicated machine that had begun to shed parts as it was pushed along faster and faster. The whole operation was rapidly approaching its capacity limits with strong signs that an utter collapse was quite probable. The advent the double-decked gallery car fleet in 1953 proved to be the basic element that saved the commuter service. Later operated in a push-pull mode with diesel power the car sets could be parked in the Chicago Passenger Terminal between runs which eliminated the backup men jobs. Hourly train service during the day between the rush hours actually was beneficial to positioning equipment while providing more service to the public. Car cleaners worked the cars in the CPT rather than in the sprawling coach yards. The need for all that high maintenance water, gas and steam pipelines at California Avenue disappeared. Another benefit was the abandonment of the Erie Street coachyard because the new cars could not clear the street viaducts crossing the entrance to the yard and had to be housed at California Avenue with attendant economies in scale. Years later a new coach maintenance shop was constructed at California Avenue to maintain the gallery car fleet. As the old suburban cars were scrapped the corresponding car shop facilities at Fortieth Street were also discontinued and cleared out.

Most of the long distance fleet was maintained at the streamliner ramp located at the Chicago Shops. In 1941 part of the yard tracks paralleling the Galena main lines were converted to a paved area with the tracks running on the walls of concrete inspection pits. The jointly owned coast streamliners and the C&NW's 400 trains were backed out of the Chicago Passenger Station to the ramp by tailhose men. The equipment backed onto a ramp track while passing through a car washer that could be positioned on all tracks as needed. An army of car cleaners and maintenance men serviced the car sets and readied the coast trains for their evening departure. After the work was completed, some of the trainsets were turned on the wye at the west end of the Chicago Shops property. There was a long tail track north of and parallel to the Galena main line west of Kenton Avenue on the elevation just to accommodate the turning of the coast fleet. In the afternoon the trains began their parade downtown in reverse. Some trains were turned on the Kedzie Avenue-Lake Street-Western Avenue (tower A-2) wye if traffic permitted. The first train to depart was the *City of Denver*, #111 at 5 p.m. The next one was #105 the *City of Portland* at 5:30 p.m. Train #111, the *City of San Francisco* departed at 7:00 p.m. and the last, #103 the *City of Los Angeles*, sailed at 7:15 p.m. The streamliners were all Galena Division trains while the 400 trains were routed on the north lines of the Wisconsin Division.

Instrumentman Art Olson asked me to accompany him to help on a job in my hometown of Elmhurst. Our task was measurement and location of the new intertrack storm drainage system recently installed through the station grounds. Water is the railroad's enemy especially when it lurks under the subgrade and destroys support for the track structure. A train operating on any track constructed with wooden ties can act like a giant pump. As cars of varying weight press down on the rails, especially at joints, the track depresses and rises in response to the passing axle loads. The faster the train, the more rapid the pumping action. At times geysers of mud may squirt up high enough to spatter the lower parts of the passing freight cars. The problem is worsened through station grounds where the platforms along the tracks block flow of storm water away from the track. Grade crossings with planking between the rails create additional barriers to normal drainage. To alleviate this problem 6 and 8 inch thin wall corrugated and perforated pipes were buried below the ballast along the tracks. Riser pipes with slotted cast iron covers were then connected to the buried mains which in turn led to collection basins away from the tracks. In later years installation of geotextile blankets that permitted water to pass through the fabric but restrain soil particles became a favored technique to improve grade crossings. Art and I located the risers and took elevations although I am not sure that was necessary as we could not get the elevations of the buried piping without removing the top screens. After we returned to the office, Art began to plot the field notes only to discover that he had overlooked half of the installation. That meant we had to return to Elmhurst on a Saturday morning and finish the job properly. The following Monday Art and I rode #703 to Lombard where a similar intertrack drainage system had been installed through the station grounds. That installation had failed to work as planned so that our field work was essential to try and determine what could be done to fix it. Of course the steel 100 foot chain decided to drop sideways into the crack between the outside running rail and the crossing plank at the Park Street crossing. We tried to wiggle it out without success. Using pocket knives we tried to cut slivers out of the crossing plank at the point where the chain was pinched without any luck. An old class R-1 ten wheeler, 4-6-0, towing a couple of freight cars and a waycar came trundling along as we were trying to free that dratted tape. Losing a tape was a great disgrace as well as fatal to finishing our work. We each grabbed an end of the tape and stretched it at a 45 degree angle away from the spot where it was caught hoping that the engine and cars did not have badly worn wheels that could lap over the outside of the rail far enough to break our measuring tape. Our luck in that regard held and the local went on west. We attacked the crossing plank again with our jack knives and finally chiseled enough wood out to free that chain. Forever after I was very careful not to allow a trailing tape get near any planking that might catch an edge.

My basic training was coming to an end. As a result of Chris' tutelage, I was feeling a lot less stupid than I did on that last day of July when I started this career. As cold weather settled into Northern Illinois, field work became more of a challenge. Our delight in being out in the field diminished in direct proportion to the rate the ground froze. The coldest place in the world is said to be out on a prairie railroad track in winter. However, in late 1947, there was one very important construction project in progress across a good part of the Galena Division that required engineering assistance. We could not wait for fair weather. Installation of Centralized Traffic Control (CTC) from West Chicago to Nelson, Illinois, about 75 miles, was a big capital investment and was absolutely needed if the C&NW was to move east-west traffic efficiently. The first part of the track work began at NI tower in West Chicago. Authorization was under AFE (Authority For Expenditure) B-1313 for the portion located in ICC valuation section 2A which ended at the DuPage County line about 3 miles west of West Chicago. The balance of the installation was in valuation section 3A which included everything west of the DuPage County line to the Mississippi River. Why that division of the railroad for the valuation accounting was established at that point in 1914 remains a mystery. Our payroll work sheets hours reflecting ICC account #1, Engineering, divided the expense according to AFE which required our estimating how much of our time was spent in each section.

The actual track changes were started in 1947 before I came on the railroad. Instrumentman John Shanklin was then the Galena Division's CTC field engineer, but with the advent of cold weather he decided that being a bridge inspector was a better career. Chris and I inherited Shanklin's assignment. We soon learned that the basic scheme called for installation of high speed #20 crossovers every five miles or so. The track work at NI just east of the West Chicago depot was a minor installation, but the next one west of West Chicago, at a place called WX, was more complicated. At WX the three track mains between Elmhurst and West Chicago converged to a double track main with the west entrance to the West Chicago yard also being located here. The high speed turnouts already had been installed when Chris and I became the CTC crew. But, the turnout joining tracks 1 and 2 caused a lot of complaint from train crews because its alignment was improperly cocked or skewed. We were to spend a lot of time running spirals into that turnout in an effort to ease the transition shock.

Our more immediate problem involved the crossover installation west of Geneva at GX and a nervous roadmaster. The standard

Centralized Traffic Control - Galena Division - 1947-1949

Control Points - crossovers and powered turnouts

Authorized by AFE B-1313 in valuation section Illinois 2A

NI	West Chicago	WX	West Chicago

Authorized by AFE B-1314 in valuation section Illinois 3A

AE	Ashton East	ME	Meredith East
AW	Ashton West	MW	Meredith West
CO	Cortland	NA	Nachusa
FX	Franklin Grove	NQ	Nelson
GX	Geneva	NY	Nelson
HX	Creston	RX	Rochelle
LX	LaFox	YD	DeKalb
MA	Malta		

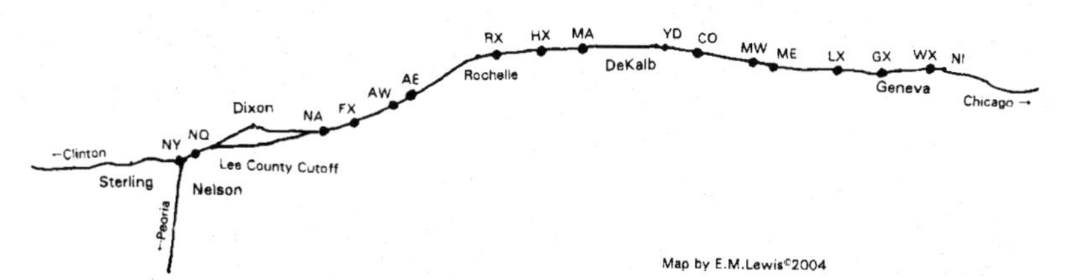

Installation of Centralized Traffic Control on 75 miles of the double-tracked Galena Division was a major capital project that involved an army of track, signal and engineering people.

number 20 turnout was over 370 feet long—given track centers of 13 feet. The frog alone weighed a ton. What made Art Benson, the roadmaster, uneasy was the set of milled switch plates that held the stock rails in position. To get the ties supporting the plates into just the right spot was Arthur's worry. So, he pleaded with the Chicago office for engineering assistance as he did not believe his veteran track men could properly read a blueprint especially for a new kind of turnout they had never seen before. Chris and I bundled up our gear and went to Geneva on a suburban train where a section foreman drove us out to the crossover site. Using the chief engineer's standard plan, Chris and I marked the location of the point of switch and the frog (149 feet apart) for all four turnouts and headed back to Geneva. Roadmaster Benson had appeared while we were working and would not allow us to leave until we had marked the exact position of every switch tie. The switch ties were all 7 × 9 inch oak varying from 9 to 20 feet in length. Using yellow keel (crayon) while being down on my knees again, I marked the measured locations as Chris read them off the standard plan. All this while a cold wind promised snow to come on that gray day. Benson insisted that all four turnouts of the double crossovers be similarly marked which meant that I was kneeling on rough stone ballast for most of that day and the next. Our plan for a quick half day job turned into a laborious two day affair. We came back on October 23rd to check with the foreman who had installed the switch ties. Since Benson was not present we were able to work directly with the steel gang foreman and quickly saw that he was competent to do the job properly. Now that the crossovers were sited, the rest of installation could proceed. Installation of the steel bungalow housing the switches and relays, digging in the underground cables and unloading the track material in the handiest locations were tasks that required great care. (After the merger with the Union Pacific Railroad in 1995, GX was renamed Randall and relocated a mile west in 2004 to better accommodate the new third main line constructed west to Elburn).

In late 1947 Chris and I kept a close eye on the work at GX because it was our first highspeed crossover. At the same time we had to start on the next crossover site, LX, near LaFox. To get there we endured a very cold ride on the section foreman's motor car. This was my first experience with a track motor car, and I found it rather unpleasant. Those 12 inch steel wheels were completely unsprung so that every rail joint jarred the vehicle and its riders. At LaFox we found the steel bungalow had already been installed. Again we marked every switch tie location which made another long, cold, day of kneeling on those ballast rocks. Making yellow keel marks every 19 inches, or so, over a 600 foot stretch of cold steel rail, was not my idea of the best way to spend a day. The signal masts were already in place so I climbed one to take a picture with my ancient Kodak camera. Two weeks later we went back to LX to check on the work and then went on west to Meredith where the existing passing track was to receive powered turnouts. These sites were named ME and MW for Meredith East and Meredith West. Meredith itself the eastern point and consisted of nothing but a short spur track serving a small stock loading pen and chute and the east end of the passing track. It had some notoriety as being the site of the last train holdup on the C&NW about 1914.

Our CTC crew was augmented by a new man, James A. Andrews. Jim was one of our co-op students from Northwestern University in

Evanston, Illinois. The railroad provided engineering entry level jobs on the Galena Division for students in hopes that they would join the railroad after graduation. Not one did. Jim was a welcome addition to our crew because I was then no longer the only donkey available for lugging our survey equipment. The weather turned colder making routine activities more difficult. Our colored crayon (keel) being made of wax did not work well on cold steel, the steel tape developed a mind of its own when being coiled up, sighting through our optical lenses caused excessive tearing of our eyes, and our wooden stakes were quickly reduced to splinters when driven into frozen ground or ballast. The next CTC site was CO at Cortland where we discovered a real problem—the mainlines were only 12½ feet on centers. Normally the two mains were 13 feet apart which may have been adequate for 1900, but any new construction would have required at least 14 foot centers. The dilemma at CO was we could not do anything about the less than desirable track centers as the winter weather had frozen the ballast which meant that any track shifting would have to wait until next year. But, we had to install the crossovers now so that the CTC project could be completed as scheduled. We elected to proceed with the tight track centers and adjusted the length of the new crossovers to accommodate the conditions we had discovered. At least we did not have to mark all those blasted switch tie locations as the gang foreman was perfectly able to install them correctly.

On December 29th Chris and I were at GX to witness the installation of Art Benson's first #20 crossover turnout. Art wanted us there in case something didn't fit. I do not know what we could have done if the carefully milled plates proved to be wrong. The 492 switch ties and electric switch machines had all been installed before the ballast froze hard. The big steel gang and a group of officials stood around in cold wind waiting for the work train carrying the switch material. An old class R-1, ten wheeler, #1353 finally showed up with its train which included material flats and a burro crane mounted on one car. Handling the one ton frogs which were 27 feet long was not difficult for the burro crane and they were quickly set in place. The big steel gang quickly cut out the main line rails, set the frogs, laid out the milled switch plates, set the 30 foot switch points, laid in the closure rails, placed the 13 foot guard rails, and fastened it all together with joint bars. Everything fit. There were no power tools in sight except for one gasoline powered rail saw. Bolt holes were drilled by a hand operated machine. Chris and I hung around until it was obvious that the work was going smoothly. We cadged a ride to town and went to Chicago on a warm train.

Intertwined with the CTC work in November-December were other engineering jobs that had to be done. One of them was a day trip with instrumentman Bob Lithgow in November to Lombard and Elburn to check on track drainage through the station grounds. Since Bob drove his own car, we did not have to scrounge transportation. This was a more efficient use of time and made a pleasant outing. The next day instrumentman Art Olson asked me to help him shoot some levels on a concrete floor in the Chicago Passenger Terminal. That little task did not even qualify as a field trip. Our new instrumentman, William F. (Bill) Polchow decided that Saturday morning (November 8th) was a good time to inventory a track on the Rockwell Street line at Taylor Street which was near Roosevelt Road. We rode down there on one of the many Chicago Surface Lines red streetcars. It appeared that a crossover track in the maze of diamonds and connecting tracks either had been overlooked in the 1914 valuation survey work or had been subsequently installed without proper authority at a later date. The local crews called it "the president's connection" for some reason long forgotten. The ownership of tracks here was a real Chinese puzzle involving the CGW, B&OCT, PRR, and C&NW. Bill was an eager new hire fresh from Navy service. Rough and ready, he enjoyed being one of the boys. After a night out with his pals, he often dragged into the office looking somewhat worse for wear. He was a good worker and performed every task assigned to him. After a few years on the Galena, he left the railroad for a job at the Rock Island Arsenal with the U. S. Corps of Engineers.

On November 12th several of us hiked over to LaSalle Street station to attend the Rock Island's open house honoring the inauguration of their new equipment for the *Golden State Rocket.* The stainless steel corrugated sides cars were open for a walk through inspection. It was a pleasure to see and smell the brand new train.

Chris and I were sent to Proviso on November 17th to stake a new sidetrack serving a wooden pole storage yard of the Illinois Bell Telephone Company. It was a simple task to set the location of a number 8 turnout and a straight track down the middle of a cinder covered area that had been a parking area for enginehouse employees. The old brick roundhouse was far more interesting. Built in 1911, it was a complete circle with 58 stalls each 90 feet long. Ten stalls were extended to 115 feet in 1929 to accommodate the new class H, 4-8-4, Northerns. The original 80 foot turntable had been replaced by a 110 footer in 1929, again to handle the new class H locomotives. I made some photographs that day of those huge Northern 4-8-4s including #3013 with its bell hung on the right side of the boiler face and #3006 and #3010 with their bells located at the top of the boiler front. There was a bevy of class Js 2-8-2 Mikes mixed in with the big boys including #2586. While the weather was overcast, the chilly air made the escaping steam quite visible. As an employee I had to be cautious when photographing around the railroad because exhibiting too much enthusiasm for trains and locomotives might label one as an FRN (railroadese for F____ Rail Nut) with decidedly negative consequences to your career.

On November 21st rodman Ed Malo and I went to Kedzie Avenue on a commuter scoot and walked a couple of blocks east toward the California Avenue coachyard. Our task, however, was located in the adjacent Griffin Wheel Company foundry complex. Constructed in the 1890s to manufacture freight car wheels, this ancient foundry was served from the south by the C&NW and from the north by the Milwaukee Road. There was a veritable maze of side tracks serving the jumble of buildings. So many of the old doorways, platforms and side clearances were in violation of Illinois Commerce Commission (ICC) regulations that the two railroads had agreed to issue a license detailing the deficiencies so that liability in the case of an accident would be the industry's problem. We spent the entire day and a Saturday morning to collect all the data.

Another nearby job was in Melrose Park where Chris and I laid out a short two car siding to serve a new grocery store just north of the depot. We arrived and found a crowd of people assembled including the city's dignitaries. Ralph Beddoes from the C&NW's industrial development department was orchestrating a highly publicized meeting to demonstrate how hard the C&NW was working to develop new customers along its lines. Looking back at this event, I can categorically say it was the most pathetic example of industrial development work I was to ever encounter. The side track required a #8 turnout set into a little used passing track that ran from JN tower at the east entrance to Proviso to Maywood. A brand new 100 pound turnout was installed and paid for by the C&NW along with the track from point of switch to the 12 foot clearance point. The rest of the little track was owned by the industry, Bonnie Bee. In the 30 years that the track existed, it saw exactly one revenue load—a car of bananas. Bonnie Bee soon burned down under mysterious circumstances and the side track just laid there. Eventually the City of Melrose Park acquired the property for construction of their new police department facility. As for Ralph Beddoes, he had been an instrumentman on the Galena Division and had transferred to the industrial department before I signed on. His tenure there was short.

A pedestrian fell and sustained injury on the sidewalk at Canal and Randolph Streets on the east side of the Chicago Passenger Terminal. The claim department requested a survey of the old concrete slab sidewalk some of which had become uneven after years of settlement. We carefully measured the old pavement and submitted a written report. On the basis of our work the claim department rejected the requested compensation. That was a good day for me, however; I received a back pay check for $157.00 which was very welcome indeed.

At the end of November Warner Frank, claim agent, drove Chris and me to Manlius, Illinois, to conduct an accident survey. A motorist contested occupation of a grade crossing with a freight train and lost the encounter with fatal consequences. Anticipating a wrongful death lawsuit, the claim department requested an immediate vision survey to establish the exact conditions at the crossing and as near the date of the accident as possible. Manlius, a small village serving an agricultural area, is located 24 miles south of Nelson on the former Southern Illinois Division. This was my first of many visits to the SI. My strongly remembered impression of the village was one of great isolation. Perhaps it was the gloomy overcast, the snow on the ground and or the chill of the day that leant a gray character to this lonesome place. The railroad was constructed from Nelson to Peoria in 1901 under the name of the Peoria and North Western Railway. There was a standard #3 depot east of the main track complete with agent as this line was dispatched from South Pekin on telegraphed train orders. The brick passenger platform was last used by revenue passengers in 1941 when the motorcar service from Sterling to Peoria was discontinued. West of the main was a small cheese factory, grain elevator and a farm machinery dealer. North of town was a locomotive water tank. We got right to work and quickly established the vision lines in the fatal quadrant of the road and railroad.

Although the weather was deteriorating in December, we had work to do at Nelson, 106 miles west of Chicago. Chris and I rode train #13 to Sterling where roadmaster Art Benson then drove us the six miles east to Nelson. And it rained—hard. We retreated to the Lincoln Hotel in Sterling for the rest of the day. Chris spent the time teaching me the intricacies of calculating railroad curves while I taught him how to play cribbage. As the next day was only overcast and threatening, but not dripping, we prevailed on Benson to take us back to Nelson. We ran a situation survey around the old coal chute and the north end of the yard that day and measured an under track coal hopper for a license agreement the next. We also inventoried a new diesel fuel pumping system which was a foretaste of the coming dieselization of the Southern Illinois subdivision.

Nelson was the north end of the Southern Illinois district of the Galena Division. The SI before 1937 had been an independent division. The old crew district rosters were still in operation ten years after the merger. The SI ran south through good agricultural land to Radnor where the line became a double track line down a stiff grade into the Illinois River valley. The double tracks ended at the foot of the grade at Limestone, continued as a single track to Kickapoo Junction (milepost 80) where the original main track ran into the Adams Street yard on the south side of Peoria. From Kickapoo the railroad was built as the St. Louis, Peoria and North Western Railway, 112 miles, to connect with an isolated little piece of the C&NW at Benld, Illinois. In 1904 the C&NW had constructed about 25 miles of railroad under the name of the Macoupin County Railway from a connection with the Chicago and Alton at Greenridge (just south of Girard, Illinois) to Benld. Named for a local entrepreneur, Ben L. Dorsey, Benld was a rough little burg that served the many soft coal mines in the surrounding area. At the turn of the century the C&NW consumed vast quantities of locomotive coal. The principal source of the coal had been the Consolidated Coal Company's mines in east central Iowa which were beginning to peter out. Wyoming's low grade coal was too far away to provide much relief for the company's general requirements. The railway created the Superior Coal Company (organized January 17, 1903) to develop locomotive fuel supplies in southern Illinois. Dependence on equitable coal rates from Girard to Chicago from the Chicago and Alton Railroad for cars of company coal to Chicago proved to be a weak link in the C&NW's strategy. To sever their dependence upon the foreign line's rate control, the C&NW resolved to build a connection from its Peoria and North Western Railway at Kickapoo to a link with its Macoupin County Railway at Girard. The plans called for a double track line at 13 foot centers on a 100 foot wide right-of-way, but only the easternmost track was constructed. (Many an unwary engineer and land agent was tripped up when locating right-of-way fences and property lines because the center of the track as built was 56.5 feet from the west property line and only 43.5 feet from the east side). The line was built in 1913 through some of the least populated part of Illinois. No towns or villages were to be found along this railroad. Near Springfield it was possible to see the state capitol dome from the track. At Benld a yard and engine servicing facilities were constructed. South of town the Macoupin County Railway had been extended 10 miles to Staunton in 1913 to serve newly opened coal mines. In 1927 another 2.61 miles to DeCamp was constructed for a connection with the Litchfield and Madison Railway. Eventually, through freights were operated by the C&NW as the coal traffic diminished and general freight traffic developed. An operating agreement was made with the L&M to crew C&NW freight trains from Benld to the Madison yard for interchange with the many railroads in the East St. Louis area.

Nelson itself did not amount to very much in 1947. Primarily a railroad junction and yard there were a few modest residences and a grocery store. The four main lines of the Galena Division ran east and west; the connection to the Southern Illinois district was double-tracked on the east wye with a single track on the west leg of the wye. A mile or more east of town the Lee County Cutoff left the Galena mains to curve gently around the hilly country of the Dixon area, rejoining the main railroad at Nachusa. Just east of the wooden depot at Nelson a huge concrete coaling tower straddled the main lines. Constructed in 1945 the structure was soon out of service as motive power changed to diesel and bunker C fired E-4 engines. Too big to take down without massive interruptions to train service, this white elephant dominated the landscape for many years. Parallel to the Galena mains were several storage tracks where condemned rolling stock was parked awaiting their turns in the electric crucibles at Northwestern Steel and Wire in Sterling. Nelson yard proper lay at right angles to the Galena mains. Containing 14 tracks, the yard was switched from the north end. A yard office was in an elevated structure on the east side of the switching lead. The lower portion was used as a locker and welfare facility by the switch crews and provided a registry point for the road crews. Several hundred feet away to the east was the infamous railroad hotel. A three story, shingle sided structure, it was operated by a private contractor to provide lodging for crews awaiting return calls to their home bases. Known as a firetrap, its operation and quality of service offered were continual sources of friction between the crews and the company. Inside the wye track were the engine servicing facilities. The brick enginehouse built in 1910 originally had 10 stalls 87 feet long. There were just four stalls left in 1947 along with a small coaling station, water and diesel fuel tanks and a cinder pit. The steam servicing facilities were in very poor condition. The lead tracks were sunk in a muddy cinder mush; oil and grease covered everything. After measuring around this area, I had to find a handful of waste (cotton threads) to get the chain clean enough to read the numbers. Steam locomotives still being handled here included class J, 2-8-2; class J4, 2-8-4 and a few "Zulus", class Z, 2-8-0. Class H Northerns, 4-8-4, were banned on the Southern Illinois because of the many class B timber trestle spans which had only three 8 × 16 inch stringers under each running rail and five piles per bent contrasted to a class A span with four stringers and seven piles. The fate of this decrepit enginehouse facility was easy to foretell. In 1948 all the steam power was withdrawn from the SI. Engineering staff cheered the demise of steam on this district. No more burnt-up timber bridges and track ties caused by cinders dropping from the locomotives, no more surface kinked rail from those unbalanced J-4s and no more repair and maintenance of coaling stations and leaking water tanks. One of those 2800 series J4s had already ruined twenty miles of brand new 115 pound rail because of dynamic pounding of unbalanced drivers. Most of these engines were sent north to serve out their lives in the iron ore country of Northern Michigan. They did not last long there as the rough track common to that country cracked the engine frames dooming them to the scrap line.

The SI was a world apart. Considered as "dark" territory because all communication between the dispatcher and his agents was by telegraph, the SI stations were throwbacks to an earlier age. Walking into a station was to enter a time machine and experiencing 30 years dropping off the calendar. Dark wainscoting and drab yellow-brown walls, wooden floors, worn wooden counters, disused train schedule boards, empty wooden ticket racks, overheated coal stoves, the telegraph table in the bay window with the train order signal handles conveniently placed, roller blinds on the dirty windows, glass insulators on each leg of the operator's chair, a Prince Albert tobacco can jammed behind the telegraph sounder, fat legged tables with bits of railroad iron piled underneath, a wooden pigeon hole desk, wooden filing cabinets, heaps of dogeared tariffs, a couple of outdated *Official Railway Guides,* perhaps an old Equipment Register, stacks of forms and waybills, an old Underwood typewriter, a sort of soft focus because of the dim lighting in the building. It also was easy to see the future for these relics. The advantage of having a live agent here to facilitate train movements and to be a link with the community was offset by the cost of maintaining his position and his workplace. The station agent was a fossil.

Despite the December weather Chris, Jim Andrews and I rode to Belvidere on train #703 headed by Atlantic #395. Our job was staking out a new side track to serve the Midwest Bottle Cap Company about a half mile west of the depot. The factory building was set back from the track a sufficient distance to accommodate a lovely gentle curved track across a brushy stretch of field. Adjacent to the C&NW

right-of-way we noted the remains of a former railroad (which would become important to us in 1963). We got the line stakes in before the early evening of December closed in. The grade stakes would have to be set the next day. We hauled our equipment back to the depot. Leaving them in the care of the agent, we boarded #706 back to Chicago. The next day we again came out on train #703 only to find the landscape covered with ice and snow. We trudged out to the site and began setting the oak hubs for the grade. Cold and wet, we decided that lunch would be a good idea and headed back to town. Not willing to leave our valuable instruments at the job site, we carried the lot back to town. Jim Andrews carried the transit but violated the basic rule of instruments: never carry an instrument any distance while still attached to the tripod. Jim, however, slung the tripod and transit over his shoulder and marched off toward town. We turned just in time to see him slip on an icy patch and land flat on his back. The lovely old 1901 Berger transit hit the ground first. Running up to him we first checked the condition of the transit. At first glance it seemed okay, but a second look saw the bright ragged edges of broken bronze. Not only was the vertical circle smashed, but a glance through the lens showed the cross hairs were also gone. That was the end of that instrument; it was not repairable out here in the field. The accident also ended our day's work. We glumly had a warm lunch and tried to reassure Jim that he would not be instantly fired when we returned to Chicago. The wait for eastbound #706 seemed endless on that cold, wet day. On the 11th our crew came back to Belvidere with a new transit acquired from the real estate department. It carried a date of 1905. Bill Wilbur did not report the accident to the chief engineer and Jim's hide was spared. The track to the bottle cap company was completed in 1948 and saw hardly any traffic—it all went by motor truck.

Belvidere in 1947 was a shadow of its former importance as a railroad junction town. The county seat of Boone County formerly had two C&NW rail lines cross here. The original Galena and Chicago Union line arrived in 1852 and built north to Caledonia in 1853. In 1885 the Northern Illinois Railway opened its line south to DeKalb and beyond to Spring Valley. At one point around the turn of the twentieth century serious thought was given to creating a Chicago bypass for Wisconsin rail traffic. The north south lines at Belvidere were abandoned in 1942 leaving the original G&CU line to Freeport and a stub of the Beloit line across the Kishwaukee River to the so-called North Yard where a casket factory provided a small amount of rail business. Green Giant Company operated a cannery east of town and a grain elevator created some business west of downtown. On December 9th a freshly painted Whitcomb diesel switch engine #403 was parked by the depot for the use of the regular switch crew still assigned to Belvidere. They did not have a lot to do but it was a contract job not easily abolished. Some years later a superintendent told me how he eliminated a regular switch job at a similar station—I think it was Antigo, Wisconsin. The regular switch crew had protested elimination of the assignment although clearly there was no switching work to be done. In order to comply with the union contract dealing with eliminating positions, the superintendent ordered the crew called every day. On reporting they were escorted to a room in the station, invited to take a seat and were locked in for eight hours except for a lunch break. Just as if they were on regular duty, no reading material was allowed. They could talk, smoke, sleep or just enjoy their company. At the end of the normal 8 hour day, the door was unlocked and they could go home. This routine lasted one week. The thoroughly bored crew agreed to abolishment of the switch job and bid in other positions on the roster.

The December weather pretty well closed down the field work away from Chicago. Just before Christmas I had an interesting trip to the construction site of a new bascule bridge spanning the Chicago River at State Street. The C&NW had a small yard operation on the north bank of the river between Dearborn Street and Wabash Avenue. The State Street yard actually occupied block number 1 of the original survey of the City of Chicago. The yard was served from the west by the North Pier line that crossed the North Branch of the Chicago River on a double-tracked lift bridge, ran underneath the Merchandise Mart, past the State Street yard and the Chicago Tribune printing plant where it split into two arms at the Ogden Slip. One arm went south went along the south side of the slip to a paper recycling plant; the north arm served a Curtiss Candy factory and the North Pier Terminal warehouse. The first Chicago railroad station was located at

Canal and Kinzie Streets where the North Pier line began. Built by the Galena and Chicago Union Rail Road in 1847, the station was just the first in a series of C&NW related depots in the area. The present day Merchandise Mart occupies the site of the third C&NW station. Called the Wells Street Station all C&NW trains had to cross that lift bridge which was always subject to boat traffic on the busy waterway. About 1908 management approved plans to build a new Chicago Passenger Terminal west of the accursed river. Acquisition of property from Noble Street east along Kinzie Street began at once. I recall that a collection of photographs showing the properties acquired was discovered in the chief engineer's vault and that no one knew what they were. After careful examination of the photos and their cryptic notations, we figured out that these pictures of the teeming mix of horse drawn wagons, carts, hordes of people in old fashioned dress amid buildings housing saloons, dry goods shops, warehouses and shacks that could only be scenes taken along the route of the proposed new railroad approach to the CPT. I have wondered what Archie Thomas, the chief engineer's chief clerk, did with this collection.

In 1911 the new Chicago Passenger Terminal was officially opened and the North Pier line remained to serve the small Wells Street and State Street yards. The old Wells Street station was pulled down soon after 1911. In 1929 the Merchandise Mart was constructed on air rights over the former station site. The railroad retained ownership of the land below elevation 31.0 (Lake Michigan datum) save for the actual land occupied by the building support caissons. Sometime after 1947 the railroad land under the entire building was sold with the railroad retaining an easement to operate two tracks under the structure.

Now in 1947 the old State Street river bridge was to be replaced by a modern bascule lift bridge. The City of Chicago's engineers determined that some railroad land was needed for supports of the north approach structure. Instrumentman Bob Lithgow and I met with the city's engineers to learn how their construction would affect the railroad's operations. Digging some of the caissons would require that some tracks had to be taken out of service or shifted. I watched the contractor drill through 80 feet of stiff blue clay overlaying the hardpan which would support the piers. The caissons were about five feet in diameter with the bottom belled out to afford greater bearing on the hardpan. The bottom of the caissons were all dug by hand as the European method of mechanical belling had not yet reached Chicago. The work at State Street was always a good excuse to leave the office to "check clearances and safety".

On December 29th Chris and I were at GX watching the installation of the crossovers. That was the last job of 1947. I now had completed 154 days of railroad service and had a credit of 6 months toward my 360 month retirement requirement. I was a railroad man, full of enthusiasm and a growing confidence in my ability to make this a career. I looked forward to 1948 and wondered what new adventures lay ahead.

BECOMING A RAILROADER, 1948

. . . The C&NW's centennial year opened with a terrific snowstorm. Word began to come in from the western reaches of the division that snow was piling up faster than it could be handled with resulting serious delays to freight movements. The old railroad response—throw more bodies at it—was invoked. Generella, a contract labor supplier for the C&NW in Chicago, scoured the skidrow areas along Madison Street and packed the men onto our commuter trains for snow shoveling work at the stations and in the yards. The engineering department was also drafted into providing supervision of this horde. I was ordered to Melrose Park on January 3rd, a Saturday morning, to supervise a snow gang at JN Tower and the east end of Proviso yards number 2 and 3.

Bundled up in thermal underwear, multiple layers of clothing and wearing high boots, I tramped west from the Melrose Park station along the main line to JN where I found my "men" huddled under the Indiana Harbor Belt Railroad overpass awaiting instructions. That was the most pathetic bunch of fellows I had ever seen. They had been dumped there by the contractor and told to wait for someone to tell them what to do. Before any work could be done, railroad pay rules had to be followed; I had to register these guys and get their social security numbers. All of this was out in the open with a cold wind whistling through the drafty underpass. At least the snow had let up and there were signs of the skies clearing. The men were mostly white Caucasians, young and old, gloved and gloveless, shod in every kind of footwear from low-cut oxfords to patched boots, wearing clothing better suited for milder weather and hardly a cap or hat to be seen. They were united in one desire—finding someplace warm. After I collected the names and numbers, true or false, the operator locked the door to his tower while the section foreman distributed brooms and long-handled shovels. He then locked the section house as he knew how these guys would try to sneak away and hide out while there was work to be done. I put them to work clearing switches on the ladder tracks leading to yards 2 and 3. Keeping them working was a chore as they invented every excuse they could to get their pay and go back to Chicago. Despite my best efforts, the gang gradually melted away during the day only to reappear on the station platform at Melrose Park waiting for the first train east. At least no one was hurt on the job. The net result was a relatively clear bunch of switches and a lot of unhappy, wet and cold men longing to get back to their warm spots on Madison Street.

The next morning, Sunday, January 4, 1948, my wife presented me with our only child, a daughter whom we named Sara. Thanks to the railroad health insurance, the hospital bill was $1.10 for eleven phone calls my wife had made. The bill from old Dr. E. W. Marquardt was $50.00 for the delivery. He had delivered my wife in 1925.

The snow problem abated. Our engineering staff had many stories to tell about their adventures as temporary snow gang foremen. The common thread to all their experiences was that it was a monumental waste of time and effort to employ bums from Madison Street as snow shovelers. We turned again to the railroad's business.

At Proviso the steel crew had completed reconstruction of the engine house turntable and requested our assistance in locating the exact center of rotation for balancing the structure. Instrumentmen Bob Lithgow and Chris Christensen were assigned and I was drafted to help. The weather was cold with an overcast. A chilly wind whipped through the gap in the circle of engine house stalls. The structural repairs had been carried out under traffic by Julius and Bill Morlock's steel gang. Heavier steel members replaced rusted iron hence the imbalance. Setting up the instrument on an infrequently used track, we ran levels on the new wide head ring rail and determined that it needed extensive shimming. Similar levels were shot on all receiving rails to make sure that the interval from pit rail to running rail was correct. Since it was well-nigh impossible to lower the pit rail, the B&B men used steel shims to raise the low spots. And so it went for two days. When the pit rail was truly level we were then able to establish the center of rotation by setting a transit line near the approximate center, rotating the table 120 degrees, reshooting the line, rotating again, reshooting the third time which yielded a triangle of lines. Repeating the procedure we were able to mark the triangle on a piece of paper from which the center point could be easily determined. The steel gang adjusted the center pivot slightly and the table was ready for normal use. The Proviso engine house was a complete circle save for the entrance tracks. Built with 58 stalls 90 feet long when the Proviso yards were first constructed in 1911, ten stalls were lengthened in 1929 to 115 feet to accommodate the new class H 4-8-4s. Stalls 1 and 2 were assigned to the B&B department for their office, locker room and material storage. During our January survey there, we often went in the B&B space to get warm. All kinds of steam power drifted in and out of the engine house. I shot pictures of M-4s #2637 and #2643, 0-8-0 heavy switchers, #491 whose 0-10-2 wheel arrangement was one of only two of that type on the railroad. A "chicken wire" diesel #4062, an F-3, also was turned while we were there, bringing with it a hint of the future for this old engine house.

The bridge construction at the State Street yard by the Chicago River continued despite the very cold weather. It wasn't cold down in the caissons where the blue clay was carved out bucket by bucket to be hauled away by a fleet of dump trucks. On January 9th Bob Lithgow and I tried to set track centerline stakes for a track shift but the 1″ × 18″ wooden stakes splintered when hammered into the frozen ground. We found a railroad lining bar in the yard office and used that heavy pointed iron bar to crack the frost for a starting hole and managed to get enough markers in for guiding the track move. Of course, with the ground frozen, that track was not going anywhere soon. On the 13th instrumentman Art Olson was pried loose from his drafting table stool to set more track stakes at State Street. Art was of the old school, no taxis for him. We boarded a Chicago Surface Lines car eastbound on Madison Street with all our surveying gear, transferred to a State Street car northbound which was detoured onto Wabash Avenue during the State Street bridge construction. Luckily, the job was minor, and we finished it quickly to return to our warm office. Recalling that streetcar ride reminded me of the parsimony of the chief bridge engineer, Arthur E. Harris. He was always reluctant to send any of his people out to look at a job preferring to have division forces do the work and report back to him with a neat situation plan. On occasion when he did send his men out in the Chicago area, he carefully handed each one 14 cents for round trip carfare. Despite his frugality, Mr. Harris was a personable man and a well regarded structural engineer who later headed the State of Illinois' Structural Licensing testing for many years.

Instrumentman Richard W. Bailey was assigned to make a vision survey at a remote public road crossing of the Lee County Cut-off south of Dixon, Illinois, Dick immediately decided to drive his own car rather than fiddle around with train #13 and a cold ride in the back of an open truck, or worse, on a track motorcar in this kind of weather. I was pleased to accompany him in his classic 12 cylinder Lincoln Zephyr. Our information about the accident indicated that a motorist had run into the side of a freight train at a public grade crossing near the Nelson end of the line. The overcast day was intensely cold as we hurried through the survey. The warm ride back to Chicago in that monumental car was a pleasure.

The Lee County Railway was the construction name used by the C&NW for the 13 mile double track main line from Nachusa to Nelson which bypassed the curves and grades of the original line through Dixon. Built in 1909 the cutoff had minimal grades and easy one degree curves. By 1947 it was down to a single track principally used by steam powered heavy drag freights. Passenger trains held the Dixon mains because their diesel power could handle the curves and grades. The double track through Dixon was maintained to higher standards than the neglected Lee County cut-off. I can remember a rather nasty freight derailment at Dixon when the cutoff turned into a welcome detour until the regular mains were restored.

Engineer of bridges, Art Harris, asked the division to take soil soundings at the Fox River bridge east of Geneva, Illinois (35 miles west of Chicago). Instrumentman M. C. Christensen took me along as the rodman. Bridge #66 was a high steel girder span that carried the double track of the Galena Division. The steel span was supported by towering concrete piers and limestone abutments forty feet high. Recent inspections showed some signs that the old stone back walls were bulging inward toward the river. To remedy that tendency, three massive concrete abutments in front of each stone pier were planned. Our job was to find out what kind of foundation support we could expect under the proposed abutments. To our regular kit of survey equipment, we added an 2 inch steel auger on a four foot pipe shaft with several four foot extensions. Loaded down with equipment we boarded the commuter train to Geneva where we then trudged down the main line to the bridge site. Here we faced our first problem: how to get down to river level. If we had come to Geneva in an automobile, we could have driven to the site by going into the downtown area, across the highway bridge and then down a park maintenance road to reach the land under the bridge. But, here we were at track level forty feet above where we needed to be and sliding down the steep and brushy embankment with all our gear was not a good choice. The limestone blocks of the abutment had been set in layers leaving big stairsteps at the ends—something like the Pyramids of Egypt. That was our road to the bottom. Since the river level was low during the cold weather, we were able to work in the dry at the foot of the abutment. Soil testing consisted of screwing the auger into the ground until we could go no further, measuring the depth reached, unscrewing the auger, pulling it up and moving to the next spot to do it all over again. We did the east side first, collected our gear and scrambled up the giant steps. Timing the trains, we scooted across the long bridge, scrambled down the west "steps" and set up to do the drilling. At this point the "Armstrong" auger ended its career. The shaft snapped off with the auger firmly wedged in a crack in the underground limestone. The good part of this accident was we had less weight to lug back to the office. The company did not replace the lost augur. In the future soil borings were to be done by professional soil engineers with proper equipment. The information we collected, however, was sufficient for the completion of the buttress design. More about this job will come in the next year.

For some reason rodman Ed Malo did not draw assignments to any of the larger survey jobs. He seemed content to putter along doing the detail work on license agreements for water lines and wire crossings, clearances and other little odd items. Deciding he needed a day out in the field, he asked me on the 23rd to go with him on #13 to DeKalb to assist him in measuring clearances at some loading docks near the depot. The roadmaster later drove us up to Sycamore to collect more clearance data, returning us to DeKalb in time for #14 eastbound. Hardly a memorable day, but so typical of Ed's low pressure approach to work.

January's bitter winter returned with blowing snow. M. C. Christensen was given the job of staking out a new siding for the Pure Oil Company at Crystal Lake, Illinois. Getting to Crystal Lake on a Wisconsin Division commuter train was a new experience, but the survey work for the siding proved to be downright dangerous. An arctic blast greeted us as we loaded a taxi from the depot with all our gear and went to the U. S. Highway 12 grade crossing southeast of the city. Our work site was about a quarter of a mile north of the crossing on the Galena Division's Elgin-Williams Bay subdivision. The fierce wind from the northwest had full sweep across the snowy stubble fields. Struggling to set up the transit and mark the centerline, proved to be almost impossible. Looking through the lens of the transit induced so much eye tearing that not only could we not see anything but there was real danger of suffering a frozen eyeball. The temperature was about 12 degrees but the wind chill factor was minus a lot more. The wind shook the transit; the plumbline waved in the breeze, the wooden stakes shattered in the frozen ground and there was no shelter from the gale. Looking back, I believe that this particular event was the most dangerous weather I ever experienced on a field trip. The track layout was easy—about a half hour job in fair weather. But, we had to retreat from this exposed site. We found an open restaurant on highway 12 and called it a day. The weather moderated a bit when Chris and I returned to Crystal Lake the next day. This time we brought along a sawed-off lining bar to crack the frozen ground enough to get the stakes set. We made cross-sections with level and rod to develop a grading and drainage plan for the contractor. When spring came, the track was built—and never saw a car of revenue freight.

Another careless motorist got in front of a commuter train at Pell Lake, Wisconsin, some miles north of Crystal Lake on the Williams Bay line, with fatal consequences for the driver. The claim agent drove Chris and me there for the required vision survey at this crossing. The situation was a bit odd as the road crossed the railroad at a very extreme angle which may have distorted the driver's judgement. At any rate the heater in the claim agent's car worked very well.

Ed Malo had another complicated license agreement exhibit to prepare to show the clearances at the Curtiss Candy Company's plant located on the North Pier line. On February 9th the weather eased enough to make a walking tour of this historic line good exercise. We started at Kinzie and Canal, crossed the drawspan which was down, wandered under the Merchandise Mart where the switch engines and cars seemed more like subway vehicles than actual railroad equipment. There were a couple of warehouses, a cold storage plant, the Chicago Tribune's printing facility, as well as the candy factory and a reclaimed paper facility served by this line. The trackage east of Michigan Avenue ran through paved streets just like a street railway's lines. Railroad operations in this district were often hampered by the public who did not realize that those shiny steel strips in the pavement were really live railroad tracks and parked their cars here and there without any thought about freight cars and locomotives. The train crews often used their engines as bulldozers on carelessly parked cars which resulted in damage claims becoming a constant headache for the claims department.

Another new instrumentman appeared on the Galena fresh from the Iowa Division at Boone, Robert D. Nelson. Bob grew-up on a western Kansas wheat farm, entered the U. S. Navy and served as a lieutenant JG. After service he went to work for the C&NW at Boone but was soon transferred to the Galena because of the large amount of construction work in progress there. Unmarried, he did not have much of a sense of humor and was quite serious in his work. Bob became the target for much good-humored ribbing because he blushed extravagantly, which, of course, brought on more efforts to rattle him. He teamed up with another of our bachelor instrumentmen, Bill Polchow, and the two of them shared many work assignments. The further and longer away from the office the better the assignment for these boys. Some mornings, however, they both seemed a bit worse for wear. While he may have lacked much imagination, Bob solidly held up his end and turned in a good work record on the Galena. He eventually returned to Boone with promotion.

Bob's first assignment was an accident survey, and I was detailed to go along to help. We rode the train to Geneva on February 11th to meet Ollie Adamson, the district claim agent. Ollie had been a claim agent on the railroad his whole working career. He drove us west of town to Evarts Road which was a Kane County road that crossed the C&NW in a zigzag fashion, an alignment leftover from horse and wagon days. The crossing was protected only with the usual advance warning signs and standard passive crossbucks. The vision survey was especially tricky because of the relationship of the parallel road and railroad. It was Bob's first accident survey, and he did it slowly and methodically. Ollie reflected that his railroad career had begun right there in 1918 when a troop train derailed just to west of the crossing. He was so helpful in succoring the injured that the railroad hired him right on the spot. Finally, Bob finished writing his notes, and we went back to Geneva for the train ride home.

When considering that I was a brand new hire only six months ago, it was amazing to see that I had now become Bob Nelson's guide dog. We went to LX west of LaFox to inspect the new crossovers on the 12th. On the 24th we had a small job on the Rockwell Street line where a failing retaining wall at Arthington Street received our attention. The Rockwell Street line had an interesting history. Also called the South Branch, it was a double track railroad that commenced at Kedzie Avenue station on the mainline and curved around 90° to a due south course at Lake Street where it passed under the Lake Street Elevated's high bridge and over the Chicago Surface Line's Lake Street trolley route. At Lake Street a single track ran back eastward to the Galena mainlines at Western Avenue forming a wye track. Within this wye were the remains of the old Western Avenue freight yard now used for track material storage and as a freight car graveyard. Western Avenue predated Proviso as the C&NW's Chicago freight yards. The tracks were laid with 72 pound rail, the ladder track was a series of overlapping 9 degree turnouts (equivalent to a number 6.3 turnout), all on cinder ballast. Nate Waterman, Freight Terminal Superintendent, claimed that Western Avenue in its prime handled cars more efficiently than any of the newer facilities and his happiest days on the railroad was when he was yardmaster there. South of Lake Street the Rockwell Street Line was a three track railroad which shared the west half of an elevated right-of-way with multiple tracks of the PRR's former Pittsburg, Fort Wayne and Chicago (Panhandle) line. The two sets of tracks crossed at Ogden Avenue in a welter of crossovers and diamonds. Here the Baltimore and Ohio Chicago Terminal (BOCT) and the Chicago Great Western (CGW) also crossed on their way downtown. A ground switchman lined the switches as there was no interlocking tower here, just a series of stop boards which all train movements had to obey. The C&NW line turned eastwards to its yard at Wood Street and beyond to the so-called Sixteenth Street Line, another multiple track elevated line that ran to a connection with the St. Charles Air Line and its big steel bridge across the Chicago River. The Air Line was the link between the LaSalle Street, Grand Central, Dearborn and Illinois Central's 12th Street stations. The St. Charles Air Line was itself a piece of Chicago's railroad history. The name came from an early railroad company that had plans to build west to St. Charles located on the Fox River about the time when the Galena and Chicago Union has pushing its own line toward Freeport. The G&CU saw its opportunity, bought the St. Charles out and folded it up. Along with the stillborn company came the right-of-way in Chicago that became the jointly owned railroad connecting all the passenger stations that facilitated transfer of through equipment. The Chicago Great Western Railway through its Minnesota and Northwestern Railroad constructed a railroad through St. Charles in 1886 along the route projected by the G&CU's rival.

From Ogden Avenue with its tangle of tracks and crossings the C&NW had operating rights over the Chicago and Western Indiana (CWI) and the Chicago Junction Railway (CJ) to 42nd Street and Halsted to directly serve the Chicago Union Stockyards.

The Rockwell Street line in 1948 was lined with empty warehouses served by a plethora of disused and rusty sidings. One of first railroad lines elevated in the City of Chicago, the many industrial spurs were crammed in to avoid spanning the many street overpasses which resulted in use of an amazing quantity of number 4 turnouts. These short little track features were just 45 feet long and contained curves over 30 degrees which is much too sharp for modern switch-engines. I have seen photographs of old C&NW 0-4-0 switch engines which would have been able to negotiate these extreme little side-tracks. The sharpest turnout authorized by the chief engineer in 1948 was a number 5 which had a 24½ degree curve. The only one of these I ever saw was installed in the hump track at Proviso where the curvature was equally divided right and left.

As luck would have it the very next day I was back at the Wood Street yard with Chris to investigate an old diner that was being used as a kitchen at the Potato Yard. The C&NW catered to the rowdiest bunch of merchants I had ever encountered. These guys dealt in carloads of produce, principally potatoes, but also onions and other root crops. The association eternally complained about the railroad's service, facilities, rates, and the lack of our recognition of their importance. Threats to move over to the Santa Fe's yard always hung in the air. Of course they had been on the Santa Fe's property and on the Burlington's before that, so that their threat to relocate was not taken too seriously. The C&NW had some doubts about the profitability of this business, but not having any kind of a cost system, could not really make a case for or against the Potato Yard. The C&NW could only believe that high volumes of cars meant profits. Chris and I measured up the crummy facility which was eventually replaced by a better restaurant arrangement. I was amused to learn that a lady clerk at Wood Street was named Olive Branch. Years later the Potato Yard and Wood Street plus the B&O's nearby coach yard became the intermodal center called Global One.

The end of February came and Bob Nelson was called on to stake out a new industry track at Proviso for the Joseph Lumber Company. On the 27th Bob and I rode the train to Bellwood. Leaving the train we climbed the rickety wooden staircase to the shaky old Mannheim Road bridge. This structure, posted for ten tons maximum weight, carried U. S. 45 across 57 "live" yard tracks and 2 mainlines. Built in 1907 as a wagon bridge when the original Proviso Yards were first built, modern day traffic on this main artery was slowly destroying the narrow two lane structure. Since repair was the railroad's responsibility, the B&B department had a never ending job of replacing the 4″ × 10″ fir planks that formed the wearing surface. The boards nailed down with 10″ boat spikes soon rattled loose under the constant pounding and often fell down on the busy tracks below. Bridge lighting was a series of iron pipe goose neck fixtures with shaded bare bulbs. They, too, soon rattled loose and hung on their electric cables along the steel trusses. It was a good thing that the public did not really understand that this rotten bridge and another just like it a mile west at Wolf Road were the railroad's responsibility. We swayed our way along the three foot wide wooden sidewalk on the east side with big trucks almost brushing our clothes. At the north end another rickety wooden stairway took us to the ground below. The switch tender's shack was right under the bridge. His job was to line the turnouts for engines going to and from the engine servicing facilities a half mile to the west. About 200 yards to the west was a gritty old wooden building known as the Welfare Facility containing the lockers and showers for the road crews who picked up, or dropped off, their engines here. The Joseph Brothers Lumber Company track we were to stake out was to begin at the Mannheim Road bridge and swing around to the northwest to their business which occupied the southwest corner of the Lake Street-Mannheim Road intersection. I briskly drove the centerline stakes into the cindery ground at 50 foot intervals along the proposed alignment. Bob was quite careful with his transit settings and seemed to know what he was doing. It was good practice for us both. When the Joseph brothers learned how much the track was going to cost them, they gave up the project and continued to receive their carload materials at a public team track.

March 1st and spring had not quite arrived when instrumentman Art Olson and I went to West Chicago to set levels for a rail welding platform. Jimmy Hicks, the welding supervisor, arranged to have us picked up at the depot and brought out to the west yard where the work was to be done. The welding line was a quarter mile long collection of tie cribs, rollers, shacks of various sorts and stacks of new rail strung out along the north side of the freight yard. The whole affair looked temporary, improvised and distinctly amateurish, but it was what Hicks had been given to work with. He was not discouraged at all and was sure that he was going to produce great welds. Setting the proper elevations along the line was not difficult as the supporting cribs could be easily adjusted with shims in the crude timber cribbing supports.

Rail, especially steel rail, is the essential item that makes a roadway into a railroad. The C&NW's use of the steel rail began after the Civil War when the North Chicago Rolling Mill began producing steel instead of iron rails. By 1900 most iron rail was gone from its tracks as the superior qualities of steel versus iron were quickly discovered. In the early days of the railroad locomotives were small, wooden freight cars had limited capacities and speeds were slow throughout agrarian C&NW territory. There was a great amount of light rail in these branch lines that survived into the 1950s. A 56 or 60 pound rail was considered perfectly good for 30 mph track. The light and flexible rail in soft cinder ballast gave a rather springy ride on some lines especially in the spring of the year. Traffic volumes, tonnages and bigger motive power required heavier rail sections. Heavier rail such as the 72, 80, 85, 90 and 100 pound sections all had their day. In 1948 the C&NW still had hundreds of miles of 60 and 72 pound rail in branch lines. Over time the lighter rails were gradually replaced by usable 90 and 100 pound sections recovered from main lines. The C&NW had a good supply of secondhand 100 pound rail that had been replaced in the main lines beginning in 1934 by the new 112 pound. Rolled to C&NW specifications and labeled as section type 10035, this sturdy rail design ended up in a lot of industrial sidings and passing tracks. The many types of rail were identified by weight and cross section design. 9030 rail was different from 9035; 11225 did not match 11228. Between the years when 100 pound rail was deemed heavy and the 112 pound weight came into use, there was an ill-fated type that weighed 110 pounds to the yard. This 11025 rail section proved to be a disaster. Corner cracking and head failure forced a basic redesign which resulted in the 112 pound section. This new rail in turn was cursed by excessive rail failures due to transverse fissures. The trouble lay in the manufacturing process where freshly rolled hot rail was allowed to cool in the open air. Variations in the rate of cooling created the irregularities within the rail that often developed into catastrophic breaks. More experimentation resulted in the concept of controlled cooling where fresh, hot rails were placed in a giant heat soaking pit and gradually cooled. The failure rate dropped drastically. Large quantities of 112 pound laid in C&NW main running tracks at first gave good service until age and accumulated tonnage over the years accelerated head cracking and fissuring. Extensive research by the AAR showed the failure could be overcome by adding more metal under the head at the web and increasing the length of the connecting joint bars. The new design was the 11525 rail that became the standard mainline rail for the next few years. In 1948 the C&NW was just beginning to receive the new 115 rail.

Rails are manufactured in rolling mills in long lengths which were hot cut to lengths suitable for shipping. Early gondola cars could accommodate 28 foot rails. As longer cars became available, 31 feet became the standard. Then 33 foot rails were used and most recently, the 39 foot rail which neatly fit in 40 foot gondolas. When placed in track the rails must be fastened together with some kind of joint bars. When railroads were new and experimentation high, the light rails were linked by fishing plates, or, as they were later styled, fish plates, that fit around the base of the rail. As rail sections got larger, more positive linkage was required and the angle bar joined the two rails with bolts through the web of the rail. The Delaware and Hudson Railway first welded rails together in 1939. Later, the Elgin, Joliet and Eastern Railway, welded their running rails into long stretches. The C&NW watched these early attempts to eliminate joints and this welding line at West Chicago was its first experiment in making long rails.

Jimmy Hicks used a butt welding technique to join secondhand 11225 and 11228 rail at West Chicago in 1948. These two types of 112 pound rail had almost identical contours and could be successfully joined. The technique used was simple. With used rail it was prudent to first saw-off the ends of the rails to eliminate any worn or battered metal and to square the faces, heat them up to a white-orange heat with gas fuel, then mash them together with hydraulic presses until the metals merged and bulged out. This was called pressure butt welding. Another technique was use of the more sophisticated electric welding. In this gas welding line the rails had to be absolutely level when they were squeezed together or a cocked weld resulted. So, Art and I had to be sure to get our levels just right. Jimmy later discovered that the squeezing process actually lifted the rails slightly so that he was creating crowned welds that gave a little lift, or bounce, to rolling stock traversing it at speed. Adjustments were made in the welding line to compensate for the unwanted crowning and good welds resulted. In these early welding efforts, the C&NW was rather timid and produced only 78 foot or 117 foot rails. The bulged metal was ground smooth after it cooled. Jimmy built quite a stock of these 78 footers; he was not yet authorized to weld long strings. One of the problems with long rails is their transportation. Invention of the connected flat cars equipped with rail racks was yet to come. The principal use for these 78 and 117 footers was in grade crossings to eliminate that hard to maintain center joint buried in the road crossing planking. Another common use was across pile bridges to eliminate joints that tended to loosen and pound the bridge timbers.

Maintaining a railroad built with jointed rail was proving to be expensive as traffic volumes and tonnages increased. Maintenance of jointed tracks required track walkers who could be identified by the two tools— a track wrench and a spike maul—which they carried on their daily trek over their assigned territory. Their task was simply to tighten loose bolts and to pound down spikes that worked up from the ties. The joint is the weakest part of the running rail. Heavy, repeated and rhythmic load and unload caused by the wheel sets work the joint and its bolted angle bars. The bolts were one inch diameter in all rail from 9035 and heavier. The bolts had special oval heads that fit snugly in slightly elongated holes in the angle or joint bars. A nutlock and a heavy duty nut wedged the whole unit together. 112 pound rail on the C&NW used a four bolt, headfree, joint bar—headfree meaning that the bearing of the top of the joint did not fully bear on the filet of the web and underside of the head of the T section rail. The joint bar area was the site for most rail failures. The four bolt, 24 inch, joint was too short to resist the bending forces and a diagonal section of the rail web broken through the bolt holes often floated within the joint. This loose triangle of rail, often referred to as a "Dutchman" would either stay in place or rattle out leaving a bad gap in the running surface. In many low speed yard tracks, these pieces of rail were just left in place rather than replacing the whole rail. In the mainline, however, end-broken rails had to be changed. Other defects could be transverse fissures emanating from engine burns or embedded impurities left in the steel when the rail was rolled. Sometimes a set of joint bars would be placed over the defect in case the rail should develop a break at that point, but this was deemed an expedient and temporary measure. I saw one rail in a Milwaukee Road side track that had five pairs of angle bars on one 39 foot rail. Jointed rail required splicing every 39 feet. That creates 270 joints in every mile and every joint is a potential problem. Even those directly involved with maintenance of jointed rail could not agree about how to handle these joints. Some advocated a program of regular oiling of each joint to facilitate slippage in the joint area; while others advocated tight joints rusted shut to promote structural integrity. Actually, neither posture was correct. Research by the American Railway Engineering Association (AREA) demonstrated the end pressure generated by temperature expanded rail was about the same value whether the welded rail string was long or short because of drag generated by tie plates, anti-creepers and ballast. Control the end pressures and rail strings of any length could be installed. Controlling the expansion-contraction of a long rail string was specified by a chief engineer's standard that called for anti-creepers (rail anchors) on every tie for 400 feet at both ends of the rail string and anticreepers on every other tie throughout the length of the string.

As the C&NW's engineering officers became more confidant with these early attempts at making the two and three rail sticks, welding of longer strings became standard. The usual length obtained by welding 34 rails together was 1,326 feet which became a production standard. Special flat cars with racks and rollers were positioned to load the freshly welded strings as they came along the welding line. Unloading a string of welded rail at a specific track site was simple. After chaining two rail strings to both existing running rails in track that were to be replaced, the rail train pulled ahead slowly with the twin welded rail strings paying out behind. Perhaps this new welding technique would lead to other major improvements in railroad maintenance.

Chicago's February weather remained unpleasant in 1948. Instrumentman M. C. Christensen was given an assignment in southern Illinois involving a new grade separation bridge near Girard and I was tapped to go with him. We rode the GM&O's day train on our annual passes to Springfield where we transferred to the Illinois Terminal's electric car. As usual we were burdened with our survey equipment. At Girard Ray Doty, our B&B supervisor for the Southern Illinois subdivision, collected us with a company truck for a short drive to the bridge site 2½ miles south of the village. The weather was cold and gloomy. Blueprints provided by the bridge engineer soon became wet and bedraggled in the sharp wet wind. Early winter darkness forced a retreat to the St. George Hotel in nearby Carlinville. There, we spread the plans out and determined how we could establish a good baseline which was not easy as the railroad at that point was in a long curve and up on a 30 foot embankment. The hotel accommodations were not four star—more like no stars. My room rate was 50 cents that night because I was on the third floor. The second floor rate was a $1.00. We could hear the rats running in the walls.

The next morning saw little improvement in the weather—windy but no rain. Illinois Highway #4 south of Springfield was one of the earliest paved roads in the state. Originally a single lane paved strip, it followed the old county roads that were aligned along the township lines and contained many sharp right angle turns to stay within the lands dedicated for public roadways. The paved road of 1920 was unsafe for 1948 traffic and the state highway engineers straightened the route to eliminate much of the sharp cornered alignment. This stretch was a new pavement that paralleled the Illinois Terminal and GM&O mainlines south from Girard. The C&NW crossed over the other railroads on a high steel truss north of Nilwood. The new state highway construction required extension of the existing truss span with a new deck plate girder. It was our job to establish the pier locations. Cold and blown, we finished establishing a baseline and marking the pier locations in time to catch the northbound ITC car and our GM&O connection at Springfield for Chicago. We would be back to this place again.

Gradually, I was being given more responsibilities. So many capital projects were underway on the Galena Division that Ed Malo could not write roadway completion reports fast enough. The list of completed work grew faster than his pencil could fly. I began to have my own list of small jobs to work on. The first one was a real daisy, illustrating a basic rule in business—leave the dirtiest work to the juniors. This little job involved installation of new rail lubricators at Western Avenue and in the lower end of the hump yard at Proviso. I soon learned why these projects had been put-off for quite a spell since their reported completion. A rail lubricator is essentially a tub of black grease with a hydraulic plunger that is actuated by the wheel of a car or locomotive hitting a cam set just a little higher alongside the running rail. The result is a glob of viscous oil jetted against the flange side of the rail head to reduce wear of the steel wheel rubbing against the steel rail. Because the lubricators authorized often ended up being placed at different locations, their installation site had to be correctly identified. In my innocence, I gleefully scheduled myself for a field trip by making a blueprint map of the general area where the new lubricators were supposed to be and took my trusty fifty foot tape. As I walked the mainline east from Kedzie Avenue to Western Avenue where the new unit was supposedly installed, I found a pile of sticky black grease on the ties heaped up to the top of the rail. That was it. Each time a wheel hit that plunger cam, a dollop of grease had been deposited on the rail, but most of it had fallen on the ties rather than being carried along the rail as planned. I measured the location of this thing to some reference point so that I could correct the station map and then discovered why Ed had given me this job—the sticky goo was on everything around the lubricator, and I soon had a good coating on my shoes, tape and pants. It reminded me of the Tar Baby that was best left alone. The section men hated these things as it made their work place a greasy mess, yet they were the ones that had to fill them with fresh grease from time to time. The B&B crews half-heartedly claimed jurisdiction over their maintenance, but were really glad that the trackmen had the nasty job of maintaining them. Hiking back to Kedzie Avenue station for a train to Proviso I managed to scrape some of that grease off my shoe bottoms, but could do nothing for my blackened trousers and grimy tape measure. At Proviso station I crossed over the mains on the pedestrian bridge and made my way to the lower end of Yard 5, the hump yard. More wary than before, I spotted the lubricators scattered on various tracks in the east end and marked their locations more or less accurately by pacing the distances from the points of switches. Even had there been no grease dispersing rail lubricators, a railroad is a dirty place. Before the general use of roller bearing equipment, car journals dripped oil along the tracks, steam locomotives dropped gobs of creamy grease from their valve gear, dirty water was dumped and sprayed by the engines, and passenger trains with their open hopper discharges left human waste scattered along the mainlines. The tapeman in charge of the chain had to be prepared to clean it thoroughly before winding it up. We learned, in those days, to avoid walking close to weedy tracks where car oil left a black haze on tall grasses growing along yard tracks.

I became a homeowner March 9th, 1948 when our family moved into its first home at 440 N. York Street in Elmhurst, Illinois, along with two mortgages.

Instrumentman Christensen and I returned to Geneva on March 10th to conduct a drainage survey about 1½ miles east of town. A culvert pipe was proposed to replace a short one-span girder bridge. Determination of the correct pipe diameter required a bridge situation survey that included a detailed topographic survey of the bridge's drainage area. Calculation of the water run-off that would pass through the pipe was based on annual rainfall records and the surcharge a 100 year event might impose. This was a tedious two day field effort which involved establishing a grid work of lines upstream from the bridge site using transit and chain. Then the elevation of each intersection point was determined. Using a plane table instrument would have been quicker using stadia sights, but we did not have one or the training to properly use it and stadia reading reductions in the office would have been too laborious with that old hand-cranked calculator. Back in the office Chris took our notes, plotted the elevations and sketched in the contours. A clever way of interpolating proportional distances between contours involved a simple rubber band marked off in units in India ink stretched across the arms of a pair of dividers. There little tricks in every trade that mark the differences between the professional and the novice. Older men in the office remarked that bridge surveys are a major pain to do and that the best ones were made from a bar stool in the nearest drinks emporium.

As the spring weather improved, the pace of the field work accelerated. Instrumentman Bailey and I tackled bridge surveys about four miles south of Green Valley on the Southern Illinois south of South Pekin. Dick's Lincoln was put to good use. Since Peoria was his hometown, he got in a brief visit with his mom before we went on to South Pekin where Ray Doty, the B&B supervisor, took us down to the bridge sites on his motor car. There were several class B timber spans crossing a low area near the Mackinaw River. These bridges acted as relief for overflow flood waters when the Mackinaw got out of its banks which it did on almost a yearly basis. The lower trusses of the big bridge #1769 at the main river were often in the water and collected a lot of debris on the upstream side. The bridges Dick and I were interested in were five little relief spans all numbered as #1769 with a fraction appended like 1769 1/5. Replacing the timber spans with galvanized iron pipes was a good idea, but the standard bridge situation survey rules could not be used as the floodplain of the Mackinaw River was immense. We made the appropriate notes, however, and our report was accepted by the bridge engineer. The next day we visited bridge #1753 about four miles south of South Pekin. Following this brief visit Dick and I went to Peoria to catch the Illinois Terminal electric car south to Benld where a vision survey was required. The trip on ITC car #284 was an adventure. Riding south we were told that violent wind storms had hit the area below Springfield and there was doubt that we could get through. Not very far out of the Springfield depot we were stopped where a big pin oak had toppled and caught the trolley wire with a limb. The car's crew broke out the emergency tool kit and hacked away at the offending limb. Dick also lent a hand to cut the branch which we then muscled off the track releasing the trolley wire which had not broken. The ITC crew inched the car past the downed tree and rolled on down the line. Arriving in Gillespie, we found that a tornado had hit the town at 6:48 that morning causing major damage. There wasn't much debris to block the electric car which ran through town in the center of the main street, but the line south to Benld was blocked. We were four miles from our job site but had to return north on the #284 car which was run around the short wye to become the next northbound train. Gillespie was a mess. One man said he was on the toilet in his basement that morning and looked up to see his house fly away. The local theater's marquee announced the showing of "Something In The Wind". Back in Springfield Dick and I caught the GM&O to Chicago.

The next Monday Dick and I returned to Springfield on the GM&O morning train and resumed our journey south on the ITC to Benld. The damage in Gillespie was shocking to see, but the rail line was open. The vision survey at Benld was routine and quickly done in time to catch the next northbound electric car which we took all the way to Peoria. The next week we were back again. This time we had a look at bridge #1734 a highway underpass that had been severely dented by a too-high load of machinery on a flatbed truck. Other than a bent lower flange on the girder, the bridge had not shifted or thrown the track out of line. That drainage area for the Mackinaw River bridge #1769 required another look, too. The next day we were north of Peoria at Akron, Illinois. On the way north we saw a grimy class J-4 Berkshire 2-8-4 and an equally dirty class J, 2-8-2 #2516 boosting a coal train up Radnor Hill. Near Akron we surveyed a timber span #1559 which was located almost underneath the

AT&SF's east-west main line overpass. We were entertained by the passage of an eastbound *Chief* passenger train, a westbound single motorcar local, and a pair of C&NW FTs southbound on our track. Completing our work, we got back to Peoria in time to catch the CRIP's afternoon *Rocket* to Chicago.

My first promotion came on March 30, 1948. On a six month temporary authority, I was appointed a rodman for the monthly sum of $291.48. The position was made permanent July 26th. I was launched on my life's first career.

The next six months for me was a period of great professional growth that I look back upon with many fond memories. A new diesel shop was to be built at the Chicago Shops. Authorized by AFE B-1366, the building with its modern diesel locomotive servicing concepts represented a huge investment and was heralded as an example of management's forward thinking. Designed by the architectural firm of DeLeuw, Cather of Chicago the general contractor was S. N. Neilson Company, also a Chicago firm. The C&NW's chief engineer had been the principal railroad officer involved in the design phase of the shop. Representing the chief's office was Maurice S. Reid, office engineer. Now the project was to be handed-off to the Galena Division for construction and finishing, and I was picked to be the division's man at the site.

Maurie Reid was a rising star in the engineering department. A graduate of Iowa State at Ames, Iowa, he had served during World War II in Assam as an Army Captain with the Railway Engineers Battalion. Afterwards, he had worked on several C&NW divisions and was now a principal engineer reporting to the chief engineer. With all these outside parties working on railroad property, he needed someone onsite to monitor activities, inspect quality of the construction and report progress to him on a regular basis. I was pleased to have been chosen despite my lack of experience and knowledge of construction practices. That lack was to be remedied over the next six months.

On this project I received a priceless education in building construction, materials testing, logistics, work scheduling, and engineering in general. I also learned to direct railroad work crews, how to deal with contractors, prepare progress charts, perform accounting audits, and to appreciate the value of quality control. Secondly, I got to poke around the many nooks and crannies of the old Chicago Shops.

The first day on the job was March 31st. Reid introduced me to DeLeuw's field man, Dan Kent (anglicized from Kvitinkas) who was to share our construction shack office. I soon got to know Nielsen's construction bosses who were to be found around the site directing the workmen. Construction was still in the excavation phase with most of the footings and foundations having been poured save for the east end where a very deep hole was being dug to accommodate a diesel locomotive truck drop pit. Most other shops being built in this early days of the diesel, used big overhead shop cranes to lift diesel car bodies off their trucks. Whiting Corporation, however, had claimed that their new method was better—support the engine car body on a steel table and drop the entire truck on an elevator device for transfer to another shop track for repair. It took a heckava deep pit to accommodate the drop and transfer bridge device.

The building itself was to be a framed steel structure with three through running tracks, one shop floor track from the east side and a short unconnected track to a degreasing vat. Five Kinnear rolling steel doors were on the east side and three on the west. The tracks were numbered from north to south; the shop track spur was number 1, the first through track, 2, *etc.* The running service tracks ran through the shop building on concrete piers which gave access to the underside of the diesels. Alongside the tracks were permanent platforms for easy access to the engine body. Just before the three running tracks reached the west entrance doors, they crossed platforms that could be elevated by electrically-driven vertical screws. The idea was to provide working and access platforms around the ends of the locomotives. In fact, these movable platforms proved to be a design mistake as they were often left in the up position when the engine hostler tried to move the engine. The mangled lift then blocked access from the shop to the west until it could be replaced. After a short time of operation, these platforms were taken out permanently. Within the shop, tracks 1, 2 and the degreaser stub were spanned by a gantry crane which facilitated removal and replacement of an entire diesel engine from the carbody. Pipelines carrying lube oil, wa-

The clean gravel ballast on the west side of the new diesel shop is a clue that this is opening day. The inactive Wisconsin coaling tower appears above the storehouse's roof on the left side while the active Galena coaling station appears over the new shop's roof in the center.

ter and gas ran under all the service platforms. Attached to the northwest corner of the shop was a two story stores and office building which in turn was attached to an old paint shop building identified in the ICC valuation inventory as M-19. (M stood for Mechanical). That is how the Accounting Department settled on the official identification of this lovely new building as M-19A since it was an "addition" to the older and much smaller M-19.

The outside finish of the shop was face brick and glass block. Fake stone letters identified the shop's owner as the "Chicago and North Western" and track numbers over each door were made of the same Tymstone material. Outside the building were concrete platforms between the service tracks. These pavements contained pipe lines for warm glycol to melt ice and snow in the winter. The piping later proved to leak badly and was disconnected. All in all, this shop project appeared to be state of the art and I was happy to be associated with it.

During April the first steel was erected in the main part of the shop on the 5th and was completed exactly a month later on May 5th. Even while that deep drop hole was still being formed and poured, the steelworkers knitted the building's frame together. Concrete floor paving began June 4th. My routine was simple. I caught the local at Elmhurst, got off at the Keeler Avenue (Chicago Shops) stop, crossed the pedestrian bridge, checked into my office in the construction shack, made my rounds by visiting with the resident engineer, the contractor and his subcontractors to see if there was any railroad involvement, and checked the jobsite and work in progress. If there was nothing in particular needing my attention, I was free to ramble around the huge sprawl of railroad shops. In these early days before the approach tracks to the shop were built, I had more opportunities to explore the Chicago Shops.

The Chicago Shops property and the Fortieth Street Yard together occupied a full square mile. The land when purchased in 1872 lay west of the City of Chicago's limits. In 1948 the boundaries were marked by Fortieth Street (later Crawford Avenue, and still later, Pulaski Road) on the east; Chicago Avenue on the north; the Belt Railway of Chicago's Cragin-Clearing railroad on the west and unpaved West Kinzie Street on the south. The Galena Division main lines ran along the southern boundary. The shop buildings extended across the lower third of the land; the Wisconsin Division's Fortieth Street Yard, occupied the northerly two thirds. By 1948 some of the prime sites along Chicago Avenue and the west side of Pulaski had been sold to companies like Hawthorn-Melody and N. Shure Electronics.

Steam locomotives were coming and going at all hours of the day and night to be serviced at the Galena Division roundhouse, M-17. This old brick circle was a gloomy, smoky cave with greasy wood block and brick floors, steam hissing, water dripping amid lights held by grimy overalled mechanics ministering to their iron charges here and there. Square timber posts held up the sooty roof through which smoke jacks protruded. Steam locomotives were to be seen in every state of readiness, or unreadiness. Smoke box doors swung open, pieces of valve gear laid out alongside, or the flash of a welding torch repairing some piping bespoke of the aging of these machines. Some were still hot from their recent trips; others were stone cold. The engine house foreman ran this dark and gloomy hell from a tiny overheated office near the gap in the ring. Engine reports on clipboards festooned the walls. While I was always welcome as a visitor, I was viewed with some uneasiness as a harbinger of a new way of life as the

Chicago Shops in July 1948. Engine house M-17 is in the background; the new diesel shop, M-19A, under construction on the left. Nestled between M-17 and the new shop was the old paint shop, M-19, that lent its official inventory number to the "addition" that dwarfed the original structure. In the right foreground an M class 0-6-0 and an E class 4-6-2 occupy one track while an E-2 simmers on the next track.

new diesel shop rose just to the south. Outside of M-17 was the main servicing area where incoming engines dumped their fires, got the ash pans cleaned, and moved to the turntable under their residual steam. Engines going to work were fueled, watered, sanded and greased on the outbound lead. Walking about this area had its perils from being slobbered-on by overfilled tenders, tripping over miscellaneous engine parts, slipping in gobs of olive-colored lubricant, or falling into the cinder pits which were located under the inbound and outbound leads. Between the leads were two tracks straddled by an overhead gantry crane. These tracks held the cinder cars which were loaded by the gantry which scooped up the dumped cinders from the watery pits. That clamshell coming up with a load of cinders was quite a sight with water spouting out in all directions as the crane positioned it over the cinder car. The pits themselves were the dangerous part. While rail people knew them, outsiders sometimes did not notice that they were indeed pits full of water and that thick layer of cinders floating on top was not meant to be stepped upon. Several of the contractor's men found out that the water was four feet deep and quite warm.

Huddled between the cinder pits and the concrete coaling tower were a couple of shacks used by the coal chute men. The Galena Division's concrete coaling structure was built in the late 1920s to replace an old coal dock whose remains we ran into when digging the new diesel shop's foundation. The coal from Macoupin County ended up here in cars on an elevated track which eased the labor of filling locomotive tenders on adjacent tracks. The twin concrete coaling station just a few feet away to the northeast belonged to the Wisconsin Division and was a monument to poor planning. Servicing of Wisconsin Division power was switched to Erie Street in Chicago after the engine terminal was torn down at Fortieth Street. The few Wisconsin Division freight and switch engines left were serviced now by the Galena while the passenger power was all handled at Erie Street. These concrete coaling towers represented an enormous cost for demolition and were allowed to remain long after their function was gone. The two tracks on the Wisconsin side were still in use in 1948 for car storage and the Lidgerwood track. This latter process involves reshaping locomotive tires without demounting them to be turned in a lathe. To reshape a tire contour, a brake shoe is removed, a tungsten carbide bit is fastened to the brake shoe mounting and the locomotive is slowly pulled by a powerful winch and cable along the track emitting great curls of blue steel peeling off the tire. Mechanics walk along the engine observing and adjusting the bit. It is easy to spot the Lidgerwood track—just look for a track ballasted with rusty steel curled chips.

Alongside the outbound engine track was a blow-off tank which looked like a cast-off locomotive boiler standing on end. Here, engines stopped and opened their steam cocks to blow sediment and water out of their cylinders causing clouds of gushing steam to shoot out of the top of the tank. Not using the tank resulted in blasting of ballast from between the ties and scalding clouds of steam endangering shop personnel in the area. The penstock for the bunker C oil used by the nine 4000 series class E-4 4-6-4s was also located in this area. The storage tank some distance away to the north required a heating unit to move the heavy black oil.

The parade of engines included class E, E-1 and E-2 Pacifics, 4-6-2s, class H Northerns, 4-8-4s, class R-1 Tenwheeler, 4-6-0s, class J,

JA and JS Mikados, 2-8-2s, class E-4 Hudsons, 4-8-4s, class M-1, M-2, M-3, 0-6-0s, and an occasional J-4 Berkshire, 2-8-4. The overall impression gained from seeing this mostly grimy crowd was that the end was coming. (And it did in 1956).

West of the new shop was the Streamliner Ramp where the jointly-owned West Coast Streamliners were serviced. Following their runs each train was tail-hosed the four miles from the Chicago Passenger Terminal to Crawford Avenue (yes, the railroad refused to change the old station name to Pulaski), and thence around the new shop site to the Ramp. The first service was washing the cars which was done by backing the consist slowly through a wash rack that could be positioned on transverse rails. During non-freezing weather a curtain of water and alkali cleaning solution was applied and rotating brushes scrubbed the car sides. Six of the ramp tracks had full length inspection pits under the trains. Usually by noon all of the Coast trains had been received, washed and inspected. An army of cleaning ladies went through the trains followed by resupply people. Minor repairs and adjustments were taken care of. At the west end of the Ramp there was a wye track, but its sharp curvature and light rail precluded its use to turn these trains of long cars. Turning the trains required a different approach. To the west of the Ramp a long stub track had been constructed parallel to the Galena Division main lines west under the BRC overpass, past the Brach Candy Company plant to a point on the elevated right-of-way far enough to hold the streamliner to be turned. The equipment was tail hosed on this extra track until clear of the switch at Kenton Avenue; then reversed direction running east to Kedzie Avenue; thence down the South Branch until the cut of cars cleared the switch at Lake Street; then it was tail hosed to Western Avenue and on to the Chicago Passenger Terminal ready for the evening departure. Curiously, most of the C&NW's 400 passenger trains and official business cars were serviced at the California Avenue Coach Yard.

Just north of the engine house was building M-16, the foundry. Here in stygian darkness, molds were formed for casting many of the car and locomotive parts used by the company. The lurid glow of hot metal being heated and poured competed with the timid light that managed to get through grimy window glass. The forming and shaping lines were illuminated by strings of suspended incandescent bulbs. The molds were blackened wood, the molding sand was black and included such esoteric things as ground up corncobs and clay. No hardhats were in use in 1948. This was a dangerous place, and I did not spend much of my spare time here. M-16 collapsed suddenly under a heavy snowload in 1949. No one was killed, but there were some injured. Another place that I did not visit very often was the powerhouse, M-4. Located north of the administration building which contained its offices and test lab, the big square brick powerhouse housed a self-contained electric generating plant that served the entire shops complex. A double track curved sharply around the south and west sides of administration building M-2 for the loads of coal that fired the boiler. The inside of the powerhouse was incredibly clean with bright painted surfaces in marked contrast to the shabby, gritty buildings surrounding it.

To the west of the powerhouse there was a row of six buildings built in the 1880s looking from above like a line of piglets feeding from their mom. The brick structures were huge old barns used by the Car Department. About 100 feet wide by 300′ long, these were the "C" buildings occupied by the Car Department. Some were open throughout; some had offices on a second level; some had storage lofts. Between some of the buildings were transfer tables, which functioned much like a longitudinal turntable, long enough to accommodate an 85 foot passenger car. The table pits were about three feet deep, but that could be really hard to tell as they were usually full of weeds. In 1948 they were rarely used, if indeed, they were operable. The first C building, labeled 1881, was logically identified as C-1, the next west was C-2, the last in the row was C-6. One of the structures had a different brick facade for some unknown reason; I believe that this one was leased to the Pullman Company for their base in maintaining the Pullman cars contained in the streamliner consists. The C buildings contained the carpentry shop which was a leftover from the days when passenger cars were made of wood, an upholstery shop which was the source of all those lovely green plush seats in the older commuter cars, a pattern shop containing a huge collection of no longer used molds, and storage of every kind of passenger car fitting that might be required. Because of the transfer tables between the buildings it was theoretically possible to roll a railroad car from the Paint Shop M-19 all the way west through the last C building. Buildings C-7 and C-8 were the most westerly Car Department operation. For some reason the axis of these structures was east-west rather than north-south like the rest of them.

The first building west of M-17, the engine house, was M-1, another barn of a structure. The east face of M-1 was really a series of double wooden doors each of which served a stub track from the transfer table pit lying between the engine house and M-1. Inside, the open area was dim and cavernous. The floor was cluttered with about every part of a steam locomotive except the boiler. There were bells on their arbors, pieces of valve gear, bearings all sizes, bins of metal pieces, pipe, number plates and headlights. Scattered around were the machines that turned, bored, shaped and welded the parts. Since there was no general lighting in the building, each machine stood in its own circle of light. Overhead, trusses supported the wooden roof, which leaked badly in rainy weather. The foreman asked me if I wanted a locomotive bell. Sure, I said. Take whichever one you can lift, was the reply. He knew that the bells weighed hundreds of pounds when still on their cast steel mounting frames. The next building west was a twin of M-1 numbered M-3.

To the north of the C buildings were more modern structures, M-14 and 15. Here heavy steam locomotive repairs were performed. Typical of most railroad back shops, the tracks coming in on the north side of the building were spanned by a gantry crane that could lift an entire locomotive if need be. In 1948 the shop was full of big power undergoing class 3 repairs, but this area was not as conducive to casual visits as some of the older places in Chicago Shops. I do recall that class H Northern, 4-8-4, were receiving one piece cast steel frames here. Tucked in between the M and C buildings were several smaller "S" buildings that belonged to the stores department. One, S-39, was an oil reclamation operation that chemically refined lubricating oils. The Testing Lab in the office building M-2 tested crankcase oil from the diesels as they had learned that trace metals discovered were valuable clues as to what engine parts were wearing and how long they might be expected to last.

Another peculiar and strictly unofficial method of identifying tracks in the west shops area were the names applied to them. There was Norway and Sweden, Iceland 1 and Iceland 2, Germany, Germany Proper and Mexico. The Wisconsin Division switch crews working from nearby Fortieth Street yard knew exactly where to place cars of incoming material provided you gave them the name of the track.

North of the Chicago Shops property was the Wisconsin Division's Fortieth Street Yard. This antique with its eleven foot track centers was one of the C&NW's original freight yards. Not only were the tracks too close together but the body tracks were curved in a big arc; the tracks at the west end were on an east-west axis but swung around to a north-south axis at the south end along Crawford Avenue. It mostly handled local Chicago traffic and stored equipment. This yard was a dangerous place to work. Because most of the switching activity was at the west end, the yard office was located near Chicago Avenue near the point where the Cragin line ran to the north.

The first phase of construction at the diesel shop, M-19A, ended when the roof was on and the brick walls began to rise on the steel frame. The second part of the project was to make it into a railroad facility by getting the running rails in place and hooking them up to the company's tracks. The running rails were produced by Jimmy Hicks at West Chicago and consisted of new 112 pound rail supplied by the mills without drilled bolt holes. The placement of the long rails onto the new concrete bit walls was a delicate job requiring careful inching and guiding of the 160 feet of rail through the cast-in steel U-bolts. The contractor's responsibility ended with placement of the concrete and the bolts; it was the railroad's job to install the rail. A problem immediately arose: both B&B and track forces claimed the work. The matter was settled when the B&B men got the rail inside the building and the track men built the track outside. Installation of the first rail taught us a lot about placing the composite steel and rubber cushion plates and shimming the rail with steel plates before tightening the clips. Exact lengths and placement were critical to accommodate the drop table and movable platform clearances at each end of the rail.

Outside of the building we had received a heap of switch parts in July that represented a special piece of track work specially designed for the east entrance of the diesel shop tracks. The engineering department called it a complex crossing—and it was. It was like a double slip switch with one curved leg. Instrumentman Dick Bailey was designated to establish the correct location of the switch as all other trackage would have to be placed to fit. Dick drafted me to help lay it out. We worked on the track plans in Chicago with a couple of interruptions for a visit to bridge #1936 near Girard and a quick accident survey at Zion, Illinois, on the Wisconsin Division at the special request of the law department. By mid-July my normal routine at the diesel shop was resumed until Dick was ready to stake the tracks. We did and redid the job several times because the contractor's site graders seemed to have a propensity for knocking over or burying our stakes.

On July 28th track foreman "Red" Surma and his extra gang stood ready to cut over a temporary bypass track so that the streamliner fleet could continue to reach the servicing ramp west of the new shop while we built the new tracks. The bypass freed-up the area we needed to build that monster complex crossing. The day started badly when the Capitol 400 broke down on the track to be moved. When it had been pulled clear, a little 0-6-0 #2629 trundled past. That was the last engine to use the old track which was scooted over and reconnected in time for an E6, #5002-A, to back through. Our work train with R-1 (4-6-0) #927 appeared with the switch parts and rails. At this point Dick and I realized we had overlooked a small detail—the new track was too close to the old Crawford Avenue station building. The building was then in use as the office of the superintendent of freight terminals, Nate Waterman. Quick action was taken by hurriedly summoning a B&B crew who sliced four feet off the overhanging eaves of the station building. A photograph taken then showed at least one engine movement past the half done project; the E6 5006B made it past, but it was close.

By August 15th the complex crossing was in place plus another #7 turnout. Ten days later Surma and his crew had the north side ladder in place. All the trackage on the east side was completed by September 10th. That ancient Crawford Avenue depot with its raw cut-off roof and an 1884 wooden passenger carboy used as a warming shed along the platform stood in sharp contrast to the shiny new shop structure just to the west. Another piece of railroad history left the scene when the locomotive blowoff tank near the Galena coal chute was loaded on a flat car for shipment to Nelson, Illinois, on August 15th. Since the tracks near the shop were to have concrete pavement next to them, it was necessary to "seat" the tracks to ensure they would not sink under traffic and leave the concrete too high. The engine house foreman sent a hostler over with a class J (2-8-2) for repeated short trips up and down the new tracks. The contractor then installed corrugated iron drain pipes along the edge of the ties, placed forms, set the glycol piping and poured the slabs. The tracks had been ballasted with pit run sand and gravel from the company's Algonquin pit so that the finished project looked very fresh and neat. The trackage west of the shop was uncomplicated and was in place by November 15th.

Going back to Nate Waterman—he was involved in one aspect of the diesel shop that marked him for what he was—a man whose vision was fixed firmly to the past. During the design of diesel shop in 1947, his input was requested as a matter of courtesy to his position as superintendent of freight terminals. He did not view the increasing fleet of diesels as a threat to his beloved steam locomotives which would live forever. He demanded that a water penstock be installed along the secondary main track that ran past his office at Crawford Avenue and behind the Keeler Avenue platform for watering eastbound steam engines. This required construction of a concrete vault with cover and extension of an 8 inch water main under the yard tracks and busy access track to the Streamliner Ramp. His rationale supporting this installation was simple: he wanted to provide water for eastbound stock train locomotives so that they could go directly to the Chicago Union Stockyards without dropping their trains to get water at the Galena engine house. To humor him the design team included the penstock in the building plans, but its installation was not high on the list. The construction was put off until a direct order was issued from the Vice President of Operation's office to the engineering department to build that penstock. Jack Goodwin, the vice president of operations, was tired of Waterman's whining. We sent the B&B forces to build the vault. It is still there. No pipeline was ever laid. Sometime later, NateWaterman, flattered me by asking me to join his staff as assistant trainmaster. I declined as that job was not my sort of life, for that matter, for anyone's normal life.

And then there was the infamous lube oil line that became my most trying experience on this project. Plans called for installation of a 4 inch lubrication oil pipeline from the new shop building to a pumphouse north of M-17, the busy engine house, a distance of some 300 feet. The steel pipe had to be protected from acids generated by the cinders expected to be found in the ground and a 2 inch steam line was to be attached to the pipe to make the lube oil flow easily in cold weather. The whole goofy pipeline was to be placed inside a treated wooden box and stuffed with insulation before being buried four feet deep. Construction was a nightmare. The whole pipe had to be in place and tested under pressure before any backfilling could be done. Since the pipeline route was right across the busy lead tracks to the engine house, we had an open excavation for sixty days through this busy ill-lit area. It was an open invitation to serious trouble. The plumbing contractor, M. J. Corboy of Chicago, did not much care if they caused any inconvenience or disruption to railroad operations. They had a contract and any delay because of need to protect the railroad would cost them money. I do not believe the pipeline could have been built in the 1960s, much less now, because of safety reasons. But, somehow, we built the miserable thing. The first problem came at the south end near the new shop building when we ran into the forgotten bluestone foundation of the old Galena coaling dock. Cutting through that took a lot of time and effort as explosives were not an option here. Going under the engine house leads was the next reason for heart failure. With the help of the B&B foreman, Nick Enders, I got some secondhand 8 × 16 inch bridge stringers that had come from the ore dock at Escanaba, Michigan, and stacked them up like mudsills to support the rails relying on the running rails to be stiff enough to bridge the narrow trench. That trench was just wide enough for the welder to crouch in it. Those iron ore dust encrusted timbers held up the class H Northerns well enough. We dealt with settlement by jacking up the track and inserting wooden shims between the stringers. After many nervous days and nights with red lanterns sprinkling the landscape, the darned thing was tested, boxed and buried by the end of September. I heard, years later, that the oil line was an immediate failure and had been soon abandoned.

On September 24, 1948 the long delayed overhead crane arrived on site loaded on three flat cars. Erecting this huge crane that spanned the entire width of the shop building was a very ticklish job because the roof had been installed in the early summer and there was not a lot of head room to work with. Two of the C&NW's track mounted American cranes were placed facing each other on track 3 in the middle of the shop. In unison they picked up the ends of the gantry, lifted it slowly as high as needed, and swung it carefully at right angles avoiding contact with the roof trusses to set it on its rails already in place high on the walls.

On September 15th, M. S. Reid was promoted from office engineer to division engineer on the CSTPM&O's Western Division. His replacement and my new boss was affable Les Deno, son of a B&B supervisor. Les was a good engineer and proved to be easy to work with. When he died in his sleep February 11, 1971, the C&NW lost a capable man.

The official dedication of the shop was December 16, 1948. The east entrance door of track 2 was covered with brown paper and sported a big C&NW logo. Among the throng of notables was president R. L. Williams and vice president Jack Goodwin plus others that I did not know then. Diesel E6 #5014A poked its nose through the paper and operation of the shop began. Except for inventorying the new track construction and correcting the station maps, my job here was finished. The very last bit of associated work was a trip to Nelson with instrumentman Bob Nelson on December 28th to establish a location for that blowoff tank we had shipped from the Shops on August 15th.

Apart from construction and engineering work, I was very interested in history and research. Stimulated by the old material I had discovered in the division office, I expanded my contacts on June 19th by boldly going to visit the company's Secretary, Mr. W. L. White, in his 14th floor office in the Daily News Building and inquire if there was some way I could borrow books or examine company records. While I am sure he was taken aback by my presumption, he invited me to come back on Saturday, and he would see if arrangements could be made. Upon the appointed day, it all came to pass. I was to have access to the vault and the records kept therein. Being careful not to abuse this privilege, I used the resources of the corporate records on several occasions to good advantage during the next thirty years. Mr. White, seeing my interest in data collection and research, asked if I would be interested in having the collection of H. W. Poor's *Manuals of Railroads* from 1871 to 1945. The Secretary's office was fac-

Opening Day November 14, 1948 for the new Diesel Shop, M-19A. E-7 5014A awaits the signal to break-through the ceremonial paper door.

ing a move to another floor in the building and did not have space for these valuable research volumes. The only stipulation was that I might be asked, from time to time, to look up certain items upon request. I agreed on the spot not having any idea of where I would house these fat green volumes. To my wife's dismay they occupied bookshelves in our living room for about thirty years. I thought they looked quite tasteful with their dark green cloth bindings and gold-leaf lettering. I used them infrequently over the ensuing years for research, and recall only one specific request from the C&NW for information—22 years later. (The books all went to the Railway and Locomotive Historical Society in 1992).

It was obvious that to succeed in this railroad business you had to not only to be interested in the craft but also to possess education credentials. I had completed about a year and a half of college work before coming to the railroad. Most of that college credit covered basic engineering requirements. Further college study would require selection of a career path and I now had my eye on exactly what I wanted—a degree as a civil engineer. The Illinois Institute of Technology (IIT) in Chicago offered evening division classes. I signed up for differential calculus in September 1948. Graduation would come in 1956.

Two more minor events in 1948 should be noted: the Brach Candy Company exploded September 7th causing some damage and interruption to Galena Division train operations at Kenton Avenue just west of the BRC overhead bridge; and the little *Pioneer* rebuilt for the Chicago Railroad Fair was parked outside one of the C buildings at Chicago Shops long enough for me to get a color photograph. Although new flues had been installed, the engine did not operate under its own steam when a company movie was later shot—it was pulled by a large track gang motorcar on a weedy spur track that ran from the Galena Division main north to the St. Charles School for Wayward Boys. A smokepot in the chimney provided enough fake smoke for the filming. The locomotive was destined to be exhibited in the Museum of Science and Industry as a permanent exhibit, but only after Frank Koval got the museum director, Major Lenox Lohr, to agree to keep it inside out of the weather.

I was beginning to feel comfortable in this new life and looked forward to the next year.

LEARNING THE TRADE, 1949 . . . My first assignment in 1949 proved to be a forecast of what the coming year held for me—paperwork and more paperwork. I was assigned to reorganize the so-called "flashlight" file. Just inside the dimly lit map vault used by the engineering staff of both divisions a set of rough wooden shelves held a dusty row of open top tin-sided box bins filled with copies of contracts and license agreements. Neatly folded and wrapped in that blue paper so beloved by the law department, each packet was about three inches wide and eight inches tall. Arrayed in no particular order the only clues to a document's identity were handwritten scrawls on the soft blue paper. These scribbles were not easy to read in the dim light of the vault so that a flashlight was required, hence the jocular term for the file. After seven days of sorting through these ancient documents, I gave up trying to devise a filing system to facilitate finding anything because it was obvious that many of these old contracts represented projects and permits either long completed or no longer of any importance. Since the originals were easily accessible and well indexed in the corporate secretary's vault just across Canal Street, I made an executive decision and put most of the dusty documents in the "round file"—the waste basket. What a collection this would have been for interested historians or railfans! There were licenses for maintaining trolley wires across the C&NW at Sterling and Belvidere (the interurban lines were long defunct by 1949), contracts for purchase of shop machinery at the end of World War I, repair of the brickwork of the tall chimney at the Chicago Passenger Station's power plant, many licenses for installation of pavements at many stations, private grade crossings on lines now abandoned, and similar odds and ends scattered all over the Galena Division. Why this crazy assemblage even existed was a mystery answered years later when an old hand recalled that a former division engineer fearing that the file clerk might lose these documents by burying them deep in some correspondence files ordered creation of this separate agreement file.

During January in Northern Illinois an inside job, even if dusty and seemingly unimportant, was preferable to working out in the cold. On one of the few nice days that month instrumentman Dick Bailey decided to make a bridge survey at North Aurora. Since he had his car with a working heater, we went to bridge #852 on the Aurora Branch. This branch line ran south from the main line connection at Geneva through Batavia to Aurora along the west side of the Fox River. The timber span was a small one requiring few measurements which were done quickly. In the first years of the twentieth century stiff competition with the new Aurora, Elgin and Fox River Electric interurban for local passengers forced the C&NW to institute frequent train service between Geneva and Aurora. From this local service a substantial Chicago suburban service developed. It was hard to believe that this now weedy railroad had seen some of the Division's fastest commuter trains before 1930. Class E light Pacifics hauled commuter trains from Aurora over this light rail and cinder ballast to Geneva and then made a fast run to Chicago in competition with the more direct route of the Chicago, Burlington and Quincy's trains. "First Aurora" and "Second Aurora" were known as ballast scorchers in the vernacular of the day. (The Aurora branch was abandoned January 29, 1982).

The lull in the January winter fooled no one, and we tried to take advantage of every good day. Our new instrumentman, Douglas E. Oakleaf, fresh from the Twin Cities, decided to inventory a rail renewal through Melrose Park and Maywood which had been completed last fall. The weather held, so we tackled another little task at Proviso on the 18th—locating a new 1¼ inch train air line at the west end of Yards 6-7-8. With the temperature hovering at 10 degrees measuring and note taking was challenging. The airline and compressor were supposed to speed-up freight train departures by precharging the train air line before the locomotives tied on their cars. Doug was another ex-naval officer with a propensity for making routine work as easy as possible. For example on a rail renewal inventory he counted several kinds of rail fittings in one pass. In addition to a regular clicker counter held in one hand, he carried stones in the other and, sometimes, even pebbles in his mouth. When ten of a thing were tallied, a stone went into his pocket. The transfer of pebbles from one side of his mouth to another accounted for another track element. At a convenient stopping place, he would spit out the pebbles, record the number, and then empty his pocket to see how many stones were collected. In this manner he was able to collect the pertinent information in one walk through. I was reminded of Demosthenes who was said to have improved his diction by declaiming to the sea with pebbles in his mouth.

My involvement with the diesel shop M-19-A was not yet finished. In the third week of January I was down in the depths of the truck transfer pit setting elevations for the B&B crews leveling the rails by shimming them with steel plates. Reading the rod required use of a flashlight. Trueing the rails was necessary to ensure that the Whiting machinery would not scrape the concrete walls or fail to match the rails in the shop floor above.

Paper, paper and more paper was the grease that moved this railroad's business. The Roadway Completion Report, form 1167, became a part of my railroad life for eight years. The railroad had suffered severe damage to its physical plant during World War II and was, in effect, worn-out. Recovery required dealing with a lot of deferred maintenance as well as improvements. Each renewal or construction required a completion report and, if applicable, map corrections. My old wire in-basket was usually stacked high with requests for completion reports. Writing one required a working knowledge of the Standard Systems of Accounts and Accounting rules mandated by the Interstate Commerce Commission. Additions, betterments, and retirements accounting was a complicated minefield of checks and balances. The railroad's accounting department was in a good position to influence the annual balance sheet through "adjustments" to the expense side of the ledger by shifting costs from capital to operating expense, or, depending upon current financial needs, shifting expenses to capital improvements. Management was not shy about adjusting the numbers. Government accountants were well aware of this creative accounting as it seemed the railroad's accounts were under perpetual audit. During bad winter weather which brought increased operating costs and usually reduced income, as much expense as possible was diverted to capital improvements. Conversely, when business was booming and operating costs stable, more roadway costs were shifted to operating expense to drive down the infamous Operating Ratio used as a measure of railroad efficiency. As an example, costs of a ballast job could be shifted to operating expense instead of to capital account #11 on the say-so of management who directed the divisions to report the labor and material costs as just restoration work rather than additional new ballast. ICC accounting rules were just so much swiss cheese when a determined management wanted to fiddle with the expense allocations. At our level on the division we stuck to our instructions and protected the reports filed by our work crews. All material used was charged out on written material slips and it was our job in the completion reports to account for all of the items reported. On a railroad things break and have to be fixed quickly often without formal authorization regardless of whether the work is capital improvement or operating expense. Our report writers learned how to argue with semi-knowledgeable accounting analysts and gimlet-eyed auditors to explain away discrepancies. We got quite adept at manipulating the system to best accommodate our urgent needs to keep the railroad running.

Although February had come, the cold winter continued unabated. In the milder climes south of Girard, Illinois, construction of the new highway underpass at Bridge #1936 was well along. The state's contractor had completed placement of the concrete piers and abutments which we had located in 1948. The steel girder spans were due to arrive soon and it was our responsibility to mark the location of the anchor bolts atop the piers. Instrumentman Dick Bailey and I met Ray Doty, the B&B Supervisor for the Southern Illinois district, in Girard when we got off the Illinois Terminal's interurban car from Springfield. Ray took us down to the bridge site, but the short winter's day did not leave much time-or light-to establish proper centerlines. Just getting up the steep embankment to the mainline track at that place was a struggle when burdened with the transit on its awkward wooden tripod as well as all our other surveyors equipment. Complicating the set-up, the main line track crossing the new bridge was on a curve. Getting out on the free-standing piers was an even greater effort. The weather overnight had not cooperated either for the next morning we found sheet ice coating everything. Ray produced an old wooden ladder he had "liberated" from somewhere and clambered up to the pier top. The ice created a very dangerous place to try and do delicate measurements much less making any kind of a mark. Ray tried pouring gasoline on the ice and igniting it. The ice was not affected in the least. We gingerly chipped at the ice with a chisel and managed to score the concrete where the anchor bolts were to be installed—all thirty feet in the air. Finally, Ray was satisfied that bolts were in the right place and arranged for a crew to drill and set them. He only relaxed later when the bridge girders all fit exactly as planned.

Returning to Chicago I helped instrumentman Doug Oakleaf chain the new tracks at the diesel shop for the completion report. The weather had not warmed so that extra layers of clothing were necessary. Even bending over to position the chain was an effort. The wax crayon "keel" we used to mark positions on the rails left only faint impressions on the cold steel rails. A few days later on the 11th I was on my own in Elmhurst doing the field work for the completion report covering installation of new switch heaters at HM Tower. HM was a mechanical "armstrong" interlocking plant that controlled the number 20 switches with iron pipe rods and bell cranks. Here the three main lines from the west merged into two passenger mains to Chicago and two freight mains to the giant Proviso Yards. In addition the towerman also manually operated the crossing gates at the Haven Road crossing that bisected the maze of tracks. The levers, rods and cables in pipes and over pulleys to line the switches and to raise and lower the gates were a classic example of mechanical ingenuity. Inside the tower on the second floor was the long row of four foot long levers whose shiny bronze tops bespoke their constant use. A diagram board hung over the lever array. Signal bells and indicators announced train arrivals from east and west. The operator had a full array of telegraph and telephone equipment at his desk. Coupled with his tower duties was a responsibility to inspect passing trains. This was old time railroading that was soon to pass. Since it was located in my own town, I often made unofficial calls at HM to absorb the atmosphere and hear about train operations on the division. (After the Union Pacific merger in 1995, HM was renamed Park).

In the previous chapter the soil testing episode at bridge #66 east of Geneva where instrumentman M. C. Christensen and I had lost a drilling auger in the limestone rock under the west abutment was related in some detail. Now the bridge department had completed plans for reinforcing the old stone abutments with concrete buttresses and had selected a contractor for the job. For some unknown reason (probably because I had been involved in sampling concrete on the M-19-A diesel shop project) the engineer of bridges, Arthur Harris, requested the chief engineer to assign me to design the concrete mix and supervise construction at the bridge. The division engineer was a bit miffed at having one of his staff assigned but was told

Bridge 66 across the Fox River at Geneva, Illinois. The tall original limestone back-walls needed concrete buttresses to curb a tendency to bulge out. Erik Borg Company of Chicago performed the work under the direction of the Galena Division Engineer in 1949. This view was taken of the east abutment May 19, 1949.

that the work was on the Galena Division and was not a full time job. I was a bit awed by the prospect of having responsibility for work on a key bridge which carried both mains across the Fox River. Saying that my knowledge of concrete design extended from nothing to zero, Art Harris handed me a Concrete Institute booklet on the design of concrete mixes and shooed me out of his office. The booklet was quite detailed and contained formulae for calculating ultimate strength based on ratios of aggregate, cement and water. In 1949 research on concrete was largely based on past practice and not on much science. The object on this job was to obtain a hard, dense concrete without making it too stiff to pour into the forms and around the reinforcing bars or using so much cement as to bankrupt the contractor. I learned the secret was in the water-to-cement ratio. I studied the handbook carefully before meeting with the contractor, Erik Borg and Son, at the job site. The old limestone abutments were huge blocks laid horizontally in tiers extending almost forty feet above the river level. There were definite signs of bulging. Failure would have created a marvelous heap of crumbled lime rock under a collapsed steel girder bridge. We were to build three concrete pylons or buttresses against both east and west abutments to hold up the old stone walls. To ensure the new work would not tip forward from the pressure, excavation to the underlying bedrock was necessary so that the new pylons could be keyed securely. The first concrete was poured on February 18th; the last in May. Concrete samples taken randomly and tested in the C&NW's laboratory at Chicago Shops routinely tested above the design specifications. One sample came in at 7,000 psi (pounds per square inch) which was far too strong, indicating that there was too much cement in the design mixture. Although I had used an on-site scale to measure weights of aggregate, and cement to produce a workable yet strong mix, the workmen didn't always get every batch just the same as there were just too many variables. Fifty years later, given the great strides in concrete technology, we would have used computer-designed mixes, transported the wet mix in transit mixers to a concrete pump and finished the work in a lot less time and, probably, at much less cost. Train crews operating across the high bridge had no idea of what we were doing down below because no slow orders were ever deemed necessary during the work. In 2003 plans were well underway to install a third track bridge on the existing south shoulders of the river piers. Modification and extension of the old limestone piers was also involved.

Just east of Bridge #66 the westbound main track descended a long grade as the railroad dropped down into the valley of the Fox River. Heavy westbound freight trains coming down that hill with their mass, heavy braking and train dynamics translated into rail creep so pronounced that a number 10 crossover joining the east and westbound mains just east of the bridge was "cocked" into unusability. The crossover provided access for switching moves to the East Batavia freight line that served both the Geneva Girls School and the former Erie Basin Ordnance plant in East Batavia two miles to the south. Nick Maas, once the section foreman at Geneva and later roadmaster at Proviso, said that the most difficult job the section faced during World War II was keeping that crossover operable. It was the section's regular routine to open a joint in the main track rail and drive the whole string of rail eastwards up the hill by taking a spare rail from the stockpile and manually swinging it like a battering ram against the open end. By 1949 installation of better rail anchors (anti-creepers) changed train movements facilitated by the new Centralized Traffic Control, elimination of most of the rail traffic to East Batavia, and the greater use of diesel locomotives eased the strain on that crossover. When the State reformatory replaced their coal-fired facilities with natural gas, there was no further need for the East Batavia line which was then abandoned along with that troublesome crossover.

March in Northern Illinois usually means winter is about over. Field jobs were saved up over the bad winter weather, the in-baskets were overflowing with requests for services and the men were anxious to get out. It is hard to chain track, measure and describe completed projects when snow and ice cover the work. Since surveying at that time was all done with optical instruments, we were vulnerable to cruel winter winds tearing-up our eyes as we vainly peered into frosty lenses trying establish sight lines while the rodmen tried to hammer wooden stakes into frozen ground. While we waited for the weather to moderate, we worked on the never-ending paper flow, corrected maps and prepared special accident drawings for the claim department. As the March days passed Chicago's false spring found us opening the office windows to get some fresh air. Unfortunately, the only windows were directly over tracks 14, 15 and 16 at the end of the train shed. Inbound commuter trains off the Milwaukee subdivision with their class E light Pacifics stopped just under our open windows belching a stream of soot, cinders and gas which stray gusts of wind deposited on our clean white drafting tables. Brushing off the offending deposit usually resulted in a black smear especially when using the back of ones hand. The remedy for cleaning a smeared drawing was simple—white gasoline. We kept a red one gallon can with a spring-loaded closure handy because the gasoline lifted the grime and graphite pencil marks but didn't affect India ink. Given that most of the men smoked in those days, the presence of volatile gasoline presented a serious fire hazard. Whenever the joint fire inspection team of railroad and City of Chicago firemen appeared, the first thing on their list was inspection of our gasoline can.

Air conditioning our engineering office in 1949, even using window units, was impossible. The power supply in the Chicago Passenger Terminal came from the coal-fired powerplant at Clinton and

Lake Streets just north end of the trainshed. Since the generators only produced direct current, the CPT had no alternating current. In 1908–1911 when the CPT was built, the C&NW had its own coal mines and the local power company was not thought to be very reliable. Therefore, the decision by the railroad to be self-contained as far as its electric power requirements may have been the correct choice, but alternating current turned out to be the more universal choice. The powerplant was maintained in spotless condition. The soaring brick chimney was periodically inspected and carefully repaired as shown by the many contracts in the "flashlight" file. A short rail spur from track #1 held two coal cars on an elevated structure inside the power plant. Switching the plant required a runaround move by the switch engine using the puzzle switches in the station's throat. By 1949 the direct current precluded usage of the most common consumer appliances like fluorescent lights, AC motors in air conditioners and refrigerators. Years later when Commonwealth Edison and the Railway agreed to convert the CPT to commercial alternating current, transformers were installed under the sidewalks on Clinton Street south of Randolph and window air conditioning units then could be used.

Throughout March I worked on a variety of little jobs that included writing the completion report for the diesel shop, a field check on clearances at the Mandel-Lear warehouse and other structures near the Ogden slip on North Pier, prepared blueprints for sanitary lines in buildings C-1 and C-2 at the Chicago Shops, made trips to the work site at bridge #66 with new rodman Jim Andrews tagging along, worked in a day trip to Belvidere, Illinois, with another new rodman, Robert R. Lawton, to determine weight of rail in the Freeport Line main track in the east end of town (we found 72 pound rail on the low side of the curve and 80 pound in the high side, just the opposite of what was expected), ran levels on the sidewalk along the west side of Canal Street for the law department which was resisting a suit concerning alleged injury from an uneven pavement, went on an emergency trip with instrumentman Christensen to check damage to a girder bridge #S-1623 when a City of Chicago ditching machine being transported on a flatbed truck struck the underside of the steel (bent but not shifted), assisted instrumentman Art Olson on an accident survey at the 11th Avenue grade crossing in Maywood, measured vertical clearances at overhead bridge #81 west of LaFox, and finally, on the 28th helped instrumentman Bill Polchow set grade stakes for the new diesel shop to be built south of the Proviso round house.

Late in the evening of the 28th of March, 1949, instrumentman Dick Bailey, new rodman William N. Lipsitz and I boarded Gulf, Mobile and Ohio's train #7, the *Midnight Special,* for Springfield, Illinois. The sleeper was available well before train departure at 11:25 p.m. and would be set out in Springfield about 4:30 a.m. allowing us some motionless sleep before getting up in time for breakfast and a connection to the Illinois Terminal's southbound train to Benld. Our project was a pipe dream of the slowly vanishing Superior Coal Company, a subsidiary company of the C&NW. Deeply concerned over the continuing loss of market for their high-sulphur Illinois bituminous coal formerly used as locomotive fuel, Superior proposed to construct a new coal grading and washing facility to serve the electric generation industry. They owned a lot of surface land in addition to their mineral holdings around Benld, Illinois. Our job was to run survey lines and make a topographical map of their land lying east of the C&NW mainline and north of the old yard at Benld. Their property had been surface mined twenty years before. The despoiled land consisted of hummocks covered by a thick growth of small trees and thick underbrush. Not realizing the tangled nature of the land over which we were to work, we had not properly equipped ourselves for hacking our way through the undergrowth. But, we did what we could and established a baseline through the site of the proposed washer and recorded elevations along the line. We stayed at an old-fashioned hotel in Gillespie, the next town north of Benld. An obliging taxi driver took us to the work area and collected us at the end of day. On Friday the 30th while waiting at the Illinois Terminal station in Gillespie for our northbound electric car, we saw the ITC's articulated train set #301 and car #284, the *City of Decatur,* rumbling south down the main track located in the middle of the main street. Our northbound car soon came along for an uneventful ride to Springfield and our connection to the GM&O back to Chicago. We worked those field notes into a set of drawings and spent a lot of hours calculating cut and fill cross sections for grading only to learn that Superior had abandoned the project.

Bill Lipsitz was one of the several Northwestern University co-op education students. The program offered participants two quarters of classroom at the Evanston campus for each quarter worked on the C&NW. They usually worked as rodmen and usually on the Galena Division. Bill was not happy with his surname and, later, legally changed it to List. Like all the other Northwestern co-op students, he chose to go somewhere else after graduation for his professional career. When last heard of, Bill was working as a salesman for Powers Regulators Company in Chicago.

When the winter's frost finally left the ground in April, pending track maintenance programs sprang into action. During the winter months mainlines were monitored daily by track walkers looking for bad joints, broken rails, misaligned track and turnouts, deteriorated grade crossings and uneven running rails. The only method for leveling track when the ballast was frozen was insertion of wooden track shims between the hanging tie plates and ties. When spring came, one of the first orders of business was getting all those shims out of the track and correcting the tie support by tamping ballast under the low ties. This was hard manual work done by strong backs using ballast forks when working in rock. Those same strong backs used square shovels when dealing with gravel or cinder ballast. The day of the hydraulic powered tamper was yet to come. In the Chicago Terminal area a few handheld power tampers could be found powered by air compressors mounted on a small four wheel tool car.

The C&NW emerging from bankruptcy near the end of World War II faced a huge backlog of deferred track maintenance. Ties, rail and ballast were the basic elements of track. Translating them into good railroad track required manpower and a lot of it. In 1949 power equipment was being developed but was not yet available so the old fashioned gangs of workmen were organized to get the work done. Gangs of 150 men were common. Out on the railroad daily track maintenance was performed by the section crews. Headed by a foreman these small five to seven man gangs operated from their section house using rail motor cars (sometimes referred to as speeders) to get to work sites in their assigned territory with their tools and supplies. Again, power tools were rare. Perhaps a gasoline powered hacksaw to cut rail could be found, but the old square shovel, ballast fork, spike mauls, lining bars, track wrenches and spike pullers were the basic tools each section depended upon. With the gradual advent of mechanical powered track maintenance equipment, the old section system was doomed. The first signs of change were expansions of the territory assignments for the individual sections which often eliminated some sections or shifted their base of operation. Then came reduction in forces which at the minimum could be just a foreman and one or two men. Since the foreman was only supposed to supervise work, the labor burden fell on his minuscule crew. These tiny crews often clubbed together to help out on track maintenance. As assigned company highway trucks became more common, the tiny section crews became more mobile, but were still an inefficient use of manpower. Eventually, the section crews were totally eliminated. Daily track patrols were then performed by an assigned track inspector who traveled the territory on his speeder at regular intervals. He did not do much track repair leaving them to a floating gang to correct. In 1949 the first section territory expansion and consolidation program began. Management slowly began to recognize that too many foremen and not enough laborers was a poor way to successfully maintain track. One negative result was the removal of a physical presence out on the line. Given the slow communications available in that day, the abrupt loss of local information was soon noticed but shrugged off as the price of progress. The regular inspection of track was severly compromised.

When the big track gangs were working, calls for engineering assistance increased. These track work programs resulted in long days in the field for our staff as we established centerline stakes for curves and spirals and some tangent track, set top-of-rail stakes for ballast programs, stringlined curves, set grade elevations for public crossings and bridges because of the raising of main line tracks on the ballasting and surfacing jobs. Our field men gained healthy tans after a summer of this very physical work. They also had the dubious opportunity of staying in every fleabag hotel on the division, eating in whatever beanery was handy and entertaining themselves in the evening as young men are wont to do. In the Galena Division engineering office staff the longest work assignments tended to depend on the marital status of each man. The young single men were most amenable to staying out all week with the track gangs while the married men tended to take the close-in work and go home at night. I was married and attending night school which were two good reasons not to spend my summer out on the line. Surprisingly, it all worked out to everyone's satisfaction with a minimum of inconvenience.

My April timebook shows field trips with instrumentman Bill Polchow to Dixon on the 5th to stringline curves, to Ashton on the 7th to stake the new #20 turnouts for the CTC project at AE (Ashton East), to Proviso on the 11th and 12th to establish the trackage to serve the new diesel shop then under construction and on the 13th to West Chicago to set line and levels for a track there.

On April 14th a new instrumentman, Ottmar W. Smith, hauled me off to Geneva to set grade stakes between bridges #65 and #67. Bill was freshly arrived from Michigan and we ran afoul of Illinois traffic laws when he ran a flashing red light at a school crossing. Flagged down by the policemen, Bill turned on his boyishly innocent face and solemnly explained that a flashing red light in Michigan merely meant slowing down. The cop accepted this explanation and waved us on with a word of warning about Illinois law. Bill later became chief engineer of the Missouri-Kansas-Texas Railroad after a long career on the C&NW, more of which will be related later in this account.

My records for April 20th showed Bill Polchow and me doing curve relocation work at that troublesome #20 turnout at WX west of West Chicago. At this CTC control point main track #1 (the southernmost track in those days) curved into a #20 turnout which was not quite correctly placed when installed in 1947. Bill and I worked out some approach curves to compensate for the original location error. On the 22nd instrumentman Bob Lithgow was pried out of his comfortable office chair by the claim department to investigate a four car derailment on the Freeport Line west of Winnebago, the first station west of Rockford. I accompanied him and the claim agent to the site of the accident. We found the boxcars were old fashioned wood and steel composite automobile cars. One of the them was upside down and full of new Chevrolets automobiles; the others were upright amid a debris field of twisted 80 pound rail, broken ties

A derailment April 22, 1949 near Winnebago, Illinois, on the venerable Freeport branch line, the 80 pound rail and indifferent tie condition failed to keep these merchandise cars rolling toward West Chicago.

and heaps of cinder ballast. Finding the cause was not difficult; it was a broken rail.

April 25th was the day for setting levels on the pit walls at the new Proviso diesel shop. The DeLeuw, Cather resident engineer was Tom Ek, a native of Alaska. The two of us set the levels on top of the concrete supports for the running rails through the shop. Just as at the Chicago Shops, the new diesel facilities were built on swampy land. At Proviso the site was just south of the round house and north of the so-called 19 and 20 mains. The railroad was its own contractor for track construction outside of the shop building and did all the grading work. Instead of removing the mucky soil, the chief engineer decided to fill the area with gravel from the company's Algonquin pit and squeeze the latent ground water out into an adjacent drainage ditch which carried the water to Mud Creek and Addison Creek at the far east end of the Proviso yards.

The best news I got in this period was a pay increase: rodmen's monthly pay went to $308.51. This was $100.00 more than my original goal in 1947 of getting a job that paid $200.

Curiously, during the good weather of May, I had few outside expeditions. But one warm day Bill Polchow called for help in staking mainline tracks in Lombard. My part of helping was to bring the field crew a supply of stakes. Our store of stakes was kept down in the bowels of the Chicago Passenger Terminal. Manufactured by the B&B department in their spare time from rough oak lumber, the piles of oak were heaped in a dungeon-like basement area that reminded me of an abandoned coal bin. Collecting the loose stakes in bundles and tieing them with coarse hempen twine was the work of a few minutes. Each bundle contained about 60 of the 1″ × 1″ × 18″ oak "hubs" weighing probably 50 pounds. They had to be humped up two flights of stairs to the train platform. Lombard was just 20 miles west which the commuter train covered in less than an hour. Since Bill didn't need me, just the stakes, I escaped back to Chicago on the next eastbound scoot. Sometimes we had to kick the bundles off a moving train when it did not stop at the work site. Insecurely tied bundles sometimes disintegrated on impact sending the oak stakes flying all over the right-of-way to the consternation of the junior party members who had to scurry around and collect them. The only other job noted in my time book was staking a curve in a new track at the sanding tower near the Proviso diesel shop on the 10th. Little trips, little jobs, but all necessary to the business of the railroad. On the 16th I was sent out for a short walking inspection across the Chicago River to the Merchandise Mart to look at the site of a proposed stairway at Wells Street. The next day my field trip was to East Elgin to examine the proposed removal of the Williams Bay branch's main track through the city. The Williams Bay line began its northward route at Foris (Fox River Switch) just south of town. The old line running through the business district of Elgin east of the Fox River offered a wonderful place for downtown automobile parking. Switch crews were often delayed because of the automobiles parked on or too close to the tracks. Sometimes the crews waited for the owners to appear and sometimes . . . On the 19th I checked the concrete work at bridge #66 at Geneva; the next day inventoried the new tracks at the Proviso diesel shop. On the 24th I rode train #13 to Clinton, Iowa, to gather eight miles of ballast soundings between Mileposts 128 and 136 (Union Grove to East Clinton). My designated digger was a section hand who labored to dig about 50 holes in the rock ballast at selected quarter mile points. This onerous work was required only because of an accounting department *diktat.*

It may be appropriate to review the *raison d'etre* of the fusty old Auditor of Capital Expenditures (ACE) department. During my early days and up to 1956 C. H. "Chick" O'Hearn was the auditor of capital expenditures. This kingdom of the bean counters was located on the 11th floor of the old Daily News Building at 400 W. Madison Street. The narrower tower section of the building extended upward from the wider 9th floor. Accounting's domain on the 11th was a huge open space for the north half; the south was occupied by the elevator bank, official offices, stairwells and lavatories. The visitor coming into the open north half of the floor was greeted by rows of desks covered with papers, ledgers, in and out baskets, filling the entire space from south to north. Ensconced behind each desk was a clerk, all men, seemingly engrossed in paperwork. A first impression recalls Ebenezer Scrooge's clerk's room. On closer inspection one could often glimpse crossword puzzles, magazines and other reading material mixed with the official work papers. One strange little fellow always erected a barricade of standing ledgers on his desk, insisting that he needed the additional privacy to concentrate on his very complicated duties which involved signal and communication equipment costs. Yes, he wore a green eyeshade and garters on his sleeves. And, he refused to share his expertise with anyone.

Each person was assigned to deal with costs associated with a particular function. One might be the Acct.20 man—Shops and Enginehouses; another, Acct.6—Bridges, Trestles and Culverts. But, why would such a department be dedicated to such a level of minutiae? Because of the Interstate Commerce Commission's insistence on a standard system of accounts for railroads.

In the 1870s midwestern farmers in particular began to organize to fight the railroad trusts which were, they felt, strangling agriculture with oppressive freight rates. Calling themselves Sons of the Grange, or Grangers, they gradually gained organization and political savvy until their collective voice was heeded by Washington. Not only had the Grangers complained about long and short haul rate inequities, but they were convinced that the railroads had systematically "watered" their common stock by issuing far more shares than the capitalized values of their properties. The Interstate Commerce Commission was established by Congress in 1887 to address these complaints and establish "reasonable and just" rates. At first the new agency was ineffective, but gained muscle in 1906 with passage of the Hepburn Act. The capitalization question was to be resolved by an inventory of all railroad properties. The ICC issued instructions to all common carriers that they would furnish a physical description of their properties. The instructions included the method to be pursued, the form of the documentation and the final maps to be presented for official acceptance. The intensive survey work was organized in 1914 and field work soon begun not to be completed until after World War I.

Description and tabulation meant *everything* had to be measured and recorded. The actual field work was carried out by teams of railroad and government men. Land for right-of-way, station grounds, yards, terminals, tunnels, quarries, tie plants, and every other railroad use had to be identified and given parcel numbers. On the final maps each parcel was identified and, in a legend in the upper left corner of the map, described in detail with original owner, size, date acquired and form of ownership claimed. All main line tracks were chained (measured) so that the exact length could be ascertained (to the fraction of a foot). The chaining could begin at one end and carry through without break, but this was often inconvenient with several chain crews operating so that meeting points were established where each parties' could begin or end their measurements. These were called equation stations. The common measuring device was a standardized one hundred foot steel tape (chain). The common quarter inch was the one we used in our work, but I own a lighter 3/16″ model that had been employed in the track elevation project in Chicago.

Not only tracks were measured. Every railroad structure was measured and recorded—stations, freight houses, coaling docks, ice houses, section houses, tools, privies, gasoline storage shanties, switch locks, switch lamps, water penstocks, water tanks, loading platforms, bridges, culverts, fences, plank crossings, signs, locomotives, shops, rolling stock, in a word, everything. The centerline of the bay window of the depot was deemed to be the actual "place" so that the town of Arlington, Nebraska, for example, was concentrated on the face of the earth as being the magic little spot in the center of the main track just opposite the middle of the station telegrapher's bay window. Talk about a geographic positioning system!

Account #3 in the ICC's Uniform Systems of Accounts was Grading. This meant that the inspection teams also had to determine the cut and fill quantities mile by mile, and record the estimated quantities of dirt, rock and other material required to build the line. Ballast (Acct.11) was also measured and recorded by type and quantity for each mile. Whatever had been an expense in construction—fencing, telegraph lines, signs, public grade crossings—was accounted for in standard formats

The Auditor of Capital Expenditure's office maintained a vault on the west side of the floor. Here the original written R&T (Roadway and Track) field survey records were kept in faded blue cloth-covered binders—shelf after dusty shelf. They were seldom needed for reference and were just a snapshot of railroad facilities in the 1915–1919 period. More important were the maps which portrayed the lands, tracks and structures. The standard valuation maps were 24″ × 56″ lithographed "vellum" (starch covered linen) with a half inch inset border leaving a drafting area of 23″× 55″. The map legend in the lower right was standardized with the name of the railroad in bold letters followed by the operating division name, the location depicted, valuation chaining station limits, scale of the drawing and the date prepared. A one inch circle in the lower right hand corner divided by a horizontal line identified theValuation Section number in the upper half and, in the lower, the sheet number. A large scale territorial map functioned as an index and showed an overlay for each of the smaller scale right-of-way and station grounds maps. When the original maps had been prepared, they were signed by railroad officers and notarized before being submitted to the ICC along with the detailed inspection reports. The map scales were either 1″= 100′ for station grounds or the larger 1″=400′ scale for trackage between stations.

Other than valuation maps, there were at least three major types of railroad maps in general use on the C&NW in 1949. One was the territorial map at a scale of 1″= 1 mile. A second type was the right-of-way maps at 1″=400′ which appeared to be copies of the ICC valuation maps. The station maps at 1″=100′ with their great detail were the most useful to engineering staff and varied greatly in lettering styles and format. The only common feature was the main line track always occupied the center horizontal axis of the drawing cloth. A very useful map was the large scale (1 inch = 1 mile, or even larger) territorial map. The main track was shown as a single bold line that cut across the many base meridians and township lines established by the U. S. Public-Land Survey in the 29 states west of the original 13 colonies, Kentucky, Tennessee and Texas. Railroads in those areas were not blessed with these rectilinear survey lines that simplified land ownership description in C&NW territory. On the 400′ scale maps main tracks were shown by a single solid line and railroad land ownerships were outlined by heavy lines interrupted about every three inches by two dots. Bridge backwalls, stations, signs, tunnels and many other features such as township lines were identified by their valuation chaining. The right-of-way maps were very handy and were kept in use on the C&NW more than half a century after their creation. The chief engineer's station maps were not as up-to-date as the divisions' maps which were maintained rigorously to reflect actual conditions. On these 100′ to the inch scale station maps, railroad tracks were shown by double lines. Structures shown were drawn to scale along with other features that might influence operation of the railroad.

Accounting had a set of the valuation maps but never made any corrections to them. They were frozen to the 1915–1919 period. The land department's set was maintained to reflect current land ownership as it was convenient to have a set of standard size drawings rather than depending on the ancient and different formats of land records created by the former railroads later to became part of the C&NW. Some of the original railroad maps were drawn on a stiff cardboard called "hard shell". To preserve these fragile works of art with their finely tinted colors and spidery lettering, they were rolled in cylinders about 6 inches in diameter and tied with string before being stored in steel drawer cabinets. Engineering received two sets of the lithographed valuation maps: one was retained by the Chief Engineer in his drafting room, another set went to the appropriate division engineers' offices. The chief engineer updated his maps as reports of completed projects were received from the divisions. The division engineering office updated their set as work was done. For the most current view of any portion of the railroad, the local division engineer's station maps and valuation maps at 1″=400′ scale were the best source.

This standardized and ordered life imposed by the ICC required that every track in each valuation section was to have a unique number. The main lines were identified to the fraction of a foot by their chained lengths. The sidetracks and industrial spurs each were measured and identified by a unique ICC track number. Accounting set up their capital books with a ledger account for every sidetrack so that each one was supposed to reflect the exact quantities contained in that unit—so many feet of rail of each weight and manufacture, so many angle bars, tieplates, insulated joints, derails, compromise joints, end of track bumpers, size and weight of turnout, length of points, kind of throwing device, hundredweight of spikes, kind of ballast, if any. Funny thing, despite the depth of detail, they were terrible records.

After all this detail work, intensive staff review, solemn pronouncements, and years of bureaucratic review, most railroads were officially declared to be under-capitalized. But, more far-reaching in effect was the order to continue capital accounting based on the 1915–1919 inventory. All improvements or changes to physical plant had to be viewed as an addition or betterment. For example, replacement of old 90 pound rail in a track with new 100 pound rail required that the 10 pounds of extra metal in every yard of track was a capital improvement while the 90 pounds was merely a replacement and, therefore, an ordinary operating expense. The additional weight of 7×10½ inch tieplates replacing 6×8s had to be capitalized because, in the accountant's opinion, tieplates never wore out when in reality they do. Rock ballast replacing cinders was a capital expense. Gravel replacing gravel was an operating expense. I learned that the only reason I had to dig those dratted holes in the main tracks in far away places on the division was only the accounting department's attempt to justify how much of ballast job was replacement (operating expense) and how much was a capital addition. When I got the chance, I fixed that little game with some imaginative work of my own.

According to the official records many side tracks were listed with four tieplates per tie when two were normal. Lightweight rails, long since melted down and made into nails now in the side of some farmer's barn, were still in service according to the auditor of capital expenditure's books. How could such closely controlled system of records get so screwed up? The very nature of a railroad conspired to undo this most carefully thought-out system. For example, who can predict a wreck or a derailment? The need to restore rail service promptly sometimes required grabbing the nearest rail supplies which might not be the same size as the destroyed material. Or, the completion report for the annual blanket AFE for tieplate replacement was usually a creative fiction made up by the division office which didn't have a clue as to where tieplates had been installed during the past year and arbitrarily charged them to tracks all over their division. When estimating a job, the prudent engineer learned to go out into the field to see for himself what was actually in place before estimating a job because the accounting department's records were so bad. Knowing the condition of their records, the auditors on the 11th floor actually welcomed our roadway completion reports. Betterment accounting for the railroads lasted until 1981 when rule changes expanded depreciation to such formerly "unchanging" items as grading and tunnels. While the new Surface Transportation Board (STB) adopted the accounting rules of the now defunct Interstate Commerce Commission, there were changes made to make them more sensible.

For the big capital expense projects an accountant from the auditor of capital expenditure's office could sometimes wheedle permission to leave the office and actually go out on the railroad to see the project he was analyzing. My completion report (form 1167) for the Centralized Traffic Control installation on the Galena Division ended up as a book of 50 pages. Covering only the details of track work (Acct. 27 Signals and Interlockers was a special case) each crossover was set-up as an individual ICC track which was completely described down to the last rail anchor. Each location required a map showing details of position and length so that the chief engineer could keep his master set up to date. Amiable (and sensible) Ray Enders from the ACE office got permission to go with me and audit some of the installations. One of the numerous Enders tribe from West Chicago, Ray had escaped his family's usual career path in the

B&B Department by going to college. He proved to be an adept student of railroad track construction. I took him out to two crossover locations and asked him count the track materials so that he could verify that my completion report was accurate. After this education and finding that the counts tallied with my report, he agreed to accept the rest of the report as being a true representation of the materials installed. That huge capital project was coopered-up in a very short order without any further discussion.

Actually, some of the ACE men were pleasant enough as individuals even though their quibbling over minor details was sometimes irritating. They loved to get away from that dismal atmosphere of the 11th floor and visit our lively division office across Canal Street. One of the characters was Joe Gaulock, a pipe smoking, greying, stolid and apparently sensible fellow. However, when he spoke of investing money with a "mechanical genius" who was building a motor that would supplant the diesel engine, I began to wonder if he had all his marbles. The machine, which never seemed to run when visited by the investors, supposedly was the size of a small washing machine. It must have been a laundry machine of some sort because it was the investors' money that was being washed. I tried several times to learn from Joe what mechanical principle this fabulous engine operated on, but he was quite guarded about details and shrugged off my efforts to point out obvious mechanical absurdities. His dreams of riches blurred his good sense.

Much of the field work Doug Oakleaf and I did that summer of 1949 was directly connected to writing roadway completion reports. There were hundreds of projects—big and small. In June I helped him do a side track inventory at Geneva Modern Kitchens in Geneva; he then helped me measure the concrete work at bridge #66. We made a good team. Doug was a former junior naval officer who came to the Galena from the Twin Cities Division. He brought with him a puckish sense of fun. One of our compatriots was that untidy old instrumentman, Art Olson, who insisted on smoking a foul pipe in the office. His engineering drawings and maps were hallmarked by small brown burn spots created by sparks from that furnace in his face. Needless to say, his clothes were often scenes of emergency slaps and brushing. Doug quietly cut-up rubber bands and mixed them into Art's tobacco humidor. The automatic knocking out of the pipe for a fresh reload, the puzzled look on Art's face at the first inhalation, the tap of the dottle into the ever present spittoon, remarks about how bad the Prince Albert was these days, and Doug's deadpan replies are just vivid memories now but in 1949 they were the cause of much suppressed mirth from those of us in on the prank. Another subtle stunt involved screwing-up Art's drafting stool one turn every day so that after a week Art would be sitting with his knees pressed against the underside of his drafting table. Finally realizing something was amiss, Art would screw it down to normal only to find a week later his chin was resting on the table. Then, of course, the minor irritant of rubber bands stretched between drawer pulls was good for a quick laugh. Poor old Arthur, he was the perfect foil.

These office pranks were pretty tame compared to one pulled in the accounting office some years before. A very pompous and self-important manager noted for his pure white J. B. Stetson hat maintained a rigorous routine. He arrived at the office promptly at 8:00 a.m. and carefully placed the prized hat in a hatbox in the anteroom before beginning his day as head of an accounting section. At exactly noon, he went to the cloakroom, retrieved his hat and went out for lunch, returning promptly at 1 p.m. After stowing the hat in the box, he again took-up the reins of power. At exactly 5 p.m. he collected his hat and left for the day. The younger clerks deeply resented this manager's attitude and his insufferable self dignity which cried out for humiliation. Collecting funds, they purchased two pure white Stetson hats. One was a size larger than their victim's own hat, the other was a size smaller. Substituting the smaller one for the "real" hat before lunch was easy. When our man put the new small hat on, it sat just a bit high on his head. He allowed he needed a haircut and went to lunch. Later, in the afternoon the larger hat went into the box. At quitting time the larger size was donned and was riding on the stuffed shirt's ears as he went home. Over the next several days the juniors substituted the hats in no particular order, even using the "real" hat from time to time so that their target could never anticipate how the fit would be. One day, he did not appear at his usual 8:00 a.m. He had gone to a doctor to complain that his head was shrinking and swelling, and he that must be going insane. Since no medical condition has found to account for this peculiar complaint, an official inquiry uncovered the deception much to the manager's relief. The culprits confessed, the ringleaders were discharged, and the manager resumed his accustomed routine.

In the middle of June I spent a hot afternoon at the company's Algonquin gravel pit taking cross-sections. The pit was operated by a contractor, Foley Brothers of St. Paul, Minnesota, who were paid according to the quantity of gravel excavated from the pit. To determine the yardage, we had to run cross-sections of the gravel deposit and calculate the yardage removed. It would have been a lot easier to count carloads shipped, but the contract was written by someone who thought that calculation by cross-section was more accurate. Later I learned that it was the accounting department who thought up that method because they could better justify depletion of the facility. This gravel pit located near the village of Algonquin on the east side of the Fox River was a wonderful source of clean gravel and coarse sand that compacted well. This was just one of several gravel pits located along the C&NW's Williams Bay line. The ancient glacier that had created its terminal moraine at this spot provided Chicago with a valuable source of good fill material for many years. The C&NW also made a lot of money hauling the gravel to Chicago until the steady improvement of highways allowed the trucking industry to capture all of the haulage.

On June 20th I took my first paid vacation since beginning work 690 days before.

An interesting field trip occurred on July 7th. Instrumentman Dick Bailey and I were sent to McGirr, Illinois, to participate in a test of wheel loading effects on light rail. McGirr is the first station south of DeKalb on the Spring Valley subdivision. This was a cooperative test involving the Association of American Railroads (AAR) test lab, General Motors Corporation's Electro-Motive Division (EMD) and the C&NW. The railroad provided the test site, a class R-1, ten-wheeler, 4-6-0 with crew, a six wheel truck Baldwin diesel with crew and Dick and me. The AAR used their test car housed in a GM diesel carbody "B" unit and a crew of technicians while EMD brought in a new 4 wheel truck diesel installed in an F-7 body.

The main track in front of the depot at McGirr was 60 pound rail, tieplated, on cinder ballast. The section crew had tamped it up so that the instrumentation would record the rail deflection rather than the sinking of the track into the soft cinders. The AAR crew in-

July 1949 at the remote McGirr, Illinois, station Electro-Motive Division tested the stresses of their new four axle diesel No. 930 on the 60 pound rail and cinder ballast of the Spring Valley branch. A C&NW R-1 and six axle diesel #1504 (DRS-6-6-1500) were also tested on this piece of track.

stalled strain gages on the rail and calibrated their instruments while Dick and I tried to figure out how we were going to read changes of elevation using our antique level and rod. Taking the head of the level off the regular tripod, Dick nailed it to a tie in the house track adjacent to the main line. Taking readings at marked places on the rail was fairly easy once Dick got used to lying on the ground in the cinders with his neck bent at an odd angle to see the tiny six inch high measuring stick through the lens. But my turn to crawl on my belly with my tiny scale came when the old R-1 eased onto the test section. Dodging dripping oily hot water and odd steam leaks, I managed to reach the marked spots with the little rod to record how much rail deflection the ten-wheeler had on the light rail. Then the Baldwin (#1504) was parked where the R-1 had been and we repeated the measurements. That diesel had various odd metal parts sticking out all along its trucks so that care had to be taken in reaching around and positioning the rod so that Dick could get the reading. I seem to recall that I had to raise and lower a yellow pencil along the scale as

Dick couldn't read the tiny characters on the scale through the level's eyepiece. That meant raising both hands to hold the measures while twisting my body in the dirt. That was the third set of records and this job was getting tiresome. The last measure involved the new EMD #930 diesel. I very well remember that hot day especially the part about lying on the ground under that panting, wheezing R-1 with my fragile fingers holding that tiny rod close to the shining treads of the drivers of a live steam engine.

That 60 pound rail in the Spring Valley line could have been the original rail laid by the predecessor Northern Illinois Railway in 1885. It was perfectly adequate for the traffic of 1949. While there was some interchange of mixed freight with the Rock Island at Spring Valley, the small amount of local traffic (mostly grain) barely justified the line's continued existence. The silica sand business in the Troy Grove area had not yet begun in 1949. The biggest attraction of the Spring Valley line was its use as a circuitous route for reconsigning lumber. Freshly cut lumber loaded in the north-western United States was often shipped eastwards on consignment without a specific customer. The load usually was sold en route and then reconsigned to a final destination from wherever it happened to be. Many circuitous routings favored by lumber brokers involved secondary mainlines and obscure interchange points because the longer a car of lumber could be kept in transit, the more time the broker had to sell it advantageously. The Spring Valley line with its Rock Island connection at Spring Valley was just the sort of slow and roundabout route lumber brokers used to ply their trade. Section men liked 60 pound rail not only because there was good supply of it—the C&NW had hundreds of miles of it, especially in the Dakotas and in Nebraska,—but also because it didn't break under traffic and laid easy on the ties. Rarely tieplated on light traffic lines, the rail didn't cut into ties because this flexible rail deformed under wheel loads and sprang back into shape like a giant linear spring. Heavier rail sections are more resistant to bending in the vertical plane and transmit wheel loads directly onto the tie and ballast. It can be less expensive to maintain a 60 pound railroad track than a stiffer 85 pound line depending on the traffic handled. But, introducing the concentrated wheel loads from a two axle diesel locomotive's truck shifts the equation. Add the technology shift to heavier cars and bigger locomotives, the limber 60 pound rail had seen its day and became a prime candidate for replacement.

Morlock's bridge steel gang that rebuilt the Proviso turntable in 1948 called again for our services at the Galena roundhouse, M-17, at the Chicago Shops. Instrumentman Bob Nelson was given the job and drafted me to help out. My experience in locating the exact center of rotation last year made this job go quickly. This project involved installation of a new circle rail which was expected to last 40 years. Great expectations of longevity were to be dashed in just seven years, however. The steam locomotives being serviced here were already doomed and would be gone in 1956.

A derailment at Proviso on July 15th required an engineering survey to settle a flaming argument. Dick Bailey and I worked in the late afternoon light to lay off the distances required and run a line of levels. This derailment was caused by a rail turning over under a departing freight train. The rail twisted over because the supporting track ties were all rotten. Calling for a survey was patently ridiculous. We were hurriedly summoned to calm the three-way shouting match and finger pointing between the car and track departments and operating. Standard instructions flatly precluded blaming a derailment on a combination of causes, just one primary reason was allowed. Of course, rotten track, bad train handling and a worn wheel often were the obvious causes, but only one department could be blamed, setting up and reinforcing the fortress mentality of the various departments. There was a definite feeling of "us against them" throughout the C&NW's departments with little regard for the company good when blame was to be assessed.

While July was filled with minor excursions like checking highway signs at Bellwood, inspecting concrete forms at the turntable at M-17, Chicago Shops, attending a public hearing in Elgin about drainage, going with Bob Lithgow to a meeting with Northern Iron and Metals at the Shops, running levels in Forest Park, conducting a sidewalk survey at Wells Street, attending a safety meeting at Proviso, I also got involved with force reduction planning that carried seeds for major changes in the way our tracks were to be maintained. L. R. Lamport, the man who had hired me in 1947, was now Engineer of

Maintenance. He sent down orders to the divisions to again extend section limits with no increase in manpower. Reallocating the Galena Division's territory among the five roadmasters and reassigning the section limits proved to be a delicate balancing act. To try and achieve parity and fairness, it was necessary to consider the miles of tangent track and curves, frequency of train movements, gradients, number of side, yard and industrial tracks, weight and condition of rail, as well as the natural break points between sections. About the only thing we didn't factor in was tradition which did cause some outcry from the field forces. Our plan, when presented to the roadmasters, was greeted with derision and fervid invective most of which revolved around the disruption and unworkability of our plan. This first step was just the beginning of a process that would lead to complete elimination of the section concept in track maintenance.

The section limit battle continued into August. The Chicago roadmaster was young John A. Wilkinson. Jack had inherited the job from his father, John E. Wilkinson, who had retired in 1948. Old John, who had almost 50 years of service with the C&NW, had begun his railroad career in 1901 when he worked on the construction of the Peoria and Northwestern Railway south from Nelson, Illinois, to Peoria. Son, Jack, had learned his father's roadmaster skills and combined them with knowledge of how to work office politics. His office was on the fourth floor of the Chicago Passenger Terminal directly above ours. A gregarious and canny man, Jack used the stairway often to visit our drafting room and determine what was going on that might affect his territory. His responsibility included all 16 tracks in the CPT, the interlocking plants in the terminal approaches up to signal bridge K on the Wisconsin-Milwaukee division and west on the Galena main to the Des Plaines River west of River Forest plus the Chicago branch lines to North Pier, the South Branch (Rockwell Street line), Wood Street, the 16th Street Line and that far away and isolated yard in South Chicago called Irondale. Somehow, his requests for new rail, ballast, bridge raising, and doubleslip switch replacement, were always couched in such reasonable terms that he probably snared more than his fair share of available funds. Under his guidance all of the complicated interlocking plant switches were rebuilt in a carefully orchestrated program that replaced the worn-out 100 pound rail section with new 115 pound material. The work was done under traffic with no delays to the commuter and long distance trains. He also obtained authority to replace the rail in the secondary mains 3, 4 and 5 from the California Avenue coach yard to the CPT with secondhand 112 pound rail arguing conclusively that these tracks were essential to the commuter operation. Jack was also smart enough to realize that heavier rail, good ties and deep ballast meant less manpower expended for future maintenance. He could see from his visits to our shop that reduction of forces was the order of the day. Reading the tea leaves, so to speak, he did not make loud complaints about the section limit expansions and was rewarded for his restraint. Some of the other roadmasters proved to be incapable of presenting coherent plans for improvements on their territories which thrust the responsibility on the division engineer to develop plans and justifications. We needed more Jack Wilkinsons. His life was short; he died suddenly August 15, 1966.

The Galena Division was blessed and cursed by a series of division engineers in the years I worked there from 1947 through 1957. There were wise and courteous gentlemen, blustery powerhouses and truculent tyrants. Each one served about two years before being promoted to greater (or lesser) positions. Most were former military men who held rank ranging from a full colonel to the oldest second lieutenant the Army ever had (he said). One had been an officer in both World Wars. Some acted as though they were still running a military unit. Outsiders were rare; most had risen through the drafting room-survey party ranks. The first order of business when taking over the division was a thorough inspection and familiarization of the territory and men responsible. (You Gotta Know the Territory). In the late 1940s and early 50s that meant either on-track trips on the unsprung and unheated Adams speeders with the roadmasters, B&B supervisors and signal men, or repeated trips over the lines on the rear of passenger trains. The best way to see track was to be close to it on the speeder. Visibility was great but every joint could be felt through the unsprung steel wheels. Public crossings were adventures because the automatic grade crossing protection was not activated by the insulated wheels of the speeder. Of course, wayside signals were also not activated by the speeder so that great care was given to getting proper clearances from the chief train dispatcher. Some motor car operators "smoked" their way on the secondary freight lines by watching for locomotive smoke ahead and behind. This proved fatal when the diesel locomotives began to appear in freight service. Ray Doty, the B&B supervisor on the Southern Illinois said he had lost a couple of speeders down the bank when surprised by a diesel powered train. After 1948 when the SI was dieselized, Ray decided to get a highway truck as his transportation. The motor cars came in several sizes. The little speeders were two man affairs and usually sported a wind screen and forward facing seats. The gang cars were bigger and heavier with benches running parallel to the track which gave one a crick in the neck when riding on the bench and trying to looking forward.

If the new division engineer survived his motor car trips, as well as the overnights in the local hotels and eating catch-as-catch-can food, he also was expected to regularly ride the varnish on the high speed main lines to monitor track quality. One would think that riding the cushions would be a great relief to riding in the wind and rain on a miserable rough-riding motor car. While more comfortable, perhaps, riding a streamliner at 90 mph across the division did not provide much information about track condition. An observation car which is a great way to see track never graced the rear of ordinary daytime trains. Since the west coast streamliner trains did not depart Chicago until 5 p.m., there was very little daylight to see the division even if a lowly division engineer could manage to get the rear seat on a posh train. The customary viewing place for the Galena division engineer was standing up in the rear vestibule of train #13 westbound with track chart handy, balancing on his toes ready to grab a handrail when the car lurched over a bad spot. Adding to the discomfort roadway dust swirled up behind the speeding train seeping into the cracks around the end door, mileposts were easily missed and senses were dulled by the noise and general commotion. As the number of trips added-up, notes became shorter and more cryptic. At last even the hardiest observer beat a retreat to the inside of the last car, making an occasional look out the back end at the track. In later years official orders came down to the division engineers and superintendents that they were expected to ride trains so many times a month. In 1956 the division engineer finally got a station wagon and conducted division business by highway until it became obvious that he was not able to inspect track that way. Near the end of my railroad career, management decreed that all suburban lines were to be traversed once a month by a special train consisting of a diesel and one suburban gallery car. It turned into a party on wheels with junior representatives from any department riding half a day at highspeed over the three suburban districts. What was accomplished in terms of maintaining track quality still is not understood.

Back to the new division engineer: after a month or so, he had developed a pretty good understanding as to where and what were the most important work sites involved. His copy of form 1136 (monthly progress report) was well worn and contained a lot of scribbled notations. He visited work in progress and began to evaluate the importance of the many requests for this and that coming from his roadmasters and B&B supervisors. Their perceived needs demanded that they become active lobbyists. New rail, more ballast cars, repairs for grade crossings and diamond crossings, new timber caps for those timber spans, trucks for their crews—were only a few items urgently needed. I accompanied one newly appointed engineer on a business car trip. Arriving back in Chicago he asked me how much he had given away as he had noticed that every supervisor leaving the car had a big smile on his face. The division engineer, however, to be successful had to keep a tight rein on expenditures of men and money or face serious problems with his budget.

An inspection trip with the top brass was the bane of a division engineer's life. These trips might occur several times a year and could be hell on wheels. On board the four or five business cars were the chief engineer, vice president of operations and their staffs along with local officers picked up at each division point. The train covered all the important main lines. Secondary lines sometimes were graced by the inspection special but many branch lines were skipped as being of no great moment or because the train running on the light rail and questionable tie conditions might have been unsafe. One survivor called the inspection train a velvet hell. While the food and service were extra special, constant pressure placed on the local supervisors was intense and unrelenting. The leisurely pace of the train allowed observations of minor items that would ordinarily not be noticed. Everything on the division could draw comments and demands for explanations. Even though every effort had been made to neatly stack, sweep, polish, or hide every loose bit of railroad gear, caustic comments and criticism flowed like water in that brass-railed business car. In retrospect I can recall that each department seemed to take intense pleasure in discomfiting the other guy's area. It was a great relief to get off that train at the end of one's territory with a pocketful of notes about things to do and not feel as though he had been skinned alive. One Galena division engineer could not handle the unending stream of criticism about work not done (because of ordered and severe budget cuts). He exploded with righteous anger at their unjust attacks and was relieved from his position on the spot. His career up the corporate ladder ended right there, and he was subsequently relegated to an office engineer's job where he shuffled papers in the chief engineer's office until the day he retired. And, it was even worse when the company president and his people traveled the road which, fortunately, was not often.

Socially, division engineers rarely mixed with their subordinates. The military nature of the organization reached down from the chief engineer's office. This individual was more like the general in charge, department heads were the colonels and majors, the division engineers were about captain grade. The assistant engineer in charge of the drafting room and survey crews was the equivalent to a first lieutenant, instrumentmen were the second lieutenants, rodmen were lowly corporals, tapemen were the privates. The chief clerk was a staff sergeant while the women file clerks were just civilians.

After a pressure-packed tour of two or three years, the incumbent division engineer was considered to have been "blooded". Promotion then could be to a more responsible position, reassignment to a more important division to garner additional experience, or being sidetracked into a deadend staff job. Many facing that last alternative picked up their marbles and went to other railroads.

The main C&NW accounting office was located in its own building 6.2 miles north of the Chicago Passenger Terminal at 4200 N. Ravenswood Avenue, two blocks north of Montrose Avenue and the Ravenswood-Wilson Avenue commuter train stop. A regular courier service was operated from Ravenswood to the general office downtown. The three story brick building housed the army of clerks needed to handle all of the station, freight, passenger, interline, payroll and other accounting. The clerical staff dealt with everything from station receipts to the tickets collected by the trainmen. In 1949 the

C&NW had a substantial battery of IBM unit record machines which fed on punched cards. Keypunching those 80 column cards was a major activity. Much of the record keeping still required manual sorting and filing creating mountains of paper. Freight waybills, timeslips, and ledgers were to be found everywhere in that beehive. I first visited Ravenswood on August 17, 1949 and got the distinct impression that I was standing at the entrance to Chaos. Our division was in need of cost information for budgeting, and I was sent to Ravenswood to try and find some help from accounting. Somewhere in that sea of paper and strange people must be the basic information we needed. In 1943–1944 while waiting for call to active duty, I had worked as a freight accountant for the Quaker Oats Company in Chicago and knew quite a bit about railroad freight bills. Ravenswood's operations were a thousand-fold larger than Quaker's. With the advent of the first main frame computers this accounting morass was partially civilized, but the clerical mind set seemed unchanged twenty years later. No, I did not find any useful information in that house of horrors.

On August 19th I was directed to accompany the district claim agent to an accident site at Sterling, Illinois. I took Ralph W. Golterman, a new rodman from the Northwestern University Co-Op program, with me to assist in the survey. It seemed that an Albert Booth and his family had driven across the main line at Avenue G in front of a train with fatal consequences. Ralph and I worked all day developing the measurements for an accident survey drawing as the area was very congested requiring a lot of detail work. Ralph proved to be a useful assistant and had a good personality. Although he sometimes appeared to be a bubble brain, there was a shrewd and focused individual under that merry facade. Later Ralph became a location engineer for Sun Oil Company and located retail outlets throughout the Chicago area. His brother, Dick, a veteran of the USAAF, had been a pilot of transport planes flying the Hump in the China-Burma-India theatre during World War II. I was to have business with him in later years. Both Golterman boys had good engineering sense inherited from their German father.

Apparently, the Sterling accident of last August had been handled to the law department's satisfaction as they requested that I do a survey for them of the uneven sidewalk at Washington and Canal Streets along the east side of the Chicago Passenger Terminal. The pavement slabs were indeed uneven after 38 years of differential settlement. The exhibit drawing was difficult to prepare because emphasizing irregularities in the pavement might unduly influence a jury should the complaint come to trial. By chosing the proper scales to preserve the accuracy of our measurements and avoid exaggerration of the dips and bumps, an accurate representation was achieved. Any concerns about the drawing, however, vanished when the complaint did not escalate into a lawsuit. Meanwhile, that nasty accident at Sterling's Avenue G crossing resulted in a plan to install automatic gates which required another trip on train #13 for a four quadrant vision survey. Another full day of measuring, then eastward on train #14 to the stop at Oak Park and a westward scoot to Elmhurst. The good news, however, was no more Saturday work for the division engineer's staff as of September 3rd.

The last quarter of the year went swiftly. A mixture of short field trips and night school Calculus classes filled the days and nights. Our instrumentmen at that period were all college graduates who encouraged the junior rodmen and tapemen to further their education. My night school classes earned me a reprieve for those overnight trips, so I turned more and more into handling reports, accident surveys, license applications, leases of railroad property, and attending the division safety meetings to represent our department. There were short trips to the North Pier, Belvidere, West Chicago, Elgin, Dundee, Geneva, Maywood, Navy Pier, Chicago Shops, State Street Yard and Wood Street Yard. A trip to Chicago's City Hall to inspect land records was my introduction to Chicago politics. Only one field trip was a real nuisance and contained elements of physical danger for Bob Lawton and me. November 1st and 2nd were cold and snowy days but Bob and I rode a Greyhound bus along with all our survey equipment to Franklin Grove, 88 miles west of Chicago, to make some vision surveys at several grade crossings proposed either for closure or improved protection. After getting the work done, we had hours to wait for the eastbound Greyhound and huddled in a shallow roadside ditch along the highway trying to stay out of the keening wind coming across the snowy fields. Neither Bob nor I owned an automobile at this time; we were totally dependant upon other peoples' courtesy or our own ingenuity to get to the work sites. Rent a car? The expense would never have been approved. We survived this event and strove to avoid such situations in the future.

At least the railroad seemed to be shaking off its postwar lethargy as the physical plant was getting better compared to two years earlier. It's antiquated personnel structure and operating policies, however, still needed overhauling if the roadway improvements were to be translated into profitability. I was content to stick around and see what the new decade would bring.

THE PACE PICKS UP, 1950

... I trust that I am not creating an impression that working in the railroad's engineering office was a boring job—it was not. Each day brought different and, sometimes, challenging tasks. I looked forward to going to the office every day as each proved to be a new experience. This new year of 1950 was a continuation of the last year of the past decade; the hours were crowded with railroad work and night school. It seemed that there was no extra time available from first light to midnight. My colleagues continued to help my education, both by sharing their engineering skills and by taking on much of the outside work and travel that could have disrupted my evening school schedule. As usual, winter weather inhibited much of our field work.

Before Valentine's Day I logged short trips to the Chicago Shops, to the Cook County Circuit Court (for a land condemnation hearing at the State Street Yard), to West Chicago (to review use of track materials in the CTC project with the roadmaster, Art Netzel), to Lombard, Wheaton, Elgin and Gilberts with instrumentman Bill Smith for completion report surveys, and to Aurora via the CA&E to inventory a new industry track at the Hill-Behan Lumber Company at North Aurora. A Mr. H. Fish sued the company when he fell while cutting across the tracks in the Chicago Passenger Terminal to an exit stairway from the crowded train platform. The law department resisted the suit, and I was drafted to make the drawings. Another crossing accident at Elida Street in Winnebago, where a Mr. Bennett disputed right-of-way with our train, brought rodman Bob Lawton and me out into the bitter cold. We rode train #703 to this first station west of Rockford where we were confronted by icy blasts coming right out of Alberta. Luckily, there was a warm restaurant nearby where we could wait for #706. Instrumentman Dick Bailey and I were hurriedly sent to another fatal grade crossing accident at Kirk Road east of Geneva on a perfectly good Saturday morning. We heard on January 12th that someone had tried to derail the Valley 400 and the Twin Cities 400 in Chicago. On February 9th I received a special request from assistant chief engineer, B. R. Meyers, to escort a Mr. Tucker, chief architect of the St. Louis-San Francisco Railway, on a tour of our new diesel shop at Chicago Shops.

Arthur Netzel, roadmaster West Chicago, was close to being more like a bear than a man. He was hewn from a solid block of wood but his stocky shape radiated a nervous energy that seemed to crackle with every motion he made. His dedication to the company and his work was monumental. Patience was not a virtue with him. One day two of us were setting mainline center stakes for Art's track crew when a freight train eased through the work zone. We all stood in the clear waiting for the train to pass when it suddenly shuddered to a stop amid the sound of air brakes applied in emergency. We all stood there and waited, and waited. Frustrated, Art picked up our eight pound maul as though it was a tack hammer and slammed it into the side of one of stationary box cars. While it didn't make the train go away any faster, he felt better and turned to us with a grin cracking his face and a mischievous glint in his eye.

On another occasion we needed a # 8 frog of 100 pound section for a new industry track in his territory. At that time purchasing and stores had embarked on a new program of inventory control for all company materials. Their frenzied counting of all the bits and pieces that make up a modern railroad seriously disrupted normal supply channels. Purchasing was in a cleft stick. They wanted to make sure that they didn't have a particular item before ordering more and yet they did not know what items they actually had in stock. The divisions could not obtain needed materials in a timely manner and had to scrounge or delay work. Complicating things further, there was no computer program involved; it was all manual data collection. I asked Art if he knew where a frog could be obtained, even if it were from a disused track where it could be "borrowed". He took me out to the east yard at West Chicago, picked up a convenient long stick and began to poke around in the dark waters of the swamp along the access road. There was all kinds of track material under that water; this was his emergency stock and none of it had been inventoried. We found the needed frog.

On the afternoon of February 13th we were hit by a terrific snow and ice storm which cut electric power to many west suburban locations from three in the afternoon until about 8 a.m. on the 15th. A long-planned test of the Budd Rail Diesel Car (RDC) had been set-up for Valentine's Day. The test run was an effort to find some alternative to those aging steam locomotives and the deteriorating suburban fleet with their five and six men crews. While the RDC had been touted as a solution for future commuter service, its performance characteristics were unknown to the C&NW. There was con-

On an icy Valentine's Day 1950 at HM tower in Elmhurst, Illinois, the single westbound RDC car is on its first test in suburban service.

siderable doubt that a single car would activate grade crossing protection, automatic train control circuits or wayside block signal systems. To monitor the test runs, division supervisory personnel were stationed in the interlocking towers along the route to observe the signal systems' reaction to the car. The corrugated stainless steel test car was identified by a stream-style of "Budd" on the letter board—I didn't note the car number located in the center of car below the windows. My assigned location was at HM Tower in Elmhurst. I persuaded my wife and daughter to ride the RDC from Elmhurst to West Chicago and return despite the awful weather. The westbound trip was uneventful if one discounts the snow, ice and lack of commercial electric power. For my part sitting around and jawing with the operator made a pleasant afternoon. No particular signal failures were noted during the test. The C&NW later acquired three of the cars numbered 9933–9935 in June 1950 and operated them in suburban service mostly on the Wisconsin and Milwaukee subdivisions until their specialized maintenance costs became an irritating nuisance. If there had been a larger fleet of these cars, maintenance could have been systematized and costs per unit kept within reason. In 1957 the railroad swapped the Budd cars to the Chesapeake and Ohio for three modern passenger cars.

This winter of 1950 proved to be unusually cold and unpleasant. Getting to the commuter train on many mornings proved to be a grueling experience requiring extra warm clothing and high boots for wading through the snowbanks. Often city streets were unplowed requiring trail breaking by the first to go to the station. The engineering force was glad to stay in the office during this bad weather and work on the ever present piles of paper and reports. Among them were the completion report of the east end of the CTC project, drawings for the accident in Winnebago and a proposition for retirement of all steam locomotive facilities on the Southern Illinois subdivision.

In the late afternoon on Friday, March 31st, I was summoned to the office of one of the railroad's top legal officers, Edward L. Warden. Sometime earlier I had prepared an accident survey drawing for the "Roulette" case. Now I was to get ready to be called as an expert witness for the trial beginning Monday. Mr. Warden asked that I prepare the necessary prints in time for the court call at 10:15 a.m. These were big drawings to duplicate and far too large for our ancient blueprint machine in the CPT. And here it was late Friday p.m. and the chief engineer's machine was always closed down at 4 p.m. The costs and difficulties in securing the necessary permissions in using a commercial print shop eliminated that option. I hurried down to the north end of the 15th floor in the Daily News Building where the reproduction department was located and caught up with the operator, Larry Schulz, as he was preparing to leave. Larry was the king of his empire, the blueprint machine room, and took no sass from anyone. He reported directly to the chief engineer, E. C. Vandenburgh, and often used that relationship to avoid performing extra service. He was an independent, arrogant and cocky little guy. Since I had never given Larry any cause to think I was some upstart punk, I now used whatever suasion I had to get my work processed first thing on Monday. Larry was true to his word and ran my drawings through the monster blueprint machine as his the first job on Monday morning. I grabbed the prints as they came out and sprinted across Chicago's

Loop to 160 N. LaSalle Street where the trial was in progress in the Illinois Commerce Commission's hearing rooms. Just like the Army's dictum, "Hurry Up and Wait", the prints were not needed until late afternoon. On April 4th I was called to appear. I had no idea what this case was about because witnesses were kept out of the courtroom during the proceedings. The group, isolated from the hearing, played pinochle until I was called at 11 a.m. Sworn in, I answered a few questions put to me by Mr. Warden, and was then dismissed. This was my first contact with the law and the courts. I never learned what this case was about nor what happened.

The law department of the C&NW was unusual among Chicago railroads. Only the C&NW and the Illinois Central maintained their own in-house staffs; other companies used outside legal counsel. In major actions, the C&NW and IC also retained outside law firms, but routine legal matters involving disputed claims, judgements, document review, commercial matters, *etc.* were all handled by the legal staff. Over the years, I was to meet and work with many of these gentlemen who were, by and large, intelligent and interesting men.

Personnel changes came in April: rodman Fred Bone went to the Madison Division, rodman Ralph Golterman returned to classes at Northwestern University, and instrumentman Dick Bailey was appointed supervisor of scales and work equipment. The open instrumentman job was given to William C. Comstock on the 28th. He had 28 years seniority on me so I was not surprised. On May 5th instrumentman Ottmar W. Smith was sent to Negaunee, Michigan, on a line relocation job for six months, which left another instrumentman position open. The division engineer, Charles E. Hise, advised me on the 9th that I was to be appointed as a temporary instrumentman effective May 11th. This was rather unusual as instrumentmen were supposed to be college graduates, and I was years from having a degree. The responsibility of the position did not overwhelm me as I had tried very hard to absorb the necessary technical skills over the past two years. My monthly pay went from $305.71 to $360.71 and I knew that the evening school schedule was going to be more of a problem as I would have to pick up my share of the out of town work.

On March 15th I spotted an IHB drag eastbound at Chicago

Shops with four dead C&NW steam locomotives. Class D #397, 4-4-2, a class M, 0-6-0 with its cab missing, a wrecked J, 2-8-2 #2477, and a class Z, 2-8-0, #1873 were in the consist headed for scrapping.

Spring came at last and I took advantage of the break in weather to get some exercise and do some company business. On April 13th I boarded #703 in Elmhurst with my bicycle and rode the train to Huntley, Illinois. Tomorrow would be the last run of this train ending passenger service on the Freeport sub when #706 came back tomorrow afternoon. Termination of this train was a loss for the engineering department as we could no longer easily get to the many small stations scattered along this line. With this inconvenience looming, I used the train one last time to clear up some odds and ends. I had a small track inventory at Huntley to do. Finishing that job I cycled east to Carpentersville on the Williams Bay sub north of East Elgin to measure some clearances, then on to Elgin and Wayne and West Chicago, each location having some minor task to be done. By early afternoon I was headed home to Elmhurst. It would have been simpler to handle this catalog of little jobs by using an automobile, but I did not then own a car and the company was not to furnish the department with one for five more years.

At a conference meeting on April 24th in assistant chief engineer B. R. Meyers' office on the 15th floor of the Daily News Building, construction of a new diesel servicing shop at Proviso was discussed in detail. The reason I was invited was explained when responsibility for managing the design and construction of the shop and approach tracks was handed to me. DeLeuw, Cather was the prime engineering and design representative and the S. N. Nielsen Company was again the contractor. We had many of the same players in this project that we had at Chicago Shops two years ago. The site of the new shop was to be on a swampy piece of company land directly south of the Proviso engine house. Modern day practice would have called for excavation of the existing peaty soil and replacing it with competent fill. Instead, the decision was made to haul in trainloads of Algonquin pitrun sand and gravel to fill in the low area. Trains of drop-bottom hoppers began moving from Algonquin to Proviso both directly from Crystal Lake and south through the congested streets of East Elgin.

The low area soon became a vast gravel dump leveled by bulldozer into a sandy, level area. The natural water ran off in a big ditch left along the northernmost yard track, the so-called Twenty Main. On May 5th rodman Bob Lawton and I chained a mile and a half of track at Proviso and on the 8th ran levels west from the Harbor Hill east of Mannheim Road to the "Middle" where several yard leads came together. After another meeting in Meyers' office on the 23rd, the track construction at the new shop site received top priority. The next day was hot and windy when Bob and I staked the centerlines. Excavation for the building proper began in early June; by the 9th it was just a muddy hole full of murky water. I was glad that I was just supervising this job and not the on-site engineer because this was a problem construction zone requiring a lot of drainage and pumping to get the concrete footings in place. I left the building construction work to the contractor. That was what he was being paid to do. My job was to make sure that the work was done competently. I knew enough to just stand back and let the professionals deal with the problems. Knowing the quality of work this contractor did, I did not have to spend a lot of time down in those muddy pits. By October the railroad crews were laying the shop approach tracks. A real problem came with the rails that were to be installed thirty feet in the air—the overhead crane rails. These rails were to be supplied by the railroad, properly marked, matched and drilled for the fastenings. I spent most of October 30th rounding up enough good 112 pound rail for these crane rails and the next day measured, punched and match marked them. Completion of this shop was scheduled for 1951. Tom Ek, DeLeuw, Cather's on-site engineer, proved to be most helpful in taking care of some of the minor railroad-related items in the shop construction.

Bernie Meyers, now assistant chief engineer and soon to become chief when Edward C. Vandenburgh retired, was an interesting person. A stuffed shirt of the old school, Bernie flattered subordinates and acted like a grand duke. Stories from former associates about his behavior when he was a bridge inspector were widely told. A bridge inspector was a sort of demi-god to division B&B men. He made judgements about steel and timber bridges often exposing himself to great physical danger as he clambered about the steel trusses inspecting rivets, dealing with rattlesnakes under the timber spans, or braving wind and wet on long track motorcar trips. Meyers carried his dignity with him at all times. At the end of a day of inspection, the B&B man carrying Mr. Meyers' valise to the hotel always walked an appropriate distance behind him. Never addressed as Bernie, he was always Mr. Meyers to anyone but his superiors. Therefore, it was a notable event when Mr. Meyers deigned to visit the dingy old Galena drafting room on May 4th. He strolled in and casually remarked that I was slated to be resident engineer on the replacement of the Mannheim Road bridge at Proviso. I am sure he enjoyed my startled reaction. The bridge job actually materialized in 1952. Despite his foibles and insufferable air of superiority, Meyers was a good engineering manager and made an excellent chief engineer of the railroad for many years.

Rodman Bob Lawton and I had a miserable little track relocation job deep underneath the Wabash Avenue viaduct on the North Pier line. We were dealing in three dimensions of difficulty with tight curves and minimum vertical clearances. The survey points fell underneath an old passenger car being used as a section tool house. It was so dark we had to use flashlights to see what we were doing. But that job had to wait as on May 15th we were hurriedly sent to Rockford, Illinois, on the Illinois Central's train #15, the *Iowan,* which left the Lakefront Station at 12th Street about 8:30 a.m. After the train wandered around the south side of the Loop, it made a direct run at Rockford 83 miles distant passing through Hillside, Elmhurst, and Pingree Grove. After we got off, the train went on to Galena and East Dubuque before crossing the Mississippi River on a long steel span to Dubuque, Iowa, and continuing onwards to Sioux City. Our job at Rockford was to make profile surveys of the 11th Avenue highway approaches and crossings of the C&NW and IC tracks. Despite heavy vehicular traffic, we managed to finish the job and catch the Greyhound bus back to Chicago. We certainly missed our old train #706 which had been discontinued a month earlier.

The next day we were sent out on an emergency at Keystone Switch south of Peoria on the Southern Illinois. To squeeze my life into this trip, I spent the afternoon clearing up the notes from the Rockford survey, went to my physics class at IIT in the evening by riding the Clark Street streetcar to 33rd Street, raced back to the Loop on the trolley, caught a Galena suburban train to Elmhurst, raced home, grabbed my traveling clothes, then went back to the Elmhurst station for the last eastbound scoot that evening, hurried up to the office to collect the surveying equipment and then hauled the whole load to the LaSalle Street station for the CRIP's Peoria night train that left at 2 a.m. The Pullman *Prairie Creek* was an old heavyweight that rode well. I woke up as we rolled through Chillicothe north of Peoria. Bob had boarded earlier in the evening. We grabbed a bite to eat at a nearby White Hut restaurant and got on a Peoria Transit Company bus to Bartonville. Keystone Switch was the junction of the industrial spur that served the Keystone Steel and Wire Company's works which sprawled across the bottomlands along the Illinois River. At the switch an old boxcar body served as an operator's shack. The division engineer, Charles E. Hise, had left instructions for us there. We were to take cross sections of the lake bottom adjacent to the railroad embankment as there had been a derailment of a gondola loaded with sand and Charlie was of the opinion that the old cinder fill had failed under the train. The opaque water gave no clue as to what was to be found on the bottom. Ray Doty, the B&B supervisor from South Pekin, soon arrived and, learning we needed a boat, set off to find one. After an encounter with a pack of aggressive dogs, Ray came back with a skiff. Bob and I chained off the stations along the main track and set up the level while Ray got the boat into the water. The lake was actually a backwater from the river and was used by the steel company as a effluent dump which changed the water into a reddish sludge. Our railroad embankment was only about five feet above the water level and quite free of vegetation as cinders do not encourage weed growth. Ray got out into the little boat with our twelve foot rod and was guided by my hand signals to the first observation point. Expecting a water depth of 8 or more feet, he stood up in the flimsy craft and vigorously thrust the rod down into the dark water. Since the water was only a foot deep, the shock of the sudden stop almost threw him over backwards into the soup. He was more cautious with the rest of the survey. Later analysis determined that the railroad embankment had not been affected by the "deep" lake alongside. Finishing up, Ray took us to the Orchard Mines grade crossing near Hollis where Illinois highway 9 met highway 24. A bus to Peoria appeared, and we made the afternoon *Rocket* to Chicago.

Bob and I went back to that miserable job at Wabash Avenue on the 19th and still couldn't make things fit. We retreated and planned another foray next week. On Saturday, our day off, instrumentman Bill Polchow, working with a lining gang in Wheaton, yelled for help in staking a curve. I rushed into town only to find he had changed his mind.

I had just arrived at a Proviso safety meeting when division engineer Charlie Hise called to report that Ed Warden of the law department had requested that I be sent to South Janesville, Wisconsin, to do an accident survey for a personal injury suit. This was not on the Galena Division and special permission had to be secured from the chief engineer, E. C. Vandenburgh, to authorize the action. The Wisconsin Division guys were somewhat unhappy that I was selected, but the grumbling was not very loud. On May 29th I caught our *Dakota 400* at 10:45. I was met in Janesville by a company stationwagon and taken to the accident site at the south end of the yard. Eugene P. Shepard, 25, a switchman from Skokie, Illinois, had lost both legs when struck by a string of cars on track number 6 on the night of September 6, 1949. I located turnouts and measured distances for the preparation of a large scale drawing of the site for use in defending the suit. I caught a ride back to town on a diesel switch engine, #1006, whose engineer just happened to have been involved in the unfortunate accident. Waiting for the Chicago trains, I got a good look at local railroad activity at Janesville. The other railroad in town was the Milwaukee Road. I watched their ten wheeler #1103 take a local out of town from their depot a couple of blocks away. C&NW train #501 westbound was powered by a class E-2 heavy Pacific 4-6-2 #2907. A C&NW motorcar on one of its last trips to Fond du Lac, Wisconsin, departed in a cloud of fumes. This car (believed to have been #9915) had been "modernized" by addition of a streamlined front end and fitted with an integral snowplow. The Milwaukee's *Hiawatha* arrived at their depot powered by an orange ALCO diesel just as the

Motor car #9915 leaving Janesville for Fond du Lac, Wisconsin, as train number 49 on May 29, 1950. This accommodation did not last much longer.

C&NW's eastbound *Dakota 400* with a Fairbanks Morse diesel unit (either 6001A/B or 6002A/B) on the point pulled into our depot. At the CPT I just managed to catch my Galena Division train west at 9:25 p.m.

I wasn't finished with Janesville yet. W. H. Kelly, general claims attorney, sent me back to use a stop watch to time how long a string of cars took to roll down the track and to time other switching moves. When the *Dakota 400* pulled in, claim agent Vince Tondryck drove me down to the yard where I timed every car and engine movement I could think of. Vince then drove to Chicago which meant I wouldn't have to wait for the eastbound train again. I made the drawings on June 1st and 2nd. On the 27th the case was called in Circuit Court. As a witness I was again isolated by being parked in another courtroom where I audited another case. The next day I was called back to court but was not called until about 2:45 p.m. to testify about my exhibits which took all of five minutes. The next day I was summoned in haste and again put on the stand. My reply to a question threw the court into an uproar with cries of "mistrial" from the plaintiff's attorney. Judge Clark retired with the attorneys leaving me on the stand with nothing to do except look at the jury and spectators. After some time, the judge and attorneys reappeared. The same question was put to me. I replied in the same way and the trial went on without a ripple. I never learned the reason for the uproar. Late that evening the jury awarded the plaintiff $125,000 of the $300,000 he had sought.

Meanwhile, the routine work went on without incident until June 8th when some of our people got into trouble. Instrumentman Bill Polchow with rodmen Willie Lipsitz and Dick Stemmerman had borrowed roadmaster Earl "Fuzzy" Pierson's personal track motor car (with permission, of course) to go out to a track raising gang where they were to set centerline stakes. They left it sitting on the main track while they did their work. Train #106, the *City of San Francisco,* smacked into it at 30 mph reducing it to flying junk. Fortunately, the surveying equipment had been removed before the impact. This was a serious incident requiring a formal investigation. A lot of people were involved in this mishap and many had to share the blame. First of all, Bill Polchow did not have a motorcar operator's permit and had never taken the prescribed rules training course. The dispatcher had given the ballast crew the wrong time for the "dead" track when no trains were supposed to be operated. Train #106 had entered the dead zone 15 minutes before the expiration time. Their orders only stated that reduced speed was required through the work area. Roadmaster Pierson should not have loaned Bill his motorcar without checking Bill's permit. The engineer of #106 escaped without blame even though witnesses said he was laughing when his train hit the fancied-up track speeder. When the dust had settled, Polchow ended up with a lot of demerits, just enough so that one more minor incident would have been cause for his dismissal.

On June 13th I took a new rodman, Vernon F. Brachman, with me to DeKalb on train #13 to teach him how to inventory track. First we tackled an easy one at YD, a CTC installation just east of the depot near the crossing of the old Northern Illinois Railway now the Sycamore-Spring Valley line. Still in DeKalb we did an accident survey at the 10th Street crossing before catching a bus to Sycamore five miles north to check vision at a grade crossing serving the Diamond Wire Company. The Greyhound took us back to Elmhurst where we both lived. Vern's dad was a banker in town, and Vern was determined not to be a banker. While he seemed a bit dense at times and was stubborn as a mule, he worked hard at learning the trade and was a willing helper.

I seemed to be a favorite expert witness for the Law Department. I was sent to Proviso on June 16th to personally measure a class M, 0-6-0, switcher #2620. A sister engine would not do—it had to be the #2620. After finding it at work and getting the crew to stand still for a few minutes, I got the dimensions requested and caught the new Budd cars east to the CPT. Bob Lawton and I then went back to Wabash Avenue with new solutions to our track problem and discovered we still did not have it right. On the night of June 19th I was on the sleeper to Marshalltown, Iowa, on train #15 at the request of the law department who needed personal measurement of an old R-1, ten wheeler, #1146 for another law suit. The engine had been all over the Iowa Division, but was now being held for me at Marshalltown until I could do my little survey. On the train I ran into Jack Datesman, one of the general engineers from the chief's office. Later, he became the Galena's division engineer. I didn't sleep well and got up as we passed through Stanwood, Iowa, about 7 a.m. After having breakfast while we were standing in Cedar Rapids, we rolled into Marshalltown where I found my locomotive east of town. My survey didn't take long and I had the whole day to rattle around town until #14 went east. I spotted the Minneapolis and St. Louis Railway's new diesel shop under construction and decided to make a visit. They were most hospitable folks and showed me all around the almost completed facility. A southbound freight headed for Peoria eased through town behind a three unit diesel set with F unit #545 on the point. The Louie numbered their diesels by the month and year received so that #545 had been received in May 1945. The new M&StL shop was brick and steel with four run through tracks. Cab floor height platforms provided easy access to units. Two of the tracks were under a high bay with overhead gantry; the other two were under a lower roof. I also discovered that Marshalltown was the home of Aunt Bessie's Black Bottom Pie and certainly had to try that. Train #14 was two hours late into Chicago so that I just caught the 10:40 p.m. commuter train back to Elmhurst.

On June 23rd the Kullman case, which involved engine #1146, was settled out of court; my trip to Marshalltown had been a waste of time.

Summer meant a break in the school routine and an increased ability to take-on a greater share of the field work. On July 5th Jim Barnes from the chief engineer's office, rodman Vern Brachman and I went to Proviso to stake the location of another crossover east of the new diesel shop. At noon we received a message that office engineer Harold Jensen, had changed his mind, and we were ordered to shift the crossover's location. There went the morning's work. The next day I had to deal with a ferret from the auditor of capital expenditures (ACE) office, Al Urkhardt, combing through a completion report on a rail renewal job in infinite detail—another morning gone. Having gotten him out of the office, I then discovered another ACE analyst, Al Butler, appearing just after lunch with more questions on some other reports. To escape this accounting department assault and to take advantage of the fine weather, I took rodman Brachman with me to one of the most isolated places on the C&NW—Irondale Yard in South Chicago. Located east of Torrance Avenue between 115th and 125th Streets, this old multitrack yard served a grain elevator complex built by the C&NW around World War I. The huge concrete silos were located next to a deep water ship channel. There was no direct C&NW trackage to this location. The one switch engine, an M class 0-6-0, went to the Chicago Shops once a month for its mandated inspections and repairs using trackage rights over the Indiana Harbor Belt and Chicago and Western Indiana Railroads to Taylor Street and the connection there with the C&NW's Rockwell Street Line. Vern and I met with Paul Disher, the assistant trainmaster, to look at the feasibility of making some track changes. By 1950 the old elevators had been purchased by Cargill of Minneapolis and new business was flowing through the facility. The usual cinder ballast in the yard tracks was covered with a residue of grain dust which made the ground feel quite spongy and filed the air with the ripe smell of fermentation. We stopped at the old brick yard office which we learned had been a school house long ago. Some time later Ray Doty, the B&B supervisor at South Pekin, remarked that as a youth he had worked with a B&B crew in Chicago. They had been sent to

Irondale following a big grain dust explosion in one of the silos. The massive structure had been lifted straight up and had settled back on its foundation. Burlap bags were found embedded in the concrete foundation at the point of separation where the ensuing in-flow of air had sucked them in.

Transportation to work sites was really getting to be a problem. On the 12th of July rodman Bob Lawton and I went to Rochelle on train #13. We had planned to catch a Greyhound bus west to Franklin Grove where #13 did not stop. At the bus station we discovered that the schedule had changed some time ago. We rushed back to the C&NW station to get back on #13 but it had pulled out minutes before we could get there. Behind #13 a work train came ambling along. We flagged it down and rode to Ashton where they stopped to unload some track material. While they worked at that, Bob and I inventoried the new crossovers at AW (Ashton West). After finishing their work, the work train set us down at FX west of Franklin Grove so that we could inventory that crossover complex. We hiked back to town to do some a vision survey at the only grade crossing in the tiny village. Then we walked over to the highway (U. S. Alt. 30) about a mile north of town and were able to wave down a westbound Greyhound for a ride to Sterling. All this on a typical July day in Northern Illinois with the temperature hovering at 93 degrees under a brassy sky.

The next day, July 13th, was cooler after a night's rain. So much cooler that we nearly froze while riding on an open track motorcar from Sterling to Morrison. I asked the foreman how he dressed for hot summer days and these cold speeder rides. He replied that he sewed on his thermal underwear August 1st and cut it off on the Fourth of July to take a bath. This remark, which was delivered with a deadpan expression, was taken with a grain of salt. Arriving at the end of a recent main line rail renewal between Agnew and Round Grove, we located the valuation stationing of the new rail for our roadway completion report. Joe Murphy, the Sterling roadmaster's truck driver, met us for a run down the Southern Illinois to Normandy. There we found an accommodating bridge crew that ran us out to bridge #1442 on their motor car for some measurements. Bob and I then walked back to town for a vision survey at the Illinois highway 92 crossing which was the only crossing in Normandy. Joe Murphy reappeared with his yellow and green truck and hauled us back to Sterling where we stayed the night. The next morning Bob woke up at 5:30 a.m. and said, "let's catch #25 to Nelson". Moving very quickly, we packed, ate and caught the train at 6:07 a.m. for the short trip to Nelson. Here we had a long hike to milepost 103 to inventory the CTC crossovers at NQ. Since there was nothing else for it, we hiked the mainline back to Nelson and on to milepost 108 to chain a mile of track. It was only another three miles to Sterling, so, of course, we walked. At the roadmaster's office we washed up and waited for eastbound #14, which was two hours late, as usual. This kind of catch-as-catch-can field trip, while productive in the number of little jobs completed, would have been much more easily accomplished with a company car.

Agnew was the scene of a rather bad rear end collision early in 1950. The cause of the accident was attributed to a locomotive engineer's circumvention of the Automatic Train Control (ATC) system. Under normal circumstances the system was activated by removal of a brass key from the relay unit located on the locomotive. The key was held by the conductor until the end of the trip. A common sight on commuter runs was the passing of the bronze key back to the engineer as the train rounded the big curve approaching the Chicago Passenger Terminal. With freight trains the key was supposed to be in the conductor's pocket back in the waycar during the trip over the Galena and Iowa Divisions. Engineers were required to acknowledge the ATC horn by toggling a four inch long brass handle within six seconds when the train was underway or the train brakes would automatically apply. Many engineers resented the constant hooting in the cab and bypassed the ATC system by illegally obtaining an extra key. In the Agnew accident the engineer had inserted an extra key into the ATC box when he noticed that the engine bell mounted in the nose of the diesel had frozen silent on that cold January day. Leaving his seat on the right-hand side while the train was rolling eastward at speed, he and the fireman went down into the inside of the diesel and tried to unfreeze the bell. Ahead at Agnew another eastbound freight had stopped on the north main (track #2). The fireman on the approaching train popped up just in time to see the waycar standing on the track ahead and yelled a warning. It was too late to stop even with an emergency brake application. The resulting collision demolished the standing waycar and sent freight cars flying in all directions, some of which destroyed the old track side grain elevator at Agnew. The conductor on the stopped train was killed in the collision. We sent out our section crews to assist in the cleanup. One of their first instructions was to look for a train control key. Despite the snow covered ground one of the men found that extra ATC key which had been thrown out of cab window by the engineer just before the impact. The investigation found the engine crew guilty, and they were discharged despite their previously unspotted records. I understand that many years later the engineer was allowed to return to duty.

The Korean conflict was raging that summer of 1950. Along with a lot of other young men I was concerned that this "police action" was going to become a full scale war and Selective Service would come looking for us. Colonel W. S. Roberts of the Military Railway Service (MRS) happened to be the C&NW's supervisor of safety. Although many of our senior railroad officials had served with the MRS, I was mildly surprised when Bill asked if I would interested in a first lieutenancy opening in his reserve battalion. Four days later, on July 21st, he gave me an application which I thought about awhile before sending it in on August 25th. I was ordered to report for a physical examination at Navy Pier in Chicago on August 30th. In September I sent in my application for a first lieutenant's commission, Transportation Corps, and never heard of it again. Roberts said the MRS had been reorganized from top to bottom and my application must have gotten lost in the shuffle. This was the second time I had volunteered for military service. I decided that twice was enough.

On July 24th I went to Rockford on the IC's 8:05 a.m. train. Our local agent, Mr. Torgerson, met the train and hauled me out to the end of the former Kenosha Division (KD) line that wandered northerly from a connection with the Freeport mainline in East Rockford through Loves Park and ending at the Larson Brothers gravel pit. The pit operators were interested in shipping gravel and needed information about the cost of certain track changes. After lunch, Torgerson and I visited with the Winnebago Service Company in Winnebago about a new side track to serve their warehouse. I went back to Chicago on the Greyhound as its normal stop in Elmhurst was a short distance from my home.

Jim Barnes again drafted me to chain track at Proviso for some project he wouldn't talk about. It was rather peculiar that a staff engineer would come out into the field and perform work that the division forces were quite capable of doing. We chained from the hump eastwards 5,800 feet plus another 5,300 feet through Yard 4. We added rodman Vern Brachman to our crew on the 31st and ran levels from the west to the hump, then down the hump lead along the south side of Yard 5 (the hump yard), down through Yard 4 to Mannheim Road and finally up the Harbor Hill, so-called because this was a connection to the Indiana Harbor Belt Railroad. James A. Barnes, a freckle-faced young man, was a gregarious and friendly person. He sometimes turned on such a naive expression during a conversation that one wondered about his qualifications. That incredulous posture was just a mask for a good engineering manager. Born in Chadron, Nebraska, to a railroad family, his father, H. D. Barnes, was a long time official who gave Jim a good grounding in the ways of the C&NW. He graduated from Iowa State and served as an officer with the U. S. Navy, rejoining the railroad after the war. He would eventually become vice president-chief engineer. Soon after retirement, Jim died suddenly on May 31, 1994. Vern Brachman and I never did find out why we ran the two miles of levels through the middle of the Proviso Yard.

On August 2nd, claim agent Warner Frank asked me to go with him to the Chicago Union Stock Yards. A C&NW brakeman had been injured there the day before. The unfortunate man was riding the side ladder of a stock car when he struck a vertical timber. The accident occurred on the Chicago Junction Railway. Warner and I could not find any kind of advance warning sign that this was a close clearance for a man on the side of a car.

With a vacation looming, I was interested in winding up my work. The last job was setting the invert elevations for a 66 inch ARMCO corrugated iron pipe drain running parallel to "20 Main" at Proviso alongside the new diesel shop. The pipeline was designed to drain the swamp water from the diesel shop site eastwards to the east to Mud Creek, also called Addison Creek.

My second paid vacation arrived but our family had little money to spend on travel. By 1950, however, I had finally earned a single round-trip pass on a foreign line railroad. With my wife and small daughter our journey began with a ride into Chicago on the CA&E from Elmhurst to Chicago. Why the electric rather than the C&NW? Because the walk to the Grand Central Station on Harrison Street was shorter. In those days a taxi ride was deemed a waste of money. We rode the C&O's *Pere Marquette* day train to Holland, Michigan. An extremely pleasant ride and a good diner, too—with no tipping permitted by official edict. A week at a rustic camp compound on the western shore of Lake Michigan passed all too quickly.

Then it was back to work and off to Proviso again to set the final centerline stakes for the new shop trackage. I had a new rodman, Byron E. Ruth, to help me with the north ladder track. We also ran levels in the drainage ditch south of the shop, just in the nick of time, as Harold Jensen came to our office the next day demanding the profile we had just run. He was impressed that we had anticipated his request. On August 23rd we learned that our good division engineer, Charles E. Hise, had been promoted to an office engineering position in the chief's office. His successor was Bruce G. Packard who was to prove to be a very difficult man indeed.

On August 25th the Federal Government seized the railroads to avert a strike.

Even though it was a Saturday morning I met Charley Cook on August 26th at the West Chicago stock yards to inspect the old railroad owned sheep barns and stock pens. Charley was the railroad's livestock agent and one of his major shippers, Armour and Company, had expressed an interest in using the secluded facility for some secret experiments. The stock yards and barns lay north of the West Chicago East Yard across some swampy ground and were rarely used. We walked the grounds noting the disrepair and the many alterations that would be required. On Monday I wrote up a proposition to fix the worst of it, but the whole idea soon vanished without a trace. Cook was an interesting hangover from a past age. His tales of the wild west, cattle drives, aggressive competition of the railroads for the livestock traffic, life in a drover car, the adventures he had when he loaded live buffalo in South Dakota for shipment east, the incredibly hard life of the cattlemen and their herders was from a time that would never return. Now in 1950 he was near the end of his working career and the whole livestock shipping business was almost dead. Even the Chicago Union Stock Yards was doomed by major changes that were reshaping the meat packing business.

In late August the railroad began replacement of the concrete sidewalks on the Madison Street side of the Chicago Passenger Terminal. The contractor who built those buttresses at bridge #66 east of Geneva, Erik Borg of Chicago, was awarded the contract, and I was tapped to oversee the work. Embedded in the old pavement were several bronze property line markers about 6 × 18 inches in size. We carefully saved these plates and stored them inside the station until we could reset them in new concrete. Some lowlife from West Madison Street stole one of them and probably sold it for booze money. On September 11th we carefully set the remaining corner and line markers and concreted them in. Don't look for them today; the sidewalks were demolished along with the CPT in 1984.

With the arrival of September, school started again for me and I settled for those jobs close to Chicago. Another new rodman appeared, George Bayer. When rodman Vern Brachman and I were working at Proviso on the west trackage for the diesel shop, we spotted the carcass of a class E, 4-6-2, #1661. This engine had been wrecked August 28th when it ran into a washout at Barrington, Illinois, caused by failure of a large water pipe under the main tracks at a grade crossing. All the valve gear was gone most likely to facilitate moving the wreck to Proviso. At this time there was another big wreck on the Galena main line east of Glen Ellyn. Instrumentman Tony Zaborowski and I went out to see if a derailment survey was necessary; thankfully, it wasn't. In the burning wreck were several cars of Anheuser Busch beer. When the firemen from Lombard and Glen Ellyn heard the cry, " There's beer over here" all the fire hoses turned in that direction. The crowd of on-lookers raided the broken cars and hauled the cases of beer away despite the efforts of our people to keep them clear of the wreckage. Much of the stolen beer was hidden in the weeds along the right-of-way. That night every hospital in the western part of the county was filled with some very sick people. When beer gets too hot in a bottle, chemical changes occur making the heated beer very unfriendly to the human digestive system. Busch representatives soon descended on the wreck and made vigorous efforts to collect every bottle for return to St. Louis.

Anthony B. Zaborowski, our new instrumentman who later became the assistant engineer in charge of the drafting room personnel, came from Milwaukee, Wisconsin, where he continued to live with his large family despite having a job in Chicago. He commuted from Milwaukee every day on an early morning train and returned home in the evening. At first he tried renting a room in a cheap hotel across the street from the Chicago Passenger Terminal, but soon the comforts of home drew him to commute 180 miles a day. An edgy man and wary as a fox, he always had a cigarette in his face. In his early days as a young engineer Tony had worked for the Tennessee Valley Authority on a big dam project. He related the story of a fellow engineer who was responsible for tunneling operations. The man kept careful records and predicted when the tunnel faces would meet. On the day predicted for break-through, there was nothing to be seen but solid rock. The engineer shot himself. Later, a recheck of his survey notes revealed an error of 100 feet where the engineer had "dropped" a station. The tunnel faces met perfectly when the last 100 feet was mined out. Tony seemed a bit haunted by that incident as he was very careful when calculating distances. People management, however, was not one of his skills. The smooth operation of the division engineering staff soon disappeared under his strange way of assigning work. He had three levels of priority for in-coming work. The in-basket on the left side of his desk were requests that had to be done. The next basket in the middle of the desk contained requests that could be stalled. The third basket containing the most files occupied the right hand side of the desk. These were the files that he hoped would just go away if not disturbed. Depending upon pressure exerted, the work would drift from right to left in direct proportion to the amount of screaming associated with the request. Nothing fazed Tony. He was immune to pressure even when the division engineer told him in no uncertain terms to move. This peculiar management style made planning field trips very difficult as it was impossible to organize requests in any kind of geographic order. Those swinging doors to the division engineer's office got a lot of action as Tony was often summoned to report why some job or other had not been done. He was his own worst enemy. I can still see him vigorously rocking back and forth in the desk chair puffing smoke like a fast-running steam engine.

On September 12th I went to Winfield on #3 to check to see about raising electric wires to permit movement of a house across the three main tracks. This was a big operation and a lot of railroad people were involved to protect the move. From Wheaton I rode the CA&E to Chicago on my annual pass arriving just in time to pack some oak stakes and other supplies for instrumentman Bill Polchow working on a mainline job in Dixon and grab train #1. After delivering the stuff to Bill, I did not hang around to help; I found a Greyhound bus going east and got on it for a late night arrival home.

We now had a serious problem with billing by the State of Illinois' contractor in connection with the construction of bridge #1936 near Girard. To settle the matter I requested that Joe Gaulock from the ACE office accompany me to Peoria for a meeting of the involved parties. On the 19th we rode Pullman *Prairie Creek* on the Rock Island's night train to Peoria, had breakfast at the White Hut and then rode a Greyhound bus to South Pekin where we met Ray Doty in his office. The contractor, Bremer, and the State of Illinois men arrived for the meeting. Joe surrounded by smoke from his billowing pipe proved to be a cagy negotiator in the four hour meeting. The allocation of costs was settled to everyone's satisfaction. Ray then drove us to Pekin to catch a bus for Peoria. We had missed the *Rocket* and settled for the local to Bureau Junction where we connected with the eastbound *Corn Belt Rocket*, train #10. The very next day a Federal auditor showed up to review the same billing we had just discussed. Project accounting was taking more time than building the bridge.

The Chicago Shops was a city in itself. With a lot of structures and a large employment, it was also the scene of a lot of thievery both by employees and outsiders. It was said that one employee over the course of many years took home enough tools and parts to build a steam locomotive. The biggest access to the property was on the south side from Kinzie Street. There were no fences, just a quick walk over the main and side tracks and there you were. Management determined that a six foot cyclone fence was the answer—one mile of it. Extending from the limestone backwall of the Pulaski Road (old Crawford Avenue) underpass west to the Belt Railway overpass near Kenton Avenue, the fence would have only two openings: a pedestrian gate at the Keeler Avenue Station and a locked vehicular gate at the access road at the west end. The City of Chicago refused to give the C&NW a construction permit for the fence. Our attorneys soon discovered that it was not the city that was holding up the permit. It was a local alderman who used an argument that the fence would establish a legal barrier to the city should they ever want to extend streets across the railroad's property. Actually, he was looking for a payoff in the good old traditional Chicago style. Legal action was being considered to force the matter when suddenly someone realized that the Chicago Shops and the Fortieth Street yard properties were acquired in 1872 before the City of Chicago's limits extended that far. The said streets to be extended by the alderman were, in fact, phantoms and had never been platted. Faced with these facts the permit was grudgingly issued and the fence was built. Of course breaching began at once with automobiles bashing it during parking while wire cutters were soon brought into use at several points along the way to make convenient pedestrian crossing points.

During this Fall of 1950 a new boiler was installed in the M-4 power house at Chicago Shops and an incinerator was built at the California Avenue Coach Yard. Although this work was supposed to be supervised by the Mechanical Department, we were called upon to set elevations for placing the boiler and to stake the location for the incinerator. Somehow, I ended up being the one who had to set-up payments to the contractors on both these jobs. These projects were not really ours, yet they afforded a good excuse for an inspection trip whenever office work began to pall.

On October 9th building M-17, the Galena engine house was damaged when a steam engine somehow knocked out a post and brought part of the roof down. The B&B crew soon put things back because the Mechanical Department needed that structure for six more years.

Accidents tended to happen. When a river drawspan is damaged, the sudden interruption to both rail and river traffic can be tremendous. The Galena Division had three drawspans in its territory: the double-tracked Kinzie Street bascule over the North Branch of the Chicago River, #1731, a lift span over the Illinois River at Pekin, Illinois; and the bridge over the Chicago River on the St. Charles Air Line south of the Chicago Loop. The latter was jointly owned as was the entire Air Line. The first one at Kinzie Street carried local freight traffic to the North Pier area. But the middle one, #1731, was a big twin tower lift span over the busy Illinois River and carried the Southern Illinois district main track. Barge operators on this stretch of river operated big tows of 15 barges or more which often hit the pier protection at the bridge knocking off bits of treated timber fenders or denting the caissons meant to ward off errant barges. This bridge was constantly under attack and commanded a lot of attention from engineering, B&B forces, law and claim departments as well as adding problems to train operations. One of the bridge department's brighter ideas resulted in construction of a sheet metal caisson filled with sand and gravel and capped with concrete on the upstream side of the bridge to encourage barge tows to stay in the channel under the open span. That kind of construction turned out to be a mistake, when soon after completion of the new caisson, it was cornered by a barge and split open allowing the sand inside to pour out into the river. Rebuilding the caisson, we filled it with concrete from bottom to top.

November in Chicago was especially unbelievable in 1950. The temperature on the 1st was 81 degrees; on the 24th the high for the day was −4 degrees. During this period we tried to complete track changes at the State Street yard office to clear the site for the city's bridge construction. We didn't quite make it and some track shifting became very difficult because of the sudden frost. In early December we went out to stake the centerline of a new track for Imperial Flooring and Waterproofing Company at Proviso. We broke more wooden stakes in the frozen earth than we managed to set. At the new diesel shop a diesel being serviced was driven right through the triple level service bridge completely destroying the screw mechanism. A blizzard roared in on Pearl Harbor Day causing vast interruptions to all train service. It was going to be a tough winter.

On the morning of December 13th W. H. "Heinie" Huffman in the chief engineer's office sent out requests for volunteer switchmen from division staff; the railroad had been hit by a wildcat strike. Rodman Bob Lawton and I were interviewed by operating officers, Charley Longman and Joe Stein (who went to the Monon as vice president-operations in 1960), to make sure we were ready, willing and able to carry out our assignments. Neither of us had ever worked as trainmen or groundmen, but we were confident we could do whatever was required. Bob and I went to our homes to collect cold weather clothing. The snow was fresh and deep with temperatures dropping to 15 degrees that night. I caught the 3:30 p.m. train at Elmhurst and reported to the yard office at the east end of Yards 6, 7 and 8 just north of the Berkeley stop at Mannheim Road. There I was presented with a brand new Adlake kerosene lantern and assigned to the 3–11 p.m. roustabout job, conductor R. St. Martin. I climbed on the Alco S3 diesel switcher and made my first mistake. I was greeted with "Get that hayburner out of here" from engineer Santore. Putting the lantern outside on the deck, I went back inside the cab to learn what we were supposed to do. The roustabout job was the maid-of-all-work that transferred cuts of cars around Proviso. The first assignment we had was pulling cars out of the south side of the hump yard (Yard 5). My training as a switchman began with trudging along a string of cars in two feet of snow to make sure they were all coupled together. Making signals with that lantern was tricky until I figured out how to make the bale rigid by properly slipping the wire locks into place. Even then, I took it slow and easy making sure I was well away from the cars as they were shoved and pulled in the darkening December evening. All together at last, the conductor and switchmen went back to the engine which then slowly pulled the cut out of yard 5, through the scissors crossovers and down the ladder track to the Middle and then to Yards 6, 7 and 8. We ran past the yard office and uphill on a connecting track to the IHB that crossed high over the passenger mains at the very southeast corner of the huge Proviso complex. There I was at midnight, up on the top of the boxcar next to the engine, on a very cold and snowy December night, trying to see a pinpoint of light that was the switchman's lantern signal that it was okay to shove the cut west into a clear track in yard 7. There were many pinpoints of light to be seen—most of them from automobile and truck traffic crossing the yards on the Mannheim Road bridge. I waited a long time estimating how long it would take a trudging switchman to line switches before I could, I thought, make out a faint circle of light that would mean to back up, or in our case, to come west with our cut of cars. I passed the signal to Santore who eased us down that long lead track into a departure track. And so the night went—slogging through snow, coupling and uncoupling cars, bleeding air from reservoirs, and relaying signals to the engineer. I thought the roustabout crew was working very slowly that night because I was a green man and the yard was full of similar volunteers. I later discovered that was the way they operated every night and always got their 16 hours in for the overtime. When the hog law limit of 16 hours was about to legally end their day, Santore made that diesel fairly fly to get to the welfare building and tie-up within the time limit. Although I actually had been on duty 22 hours and 40 minutes, I put down 16 just as the rest of the roustabout crew had done. I caught a westbound bus on Lake Street to Elmhurst, got to bed at 7 a.m. and slept well.

The next day I worked the west end of Yard 7 as a switchman from noon until 8 p.m. The temperature was plus five degrees. This was a different kind of work from last night. We were actually making up trains by pulling and sorting cars and cuts of cars. The day was an endless succession of pulling cuts out of snowy yard tracks, quick accelerations, grinding stops, the crash of couplers and slack running in and out, men running along side cuts holding uncoupling levers, lining switches and trying to remember the proper signals relative to the way our diesel #1090 (an ALCO S-2) was facing. All C&NW diesels had an "F" painted on both sides at one end indicating which end of the locomotive was the front. The groundcrew had to know which was the front in order to give the correct signals. We made a few mistakes by putting cars into the wrong tracks, but no one was injured or any damage inflicted—that we knew of, anyway.

The third tour of duty was a switching assignment again—at night. Temperatures were -5 degrees following another storm that put a lot of snow down in the yard. The work was a repeat of the previous day, pulling and sorting cuts of cars with me the only green man in the crew. This night we had another diesel engine #1072. Frequent visits to the yard office with its glowing stove were welcome relief from the bitter cold. We made a mistake this time that upset a lot of trainmen. Somehow the switch to the waycar track was opened by the field man. I was the middle man half way up the ladder and the conductor was at the head end with the switch list. Just as an old tank car was kicked down the lead, the conductor spotted the open switch to the caboose (waycar on the C&NW) track and yelled at me to catch that car. I swung aboard and began twisting the vertical staff handbrake wheel. Nothing happened; ice must have frozen the locking dog open. I turned that staff as hard as I could with no effect. I rode the slowing car into the waycar track; it was not going to stop before hitting the line of standing waycars. Meanwhile the conductor had sprinted past me while I struggled with that useless brake and ran along the line of waycars yelling, "Runaway! Runaway!" while pounding on the sides of those wooden cars. The sight of the sleeping trainmen flying out of those waycars wearing little else but long johns and landing in the snow banks along the track was something like a Mack Sennett silent movie scene. I bailed off the tank car just before it hit the first waycar. The collision was about like a hard coupling and everything stayed on the track. I was surprised at the amount of dust that flew up in the clear early morning light and hovered there as a cloud in the still air. There were comments about greenhorns endangering men's lives and other unprintable language, but the cold air and their underwear uniforms quickly chased the discommoded trainmen back aboard their warm cars. My shift was over at 7 a.m. and I went home to sleep.

The strike was over on the 16th. I was glad I did not have a fourth tour as the temperature was again at −5 degrees. Since there had been no interchange with other railroads, the C&NW volunteers could only switch cars on hand for departure until the strike should end. Proviso, for once, was all sorted out and had, it was said, 25 trains ready for departure. The cost was one locomotive banged-up when a cut of cars drifted out the end of a track and one man suffered a hernia throwing a snow-clogged switch. It could have been worse. For some reason the Federal Attorney General's office took all our names.

President R. L. "Bud" Williams sent personal letters to all the volunteer participants along with checks of $25.00 for each work assignment. Dated January 12, 1951, the letters said: "I am certain that you must have something of the same feeling of satisfaction that I have, when trains continue to roll and yard operations are kept in motion, in spite of the most discouraging circumstances. The fact that you volunteered and reported for duty in train and yard service during the unauthorized strike of trainmen December 13 to 16, 1950, a line of work well outside the scope of your regular duties, makes this entirely plain. Speaking not only for myself, but for the Railway Company, we are grateful for this service, and very appreciative of the spirit of loyalty to the 'North Western', which prompted you. It is most heartening." My check was for $75.00.

1950 was almost over. I made a quick trip with Bob Lawton to Wheaton to check a freight platform, to West Chicago to locate a new truck garage (the roadmaster finally got a place to park his new truck), and to Melrose Park to chain some trackage for a completion report. Instrumentman Bob Nelson transferred December 27th to the Iowa Division located at Boone, Iowa, and I applied for the permanent position then opened on the Galena. The day after Christmas I was again pressed into service to run a snow gang at Proviso. The same problem of keeping them shoveling snow instead of hiding out in some warm place kept me constantly on their necks. These poor brokendown specimens of humanity were not fit to do this kind of labor and should not have been sent out by the labor contractor. The year's weather summary said that this winter was the hardest since 1934.

Far to the north, on the west end of the Madison Division, the last passenger train to Lancaster, Wisconsin, ran on December 30th behind class R-1, 4-6-0, #344. Through the courtesy of my good friend, William F. Armstrong, who worked for the chief architect, I received a nice set of pictures recording that last run. Bill inherited his interest in railroads from his grandfather who was a C&NW superintendent in Iowa around the turn of the century. Bill, born April 9, 1921, graduated from Iowa State at Ames with a degree in architectural engineering and served as an officer with the Navy in the South Pacific on a transport vessel during the war. Joining the C&NW in 1946 he worked for L. C. Winkelhouse, chief architect, who reported directly to the chief engineer. Counting the draftsman, the tiny department of three was responsible for design, construction and remodeling of all buildings. Using his responsibility for structures, Bill traveled extensively across the entire railroad and was able to record many historic events and structures with his trusty Argus C-3. His pictures have been widely reproduced in railroad interest publications.

BRANCHING OUT, 1951. . . A minor derailment at Geneva on January 2nd opened the new year for the Galena Division engineering department. The fine tuning of the morning commuter fleet's operation became a raucous squawk that day eliciting a lot of bad public comments. Management told the division engineer to get out there and find the cause of the mishap. Bruce Packard personally led the investigation and conducted a standard derailment survey on the curved St. Charles branch connection just west of the Geneva depot. Vern Brachman and I were probably selected as his assistants because we also lived in Elmhurst and were handy. Perhaps, Bruce just wanted to see how we functioned in the field as we were new to him. Apparently, we did all right for the very next day I got my permanent instrumentman position. The weather continued cold with ice and more snow. Indoor railroading was a lot more attractive even though we had those dratted completion reports to write along with the usual estimates, map corrections, leases and licenses. *Forbes Magazine* reported that C&NW managers were the country's least efficient. We sensed the truth of that remark but had no basis for comparison.

The office force was dwindling. Rodman Dick Stemmerman took his alcohol problems with him to Boone, Iowa on January 5th leaving the Galena Division four persons short. On the 10th we had an assistant engineer, three instrumentmen, two rodmen and a tapeman who announced he was leaving on the 12th. Our shortage got to be a joke as requests for engineering services poured in. On the 12th there was a major building fire at 320 N. W ells Street. I was sent over to assess its effects on the North Pier line. In mid January the weather eased up a bit only to clamp down again toward the end of the month. During that brief mild spell, we managed to get several little jobs done: vertical clearances at bridge 16-B at Proviso, set some elevations for a B&B crew working on the transfer table at building C-1, Chicago Shops and conduct a situation survey at the Wells Street fire site. Despite –5 degree weather, Bob Lawton and I rode the CA&E to Aurora on January 31st to inventory a new industry track serving the Hill-Behan Lumber Company at North Aurora. There was a bit of urgency about this little job as it was billable and the railroad needed the cash.

That same day the switchmen went out on strike again. This time I declined an invitation to go to Proviso and switch cars because I was leaving on a long-planned vacation to Texas the next day. Our ride south in my father-in-law's automobile was scheduled for February 1st.. Our return was to be by rail, and I had my half fare tickets all ready. Severe winter weather with minus temperatures still gripped the Midwest as the switchmen's strike spread to other railroads. The CA&E, which mostly paralleled the C&NW's Galena Division, had also shut down (January 29th) in a labor dispute which sent a huge number of their riders to the C&NW trains. The C&NW put on an extra morning and evening train to reduce the commuter crush. By February 8th the switchmen were back at work, but the CA&E was still out. (They resumed operations March 11, 1951). I missed most of this turmoil by being on vacation, but on my return from Texas, Warner Frank, the Chicago area claim agent, immediately called for an accident survey at Roosevelt Road on the South Branch. Luckily, Vern Brachman and I benefited by a break in the cold spell to do our work there. The best news of this period was a pay increase; my new rate was to be $396.78.

On January 27th I was officially notified that the long overdue replacement of the Mannheim Road viaduct at Proviso was to begin and that I was indeed to be the railroad's resident engineer. Although the project took three years to complete, I did not spend all my time on-site as there were interludes in the construction where I could make myself available for the normal division work. Being in the Chicago area more or less constantly, however, did greatly facilitate my night school attendance. On February 23rd I met with C. O. Weis, State of Illinois highway engineer and Leroy "Roy" Stift, the C&NW superintendent at Proviso. It was going to be necessary to open a clear sight along the old bridge across the railroad yards. I counted 57 live tracks to be crossed as well as the busy switching leads to Yards 6-7-8. The west end of old yards 2 and 3 did not seem to be a problem. Yard 4 was the biggest problem for it was mostly a storage yard for crippled and retired freight cars. Roy didn't see much problem with Yards 2 and 3, but 4 was a monumental pain as many of the cars were wrecks or were missing drawbars. Roy felt that replacing the bridge was going to be an enormous headache and major interference with his operation of Proviso. I assured him that I would do my best to keep his railroad fluid but I hoped he would issue an order to his yardmasters to give me some cooperation when I needed it. Yes, he issued an order, but it got buried on the bulletin boards, and I was to spend a lot of time cajoling bristly yardmasters to send switching crews to open the yards. The State's survey crew showed up on February 26th to reconnoiter the survey site and scheduled the next visit two days later. I got there on the 28th and couldn't find any surveyors. I later learned they had gone to another job on the Congress Street Expressway then under construction. Packard was irate (as he usually was) and chewed out the state engineer for wasting our valuable time. Not the way to win friends and influence people.

The survey actually started on March 5th and I had my first rough experience working with the yardmasters to get the sight line opened across those many tracks. In Yard 4 a switch crew worked an entire shift to open 13 tracks. Many derelict cars had to be chained-up to be moved a few hundred feet. Essentially Yard 4 was bowl shaped with the overhead bridge at the lowest point. Quite often, every night it seemed, a careless or thoughtless crew would slam cars into standing cuts that would then drift slowly to the bottom of the bowl thus blocking our sight lines. Every morning my first job was to track down the yardmaster at the East 5 to get a crew down in Yard 4 and pull it clear. Finally, I decided to fix this daily agony. After getting the cars moved one last time, I had the sectionmen spike the switches shut despite the yowls from the yardmasters about loss of track space. While Yards 2 and 3 were not so bad, they could be a headache as the night transfer jobs either hadn't read the general order about leaving cars well clear of the bridge or just didn't care. Many a morning I bled the air off those smelly stock cars while waiting for a crew from Yard 8 to shove them clear or to take them away. Somehow, the survey was finished by March 16th. The next event was soil sampling along the centerline of the new bridge. While not a problem on land between the tracks, crossing into the active track zone was another story. The contractor's drilling rig was mounted on a caterpillar tread crane which was able to cross tracks on timber blocking placed by the rails. They did not have to drill very far as there was a solid limestone ledge about 26 feet down. The entire area was underlain with solid limestone which formed an underground layer that ran from east of the bridge site about four miles to the East Branch of the DuPage River just west of Elmhurst. Commercial quarrying of that limestone layer had been carried on in Elmhurst since the 1870s. That quarry site is still there looking to the airline passenger landing at O'Hare like a gaping hole where a tooth has been pulled. Back at Mannheim Road the final soil sampling was done by the state's contractor, Gordon Benson, on April 23rd and 24th. That finished the field work for now. The next stage was drafting plans and preparing bidding documents. I went on to other projects. Actual construction did not commence until May 6, 1953.

Turning back to February my journal showed that Bob Lawton and I were out at Bob's new project at Chicago Shops on the 19th. The Santucci Construction Company was installing a 14 foot diameter sewer line parallel to the Belt Railway tracks across the western part of the Shops complex from Kinzie Street on the south to Chicago Avenue on the north. Tunneling of the main tracks and open cut for the balance of the job made this work a tremendous disruption to the railroad's operations. Removal of spoil by trucking required flagman protection at the only grade crossing of the Galena main lines at Kilbourn Avenue. The network of old shop tracks with their funny names (Iceland, Sweden, Germany 1, *etc.*) had to be removed one by one and replaced. Since many of these tracks were old and obviously disused, their replacement seemed like a waste of money. But the City of Chicago was footing the bill, so we restored them. At Proviso on the 21st rodman Vern Brachman and I chained a rail renewal from Grand Avenue to North Avenue at yard 9 (North Yard), inventoried the material, hiked over to yard 7 and chained another mile of track to Mannheim Road. There we inventoried another new track at Joseph Lumber Company. It was a very full day with much bending over at 100 foot intervals. I was lean, thin and limber in those days.

Winter returned in March. A storm dropped five feet of snow in Iowa on the 19th causing huge backups of freight cars at Proviso. The same cold front dropped temperatures in Illinois making our job of staking a curve in the number 3 main at County Line (Proviso) a numbing and uncomfortable experience. Near the end of the month, on the 26th, a new rodman, Byron Ruth, reported for work. Byron was an energetic young fellow who may have lacked sophistication, but made up for that with his generous and candid character. His uncle, Hal Ruth, was then the assistant engineer on the Wisconsin Division. Another new rodman showed up April 2nd—John Zima. Our labor shortage was somewhat eased by these rookies. The next day I took rodman Bill Brown with me to Rockford on the Illinois Central's *Hawkeye* at 11 p.m. After we laid out a track to serve Cellusuede Company on the KD line, Bruce Packard, the division engineer, accompanied us to Freeport for a meeting with J. G. Gokey, the owner of the Structo Manufacturing Company. The Rockford agent, Williams, drove us up there and back which greatly relieved our transportation poverty. I suspect that if Packard had not been there, we would have had to scrounge some other way to get to the job site. The meeting was held on the ground where a new factory was planned east of Freeport. We laid in a neat sidetrack to serve the proposed loading platform and went back to Rockford where the next day we did some work on the Churchill Spur and conducted a survey on bridge #474 near the old J. I. Case plant. Returning to Elmhurst on the Greyhound, I was glad to know that the next field trip was to be a positive vacation—we would have a car! Rodmen Ruth and Zima picked me up at home for a field trip to Nelson where we staked out a long track to a gravel pit operated by D. F. Butler on the south bank of the Rock River. The track began east of the Nelson tower known as NY and ran north of the main lines to the pit. On the homeward trip we inventoried the CTC crossovers at RX (Rochelle) and HX (Creston). This certainly was a more efficient use of our time.

In April at the State Street yard, our paving contractor was preparing to place the concrete at the new yard office building and requested that we set the elevations. On the 25th the claim department hustled us out to Halsted Street on the 16th Street Line where Howard F. Rathbun, a brakeman, had suffered a fractured hip and loss of two fingers in a 1946 switching accident. We measured the accident scene as directed and thought no more of it. Four days later while working with Ralph Golterman at Proviso staking a track for Imperial Flooring and Waterproofing's concrete mixing plant, an urgent message was delivered to us on the site. I had to be at court at 2 p.m. With some lucky bus connections I managed to get home to Elmhurst, changed into my city clothes and got there on time. I testified about 35 minutes. John Danielson, our attorney, said I had done well. The jury awarded Rathbun $111,000. While visiting John in his law office on the 27th, I took my copy of *Pioneer Railroad* and arranged to get it signed by Lowell Hastings, Bradford W. Carlton and Frank V. Koval. (The book went to the C&NW Historical Society for their archives in 1992).

Instrumentman O. W. Smith returned to the Negaunee track relocation project on the 27th. Since we were short of help again, division engineer Bruce Packard "helped" Bob Lawton, Ralph Golterman, John Zima and me layout a new bridge number 16-B on Proviso's Harbor Hill. After he left, we got along much better. I took rodman Bill Brown with me on several trips because he had a car. On May 11th we went to Algonquin and then to Garden Prairie on the Freeport Line where we had to cram a tight little industrial sidetrack to the Northern Seed Company's new warehouse. As I set the instrument up at the point of intersection of the curve, I was conscious of a bad smell in the immediate neighborhood. Looking more carefully, I saw that I was standing next to a pile of rotting muskrat carcasses. Holding my nose and being careful where I stepped while running the transit, I gave the angles to Bill who was wondering why I was in such a hurry to finish. We investigated a drainage complaint at bridge #28 east of Winfield on May 15th. Bill and I listened to a property owner north of the main line claim that the railroad embankment was blocking surface drainage causing water to back-up into his greenhouse. Later investigation showed there had been an opening under the tracks before 1923, but during construction of the third main, it had been filled. Wet weather always brought drainage complaints. On the 22nd I went out to check on one at Lombard. It was easy to figure that one out—the Catholic church had paved its parking lot and concentrated run-off overloaded the normal drainage ditch. Could this be an Act of God? On May 23rd I rode train #111, the *City of Denver* to Sterling where rodman Bill Brown was supposed to meet me. However, Bill missed train #23, and had to ride #15 which got in at 4 a.m. Early that morning (Bill didn't get much sleep) we grabbed train #26 to Nelson where we staked out the D. F. Butler gravel pit track. Joe Murphy, roadmaster Art Benson's truck driver hauled us to Sterling in time to catch #14 which was ahead of schedule so much that the conductor graciously stopped at Wheaton to let me off to catch a local to Elmhurst. On June 6th, Harold Jensen sent me to Proviso on an emergency trip to investigate a rumor that defective rail was being installed in the new shop—the rail was fine, minor surface defects in it had been greatly exaggerated

I was accepted as a junior member of the American Railway Engineering Association (AREA) April 20, 1951 and became a full member March 3, 1957. On June 29, 1951, chief engineer, E. C. Vandenburgh, wrote to me commending my attendance at the evening division of the Illinois Institute of Technology. Probably just a form letter, but it was nice to have some recognition of my efforts. On December 22, 1952 Jack Goodwin, then vice president-operations, sent me a similar letter on behalf of president Paul E. Feucht. Another letter dated September 18, 1953 came from M. S. Reid, my division engineer. The chief engineer B. R. Meyers, (Vandenburgh's successor) wrote on August 31, 1954 and the next year on September 16th. Was there a rotating duty list for writing these congratulatory letters?

TIME FREIGHT 381. It would be impossible to recount the number of times Robert Lawton and I had heard that number mentioned by train dispatchers, superintendents, trainmasters and crews, roadmasters, section men, tower operators and anyone associated with train operations on the Galena Division of the C&NW. The number referred to a freight train that operated from Proviso Yard west of Chicago to Madison, Illinois, across the Mississippi River from St. Louis, Missouri. The train was usually long and heavy and ran west from Proviso to Nelson, Illinois, where it turned south to South Pekin and Madison. Three Eighty One had an aura about it that made it special from all the other C&NW freights. Two railroads handled 381 with four train crews, a foretaste of the run-through concept that was to become common practice as diesel locomotives became standard on all of the major interchange railroads. Bob and I decided to ride 381 all the way. That was one way we could inspect the railroad and structures in places we rarely went in our normal engineering work, and we could experience train operations at first hand. Saturday, June 9, 1951 was day we picked for our adventure. Armed with our annual passes and locomotive permits we went to the Bellwood commuter stop loaded down with lunches, timecards, track profiles and cameras. While the weather was overcast, there was a good chance it would improve as we went west. There was a lot of switching activity on the east leads to yards 6-7-8 where outbound trains were being assembled. About 8:30 the roustabout crew with diesel switcher #1018 pulled a long string of cars from the hump classification Yard 5 through the "Middle" to the old third main track in Yard 3 where we were waiting by some reefers gently dripping brine. Yard 3 was a relic of the original Proviso Yard before the big expansion of

1926–28. Switch engine #1018 had the first piece of today's 381. These were cars that had come down from Wisconsin the previous night. One flat of farm machinery was marked bad order; the roustabout crew cut it out to take back to the rip track. This cut of cars was not yet a train. We walked the length of the train and noted flatcars with big industrial cranes from South Milwaukee, cars of Wisconsin cheese, cars of furniture, malt, beer, paper products, gasoline engines from Waukesha, lath, farm equipment and an export load of bottling machinery destined to Montevideo, Uruguay. As we hiked back to head end, there were muted hisses, creaks of thermal expansion, the sounds of heavy things settling, all of which added to an atmosphere of something about to happen.

Crossing over the rickety old wagon bridge that carried U. S. 45 over the yard tracks, Bob and I came to the north side of Yard 4. From the switch tender stationed under the noisy bridge we learned that we were to have the customary three unit diesel set which had supplanted steam power on the Southern Illinois in the past three years. As engineering staff responsible for track maintenance we were glad that those dirty, loose-jointed, unbalanced old steam engines were gone. Diesel power was a lot easier on our track. Steam locomotives tended to pound joints, chew-up rail, and drop hot cinders on our treated timber trestles and track ties. At the Welfare Building to the west, we could see the yellow, black and green diesels waiting for their road crew. They had just been brought down from the new diesel shop by the hostler and were called for 9:30 a.m. As we chatted with the switchtender, the squawk box advised the yardmaster of Yard 6 that he had better "get the SIs on 381". The SI was, of course, the Southern Illinois, which had been its own division before 1937. Bob and I walked to the murmuring diesels and threw our gear into the cab of F7 unit #4079-A as engineer "Wild Bill" Alexander and fireman Ray Kimes climbed up and checked the units over. At 9:30 a.m. Alexander kicked off the engine brakes, and the three units ambled east up the Harbor Hill to cross over to the lead track that would take us to Yard 3 where our train was being assembled. Alexander hurriedly set the brakes as a grimy B&O mike (2-8-2) churned past on the next running lead under a cloud of black real estate trying to make the top of the hill with a heavy drag from Yard 5. He slowly made it onto the IHB interchange. When the faded red wooden caboose crept past, we got the signal to pull up and then reverse back down the hill to Yard 3. We coupled onto the cars, tested the air and became train #381. Another delay while the roustabout crew blocked the departure track cost us 42 minutes. Kimes speared our orders on the fly as we rolled past Wolf Road under the gaze of Frank Krish perched in the general yardmaster's high tower on the old ice house. The F7s eased through the switches at County Line at the west end of Proviso. "Red board" sang out the fireman. "Red board", replied the engineer as he made a brake application. The reason for the stop was soon clear: The *City of Portland* (#106) appeared eastbound at HM tower just ahead of us in Elmhurst. The streamliner swiftly rolled east with its string of travel-stained yellow cars as "Green board"was heard from fireman Kimes; "green board" replied the engineer as the ammeter needle climbed to 625. Every 50 seconds the automatic train control horn sounded which Alexander had to acknowledge within six seconds by toggling a brass lever on his right side. Bob and I walked back through the straining F7s and experienced the noise and vibration of diesel engines under heavy load.

Through the Tree Towns of Elmhurst, Villa, Park, Lombard and Glen Ellyn 381 accelerated its substantial train of 99 cars. At Wheaton the ammeter began to drop as Alexander notched the throttle down. A glance at the track chart profile provided the answer—we had crossed the divide between the Lake Michigan and Fox River drainages. These three main tracks from Proviso to West Chicago with their heavy rail and deep rock ballast were racetracks for the many suburban trains. This was Saturday and just one eastbound scoot was spotted loading passengers at Wheaton. South of the mainline at Wheaton we could see the CA&E fleet of interurban cars in their new scarlet and gray paint; the older livery of blue and gray was fast disappearing. Winfield was passed. Cruising around a curve and through NI tower at West Chicago we entered CTC territory at 11:30 a.m. We rolled across the EJ&E crossing, past the West Chicago yards filled with the commuter fleet, eased through the high speed equilateral #20 turnout at WX into the double track that extended all the way to Council Bluffs, Iowa, and crossed Kress Road and the long diagonal crossing at Roosevelt Road, U. S. Highway Alternate 30. Alexander advanced the throttle for a short upgrade haul before the train crested the ridge marking the east edge of the Fox River Valley. 381 dropped down the grade to the river crossing at bridge #66 where I had been involved with the buttress reinforcement, through the Geneva station grounds, around a sweeping curve to GX crossovers and then the long upgrade of the west slope of the river valley. In steam days the LaFox Hill was a difficult slog; to the diesels it merely meant more amps to the motors. Cresting the hill, Alexander applied the dynamic brakes bringing the train to a smooth stop at Meredith siding for a mandatory standing inspection of the train. The 74 loads and 25 empties were okay and the speedometer soon read 60 again. The June blossoms of mustard, blue lupine and the scattered pink of prairie rose mingled with the long stemmed grasses that bowed as the pressure wave from the rounded nose of the 4079A spread outward to the right-of-way fences.

Galena Division timecard No.9 listed the speed limit in this territory as 90 for streamliners, 80 for conventional passenger equipment and 60 for freights. 381 was operating close to the 60 limit as it passed Cortland but slowed through DeKalb, recovered through Malta and slowed again on the long upgrade to Creston. Dropping down toward Rochelle, 381 picked up the pace and crashed with a shuddering roar of many wheels across the CB&Q Aurora-Savanna mainline's diamonds on the west edge of town. The F7s held a steady 60 past the asparagus fields in the ancient lake bed to Flagg station with its grain elevator monument, through the village of Ashton and past Franklin Grove with its infamous drawbar gulch just to the west. Looking back along the train one could see only a dozen car lengths or so because of the shimmer of heat and clouds of dust.

Knowing that train #1, the *Los Angeles Limited* had left Chicago at 12:01 p.m., engineer Alexander tried to anticipate the train dispatcher's next move for us. *Madame X*, an eastbound freight from Iowa passed us on the other main track, Alexander announced that we would be going down the Lee County Cut-off south of Dixon to get out of way of #1. A yellow approach signal soon confirmed his guess and began throttling back to 40 mph three miles east of Nachusa, the entrance to the low grade cut-off. It seemed to be much too soon to cut back the throttles when still so far from the diverging route and another sharp hill yet to climb; yet we watched the speedometer go from 45 to 55 as the 99 cars behind shoved the diesels toward Nachusa. Swinging to the left at NA, I noted that the speed was exactly the prescribed 40 mph. Experience with the road and its hills and curves was the mark of a good engineman. The Lee County Cutoff was constructed by the C&NW 1909–1910 to ease operating problems through Dixon, Illinois. Stiff grades and curves marked the original mainline. With the advent of heavier power and trains, the cut-off with its low grade double tracked line with easy one degree curves was a welcome alternative route to the operating department. The second main was removed in 1941 regrettably just before World War II. This was to be our last ride over this piece of railroad as the autumn floods of 1954 were to wash it out and the decision was made not to replace it.

381 approached Nelson and eased to a stop at the huge concrete coal chute that spanned the mainlines. There we took on water as #1 swirled past on track three. 381 pulled on past the NY tower and turned to the left stopping at the south end of Nelson yard. Wild Bill and Ray climbed down from the engine—their day was finished. The warm summer sun came out as we watched a grimy class H Northern #3032 pull in on an adjacent track with a drag from Clinton, Iowa, some 30 miles to the west. They had too many cars for the track so a cut and double move was necessary. When we spotted our new crew climbing aboard, we hastened back to the 4079A. Our new crew was engineer John Roman, fireman, G. E. Walters and head brakeman "Lightning" Parks. They were anxious to get rolling to make up the time lost from our late Proviso departure. We went south with 98 cars leaving Nelson at 2:43 p.m. That first fourteen mile tangent was the perfect place for #4079A to get the train rolling. Engineer Roman gave Bob and me a guided tour, pointing out scenes of interest and recalling the old SI. days when they were an independent division. There were stories of blinding fogs, washouts, steam locomotive adventures, the 1938 tornado that destroyed the South Pekin engine house and other vignettes of life on the railroad. Roman knew this part of the railroad had been constructed in 1901 as the Peoria and North Western, but surprised us by saying that passenger service had been maintained into the 1930s.

"Green board" from fireman Walters. This for the approach to an automatic interlocking where the SI crossed the CB&Q's Mendota-East Clinton branch line. We were now rolling on the new 115 pound rail underpinned with slag ballast. The railroad left the level lowlands to climb into gently rolling uplands that marked the beginning of the great Peoria moraine deposited by the continental ice sheet thousands of years ago. A hollow roar from our train crossing a steel bridge marked the crossing of the Rock Island's main line at Langley. Bob and I noticed that the "stations" marked on our track profile were mostly just a signboard identifying a grain elevator or sidetrack. Van Petten, Langley, Morse, Storage, Camp Grove and Akron were not communities; just railroad names and, perhaps, a house or two out on the great Illinois prairie. At 3:38 p.m. we passed under the CB&Q's main line west at Buda, which was a small town. However, no silver-sided *Zephyrs* were to be seen. Our 4079A flowed smoothly along into the spirals and curves and straightened-up gracefully upon entering the tangents. This was a good piece of railroad. The right-of-way was lined with willow, milkweed and wild mustard while beyond the fences the corn rows stretched away to the horizon. Dynamic brakes brought us down to a crawl as 381 eased through the spring switch at Storage. We dropped the head brakeman to make a rolling inspection. The train eased up to the depot at Broadmoor standing in the shadow of a bedraggled wooden coaling tower unused since 1949. We were supposed to meet and pass our counterpart northbound #380 here, but they were not in sight. The crew all detrained to find and eat the delicious wild strawberries growing in profusion on the right-of-way. With some regret we saw 380 approaching. Passing at low speed our guys volunteered to teach 380's crew how to get a train over the road. 380 also had three F7's on the point plus two new GP7s about 40 cars deep in their train acting as helpers. 380 was a big train with 140 cars and had needed that extra power to climb the Radnor hill which we were soon to descend. 380 slowly moved north clearing the single track main. Then 381 moved out the south switch with motors whooping it up. Engineer Roman pointed out the old grade of the railroad some sixteen feet over our heads at Camp Grove. The original railroad had been built over a sharp hill. About 1917 the hill was removed to ease the grade; the dirt being used to fill a low spot nearby. The grain elevator built at the old grade level looked like a medieval tower high above the track.

Thoughts of getting back on schedule went out the window when Walters grabbed our orders at Radnor, the next open station. Roman read the bad news. There had been a derailment thirty miles ahead—an ore hopper had jumped off the track on the Illinois River bridge #1731 near Pekin. We were instructed to take our train into Adams Street yard at Peoria, drop the Peoria cars and go to beans (take dinner) after which we were to back the rest of our train to Kickapoo Junction and then pull our remaining cars south to the point of derailment and wait for further instructions. The ore car, we learned later, had suffered a burned-off journal and that the car had not derailed on the river draw span but on another nearby bridge spanning a highway where it narrowly avoided falling onto the pavement below.

Radnor (renamed Pioneer in 1962) was located at the top of the ruling northward grade. Here the double track line dropped into the Illinois River Valley, twisting and turning through a ravine-scored landscape. In 1951 management refused to install new rail in this section of track because they thought the heavy uphill trains using lots of sand would soon ruin the new rail. Consequently, both mains were laid with secondhand 100 pound rail on cinder ballast. In the later days of steam class J4, 2-8-4s, boosted coal trains and heavy freight from the Illinois River bottoms to the top amid clouds of coal smoke and showers of cinders. Kickapoo Creek, which had carved out the drainage which the railroad followed, had a nasty habit of taking parts of the railroad embankment out during the heavy rains common in this region. Halfway down the hill, rail lines of the CB&Q and the M&StL joined the C&NW through the narrow valley of the Kickapoo. On one memorable occasion an empty coal hopper escaped from a "Louie" crew at the top of what they called Maxwell Hill. The loose car ran wild down the grade passing through several junctions and switches, and came to a gentle halt at the end of a track in the old Peoria Union Station. Now, it was late in the evening as we edged down the grade, the whine of the dynamic brakes increasing as the weight of the train pushed against the locomotives. In places sheer limestone bluffs crowded the railroad. Leveling out at Pottstown a short distance northwest of Limestone Station and the Farmington Road grade crossing, 381 rumbled through a couple of steel truss bridges spanning the creek bed before the familiar shape of the tower at Kickapoo Junction appeared in our headlight. The tower controlled one switch—the junction of the Adams Street line with the main. The signal indicated our route was set for the Low line to Adams Street. 381 ground to a stop at 7:06 p.m. But its night's work was not yet finished. Bob and I accompanied the crew to a nearby hash house and were barely into our supper when conductor Carroll rushed in with orders to run light back to Broadmoor with our F7 units as an extra. That northbound 380 we had passed there had tried to set out a bad order car. In the process the two train crews managed to jackknife the two GP7 helper units and block the main line. Our job was to run up there and pull 380's cars south into the clear. Our crew and their diesels at Adams Street had the only available C&NW locomotives between the derailments at Pekin and Broadmoor.

Renamed Extra 4074C North we ran back up Radnor Hill forty miles to Broadmoor, pulled the 95 car tail end of 380 clear of the derailed GP7s and set them over into the passing track. Changing directions and name, we ran back to Adams Street as Extra 4079A South. Retrieving our train at Adams Street, we backed through Kickapoo Junction onto the main in the dark humid night and headed south again on the High line under our original name as 381. Approaching the high steel truss spans across the Illinois River, we saw with relief that the track was now clear of the wayward ore car and rolled on to the yard at South Pekin at 12:59 a.m. Before Bob and I ever started this trip, we had monitored 381's daily progress and found that it rarely was more than an hour late. So much for predictability on a railroad.

We got off the engine with the crew. They went home while we waited for the next part of our trip. The diesels went to the fueling station near the engine house for a drink of oil as the operating plan called for a round trip to Madison and return on this fueling. In the yard 381's train was pulled apart and reassembled. The new 381 had 62 cars and 35 empties for a tally of 4123 trailing tons. The new crew appeared: engineer Percy Black, fireman William B. Earhardt. The units came back, checked out to the crew's satisfaction, were driven to the south end of the yard and tied onto the cut of cars that then became 381. The highball came from the rear end at 2:15 a.m.

This part of the SI was the old railroad with worn 100 pound jointed rail and skimpy ballast. The low joints and 25 mph speed limit combined to make the big diesels into giant rocking chairs. Bob and I nodded off from time to time as the swaying diesels rolled south. We traded turns in the brakeman's "third" seat in the engine cab. Somewhat hypnotized by the oscillating headlight, Bob suddenly started from a doze saying that he had seen wild turkeys on the track. To clear our heads we tried putting our faces into the cool night air rushing past the side window of the diesel. To say we saw much of the SI south of South Pekin to about Barr would be a stretch. This part of the railroad was built in 1914 as the St. Louis, Peoria and North Western. Designed for eventual double tracking, only the easterly or southbound main was ever built. (The history of construction of the StLP&NW was discussed at length—see 1947). After World War II when locomotive coal demand tapered-off, the value of the SI for merchandise freight service became more apparent. Traffic from Wisconsin factories and grain from Iowa destined for southern points were among the traffic mix that reached substantial levels on the SI. Since the SI avoided all developed urban areas until the trains were handed off to the Litchfield and Madison Railway at Benld for delivery to connections at Madison, the route developed a remarkable amount of business over the single-tracked line.

There was another running inspection at Barr where we crossed the Chicago and Illinois Midland Railway (CIM). At the CIM's neat brick tower we picked up new orders because we were so far off schedule. At 3:33 a.m. we saw the first rosy glow of predawn. The pale light began to show the country details at Sweetwater hill, Lick hill and the Sangamon River crossing. Near Lick we looked down on the Wabash Railroad's main lines marked by a regular pattern of green signal lights stretching toward the horizon. A fox loped along in the half light and turned into the brush as the headlight beam picked him out in stark relief. We could see the Illinois State Capitol building dome and lights of the city over in the east. Further south near Gillespie the automatic interlocking at ON tower beckoned us on with a green eye. 381 clumped over the Big Four crossing there, swung around a sweeping curve to the right past high slag piles from abandoned coal mines at Henderson, clumped across the Bear Creek trestle and gently stopped at the Benld depot. It was 5:15 a.m.

Our North Western crew said goodbye and a Litchfield and Madison engine crew came aboard for the last stage into Madison.

Engineer George Osdendorf and fireman Julius Werner remarked that we looked like we had been up all night. Osdendorf glanced at our credentials and kicked off the brakes at 6:11 a.m. Our train was renamed in the L&M Timetable No.2 as "381-Ozark Freight". Now that the sun was up we could see that the country south of Benld was quite different from the rolling hills we had traversed. We saw our first oil pumps bobbing their heads in this flat land cut with deep brushy ravines. There were more scattered towns and villages as we approached the metropolitan area, crops were further advanced and more rail lines seemed to close in. We crossed the Illinois Terminal, an electric interurban line, three times: the first just south of Benld, the second at DeCamp and the third near the end of run in Madison. At DeCamp 381 also crossed the Wabash's main line to reach the L&M's main track.

The L&M was a surviving piece of the long gone Chicago, Peoria and St. Louis Railway, often derisively called "the Wooden Axle". 381 rolled smoothly along at 40 mph, and we got a good look at this property. Coming through their headquarters city of Edwardsville, we spotted the engine house with five light mikados (2-8-2) mounted on very small diameter wheels steaming quietly in the Sunday air. Our cameras were not ready and we missed the shot. The engineer remarked that was the entire fleet parked there except for a diesel switcher at Madison. South of Edwardsville the L&M paralleled the Nickel Plate main line and saw one of their famous Berkshire 2-8-4s with a long string of high autoparts cars wreathed in a gray smoke cloud. Coming around the curve and onto the Mississippi River bottoms at Glen Carbon, Bob and I could see the skyline of St. Louis. This stretch of L&M was a jointly owned section with the Illinois Central and built to higher standards for passenger train speeds. (Curiously, the L&M owned one rail and the IC owned the other, a fact I was to learn in 1958). Too bad that we were almost into Madison as increasing numbers of crossings and industrial trackage appeared. Easing through this tangle, 381 stopped at the yard office at 7:46 a.m., 6 hours and 41 minutes off the advertised. Bidding the crew goodbye as they prepared to double back to Benld, Bob and I hiked over to the ITC's Granite City electric line and rode a green and cream colored PCC car over the McKinley Bridge to St. Louis. After breakfast, we explored the cavernous St. Louis Union Station and climbed on the GM&O's *Alton Limited* for Chicago. We slept all the way back.

The fair June weather and the end of my school term meant that field trips multiplied With rodman Byron Ruth and his available automobile, we made a one day trip to LaFox, Elburn, DeKalb, Creston, Rochelle, Ashton, Franklin Grove and Dixon and the next day, the 13th, went on to Rockford and Belvidere sweeping up a lot of little jobs that had been waiting for our attention. East of Belvidere we ran two miles of levels for a drainage survey. In the office a miracle occurred: a new (used) blueprint machine was installed replacing that carbon arc horror that had terrified me on my first day on the job. On the same day it began working, the 18th, a new rodman, Len Anhalt, reported for duty. Than I made a mistake: I signed up for summer school using my GI Bill credits just in time for the division engineer, Bruce Packard, to send me to the SI for a three week job. This presented me with a crisis because dropping a class meant termination of future government payments for tuition. Eventually, the matter was settled by some official letters, but I did not gain much respect for Packard's insensitivity to personal needs. It was clear that he thought he was still an Army captain and his troops had to obey every command.

I was accepted as a "military" member of the Society of American Military Engineers (SAME). Charley Hise, the former division engineer, had sponsored me. The organization had a Chicago chapter, held regular meetings with guest speakers and put out an informative magazine. Over several years I made some good contacts through this group, but I did feel a bit out of place as my military experience in the Army Air Force had been minimal and certainly not as an engineer.

Sunday night, June 24th, found me on the Rock Island's Peoria sleeper with two new rodmen, Len Anhalt and Sid Kahn. We were met the next morning by Ray Doty, our B&B Supervisor, who drove us to milepost 14.5 south of South Pekin where we were to run top of rail levels to the Salt Creek bridge south of Luther, Illinois. This part of the SI which Bob and I had ridden over a few weeks ago was to be reballasted as the cinders and dirt were not holding up very well under the heavy trains now operating on this line. Part of our task was to try and find as many of the original 1912 benchmarks as we could. Since most of those survey markers from forty years ago were merely nails in tree roots, we failed to find a single one because the trees were long gone. We ran the levels with occasional back sights to reduce systematic and accumulated error, not that such accuracy was really needed on this type of project. Letting the boys run the chain, I recorded the notes as we went down the line to milepost 20.5 under a blazing summer sun. By 3:30 p.m. we were out of energy and water. Ray Doty picked us up at a road crossing and took us to a motel in Pekin. The next day, the 26th, we made good progress to milepost 28.2 under a hazy sky. Rain the night before elevated humidity levels causing us to perspire extravagantly. Our canteens were empty by noon. The next day Ray took us out to the job, but decided to hang around as he felt it was going to rain—and it did when a severe thunderstorm cell moved through. There was no chaining done that day and the next was no better. The storms were so intense that there was significant property damage to structures in the area. Giving up, we took the Rock Island's afternoon *Rocket* to Chicago. The next week was a repeat of the former. We were on the Rock Island sleeper on July 1st.. Ray again took us down to where we had left off and we chained rapidly down to the Illinois Central crossing at Luther. Ray picked us up there and took us on a slight side trip to Mason City where a bridge under Illinois highway #4 had washed out and an unsuspecting motorist had driven into the huge hole gouged by the flood waters with fatal consequences.

I was getting to be known as the Peoria Kid. When there was some engineering job on the SI it seemed to end up with me. We had a report of bank slippage on Radnor hill. Ray Doty ferried us out to the general area, and we hiked to the spot where the eastbound track had slumped. Poking around in the brush we discovered some old French drains in the side slope. A French drain is simply a trench filled with large stones, cobbles and crushed rock to provide a drain for surface water. These old drains were silted up and not functioning as intended. Although Packard wanted borings made, the next day was the Fourth of July and organizing anyone to make soil borings was impossible. We caught the *Rocket* and went home. On the night of the 4th Len, Sid and I were on the Peoria sleeper again. I sent Len and

Sid to run levels on the ballast job south of South Pekin while Ray Doty and I went to Limestone to collect some section men to help drill holes in that failing embankment. No one was there because the track supervisor had forgotten to tell the section foreman. Ray and I then went down to South Pekin where I caught up with the leveling crew to give them a hand. On the 6th Ray took the boys out to run levels around Green Valley and Allen while I took a bus to Peoria and hired a taxi to run me out to the Farmington Road crossing where Limestone station is located. With the section men doing the heavy work, we managed to get three decent holes and found nothing but cinders and clay. I went back to Chicago on the afternoon *Rocket.*

Sunday night, July 8th, and I was back on that Peoria sleeper again. Because of a train wreck at Utica on the 7th, our train was delayed three hours. Monday morning was another hot and humid day. We started our levels at milepost 20.4 and got down to 23.6 before a work train picked us up for a ride to South Pekin. The next day I went back to Peoria with section foreman Austin and got three more holes drilled at the slump site which was enough to get a fix on the slippage problem. There was a wet and slippery clay bank under the cinder fill causing instability under the outer track. On Wednesday the 11th Len, Sid and I got up at 4:30 a.m. to catch the work train for a ride down to milepost 23.6. We got another 5.25 miles done before a fast moving rainstorm hit forcing us to seek cover. We were so close to being finished. The work train came north early because the rain also had stopped their work. The next morning we caught the train again and finished our line of levels. The roadmaster's truck driver, without his truck, picked us up on his track motorcar. He couldn't use the highway because of another washout. After lunch at South Pekin, we grabbed a bus to Peoria and had enough time to stake out a short industry track to serve Heller and Company before boarding the *Rocket.*

The summer of 1951 was notable for heat and humidity. There were no more lengthy field jobs for me in sight, just a bunch of little excursions to take care of minor items. The other instrumentmen worked on the main line rail and ballast jobs in the western reaches of the Galena division setting centers and stringlining curves. On the 31st I took Sid with me to make cross sections east of Poplar Avenue in Elmhurst where Bruce Packard had made a deal with a local contractor to excavate some surplus dirt on the right-of-way. Fred Houx, Sr., the superintendent of freight terminals, had several little tasks for us at Wood Street yard. Gus Johnson, the B&B supervisor, went over plans for a new oil burner installation in the furnace at the Glen Ellyn depot with me. Why me? I had no expertise with heating equipment.

On August 21st the Land Department requested division assistance for a survey at Proviso. Sid Kahn and I were selected to help Bill Rundell, their old time surveyor, rerun the 1910 centerline of Lake Street which bordered the yards property on the north. The object of the survey was to locate the Indian Boundary treaty line that sliced in a northeast-southwest direction across the freight yards. Given the substantial changes in the area since 1910, it was a tough job and too much for a one day affair. We tackled the survey again on the 23rd. At one point we were cutting down 8 inch diameter trees with a hatchet to obtain sighting lines, but on the next day we were able to close the survey. This was property in the northeast corner of the yard sold to Henry Crown's Material Service Company which planned to build a huge ready-mix concrete plant.

On August 28th I met Sid Kahn on a westbound CA&E train in Elmhurst. We went to Wheaton and rode their Aurora car to Batavia Junction where we transferred to the Batavia Branch car—all of this to see a man about a fence problem. As long as we were in the area, we took a bus to Aurora and measured trackage at the old National Brush Company plant and checked on an illegal pier built on C&NW property along the Fox River. Returning to Chicago on a CB&Q air-conditioned car was a marked contrast to the hot, steamy morning we had spent out in the sun.

On July 25th Sid and I were in Glen Ellyn making an accident survey at the Park Street crossing where Wilton F. Swatek and his two daughters, Marilla and Jean, were killed by a train November 25th. Attorney John Davidson wanted poster card exhibits for the Glen Ellyn case showing the crossing gates in various positions. I was a bit at a loss in understanding just what John wanted. He called a conference with his boss Drennan Slater and colleague Ed Warden. They thought the exhibit I had prepared was satisfactory but agreed that I was an engineer not a graphic artist. To strengthen my testimony in the Swatek case, the attorneys decided I should ride the cab of the same train responsible for the fatalities, #104 the *City of Los Angeles,* through the accident scene. I made an adventure of it. To catch #104 eastbound, it would necessary to be in Clinton, Iowa, its last stop before Chicago, early in the morning. Of course, the easy way was going to Clinton on a C&NW train, stay the night in a hotel and get up very early to catch #104. Instead I rode to Chicago from Elmhurst on the CA&E at 5:30 pm. The second leg was on the Illinois Central's *Hawkeye* train #11 to Dubuque, Iowa. At 3 a.m. I walked across the Mississippi River bridge to the CB&Q depot in East Dubuque, Illinois, where I watched four streamliners pass while waiting for the CB&Q-GN's eastbound *Western Star.* Boarding it, I rode to its first stop at Savanna, Illinois. After breakfast in an all night cafe by the depot, I finished my trip on the CB&Q's two car local #48 from Savanna to Clinton, Iowa, where I arrived in plenty of time for C&NW #104 only to discover it was running late. When it did limp into town with two engines down, I climbed on and displayed my permit to engineer Jacoby and fireman Klingberg. Despite its lack of power the train made up 13 minutes of the delay by running in the 70–85 mph range. As we approached the Glen Ellyn accident site, I positioned myself by the lefthand side window and threw a small sack of flour out at the instant the engineer said he could see the Park Street crossing. Later, I went back and measured the distance to the flour splat which gave a clear idea as to how much time an engineer would have had to react to an automobile on the crossing—not very much.

Just for the fun of it, I rode the cab of the *Flambeau 400* #153 to Green Bay, Wisconsin and got my first look at Wisconsin and Lake Shore Division tracks. Coming back on #216 I rode in the diner and enjoyed a good meal.

On September 6th I had to face the mysteries of string lining curves. Basically, string lining consists of measuring 62 foot chords and their 31 foot midpoints. Using the fact that a railroad track moves 2 inches inward at one midchord point while moving move one inch outward at the adjacent midchord points the trick is to develop a pattern of shifting the curve until the movements inward and outward balance. Why 62 foot chords? Because the midpoint chord measurement in inches to the inside gage face of the rail gives the curvature of the track directly in degrees, *e. g.*, a 3 inch measure from the center of the string to the face of the rail meant the track was a three degree curve. A wonderful computer problem, but personal computers did not yet exist. We had just a length of chain, a six foot rule, marking crayon (keel) and field books for recording the readings. My training ground in this arcane art was a curve in track 3 west of Villa Park quite near my parents' home. After measuring the chords and recording the mid-chord readings, I spent the entire afternoon trying to balance the throws. Finally, I got a good solution that kept the track on the "dump" (track bed). Sid and I set the centerline stakes the next day and nailed down the stainless steel superelevation tags on the track ties at the appropriate points. These tags were stamped with fractional inches so that a track crew could read the tag of say "1½" inch and raise the outer rail that much over the lower rail to have the correct superelevation for that particular place in the spiral leading to the curve. As long as we were in the area Sid and I walked down the main line to the CGW overhead bridge (#24½) and measured the vertical clearance for the Bridge Department before calling it a day.

Trainmaster Elmer Kohler was having a lot of problems with automobile parking in the North Pier district and asked for a walking inspection. I was glad to revisit the area and took notes as we went along. The track structures were sunk deeply into the mud leaving the rails, at best, even with the ground. Motorists often had no idea that there was a railroad track there. This problem was insoluble until the line was relayed some years later with usable 112 pound rail which stuck up so high that automobiles could not easily park on the tracks.

I made another trip with claim agent Ollie Adamson (and mayor of Geneva, Illinois) to Cherry Valley, the next station east of Rockford, on September 13th to make an accident survey at the Illinois highway #5 grade crossing of the C&NW Freeport line. Finishing that one, he then took us over to Big Timber Road crossing west of Elgin where there had been another accident requiring a survey. These jobs were so common that we became quite adept in running the surveys.

On the 17th I inspected a new welfare building at West Chicago after which I made a call on the nearby pickle works and explained to them that they needed a license from the railroad to maintain their private driveway on our property. I met with Seth Steiner, the

CA&E's maintenance engineer, to work out improvements to the Main Street crossing in Wheaton. Our railroads ran side by side there and the grade crossing was a joint problem. We measured it up, ran a line of levels, and then I caught the 11:01 a.m. C&NW train to Chicago. On the 25th Sid Kahn and I rode the Illinois Central's *Iowan* to Rockford where we grabbed a taxi out to the west side of town to reset the stakes for a new industry track. Luckily, most of last year's stakes were still in place; we added some grade stakes and hiked back to the depot where we talked Russ Longdon, Mr. Williams' chief clerk, into driving us across country to Oregon, Illinois, where we were able to catch the CB&Q's *Zephyr* to Chicago. I sure missed our old train #706 when working on the Freeport Line.

Sid and I rode #13 west to Morrison on September 26th to finish up a survey on drainage through the station grounds. We were interested spectators that afternoon as we watched a couple of traveling mechanics install a new wrist pin in a disabled class J-1, 2-8-2, #2530. The dispatcher in Chicago was most accommodating by instructing eastbound #14 to stop at Morrison and pick us up. We only had to ask, he said later. The next day Sid and I went down the 16th Street line in Chicago before inspecting a distressed retaining wall at Washington and Rockwell Streets. The big event of the day was watching a 600 foot long lake freighter with a 72 foot beam squeeze through the Chicago River past the Daily News Building.

On October 2nd I grabbed our survey equipment and boarded the *Commuter 400* to Evanston where Byron Ruth waited with his car. We drove to Freeport and staked a track for the Structo Company's new factory under construction east of town. There also was some work in Rockford on River Lane for the Newkirk Company which we took care of before heading home with a stop for dinner in Marengo. The next day in the office assistant engineer Tony Zaborowski was quite annoyed that we had not stayed in Rockford overnight but forgot what he was upset about. Shrugging our shoulders, Byron and I went to Proviso to survey for the trackage required for the new Material Service plant, staking it out on October 4th..

Then we were handed a new assignment—locate the 14 inch waterline at Proviso so that it wasn't damaged when the new Material Service tracks were built over it. The waterline was built about 1935 to bring Lake Michigan water to the Proviso engine house. Local well water was heavily mineralized which made it unsuitable for use in locomotive boilers. A deal was made with the City of Chicago to tap their system at a pumping station located at the west city limits at Austin Boulevard near Lake Street. The water line made of 14 inch transite (an asbestos-cement compound) was laid in the center of the track elevation from Austin west to the Des Plaines River, carried across next to the railroad span in an insulated box and then westward buried in the north right-of-way line through Maywood and Melrose Park, under the IHB overhead bridge and then making a big looping curve to the northwest where it terminated at storage tanks west of the engine house. Rodmen Bill Brown and Len Anhalt and I cut brush and gently probed the sandy ground until we thought we had a fix on the location so that proper encasement could be installed to carry the track load over this fragile pipeline. The water in this pipe was a temptation to all the cities through which it passed. Melrose Park managed to negotiate a tap on the line near 22nd Avenue for "emergencies", but the City of Chicago was adamant that the western suburbs should not have access to the water. Many years later a compact was formed with several of the suburban cities to obtain Lake Michigan water through a vast system of tunnels and large diameter pipelines.

Instead of Peoria, I was now spending a lot of time in Rockford. On the 16th I took the Illinois Central morning train to that city for a meeting of the three railroads that owned bridge #474 at Kent Creek west of our depot. The Milwaukee sent two people, an attorney, Mr. Simmonds, and assistant division engineer, Campbell. The IC was represented by their assistant engineer from Waterloo, Iowa, Mr. David. Repairs to the bridge were discussed at length, but it seemed to me that there were a lot of people attending a meeting in Rockford that could just as well have stayed home and talked on the telephone or written a letter or two.

These personal injury accident cases were always unique. A passenger named Fiedler claimed injury when boarding a train at the Austin station in Oak Park. Rodman Bill Brown and I measured the platform and surrounding area for the case in October 1951. One year later the Law Department requested that exact measurements be taken from the platform to the bottom step using the same passenger car in the same position as at the time of the incident. On October 1, 1952 I was at Austin waiting for an eastbound local that I knew had the 1910 steel car #3120 in its regular consist. I waited and waited; the train was 50 minutes late because of rail laying and ballast work with resultant single tracking between HM at Elmhurst and JN at Melrose Park. The train finally arrived and waited for me to do my measurements; the crew had been alerted to my being there. I was quick about it and the train did not suffer much further delay. Later, I drew a cross section of the car and showed its relation to the platform along with those critical dimensions. In October 1952 the law department sent out an emergency call for me to appear in court. I was working on a grading job at Wood Street when I got the message. I grabbed a streetcar on Western Avenue to Madison Street and went directly to Judge Igoe's Federal court still dressed in my rough work clothes. The C&NW's attorney, Bob Russell, blenched when he saw me in that bright red plaid shirt, but put me on the stand anyway. Later he explained that defense witnesses in personal injury cases should always dress in neutral shades and, especially not in red, as it made the juries think of blood. But, they had to take me the way I was that day.

The last quarter of the year brought the annual budget scramble. The flood of requests for estimates of this and that overwhelmed Tony Zaborowski's unique system for dodging work. His inherent nervousness made him increasingly erratic and hard to reason with. The division engineer, Bruce Packard, did not help with his grating personality and sarcastic manner. Confrontational management kept everyone at arm's length. With his defensive cigar in his face and truculent attitude, Bruce seemed to go out of his way to take the opposite view in every argument just for the sake of arguing. These two men did not provide much help or guidance as we struggled to deal with the growing stack of work. Our own modest request for a station wagon, which would have been an enormous help in dealing with the many far-flung requests, was summarily rejected. We all got a smile, however, when I found in the local paper that Bruce had been hauled up before the local Elmhurst magistrate and fined for letting his dog run at large.

That new industry track at the Structo Manufacturing plant east of Freeport became a monument to my bad relations with the division engineer, Bruce G. Packard. I was angry enough that I seriously considered resignation. The whole business started in April as related previously about the first meeting with Structo's president. At that time we were escorted through the toy manufacturing line to see the creation of the famous Structo line of pressed metal trucks. This initial visit for determining the location of the new side track and subsequent field inspections were absolutely normal. Construction was carried out by Structo's contractor during the summer so that the new railroad grade was ready for centerline stakes and final grade stakes on October 23rd. But winter came early and the ground froze halting completion of the grade.

After the Spring thaw came in 1952 and the ground had dried out, Structo's contractor completed the grade and called for the final location stakes to be placed. President Gokey, called Packard on Friday June 6, 1952 and demanded instant engineering in the form of finish grade and centerline stakes. Bruce told him (without asking me) that I would be there without fail the next day. Although the timing was somewhat inconvenient for me, I made arrangements with rodman Byron Ruth to meet me early Monday morning at the hotel in Freeport as I would take the Illinois Central night train Sunday evening to get there. Byron was spending the weekend with relatives in Freeport. Our plans then went all to hell. On Sunday evening I went to the Elmhurst station to catch the last commuter train to Chicago to make connection with the late evening Illinois Central train to Freeport. At the scheduled time in Elmhurst, no train showed up which was quite out of the ordinary. There was no agent on duty there late Sunday night and no handy public telephone in sight. The drugstore across the way was open and I finally got through to the train dispatcher in Chicago who advised that the wayward eastbound was hauling a dead steam locomotive from West Chicago to the Chicago Shops. By the time the hospital train arrived with its cripple in tow, limped toward Chicago at 20 mph, set out the disabled engine at the Shops, and then raced at high speed to the Chicago Passenger Terminal, my connection to the Illinois Central across Chicago's Loop was lost. I did not know where Byron was staying much less have a telephone number to reach him. I wired the Freeport hotel with a message that I was delayed and went looking for a Greyhound bus to Freeport. No busses. I went home on the last westbound train, slept a little and caught a westbound Greyhound from Elmhurst early in the morning after calling the Freeport hotel again with a message that Byron was to meet the bus. When I did not show up with my survey crew promptly at 8:00 a.m., the irate president of Structo, Gokey, got on the telephone and called the C&NW's president. Annoyed, he directed the complaint down the chain of command to the chief engineer who landed on my boss, Bruce Packard. The first thing Packard said to me when I came into the office was: "You're fired as of right now." Turning away, he did not want to hear anything I had to say, but, after listening to my recitation of the events, very reluctantly, walked down the hall to the chief dispatcher's office and found the notation about the crippled steam engine on the train sheet of June 8th. Sure enough, the circumstances were exactly as I had related them. His apology was grudgingly given. I accepted reinstatement and life went on as before. The fact that our work for the hair-triggered Gokey had been performed at 10:30 instead of 8:00 had certainly stirred up a hornet's nest. Gokey probably was just upset that he was paying for a grading crew that was standing around waiting for us. This incident underscored the inconveniences that lack of a company car was imposing on our division's work, but we still were not to get a station wagon for three more years. (As further information about Structo Manufacturing: J. G. Gokey had purchased the toy business in 1935 but was now expanding the product line to include outdoor grills. The company was sold to King-Sealy Thermos in 1965 and eventually ended up as part of Household International. Gokey died in 1975).

Back to 1951 and a problem we had with ten car commuter trains. The old cars then operated were so crowded that employees riding on passes were required to stand for revenue riders. The 1910, 1914 and the light weight 1927 cars were open invitations to accidents. When the train came into a station, the trainmen had to open the vestibule traps at each end of the cars and detrain before the passengers to be able to assist on the steep stairs. In multiple track territory the homeward crush was so heavy that trainmen intent on doing their jobs on the platform side were unable to restrain impatient passengers who opened the vestibule traps on the other side of the train and jumped off onto the open track alongside. To forestall this activity low wooden picket intertrack fences were installed through station grounds. Over the years steel posts and barbed wire replaced the wooden fence despite many municipal codes barring the use of barbed wire within city limits. Our Law Department resisted many communities' attempts to force the railroad to give up the use of that fence wire. Most of the passenger platforms were too short for the ten car trains so that the end cars often stopped on grade crossings beyond the fence protection. The obvious danger of opening the track side trap and jumping out, perhaps, into the path on an oncoming train never entered the mind of the homeward bound commuter. Some employees who were not train crew members tried to help the overwhelmed trainmen by standing firmly on the track side vestibule traps and directing the commuters to detrain on the platform side. The loud, colorful, and, often, offensive abuse heaped on these safety-minded officers was unbelievable. I can still recall the evening when the longest westbound commuter stopped and the usual gang of commuters rushed to get off the train at the Cottage Hill grade crossing in Elmhurst from the wrong side of the stopped train when a class H Northern 4-8-4 came screaming by eastbound on the center track. This near-miss incident sobered the crowd for awhile, but gradually, the dangerous situation returned. The trainmen were powerless to control the crowd because there were at most only five of them on these ten car trains and there were twenty traps to open and close. Eventually, when the new commuter cars with remote control pneumatic doors came into use, the problem was temporarily solved until the observant commuters learned to press the door buttons on the wrong side of the train. The railroad then retrofitted the cars with a different button system that required a plastic stick to operate the doors. That helped a lot. While the average commuter may press a clearly marked button, the thought of using a special tool, stopped the passengers' bad habit of playing trainmen. (A pencil, of course, does the trick quite well).

Chief Engineer B. R. Meyers requested a special engineering survey at Proviso, On October 29th rodman Byron Ruth and I chained eastwards through the hump Yard 5 past the Middle and east on track 25 in Yard 7. This was the same job that we had done with James Barnes the year before. The next day we ran levels over the route chained the day before despite the raw and chilly weather. Winter came early in 1951; by November 1st we had snow. Office work was now the norm and drawing up Meyer's profile was a good excuse to stay in and keep warm. On November 6th the locomotive firemen announced a strike against the B&O, C&NW, TRRA and L&N, but a Presidential order set the strike back 60 days. In the middle of November Bill Armstrong from the chief architect's office and I made a trip to the western part of the division together in his car which was a great help in taking care of many minor tasks scattered along the line. We inspected the Morrison depot where the B&B gang had done some remodeling work including a new fascia board on the outside of the building. Their hammer marks were very clearly visible in the smooth wood. Bill's comment: "wood butchers". After all, these guys were used to working with 12 × 12s and were not fine craftsmen. On the 23rd Len Anhalt and I made a quick trip to Chicago Shops to measure new pavement at the old transfer table and inventoried a double slip switch at Harding Avenue. On the way back to the office, we saw a class H #3028 sitting on the interlocking plant at Western Avenue with a thrown driver tire which tied-up the C&NW west line and the Milwaukee Road's main lines.

On the evening of November 23rd Bill Armstrong and I rode the Illinois Central's *Southern Express* (train #25) to Centralia, Illinois for a day of railroad visiting. Our principal objective was the CB&Q engine house which was the last stronghold of their big 2-10-4 steam locomotives. Mr. Libber, the roundhouse foreman, without any hesitation gave us permission to tour the premises. We also saw a Southern Railway local headed by a Pacific #1234 and the Missouri and Illinois (Mike and Ike) motorcar #625. We rode to Herrin, Illinois, on the CB&Q's motorcar #9844 and stayed with it to Litchfield hoping to catch a Wabash local to St. Louis, but our meandering motorcar was too slow. A bus ride west to Alton, Illinois, brought us to the Illinois Terminal's Granite City electric car line and the connection to St. Louis on car #100, a drop center entrance car. After a night in the old hotel in the

St. Louis Union Station, we rode the CB&Q local train #47 headed by diesel #9928A from St. Louis to Galesburg, Illinois, all on our passes of course. Leaving St. Louis we spotted an SSW 4-6-0 #664 wandering among the puzzle switches. After our train crossed the Mississippi on the Eads Bridge, the brakeman came through just behind the conductor and asked each passenger if they wanted lunch and handed them a menu card. After collecting their requests, he dropped a message off for the operator at Whitehall who wired Beardstown with the orders. The box lunches came aboard there at 12:45 p.m. When we arrived in Galesburg at 4:30 p.m., we connected to the eastbound *Coloradan* #6 to Aurora (diesel # 9934B and 14 cars) where we transferred to the CA&E electric car #424 to Wheaton. A short walk took us to the nearby C&NW station and the short final leg of our trip to Elmhurst where we both lived. Railroad men on holiday!

The next Monday, November 26th, began with an engine failure at West Chicago which threw the morning commute into a snarl. The dispatcher hoping to ease the crush of passengers waiting for trains to work stopped #24 in Elmhurst much to the discomfort of the through passengers who were inundated by pushing crowds of angry commuters that filled every seat, aisle and vestibule. On Tuesday I took our new tapeman, Frank Trimberger, with me to the Chicago Shops to give him a little seasoning. Later that week, Bob Lawton and I were in the office later than normal when a fluorescent light fixture over Wisconsin Division assistant engineer Hal Ruth's desk burst into flame. We called the Chicago Fire Department who responded promptly as they had a station on Washington Street just across the street from the CPT. Damage was minor.

On December 3rd Frank and I chained tracks at State Street and North Pier in showery springlike weather. Willie List (formerly Lipsitz) dropped in for a visit before going to Little Creek, Virginia, for active duty as a naval ensign. On the last day of the warm spell, Bill Armstrong and I made a trip to Williams Bay, Wisconsin, before returning via Milwaukee and Racine. Winter arrived on December 12th with the cold and snow expected in Chicago. I hiked out to the Clinton Street and Noble Street interlocking towers to check on the new oil burners installed a few months previously. The tower men were very happy to have the automatic heat instead of having to deal with the old coal stoves to stay warm.

Because of the school year our family vacations usually were taken in the last two weeks of December. This year we went west and south before returning to Chicago. Our railroad trip passes could include a round trip anywhere on any combination of railroads so long as no part of the journey went through the origination city. We rode AT&SF #23 west to La Junta, Colorado (three hours late in bitter −22° weather with frozen air and steam lines) to Lamy, New Mexico, where we bussed the 16 miles into Santa Fe. Renting a car we visited Bandolier Canyon and Taos, New Mexico before resuming our rail trip at Albuquerque in the Pullman *Lake Sedgwick* attached to AT&SF #123 south to El Paso, Texas. I rode the streetcar line that ran in a loop through El Paso, across the International Bridge into Juarez, and back to El Paso. From El Paso we traveled east on the Southern Pacific #6 to New Orleans, Louisiana, via Del Rio, San Antonio, Houston and Beaumont. At Del Rio I spotted a 1930 Baldwin 4-8-4 T&NO #706. The final leg north was on the L&N's #6 *Hummingbird* which arrived in Chicago on the C&EI three hours late amid snow and cold. When I returned to the office, I was greeted with many grim tales about fighting the heavy snowfall which made this vacation trip even more appreciated.

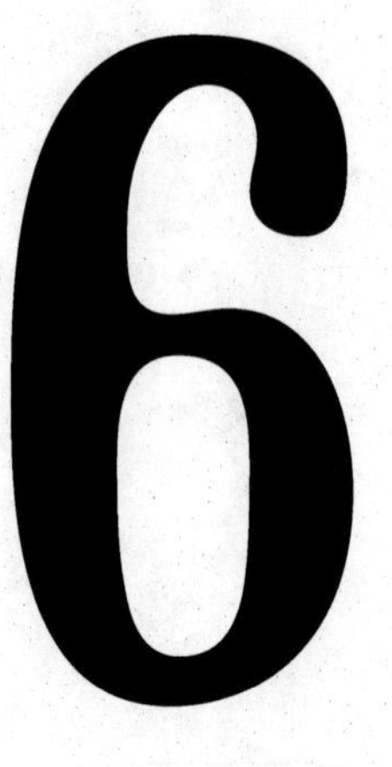

HARD WORK AND OPPORTUNITIES, 1952 . . . I should have stayed on vacation. Walking into the office of the first working day of 1952 I received a stiff dose of reality. There was a huge stack of files piled on my desk and a preemptory order to go to Sterling at once and conduct an accident survey. Assistant engineer Tony Zaborowski said that #101 the *City of Denver* had been sideswiped there on December 28th and that a profile of the track at the accident site was urgently needed. Since I had school that night, rodman Anhalt volunteered to go out on #13 to start the survey and I would follow the next day with rodman Kahn to finish the work. When Sid and I got to Sterling, we ran the levels on the marks Len had made and also inventoried some new yard tracks west of town. While there I noticed the CB&Q was grading for new trackage on the north side of our mainline and I reported the matter to the office. We were instructed to stay in Sterling and rodman Bill Brown would join us the next day with a transit to run some additional lines for the accident survey. Bill arrived about noon with the transit. Unfortunately, he brought the wrong tripod and the transit did not fit which made doing the survey a bit awkward. To add to our problems, it began snowing. After doing what we could, we grabbed an eastbound Greyhound bus rather than wait for #14. On the way into Chicago the bus had to detour around a major train wreck on the Illinois Central's grade crossing at Illinois highway 64 (North Avenue) near Villa Park.

That Sterling trip triggered my annual bad cold which I always seemed to develop on return from vacation. Nursing it, I stayed in the office to update the Sterling map. The survey we had prepared cleared the Engineering Department from blame in the sideswipe accident which later was thought to have been caused by a shifted load of scrap on the side track next to the main. On January 16th we heard that general manager Charley Longman had been canned. His departure started a ripple of changes throughout the operating department. The winter in the West was bad; train #102 the *City of San Francisco* was stuck in the snows of the Sierra Nevada in California. Instrumentman Ottmar W. Smith came back to us from his line relocation project at Negaunee, Michigan. We did another accident survey at College Avenue in Wheaton on the 18th. Then, still another bad crossing accident in Glen Ellyn ruined my weekend as I had to make an emergency survey and drawing ready for the first thing Monday morning.

The railroad is a dangerous place and our newest employee, tragically, was a victim. Rodman Frank Trimberger was struck and killed by westbound train #35 just west of the IHB overhead bridge on January 23rd. His best friend, Len Anhalt, held him in his arms and watched him die from massive head injuries. Les and Frank had been running levels on the mainline on a cold day; Len was running the instrument while Frank held the survey rod on top of the outside rail **while standing with his back to oncoming traffic.** The steam chest on the locomotive struck the bundled-up Frank and threw his body clear of the train. It was said later that had he been struck by a diesel, the result would have been different as diesels, for some unexplained reason, tend to suck struck objects underneath the unit while steam locomotives tend to bounce, or throw, objects away from the impact point. This particular spot on the railroad was already notorious as being a dangerous place. The short curve under the overhead bridge masked trains coming from the east. Some years earlier a woman had been killed when fishing from the Addison Creek bridge a short distance from the IHB overpass. "Dinty" Moore, a yardmaster, had been killed here at some unremembered date while passing signals for a crew switching a circus train at JN tower. The shock of this tragedy sobered our office force and made us a lot more aware of our surroundings when out on the road.

January continued cold and unpleasant and we made every effort to stick to our desks. I was appointed to represent engineering on the division scrap committee and attended a meeting at Chicago Shops on the 24th. Four days later rodman Anhalt and I did an accident survey at Kedzie Avenue. Rumors circulated that DeLeuw, Cather Company was engineering a major construction project to depress the three C&NW main tracks through Elmhurst in response to growing public concern about increasing delay at the six grade crossings in that city. The rumors were just that.

At last February brought relief from the bitter cold. On the 1st I went with instrumentman Bob Lawton to help set grade stakes near North Avenue in Northlake for the new Roberta Avenue sewer line being built by the C&NW's contractor. Since before World War I the railroad had been dumping human waste into surface ditches at Proviso. That practice had to stop as the area was being built up and people were a lot more conscious of what was flowing down the creek. Bob's project was a mile long 24 inch concrete sewer main running from the new Proviso diesel shop past the Proviso engine house and thence north along Roberta Avenue to a connection with a City of

Chicago Sanitary District trunk at North Avenue. Sewer construction always starts at the lowest point on the line to facilitate drainage which meant that the first work was done at North Avenue and proceeded south toward the Proviso yards. The contractor was the same Santucci Company which had recently constructed the giant storm water drain across the C&NW's Chicago Shops property. Santucci was a wizened little Italian who liked to personally inspect his jobs at unexpected and irregular intervals just to see what was going on. The Roberta Avenue job was only a couple of miles from my home so that helping Bob was always a good excuse for an early quit. Bob and I had established the route and grade line by the 19th. Just six days later Santucci had 2,000 feet of the line laid. By the 27th the sewer had reached Lake Street. March snow storms put a stop to sewer construction and to most of our other survey work. We were also suffering from a manpower shortage with only three instrumentmen and one rodman on the staff. Helping each other was the only way to get anything done. On March 13th I assisted Bob Lawton to set grade stakes for 12 inch connector lines to the new sewer. Two weeks later on the 26th I got my reward for being so helpful. We were measuring distances between the newly installed manholes. As I held the tape taut while perched on a manhole cover, Bob let go of his end of the tape to write a note and over I went backwards into the big icy puddle surrounding the manhole. Wading out, I quickly did up the chain while briskly hiking through the cold afternoon to the nearby Proviso engine house where the B&B's steam heated locker room was a good spot to dry out. I was soaked through; even the inked signatures on the railroad passes in my wallet were smeared.

I could not shake Peoria. On February 11th I was on the Rock Island sleeper *New Yama* with rodman Bill Brown in tow. On time arrival allowed us to get breakfast and grab a South Adams bus. At the Adams Street yard we chained and inventoried a rail renewal. Track inspector Bill Campbell showed up with an ancient Dodge truck and drove us north to Akron where we located the end of a recent 115 pound rail renewal and then hauled us down to Radnor where Bill Brown and I set off to locate all the stretches of usable 112 pound rail that had been installed in the curves down the hill from Radnor to Kickapoo. That was a stiff three mile hike especially when having to bend over every hundred feet to mark a chaining station.

On March 28th instrumentman Bill Smith and I tried to run the property line along the Belt Railway of Chicago at Chicago Shops. Nothing checked with the old surveys. That remained Bill's problem to puzzle out because I was off to Proviso where Bob Lawton and I measured a long track in yard 6, checked on his sewer job and staked three new industrial tracks for the Material Service Company all in one day. Improving weather in April really brought on the field requests for engineering services. Rodman Sid Kahn departed on April 4th leaving us short another body. On the 8th Byron Ruth came to work just in time to help chain five miles of track at Proviso. All this chaining of yard tracks was in response to operating department complaints that the tracks were not as long as they thought they ought to be. I was getting a little cynical about engineering staff always having to carry the sack when an argument raged between departments. On the 11th Byron and I caught #13 west to Geneva where our claim agent, the venerable Ollie Adamson, collected us for a crossing survey at Elburn. After we finished, he drove us north to Richmond on the Williams Bay line for another vision survey at a grade crossing. On the 15th the Byron Ruth—Gene Lewis team did another accident survey at the Crawford Avenue (Pulaski Road) viaduct followed by another little measuring job at the California Avenue coach yard. Then it was back to Irondale with division engineer Bruce G. Packard and Jack Wilkinson, the roadmaster, to inventory track at 120th Street and Torrance Avenue to be leased to the Indiana Harbor Belt RR (IHB). On the 17th, a beautiful warm spring day, I worked near the Bellwood commuter stop on track changes proposed for the east end of yard 6. Since operating couldn't make the yard tracks hold more cars by having the engineers measure them, they thought by rearranging the ladder tracks they could increase capacity and flexibility. A magician I was not. The next day saw a midday emergency trip to Geneva to measure a signal box east of bridge #66. On the 22nd I made an accident survey in Elmhurst at the Myrtle Avenue crossing only to learn that as I was making the drawing for the law department the case had been settled. Disgusted, I scrapped that piece of paper and walked up Canal Street to measure a rail renewal at the diamond crossing just west of the Kinzie Street drawspan. This strange track connection belonging to the Pennsylvania Railroad ran north from the Union Station's tracks along the west bank of the Chicago River to Grand Avenue and the old Grand Avenue freightyard of a C&NW predecessor company. The track also provided access by the Milwaukee Road to their swingspan that crossed the North Branch of the river a short distance to the north. This line which wandered north to the vicinity of Wrigley Field had not seen much traffic for years, but the C&NW's traffic to the North Pier had worn out the light rail on its own running rails. Unfortunately, relaying half of a diamond crossing is impossible; it all has be renewed at once. Since this was a billable project, the C&NW got a new improved crossing at a bargain price because the Pennsylvania Railroad and the Milwaukee Road, the other owners, had to pay two thirds of the cost.

Rodman Byron Ruth and I made an accident survey under the Merchandise Mart on April 25th. It was sobering to realize that human beings had been hurt at these accident scenes. The Trimberger tragedy was still fresh in my mind. The good weather of April even reached into the musty offices of the auditor of capital expenditures. Ray Enders was the big fellow who should have followed a career in a Bridge and Building gang like his father Nick and his many West Chicago cousins. But, Ray had college time under his belt and determined that accounting was more to his taste than wrestling treated timbers in some bridge far away. Using the excuse to his superiors that he had to field check the division's reports, he arranged with me to go to the CTC crossover installation at Nachusa (NA), Creston (HX) and Malta (MA) on the 28th. He enjoyed the experience so much that he pulled the same reason for the next day for a trip to Meredith (ME and MW). I made sure that he earned his outing because I had written the completion reports and was quite sure of their accuracy. I introduced Ray to the joys of counting and measuring track components especially the minor pieces like rail anchors. Later, we stopped in DeKalb to visit the roadmaster, Earl "Fuzzy" Pierson and his clerk Florence. Sitting in their office in the old brick depot, Fuzzy was very depressed about the decline of track maintenance. I was used to his never-changing tale of gloom, but Ray was all ears and nodded sympathetically at the right moments because his father and brothers told much the same kind of story. At the end of the month Len Anhalt left for Iron River, Michigan, leaving our staff with four instrumentmen and two rodmen.

I was determined to finish that huge completion report for the Centralized Traffic Control installation from West Chicago to Nelson. To avoid repetitive copying, I designed a form that detailed all the types of materials installed so that I could merely add the place name and note the number of units of each item that were installed in that location. The report was almost finished when Ray Enders questioned the location of the switch stands at each installation. To settle the matter and to avoid catch-as-catch can trips to each spot, I rode train #381 from Proviso to Nelson. The trip was not very swift across the division as we kept losing the train air because of one car with a tricky air brake. These bad brake cars were called "dynamiters" because of the effect they had on stopping a run. The dispatcher didn't help either by sticking us with a lot of yellow (caution) signals. However, in the five hours it took to get to Nelson, I spotted and recorded all those switch stands for Mr. Enders. Since I was at Nelson for a time waiting for #380 eastbound, I measured new concrete foundations for a couple of buildings for another report. We made good time on #380 by passing two other slower eastbound freights and dodging the westbound streamliner fleet, a good demonstration of how flexible an instrument the CTC installation could be. By June I had the voluminous completion report all finished complete with blueprint maps of each crossover location and turned the hefty pile over to the auditor of capital expenditures. AFE B-1314 was finished.

On the 15th I ran a ditch profile along the north side of the main line at Elmhurst from West Avenue to Salt Creek west of town for Bruce Packard. Bruce happened to live in Elmhurst not far from the railroad and strolled over to supervise our work. My rodman, Byron Ruth, and I felt that we could have gotten along much better without this supervision as Bruce had a talent for obstructing work. He left about 10 a.m.; we finished the job in short order. Packard wanted to stand around and argue about how the work should be done instead of just doing it. He did not seem to understand that time is important, and that we, younger people, had a lot yet to do with our lives besides standing around debating with him.

On May 19th instrumentman Bill Smith and I were headed for Geneva until Bill's car jammed in first gear in the middle of Elmhurst's main street. After a struggle, the recalcitrant gear box grudgingly loosened and we limped our way to bridge #66 to check the high water level. Of course, there were always more tracks to measure; one at Ashton and one at Malta. In June we received direct orders from the chief engineer that the completion report for the new Proviso diesel shop would be finished at once. Management needed the capitalization for the company books to support some kind of financial maneuver transparent to our lowly division eyes. I did the track work portion of the 1167 report. Harold Jensen, now an office engineer in the chief's office and formerly my first division engineer in 1947, was directed to write the building description portion. He came over to our office with papers in his hand, threw his arm around my slim shoulders and said I could just as well write the building description. Dropping the mostly undone report on my desk, he gave me a pat on the back and said he would be glad to read it over, if I wished him to. Harold Jensen was still Major Jensen and this was the army, Mr. Jones. It was not a big problem. I let him impose on me because I really respected him. He had earned my loyalty a long time ago.

When I was in town I would often spend part of my lunch hour in the Galena Division's chief dispatcher's office. A table behind the dispatcher was a perfect place to perch while watching him move trains across the division using the CTC display board and long train sheet. Chauncey Grimm played that CTC panel like a musical instrument using the visual displays, telephone and his profound knowledge of the railroad and its people. Keeping the railroad fluid was an art in these pre-radio days. Always courteous to everyone, he was the perfect gentleman who could be depended upon to be a good humored guide when I brought visitors in to see the operation. The dispatcher's offices were located on the third floor of the CPT along the south corridor. The Wisconsin and Galena Division offices were in adjacent but separate offices. Despite their southern exposure, the window blinds were always drawn and the rooms kept quite dim. There was a striking contrast between the Galena's room with its new green and black CTC cabinets with their sparkling tiny lights and the Wisconsin's somber atmosphere populated by low-spoken eye-shaded dispatchers at their special long desks made to accommodate the long train sheets. From the next office could be heard the muted chatter of telegraph repeaters as many routine reports still came in from on line stations by wire. In these gloomy caverns one could almost imagine bats flying around the high ceilings. This was the operating heart of the railroad.

After a series of very hot June days, two bridge fires at Proviso hit the news. On the 25th the timber trestle over Lake Street (U. S. 20) carrying the freight mains from the Galena Division to yard 9 (North Yard) caught fire blocking inbound trains from the west. Of more importance to the commuting public was the fire at bridge 15¼ which burned the timber part of the span carrying the passenger mains across Addison Creek. (And where Frank Trimberger had been killed in January). That evening it was necessary to detour the westbound commuter fleet through the Proviso freight yards with resultant slow speeds and rough rides. The grumbling patrons did not appreciate how lucky they were to get home merely late instead of possibly very much later. Freight operations had to cease, of course, which compounded the railroad's problems. During this period in late June the weather presented us with a spell of intense heat which caused rail in the main lines to expand and move out of alignment. It was fortunate that welded ribbon rail was not yet installed in the main lines as the art of rail anchoring was still imperfectly understood. Track inspection was increased and slow orders placed which resulted in more serious disruptions to normal train operations.

Ollie Adamson accompanied me to the Chicago Shops to measure a diesel locomotive involved in a personal injury accident. I was certain that the manufacturer's construction drawings would have been more accurate, but he was sure that in presenting testimony in a court of law, my personal measurements would carry much more weight before a jury. We rode back to town in air-conditioned comfort of the lounge of a 400 streamliner backing to the CPT from the Streamliner Ramp. Late that same day a harried Ray Schaffer from public affairs rushed into the office and breathlessly asked that levels had to be run at once on the steel viaduct spans along the west and south side of the Galena tracks at Clinton Street. It seems that the artist doing the calendar painting for 1953 had just displayed his work to president R. L. Williams and his staff. The 400 streamliner depicted seemed to the president to be running uphill. To settle the matter poor Ray was dispatched to the division to determine if the steel girder spans were level. We dropped everything and went out in the blazing afternoon heat to run a line of levels around the sweeping curve. The steel spans were indeed level.

A railroad mishap in full view of the traveling public occurred at 11:00 a.m. July 7th when the cars for train #153 backed into the bumping post, derailing two cars. Jacking up and re-railing the cars was a delicate task given its location under the train shed roof and on the undertrack girder structure. Soon, thereafter, broad white lines were painted on the platforms ten feet from the bumping post at each track. Woe betide the trainman or locomotive engineer that violated that sacred space beyond the "dead" line. It has been said that the best protection for the end of a track was just a yellow mark and a ten day suspension for passing it. While #153's mishap was not quite what the term "wabashing" meant, the descriptive expression of cramming too many railroad cars into too little track had an interesting origin. About the turn of the twentieth century in the East St. Louis, Illinois, area with its many railroad yards and inter-road movements, a Wabash transfer crew arrived at a CB&Q yard to deliver a cut of cars. Since they were on short time and it was nearly the magic midnight hour when railroads make their daily car counts, the Wabash crew shoved their cut of freight cars into an "empty" track in the Burlington's yard. The track, however, was not empty. The Wabash engineman thought his cut of cars was rather hard to shove, but by opening the throttle more, he leaned into the string until it cleared the ladder, cut off and went home satisfied that the interchange was completed before midnight. The next day the rising sun disclosed a heap of demolished wooden freight cars at the other end of the yard. The Burlington yardmaster could only say he had been "wabashed". I believe this story can be found in Freeman Hubbard's classic book *Railroad Avenue.*

We were short of engineering staff all summer. Bernie Meyers, assistant chief engineer, did not help matters by making special requests for surveys to be done right away. He issued direct orders to our assistant engineer, Tony Zaborowski, to get out of his office chair and conduct a field survey of the Material Service Company's recent purchase of railroad land at Proviso. To get the cross section survey he thought he needed, Meyers dragooned his architectural engineer, Bill Armstrong into helping with the field work even though Bill had no experience with survey work. The next day we plotted the notes and began frenzied calculation of yardage, all with a manual calculator and slide rules. I never learned why we had to figure the grading quantities on this sale of land. In my view, however, Meyers owed the division a favor. I asked for and received permission for Bill to accompany me on a trip to South Pekin. On July 17th Bill and I rode the Rock Island's Peoria sleeper *Prairie Point.* I slept well in lower 7. In Peoria we found the bus to Pekin and met Ray Doty, the B&B supervisor, for the five mile run to South Pekin. After Bill and I went over the station plans to update them, a signal department truck took us back to Pekin where we happened on the C&IM's famous American type 4-4-0 #502. Going on to Peoria we boarded the Illinois Terminal's Decatur car and went to the east end of the line at Watkins near the Indiana state line. The car then reversed and went to Springfield. This little trip completed my personal coverage of the entire ITC line some of which was soon to be abandoned. Bill and I came back to Chicago on the GM&O night train in sleeper *McAllen.*

Rodman Bill Brown worked through the hot weather on the ballast job south of South Pekin where Len Anhalt, Sid Kahn and I had run levels last year. The ballast being placed was our old friend, Algonquin pitrun sand and gravel, which did not qualify as very good ballast. Each day Bill conscientiously set the 30 inch long $1\frac{1}{2} \times 1\frac{1}{2}$ inch oak ballast stakes with their blue tops mile after mile following the profiles we had drawn in the office last Fall. Much later, I learned from the roadmaster that Bill's strenuous efforts were mostly wasted. The roadmaster had watched the performance for some time. An old section foreman working with ballast gang would wait until Bill was out of sight and then walk along the line of work with a spike maul "adjusting" the grade stakes to achieve what he believed to be a better grade line.

On July 18, 1952 torrential rain in the Rockford area washed out the Freeport line. July recorded 13 days of +90 degree days ending up with a surplus of 905 degree days above normal. The engineering

Noontime railroaders in the Galena Division engineering office. Bob Lawton, rodman, Chuck Jevne, Land Department, and Gene Lewis, instrumentman, observe Bob's HO switch engine 801 pull a cut of cars from the "yard" on July 29, 1952.

staff tried to minimize days in the sun during this heat outburst. On the 29th Joe Zack, a photographer from the *Chicago Daily News,* came to our division office to make a photo story around our portable HO model railroad which we often set up at lunch time across our drafting tables. Several weeks later the spread appeared in the rotogravure section. On the last day of July, beginning my sixth year with the railroad, I consulted with C. P. Nelson, assistant superintendent of the car shops, about improving ventilation of a shop in building C-9. The next day I was back at Chicago Shops measuring rail at the engine house and trying to get information about the new boiler in M-4 to write the completion report. That was another job Harold Jensen was supposed to have done, but my reward for being good at writing that 1167 for the Proviso diesel shop turned into doing another favor for Harold. What could I say? I suppose, I could have tried to send in a report that stated "Steam generation boiler, each 1" but it was much more complicated than that.

On August 12th I wrote up a retirement proposition to eliminate all steam locomotive facilities on the Freeport line. On August 19th instrumentman Bob Lawton, new rodman John Cussen and I were at the east end of yard 7 at Proviso staking out a new half ladder track in the existing ladder. John Cantwell, the Galena Division superintendent, had an idea that two switch engines could work side by side on separate ladder tracks and do twice as much work providing the two crews didn't kick cars into the common track at the same time. Later that day roadmaster Nick Maas went with us to yard 9 to work out a track change that would eliminate a failing diamond crossing. On the 28th I was back at Proviso with Maas to work out changes to the east end of yard 6. On September 9th our crew was summoned to Proviso to explain why the new tracks were not straight. Of course they were not straight because the section foreman had not followed our stakes, relying instead on his "eye". There was always a battle between the optical view of the engineer and eyes of "experienced" old timers. While changes were being made at Proviso, Bob Heron in the operating department downtown was complaining about the general bad condition of main line tracks around the system. The chief draftsman, Ralph Walthers, another old C&NW veteran, retired August 29th. Ralph's head was full of the railroad's former days, the place names that used to be and his beloved maps and records which he was sure would never receive the tender care he had given them.

Accidents continued to happen. On September 5th I rode the CA&E to Lakewood station north of West Chicago to meet Ollie Adamson and a court reporter to survey the Pecatonica Street grade crossing in Winnebago. A photographer, Gene LeSeur, joined our party which was helpful as he later gave me a ride home during which he regaled me with tales of auto racing with the famous Ralph DiPalma in 1921. Other jobs in that month included more track changes at State Street Yard because of widening of the overhead street viaduct. Paving work at the Railway Express Terminal at Kinzie and Halsted Street required periodic taking of concrete samples for the test lab. I noted in my diary that the unburned portion of the Wolf Road timber bridge across Proviso yard was being removed this month. Part of the deal that Bernie Meyers cooked up with the State of Illinois on replacing the Mannheim Road bridge a mile east of Wolf Road was that the railroad by paying half the cost of the new

Mannheim structure would be relieved of any obligation to provide a new bridge at Wolf Road. Of course the public would not learn of this contractual wrinkle for many years when the outcry for a new Wolf Road bridge would be squelched when the railroad revealed the agreement with the Illinois highway department.

We lost another rodman on the 15th when Don Deneen left. A week later Edward E. Coover, Galena Division superintendent, was relieved of duty and replaced by H. B. Smith. Although Ed Coover had once been superintendent of the Nebraska Division, he failed to manage the turbulent Galena Division. On the 16th I had the privilege of hearing a speech given by Wayne Johnston, president of the Illinois Central, at a convention of the American Railway Engineering Association. Movies of construction of the Quebec, North Shore and Labrador Railway under extreme weather conditions were also shown. I was beginning to sense that there was a lot more to this railroad engineering world than my little corner on the Galena Division of the C&NW.

Wood Street's problems with the potato dealers came up again October 3rd. In company with my ex-boss Charley Hise, now an office engineer in the chief's office, Gus C. Johnson, B&B supervisor, and Jack Wilkinson, the roadmaster, I went to the Potato yard office to establish the limits of a new leased area and parking lot. Later in the day, B. R. Meyers, assistant chief engineer, J. E. "Jack" Goodwin, vice president operations, J. J. "Joe" Stein, general manager, N. L. "Nate" Waterman, superintendent of freight terminals, and T. A. "Tom" O'Donnell, assistant freight traffic manager-sales, straggled in to impress the volatile dealer's association that they could indeed attract the railroad's top brass who would surely lend their august positions to solving their space problems. On October 8th I rode the Western Avenue streetcar down to Wood Street to check on progress being made on the new arrangement. All these titles and the associated names bring up the point that on the C&NW signatures on letters rarely included given names of individuals—just initials and last name. The convention began in the telegraph days when sets of initials sufficed to identify the sender and the receipient. The foregoing list of people who attended the Wood Street meeting, for example, could have been expressed as: CEH, GCJ, JAW, EML met BRM, JEG, JJS, NLW and TAO at Wood Street October 3.

October was more hectic than usual. I had the paving job at the Express Terminal to monitor, the grading work at Wood Street needed attention, and a lot of other little jobs demanded some attention. Fortunately, the usual autumn shortage of cement slowed down the paving job so that the Wood Street grading work for which I was the official reporter to management could be juggled to fit in other urgent demands like an accident survey at the California Avenue coach yard on the 21st, Proviso track changes on the 22nd, a visual inspection of all the Elmhurst grade crossings for Charley Hise, helping Bob Lawton run cross sections at the Algonquin gravel pit on the 23rd, trips to St. Charles and Geneva inspecting pipeline installations, and checking the vertical clearance at overhead roadway bridge #81 at LaFox on the 31st. With Byron Ruth and his handy car we also inventoried the reconstructed Elburn stockyards (now there is a daunting job, measuring and detailing all the parts that go into a stockyard), and that almost fatal (to me) new side track at the Structo Manufacturing Company plant east of Freeport. Byron stayed in Freeport to visit his relatives while I caught the Illinois Central eastbound local at Cedarville. The IC crew graciously made an unscheduled stop in Elmhurst for me to get off close to home. A mild spell in the beginning of November allowed more quick field trips to collect bits of information needed to clear away those ever-accumulating completion report requests. Things as minor as a serial number on the new oil burner in the furnace at the Belvidere depot, or the location of the end of a rail renewal made field trips necessary. Byron and I made a trip on the 6th to RX east of Rochelle, inventoried turnouts in town and then hiked two miles to milepost 77 just to chain the location of the end of a rail renewal. He went to Freeport while I rode the Greyhound to Chicago. On the 7th I did my first job for M. B. "Bob" Lithgow, our erstwhile instrumentman, who had transferred to the industrial development department. On the 10th I went to see the manager of the Fox Valley Ordnance Works (a former munitions plant) at East Batavia, Illinois about some track relocations (which were never done). I rode the CA&E Batavia branch car east to the junction with the Chicago bound car. The completion report backlog grew so large that assistant engineer Tony Zaborowski stopped rocking in his office chair and offered to help. During all this commotion I was still attending engineering classes at IIT in the evening. The weather continued mild which meant that building contractors were pushing their jobs as hard as possible. On November 13th I went to Wheaton to check on the city's construction crew which was pushing sewer lines under the triple track main line at College Avenue. The next day saw the end of the paving project at the express house on Halsted Street. Byron Ruth was a big help during this hectic period. Not only a good rodman, he was outstandingly helpful and innovative in carrying out our routine work. He worked hard at his college work at Northwestern and served several tours with the C&NW as rodman and instrumentman. In later life he became a professor at the University of Florida, Gainesville. I can still hear his strongest expletive,"Gol' lee".

On November 20th everyone was gone leaving me alone in the office—I was the acting assistant engineer, a job that I would officially fill in another four years. On this day in 1952 I had to deal with an emergency when Blommer Chocolate Company's contractor accidentally severed the Western Union cable along the south side of our retaining wall at Clinton Street, deal with a report that a grocery store under construction in Melrose Park was encroaching on our right-of-way, arrange slow orders to protect the under crossing work at College Avenue, Wheaton, plan revisions for the State Street widening project and field three calls from the Fox River Ordnance Plant about their proposed track changes. Tony Zaborowski was back the next day, so I escaped on a train to Winnetka to meet Byron and his useful car for a trip to Williams Bay, Wisconsin, to do a track inventory with additional stops at Harvard and Crystal Lake on the return. Winter suddenly came on November 26th with high winds and cold. The railroad always went to pieces at the first cold. Commuter trains ran late sometimes without lights or heat as train lines froze up. The snow came on December 1st and another Chicago winter season was underway. Before Christmas there were short spells of warmer weather between the cold blasts. Despite the increasing cold Bob Lawton had to oversee the Santucci Company which was pushing ahead with the big City of Chicago tunnel across the west end of the Chicago Shops. The excavators were so deep in the ground that the ground did not freeze solid which meant that the surface had to be cleared in advance of the dig. The little used network of tracks serving the westernmost shop buildings were lifted—many were never reinstalled next year. My school break had come. Armed with a handful of trip passes I took my family on our annual Christmas vacation.

The first leg of our trip was on the Big Four's (NYC) train #438 from the 12th Street Illinois Central station to Cincinnati, Ohio. The train was steam powered. Arriving in a strange city at 5:10 a.m. (Because of the change in time zone) is a bit hard to bear because it was too early for the restaurant to be open and our five year old was hungry. We admired the Cincinnati Union Depot until the cafe opened for breakfast. A CUT 0-6-0 #9 quietly simmered among the passenger coaches from many railroads. We rode the Southern Railway's *New Royal Palm* south 338 miles to Chattanooga, Tennessee, on a half fare rate. There in the station we saw the Western and Atlantic's famous locomotive "General" and a replica of the South Carolina RR's " Best Friend of Charleston" before having dinner. We then rode the Southern's #41, the *Pelican,* in Pullman *General Longstreet* to New Orleans, Louisiana. That part of the trip was on a full pass. En route we woke up near Hattiesburg, Mississippi, and had an interesting rear view from the back of the train across Lake Pontchartrain before arriving in the Crescent City at 9:45 a.m. After a few days in this river city which included riding the St. Charles streetcar line, we boarded Texas Pacific's *Louisiana Eagle* on the 21st to Dallas, Texas, another free pass. It was a rough ride in a beautiful coach. After a day of sightseeing, shopping and riding the Oaklawn and Forney cars, we boarded a coach on the Fort Worth and Denver's #8 but soon decided to negotiate for Pullman space in *Lariat Crest.* The weather got colder as we went north. I got a glimpse from the train of my old USAAF station at Childress, Texas. There was an engine change at Amarillo where a pacific 4-6-2, # 552, became our new power. At Dalhart, Texas, we got off a very late train to meet my wife's parents who had driven from Illinois through incredibly bad weather across Kansas. Together at last, we drove west to Taos, New Mexico, for Christmas. After the holiday we drove south past Magdalena, New Mexico, where we spotted an ATSF 2-10-0 #2565 plodding eastwards. After a cold visit in Albuquerque, we went on west to Globe and Phoenix, Arizona. There we boarded a westbound AT&SF train while her parents continued south into Mexico. Our Pullman ride in *Kiowa Tribe* on the Santa Fe from Phoenix to Los Angeles, California, was uneventful except for a severe shaking at Cadiz, California, where our train was consolidated with another in the night. We visited some relatives and were treated to motor tours around that vast city. The Pacific Electric lines around the Los Angeles area were being dismantled at this time. On New Year's Day we took Union Pacific #10, the *City of St. Louis* from Los Angeles to Salt Lake City. Again, arrival in the dark at 4:40 a.m. in a big railroad station with nothing open is daunting, but we survived. The winter morning light finally came about 7:30 a.m. During our bus tour of the city we found a D&RGW 2-8-0 #223 in good condition on display in Liberty Park. That afternoon we left Salt Lake City on UP #38 and #24 eastbound in the Pullman *Clan Campbell.* Leaving town we saw three old Bamberger Railroad cars and an electric freight jack parked in the station. All the overhead wires were down. We had a Bigboy 4-8-8-4 #4005 on the head end from Ogden to Cheyenne. The next day in Cheyenne, Wyoming, a large fleet of old Union Pacific steam locomotives on the scrap line including Consolidation 2-8-0s and grimy 2-10-2s with OWR&N on the rear of some of the tenders caught my eye. At Omaha we got back on home rails for the trip to Oak Park 30 minutes ahead of schedule on my daughter's birthday—January 4, 1953.

While I was gone, president Rowland L. Williams, had become chairman of the board on December 31st. Paul E. Feucht, who had joined the C&NW August 1, 1951 as executive vice president, was elected president. This shuffle at the top was to be of no help to anyone much less for the railroad.

SPANNING PROVISO YARDS, 1953 . . . Any thought that the work load might have eased in my absence was instantly dispelled when I walked into the office on my first day back from vacation. I first asked about our staffing and was told there was no improvement. Despite the below zero temperature I had to check on that sewer job at Wheaton. I boarded #3 in Elmhurst on the 7th for the short ride to Wheaton. I took one look at the job and stopped the work at once. After a long and serious discussion with MacKay, the city's engineer, I watched while corrective measures were swiftly made to make the work site safe before returning to Chicago on the 11:00 scoot. Despite this first cold snap, the winter was to one of the warmest on record. Instrumentman William F. Polchow left to become chief engineer of the Toledo Terminal Railroad, a former electric interurban line, in Ohio. Now we were down to the assistant engineer Zaborowski, two instrumentmen: Bob Lawton and myself and two rodmen: Bill Brown and John Cussen, a far cry from the thirteen man staff we had when I started here in 1947. Bruce G. Packard was still division engineer. In March two more rodmen joined our group (I have forgotten their names). Byron Ruth came back from Northwestern and was put on the payroll as instrumentman, a well deserved promotion and a welcome addition to the crew. On April 1st two more rodmen joined: Carl Wojan and Bruce Meinders. Although we had a lot of rookies, at least we had a full crew. Taking advantage of a January thaw, I took John Cussen with me to Proviso to locate the end of a rail renewal in track 2 of the passenger main, walked to Wolf Road to check a signal bridge, and then went on into Elmhurst to measure the crossovers at HM tower and west of Cottage Hill Avenue. We then took the CA&E to Glen Ellyn because the C&NW's midday service in those days was not very frequent. We had to find and record the end of another rail renewal job near Taylor Street on track 1. We came back to Chicago on the CA&E having covered a lot of ground and getting several items cleared off our long to-do list. The next day we rode to Geneva and measured vertical clearances at bridge #65 in the rain before coming back to the office to work on an accident survey drawing for the law department.

Restlessness was endemic on the railroad at this time; I heard of several people planning to leave. On February 4th our faithful chief clerk, Tony Daleiden, announced that he was taking a job with the engineer of bridges. On February 5th rodman Bill Brown said he was leaving and was gone by the 13th. Our stenographer, Betty Jane Coe, quit on the 6th. I even flirted with change by interviewing the Railway Underwriters Insurance group for a very interesting position. Weighing the large amount of travel that their insurance investigational work required against my desire to finish my engineering education, I turned down their offer. I mentioned it to Bob Witt, a long time instrumentman on the Wisconsin Division. He interviewed Underwriters and was accepted in late March. Bill Armstrong, the assistant architect, had an offer from the Milwaukee Road. Knowing that Bill was getting restive and that he himself was to retire soon, L.C. Winkelhouse, offered to swap jobs with Bill. The change was approved and Bill became chief architectural engineer for the C&NW. On February 11th another new rodman appeared: Carl K. Oleson. The fact that Carl had no technical background or training underscored the difficulty of filling our needs. Carl was willing in that dogged Norwegian way of his and held on for several periods of employment eventually becoming a B&B foreman. He was fatally injured in an accident in the Chicago Passenger Terminal some years later while supervising platform repairs.

I continued my short field trips to collect information about new crossovers at Wheaton, locate the new depot at Winfield for proper map corrections, do a rail and switch inventory, locate crossing protection gates at Washington Street and describe the new sanitary facilities at NI tower, all in West Chicago. Rodman John Cussen drove us south to the lonesome station of Dimmick far down the Spring Valley line to check the U.S. highway 51 grade crossing. As we went west toward Langley on the Southern Illinois line to chain and inventory a sidetrack, the rain changed to heavy snow making driving very treacherous.

On January 27th I was in lower 4 of Pullman *McFaddin* on the Illinois Central's night train to Springfield. I rented a car and drove to Benld to look at a bridge on the Mine 4 spur. Coming north to Virden I inventoried a newly constructed passing track extension, inspected bridge #1910 and, finally, made a station ground survey at Culver, Illinois. Dropping the car, I rode the IC *Daylight* back to Chicago. At least I could now rent an automobile and get it through on my expense account. Bob Lawton and John Cussen were spending a lot of time in the Clinton Street interlocking plant stringlining curves for the new chief draftsman, Tom Beebe. His office was responsible for preparation of the track standards used throughout the C&NW. Given the wild variety of rail sizes and sections and the varying climatic conditions the railroad operated in, preparation of general standards that fit all conditions was a difficult task. At the moment he was concerned with adapting the old standards for 100 pound rail doubleslip switches to the new 115 pound rail. The geometry of these complex switches changed because of the greater height and thickness of the rail. The Clinton Street and Lake Street plants badly needed replacement. On the 13th John Cussen and I went to Sterling on #13 to locate switches between mileposts 110 and 111. I came back in the cab of #14 to Oak Park. The weather held and I rode #3 to West Chicago to see the roadmaster, Art Netzel, and to check out the new welfare building. The balmy weather was replaced by a sudden cold snap which created a plague of broken track bolts and consequent slow orders for the Iowa Division's main tracks.

As the weather eased in March, John Cussen and I ventured to Lombard where the gravel driveway behind the depot was washing out into Main Street. This was just the beginning of a messy job that required agreement of the railroad's engineering staff, the village of Lombard and adjacent land owners for placing a concrete pavement. It took years to accomplish. There were some jobs that had to be done "rain or shine". John Cussen and I got very wet March 11th staking out a must-do new trackage layout to serve Contracting and Materials Company at Proviso.

On March 16th I slipped out to photograph the test run of the Budd rail diesel cars on the North Pier trackage. The three car set was packed with local dignitaries from the city, business community and railroad. The object was to see if a successful commuter run could be established to Michigan Avenue north of the Chicago River. The run was made at a very slow speed as the vertical clearances at Dearborn and Clark Streets were tight. The test train reversed at Michigan Avenue and went west on the old original Galena and Chicago Union line to Noble Street where they again reversed and ran to the CPT. While a cute public relations stunt, this train service was not practical and was never instituted.

On March 17th I rode Wabash #17, in Pullman *Dorchester,* to Litchfield. Even though the train arrived at 5:25 a.m. the Benld section foreman was there to take us to Eagerville (Mine 1) for a day of chaining and inventorying tracks. Because the tracks were sunken into a cinder-coal crust, it was not an easy task. As we were doing our measuring an older man came strolling up to see what we were up to.

On March 16, 1953 a big promotion by the C&NW involved operation of its three RDC cars on the North Pier Line to test the practicality of suburban service directly to glittering North Michigan Avenue. Here the train loaded with civic officials and local dignitaries moves slowly through the State Street yard which was located in block Number One of the original City of Chicago plat of survey. Tight clearances, poor track, lack of passenger handling facilities and obvious complications to the regular commuter fleet operations relegated this experiment to just a public relations stunt.

Although he turned out to be the president of a short line railroad, the Columbia and Millstadt, he looked as though he had just gotten off a tractor plowing in a nearby field. You can't tell a book by the cover, as they say. We finished up our work and were driven back to Litchfield for Wabash #10 which was running an hour late.

There was some good news; instrumentman's pay went to $412.65 from $396.78 retroactive to October 1, 1952. That was almost an extra two hundred bucks a year and much appreciated. After Easter the field work began in earnest. There were a lot of new industrial tracks to be built, and we had to stake their centerlines. On March 20th John Cussen and I did the Allied Chemical track at Proviso's Yard 9. On April 6th with rodmen Milt (?) and Carl Wojan we staked the Contracting and Materials track near the engine house at Proviso. The next day instrumentman Bob Lawton and I inventoried the Hoyne Iron and Steel Company tracks near Kedzie Avenue. Rodman Carl Oleson and I went to Morrison on train #25 to make a four quadrant vision survey at a grade crossing. That was an all-day job complicated by rain, hail and sunshine. We caught #14 eastbound and rode in the cab of #5018A with engineer Striebel and a fireman who was introduced only as "Wartnose". They were kind enough to stop at Wheaton so that we could catch a local train to Elmhurst. April continued its hectic pace as we strove to catch up on the winter's accumulated work. I had an accident survey to make east of the Fox River bridge #66 at Geneva. The next day I went to the Chicago Shops with Jack Reese, chief lineman, to figure out a way to get some Western Union wires moved at the diesel shop. There was an emergency call to realign the third main at the Shops on the 20th, so I grabbed Carl Wojan and Chris as rodmen. Two days later Bill Armstrong and I drove to DeKalb to deliver a package to Earl K."Fuzzy" Pierson, the roadmaster, to check on a drainage complaint at the Dixon stockyards, to help Bill measure the interior details of the Nelson Hotel where our trainmen stayed between runs and, finally, to measure and inventory the passing track extension at Manlius on the Southern Illinois subdivision. April 23rd was safety meeting day at Proviso. I met the new superintendent, Walter Johnson, whose help I would need in the very near future.

There were other personnel changes on the railroad: The division engineer at Sioux City resigned to go to the D&RGW, his place was taken by Les Deno who moved from Escanaba, Michigan, and Wagner, the division engineer at St. Paul, had resigned. On the 28th I took Chris with me to Proviso to reset those pesky centerline stakes at Contracting and Materials because their grading contractor seemed to have a penchant for knocking them out. My third visit to Irondale was on the 30th. when Jack McCord, the assistant roadmaster, drove John Cussen and me to South Chicago to chain and inventory trackage to be leased to the Indiana Harbor Belt. At this point my engineering duties changed dramatically when the reconstruction of the Mannheim Road bridge began in May.

Assistant chief engineer B.R. Meyers had mentioned to me some three years ago (May 4, 1950) that I was to be the railroad's resident engineer on this project. Looking back now, I can see that I should have been more flattered by being selected so long ahead of the ac-

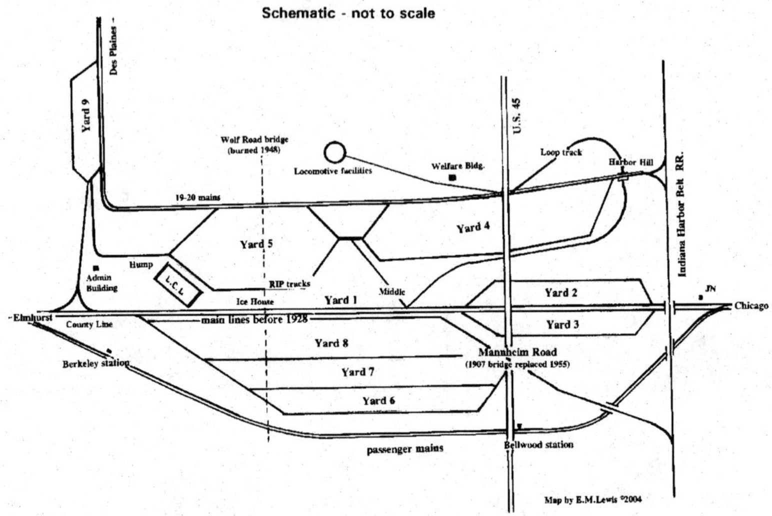

Once styled the biggest freight yard in the world, the collection of yards comprising the Proviso complex grew from the original yards 1, 2 and 3 built 1911–1917 into a sprawling sea of tracks that required relocation of the passenger mains in the 1923–1927 project.

tual job start. Meyers must have been confidant that I would still be around when the project began. May 6, 1953 marked the first day of the Mannheim Road viaduct replacement project. The State of Illinois had completed the construction plans, put the job out to bid and awarded the grading, foundation and old bridge removal to the Thomas McQueen Company of Forest Park, Illinois. The structural steel manufacture went to the Mississippi Valley Steel Company of Melrose Park, Illinois, and its erection to the Overland Construction Company. The C&NW paid one half the cost of this structure and was obliged to provide access to railroad property as well as dedicating an engineering representative to facilitate the job. In return for footing half the bill the Railway was relieved of any requirement to restore the fire-destroyed crossing of the Proviso yards at Wolf Road one mile to the west of Mannheim Road. The new four lane steel and concrete highway bridge was to replace a spindly two lane pony truss girder and wood wagon bridge constructed in 1907 that crossed 57 active railroad tracks. My first problem was the same one I had experienced in 1951—get the yards opened up so that survey lines could be made for construction layout. After a lot of chasing around Proviso, I finally caught up with Walter Johnson, the new superintendent, and explained that the derelict cars in yard 4 must be cleared away from the bridge. After that I got on an eastbound train and went to see my old boss Charley Hise to explain that Meyer's and Packard's idea that I should work in the office in Chicago and also manage the field problems associated with the construction at the bridge was not going to work—an onsite presence was absolutely necessary. He agreed and cleared the matter with Mr. Meyers. There were to be periods when I did not go near the Galena office for months at a time. Bruce Packard, the division engineer, and Nick Maas, the Proviso roadmaster, met with me on the site the next day to go over the plans. Yard 4 was now clear, Johnson was as good as his word. I now went after the lesser problems in yards 2 and 3. On May 8th the entire bridge survey line was open and ready for the state crews. One yardmaster said that was only the second time in 25 years he had seen from one side to the other. On Monday, May 11th, four tracks in yard 4 had again drifted shut. It took an entire morning to cajole the yardmaster at the east end of 5, to clear the fouled tracks. On another occasion, Johnny Larson, the yardmaster at Yard 6 was so exasperated at my pleas for a switch crew that he told me get some iodine out of the emergency medical kit, paint a number on my forehead, insert a drawbar in the appropriate part of my anatomy, and call myself a switch engine. Realizing he had overstepped the bounds, he ordered a switch foreman named "Mississip" to take his crew over to yard 2 after he finished the cut he was currently working. In addition to the constant strain of keeping the line open, I had to round up a section crew to take out a single track tie that just happened to be where a survey monument had to be installed. As I grew into the job, I learned that just doing small things like spiking switches, bleeding air on cars to be moved, or hanging red lanterns to protect the work was a lot easier than waiting for someone to perform the work. A post in the intertrack fence at the Bellwood station had to be moved. That was one job I would not tackle as it was between the high speed passenger mains, a territory that I treated with great respect. It was just a quarter mile to the east where our tapeman, Frank Trimberger had been killed by a train in January last year.

On May 15th the State's test pile contractor arrived on site to learn where the access roads were and how we were going to work in the midst of all those tracks. At his request I arranged for three railroad flat cars to carry his rig. My notes for the 17th showed that the engine house at West Chicago had burned down and at Proviso, the general contractor, Thomas McQueen, was installing the construction shacks where we were to work for the next two years. Of course, I was having to deal with that Yard 4 problem again. On the 19th I went downtown and attended a meeting in the State of Illinois building with Phil Sawicki of the State's Bridge Division, the general contractor and the structural steel supplier, Mississippi Valley Steel. On May 20th the drilling contractor loaded his caterpillar tread crane onto an SP 52 foot flatcar on the IHB lead from Yard 6 east of the bridge. His method was simple. The crane on the ground lifted one end of the flat, the crew rolled the truck clear, the end of the car was lowered to the rail, the crane rolled up the flat car until it balanced over the remaining truck, turned and lifted the end of the car resting on the rail high enough to allow the truck to be rolled back in place, and then positioned the crane in the center of the car. We were ready to roll. I ordered a work train for the 21st. The three car train was led by a loose-jointed old class J Mike, 2-8-2, #2512. We switched the consist to various spots in yards 4, 3 and 2 where steel pilings were banged into the ground until they met refusal. The pilings did not go very far because there was a solid limestone ledge just a few feet down. The work went quickly and we finished the test pilings in the railroad yard area that same afternoon. Taking the crane off the car was simply a reverse of the loading process. The car had to go to the rip track to fix the disconnected air brake piping caused by the truck removal and replacement. More test piles were needed in open areas away from the railroad operations which was a relief to me because I didn't have to supervise a work train. I arranged to have a temporary grade crossing installed east of the Bellwood passenger platforms across the double tracked main line and the contractor moved his equipment across without incident. The crossing was protected with a chain and a railroad switch lock; I, of course, had a key.

The contractor's boss on the job was a tough and canny Swiss named Ernest Sporry. Aggressive and innovative, Ernie was to become a very good friend from whom I learned a great deal about construction methods and organization. The construction trades respected his leadership abilities despite his sometimes brusque, but very fair, relations with them. The men obeyed him without question and went out of their way to make the job go smoothly. The owner of the Thomas McQueen Company was a dignified ruddy faced old Irishman. The caricature of the selfmade man, Tom McQueen was often found out on the job watching the work, always prepared to roll up his sleeves and step right in, if necessary. He was close with his money and had a lot of it.

I was settling in as the C&NW's Proviso engineer. Every day I first went to the bridge site to make sure the track corridor was clear. If everything was normal, I often went to the roadmaster's office to help his clerk sort labor reports as there were so many projects under way in the yard that allocation of labor costs was easily muddled. If nothing in particular required my attention, I would hop a commuter train and go into the office and help with the paper blizzard. I could not shake the ongoing job of writing completion reports. On June 4th Bruce Packard suggested that I might want to bid the job of estimator in the bridge department as it paid $458.00. I decided to stick with the present arrangement. On June 5th I prepared a plan for a small temporary bridge at the Mannheim site. The engineer of bridges, Art Harris, did not approve of it, but Meyers overruled him. Any construction job at Proviso was automatically shunted over to me—like paving replacement at the Administration Building. On June 11th I went over this work with Jim Gordon of Sackley and Company, an old established corporation in the area that had gotten started by building sidewalks in the new suburbs around Chicago in the 1920s.

The State assigned Pat Mair to be their field engineer at the Mannheim job. A lanky, slow spoken fellow, he gave the impression that he was a country boy just off the farm, but that image was soon modified by his reasonable, measured reaction to problems that arose on this big job. We shared an office in one of the construction shacks which were supplied by the contractor. Certainly not palatial, the rough buildings were about the size of a house trailer and were equipped with running water and toilets. The power and water lines were connected to nearby railroad lines. McQueen and Company learned at first hand about the infamous buried slag on which Proviso had been built when they tried digging ditches for the water and sewer connections. When the first Proviso yard was built around 1911 the company obtained a huge quantity of waste material from the steel mills for little but the cost of hauling. Called popcorn slag because of its appearance when fresh, it was dumped in great quantities into the swampy land as a base for the new railroad yard. What was not under-stood, however, were the chemical changes that occurred when the slag was buried in a wet environment. It changed into a vitreous material much like glass. The contractor's ditcher hit this petrified slag while digging the water and sewer lines for our construction offices. The machine juttered and shuddered as it attacked this hard material and soon began to shed pieces of itself. The only way to deal with the slag was use of jack hammers. That slag problem would come back to haunt me some years later. Before the digging started I had my first experience with the strange art of dowsing. Using a bent coat hanger, Sporry's labor foreman walked slowly over the area where we believed there was a 1 inch waterline. The hanger dipped repeatedly in one location. Sure enough when we had chopped a hole there, the waterline was found. I tried my hand at dowsing with no success at all. Oh, ye of little faith!

Freight cars set-off by transfer crews continued to be a problem, especially in yards 2 and 3. Despite standing orders, night crews would drop their empties under the bridge and go on their merry ways leaving me to plead with the yardmaster the next morning to move them into the clear. It got to be a routine, but the yardmaster would never send a crew over to clear the area on his own initiative; he waited for me to come around yowling at him before he took action. I think he enjoyed that little game. Finally, the surveyors had established their sight lines and had the piers monumented so I took a short vacation with my family—we rode trains.

On June 25th we went to Omaha on #1 arriving 1¼ hours late because of a sick diesel. That evening we boarded Pullman car *Shorewood* on train #13 for Chadron, Nebraska. Train #13 that evening had a GP7 #1638, five baggage and express cars, one coach and our Pullman. We had a breakfast stop at Long Pine, Nebraska, and a leisurely amble to Chadron where I was surprised to see that the consist was actually washed before returning east as train #14. These two trains would continue their lonesome journeys until July 1958 when, amid

Motorcar #9926 at Chadron, Nebraska, on June 26, 1953 being loaded with lcl freight. This was train #1 to Rapid City, South Dakota.

a storm of local protest, the ICC authorized their discontinuance. Especially damning at one hearing session was the response to the law judge's question of how the protestors had come to the hearing in Omaha—no one had come to the hearing on the train. At Chadron I saw my old colleague from Galena Division days, Doug Oakleaf. He said he liked living out here in the sandhill country. In the afternoon we boarded motorcar #9926 to Rapid City, South Dakota. After a day of sightseeing there, we came back to Chicago on #518, the *Dakota 400*. #518 that evening was headed by GP7 #1619 hauling four mail and express cars, our single coach and a Pullman *Clarkes Gap*. The light rail and sketchily ballasted track made for a bumpy ride. We also found the coach was uncomfortable especially when the trainmen came through the car collecting our pillows at 5:00 a.m. just before arriving at the division point in Huron.

Since the Mannheim job was inactive while awaiting delivery of steel piling, I went back to the Chicago office which was just as well as we were enduring a summer of record heat and humidity. On July 1st all three main lines were blocked at West Avenue Elmhurst by a 23 car derailment. I took some pictures and then rode to town in one of the then rare air-conditioned car on an emergency commuter train. By evening two tracks were open but the eastbound was still blocked muddling the evening commuter trains which were compelled to use the hand throw crossovers at Wheaton. The eastbound streamliner fleet was rerouted at West Chicago northwards on the old Freeport line to Foris (Fox River Switch) at Elgin and then north through the city on the old Williams Bay line to Crystal Lake where they turned east to Chicago. That was a slow trip over a line that had seen its last passenger train February 15, 1932.

There were more personnel changes announced July 6th: James A. Barnes became division engineer at Huron, South Dakota, Culbertson went to Madison, Wisconsin, M. S. Reid came from St. Paul to the Galena Division replacing Bruce G. Packard who was sidetracked to an office engineer's job on the chief engineer's staff. (His temper had gotten the best of him during an inspection trip). Since we were still waiting for the piling at Mannheim Road, I busied myself with other Proviso jobs like measuring rail renewals, inventorying tracks, checking on other bridge jobs, *etc.* Meanwhile, the Chicago office was falling apart as the division engineer was on his way out (last day July 15th) and his assistant was ill with a stomach ulcer. I had to go back into town and run the office. The next day was Maurie Reid's first day as division engineer. Tony, somewhat recovered, finally crept back to work, but was not well enough to do anything useful. I did what I could with the paperwork and got in a couple of short trips to Wood Street before the news came that the steel piling had arrived and that I was needed at Proviso. Jack Goodwin, VPO, authorized a company telephone for me at the jobsite (extension 257, installed September 11th) which proved to be a valued tool for chasing those yardmasters and for ordering work trains. On July 22nd I was once more summoned to run the office when the entire staff was away. There was a derailment in the Clinton Street interlocking. Knowing that a derailment there would be a hot subject, I walked out there to inspect and found that it was just one set of wheels off. I reported the matter to the chief engineer's office. Leonard R. Lamport, the engineer of maintenance (the man who hired me in 1947) called to express his satisfaction at the way I had handled the incident. Of course the next day I had to conduct a regular derailment survey just to close the matter properly. During the mid day lull in commuter train operations, I took rodman John Cussen with me to do the survey and also inventory a rail renewal including four double slip switches at Lake Street. My day was not over; I attended a special assessment meeting in Wheaton that evening for the law department.

At Proviso on July 24th grading of the new highway approaches was underway. The contractor, Thomas McQueen, had scored a real coup. They had successfully bid the excavation of a new highway interchange at 25th Avenue, Bellwood, and the Congress Street Expressway. Disposing of the excavated dirt was no problem, since they also had the contract at the Mannheim Road job. On top of being paid to dig the earth out at 25th Avenue, they also being were paid to place it as fill—a golden deal. McQueen, a canny Irishman, also took this opportunity to squeeze the last bit of use out of a fleet of ancient Mack trucks. These chain drive museum pieces ran on solid rubber tires. They looked like World War I army surplus and were probably the first trucks acquired by the young Thomas McQueen when he started his company. Use of this equipment on public roads was questionable, but in 1953 he got away with it.

On August 21st pile driving began at the north abutment. Mannheim Road, which carried U. S. 45, was a major north-south artery with much heavy truck traffic. We had to close the road for two weeks to permit installation of a 66 inch diameter culvert through the north approach. Taking advantage of the road closure the entire north abutment for the four lane highway bridge was also built. The detour was one mile to the east on 25th Avenue which crossed the railroad's main lines at grade. There was a tremendous amount of highway traffic inconvenienced by this diversion and the risks of rail and road crossing was very high, yet no accidents were recorded to my knowledge. On August 26th work at the south abutment began. For the contractor's crews these first pieces of building provided an opportunity for seeing how they would work together as a team. Happily, they were far away from active railroad tracks. No major problems arose as the crews learned and improved upon the routine. The new bridge was to be a four lane concrete highway with sidewalks on either side. The basic structure consisted of a series of simple steel girder spans set on solid slab upright concrete piers founded on steel H piles. Construction of the easterly two lanes adjacent to the 1907 wooden bridge was to be completed and put into use before demolition of the old bridge could proceed. The western half of the new bridge would then take the place of the old bridge. The new piers were numbered from north to south. The construction procedure for the piers consisted of driving steel sheet piling to form an enclosed area like a caisson between the railroad tracks. The inside of the cofferdam was then excavated about six feet deep using a crane with a clamshell bucket. Then a pattern of H piles were driven to refusal by a pile driver equipped crane. After cutting-off the piles about a foot above the bottom of the grade, reinforcing steel was placed and the caisson was filled with concrete. After the concrete had set, the sheet piling was extracted for reuse by a compressed air-driven pile extractor mounted on a crane. The second operation involved setting forms for six foot high concrete collision walls on the footings, placing reinforcing steel, pouring the walls and stripping the forms after the concrete had set a sufficient time. The third operation required erection of forms for the bridge pier itself. These were about 20 feet high and presented a real problem in handling within the confines of the railroad yard. The thin-walled piers required setting the reinforcing steel, pouring concrete, stripping the forms and finishing the surface in areas that often were nothing more than the land between two parallel tracks. To complicate matters further, the construction sequence did not include just one pier location at a time; each of the above operations could be observed in progress at several piers simultaneously. Doing all this digging, pile driving, concrete pouring and form handling in the middle of a busy railroad yard took some clever planning and the railroad's cooperation. Sporry was the genius that figured how to get to and work at the pier sites amid all those railroad tracks. At his request I ordered some 70 ton capacity flatcars for the initial sheet and H Beam pile driving. He loaded a big Lorain crane on one of the cars and the steel piling on the other cars. On September 1st I arranged to have the rig switched to the pier 7 location. The scheme worked so well that the contractor fixed up another crane set on flatcars to double the amount of pile driving. I was able to establish a noon switch which meant that we could get two operations a day with each crane set. After getting the equipment into position, I personally spiked the switches shut and set warning lanterns. We worked six days a week for we knew that this work could not be done after cold weather set in. Meanwhile, chief engineer, B. R. Meyers, told division engineer Reid to quit complaining about my non-availability because I was deeply involved on a project in which he had a special interest.

All through September the mad shuffle of equipment went on. On the 16th the last of the pile driving was completed much to everyone's relief. I was glad to get those big (and expensive) cranes off the flatcars and out of the railroad yard. The State's engineers had to count the blows during pile driving to calculate bearing capacity and complained that their hearing was being affected. The railroad people were also heartily sick of the constant booming sound. By October 6th the underground work was finished and there was a nice, neat line of concrete slabs six feet high marching across the yards between the tracks. During this hectic period I had several problems with the switching arrangements and had to step on a couple of yardmasters when construction crews were standing idle just because someone "forgot" to tell the switch crew or just plain got stubborn. Maurie Reid was also upset, not only at my absence, but also at my nerve in submitting overtime hours on my timesheet for Saturday work. He came out on the job and said flatly that my monthly pay was exactly that. Mr. Sporry was present at that conversation and said that "if the railroad was too cheap to pay overtime, then Thomas McQueen Company would pay the difference." Reid left. I then got an official letter from him October 6th telling me not to work Saturdays anymore. As much as I could I obeyed his instructions. In those days there was no official railroad policy defining what monthly salaries covered. The engineering staff was not unionized and management felt that we were paid for 25 hours a day 8 days in the week (24/7 in modern parlance). Yet, I knew of numerous examples of overtime being paid for special work. This was a small potatoes argument to me, and I concentrated on doing my best to keep the railroad and contractor out of each other's way.

Charley Hise sent me down to see about a problem in connection with construction of the Congress Street Expressway at the Rockwell Street line underpass. I met with Mr. Haupt, president of Strobel Construction Company but quickly determined that the decision making required was far above my level. I threw the problem back to Charley and went back to Proviso. When at the Rockwell site, I noticed that demolition of the Garfield Park elevated line was beginning. The CA&E's access to Chicago was gone and the railroad itself was doomed.

The next phase of construction at Mannheim Road involved those high, thin, concrete slab piers. Again, Ernest Sporry figured out an elegant and practical way of building and handling these tall concrete forms so that there was no interference with railroad operations. He asked me to secure some old wooden boxcars and flats. Spotting them on the Material Service lead tracks (with their permission) the wooden forms for the piers were prefabricated and loaded on the flats along with a crane. Sporry's men cut holes in the wooden roof of the condemned boxcars before I had the consist switched into the track next to a new collision wall. The box car next to the construction site was immobilized by setting the hand brake and jacking it sideways onto its side bearings. The crane on the flatcar picked up one of the prefabricated forms and held it in position while the carpenters set timber bracing to the now solidly planted box car. In went the reinforcing steel and then the rest of the prefabricated wooden forms. Concrete was poured from mixing trucks parked on the old bridge above the construction site. Leaving the anchored box car in place until the concrete had set, I arranged to switch the rest of the work set to another track with another condemned box car and the process was repeated. It was an orchestrated ballet of switching the work train around to the various spots.

While involved in this complicated equipment shuffle, I learned that the assistant engineer, Tony Zaborowski, had been found unconscious in the Chicago Passenger Terminal and had been rushed to the hospital. He had suffered a ruptured ulcer and also had tuberculosis. That was on October 8th. The next day division engineer M. S. Reid called me at Proviso and asked me to come in to the office on Monday the 12th as I was now the senior man in the engineering office. I reported only to find him tied up in a safety meeting. Since I had an equipment move to supervise at Proviso, I started to go back but was cornered by Nate Waterman, superintendent of freight terminals, John Cantwell, Galena Division superintendent and Walter Johnson, the Proviso superintendent, who wanted to talk about substantial track changes at Proviso. Reid finally appeared from the meeting but was undecided about what I was to do. I went back to Proviso to get the next set of equipment placed. It was to be a race against the weather to get these tall piers poured before the winter cold arrived. By October 29th 5 of the 17 piers were completed. The State's engineers were getting very nervous as Sporry kept his crews going on overtime. We shuffled the work cars from site to site with very little slippage—the yard crews seemed to be getting with the routine. Phil Sawicki the State of Illinois's bridge engineer was replaced by Richard Golterman, the brother of a rodman, Ralph W. Golterman, with whom I had worked several years ago. The downtown highway engineers said they were getting complaints from other work sites that we were not protecting the concrete from cold as other contractors were required to do on other jobs. On November 16th the temperature was 71 degrees when we poured concrete. The last pier, #17, was poured on November 20th when the day's high temperature was 66 degrees. Sporry's Swiss luck had held up. That finished the bridge work for the winter. It was now ready for steel erection. In early November I was invited to visit the fabrication plant at Mississippi Valley Steel in Melrose Park to see our steel being processed. I was surprised to see that the girders were assembled on the shop floor to make sure that every piece fit.

Now that the bridge job no longer required by daily attention, I was drawn into the rearrangement of the east yard leads for yards 6-7-8. That unofficial title I had acquired as the Proviso engineer resulted in Chicago happily dumping everything associated with Proviso into my lap. Rodman John Cussen took a job in the chief draftsman office leaving me with those dratted completion reports again. Returning to my old desk on November 30th I picked up small jobs around Chicago such as helping assistant roadmaster Jack McCord with an inventory at Hoyne Iron and Steel near Kedzie station; conferring with the law department about an accident case at McHenry; and preparing a drawing for an accident at McCullom Road south of McHenry. Seeing that rodman Carl Oleson was a likely candidate, I began training him in the mysteries of doing completion reports which involved several expeditions to LaFox, Elburn, Maple Park and Cortland for a rail renewal. I now owned a car, my very first, and it made a whale of a difference in collecting information from scattered points on the division. There were more stops at Geneva, West Chicago, Batavia and Winfield where I showed Carl what was required to properly write a report. We trimmed the delinquent list from 100 to 68 by the time I took my annual Christmas vacation. While I was gone, the railroad had a severe financial crisis that involved layoffs. I lost two day's pay but was grateful that I had gotten that extra overtime pay last summer. The future of the railroad seemed bleak.

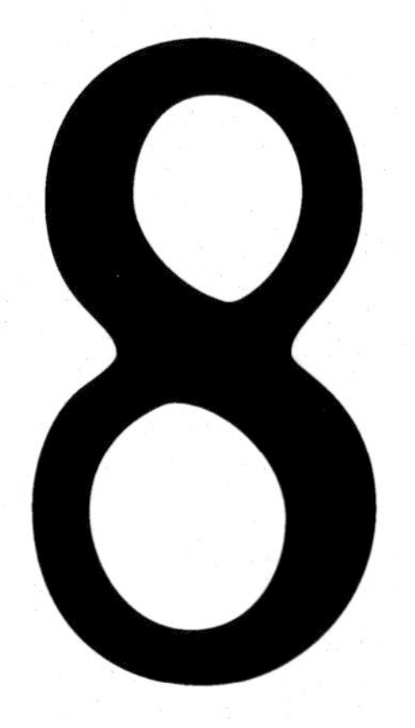

FOUR LANES ACROSS PROVISO, 1954 . . . The usual heap of work to be done greeted me upon my return from a Florida vacation. The new year started quietly with the Mannheim bridge job inactive and last year's hot jobs asleep. Nothing seem to be happening anywhere; it was if the cash crisis of December 1953 had cast a pall over the railroad. The extreme cold weather didn't help, either. I ventured out to see what Byron Ruth had done to the Cleveland Chair Company track at Proviso and determined that a few adjustments were necessary to make it right. Then the first shipment of bright orange structural steel arrived at Proviso in the morning of January 19th which opened the floodgates on that job. Unfortunately, the State's engineers had made a slight error in the placement of the anchor bolts atop the piers and some tedious adjustments were necessary before erection could commence. Overland, the steel erector, set the first 30 inch deep girder at 2:30 p.m. that afternoon and the race was on. I had ordered flagmen and slow orders for protecting the main lines so that the work moved safely and quickly. By late afternoon the girders over the main lines were all set in place. The next day M. S. Reid, the division engineer, asked me if I wanted to take the assistant roadmaster's job at Proviso. I declined as the work would have interfered with my night school schedule and my G. I. Bill credits would be terminated if I dropped any courses. With only two more years to go, I wanted to complete my degree work. On the 22nd Overland's steel foreman, Ken Blakey, said the gang had been called to the Griffin Wheel foundry at California Avenue to fix an emergency there. He invited me to accompany him to see the work. The foundry, built in 1890, had only one clean and neat area where a new casting line had been installed. Griffin had figured out how to cast steel freight car wheels, something the industry desperately needed. The process was still a deep secret when we visited, but then, who would we tell? Saturday the 23rd was a makeup day for the layoff of December 24th. Instead of going to the Chicago office, I went to Proviso to help Pat Mair, the State's resident engineer at Mannheim Road, update the construction records. Steel erection continued despite the cold. Snow or ice on the steel, however, kept the steel gang grounded. The erection crew were Mohawk Indians from northern New York State. Absolutely fearless when working high in the air, they were a close-mouthed, clannish bunch; not unfriendly, just withdrawn. When working on the high iron, they moved easily and surely with their drift pins, wrenches, hammers and buckers amid hot rivets flying from the forge to the catch bucket. I never saw one rivet fall to the ground.

Elsewhere at Proviso I saw C&NW steam locomotives being dismantled at Yard 9. On January 8th a class J-4, 2-8-4, #2808 was reduced to a naked boiler and running gear along with a class Z, "Zulu"

2-8-0 #1735 and a class M-4 0-8-0 switcher #2643 on the line next to succumb to the torches.

Back in the Chicago office, the Bridge and Building paint crew actually did our walls. Dull yellow-brown changed to institutional green with a white ceiling. At Mannheim Road steel erection went swiftly after getting a new crane operator—the erecting crew did not trust the first one. The big girder spans went across the yard 7 and 8 switching leads without difficulty but could have been done sooner save for an uncooperative yardmaster, Sparky Adams. The steel was getting deeper and longer as the crew moved from south to north. The 30 inch girders changed to 36 inch which then became built-up plate girders across yard 4. On February 4th I learned a valuable lesson, do not trust a section man's assurances. I found that the "dead" tracks where the crane was working were not dead at all; I spiked them shut myself. On the 4th and 5th the blast plates over the switching leads were put up. The latter were floppy, awkward plates that were finally wrestled into place. Ironically, they were designed to deflect steam locomotive stack blasts under the new bridge even though the steam era was almost dead. The blast plates over the main lines were set by operating the crane from atop the old bridge. On February 8th I had that old problem of derelict freight cars blocking access to the bridge work site in yard 4. This time I went to the general yard superintendent who could see we had a lot of big loads of steel girders waiting for the railroad to clear that yard. The first big girder went up on the 9th. The warm spell we were having should have made the steel erection go quicker, but now there was a problem with some of the new girders coming from the fabricator were slightly crooked. The defective girders had to be returned which was a larger problem than it should have been. Mississippi Valley Steel's plant could almost be seen from Harbor Hill, yet the Indiana Harbor Belt couldn't seem to get traffic through their nearby Norpaul yard. At least the riveting gang could continue their work.

At the Proviso safety meeting February 25th Galena Division superintendent John R. Cantwell announced that Proviso was to become a storage and transfer point because of the high cost of lading damage caused by the loose car switching. What that meant was not clear. By March 1st changes in operation at Proviso were instituted, but Walter Johnson, the local superintendent, said he felt that he was just going around in circles. In news from the Chicago office I learned that the Peninsula Division had been abolished and that Bill Comstock had bumped our instrumentman, Ralph W. Golterman, back to rodman. Early in March Bill was appointed assistant engineer to replace the incapacitated Tony Zaborowski. Willard B. Comstock was a full-blooded Southern Cheyenne Indian educated at the Carlisle (Pennsylvania) Academy and was in school there when the famous athlete, Jim Thorpe, attended the school. Because of his Indian heritage, Bill always had a struggle with the railroad's engineering hierarchy. Any mistake was magnified beyond its importance. Such a simple error as the calculation of the MBM (thousand board measure) for replacement of an engine house roof for 1 inch material instead of the 2 inch thickness required, set a black cloud over his head that followed him from job to job. Bill was to perform very well in his new position with his calm demeanor and dignified approach to the many problems facing the office. A newly hired tapeman, Arnold Thomas Stone, turned out to be a fugitive wanted in six states and by the FBI for passing bad checks. He had just been released from the Chillicothe (Ohio) penitentiary before he took the railroad job. He vanished on February 26th when he felt the heat, but was soon apprehended in Evanston, Wyoming, according to a newspaper item on March 10th. The railroad's lack of a central employment office and failure to check references was proving to be a serious problem.

I was called to Judge Sullivan's court March 5, 1954 to testify in a personal injury suit involving an accident at Elburn, Illinois. John Gobel was our attorney. Afterwards, I raced back to Proviso only to find the erecting crane had broken down and was being fixed. I needn't have hurried. The good news, however, was my instrumentman's pay went to $429.34. On March 11th Reid called me at Proviso and said that there had been a report of cables stretched across the main line near Proviso and that #103 *the City of Los Angeles* had struck them last night. I checked out the reported incident but could find nothing amiss. I wondered where these stories came from. By the 18th of March all the steel was up and being riveted. I had no reason to stay at Proviso and resumed my duties in Chicago.

On April 1st I visited the catacombs of the Ravenswood accounting office again seeking records on ballast installed in main tracks. It was hopeless. Since the weather had moderated, I gathered up Carl Oleson in Batavia and ran through our laundry list of things to do: check the Railway Express facility in Geneva, a low clearance at South Elgin, two bridges at Carpentersville, chain a track at Huntley, see to an outstanding bill against the City of Belvidere, check two bridges in Rockford, chain and inventory two tracks on the Album Street spur in East Rockford, and inspect the depot at Freeport. We stayed the night in Beloit, Wisconsin. The next day we went to Lake Geneva, Wisconsin, inspected the depot, measured a sidewalk at Sage Street, ran levels at McCullom Road crossing near Richmond, checked on a bridge there and located some crossing signals at Crystal Lake. All of this piddling work cleared up 18 requests for information. But the list of outstanding roadway completion reports was too long to ignore. On the 14th and 15th I hit the road again to inspect an undertrack coal hopper in Sterling. I then cut across country to Churchill on the Spring Valley line to check some bridges and traveled through Ladd, Seatonville, Spring Valley and Triumph. Coming through Dixon I made a picture of a shabby IC #341 (0-6-0) working the interchange north of our depot. (Their line from Freeport to Clinton, Illinois, called the "Gruber Line", was to see its last trip December 21, 1985). A public hearing regarding a drainage assessement in the Township of Ophir had to be covered to make sure that some sneaky tax would not be levied against the railroad. I came home via Elva and DeKalb. This personal transportation was certainly a great improvement to the chaotic travel arrangements we used to endure. And the company paid me a mileage allowance.

On April 19th instrumentman Byron Ruth and I were called out to the Clinton Street tower to help determine why the interlocking machine was not working well. Some differential settlement seemed to have twisted the frame slightly causing the interlocking levers and bars to seize up. We carefully ran levels on the corners and recommended some shims at the critical spots. The interlocking machine itself was a General Railway Signal product containing 168 pistol-grip levers. It controlled 29 upper quadrant semaphores, 33 dwarf signals, 26 turnouts, 12 double slips plus a number of locking levers to protect train movements. With 153 working levers the inside of the machine was a marvel of machined metal pieces that worked together like a Swiss watch. There was nothing digital about this machine—it was pure analog and worked very well in concert with the Lake Street plant to protect train movements in and out of the Chicago Passenger Terminal's 16 tracks. An electronic replacement was many years away. (The interlocking was replaced by electronic controls and the old brick structure was demolished November 26, 2002).

During May instrumentman Bob Lawton and I made a trip to Southern Illinois for a bridge check at Girard, a rail renewal at Womac and several other minor items at the far south end of the division. By the 12th we had the completion report list down to 23 items. Seven days later, there were 7 left. By May 27th there were 4 left to do. On June 2nd for the first time in memory, the office paperwork was caught up. Bill Comstock ran a good operation and his men responded.

At Mannheim Road the last concrete was poured on the east half of the new viaduct on May 21st. I had not been involved in this work as there was no interference with railroad operations. While I regretted not being with the construc-tion through the entire process, there was other railroad work to be done. A weekend trip with the family to central Illinois yielded a nice view of the Wabash's tiny 2-6-0 #573 in all its dingy glory parked outside its engine house at Bluffs, Illinois. This survivor, and a companion, lived only to provide power across the nearby Meredosia rail bridge spanning the Illinois River. Four years later I was to met #573 again at the St. Louis Transportation Museum at Barrett's Station.

My first contact with the Iowa traffic department was May 26th when I met W. F. Winkrantz, the district freight and passenger agent from Des Moines, Iowa, in Clinton, Iowa, to see what his customer, AGRICO and their storage facility at Fulton, Illinois, needed. Their business had expanded and they needed more trackage. This was the facility built on top of the former Galena and Chicago Union Rail Road's engine house where M. C. Christensen, Art Humburg and I had surveyed their original track layout in late 1947. I called their engineer in New York and was promised a drawing of the proposed layout. On the way back to Chicago I stopped at Morrison to check the grade crossing at Cherry Street and picked up the roadmaster,

Arthur E. Benson, to take him home to Sterling. On the way home I visited Earl "Fuzzy" Pierson, the roadmaster, in De Kalb where he asked me to meet with a Noel Baker who was interested in purchasing a piece of railroad property.

Division engineer Reid asked for a volunteer to accompany the weed sprayer from Nelson south to Benld. Because instrumentman Bob Lawton was a new father and Ralph Golterman had another commitment, I agreed to go. I rushed home to grab my traveling kit and got back to Chicago in time to catch #105 the *City of San Francisco.* I would rather have caught #111, the *City of Denver,* because it stopped in Sterling but my timing was bad. Riding the cab of #105 in the long summer evening was a good way to look at the division's main tracks. Our first stop was Clinton, Iowa, where I switched to #16 eastbound, which was running an hour late, for the short ride to Sterling. Early the next morning I reported to roadmaster Art Benson's office. He drove me to Nelson to catchup with the weed spraying extra. The man in charge was Dick Bailey former instrumentman on the Galena. We were stuck at Nelson because there was no locomotive power for our train. When #386 pulled in from South Pekin, there was a GP7 #1631 in the consist—that was our power. Finally, at 11:00 a.m. we commenced spraying aromatic oil on all the tracks from Nelson south to Speer. The aromatic oil was a light petroleum distillate which coated the weeds and smothered them. There was a constant mist of the stuff hovering around the spray car. It permeated our clothes and coated our skin. Years later I learned that this stuff was a carcinogen and had been banned from this type of use. In 1953, however, the smelly oil did kill weeds. Benson, the north end roadmaster, and Bill Baker, the south end roadmaster also rode the train, mostly in the waycar. My job was to make notes about the number of gallons distributed per mile. After a lunch stop at Manlius, we sprayed to Speer and then ran light to South Pekin, tieing-up at 7:50 p.m. We sprayed the South Pekin yard on the 19th and did the mainline from Speer to the Adams Street yard at Peoria. Other than recording gallonage I was essentially useless on this trip except when helping to switch in fresh oil tanks or set-out the empties by coupling or uncoupling the tank cars. On June 20th we got up at 4:30 a.m. and were out of South Pekin at 5:53 a.m. We used less oil on the south end and consequently arrived in Benld at 1 p.m., earlier than planned After a bite to eat, I hopped on the Illinois Terminal electric car to Springfield and waited there 2½ hours in my smelly clothes for the GM&O's *Ann Rutledge* to Chicago. The 21st day of June was the eleventh in a row of plus 90 degree temperatures. I was glad to stay in the office to write up a report on spraying and rid myself of the oil.

After that trip I was designated as the man in charge of the weed spraying. My attempts to line up a work train to spray the Freeport line were futile as the spray car had been bad ordered at Nelson because of a hot journal. I was not unhappy at the prospect of another day out of the heat. On the 24th roadmaster Jack McCord picked me up in Elmhurst at 5:15 a.m. for the short trip to West Chicago. It was to be another scorcher. We found the spray train was all mixed-up and the yardmaster unwilling to switch the cars around. After firmly leaning on this scissorbill, his switch crew lined up our train. We got started at 8:13 a.m. with an ALCO S-2 diesel #1213, conductor Les Weltman, engineer E. M."Phonograph" Clarke and two brakemen, Joe and John. We sprayed every track on the Freeport Line to Rockford, 62 miles, and tied up at 8 p.m. We had no particular problems other than sliding on oily rails through a scrap company's closed gate at South Elgin and having to deal with a hot journal on the spray car. Since the temperature was predicted for 100 degrees the next day, we commenced spraying the Rockford yard tracks at 5:00 a.m. including a run up the KD line to Loves Park, spraying all the way. On the final pass through the Rockford yard, we backed the waycar and the engine over a weed-hidden derail on the CB&Q interchange, dropping the waycar and one truck of the engine on the ties in a cloud of dust. When asked, the dispatcher in Chicago said it was too risky for us to try and re-rail the engine and that he would order the mechanical department's "400" truck at Proviso to come to our rescue. He was afraid that if we used the rerailing frogs and something slipped, we might crack the diesel's gear case on the rail. While waiting for the repair truck from Chicago, we looked over the situation again and decided that the dirt filled track with its light rail might help the re-railing. We carefully placed the frogs and eased #1213 back on the rail with a thump. Getting the waycar back on track was simple and off we went with a total delay of only 1½ hours. We ran back to West Chicago at a good clip. I made the mistake of riding the spray car which bounced extravagantly along at the rear of the train with me hanging on for dear life. It was a good thing another hot box developed on one of the oil tankers, because we had to stop and cool it down using the number one emergency treatment—urination.

On the 26th Jack McCord took the weed train up the Williams Bay line and I took the day off, my first in two weeks, so that I could attend the dedication of the first half of the Mannheim Road bridge. The C&NW was represented by J. J. "Joe" Stein, general manager, who knew nothing about the project, but loved being in the public eye. Festivities included a lot of hot people under a bright sun surrounded by fire trucks and dignataries from several nearby villages.

The second phase of the Mannheim Road bridge construction began June 28th, and I was back in business at Proviso. The first thing to be done was to wreck the old structure. Ernest Sporry, McQueen's superintendent, first tried a crew using manual tools to yank the ½ inch square 10 and 14 inch long boat spikes from the 4 × 10 inch planks spiked to the underlying stringers. That technique did not work at all. Sporry then used his innovative genius and invented a tool to remove that road surface quickly. On the 29th he tried out the new device and, after a minor adjustment or two, it was used to strip the roadway planks from the deck from the north to the south end. About 12 feet high the contraption looked like a giant monkey wrench or an upside down letter "F". The jaws were about 6 inches wide, just enough to fit over the edge of the 4 inch decking. A caterpillar tread crane lifted the device, slipped the jaws over the planking, lifted it tight with a cable and then reeled in the cable attached to the top of the wrench which levered the plank, boat spikes and all right off the stringers. The old timbers came loose in great clouds of dust, splinters, spikes and the sound of snapping fir lumber. By July 6th the deck removal had been completed for 26 panels of the 55 total. I was greatly relieved on July 15th when the deck removal had been finished over the live tracks below. Lifting the 8 × 16 inch stringers off the old steel and into trucks was a straight forward task.

Placing the first girder of the new Mannheim Road (U.S. 45) viaduct across the Galena Division passenger mains at Bellwood station on January 21, 1954. The old trestle in the background was constructed by the C&NW as a two-lane wagon bridge in 1907. It became U.S. Route 45. In 1953 the rickety structure was posted with a 10-ton limit warning that was always ignored by the truckers using this busy highway. The creaking and quaking of the ancient structure under modern traffic was especially unnerving for pedestrians which often included engine crews who used it to cross the railroad yards from the suburban train station on the south side to the Welfare Facility on the north. The first half of the new structure was opened to traffic June 26, 1954 amid great fanfare.

Overland Construction cut the old steel pony trusses with oxygen torches starting at the north end over yard 4. As the trusses were cut loose, a crane lowered them into the waiting gondola cars I had ordered spotted on selected yard tracks. The loads were shipped out as commercial freight to scrappers. The span over the mainline was demolished July 15th, after flagmen and slow orders were established to protect regular train movements. The trestle work between the main tracks and the yard 6 switching lead was pulled over by a tractor and cable in a great cloud of dust and splintered timbers. Yet, we still had to deal with the old aboveground concrete piers. These old foundations were hard to break up because concrete continues to harden over time and these had been curing for 47 years. After the crews had drilled holes in them, Sporry and I blasted them with 30%

Pulling down the old wooden trestle at the Bellwood suburban station on July 15, 1954 created a cloud of dust and splinters from the old fir timber mixed with years of embedded road grit.

The final gap in the Mannheim Road viaduct across Proviso Yards is closed March 18, 1954 while the steel crew rivets the girder section spanning Yard 4. The old bridge to the west is still in service.

The stripped steel truss sections of the old Mannheim Road bridge are ready for removal on July 19, 1954. The old steel was torched and lowered piece by piece into gondolas spotted on the tracks below. This scrap was revenue traffic for the C&NW. Engine 1048 standing on the Yard 3 ladder is a Fairbanks-Morse H10-44 constructed in September, 1946.

dynamite. Sporry was a good instructor on the use of explosives and was careful to teach me the proper handing of the dynamite. My grandad used to blast stumps and congregations of crows in his woods in southeast Missouri, but I didn't learn very much as he always got a terrific headache after just one blast. On July 23rd Sporry and I carefully tried the first shot using a blasting blanket made of woven hemp rope, after warning the railroad crews. It just went "whump" and did not throw any debris. The old concrete was very hard, but was not reinforced, so it acted like solid stone, cracking along predictable lines. The pieces were swiftly picked up by crane and hauled away by trucks working off the new bridge structure. By the 27th the old piers were all broken up and removed. Rules to remember when blasting: always wire charges in series and use a high voltage source. On July 28th the last two cars of scrap were loaded, Overland moved out and every aboveground vestige of the 1907 wagon bridge was gone.

The second half of the new overhead structure at Proviso stands ready for installation of the roadway October 6, 1954. The four-lane structure was opened for traffic in May 1955. GP7 1534 stands on the Yard 2 ladder. This engine was one of a batch of 30 dating from March 1951.

The old bridge was not quite gone underground as we still had the old foundations between tracks to remove. Early in August Sporry tried to drive the steel sheet piling alongside the old buried concrete. That tough old concrete twice broke the steam pile driver. That underground concrete had not been exposed to any weathering and had grown stronger with age. Sporry dealt with them by drilling and blasting down to their very bottom which was about ten feet below track level. The dust from an explosion had hardly drifted away when the cleanout began as the crane yanked up the broken pieces of rubble. Intensity of construction on this west half of the bridge was even greater than on the completed eastern half. The now familiar ritual of sheet piling, excavation, pile driving, forming, pouring concrete, erection of high forms, with the attendant shuffling of cranes on flat cars and the old condemned bracing boxcars worked like a orchestrated ballet despite the constant battle with recalcitrant yardmasters to get rail equipment positioned promptly. Although the concrete removal had slowed the work down, backfilling the holes was easier. Loaded dump trucks operating on the new bridge structure merely dumped great Niagaras of sand and gravel into the holes 30 feet below. Seeing to the safety of this operation required my constant attendance despite Reid's unhappiness with my long absences from the division office. I wasn't forgotten by the division, however, when I did go in for some reason or other, there was always a job or two handed me with my paycheck for "my spare time". One of these little projects was a stadia survey for a new track to serve the Green Giant plant at Belvidere, Illinois, some 60 miles west of the Proviso job. During a break in the Proviso work, I took new rodman Norman W. Hillegass with me to do the survey. We found the site on the east side of town covered in giant ragweed. I asked Norm if he was allergic to ragweed; he said "no" and grabbed the stadia rod. I was not expert in the use of stadia, but knew how to take the four readings at each spot which made a lot of notes to translate for plotting. We also inventoried the new Sundstrand Corporation track west of town as long as we were in the vicinity. I gave the notes to Norm to work up and went back to Proviso to arrange a work train movement. Norm learned that he WAS allergic to ragweed pollen and suffered a long siege of sniffles and watering eyes.

At Proviso a disaffected freight conductor named Koepke flooded the timekeeper's office with claims for penalty pay because he felt the bridge job was not protected sufficiently and that a lot of job opportunities for flagmen were being lost. To make the railroad safe I had personally taken responsibility by spiking the switches of the yard tracks to create "dead" tracks that did not need flagmen and all work at the main lines was protected properly by train orders and flagmen. Every one of his time slips was rejected foiling his efforts to extort money from the project. He really had it in for me and let me know his opinion in no uncertain terms on several occasions. I was firmly polite and ignored him on advice of the superintendent.

The 1955 capital budget season came, and I went into Chicago on Saturdays to help out with rail renewal estimates at Radnor, Crystal Lake, passing tracks at Manlius, Virden and Allen, and track 2 in the CPT. Only Bill Comstock, Bob Lawton and I were qualified to deal with these complicated estimates.

Back at the bridge Overland Construction began erecting the first steel girders for the west half. As before, they started on the south end with the 30 inch girders. I set up slow orders for the main line on September 28th and was able to annul them at 2 p.m. well ahead of the evening commuter fleet. On October 1st the girders went up over the yard 6 switching lead and caused a 15 minute delay to switching operations, probably the longest interruption on the entire job. Riveting followed on the 4th. The next day McQueen unloaded all the rail equipment, and I was finally free to open all the yard tracks to normal operation until the last steel erection a month later. What a relief! We had used a total of 742 car days, a bill the contractor was happy to pay. My role in the bridge construction was almost over. We only had a couple of incidents that marred our safety record. One involved a passenger crossing the tracks on the contractor's private crossing at Bellwood station to the eastbound platform. He fell over the guard chain and broke an arm. Since there was a license agreement with the C&NW permitting the crossing and limiting the railroad's liability, McQueen's insurance carrier had to settle with the claimant. In another accident a motorist crashed the construction barrier on the new bridge. He tried to drive over a bridge that wasn't there yet. The last phase of the steel erection began November 14th after I had made arrangements to clear yard 4 one more time. A spate of wet weather stalled steel erection for a few days but the last steel of 1954 went up on December 3rd. The final steel unit would have to wait until 1955.

Returning to October 1954 the weather turned from very hot to very wet. A record rainfall on the 10th caused the Chicago River to back up and flood the Daily News Building's subbasements. Some 42 feet of water drowned the printing plant and endangered the building as the water soaked paper stock began to swell. The C&NW offices on floors 10 through 16 were not affected. The Milwaukee Road commuter trains could not enter Union Station because of the high water although some long distance trains could use a few of the outer tracks. Proviso was flooded. Built on a swamp, the main drain, Addison Creek, became a raging torrent in which a couple of women drowned in nearby Stone Park. The Des Plaines River into which Addison Creek flowed was also backed up as was the downstream Illinois River outlet many miles to the south. By mid October the waters had calmed down, and we set stakes for raising the yard 7 switching lead to increase the speed of cars being switched. Back in the office we were faced with requests for 86 budget items that required engineering estimates and supporting blueprints. By October 22nd we had pounded out 58 of the items which included steel garage buildings, rail renewals, ballast jobs, machinery purchases, and everything else needed to improve the railroad's track and right-of-way. The requests kept flowing in as though there was no limit on the amount of capital funds available. Since we had no way of judging what was hot and what was not, we threw them all together as best we could knowing that not all would make the cut. It was with some relief that I was sent to Proviso to set the pile cutoffs on bridge 16¼A. This was the bridge spanning the famous loop track that was used to turn the class H Northerns before the turntable at Proviso was made large enough to handle them. Although the big steam engines were near to death, the bridge was redriven because it carried the main connection to the Indiana Harbor Belt Railroad from Proviso.

After the budget rush was over I settled into the other huge piles of files. Measuring the duct work installed for the air-conditioning on the third floor of the CPT last summer was an example of a tedious job that had to be done. On December 13th I took rodman Norm Hillegass to Earlville on the Spring Valley line for a vision survey and then down to lonesome Dimmick for another survey at the U. S. highway 51 grade crossing. Automatic protection was proposed for both locations and justification was supported by the surveys. On the 14th I drove to Harmon Township (south of Dixon) to attend an evening hearing of a local drainage district and make certain that a proposed tax assessment didn't unfairly burden the railroad. Our law department kept an eagle eye out for all these assessment districts, but they did not send lawyers to these meetings when they could order division staff to cover for them. On the 16th Norm and I measured clearances at the American Can Company in Maywood, went to Proviso to inventory rail renewals, and inspected the Proviso engine house in connection with a proposed waterline to serve the Western Contracting and Supply Company, lessees of railroad property nearby. That was the end of 1954 for me, and I prepared to go on my usual Christmas break vacation. Before I left town, instrumentman Bob Lawton called me at home to say that the merger work for the proposed consolidation with the Milwaukee Road had been dumped on the division engineering staff. That was the 23rd of December. The news did not change our plans. On the 26th we left Chicago on B&O train #8 the *Shenandoah* in a new Pullman car *Mahoning* en route to Washington, DC. All along the way we saw that there was still a lot of B&O steam locomotive power still in service. We also rode the Pennsylvania Railroad's #120 *The President* to New York, the New Haven's *Bay State* to Boston and came back to Chicago on the New York Central's *Wolverine* in Pullman *Island Nymph* through Buffalo, across Ontario and the St. Clair tunnel. It was on time.

THE CHALLENGE BEGINS TO FADE, 1955 . . . My colleague, Richard W. Bailey, was appointed acting division engineer on the Wisconsin Division at the beginning of 1955. Something was happening with the railroad's ownership; someone was buying the preferred stock whose price had doubled in the past month. Anthony B. Zaborowski suddenly reappeared on the 28th of January and began working 30

hours per week as an instrumentman, bumping Carl Oleson who had been holding the job on a temporary basis. The inconvenience to Carl lasted until March 31st when Tony bid the open assistant engineer's job on the Madison Division. (Tony was to die of heart failure December 26, 1964; I attended his service on the 29th).

The last bit of the Mannheim Road bridge project began January 11th with the arrival of nine cars of structural steel. The switch crew managed to get four of the loads in the wrong place. Then on the 13th erection stopped because the base plates were missing. We found that they were in gondola car bad ordered on the IHB at Norpaul yard. Pat Mair and I drove to Norpaul in my Ford station wagon and unloaded the missing plates from the stranded freight car. They made a heavy load for my modest car. Then a splice between two girders could not be made because a stray bolt had been left between the two plates during shop assembly. It's tough to take heavy plates off a girder thirty feet in the air to extract an errant bolt. On January 18th the last steel was up and the contractor was out of the railroad's active zone. My job of protecting equipment and personnel was over. I ordered removal of my telephone and said goodbye to my desk in the construction shanty. The bridge was completed and opened to traffic May 16th.

Just because my involvement with the big highway bridge was finished did not mean that were no more Proviso projects for me. On January 20th division engineer Maurie Reid sent me back to Proviso to run a topographic survey on railroad land west of Wolf Road and south of Lake Street. The drawings had to be ready on the 24th. I took a new rodman with me, John Sierzga, and set off into abominable weather. The snow and bitter cold kept us from finishing the survey. We let the weekend intervene hoping for a break in the weather. On Monday we finished the survey despite two more inches of falling snow. I learned that this was to be the site of a proposed warehouse financed by the railroad and leased to Del Monte Foods on a buy-back plan. I was to become more involved in this project later in the year. Meanwhile, other work had to be handled. At the insistence of Gene Farrell of the claim department, I made 49 elevation observations on a two foot square piece of concrete pavement in the westbound passenger platform at Elmhurst—in a driving snowstorm. I never found out why that was necessary. During this extreme cold and snow the railroad's commuter operations were quite ragged. The 5:11 p.m. westbound from Chicago died west of Melrose Park at milepost 12 on January 31st forcing transfer of passengers to a following train along the open track without benefit of any platform. My journal noted that on March 1st the 5:08 p.m. broke down at Keeler Station causing a 40 minute delay to cut out a bad order passenger coach. On St. Patrick's day eastbound #6 broke a knuckle at Oak Park, blocking the local track and screwing up the morning rush.

Back in February instrumentman Bob Lawton and I cross sectioned land east of York Street, Elmhurst, and picked up small jobs at Dixon, Sterling, Morrison and southeast at Dimmick. I was beginning to think that this office work was getting to be a crashing bore compared to the excitement of a big construction project. It was time for some of these new "kids" to start taking over the roadway completion reports and associated paperwork. On February 21st assistant chief engineer P. V. Thelander, who had once been the Wisconsin Division engineer, dropped a hot potato in all the divisions' laps. All outstanding contracts were to be examined to determine if the railroad was billing and collecting monies due. The nasty assignment ended up on my desk. It turned out to be fairly simple once I had made a list of the contracts and compared them with accounting's records of billings and receipts. We were clean as I reported to Thelander on the 24th. Our rapid response, however, was a bad move because he then found more things for us to do. The next messy job was a physical check of all the industries along the jointly owned Dixon River track. Maurie Reid, the division engineer, then stepped in and told Peter Victor Thelander to quit interfering with his engineering staff.

Anticipating that new commuter cars were coming, management asked the Galena and Wisconsin divisions to measure all their passenger platforms. On March 9th rodman Norm Hillegass and I did Lathrop Avenue, River Forest, Maywood, Melrose Park, Bellwood, and Proviso. When we got to Proviso we also chained the new piggyback tracks. The next day we went down the South Branch and found another track at Roosevelt Road that had either escaped the 1914 valuation or had been built without proper authority. The track appeared to be different from the crossover Bill Polchow and I had discovered there in 1947. I wonder how many other short bits and pieces lurked in that tangle of steel crossings and paper contracts. On the

14th Norm and I measured Oak Park, Ridgeland, Austin Boulevard, Austin and Keeler, Elmhurst, Villa Park, Lombard and Glen Ellyn. Luckily, Norm and I had already measured the platforms at Winfield and Wheaton because the next day a March storm hit with winds so high that the gas fired switch heaters in the approach tracks to the Chicago Passenger Terminal blew out allowing the driving snow to plug the switch points. It was almost winter's last gasp. We finished the passenger platforms on the 28th with West Chicago, Geneva, DeKalb and Rochelle. Before the big snow came, Norm and I had staked a tight little side track to serve the Configured Tube Company at 25th Avenue in Melrose Park.

Being a junior member of the American Railway Engineering Association (AREA), I was encouraged to attend the trade shows and meetings. Bob Lawton, Bill Reed and I attended the big suppliers' show at the Coliseum March 15th. These shows were important to engineering personnel. New products and techniques for track maintenance and construction were displayed here plus valuable contacts could be made with other railroads. It was so easy to think that your own company was the center of the universe. These shows quickly dispelled that notion. Bill Reed worked as draftsman for Bill Armstrong in the architect's office until he relocated to San Francisco, California, to work for the Western Pacific Railroad.

The land department sent their surveyor, Charles "Chuck" Jevne to the division requesting help for establishing the old south line of Lake Street at Proviso. This boundary was the north property line of the railroad's Proviso yards. The big problem was finding the old monuments which had largely disappeared over the years. We had almost finished the job on March 24, 1955 and decided to resume the job the next day. A mistake, as it turned out because another snowstorm reduced visibilities to zero. In those days weather forecasting was pretty much a case of looking out of a window. By chance I had met the local NBC TV forecaster, Clint Yule, when riding that late night train after school. He confirmed my opinion that it was pretty much guesswork. Clint later got out of the weather business and moved to Galena, Illinois, where he set up a business as an investment counselor. From guessing at weather to guessing at what the market would do—seems like much the same level of risk. On the 30th I went to Chicago Shops to measure some new trackage serving building

M-50, formerly the heavy steam locomotive repair shop. What a change from 1948. No engines and no repairs were in progress; the empty building was being converted for industrial use. That afternoon I got a look at a new doubledeck gallery car from the St. Louis Car Company—car #1.

On April Fools' Day while Norm Hillegass and I were out in the middle of the busy Clinton Street interlocking plant measuring the new rail installed, the rest of the boys back in the office were busy nailing my desk drawer shut, fastening my drafting stool to the desk with rubber bands and secreting a big hunk of rail in my briefcase. Overcoming these good-natured pranks, I went to Proviso the next day to measure new rail at bridge 16-C, dropped in on the Mannheim Road job, located a new truck garage and roadway for map corrections and inspected the new piggyback ramp facility south of the administration building. That evening I rode home in car #2, the second of our new St. Louis Car Company gallery cars. The advent of these new cars made some strange looking commuter trains—shiny new yellow and green cars mixed into consists of unwashed olive green "low" cars and the standard heavyweights of ancient (1910, 1914) coaches. On the 13th I met with Mr. Vorel of Griffin Wheel Company at California Avenue and got a look at their new steel wheel casting line in operation. Vorel wanted some track changes made at the plant involving some tricky realignments and some very sharp curves. Bob Lawton and I agreed to do the engineering and staked out the changes on a Sunday morning. Vorel then had a private track contractor come in and do the actual track work.

Construction started on the new Del Monte warehouse building at Proviso April 12th. The general contractor was Ragnar Benson Company of Chicago. After a quick look at it on the 15th, I attended a meeting in chief engineer Meyers' office where we were told in no uncertain terms that the facility had to be finished and ready for occupancy by August 1st. I was made responsible for getting the tracks built. In addition to our normal business our division office was struggling to deal with the demands of the C&NW-MILW coordinating committee. One of their requests involved making huge blueprints of the entire Proviso yard complex by April 27th. With Mr. Meyers' words ringing in my ears, I left the map making mess to the rodmen in the office and drove my car to Proviso. I had asked Norm Hillegass to follow on the train with the transit and tripod. Not only did he fail to bring stakes or a maul, he brought the wrong tripod for the transit. That was the second time a rodman had screwed up a field trip by thoughtlessly grabbing the wrong equipment. I drove to the Mannheim bridge office and borrowed the State of Illinois' transit from field engineer Pat Mair. By 4 p.m. Norm and I had a preliminary line established for the tracks to serve the future warehouse. The next day, with our own proper tools, we reset the stakes and grade lines. On May 9th we again checked the track alignment to ensure that the tracks were the correct legal distance from the warehouse platforms which were now far enough along to permit accurate measurements. In Illinois the distance from the centerline of track to the edge of a freight platform had to be exactly 5 feet 8 inches according to rules laid down by the Illinois Commerce Commission; other states had different requirements. Bill Armstrong and I inspected the concrete work in progress and were appalled at its poor quality. A commercial product, Sonotube, was being used for forming the column footings. This product was essentially a thick resin-impregnated cardboard tube. I ripped open one of these tubes with my pocket knife and found fist-size voids in the concrete. Obviously, proper vibration to attain density around the reinforcing steel had not been done. The contractor's excuse "you guys were in a hurry, so we were in a hurry." And then there was the brick dust used in the flooring concrete that interfered with the proper setting and finishing of parts of the floor. Ragnar Benson, himself was there, but the tough old Swede did nothing about our complaints. I resolved at that point that I would never hire that company to build anything for me—even a dog house. These quality lapses also pointed out the need for on-site inspection; this remote supervision supplemented by sporadic visits, was not producing a quality product. Construction continued, however, with or without our input. On June 3rd M. S. Reid left town after ordering Nick Maas, the Proviso roadmaster to commence construction of the warehouse tracks. Since the assistant engineer, Bill Comstock, was on vacation, I was the man in charge of the engineering office when an urgent call came from the Ragnar Benson Company. The frantic caller wanted the railroad construction crews to stop working on the tracks at once. His AFL union representative was standing at his elbow demanding that the Laborers Union perform that work. I explained that the railroad trackmen were also AFL, that the building was owned by the railroad, that I had no authority to stop the trackmen, and that the only person that could stop them was out of town and unavailable. The railroad crews stayed at work, the contractor's men stayed on their jobs and the AFL union official had performed the charade that he was trying to protect his membership by claiming the work. We stood firm on our premise that the track work was a railroad job and that the AFL unions needed to sort out their own jurisdictional disputes. We had a time line to meet and certainly wanted to avoid an intra union snaffle. Some performance—everybody won. By July 15th Nick Maas' bulldozer operator was grading for the new tracks by the building. There were some growls by the contractor's laborers, but they were told firmly that they were on railroad property. The weather continued uncommonly warm throughout July; on the 27th the high for the day exceeded 100 degrees when Meyers, Reid and Ragnar Benson decided to tour the work in progress. After the weather cooled, the roofers went on strike. The August 1st deadline went by without comment. On the 29th I made an inspection with Meyers and Reid. Two days later, Tony Shactner, the railroad's insurance supervisor, came to my home in Elmhurst in the evening reporting that the warehouse was on fire. A quick trip the two miles down Lake Street to the warehouse site proved his information was erroneous.

I had gone to Rockford on the Illinois Central's train #15 on May 3rd to meet Fred Shappert, owner of a construction company that had secured the contract to build a modern overhead highway bridge to replace our old wooden structure #450 near Cherry Valley. We went over the ground rules involving the railroad's requirements for safe conduct of the job and protection of our trains. He graciously drove me south to Dixon where I inspected bridge 190½ while waiting 3½ hours for a cab ride on train #14. On the 13th rodman Norm Hillegass and I went to Rockford for another conference with Shappert and the Illinois Central's engineering representative. I was to work with Shappert again in 1956 and again in 1963 when he would become an important factor in a big railroad project in Belvidere.

On May 5th construction began on new escalators leading from the train floor to Canal Street. The racket of jackhammers was right under our office. Conversations in the office became shouting

Bridge 450 near Cherry Valley, Illinois, May 3, 1955. The view is east on the Freeport branch. The railway-owned wagon bridge was to be replaced by a modern structure for the new Rockford bypass of U.S. 20. The successful bidder for the bridge contract was Fred W. Shappert shown holding a set of plans. He was to play an important role in a huge C&NW industrial project in nearby Belvidere eight years later. There were many of these wooden overhead structures built by the railroad to accommodate the public. Note in the background the whipguard designed to warn crewmen "decorating" the tops of rail cars that a low clearance was nigh.

matches. On May 17th I was summoned to Judge Sam Perry's Federal court to testify about an accident survey in Elburn I had made some years ago. The testimony was routine and short. That day I heard there was a C&NW stockholders' meeting and that new interests were coming into the railroad's board of directors.

Wet weather often typical of late spring in Illinois caused much damage to property this year. I had a meeting on June 8th with an engineer from ARMCO who was in charge of installing a sewer line under crossings of our main line at Elburn. On the 14th I audited another of those evening meetings of a drainage district near McGirr on the Spring Valley line. I picked up a new rodman at his Batavia home for a trip to Peoria on June 20th. Stanley H. Johnston, was the son-in-law of P. V. Thelander, assistant chief engineer. Stan was a strong, intelligent lad with a very direct and positive character. We were to remain friends and colleagues for many years. We did a survey on the Farmington Road grade crossing at Limestone station near Peoria, returning home the same day. I took Norm Hillegass with me to Rockford on the 28th to survey tracks in the west yard where we planned to install simple turnouts to replace a worn-out doubleslip switch. We also inventoried the new Texas Company trackage at Winnebago and stopped in St. Charles to see the Howell Company about proposed track changes.

At the end of June the engineering department was ordered to cut payrolls by $110,000. The bad news was delivered at 5 p.m. The Galena Division engineering office lost one instrumentman, one tapeman, and, in the field, one roadmaster. Because of an agreement with the operating unions, it was necessary that official positions also be cut when union jobs were reduced. Old Tom Beebe, the chief draftsman, L. C. Winkelhouse, Bill Armstrong's draftsman, H. D. Barnes, comptroller (Jim Barnes' father) and Bernie Whitehouse, the fire inspector, were among those taking retirement- they were all past 65. The Galena lost a roadmaster position; Earl "Fuzzy" Pierson at DeKalb was made assistant to roadmaster Arthur E. Benson at Sterling. The musical chairs of bumping continued. On July 1st Wisconsin Division instrumentman Art Humburg bumped the Galena's Ralph W. Golterman. Galena instrumentman M. C. Christensen planned to bump Bill Comstock who would then take Wisconsin Division Jeff Edee's place. One more instrumentman in the chain and I would have to go back to being a rodman. Rumors flew like fleas on a lost dog. On the 13th I had an opportunity to have a chat with L. R. Lamport, the man who had hired me in 1947. He assured me that everything would work out eventually. Since his office monitored the seniority rosters and determined individual qualifications for job assignments, I felt he had a pretty good handle on the changes. The key to this job shuffle was M. C. Christensen. Chris was offered, and accepted, the job of bridge inspector which took the strain off the bumping chain. This left our assistant engineer, Bill Comstock, rodman Hillegass and me in place. Byron Ruth returned to school on the 11th and rodman Carl Oleson resigned to take a job with the State of Illinois. It was tough to work for a poor railroad.

Lack of manpower cut into our ability to provide engineering services. That and the brutal August heat gave office work an inviting priority. By August 5th there had been 11 days in a row of temperatures over 90 degrees. On the 9th the heat was broken by heavy rain showers. I helped Jack McCord, roadmaster at West Chicago, inventory the tracks of Material Service Company at Algonquin as well as Consumers Company's tracks at their pit south of Crystal Lake before the extreme August heat returned in the third week of the month. Another new rodman appeared from the Northwestern University Co-Op program, Alfred J. Kuhn. Al was a champion swimmer, tall, athletic, confidant, and a somewhat arrogant young man. In later years he would work for the American Red Cross in Hammond, Indiana, become their executive director and then take a teaching post in the Sudan. But now he learned to color blueprints and hold a surveyor's rod which he did with that air of faint disdain he affected. Al went back to school on December 14th.

Good old Charley Hise, office engineer and former Galena Division engineer, had a talent for giving me awful jobs. The City of Chicago water consumption report was a nightmare for the railroad. Chicago got its water from Lake Michigan and because the households and businesses in the city were unmetered, consumption in the hot summer of 1955 soared. The railroad carried lake water to Proviso through a 14 inch main from connection to Chicago mains at an Austin Boulevard pumping station. My job was to find out where our meters were, who read them, how often and how the billing was handled. I found the reporting was ragged, spasmodic, compromised by a lot of unsubstantiated adjustments in the billings and that metering was a farce. I gave the information to Charley to let him figure out what to do about the mess.

The law department kept finding notices of assessment hearings in the papers—they had a clipping service. On August 25th I was sent to Hamilton Township south of Dixon. On Halloween there was another meeting at Walnut south of Nelson. In this case I had to find a farmer near Normandy, the town clerk at New Bedford and a lawyer at Princeton. I think that it would have been beneficial for our young lawyers to leave their comfortable offices and do a little field work instead of dumping these tax assessment investigations on division staff.

Returning to that hot August, it was 94 degrees when Stan Johnston and I were sent down the 16th Street line on a priority job involving track changes. In later years we would have needed a police escort to go into that neighborhood on the near south side of Chicago. Out west, Douglas E. Oakleaf transferred from Chadron, Nebraska, to the Twin Cities as assistant engineer on August 30th. Rudy Kassel, rodman on the Wisconsin, left the railroad September 9th. Our rodman Norman Hillegass was the only bidder on the open job on the Black Hills Division at Chadron. This personnel shifting was really hurting our ability to do our work. Rudy Kassel, a big gangly kid, was involved in a strange incident during his brief career as a rodman on the Wisconsin Division. "Heat numbers" was a term that could bring terror to a rodman's job. Every rail has a unique number describing the heat (batch) and position in the ingot pressed into the web of the rail at the time it is rolled in the mill. Before there was understanding of the reasons for failure of steel rails—transverse fissures, piping, head crushing, corner breaks—the heat numbers of all rails laid in main tracks were recorded in giant ledgers by milepost location. Instructions from the engineer of maintenance required that failure of three rails from the same heat number called for removal of the rest of that heat from the track because of a high probability of failure. The rodman's job was finding those heat numbers in the ledgers and sending a list of the rails to be replaced and their location to the affected roadmaster. The rail sections most liable to breakage were 9030, 9035, 10030, 10035, 11025, 11225 and 11228 pound. To find these elusive numbers the rodman had to go through many ledgers. It was convenient to open them out on desks, open desk drawers, waste baskets and whatever flat surface was available. Instead of tidying up at the end of one day, Rudy merely walked out the door at 5 p.m. The cleaning staff took away the books left open on top of his waste basket. The next day a frantic search did not recover the lost books and Rudy's name was mud. Actually, he did us all a favor. An attempt was made to rebuild the missing data from roadmaster records, but it was all in vain. Soon afterwards when confronted with this breakdown in the system, the entire heat number program was scrapped to the great joy of division staff. Technology, however, was really the reason why it was no longer necessary to maintain the system. The newer 11228 and 11525 pound rail was all controlled-cooled after 1948 which eliminated most of the internal defects that had given birth to the heat number program in the first place. The railroad was also actively using the Sperry Rail inspection service that employed an electronic scan mounted in a motorized coach to detect rail defects. These two advances gave a greater and more positive measure of safety than the cumbersome heat number system ever could provide.

Although I had inspected the new piggyback ramps built at Proviso, I did not realize at the time that a new era in freight transportation had begun—trailers on flat cars (TOFC). Norm and I had done some surveying for the new trackage at Proviso on August 31st but were unaware of the rapid growth and swift technological changes that would be the hallmark of the world of intermodal transportation. The modest ramps at Proviso were simply circus-style loading facilities—a style of operation doomed by its inflexibility and slowness.

August was over at last. It had been the hottest summer on record since 1871, averaging 76.4 degrees which broke the 1921 record of 75 degrees. There had been 40 days in 1955 exceeding 90 degrees. The new month of September may have been cooler but some unpleasant events occurred. The 7th of the month was a bad day for the railroad. First, wooden trestle bridge 200-D carrying all ten tracks of the Nelson Yard caught fire and was totally destroyed. That same morning the C&NW's general solicitor, Drennen J. Slater, tipped his hat to the locomotive engineer of speeding train #151 at Evanston just before stepping into its path.

During the last quarter of 1955 my life in the engineering department was definitely getting stale. Construction projects were few and far between. I was bored. If it were not for the nearing completion of my degree work, I sometimes wonder if I would have stayed with the railroad through this gray period. To accelerate completion of my required college courses, I began to take on additional classes. On September 15th Stan Johnston and I did a little stadia survey at Adams Street in Peoria. We also met Fred C. Jones, our local district freight and passenger agent, someone I would get to know very well in future years.

On September 16th there was a rumor that the C&NW was losing the coast streamliners to the Milwaukee Road. Our reaction was one of shocked disbelief and worries about adverse effects on future maintenance funds for our high speed main line west. By the 21st the news was all over town. At the same time we learned that the general offices of the Omaha Road (CStPM&O) were to be moved to Chicago. Our operating unions became very vocal about the streamliner loss on the 23rd. I had a chat with Frank V. Koval, public affairs, on the 26th about the coming change. He said the usual things about how good it would be to get rid of those costly and declining passenger trains. Although he was correct, it was a blow to our pride to lose those beautiful but aging trains. The new escalators were now in operation in the Chicago Passenger Terminal, but they would now serve only the commuter passengers. By October 5th the job loss because of the streamliner loss was said to total 707. The actual changeover occurred October 30th.

Major changes proposed for the Proviso yard tracks bumbled along. After a talk with Nate Waterman and Walter Johnson, I had to rework the plans and estimates for the yard 7 and 8 ladder tracks. Little jobs dribbled in: track extensions at California Packing's DeKalb plant, specifications for that tricky paving at the Lombard depot, low steel clearances at signal bridge A at Clinton Street, track inventories at the Proviso piggyback ramp, a revival of the cramped little siding for Configured Tube in Melrose Park, a track for scrap dealer M. S. Kaplan north of yard 2 at Proviso, and an accident survey in Maywood requested by Warner Frank of the claim department that resulted in a long afternoon with attorney Bob Russell going over the details. On October 17th Stan Johnston and I staked out 1,600 feet of track for the Kawneer Company just east of the DuPage County airport north of West Chicago. We should have had a third man on our survey crew to properly do the job but I was forced to use clever mathematics to calculate our way around that long curve. This new track was to carry shipments of aluminum shapes for the company's storefront manufacturing business. On the 26th Stan and I staked a new track for the *Elgin Courier-News*. That day we also got a look at an eight car wreck of gravel cars at Carpentersville. Then on the 29th it was back to State Street yard in Chicago with Al Kuhn as my rodman to run a series of track levels. November 1st found Stan and me doing a rush job at Mine #3 near Benld. We had ridden the Rock Island's Peoria *Rocket* and were met by Bill Baker, our roadmaster for the south end of the Southern Illinois. After a meeting with Fred Jones, DF&PA, we staked out a track for the Meadow Brook Dairy at Limestone and spent the night in Pekin. During Halloween night vandals threw the siding switch at Girard and sidetracked train #381 at 40 mph. Luckily, it stayed on the rails. The next day with Bill Baker driving as only Bill Baker could, we went south very fast with only a brief pause at a surfacing gang working south of Allen before meeting a Mr. Kiss and Mr. O'Gorman of the Superior Coal Company. Mine #3 was no longer in operation; the area looked abandoned. Kiss and O'Gorman had a prospective user of the old rail yard and needed some track changes, Stan and I developed a plan and cost estimate for them on the spot. We went back to Chicago on the Wabash from Litchfield. A word about Bill Baker—he drove fast because he had once been a race driver and knew how to put his foot through the floor boards and still stay on the course. In 1955 Illinois did not have a speed limit and everyone drove as fast as they felt like. Bill gave us lessons on how to enter a curve, when to accelerate out of a corner, always emphasizing that one had to look far down the road and anticipate changes in conditions. Bill lived like he drove and died a young man from a sudden heart attack. He was also a good roadmaster.

On November 9th we learned that the board of directors of the Milwaukee Road had voted to pursue possible consolidation with the C&NW based on a consultant's (Wyer) report of October 26, 1955. On the 11th the C&NW board reluctantly agreed to further merger studies. Rumors flew again on December 2nd; someone was buying C&NW stock as it had risen $3.00 in two days. R. W. "Dick" Bailey's pending transfer to Huron, South Dakota, was held up because there was some concern that a merger would affect the position (it didn't). November ebbed away, nibbled to death by little jobs here and there: a minor track change at Geneva Modern Kitchens, water problems at the KPA warehouse at Proviso, measurement of a new diesel fueling station at DeKalb, a trip to Marengo with roadmaster Jack McCord, and lots of roadway completion reports to write. November 17th and 18th were more bad days. Derailments and collisions in the Clinton Street interlocking late on the 17th caused massive delays to the commuter operation into the next morning. A class E Pacific, 4-6-2, #577 blew-up at West Chicago on the 18th killing the fireman. Another locomotive suffered a burned crown sheet at Geneva, and there was a freight train wreck at Round Grove that blocked the main line west.

At Sterling the CB&Q was still trying to figure out a way to get direct service into the Northwestern Steel and Wire Company mill. The Q had operating rights on the C&NW from Sterling west to Agnew but no right to serve local industry from the C&NW main. Their station in Sterling was north of the C&NW and the steel company was south of the C&NW mainlines. To forestall and impede any CB&Q's aspirations, the C&NW decided to construct a yard track south of the main line from Sterling to its next station west, Galt. On November 29th we were told to get going on the project. The reason for the construction was kept secret at the division level: only the superintendent, H. B. Smith, the division engineer, M. S. Reid and I knew the reason why this yard track was suddenly needed. Later I learned that Harry Murphy, president of the CB&Q had developed a personal friendship with Paul W. Dillon, the eccentric owner of the steel works. Dillon told Murphy that if they could get the Q into the mill, they would get all of NSW's business. This struggle was to continue for several years. While the C&NW won in its efforts to block the CB&Q, Northwestern Steel & Wire eventually went bankrupt along with much of the Nation's steel industry and ceased operations.

The doom of the steam locomotive came nearer when we began a study on December 8th to dieselize the entire Galena Division. When the report was completed on the 14th, all facilities associated with operating steam power had been identified, costs estimated and salvage expected. I then went on my annual Christmas vacation with my family.

On December 16th we rode the CB&Q's *Blackhawk* to St. Paul, Minnesota, connected to the Great Northern's *Western Star* to Spokane, Washington, where the SP&S carried us to Portland, Oregon. The train was six hours late into Whitefish, Montana. I negotiated with the Pullman conductor for sleeping car space because of the missed connection at Pasco. On the 21st of December we rode Southern Pacific's *Shasta Daylight* to Oakland through flooded country and came across San Francisco Bay at midnight on the ferry. California was in the grip of an historic rain event—too much water. On the 23rd we took the SP's *Del Monte* from San Francisco to Monterey over flooded tracks where trackmen walked the train through the high waters. On Christmas Eve all rail lines north and east were washed out. By the 27th the SP's line from Monterey to San Francisco had been returned to service. Despite some financial anguish I had purchased half fare tickets for the WP's *California Zephyr* to see the famed Feather River Canyon. Because of the floods, we found our train was to be detoured over Donner Pass on the SP. On the 31st we started out in Oakland on the WP to Sacramento where the train was carefully switched to the Southern Pacific on a freight transfer track. The trip on the SP was slow because of soft track and landslides. By New Year's Eve the train had made up much of the lost time across Nevada and was only a half hour late into Salt Lake City. The D&RGW rolled us across the Colorado plateau and over the Rockies to Denver where the CB&Q took over for the rest of the eastward trip. We got off in Aurora and rode the CA&E electric home to Elmhurst.

10

REWARDS AND ACCOMPLISHMENTS, 1956 . . . By January 1956 our office force was down to an assistant engineer (Bill Comstock), three instrumentmen (Bob Lawton, Art Humburg and myself) and one rodman, Stanley H. Johnston. Art went to his father's funeral in California and decided to move there himself later in the year. Another sometime intrumentman, Ralph Golterman, announced his resignation February 27th to accept a position with the Sun Oil Company and Stan

Johnston left on Leap Year's Day. On January 31st the C&NW offered its unions an increase in wages along with a revision of certain work rules. If the unions did not agree to the changes, the pay increase was halved. My monthly instrumentman's salary went up $43.00 to $472.34.

On the Galena division the Sterling yard track construction was our top priority. Stan and I staked the new lead track on the 16th of January. Other jobs at that time included track inventories at West Chicago and Elgin, a survey at Rochelle, and a big survey on the former Westward Ho golf course adjacent to Yard 9 at Proviso where the Automatic Electric Company planned a huge new manufacturing facility. Pullman Company forces moved out of the Chicago Shops to the California Avenue coach yard. When the move was completed February 7th, the Streamliner Ramp servicing operation was shut down.

On February 3, 1956 we got word that Ben Walter Heineman had demanded the chairmanship of the C&NW Board of Directors along with 10 of the 18 director positions. There was a lot of speculation at this sudden news and wondered what it would mean if it succeeded. By the 13th Heineman had made his point and on the 21st took control. On March 2nd he announced that the new president would be Clyde J. Fitzpatrick, former operating vice president of the Illinois Central. Ben Heineman, born February 10, 1914 in Wausau, Wisconsin, to a family whose fortunes were founded in lumber and banking, was an attorney with the firm of Swiren and Heineman. We gradually picked up more details on his involvements with railroads. He had represented CGW stockholders and had secured control of the M&StL in the early 1950s. As time went by we were to see a lot more of this powerful and talented man. By March 7th the C&NW-MILW consolidation study was a dead issue. Heineman represented a new chance to get this valuable property moving and many of us welcomed the coming changes with curious anticipation. From our lowly point of view, things could only get better. The ripple of change affected many departments even though the Heineman forces had not yet taken full control.

February 9th was the last day for the old president, Paul E. Feucht, who hospitalized himself. Charles C. Shannon, protecting his erstwhile boss, later said that the heart attack was Feucht's own invention. Considerable mystery still hangs around this incident. The next day a rumor cropped up that B. R. Meyers was to become assistant vice president of operations (that did not happen). On the 28th we heard that the Southern Pacific was very unhappy with the Milwaukee Road's handling of the streamliner fleet and would welcome a different "Ogden" connection. That started wild speculations that the C&NW might consider extending the Lander branch across the mountains to Ogden. That spring Bernie Meyers and Jack Perrier went off on a secret expedition to the South Pass area in Wyoming. Using an aneroid barometer they drove around through the country measuring altitudes seeking a rail pass from Riverton or Lander toward Ogden. Meyers never said anything about this trip, but Jack Perrier gave me a short synopsis when we were briefly involved in 1965 with PPG about extending rail service to a mine site in that area, of which more will be related.

Meanwhile, our routine work went on. Stan Johnston and I staked out a new track to serve the welding line at the West Chicago yard and set levels for welding supervisor Jimmie Hicks. Later, we ran more levels on station platforms at Winfield and Glen Ellyn. Instrumentman Art Humburg drafted me and new rodman Larry Hoffman to help him run a survey at Navy Pier on February 27th. The old Chicago Surface Lines trolley car tracks running down the center of the pier were to be transformed into freight railroad tracks. The electric street railway had built their double track line with eleven foot centers whereas the railroad required a minimum of thirteen. One of the trolley tracks had to be shifted two feet.

Our staffing situation remained grim; the work stacked up. On March 9th I met with the mayor and city engineer of DeKalb to discuss a new storm drain system. Of course, they expected the railroad to pay its share and, of course, the railroad would resist paying anything. Back at the office Art Humburg was increasingly absent as the siren call of California was in his blood. Wisconsin rodman Martin Goers resigned after 15 years of service and the assistant engineer, Hal Ruth, who recently had been in ill health, announced his retirement in June. The washed-out Lee County Railway was removed in March 1956 from Nachusa to Nelson.

On March 19th I again met with Fred Shappert of Belvidere along with Charles Ind, Charles Richards and Bill Howard in Rockford to plan the construction of the Auburn Street overhead bridge which was to cross our old KD branch line at Spring Creek Road.

While in town, I made the acquaintance of our sales agent Bill Plummer who had taken Williams' place. The next day turned into a Chinese fire drill. Bruce Packard, our former division engineer, who had been relegated to a staff job in the chief engineer's office, suddenly appeared in our drafting room to borrow two men to help him do a survey at River Forest. He had been told to determine if the Soo Line and the C&NW could be connected there. The Soo passed over the C&NW on an overhead steel span approached by high embankments from either side through a built-up area. (Nothing ever came from this wild goose chase.) Meanwhile, Art Humburg and rodman Larry Hoffman went off to Lombard on some job or other overlooking a planned meeting with California Packing Company officials in DeKalb. The CalPak people were exceedingly annoyed at being stood up and complained all the way to the top (since Feucht was gone and Fitzpatrick had not taken over, I am not sure what the "top" was). I then arranged to go out and meet with them on the 21st, but the law department put a hold on me for a pending court hearing. Division engineer M. S. Reid now wanted us to male a photo record of company structures so I dusted off my old Rolleiflex and loaded up with a lot of black and white film for an expedition up the Freeport line with Bob Lawton, snapping pictures of every above ground building. At Rockford we staked out the track shift required at the Spring Creek Road construction site. On the 26th I helped Art Humburg and Larry Hoffman establish the centerlines and elevations for tracks to serve the new KPA warehouse at Proviso. On March 28th I began work on an estimate to bring the Chicago Transit Authority's Lake Street Elevated line up on our elevated right-of-way through Oak Park and Forest Park.

On April 1st Ben Heineman and his colleagues formally took over operation of the C&NW. The next day the first inspection train left to give the new management team a look at what they now controlled. Resignations streamed in, among them C. H. O'Hearn the auditor of capital expenditures who departed April 5th. Remaining officers made extra efforts to demonstrate their value to the company. Leonard Lamport, chief engineer of maintenance, Harold Jensen, office engineer and Maurie Reid, Galena Division engineer combined as a group April 7th to prepare a huge re-make of Proviso yard and abandonment of the West Chicago yard. This secret presentation required preparation of many large blueprints. Since they couldn't let a lowly rodman in on the plan, Wisconsin Division engineer Jack L. Perrier was pressed into service to make the copies. Their plan estimated a cost of $603,000 for rehabbing yards 2, 3 and 4 which, at the time were mostly light rail with few good ties and completely useless for good freight car handling. The plan was a doomed attempt to revive past operations without a thought that there might be better ways to move freight cars through terminals rather than constantly making and breaking trains. On April 24th big mouthed John E. "Jack" Goodwin, vice president of operations was gone along with his endless stories about how he used to run the car shops of the Missouri Pacific in DeSoto, Missouri. A bright young acquaintance, Nelson Budd Wilder, was cut from the power board, but Jack Perrier gave him a job as rodman on the Wisconsin Division April 26th.

Our work at Proviso during April included monitoring the grading work at the KPA warehouse site on the 10th; three days of laying out tracks for the Automatic Electric plant at yard 9; and staking a #14 turnout at Grand Avenue which would allow faster entry to the receiving tracks in yard 9 for incoming trains. While at Proviso on May 18th I spotted some Fort Dodge, Des Moines and Southern electric locomotives and a Pere Marquette 2-8-4 awaiting scrapping in yard 9. Last April 20th Bill Comstock had said he was going to stick with his assistant engineer's position on the Galena and, if I wanted to bid for that level of job, I could try for Hal Ruth's vacated spot on the Wisconsin Division when he retired in June. A month later on May 25th Bill Comstock changed his mind and bid on Hal Ruth's position. Bill's seniority as instrumentman predated mine by many years. Galena Division engineer M. S. Reid was visibly annoyed by his good office manager bidding off the division. Bill got the job effective June 11th. Hal's last day was on my birthday, June 5th. This job switching was to turn out to be a lot more complicated as we shall see.

The Sterling yard track project came off the back burner with a roar when Paul Dillon, the sole owner of Northwestern Steel and Wire Company, discovered the C&NW had purchased an extensive tract of farmland land west of town and south of Galt, the next station west, which would effectively stop the CB&Q from ever gaining entry to his steel mill. Paul Dillon was a colorful character and a self-made man. The steel mill was his creation and filled a niche market in the industry. The mill used scrap metal in their electric furnaces to produce plate, bar, fence posts, wire and other steel products. The impressive plant was about a half mile long lying west of Avenue G and north of the Rock River. The many yard tracks were full of derelict steam locomotives waiting their turn in the furnaces along with many gondola cars heaped high with every sort of scrap steel. Paul was a careful operator and close with his money. He and his son personally inspected the cars of scrap to make sure there was no timber mixed in the loads as wood dumped into the electric crucibles was big trouble. Curiously, wood did not burn in the oxygen-poor melt but floated around on the liquid metal and could, possibly, break the expensive graphite electrodes. Paul Dillon also had a habit of rearranging his yard tracks simply by telling his section man. These verbal instructions accompanied by a lot of hand motions describing what was desired were often superseded the next day by a new set of arm motions and words. Of course, Paul continued to operate the retired steam engines if there was any life left in them. He kept a fleet of Grand Trunk Western 0-8-0s shuffling cars of scrap steel and finished product in his yard from 1963 into the 1980s before the worn-out locomotives themselves ended up in the melting pot. The C&NW readily agreed to amend their switching agreement with NWS and construct additional yard tracks south of the C&NW main line and the steel company's property because that meant even more physical barriers to CB&Q entry to the steel mill. The Q's trackage all laid on the north side of the C&NW mains. While the CB&Q had those trackage rights on the C&NW west to Agnew, the old contract specifically barred them from serving any local industries along the way. I heard later that their official family was really annoyed at their inability to get direct service into the mill yard. At one point the Burlington even considered bridging the Rock River with a major steel span to gain entry to the mill from the south, but that was too costly. The CB&Q legal eagles tried and failed to break that old trackage agreement. The C&NW land purchase at Galt effectively sealed off entry from the west and south by the Burlington. During the course of this long running controversy the C&NW spiked down an iron spine through its ring of farmland in 1962 by building an industrial lead track to serve a new Armour facility which made doubly certain that the CB&Q would not be able to throw a railroad line across the C&NW

land. In December 2000 Northwestern Steel and Wire declared bankruptcy mostly because of institutional changes in American steel manufacture and the flood of imported steel. The mill was shut down and 1,400 employees were laid off in May of 2001.

Back to 1956: Bob Lawton and I ran cross sections along the steel mill property on April 30th preparatory to construction of the new yard tracks. That survey work was the opening gun in a long battle to keep the steel company as a captive industry. Bob and I were hard pressed that summer to finish the drawings and endless yardage calculations required. By August, however, the grading was essentially finished, and the roadmaster was demanding that we provide engineering services to stake the new track centerlines.

On April 25, 1956 freight service on the Galena Division was dieselized. All steam facilities were to be scrapped. The final steam suburban operation occurred May 9th when a class E, 4-6-2, #614 took the last scheduled steam powered train out of the Chicago Passenger Terminal with Chicago Mayor Richard J. Daley, Ben Heineman and Clyde Fitzpatrick (and his cigar) beaming on the platform. Full dieselization of the entire railroad was announced May 11, 1956.

Personnel changes continued. Leo G. Tieman became assistant engineer in the general office on May 3rd. H. B. Smith, Galena Division superintendent was relieved and left on May 9th along with three other division superintendents. Ottmar W. (Bill) Smith went to Madison on May 10th to replace James A. Barnes, detached for special duty. One of Jim's first assignments was to visit other railroads and find out how they were organized. His first report was on the AT&SF; he concluded that the Santa Fe was over staffed and inefficient when compared to the C&NW. Bill Reed, Bill Armstrong's draftsman, resigned May 28th to take a job with the Western Pacific in San Francisco effective June 4th. Bill Armstrong had also announced his resignation May 26th because he wanted to work for a licensed architect for the three years needed to qualify for his own licensing. With Bill Reed leaving, however, he reconsidered and withdrew the resignation.

Personally, I was somewhat exhilarated by this new management's action when I saw many of the old dinosaurs being put out. Perhaps the excitement in my own life had heightened my perceptions. May 25th was my last night school class before graduation. I had finally earned a bachelor of science degree in civil engineering from the Illinois Institute of Technology. Along the long eight year trek I had made several career decisions that may have led to different opportunities. Completing the degree work, however, was my prime objective. I mentioned the achievement of my goal to Frank Koval; he arranged with his assistant Hal Lenske for a story and photo session. Hal was a delightful guy to work with. He was later to become director of commuter and passenger service with the C&NW and played a major role in creation of the Chicago Regional Transportation Agency (RTA). (He died in Panama City, Florida, on July 20, 2002 at the age of 83).

On May 29th Charley Hise and I traveled to Cherry Valley and Rockford to investigate the possibility of a joint track operation with the parallel Illinois Central; it came to nothing, however. After lunch in Rockford, we met with Glenn Kasdorf and Bob Wheeland of the Illlinois State Highway Department on the pending start of the Auburn Street job and to develop a plan for flashing light signals to protect the at grade crossing of Spring Creek Road. On the first day of June I met with the our Chicago sales office man, Walter Hoffman and the manager of Booth Cold Storage Company whose warehouse was located on our North Pier line east of the Merchandise Mart. The immediate problem was the vertical clearance at the Clark Street overpass. The underside of this old structure was scored and scraped by too-high refrigerator cars. Booth needed service which required reefers, but the newer mechanical cars were taller which meant they could not get to the Booth sidetrack. The easiest solution at first glance would be dropping the sidetrack a foot or so. But, the base of rail already rested on top of an old brick sewer owned by the City of Chicago. Plans for reconstructing the Clark Street overpass were not drawn yet—it was to be many years before it was rebuilt. Booth moved away from this site and their business was lost to the C&NW. The old North Pier line lasted into the 21st century, but eventually succumbed when the inbound newsprint paper traffic of the *Chicago Sun Times'* was diverted to a new printing facility on the south side.

Major events in my life coalesced suddenly on June 8, 1956. At 12:03 a.m. eastbound train #252 derailed into the side of #384, also east-

The infamous burnt-off journal—the cause of a major wreck at Wheaton, Illinois, just past midnight on June 8, 1956. Two freight trains heading for Proviso on parallel mains contributed 56 cars to the pile-up.

bound, in the triple track territory about one mile west of the Wheaton depot. The journal on an old tank car had burned off. When the noise and dust subsided, there were 56 wrecked freight cars heaped up six deep on all three main lines and down the embankment. The pole line on the south side was down taking with it Illinois Bell Telephone's cables. Willard "Bill" Comstock lived a short distance from the scene of the accident. Hearing the racket, he left his wife Marjorie in bed while he went to investigate. He soon returned and, luckily, was able to get through to the chief dispatcher and tell him about the wreck. Getting into his work clothes, he headed for the site—there was no more sleep for him that night. I was peacefully asleep in Elmhurst ten miles away until Bill Armstrong, an

The Clinton, Iowa, and the California Avenue hooks working at Wheaton on June 8, 1956 to clear the three-track Galena Division main line.

early riser, called me before 7 a.m. to say that there were no commuter trains that morning because of a wreck at Wheaton. My wife and I picked him up in our car and drove there at once. The commuter stations along the C&NW were packed with disgruntled customers. The railroad was unable to provide much commuter service at all because most of the west line's morning fleet was trapped beyond the wreck. The CA&E, CB&Q and Milwaukee got a lot of extra riders that morning.

The pile of wrecked freight cars was impressive. Crews from Clinton worked on the west end of the heap; the California Avenue crew, on the east. It was a giant game of jackstraws as the wreckers carefully pulled cars away from the edges of the pile, set them upright on whatever trucks were handy and moved them away. I found Bill Comstock there directing some of the trackmen who were waiting to reconstruct the main lines when enough of the wreckage had been cleared. Not knowing then what the cause of the derailment had been, we decided that a standard derailment survey should be started. We began making a location survey of the wrecked cars and recording details of the undamaged track. About 10 a.m. Bill decided he was hungry and went home to get some breakfast as he had been up all night. He went into the house, carefully hung up his coat and fell dead on the living room carpet.

We didn't notice Bill's absence for a long period and then thought that he must have had something else to do. Our immediate problem with the wreck was trying to identify the twisted ore cars, broken wooden refrigerators spilled loads of oranges and tangerines, steel box cars full of canned goods and tumbled loads of lumber. The day grew warmer and the lack of drinking water was giving us dry mouths although an orange or two tasted pretty good after wiping off the iron ore dust. Shortly after noon a car was lifted off the pile revealing the body of a hobo who had been riding one of the trains. Everything changed at that point because a fatality caused by a train accident automatically involved the Interstate Commerce Commission and other legal bodies. Now we had to make extra efforts to accurately locate the various wrecked cars so that exhibits could be prepared for the inevitable hearings and reports. A burned-off journal wheel set had been identified as the cause of the wreck which eliminated the need for us to run a mile of levels on both mains to the west of the accident site. Our task to locate wrecked cars was hampered by the fact that many were now on wheels and had been moved out of the way to Wheaton. The *Chicago Tribune* had photographed the wreck from the air which proved to be our salvation. Using the photos later we were able to recreate the original scene fairly accurately. Instrumentman Art Humburg and rodman Norm Hillegass worked very hard to collect the important data we needed on that hot, clear summer day.

The first objective at any wreck is getting a through track restored. The south side of the pile was thinnest so the Clinton and California Avenue hooks concentrated there. Westbound train #13 with E8 #5029 waited while the wrecking crews cabled, chained, pulled, and pried cars out of the heap. Meanwhile trackmen had assembled rail and ties and waited as a bulldozer shoved pieces of freight cars down the bank and leveled off the track bed. The wreckmaster, a car department foreman, was in charge of the whole operation—not even the chief engineer could give an instruction at a wreck site without the wreckmaster's approval. About noon track #1 was opened on a skeleton track with just enough clearance for #13 to creep past the remaining debris. The work continued all afternoon and into the night. Marjorie Comstock, who had a clerical job with the railroad, had somehow found her way into Chicago to do her day's work. Returning home that evening, she entered the house and found her husband, Bill, dead on the floor. I did not learn of Bill's death until later that evening because I had left the wreck scene about 5 p.m. to attend my graduation ceremony at the Illinois Institute of Technology in Chicago. I had my engineering diploma and a dead boss.

The clearing work went on through June 9th enabling trains #3 and #4 to squeak through the wreck site. It was to be several more days before the main lines were clear and back to normal speeds. Untangling the communication lines took even longer. We buried Bill Comstock on Monday the 11th. That same day I was notified that my bid for the assistant engineer's position of the Wisconsin Division was successful and that I was to report for duty on the 14th. My last day on the Galena was the 13th marking the end of continuous service that had began July 31, 1947. Just to cover my bases, I used my new seniority as assistant engineer from the Wisconsin Division assignment to bid Bill Comstock's vacated job on the Galena thus pre-empting any bids by instrumentman seeking promotion.

My new boss was Jack L. Perrier, division engineer. On the 14th I sat down at Hal Ruth's old desk and began to clear it out. Apparently, Hal had not understood what a waste baskets was for, but I did. Because the engineering work backlog on this division was a lot lighter than on the Galena, I had little to do. On the 18th Leonard Lamport, chief engineer of maintenance, called and asked if I still wanted the Galena job. I said "yes" and was appointed the next day, effective June 25th. So, I was right back "home" again and was now officially in charge. Galena instrumentman Art Humburg was appointed on July 3rd to the Wisconsin job I had just vacated despite concerns about his sporadic bouts with booze. On the Galena I had my own personnel problems now. Rodman Larry Hoffman left on

June 29th; Art Humburg, as mentioned above, went to the Wisconsin and was replaced by Norm Hillegass so that my staff consisted of two instrumentmen: Robert Lawton and Norm; rodmen John Colloton and Dillis Allen. On July 17th Al Kuhn rejoined us from Northwestern University as a rodman and, on August 6th, a rodman from Boone, Iowa, reported. Raymond E. Snyder, a very young wise guy who was to grow-up on the Galena and have a long career in the engineering department—we will meet him at several points later in this account. On August 10th I learned that our division engineer, M. S. Reid, was soon to be promoted to assistant chief engineer-maintenance and William Wilbur from Green Bay was to be our new division engineer. What goes around, comes around. This was the same Bill Wilbur to whom I had reported on my first day on the railroad, July 31, 1947. Reid's last day on the division was August 15th. I received another invitation to be the chief engineer of the Fort Dodge, Des Moines and Southern Railroad on June 30, 1956, but declined just as I had done some six years earlier. B. R. Meyers, chief engineer of the C&NW, sent another letter of congratulations on August 17th giving me a boost in confidence that my best career path was with the North Western.

The senior bridge inspector was Leo D. Garis. An ardent union organizer, he had long sought to establish a mandatory union shop for all technical workers from division assistant engineer on down including chemists and laboratory technicians at the Chicago Shops and draftsmen in the chief engineer's office and the land department. He had waged a long, and futile, battle to sign-up the Galena Division office. He could not muster one vote on the obdurant Galena. His frustration with us grew apace. At one point Leo had the temerity to publish an item in their union rag that I had joined the union. When confronted with this blatant untruth, he had to recant and apologize which added more bile to his discontent. There was no love lost between us. In a twist of fate, Leo Garis was assigned to my office as an instrumentman in August 1956. Nearing retirement, Leo was just filling-in time until he could qualify for a pension. His labor position and dirty-tricks did not sit easy with me; I was now the master and determined that Leo would work, not sit out his time. His

Galena Division career began with assignment to writing roadway completion reports as we had a huge backlog of them. When B. R. Meyers urged us to get that Sterling yard track extension completed, I assigned the job to Leo. He spent a lot of time out in the hot summer sun clambering up and down the new grading work in progress running levels for cross sections. Very soon, he had enough. On August 22nd Leo announced he was taking his retirement in September. Leo had less than a year of retirement before he died July 14, 1957.

As expected Art Humburg resigned as the Wisconsin Division assistant engineer August 20th and headed for California. My old friend, Robert R. Lawton, bid the job and was appointed assistant engineer of the Wisconsin Division on September 10th. Rodman John Colloton's last day was August 31st. My staff was getting thin again as I had only a newish instrumentman, Norm Hillegass and two green rodmen. Then mercurial Norm decided to take a bridge inspector's job effective September 14th leaving me with no instrumentmen at all. I hired a young Sioux Indian, Carl, from the Rosebud Reservation in South Dakota, as a rodman because he seemed to possess an aptitude for learning the rodman's trade. On September 12th I received a direct order from Mr. Lamport to dismiss him at once—because he was an Indian. That was one of the hardest things I ever had to do. Al Kuhn, a champion swimmer, came back from the Los Angeles Olympics on September 11th and I got him an appointment as a temporary instrumentman which lasted until he went back to school on the 27th. Ray Snyder, my brash young rodman from Boone, bid on a rodman's job at Norfolk, Nebraska, on the Nebraska Division. I must have been a tough boss as I couldn't keep a staff together. Things began to improve when Carl K. Oleson, a former rodman who had gone to work for the highway department, came back to the railroad as an instrumentman on the 25th. Two days later Russ Padgett from Green Bay came on board as a rodman. Mostly just a willing warm body, Russ was not very knowledgable about engineering matters. Then on October 1st Fred Kruse came from Sioux City as instrumentman along with another new rodman from the Northwestern University program. We had the nucleous for a good staff—after they had gained some experience. Ed Marquardt, the longtime chief clerk on the Wisconsin, moved over to the corporate secretary's office in the Daily News Building. Our chief clerk, Al Simandl, was angling for Ed's vacated job, but was encouraged to stay where he was.

Amid all this personnel shuffling, we actually had engineering work to do despite the large number of green hands available. In early July the hot project was a big repair project to fix railroad structures at Proviso and the Chicago Shops. Maurie Reid, the division engineer, and Leo Tieman from the chief engineer's office, collared our B&B supervisor, Orville Olson, and his foremen, Mel Barsema and Bill Enders, to help estimate a massive rebuilding and repair estimate. The summer of 1956 was another extraordinary season for heat. And John R. Cantwell, Galena Division superintendent had hot ideas about making major yard trackage changes at the west end of yard 4. Yard 4 had been a dead storage yard for derelict cars for many years and was now to be converted into an active switching facility. We were to redesign the ladder tracks on the west end so that loose car switching could be done with greater facility. Given the down grade of this yard the free-rolling rolling cars accelerated all the way to the bottom of the yard at the Mannheim Road bridge. Impacts with standing cuts of cars were excessive. Of course, the whole concept was flawed—loose car switching itself was the problem. Instead of enlarging that operation, figuring out ways to minimize car shuffling would have been smarter. On August 2nd Cantwell suddenly woke up to the problems he was creating and had us radically change the whole track arrangement. Three days of very intense work was done by Norm Hillegass, rodmen Dill Allen and John Colloton, to alter the plans on the ground which had to be right the first time as the roadmaster's track laying crews were building switches and tracks right down their necks. Cantwell was hell bent to get the work completed as he had big operational plans that were to have commenced July 16th (but didn't). And then the cash drawer was slammed shut August 1st. All main line rail and ballast gangs were shut down, no work trains were to be ordered as there just was no more cash available. With the abrupt cessation of work, I took a weeks' camping vacation in late August with my family. In Sturgeon Bay, Wisconsin, I spotted the rarely seen Ahnapee and Western and its #166 diesel.

Seemingly as a counterbalance the heat of August was replaced by a cold September which saw on the 20th the earliest freeze on record. In the division office we were well along with the annual budget and had a pretty good handle on the engineering jobs around our territory. The 1957 program work was all submitted by October 5th. A new budget system was to be instituted and a system wide inventory of materials was planned. These "new " systems were designed to use the existing unit record monsters crouching in their lair at Ravenswood. The nascent computer science technology just beginning to unfold would supplant these old mechanical crocks—someday.

On July 20th we took delivery of a new dark green Chevrolet station wagon for the Galena Division. Division engineer Reid promptly glommed onto it and went sailing around the division with the superintendent. We finally got it back when he left for his new job three weeks later and the new boss, Bill Wilbur, realized that the field crews needed it more than he did. On October 4th Bill Armstrong and I drove to Clinton to inspect the construction of the new freight car shop there. This was to be an important addition to the C&NW for many years. We ran into Harold Barr, the Iowa's division engineer and Al Peagan, the division superintendent. Al was a strange fellow. He had engineering training on the C&NW but had gone into the operating end. After a checkered career on the railroad, he ended up as a vice president of Ford, Bacon and Davis, a New York engineering firm. Al was not liked very much by the men he supervised on the Iowa Division. His eccentricities were notorious and viewed with some bewilderment, if not laughter. For example, in the first days of the Heineman-Fitzpatrick regime, their official inspection train rolled into Clinton early one morning and stopped to change crews. Al Peagan came stamping down from his office in the station without a single word of greeting to the cluster of men waiting for the train to be readied. He walked up to the Iowa Division conductor, bruskly took the train orders protruding from the man's lapel pocket, turned his back to the surprised observers and leafed through the orders. When finished, he jammed the orders back into the conductor's pocket, turned on his heel and marched off, still without a word to anyone. The conductor then mildly observed to Mr. Heineman and Mr. Fitz. that if Mr. Peagan really had wanted to see today's orders, he should have asked for them. The conductor then reached inside his uniform coat and produced the correct set from an interior pocket. Apparently, Al had played that stunt with his train crews one time too often.

The autumn was filled with the paper work that made the railroad run—at least the people who wrote and read the stuff had work to do. I was getting rather stale at not doing as much field work as in past years. Somehow, sending crews out and hoping they were getting the right results, was making me uneasy. Yet, I knew that going along with them to check their work, could be inhibiting. The railroad continued to operate under its new management team despite little incidents like the 26 freight cars piling up in the center of Morrison on the 22nd of September. This incident took Bill Wilbur away from the office for a few days allowing us to get more of our work done without his interruptions.

M. B. "Bob" Lithgow, with whom I had worked when he was an instrumentman on the division many years ago, had been with the industrial development department for several years. He asked me to accompany him on a tour to look at railroad and private properties in Elgin, Rockford and Caledonia on October 30th. While on the trip, he tried to sell me on taking a job in their department as an industrial agent. The job paid a bit more than my current pay ($560.00 versus $520.40). The change, however, would be a huge one for me. Moving from the known engineering world to the unknown traffic department was a sea change. His offer was tempting as it came from an old friend. I said I would think about it. The next day Jim Barnes said there was a big change coming in engineering and that I could be a division engineer in three or four months. Personnel problems continued to plague the orderly conduct of business. Norm Hillegass came back to the Galena in November—he had experienced enough bridge inspecting. Antsy Norm was never to find his true niche at the C&NW; he eventually ended up as a salesman for Pettibone-Mulliken, a railroad work equipment manufacturer. On November 9th I was introduced to the head of the industrial development department, Gene F. Cermak, and his senior aide, John F. Daeschler, when we went to Proviso together to inspect a proposed site for the new warehouse and manufacturing plant of the Inland Container Company on Grand Avenue. Cermak offered me the same job that Lithgow had dangled in front of me. The offer forced me to stand aside and coldly assess where I stood on my career path. On the engineering side, there was the prospect of becoming a division engineer, but somewhere else, probably at Chadron, Nebraska, or Huron, South Dakota, or Norfolk, Nebraska, or some other far off spot, where I would be "seasoned". My family would not like that because my wife's ties to her parents were very close and her art career was here. I was loath to take on a mortgage as my home in Illinois was free and clear. Moving meant expense and debt. However, on the traffic side, the industrial development department was becoming a jewel in Heineman's crown, so that future prospects could be alluring along with the appeal of learning new skills. I tested the waters with my close friends and even interviewed the chief engineer. While not expecting any firm commitments from "Mr. Meyers", as he insisted on being addressed by juniors, I let him know that I was receiving offers of job opportunities. On balance, engineering won for the time being and I so advised Gene Cermak on the 23rd. While Gene was not happy at my choice, I heard from the grapevine that he had not given up and was seeking authority for more pay for the job hoping to lure me with dollars. By November 26th my division personnel roster was complete, at last. I had seven men: 3 instrumentmen and 4 rodmen. Life was too easy.

In November purchasing and stores departments were combined. This was one of the first major changes in the railroad's organization, the first of many to come. All in all, December was pretty quiet. My staff of willing lads tackled the many tasks at hand and really did a good job with a little guidance here and there. By now we had pried the new station wagon away from Bill Wilbur and fitted it out with extra shelving and boxes for our survey equipment. Accidents still happened, however; Norm managed to smash our good (1910) Berger transit while working at HM tower in Elmhurst. On December 12th I accompanied Bill Wilbur to Elgin for a meeting with the State of Illinois Highway Department concerning a new highway bridge to be constructed across the Fox River just outside their office windows. The bridge would carry the U. S. 20 by-pass around downtown Elgin and cross over both our Williams Bay and Freeport main lines. This month saw a few field jobs like changes to the West Chicago coach yard and those sidetracks at Sterling in progress. Given the calm atmosphere, I took my annual winter vacation as usual. This was the first year we did not ride trains somewhere.

Instead we made a long automobile trip south to New Orleans by way of Mammoth Cave, Kentucky and then west to Laredo, Texas before returning via San Antonio, Texas and Anadarko, Oklahoma.

One bright spot in this pivotal year of 1956 came from the Veterans Administration on May 1st when they advised me that Public Law 346 of the 78th Congress stated that training and education benefits would terminate nine years after the end of the war which had been determined by the 80th Congress as July 25, 1947. That meant that July 25, 1956 saw the end of that part of the G. I Bill which had covered my college education. I graduated June 8, 1956—just in time.

11

COMING TO THE END OF TRACK, 1957 . . . January 1957 gave us a lot of snow which caused the usual operating problems for the railroad. When I returned to the office after my vacation in the Southwest, I found one of my survey crews was in South Pekin and another at Proviso. At least I still had men on my staff. On the 11th I met with a vice president of Medusa Portland Cement who wanted to purchase an old shale heap on C&NW land at Spring Valley. The shale was waste material from a third vein coal mine that had been closed since the mid 1920s. In a twenty year period after the turn of the century there had been a vigorous bituminous mining industry in the region. A great amount of locomotive coal came from the underground mines in this area on both sides of the Illinois River. One of the great mining disasters of Illinois occurred nearby at Cherry, Illinois, some 8 miles from Spring Valley. On November 13, 1909 the St. Paul Coal Company's mine caught fire killing 259 miners. The St. Paul Coal Company was owned by the Chicago, Milwaukee and St. Paul Railroad. The coal in that section of the Illinois River valley was mined out quickly as the seams were not very thick. Each little town that had a mine was monumented by a towering pyramid of reddish shale over a hundred feet high and on which no vegetation would grow. Oglesby, Spring Valley, Granville, Ladd, and Seatonville were among these former mining towns. Medusa wanted the waste shale for manufacture of cement at their Dixon, Illinois, plant. I took him to the land department as this was strictly a land sale.

New construction projects in Chicago were now coming to life in 1957. One of the biggest was the expressway planned to connect Chicago with its new airport site. A multi-lane roadway was planned to pass underneath the Railway Express Terminal building at Halsted and Kinzie Streets as well as under the Galena Division's elevated main tracks. That involved digging a very deep hole. Despite a severe cold snap, another track construction job came in from Rockford—a 2,000 foot extension to the old KD line north of Loves Park. On the 22nd seven freight cars derailed west of the Rochelle depot wiping out both main lines. Bill Wilbur, the division engineer, invited me to go along and see how bad it was. The next day came requests for three more track construction projects including an industrial siding at Proviso north of Grand Avenue to serve the Inland Container plant then under construction, a track extension at Cargill's chemical plant that was tucked away in an old gravel pit at Carpentersville and a really hot project: doubletracking the Proviso hump. This last one turned into a real challenge.

The impetus to double track the Proviso hump came right from the top. In approving the construction Mr. Fitzpatrick made it clear that the project could cost no more than $25,000, the limit of his authority. If the project were to cost more, a special request for funding would have to go before the board of directors. As the project went along we did some very creative accounting to stay within that cap on costs. We began a detailed situation survey on January 30th. Instrumentman Fred Kruse and his team developed the notes and then worked furiously on drawings. The second hump was to be located north and east of the existing 1926 hump. The design of the new facility required careful adjustments of its location to control the steepness of the gradient. The proposed crest was moved west, then east, then west again as Bill Wilbur tinkered with the design. Each new location caused recalculation of vertical curves, grading yardages and alignment. I was uncomfortable with this shifting and fiddling because Wilbur was working without any real information on car rollability, the effects of acceleration grades, or even statistics for prevailing winds. All he knew was that the old hump was felt to be too steep for modern cars with roller bearings. At last a configuration was adopted that his intuition deemed acceptable and we "froze" the design. From this base we developed grading plans and the connections to the existing bowl tracks. Actually three designs were presented to the chief engineer on February 1st. Mr. Meyers rejected all three. A new modification shoved the proposed hump crest further to the east and set it much lower than the old hump. Meyers approved this plan and told us to get it built in two weeks. On February 6th and 7th Fred Kruse staked out the new plan and a grading contractor was engaged to shape the new hill and approach. It looked easy—just add fill to the north side of the existing hump lead, smooth it out and cut through the old fill east of the crest to connect to the bowl tracks. On the 13th Wilbur and Meyers inspected the work just before the curse of Proviso struck in the form of that hidden old devil, popcorn slag. As related previously, the waste slag material upon which a great part of Proviso was founded consisted of steel mill refuse called popcorn slag which chemically changed when in contact with ground water to form an obsidian-like glassy material. Our grading contractor brought in big equipment to break-up this iron-hard layer. And so he did, very slowly. Meanwhile, the old hump was kept in operation right alongside of the new construction—a safety nightmare with flagmen present at all times. At least that hardshell layer of slag under the old hump eliminated the necessity of installing cribbing walls to hold the bank up as the remaining slag stood in vertical faces like formed concrete. Although the weather had moderated, by February 25th Meyers was getting very impatient with our progress, but breaking through that shield had seriously disrupted our work schedule. We were still having problems with track alignments the next day, but construction went ahead anyway. On the 27th gravel ballast was being dumped on the second hump lead. To make sure the project was being built according to plan, I went out on the 28th to double check the work. It went into service March 1st—four weeks after we began construction. Knowing that the first day would determine if it worked or not and not wishing to be there if it didn't, I waited until the next day to see it in operation. The new vice president of operations, S. C. Jones, was also there along with a haggard-looking John Cantwell. (John was soon to be fired as superintendent of freight terminals). No serious difficulties were experienced with this new facility. In times of slow business, however, it was the old hump that was used because the retarder operators were more familiar with the idiosyncracies of the old system.

S. Chasey Jones, Mississippi born, was a smallish gentleman who stood stiff as a ramrod, with a fedora tilted forward over his brow and holding an ever-present cigar. Always addressed as "Mr. Jones", he had most recently been the Illinois Central's Iowa Division superintendent. A stern disciplinarian and a firm believer in the concept that appearances made the man, his very demeanor bespoke control. At a derailment on one occasion, he was observing the tumultuous activity when he saw a young assistant trainmaster crawl out from under a smashed boxcar dragging the end of chain to be used to pull the wreck loose. Mr. Jones looked at the disheveled young man and said in his Southern drawl," Wha's yo' 'thority?" "Sir?" replied the dirty, shirt sleeved junior officer. " I said, where's yo authorty?" " My authority?" "Yas Sah, wha's yo hat?"

In January Bill Martin, Galena Division superintendent, was relieved from duty because of excessive overtime of crews on the Proviso-Clinton trains. The severe weather and poor condition of the railroad, of course, had nothing to do with the dismissal—indeed! Master mechanics, superintendents and 200 section men also were also discharged. On January 5th 59 more section men were laid off. The new Galena superintendent was Del Perrin from the Illinois Central. Along with president Clyde Fitzpatrick the flood of his former Illinois Central pals into the C&NW ranks was styled by some old

C&NW heads as an Illinois Central Mafia invasion. Perrin was an Iowan and had some fame as a college football player. Called "June-bug" mostly because he looked like one, he did not waste any time when he told me to prepare a proposition for new crossovers at Sterling. On February 20th we added another rodman to our staff, David J. Blutt. David was instrumentman Fred Kruse's brother-in-law. A very young man, David's last job had been sacking groceries in Sioux City, Iowa. To his credit David had a sober head on his shoulders, a good work attitude and a desire to apply himself to mastering our craft. He eventually attained the rank of assistant engineer on the Central Division at Mason City, Iowa. I reflected that just a few years ago a rodman starting on the railroad had to have some experience in land survey or other technical skill if not a college degree. *Sic transit gloria mundi.*

Our venerable B&B supervisor at South Pekin, Ray Doty, retired February 28th as did our starchy and pompous roadmaster at Sterling, Arthur E. Benson. The new roadmaster at Sterling was Al Johnson who was to prove many times to be a practical and steady man. There was some good news in the paycheck: my rate went to $547.62 from the former $472.34 on February 1st.

Rumors were on the loose again. The Wisconsin Division was to be moved to Milwaukee, Wisconsin, and a new Terminal Division was to be created. The rumors of February 18th turned into reality and on the 28th Jack L. Perrier was named the new Terminal Division's engineer. The new division was formed out of big chunks of the Galena and the Wisconsin including that big stewpot at Proviso. My Galena division office had achieved a very smooth level of operation as each of my new boys grew into their responsibilities. They were all youngsters—I was the old man at 30. The team worked together well, each striving to do his part which greatly simplified my job and gave me considerable latitude in running the office. Bill Wilbur seemed to think that everything was running along well and absented himself a lot from the office by riding passenger trains and going on the annual bridge inspections which he so loved to do. The smooth tenor of life changed abruptly on March 11th when we learned how much of the old Galena Division we were to lose on April 1st. My staff would be cut to four positions: two instrumentmen and two rodmen. What a comedown from the twelve we had in 1947! A look at the capital budget showed that most of the approved work for 1957 labeled for the Galena Division was actually going to become the Terminal's responsibility. April 1st came and went; the actual transfer occurred June 1st. Until that change took place, we had work to do. There was a new interchange track with the Toledo, Peoria and Western Railroad at Hollis to be built near the highway #9 crossing at Orchard Mines. There were hearings on a new overpass of the North Pier line at Dearborn Street in Chicago on March 5th. At the hearings I met the city's bridge engineer, Michuda, and Col. Corey of the U. S. Corps of Engineers. I knew Corey from my membership in the SAME (Society of American Military Engineers). In Elgin a boxcar derailed and fell off the Chicago Avenue (U. S. 20) underpass landing on the roadway below—no one was hurt. I got a message to Fred Kruse who was working on a cross-sectioning job at the Algonquin gravel pit, to detour to Elgin and make the accident survey.

My social relations with the railroad broadened when I was invited to join the C&NW Toastmasters' Club on March 20th. The first meeting of the group was held at the Cafe Bohemia at Adams and Clinton. Frank V. Koval from public affairs, Gene F. Cermak, industrial development, Stanley B. Boardman, traffic, were among the group. I was the only invited member from engineering or the division level. The so-called Pioneer Club met monthly for a time until it gradually petered out. Sobering news came down that Leonard R. Lamport, chief engineer of maintenance, was retiring early at age 58 effective April 1st. He said he could no longer be responsible for the railroad's safe operations given the unacceptably low maintenance budget. He later took a position with the Association of American Railroads' test laboratory in Chicago. P. V. Thelander also retired April 1st. That March 22nd was what my journal described as a "stemwinder". I attended a very loud meeting in B. R. Meyers' office with F. C."Carter" Harrison, superintendent of freight terminals, and Carl Hussey, the new assistant general manager, a recruit from the Illinois Central. The subject was track changes at the east end of Yard 5 at Proviso. I had to call Fred Kruse and his crew off a topographic survey at the Inland Container plant site north of Grand Avenue, Proviso, to rush over to Yard 5. Gene Cermak, director of industrial development, had promised the topographic work for Inland who needed it for their building construction plans. Gene was livid when he heard that I had pulled Fred off the project. He called Meyers and complained that "his" work was being interrupted and that "his" client was being discommoded. The call was inopportune as the dignified Mr. Meyers was not about to be dictated to by this young, redheaded newcomer who did not understand that, he, the chief engineer would decide where and when his engineers could best serve the company's interests. Cermak would regret ruffling the chief's feathers. We went back to the topo work on March 29th. Meanwhile, back on the 22nd, the suggested track changes at the east end of Yard 5 had to be seen on the ground to be understood, thought Mr. Harrison, so off we went. "Carter" Harrison was another puffed-up windbag promoted far above his capabilities. That entire afternoon while I was with them, Carter belied his self-important role as a fearless leader of men when he groveled like a fawning lackey before Carl Hussey, the new and unknown "lord" from the IC. I could hardly believe what I was seeing. Years later, after I had known Carl Hussey for some time, I learned that a good part of his secretive and aggressive furtiveness hinged on a deliberately cultivated persona which suggested to others that he always had some private information that only he knew. Once, when in his cups in Peoria, his mask slipped, and I got a glimpse of the real Carl Hussey. On this visit to East 5, however, Carl had Harrison's measure and calmly let him hang himself. F. E. Harrison was soon an ex-employee. An excellent and capable negotiator, Carl Hussey protected C&NW interests on several boards of jointly owned railroad terminal properties. As far as the track alterations at East 5 were concerned, approval was secured and my crews worked the rest of the month on this hot job plus the proposed extension of the old main lines through Proviso.

More organizational changes occurred. The Sioux City Division was abolished and merged with the Iowa on April 1st. M. S. Reid became Engineer of Maintenance succeeding Leonard R. Lamport. Harold W. Jensen became maintenance engineer, a title which he detested as he said, " every building in Chicago has a maintenance engineer which is just another name for the janitor." W. H. "Heinie" Huffman became an assistant engineer on Meyers' staff and Vern Mitchell was appointed as engineer of signals. Bob Nelson, an ex-

Galena instrumentman, transferred to Boone as assistant division engineer. On April 3rd there were strong rumors that the Galena Division was to be extended to Cedar Rapids, Iowa, to compensate for the loss of the Chicago Terminal territory. I lost another instrumentman—the peripatetic Norman Hillegass took a job with the State of Illinois highway department effective May 16, 1957. I heard on April 12th that Joe Gaulock, my long-time accounting contact, had taken a job with the Milwaukee Road.

On April 4th the Chicago, Aurora and Elgin Railway received authority from the Illinois Commerce Commission to shut down its passenger service effective April 28th. This heavy electric interurban line had operated a high speed third rail line from Chicago to Wheaton, Aurora and Elgin since 1902. The almost hourly trains into downtown Chicago over the elevated lines of the Chicago Rapid Transit's Garfield Park line was a well patronized service. The construction of the new Congress Street Expressway wiped out the elevated track structure and the transit authority's trains were detoured to old streetcar tracks in Van Buren Street parallel to the new highway construction. The CA&E chose not to operate on this temporary trackage and terminated their trains at Forest Park requiring their riders to transfer to the "El" trains there. The slower service, need to transfer and the two seat ride caused the passenger count to plummet as patrons found new ways to go to Chicago. Kansas investors of the CA&E saw it as more valuable as scrap than as an operating railroad. After a series of legal delays, a final court ruling came down on the morning of July 3rd. The railroad quit at noon, stranding many commuters in the city. The C&NW west line trains were jammed that evening.

Bill Armstrong, chief architect, and I went to Elburn on April 9th to review plans for a new depot there. The agent had an office in the old wooden freight house which quite probably was an original Galena and Chicago Union Rail Road building. While we were out west, Bill and I also inspected the construction underway at the Clinton car shop. On April 25th the railroad felt constrained to put on a demonstration of cooperation at the potato dealers' building at Wood Street. Walt Foutts, our perishables sales agent, put up with a lot from these free-wheeling produce shippers who seemed to thrive best in a confrontational atmosphere. Knowing that the railroad earned a lot of money from this volatile business, a new 15,000 square foot office building complete with air-conditioning was built for the Chicago Carlot Potato Association at 1421 S. Western Avenue. On April 30th I attended a meeting with superintendent Les Bean to discuss a plan to remodel the unused streamliner ramp at Chicago Shops into a coach servicing yard. Now that I was short an instrumentman, I took on more of the field work to help out. Fred Kruse and I ran a mile of levels under the hot sun of southern Illinois at milepost 115 where a new highway crossing was planned. The next day we went to Peoria and laid out track changes at Adams Street where the old concrete highway viaduct was being knocked down. On the 13th I made a swing through West Chicago to see roadmaster Jack McCord, called on the state highway office in Dixon, met with Manfield, a scrap dealer in Sterling, about a new track for his yard and listened to assistant roadmaster Earl "Fuzzy" Piersen complain about Jack P. Datesman and James A. Barnes telling him from Chicago how to maintain track. I drew the line (pardon the pun) at stringlining curves. Fred Kruse, Jim Simons and Davie Blutt had younger backs for that kind of work on the three main lines through Glen Ellyn May 23rd and 24th. This year the accounting department got ballast section reports that were pure fiction. I had started my railroad career digging those miserable holes in the main track ballast (that is, the section men did the actual digging). Being so short of help now and knowing that the only purpose of the information was allocation of monies between capital and operating accounts, Norm Hillegass and I sat in the comfort of the office and drew ballast sections quickly by reviewing past year records and applying our engineering imagination. We were deliberately slow about sending accounting our "work" to allay any suspicions about how they were developed.

On May 22nd we heard that the C&NW was to purchase the Litchfield and Madison Railway which operated our freight trains from Benld to Madison, Illinois. Division engineer Bill Wilbur had been spending quite a bit of time on the south end but had not revealed why. Of course, Bill never did communicate with us much anyway, especially verbally. Most of his instructions came in minuscule script on bits of papers or on the corners of files sent to the drafting room and signed "ww". On June 3rd W. H. Huffman confirmed the coming purchase of the L&M. It wasn't until June 25th that I got my hands on a L&M track chart and saw what the hills and curves of our new south end looked like. The actual purchase and merger occurred January 2, 1958.

The pending transfer of part of our Galena division to the new Terminal Division involved intense scrutiny of all files and maps. We cleaned out the appropriate items and dumped them onto the new division. In the process we scrapped a lot of obsolete material all over the Galena. The new jobs were bulletined. The last day of the old Galena was May 31st. It was like major surgery setting up the new entity. The new Terminal Division was defined as: the Chicago Passenger Terminal and Galena mains west of Elmhurst to milepost 16.5 which included the California Avenue coach yard, the Chicago Shops, Proviso yards, Rockwell Street and 16th Street lines, Irondale, and North Pier line. They also got part of the Wisconsin Division lines north and northwest. I believe the northwest line extended to Crystal Lake and the north line went to Kenosha, Wisconsin. Robert R. Lawton, the assistant engineer, was not only faced with the task of dealing with a huge heap of maps and files from the old Galena, but also had to sort through his collection of the same stuff from the truncated Wisconsin Division. The Wisconsin Division headquarters was shifted to new offices created in the lakefront passenger depot at Milwaukee. Not much field work was accomplished for a time. We had given away so much territory and the work that went with it that we were actually able to catch-up on our work backlog for the first time in memory. Rodman Russ Padgett transferred to Green Bay June 10th as it was closer to his former home. On June 6th I went to DeKalb to see attorney Harold Rissman about a drainage district assessment. In June our rail and ballast jobs were suddenly reinstated which changed our slow period into one of intense activity. Then the section limits were stretched again causing another round of reallocation and balancing. On one of our rare slow days, I set the lads to cleaning out the map vault especially the top of a ceiling-high steel cabinet where rolls of ancient construction projects lay in dusty solitude. We scrapped 90% of these old rolls of blueprints. Many of these plans covered repairs to structures that had long since been demolished. There was no historical society active then that would have been a good repository for these items. The one cabinet we did not disturb was the huge collection of original field books. There were over 1500 of them, all carefully numbered with their contents sort-of indexed. While finding notes for a specific old survey would have been a challenge, these leather and clothbound books were considered legal documents and had to be preserved. These old books contained a lot of history on their sometimes rain-soaked pages.

Western Electric Company announced construction of a huge distribution warehouse and manufacturing plant at the corner of Hawthorne and Kress Roads west of West Chicago. The facility was to be served by a long track running north from the west yard at West Chicago. This was a job I would have liked to have done myself, but I sent Fred Kruse and Jim Simons. On July 9th they were shorthanded and asked for help; I was pleased to oblige. Fred Kruse, Davie Blutt and I set the alignment on this long straight track. To me there was a lot of pleasure in laying-out a long, new track; it seemed that all we had been doing recently was tearing-up old rails and structures. Speaking of which, rodman John Colloton on the Terminal somehow managed to break the plate glass cylinder on that old horror of a blueprint machine June 28th. It was a blessing to get rid of that thing. Instrumentman Carl K. Oleson resigned again on June 29th effective July 5th leaving me with one instrumentman, Fred Kruse, and two rodmen, Jim Simons and Dave Blutt. On the first of July Jim was stringlining curves at Glen Ellyn, while Carl and Dave were helping Charles P. "Chuck" Jevne of the land department in a survey at Sterling. That left an office force of me. On July 15th I managed to get James L. Simons promoted to instrumentman. Jim was a skinny kid who had served with the U. S. Marines and had begun his railroad career in Boone, Iowa, before coming to the Galena Division. Although he had trained in assaults on beaches and in troop exercises in desolate places, he was deathly afraid of snakes. His mistake was letting the other lads discover that phobia. They often had a bit of fun at his expense when working out in a weedy field. At the appropriate moment, someone would yell "snake", and Jim would drop everything and fly out of the area. Years later, when he was division engineer on the Black Hills Division at Chadron, Nebraska, the reality of rattlesnakes lurking under the timber bridges and on the sandhill prairies kept him very alert indeed. Jimmy had a temper that was very close to the surface. He was an intense individual burdened with a severe cigarette habit. Despite his hair trigger nerves, he responded well to the responsibilities assigned to him. I wished I had more like him on my staff.

Terrific rainstorms hit the area in mid summer; one yielded 6.28 inches in one day—the worst since 1871. On July 15th Bob Lawton was working in my home town, Elmhurst, stringlining curves. It didn't seem right to have another division working my old territory especially in my own home city. Anyway, we didn't get any part of the Iowa Division to Cedar Rapids as had been suggested.

On July 11th I received instructions to prepare suitable quarters for new IBM (International Business Machines) equipment at various designated places on the railroad. A new punch card based system called CAR-FAX was to be installed for car tracing and train operations. In the rush to get air-conditioned rooms established in the many crummy old yard offices, I got into hot water by commenting too loudly about the ridiculous use of old fashioned punch cards when newer technologies were available. The CAR-FAX system had been concocted by Fred Baldauf, an old time car accountant, from unit record equipment and millions of 80 column punch cards. These IBM machines had been around a long time. My father was the head statistician at the Quaker Oats Company in Chicago since 1930 and had installed this very same type of equipment there in those early days. I could well remember as a small boy being taken to his office on a Saturday morning and allowed to randomly punch cards, run them through the sorter and collator to admire the nonsense printed out. Here we were in 1957 using that old technology to create new mountains of punched cards. IBM laughed all the way to the bank. CAR-FAX was based on the idea that a card deck represented a freight train with every card (or two) being a freight car, waycar or a locomotive. The deck was created at the departing yard, run through their IBM equipment for transmission to the next yard with a separate copy going to the computer center at Ravenswood. The receiving yard upon receipt of the transmission, punched out a new train deck so that the cards (or cars) could be rearranged (switched), if necessary, and sent along to the next yard(s). Printouts were created at receiving yards for yardmasters and switching crews. It was loose-car railroading at its worst. My mutterings reached Fred Baldauf's ears and he complained to B. R. Meyers. I was hauled on the carpet and told to keep my comments to myself. Over the next few years while the CAR-FAX system existed, the enormous expense of transmission lines, IBM equipment rental (or purchase), millions and millions of punched cards and the storage required for them, a huge increase in use of printer paper plus the air-conditioning costs, all made it a very expensive (and awkward) way to move information across the railroad. On the plus side, however, there was the benefit of getting people used to electronic technology in the workplace. The horror stories of those famous card (train) decks became the stuff of legend. There were stories of the chaos that ensued when an arriving train deck was dropped on a freshly mopped floor, of card punches jamming, communication links failing, and the growing problem of what to do with all those used cards which had to be stored for auditing purposes. Soon after the system began, the flood of repetitious data transmissions totally clogged the central office computers at Ravenswood. Another useful benefit from this primitive system was an eventual understanding that centralized control was achievable and that the current yards which operated like independent little kingdoms, were not going to have things their own way much longer. Management, too, learned that the old ways are not necessarily the best ways and that control over the property depended upon improved communications.

At this time another new program was instituted—Responsibility Accounting. Now this seemed to have possibilities. On July 25th the new accounting, budget and cost control functions were introduced to us at a big staff meeting attended by all division engineers and selected assistants. The proposals promised to reduce paper barriers—by creating a different variety of paper. At the end of July the new budget procedure (accompanied by severe local rainstorms) came into being. The deadline for the 1958 budget was August 2nd; we were finished the day before. Poor overwhelmed Bob Lawton on the new Terminal Division did not even start his budgeting process until the day before it was due.

A red letter day for me came July 31, 1957—I had completed ten full years of active railroad service with the C&NW which established my right for a pension under the Railroad Retirement Act when I should retire. That was so far in the future, I hardly thought it important at the moment. Was I not going to live forever? A more important benefit obtained with the ten years of service was entitlement to **two** roundtrip foreign line trip passes per year.

The strange A. T. "Al" Peagan was relieved of duty as the Iowa Division superintendent and moved to Chicago to be an assistant to the engineer of maintenance on August 15th. Peagan's replacement the next day was Del Perrin from the Galena. On August 15th the 1958 budget was returned to the division with suggested alterations. Recasting some of the program work for additional rail anchors, ties and tie plates was a time consuming task. On July 23rd a huge industrial complex, CENTEX, was announced for the flat onion fields west of Chicago's new O'Hare Field. Served by industrial lead tracks from the Des Plaines Valley line that connected Proviso and Butler, Wisconsin. This was all Terminal Division territory now. Bob Lawton's crews were to spend a lot of engineering time in this new development. One of Bob's new rodmen at this time was Varis Perkalitis, a tall blond Lithuanian, who later became a flight officer in the USAF serving as a navigator on B-52 bombers. On August 28th word came that train #386 was all over the ground at Radnor.

By the end of August it was very obvious to me that I was bored stiff. The open files were few, there was no backlog of field work to do, the boys in the field were going along in good style, the annual budget crisis was past, and next year's budget was already neatly systematized. I felt uneasy and out of touch with the railroad. I was ready for a new challenge, but feared it meant the relocation railroad life demanded and my family did not want. In early September I got into a big program of grade crossing elimination in Wheaton. In return for installation of automatic gate protection, the railroad bargained for closure of many of the public grade crossings. Each crossing required a twenty-four hour traffic count which meant that a team of section men had to physically observe and tally all the cars, trucks, pedestrians, bicycles and horses that crossed each hour. Our survey crews ran vision surveys in all four quadrants at College Avenue, Chase, President, and Cross Streets which entailed dodging a lot of traffic. In those days we never thought a flagman was necessary to protect our surveyors especially on these slow speed city streets. On September 10th we did Hale, Main, Wheaton and West Streets in Wheaton. On the 11th the boys did the crossings in the next town east, Glen Ellyn. Back in the office I found a mound of paper to be sorted. These were all the new procedures for payroll and material handling. Just sorting it out took most of a day. The itch to change was getting stronger. I went to Spring Valley on the 19th with Vernon Giegold of the industrial development department to meet with the manager of the Stewart-Warner plant about rail service from an old mine track running near his facility. I would get to know Vern very well in coming years. The Mannheim Road bridge project came back to life briefly when I was served with a *sub poena* September 20th. There had been some accident there in 1955, but I had no memory or record of the incident. I took the summons to attorney Bob Russell and met with him and Ben Allison, the chief electrical engineer (who happened to also be the mayor of Elmhurst) to develop an exhibit for the suit. It was settled out of court on the 27th. That day John Daeschler, the senior agent from industrial development, offered me a job with them that paid in the range of $625–$650. I treated it as an off-hand remark and did not pursue the offer. On September 30th Gene F. Cermak inquired if I were interested in a job. That was more substantive coming from the head of the department. What with being bored with track maintenance on the division, reluctant to go through that disruptive cycle of relocation so necessary to advancement in engineering, a desire to see more of the railroad's operations, and the chance of earning more money, I took the offer under careful consideration. During the summer I had done a small job for Burgess Norton Company in Geneva; their chief engineer called B. R. Meyers on the 8th to compliment him for my work. Caught by surprise, Meyers said I would be a division engineer in two or three years. (The circumstances were related to me by Bill Armstrong who happened to be in Meyers' office when the call came in).

On October 7th L. H. Cather and John Edward 'Stinky" Linden of the prestigious engineering firm of DeLeuw, Cather approached me about taking a job with their firm. (Ed Linden later became president and chairman of DeLeuw, Cather International and died May 17, 1991). This offer to join the well known engineering firm opened up the prospect of a lifetime of construction work which recalled a sobering remark made by an acquaintance some years before. "Before you know it, you will be 55 years old, standing in a drafty construction shack reading blueprints while a cold wind whistles up your pants." That was not the kind of non-railroad future I wanted to experience. Meanwhile, Gene Cermak was seeking authority for a job in the industrial development department that would pay $615. By November 14th I heard that Cermak had approached Meyers for permission to offer me a job in his department and that Meyers had refused to cooperate. While president Clyde Fitzpatrick gave Cermak the okay to offer me a position, he decreed that Meyers had the right to compete. Thanks to friends here and there, I was fully aware of what was going on. Actually, my mind was pretty well made up by this time, but I kept quiet to see what would develop.

Heineman and Fitzpatrick left Chicago October 3rd in their special train for a big inspection trip. Since division engineer Bill Wilbur had to go with them over the Galena, I spent several days gathering information about work in progress on the division in case they should ask Bill. On the 4th I drove to Creston and Sterling to check vertical clearances for a special high steel load. I met the Clinton man and told him the load was too tall for our bridge #117½ near Creston. This was an old wooden frame bridge overpass consisting of a flat top and two steep approaches that was built to accommodate horses and wagons, not modern automobiles. It should have been replaced long ago, but the local county highway authorities were loathe to spend money on modern bridges because future maintenance would be on their budget and no longer the responsibility of the railroad. There were cases where the railroad would contribute to the cost of rebuilding these lightly used country road bridges when the structures neared critical maintenance status or a mishap had damaged it. On the 9th I drove to Peoria with Jack Donovan of Sumner Solitt Company, contractors, and Vern Giegold of industrial development, to meet with Randy Egbert, chief engineer, and Bruce Gifford, assistant to the president, of the Toledo, Peoria and Western about a proposed Gulf Oil facility at Hollis. A new jointly owned and operated track would serve this receiving station for barge loads of anhydrous ammonia which would be distributed throughout the corn growing areas of the Midwest by rail. I had no premonition of the big squabble that would erupt with the TP&W over train service in this supposedly joint area.

Thoroughly tired of wrestling with the function budget which would govern responsibility accounting in 1958, I again went with Vern Giegold of industrial development on October 14th to visit with the Borden Company at Elgin. They needed more track room, but the physical layout precluded installation of any new track short of heroic (and expensive) measures. Borden's old cheese plant was located on the Freeport line west of the West Elgin depot. The main line ran along a shelf cut out of a natural bluff so that a steep wall was along the west side of the track and a sharp drop existed on the east side to the street below. The ground floor of the two story Borden plant was at street level and the railroad side track was at the second floor level. Street viaduct underpasses hemmed in the plant on each end. Any new sidetrack would have required bridging a city street with new steel structures. The numbers just would not work especially since the railroad wasn't picking up any of the costs. Their business was not growing, the facility was old and inefficient, and the matter was shelved. On October 15th I went to South Fulton on a wet day to lay out a #14 turnout for a runaround track at bridge 252¼. Luckily, Ralph Perino, the track supervisor, happened to come by and give me a hand measuring the location of the frog and switch points. The next day I met with an Illinois Commerce Commission inspector at bridge #345 near Elgin despite having to cope with a flat tire on the company station wagon. On the 18th my only field survey team went off to Troy Grove to stake new tracks for a silica sand mining company.

I went back to Peoria on October 22nd with Bill Wilbur to meet with the TP&W's Randy Egbert about tracks to serve a new power plant for the Central Illinois Light Company (CILCO) at Hollis. This was getting to be a popular spot as we had just developed trackage there for Gulf Oil. TP&W's president, J. Russel Coulter, and Herb Davis, a vice president of CILCO, were also present at the celebration party held in the TP&W's private car. On the 24th I sent Fred Kruse and Jimmy Simons down there to stake the power plant trackage. The new facility was to be a gas burner so that freight revenues would be confined to inbound power plant machinery. The generating station was located in the flood plain of the Illinois River and was surrounded by a substantial levee. If natural gas were to be in short supply, the water location, theoretically, would permit use of barge coal to fire modified boilers. This scenario was unlikely, however, as CILCO was heavily invested in gas fields and transmission lines. Halloween night was an adventure for Terminal Division passenger train #518 when it struck an outhouse—unoccupied—placed on the track at Cary, Illinois. On November 18th, the C&NW spilled a freight train on the diamond crossings of the CGW in the middle of Marshalltown, Iowa. Bill Ward, the B&B supervisor at West Chicago, suddenly died November 1st; he was replaced December 9th by Gebhard Olson.

November brought another round of job cuts as car loadings plunged along with the stock market. Fred Kruse, my mainstay instrumentman, bid on an open job at Boone, Iowa, because of the high cost of living in Chicago. On November 7th the railroad got a new vice president of traffic—Edward A. Olson. Ed had been an official of Libby, McNeil and Libby, a big food processor. He did not have a railroad background which may have been a plus factor. Because of the deep drop in revenues we got the 1958 function budget back on November 12th for severe revision

My boss, Bill Wilbur, had been acting strangely all that Fall. He was out of the office at every opportunity. He spent weeks with the bridge inspectors on their annual tour of the division or riding train #3 west to monitor track quality. He made himself very hard to find leaving me to run the office and field all the phone calls. On November 26th Chief engineer Meyers called me to his office and devoted a half hour to the subject of my promising career in the engineering department. No particular sequence of events was forecasted, of course, but he promoted the bright future that was to be mine. The fly-in-the-ointment was the unspoken requirement that I would need to relocate from time to time at the whim of management to prepare me for greater responsibilities as I learned the railroad. I listened knowing full well I had already made up my mind on a different strategy to promote my railroad career and it was quite different from the traditional roadmap he was unrolling. I had decided that I could get more railroad experience by changing jobs rather than my residence. I was tired of engineering and its repetitious office routine. The jobs and personnel would change, but the narrow view of the railroad from the engineering department was not enough to satisfy my interest in the wider subject of railroading. I didn't say anything to Meyers then, but did mention it to Bill Wilbur on one of his rare visits to his office. That man then surprised me, he became all smiles, bade me to sit down, congratulated my choice, and proceeded to tell me about his life in the army and his railroad career. I had never heard him use so many words and complete sentences before in all the years I had known him. We remained good friends for the next thirty years. On December 3rd Meyers finally allowed Gene Cermak to officially address me about the job in the traffic department but allowed that he thought I would not take it. He was wrong. I told him on December 10th that I was moving on. I could not read his reaction other than a shade of disappointment. I sensed that any attempt to return to engineering was probably closed to me forever. W. H. Huffman called me on the 17th to bulletin my Galena Division job—my last bridge to engineering had been burned.

On the 19th Clyde J. Fitzpatrick signed my payroll authority for a $622.15 monthly salary. I wrote to B. R. Meyers thanking him for his many considerations over the years, *etc.* Meanwhile, Bill Wilbur was gone again—up the Freeport line with Robert C. Conley, the new Galena Division superintendent, another of the Illinois Central mafia. On December 14th William F. Armstrong was appointed engineer of buildings.

My last days on the Galena Division were devoted to clearing away files and transferring duties. On the 23rd I finished redoing the section limits—again. There were only 27 sections left for the entire division, and they would soon disappear as new mechanized work gangs were organized in a revolution of track maintenance practices. My last day was December 24th and it was a purely social half day. The transfer had come at an ideal time as there were few outstanding commitments that my replacement, Fred Kruse, would have to fulfill. With this change in departments came a major step on the railroad's hierarchy, I was no longer a contract employee. I was now on the lowest rung of the management ladder—an official whose name would be published in the *Official Railway Guide.* Based in Chicago, I would stay firmly in my mortgage-free Elmhurst home and get my "seasoning" by serving as an engineer in different departments rather than on different divisions. I was blessed with a degree of independence rarely found in the railroad industry and was not burdened with family, house and money constraints. The luxury of being able to choose my own career path was a satisfaction.

Although not officially in the department, I went with Vern Giegold and another industrial agent, John Foxen, to the east side of Geneva on the 29th to meet with a developer of property at Randall Road and Roosevelt Road. He was planning an industrial park and wanted information on construction of railroad sidings. The land which sloped upwards to the north was not too well suited to industrial development and eventually became a mixed use area with some light industrial plants like Globe Battery and a road frontage of homes. That evening my family and I left for our annual Christmas train ride. We went to Detroit on the NYC's *Wolverine* train #8 and transferred there to the NYC *Detroiter* to New York returning the next evening on the NYC's *Chicagoan*—a lightning one day visit to the Big Apple. The second phase of my railroad career was about to open.

12

A STRANGER IN THE TRAFFIC DEPARTMENT, 1958

. . . I reported to room #1018 in the Daily News Building on January 2, 1958 as the newest of the C&NW's industrial development agents. Among my new colleagues was M. B."Bob" Lithgow formerly an instrumentman on the Galena Division. Others were already known to me: Gene F. Cermak, director; John Daeschler, Vernon C. Giegold and John J. Foxen, industrial agents. The chief clerk, Donald E. Guenther, secretaries Naomi Scharenberg and Helen Johnson were new to me. My first impression of this new place was not favorable—a gray pall of cigarette smoke hung in the air. So, I was to be a non-smoker in a smoker's world. Already, I had set myself apart, but my initial reaction was to swallow the vague disappointment and give myself time to adapt. I sensed that this department would be a good place to learn more about the railroad as it carried a heavy responsibility from management to locate more revenue producing industries along our far-flung lines. The stress level here was to prove to be much higher than in engineering.

The industrial development department was a reorganization in July 1956 of a cozy little gentlemen's club of the same name run by A. O. Olson. Al Olson had inherited the function from R. E. Case, whose shadowy presence had left little trace. The department's principal job appears to have been soothing irritated shippers. Should a customer happen to walk in the door with plans for a new industry to be served by the C&NW, he may have received polite encouragement and little else. In those early days the "gentlemen" were not motivated to get into the details that a new location often requires. Al Olson's little aggregation appeared to be a quiet little backwater staffed by under performing individuals eased away from traffic and engineering. One of the gang was dapper Ralph Beddoes, once an instrumentman on the Galena Division. John Daeschler, in his twilight years, was a former traffic man accustomed to a stately and leisurely pace in life. My friend, Bob Lithgow, who joined Olson's group about 1954, enjoyed his comfortable surroundings and enthusiastically waited for anything to happen. Harvey Buchholz had been a freight sales manager in Milwaukee until the traffic department moved him to the new organization in 1956. Their passive lifestyle abruptly changed when Gene Ferdinand Cermak was brought in by the Heineman regime from the Chicago and Eastern Illinois Railroad (C&EI) to replace A. O. Olson.

At the little C&EI Gene had been their director of industrial development and freight traffic manager, a pretty grand position for a former yard clerk without a high school education. When he first started work on a railroad as a teenaged mudhop (yard clerk), he had caught the eye of a railroad officer whose ability to spot coming talent was excellent. Gene was the son of a baker who had fought with the Czech Legion in Russia during the First World War. After the collapse of Russia forces in 1918, the Legion joined the royalist White

Army against the communist Red army. Eventually the White forces were defeated and the Czech Legion escaped capture by crossing all of Siberia to the Pacific port of Vladivostok where Gene's father managed to board a ship. He ended up in St. Louis, Missouri, where he married. His son, Gene, was born January 26, 1927. Gene grew-up in the Dago Hill section of south St. Louis as a street savvy skinny kid with a crop of red hair above his sharp featured freckled face. After Gene was taken in hand by his mentor, a veneer of civilization had been applied and his social graces were polished to the point where Gene could charm his way through society and the business community with ease. His energetic approach to every activity, earned their attention and respect. A sharp dresser, he learned that appearing to listen and exhibiting a willingness to be of help enabled him to swim in any environment. In a word, he became a super salesman. I was charmed too—at first. After long years of association with Gene Cermak my initial admiration faded to understanding. I found him to be not unlike Nebraska's Platte River—a mile wide and an inch deep. He knew all the right buttons to push in making the sale, but he was never around when problems arose. Golf was his mania (he shot in the high 70s). This addiction developed into a very real impediment to the conduct of the railroad's business. In 1957 Gene F. Cermak was just what the C&NW needed—a super salesman. It proved to be an interesting journey with him as he rose like a fireworks rocket only to quickly peter out.

Cermak's concept of aggressive industrial development was to provide a one-stop service for new railroad customers. ID agents were trained to oversee and arrange all the elementary tasks involved in getting a new industrial customer established and start producing new revenue for the railroad. Build a siding? Get engineering involved. How about a track agreement? Set it up with the Law Department. How to establish rates for products? Introduce them to the rate officers. Need a lease or purchase of railroad property? Call on the real estate department. How to get switching service? Bring in the operating department. Under Cermak's program every question from the client was to be handled by the industrial agent. Because every plant location involved different physical requirements and a cast of personalities, the industrial agent had to appear to know exactly which buttons on the railroad to push for solution of the client's problems, to make the process seem routine, and to do it so promptly that the customer felt like they were receiving personal attention. The successful industrial agent learned to closely monitor client problems through railroad's departments and to make darned sure that the results were responsive and accurate. In a word, the industrial agent had to turn himself inside out to properly serve the client. Behind the scenes when interacting with railroad departments, another strategy was often necessary. The industrial development agent quickly learned that getting good cooperation from busy railroad departments was often like pushing on a string. Ranting, raving or scolding could backfire. Passive contacts were also prone to failure in securing answers to problems. The successful agent learned to understand the business of each element of the railroad and become a friend to all. You could catch more flies with honey than with vinegar, as the saying goes. Over the next ten years I was to see the above scenario acted out many times. Each industrial agent had his own method of dealing with the public and with the railroad. The successful agents never tried to use muscle on other railroad departments. Trying to force a harassed division engineer to toss his work schedules aside to build a new industry track usually failed and left a bitter taste in everyone's mouth. Force the operating department to establish switching services that interfered with normal operations and the next time the agent really needed a favor, he could whistle for it in vain.

The culture Cermak brought to the industrial development program was a revolution to the railroad with its militaristic structure. Cermak's men slowly learned that cooperation was gained by building relationships. Successful plant locations like the Inland Container plant at Northlake or expansion of an Iowa grain elevator were quickly brought to Heineman's attention who responded by lavishing praise on the new department. Over time as the industrial development department began to show results, its prestige rose in the hierarchy and other railroad officers began to appreciate the value of the effort, easing sometimes prickly relationships. That is not to say frictions totally disappeared; there are politics in everything.

My education as an industrial development agent began by sending me out for a couple of months to learn the ropes by tagging along with other agents. They all had assigned specific territories to eliminate turf battles over prospects. In 1958 there were two departmental offices, the main one in Chicago and a satellite in Milwaukee, Wisconsin where Harvey B. Buchholz was the manager. My apprenticeship began January 6th when I rode a sleeper on train #405 to St. Paul, Minnesota. Bob Lithgow was the guide on my first trip. The next day we met Ray J. Steiner, assistant passenger traffic manager, who took us on a tour of Bayport, Minnesota, and the old Omaha Road shops at Hudson, Wisconsin. I learned that lunch was always a big event when on the road. Today's was with the vice presidents of the St. Croix Transfer Company. Dinner could be an even bigger event depending on whom one dined with. This evening it was at the St. Paul Athletic Club where we were staying. After the Spartan life we had led on the division engineering staff, the Lucullan feast was a culture shock. Expense account took on new meaning. On the 8th Bob and I went with Guy K. Rossman, general agent at St. Paul, for a tour of the Westminster yard and the Hazel Park area east of the city. After another great lunch at the Minneapolis Athletic Club, we called on Northern States Power Company. I was beginning to see that one of the principal means of being a success in this business was development and maintenance of contacts. Later that afternoon we ran into Frank Tribbey, director of the new piggyback department. On the 9th Bob Lithgow and I drove south to Shakopee, Minnesota, via Nicols to look at a new 2200 acre industrial park planned for a gentle valley along the Minnesota River. Then it was back to Chicago on train #406. This trip was instructive in one critical regard; I needed a new wardrobe as I was no longer the grubby engineer just back from a field survey out on the line. Elsewhere on the railroad at this time a series of disturbing accidents occurred, the worst happening on January 13th when the eastbound Williams Bay commuter train running in a dense fog slammed into the rear of stopped passenger train #514 at Cary, Illinois. A woman pedestrian had been struck by #514 at a grade crossing causing the train to stop while the police and coroner were summoned. The rear Pullman, which had been added at Rochester, Minnesota, was badly crushed with several passengers injured.

On January 13th Bob Lithgow and I were on train #405 again headed for St. Paul. The next day we took CMO train #203 south to Mankato, Minnesota. After lunch with Bob Drengler, the local trainmaster (whom I was to know better in Chicago years later), we drove to St. James and Fairmont, Minnesota, for a Chamber of Commerce dinner. A snowstorm on the 15th changed our travel plans, so we stayed in Fairmont and filled out Cermak's new-fangled community work sheets. Gene wanted a completed set for every town on the railroad. We managed to drive back to Mankato in time to get train #204 back to St. Paul. The community profiles were a great idea, but a big problem. We were creating a monster that could tie up an army of clerks and data collectors just to keep the files current. As we got used to doing these profiles, we soon learned that the best way to utilize this tool was not to bury old data in a file. It was a lot more efficient to just drop everything and make a mad dash to a targeted community to collect fresh data when a hot prospect showed interest in a specific location. We were late going north on train #204 as they had dropped a brake beam en route from Omaha. On the 16th Bob Lithgow and Guy Rossman with me still just tagging along called on Rahr Malting to run down news that they were planning a new 1½ million bushel barley storage facility at Shakopee. Later, we visited the State of Minnesota's business development director. Bob was invited to dine with Governor Orville Freeman; I had to fend for myself. Late that night we returned to Chicago on train #406.

January 21st I was alone on train #401 en route to Wyeville, Wisconsin, and Rochester, Minnesota. Bob Lithgow had established a comfortable arrangement for covering his Minnesota territory. He lived on a farm east of Barrington, Illinois. On Sunday evening he would take the sleeper on #405 to St. Paul, spend Monday through Thursday in St. Paul-Minneapolis, ride the sleeper on #406 to Chicago, spend Friday in the office and then have the weekend at home with wife Mary. The more logical plan of having a resident agent in the Twin Cities would not happen during Bob's tenure. On this trip I met Bob and Walter J. Hitzman, district traffic representative, in Rochester to make a call on a prospective customer interested in setting up a propane distribution business at Lewiston, Minnesota. We also made a courtesy call on a concrete block manufacturing plant to see if the railroad could use their private siding for a paving contractor who needed a spot to receive cars of cement and gravel for a nearby highway job. I can still recall the agony of that trip—riding in Walt Hitzman's car with him driving was an unforgettable experience.

Those Minnesota highways ran straight up and down the hilly southern part of the state. Walt would hit the gas going upgrade and brake on the descents so that Bob and I first rocked backwards uphill and nodded forward on the downside as the momentum changed. Poor old Walt was totally unaware of how uncomfortable his passengers were. On the 23rd we visited Dodge Center and Austin, Minnesota, where Bob and I developed a track layout for a new lumber yard at Worthington, Minnesota. Now, that was something I knew how to do. We caught train #204 north at Mankato. (This train's last run was to be October 24, 1959). On the train we found Miles Koenig, the division engineer, who had dinner with us at the Athletic Club later that evening before Bob and I went back to Chicago on #406. We had a good shaking at Altoona, Wisconsin, where the switchmen seemed to enjoy slamming the sleepers together.

On January 29th Ray Steiner conducted a special business car trip for off-line sales representatives from Chicago to Clinton, Iowa. I was invited to be a tour guide since our car was attached to train #3 westbound over the Galena division, my old home territory. The group also got a tour of the new car shops in Clinton before we returned on #4 late in the afternoon.

I rode train #149 to Milwaukee on January 31st. This trip was taken at the request of Harvey B. Buchholz, our Milwaukee manager. I was tapped to make site drawings of potential industrial properties south of the city around Oak Creek. Another Cermak innovation, the site drawings were letter size yellow sheets that described the principal facts about a property on the front side and a drawing (more or less to scale) on the reverse showing railroad lines, utilities, highways, drainage and other important features. I eventually became the sole draftsman for all our agents of these valuable and useful documents which were a lot easier to maintain than those short shelf-life community profiles. Harvey was rather new to this industrial development program. As mentioned above, Harvey had been the C&NW's division freight and passenger agent in Milwaukee for many years and was assigned to Cermak to serve out the few years remaining until his retirement. Since he was so deeply embedded in Milwaukee, Cermak just let him stay there. Harvey was as placid a man as I had ever seen. He was also a stamp collector, and we got along well from our first meeting.

Winter's snow was on the ground from New Year's Day to the end of February. On February 4th I toured Boone County, Illinois, with Arthur Fitzgerald, a consultant from Glenview, Illinois, who had been retained by the county to establish land uses for a zoning ordinance. Little did I realize that Boone County and its zoning would become very important to us in 1963. The next evening I was in the sleeper on train #209 headed north to Marinette, Wisconsin. The ride was very rough. I was coming down with influenza or something else, that made me feel pretty rotten. Despite my acute physical discomfort I managed to accompany Tom Bailey, the railroad's local sales representative, to a joint meeting with city dignitaries of the twin cities of Marinette and Menoninee, toured their industrial areas, collected information about their cities and escaped to Chicago with my dripping nose on train #162. (Tom Bailey was killed near the end of February when he drove into a Milwaukee Road switch engine in Marinette.)

I had recovered enough by the 11th to catch #149 to Milwaukee and accompany Harvey Buchholz on a trip to Oshkosh and Fond du Lac, Wisconsin. We attended the traffic club dinner in Neenah-Menasha and glad handed the shipper community most of whom were quite relaxed by liberal oiling with free booze. This kind of meeting was as alien an affair as I had ever experienced. I had to get used to it, however, as this was the social environment in which the railroad's traffic department operated. While being an interested observer of the scene, I was really never an eager participant. Not a teetotaler, a little alcohol went a long way with me as I was very uncomfortable with any loss of control. My limit was two mixed drinks in an evening. I learned to mingle and be affable and avoided being labeled a "stick". Adding ice cubes to a tall flavored-water drink made it last a long time. Over time I made a lot of valued friendships not based on alcohol. The following day we went to Appleton, Wisconsin, and met with traffic managers at several paper mills before going on to Green Bay. There we met Ray J. MacCarthy, district freight and passenger agent (DF&PA), who had recently come from Minneapolis, and Oscar T. Schwein, a traveling agent. On the 13th I ran into an old friend from the diesel shop construction days in 1948, Les Deno, now division engineer. After a quick tour of the Ashwaubenon area near Green Bay, we caught train #216 south to Milwaukee and Chicago.

On a bitter cold February 17th, Gene Cermak, John Foxen and I rode train #401 to Milwaukee where Harvey Buchholz joined us. Gene and Harvey got off at LaCrosse, Wisconsin, while John and I continued on to Rapid City, South Dakota. It was a long ride. The next morning we woke up in the Pullman in Midland, South Dakota. The train was very late so the breakfast stop at Philip was at 10 a.m. We hiked up the main street from the depot to the cafe—a throw back to the days before dining cars. Rapid City was warm compared to the country to the east as the chinook wind was blowing. We did our community data collection duty, met with local dignitaries and looked at their industrial sites. The next day we drove north to Belle Fourche and out to the end of Bentonite Spur constructed in 1949. John and I had never been out here before. I had a hunch that someday someone was going to be interested in extending that line further northwest into Montana and convinced John that we should drive a few miles along the highway to Alzada just across the Montana line to take a look at the intervening territory. The next day we inspected the engine house in Rapid City before boarding the sleeper on train #519 eastbound. The ride was very rough on that light rail. This line had been constructed by the C&NW under the title of the Pierre, Rapid City and Northwestern Railway in 1906–07. Some locals called the line the "Pretty Rough Country and No Water" as that succinctly described the South Dakota country west of the Missouri River. When we arrived at Wyeville, Wisconsin, John and I switched to train #400 which got us into Chicago an hour sooner.

The western part of the C&NW held a fascination for me that was to last to this very day. I began to research the history and development of the bentonite deposits northwest of Belle Fourche. This useful clay was scattered in deposits along the Little Missouri River from South Dakota into Wyoming. At Harvey's request I also researched barite deposits in southwestern Wisconsin, but that never came to anything. That part of railroad west of Madison, Wisconsin, and the branch south toward Galena was slowly sinking into oblivion; there just was not enough business to justify its continued existence.

My colleagues, who were mostly promoted clerks and traffic men now magically transformed into industrial development agents, seemed to thrive on thinking up nutty ideas which faded as soon as the bright light of practicality hit them. The dying CA&E electric line's Aurora Branch attracted John Daeschler's attention. He asked me to make a study of its potential for industrial development and went with me on a tour in late February. I was very familiar with the area and knew the County of DuPage's planning and zoning projections were in favor of residential housing not heavy industry. I prepared a report in March 1958 which put that matter to rest. John and I also made a trip to the Calumet Industrial District on the far south side of Chicago. The ride on the Illinois Central suburban train to 95th Street was the most interesting part of this jaunt. We went to visit a man named Addison Brown who had invented a new type of container on a flatcar. The car looked like a boxcar that was all roll-up doors. Brown was enthusiastic about the car and its prospects; we looked at the concept and decided that this guy would never be a good prospect for construction of a manufacturing facility on the C&NW. John and I spent the 26th of February driving around the triangle of Harvard, Illinois, Genoa City, Wisconsin and Zion, Illinois, looking for potential sites for a proposed new retail lumberyard. John's prospect rejected everything we had turned up. Then they changed their area of interest so we had to go out again and look for land in the Des Plaines, Evanston, Skokie areas in early March. That prospect never established a new lumber retail outlet to my knowledge. John Daeschler, in the twilight of his long railroad career, was the old time traffic man put out to pasture until his time for retirement would come. An easy going, likable man, he had a wide circle of friends with whom he was comfortable. Cermak had inherited him along with his ever present cigar knowing full well that John was not the kind of hot shot industrial agent he needed. John was our guy, however, and we tolerated his slow and easy pace. The Chicago area was his assigned territory, but he never really got involved in the new style of aggressive representation that Cermak espoused.

March 10th I was on the Rock Island's train #9 to Des Moines, Iowa, with Vern Giegold, who was to be my instructor on this trip. After a tour around Des Moines, we headed north to Algona, Ames, Boone, Fort Dodge and Mason City. Vern had a lead foot so that we hit 105 mph at one point—out in the country, of course. Iowa had no speed limits then but the little towns along the way derived part of their municipal incomes from the 25 mph speed zones that appeared without warning at the town limits. In Mason City I made pictures of the electric juice jacks of the Mason City and Clear Lake Railroad (#52 and #107). Vern drove back to Des Moines at his usual frantic pace which was completely unnecessary as we had a long wait for eastbound Rock Island train #2 which rolled into Chicago at 5 a.m. the next day. Vern also had responsibility for that part of Illinois outside of Chicago and insisted on escorting me around the territory I already knew so well. Geneva, Sycamore and Rochelle were his chosen spots; I showed him a thing or two that he didn't know about these places. On March 20th I went to Milwaukee with Vern and John Foxen on train #149 for a luncheon sponsored by the industrial development council there. Ben W. Heineman was the featured speaker with Mr. Fitzpatrick also attending the affair. We all came back to Chicago on the *400*.

The Litchfield and Madison Railway became part of the C&NW's Galena Division January 2, 1958. Vern and I decided we had better go and see what industrial potential the property had. On March 24th we were the only two passengers in the parlor car of southbound Wabash train #11. Jim Simons and Ray Snyder from the Galena engineering office were on board too—back in coach. The next day we met Edward E. Harney, our city agent from St. Louis, to make calls on LaClede Steel, Gaylord Container Company and the chamber of commerce. That afternoon we toured the Madison, Illinois, area which seemed to be mostly railroad tracks going in all directions. We met Mr. Moudry, the terminal manager of the ex-Litchfield and Madison Railway and then headed north to Edwardsville, the former headquarters, looking for potential industrial sites. We did not find very much property that would be attractive to manufacturing plants. Yes, there was plenty of land; but there was no population base and few communities with public utilities. Vern had never seen the Southern Illinois so I drove him north from Edwardsville through Stallings, DeCamp, Benld, Lick, Springfield, Luther and South Pekin to Peoria, Just like old home week to me, but Vern was bored out of his mind. Pressing my small advantage I introduced Vern to J. Russel Coulter, president of the TP&W and also let him know I was now in a different role with the C&NW. Around Mapleton and Hollis southwest of Peoria the C&NW and TP&W had agreed in 1957 to creation of a joint service area. At that time the idea of two railroads jointly serving an industrial area was a new concept. The agreement was based on a personal relationship of Coulter and Fitzpatrick, the two railroad presidents. As yet, there had not been a lot of demand for this joint railroad service as outlined in the document, but, when it was needed in later years, a bitter legal feud ensued. Vern and I caught the afternoon *Rocket* home. The news of the day was a coming Interstate Commerce Commission hearing concerning the Peoria and Pekin Union Railway's attempt to crash the C&NW-TP&W's joint service area. The P&PU claimed an old abandoned right-of-way that would provide access to the area, but the commission ruled on September 2nd that the area was to be solely a CNW-TPW affair.

My apprenticeship moved up a notch. On April 2nd I was assigned to work with Vern Giegold in Illinois, Iowa and Omaha, Nebraska, but, I was also given the rest of Nebraska and Wyoming as my very own territory to administer and promote. The last part was fine, but I sensed that working for Vern was going to be a bad arrangement for both of us. On April 9th I rode Rock Island's train #9 to Des Moines, Iowa. The next day I met Arthur E. Knight, our local sales agent, to make a call on Bob Sargent of Sargent Grain Company. Bob was interested in a piece of railroad land at Hull Avenue in northeast Des Moines for a new grain handling facility. Later that day Art and I saw a Mr. Holman of the Hiland Potato Chip Company who also wanted a piece of that Hull Avenue property for a new chip manufacturing plant. It was a very busy and, potentially, productive morning. Art had things to do so I hopped on the eastbound *Rocket* to Chicago. Slowly, I was beginning to learn who wielded the levers of power in the traffic department. One man I learned about quite early was Iver S. Olsen, freight traffic manager (sales), who was the smartest player in the department. He and Cermak squared off and had many pushing and shoving matches to see who had the vice president's ear. Iver was a good and sensible person, but one not to be crossed.

With advent of Spring construction problems dormant during the cold weather came to life and clamored for attention. Our department developed a reputation for getting projects unstuck in addition to serving the immediate needs of new clients. Bob Lithgow stepped in to get Material Service Company's new trackage at their Proviso concrete mixing plant built in a hurry because there were 8,000 cars of gravel aggregate due to come as soon as the frost came out of the ground. The rock was slated for paving on the new Congress Street expressway (later named the Eisenhower). Since the Terminal Division was responsible for building the track, Bob camped on them until the job was done. Construction of the long lead track at Western Electric's new facility in West Chicago finally got underway on April 21st with a little nudge from our department. On the 22nd I met Donald L. Gunvalson, the district freight and passenger agent at Rockford, Illinois. Known as "Gunnie" or "Sarge", Don was an alert and energetic salesman with good instincts about customer service. A veteran of jungle fighting on Bougainville during World War II, Don was to become an important member of our team in a big project in 1963. On April 30th I got a call from Edward "Stinky" Linden at DeLeuw, Cather. He had a railroad engineer's job open for the Chicago Consolidated Terminal Study his firm had landed. Flattered to be contacted again by this prestigious company, I said that I had just started a new career and thanks but no thanks. On May 12th the bentonite mining prospect came up again; I got my papers out for review. My engineering background, while alien to the traffic department, was becoming useful in many areas when dealing with clients. Many of our field representatives began to send people to me for special projects and began to ask more technical questions. Perhaps, I could fit into this strange world.

On May 12th Vern and I went to St. Louis again on that comfortable old-fashioned Wabash train #11. A railroad tour sponsored by the St. Louis Chamber of Commerce was the next day's event. A big crowd rode a passenger special that included several gondola cars outfitted with benches. We boarded the odd collection of equipment in the St. Louis Union Station, crossed one of the river bridges and rambled all over the east side visiting industrial sites and railroad yards on the Terminal Railroad of St. Louis (TRRA) and the Alton and Southern (AS). The weather was fine and the view of the East St. Louis was, well, interesting. I did catch a picture of an Illinois Terminal center-entrance electric car #473 en route to Granite City. On the 14th Vern and I headed north in a rental car to a meeting with Randy Egbert and Roger Fischer of the TP&W regarding a proposed grain elevator in our joint industrial area. Dropping the rental car, we rode the afternoon *Rocket* home. But on a Sunday in late May we were on a train again—Rock Island's train #9 to Des Moines for what was to me a new experience—a golf outing. I had never played golf, was not interested in golf, would never become involved with the game nor care to join the clannish social climate surrounding it. In May 1958 I had not yet made any judgement about golf, but this outing made it clear that it was not my kind of game. Our party on train #9 included Cermak and Giegold from our department, Raymond E. Degnan from the newly organized foreign freight department (in September 1956), and myself. It looked like most of Des Moines and central Iowa's traffic people were at the Hyperion Club north of the city at Camp Dodge on this very hot day. Since I was neither skilled nor equipped for golf, I just walked along with Cermak's foursome and observed the rituals of the game such as the freshly lit cigarette casually tossed on the green while the smoker lined up his shot, the coin placed as a marker, the "practice" swings before the actual game stroke, the questions of "whose away?" I was very happy to be borrowed by Don Fromknecht, the Des Moines district freight and passenger agent, to visit with a shipper about his proposed new elevator construction. The evening was another event involving plenty of booze before the evening banquet about which I only remember being very ill at ease because I again was not properly dressed (I had no idea beforehand that one carried an evening ensemble on these outings) and it was a very hot evening. Missing golf outings was to become an art form on my part. The next day Don Fromknecht and his traveling agent, Vernon D. Buckman took Giegold and me west to Jefferson, Ralston and Carroll along the Iowa Division main line, making calls on shippers and showing industrial properties to us. These Iowa towns all looked alike to me. Even now, I would be hard pressed to tell one from another. May 28th was a continuation of the tour to Denison, Logan and Missouri Valley. After another night of eating too much and being very hot the next day, we crossed the Missouri River to Blair, Nebraska, before returning to the Iowa side near Loveland and catching our train #14 eastbound at Missouri Valley. That Iowa division main line train even made a lunch stop at Boone. Vern Giegold proved to be a terrible gin rummy player, but as he never carried a book to read, we passed the time playing cards. It was a long train ride to Chicago.

In the summer of 1958 our family took an automobile vacation trip to the West. On June 24th in Casper, Wyoming, I combined some business with pleasure by calling on our DF&PA John D. Boyles who had recently been assigned to this on-line post from the off-line office in Kansas City, Missouri. This was the first of many trips to Casper and Wyoming which was to become a very important region in the latter stages of my railroad life. At lunch in Lusk the next day, we had no premonition of the importance to my work that this town would assume in later years. On the way home we stayed overnight at Valentine, Nebraska, and saw train #14 make its station stop. It's last run was to be July 6th.

After a cool and dry June, July came on wet and hot. Bill Armstrong and I made a tour together to Crystal Lake, McHenry, Genoa City, Wisconsin, Hebron, Woodstock and Carpentersville—just like old times. On Bastille Day I rode the CB&Q's *Nebraska Zephyr* to Omaha, Nebraska, through threatening weather. Tornadoes were reported all over the area amidst news that the revolt in Iraq had resulted in their king's death. After a courtesy call at the Omaha sales office, I rented a car and drove west into my own Nebraska territory visiting Wahoo and Fremont. Wahoo was the boyhood home of both Darryl F. Zanuck, movie director, and Howard Hanson, an important American musician. The next day I drove west to Norfolk, Nebraska, (youthful home of Johnny Carson, a TV late night host in the 1980s) and called on the division office there. Holman F. "Shorty" Braden was the assistant engineer. Shorty was a stubby little guy with a cross-grained disposition. He had worked in the South Pekin, Illinois, engineering office when it was an independent division. A pretty good photographer, his pictures of the Southern Illinois' steam locomotives at work in Illinois were later published in the *Locomotive Quarterly*. Holman was to be both a friend and a pain in the butt. I drove north to Wayne, Nebraska, and collected information for one of Cermak's community surveys despite misgivings about the value of time sensitive material residing in a file. That was the last one of those things I did. After an overnight in Sioux City, Iowa, I called on Wayne A. Anderson, the DF&PA, to let him know I was in town as these local guys were often very upset if we should come to their territory unannounced. A quick tour of the industrial lands at South Sioux City, Nebraska, and the lands around Sergeant Bluff, Iowa, and I was off south through Onawa (widest main street in the world), Iowa, to Blair, Nebraska, and into Omaha where I caught the Rock Island eastbound at 11:40 p.m. The train was very late because of slow orders—soft track. The Pullman *Golden Banner* was a smooth ride to Chicago. On July 18th a Mr. Walker came into the office to inquire about company lands along the west bank of the Missouri River at Blair, Nebraska. There was quite a bit of property along the river left over from the days of John I. Blair's Sioux City and Pacific Railroad and its ferry operation across the river at that point. I arranged a meeting with Kendall Cady, our newly appointed vice president of real estate, but the first development at Blair was to be one of my own contacts later in the year. I believe it was Mr. Cady who changed the name of the land department to the more modern one of real estate.

July 30, 1958 saw the opening gun in the Kroger campaign. A representative from the Cincinnati, Ohio, headquarters came to Chicago to inspect four potential sites for a new grocery distribution warehouse. Kroger was going to invade the Chicago market. By September 5th they were down to two sites—both served by the C&NW. It is hard to recall how slow communications could be in the 1950s. The long distance telephone was the quickest method, Western Union telegraph was still in use, there was no facsimile (FAX), no express package service or priority mail offered by the U. S. Post Office, FED EX or UPS were not invented yet. Personal computers, internet, e-mail were science fiction or something from a Dick Tracy comic strip. The face-to-face meeting was the accepted method for communicating. Even the biggest project seemed to go through a sort of languid theatrical performance with the principals appearing now and again on the stage at irregular intervals, say their parts and then exit left or right. Actions occurred in bursts of energy followed by calm periods which were ideal for plotting strategy, getting in more golf or, sometimes, actually developing information for the client. So, it was well into October before the basic question was asked by Kroger, "How much was the land?" By using estimated costs for some creative grading to bring the site up to a suitable elevation for building, Cermak and I developed a base price for the land. We went to Cincinnati on the NYC train for an all day meeting with Kroger's engineers and real estate manager. Ken Cady, our vice president of real estate, had surrounded the land sale with some onerous conditions, but we did the best we could to land this big revenue producer. The westbound NYC train at 3:35 p.m. was an hour late into Chicago. On October 30th Cermak and I attended the Chicago Association of Commerce and Industry's annual luncheon meeting at the Sherman Hotel to hear our boss, Ben W. Heineman make the main address using Cermak as a ghost writer. After greeting Mr. Fitzpatrick, Gene and I ducked out to work on the "Kay Roger" project as we styled it—a pretty lame attempt to mask the client's identity. Despite Ken Cady's objections, Mr. Fitz okayed Cermak's proposal to Kroger on October 31st. The project sped up when Scott and Schenk of Kroger came to Chicago on November 3rd to review the proposed building sites including a third one located on the Milwaukee Road in Franklin Park. With the help of Terminal Division engineer, Jack L. Perrier, and my old contractor friend Thomas McQueen, we arranged to get fill dirt for the Kroger site for 75 cents a yard, delivered. Since McQueen had several expressway construction sites going on nearby and was looking for somewhere to get rid of extra dirt, the grading price was a rock bottom steal. A week later Howard Davis of Kroger said our site was looking better and better. On the 24th the whole thing blew-up again over real estate taxes. Yet, the project did not die; it just went to sleep until coming back to life in 1959.

And then there was the futile exercise we went through to find a new home for a waste paper reprocessing facility for Container Corporation of America (CCA). Reprocessing scrap or waste paper wasn't called recycling in those days. CCA operated a big reprocessing plant on the south side of the Ogden Slip at the end of the North Pier branch. They were using space under the Outer Drive viaduct structure for their waste paper storage area, and the City of Chicago was getting very testy about the mess. Vern Giegold and I went looking all over northern Illinois in August for a new location for this paper making operation. The key element was availability of a good supply of water. After Vern and I had wasted a lot of time driving around the area checking on places like Rochelle, Oregon, Crystal Lake, Woodstock and Elgin which did not have bounteous quantities of excess industrial water, I decided that checking the aquifer and well records of the State of Illinois' geology section might be a better approach. On August 18th I went to the University of Illinois at Champaign-Urbana, Illinois and collected appropriate materials before going on to Springfield where I met with Jim Cannon of the Industrial Development Board, Clarence Klassen, the state's sanitary engineer and Mr. Casey, chief of the Waterways Board. Two days later Vern and I took George Russell of Container Corporation on a tour to look at possible sites in Rochelle, Sycamore and Geneva. On September 5th Cermak heard rumors from his network of contacts that we had lost the plant, but Russell would not confirm that story. With some surprise we learned on the 15th that CCA had authorized an engineering study on two sites. One was at Genoa, Illinois, on property that could be served by either the Illinois Central or the CMStP&P. The other site was at Carpentersville on the Fox River near the C&NW. The Carpentersville site had problems said the CCA engineers on November 11th. I went out there with Fred Kruse from the division office two days later and tramped around the muddy property. The problem, however, wasn't physical; it was monetary. The owner was a land developer named Fred Besinger who was constructing an enormous complex of homes and commercial properties nearby. Besinger's idea of land values was much higher than CCA's. Just before Christmas we learned that CCA had put all their plans on indefinite hold. Eventually the North Pier plant was closed and its replacement was constructed in another state.

On August 13th I went to Beloit, Wisconsin, to see the owner of a falling-down grain elevator because one of our salesmen had mentioned that the man was thinking of building a new one outside of the city limits. The existing elevator was a relic and so was the owner. Cantankerous would be a pleasant adjective for him. I explained to him that a site on the C&NW at South Beloit, Illinois, would be advantageous for his business as grain rates were lower across the state line in Illinois. It was like talking to a post. His elevator business actually was at the end of its useful life. Little country elevators like his were being phased out one by one as large modern facilities developed grain storage into a wholesale business dependent upon high volumes of throughput. His old elevator would soon disappear without a trace. This was one project I wrote-off very quickly. I was learning how to determine what prospects had potential from the many wild ideas that floated into our office.

The railroad was threatened by a telegraphers' strike in 1958. Although the strike was stalled August 21st by an injunction, the telegraphers' job insecurity fears triggered by the mass closing of small stations were not allayed. Their craft was no longer needed to run the railroads. The men, their equipment and code were obsolete. They had to go, but it was 1962 before the matter was finally settled.

The temporary "shoo-fly" trackage at the old Railway Express building at Halsted Street in Chicago was first used by the commuter fleet on August 25th. Putting a big temporary bulge in the Galena Division main lines allowed excavation of the massive hole for the six lane expressway undercrossing. It seemed a shame that the express house tracks had to be maintained as temporary tracks when the express business was gone and the tracks were idle. But, planning for the construction was done when the facilities were in use and the contractor got paid for the work specified in his contract. The new undercrossing was a three level affair with the elevated C&NW mains on top, Halsted Street underneath the railroad and the expressway to O'Hare Field at the bottom. Hubbard Street running parallel to the railroad elevation crossed Halsted Street at the same grade. Local radio traffic announcers soon named this deep spot the Hubbard Tunnel. Construction went along swiftly; the new bridge for the railroad went into service October 26, 1959.

Vern and I continued to chase prospects that should have been politely ignored. On September 8th we drove to Silvis, Illinois, to meet a man named Arnold about a cattle concentration point near Morrison. We should have known better—the business of moving cattle by rail was about over. I suspect that Vern just wanted an excuse to go west to justify attending a social affair at Cedar Rapids, Iowa—dinner with the Chamber of Commerce and a golf outing the next day at the Elmcrest Club. Gene Cermak and Ray Degnan showed up for the golf. Again, I was a fifth wheel and endured the tedious awarding of golf trophies and the phony jollity. In the evening Ed Olson, our vice president of traffic, entertained local dignitaries on his official car #401 parked at the Cedar Rapids depot; we were also invited to share in the festivities. The next day, September 10th, was another golf outing in Clinton, Iowa, followed by a boozy dinner and evening. At Clinton, I "disappeared" after nine holes, borrowed a car and went on a tour to look at industrial properties, later reappearing for the dinner and the subsequent fun and games. I was having difficulty fitting into this aspect of "business".

As summer faded into autumn, better quality prospects appeared. On September 18th a telegram came from International Paper Company inquiring about C&NW properties in Omaha, Nebraska, for a paper products warehouse. That same day the officers of the Davenport Elevator Company came into our Chicago office to discuss locating a new facility on our line at Des Moines, Iowa. Vern decided that the paper company was the more important prospect and hustled us off to Omaha on the CB&Q's *California Zephyr* on the 23rd. Finding a suitable piece of property served by the C&NW in Omaha proper was a tough proposition as this was a Union Pacific town. The CB&Q and Missouri Pacific also served the city, but the C&NW was a distant cousin there. The two C&NW lines went west and north from Omaha. The old Omaha-Irvington-Arlington line wandered west from the UP's station downtown turning northwesterly along Little Papillion Creek near 72nd Street. This railroad looked like an industrial lead track with light rail and no visible ballast showing through the mud-caked ties. To think that it once saw passenger trains, and not very long ago, was a stretch of the imagination. There was a grain elevator near Dodge Street, Omaha's main east-west artery. The other C&NW line was the old Chicago, St. Paul, Minneapolis and Omaha (CMO) line that ran straight north to the suburb of Florence and thence to Fort Calhoun and Blair, Nebraska. In 1958 this railroad saw a wayfreight now and again, a far cry from the Omaha Road's early days when passenger trains between the Twin Cities and Omaha ran frequently. Property along the line in this area was not flat because the river hills along the Missouri River had required some rather heroic track construction by the CMO's builders. At one steep grade freight trains were regularly pushed to the summit. Another old C&NW line went west from Florence to Irvington slicing through the hilly terrain. This railroad was really out in the undeveloped area north of Omaha in 1958. There was little but bare grassy hillsides to be seen near the famous Boys' Home north of the city. Realizing that we were not having much success finding a piece of land that could be utilized and serviced on the Nebraska side, Vern and I prowled around Council Bluffs, Iowa, where the C&NW had a better presence and the land was level. This was my first experience with the bewildering tangle of railroads in that east bank city. Luckily, I had gotten a print from the chief engineer's map so that the ownerships of CGW, Wabash, Rock Island, Burlington, UP, and C&NW could be puzzled-out. That same day Vern and I called on Harold Kelberg, the division freight agent, and city agents Bud Campbell and Joe Gurnon who had a prospect that wanted to locate in the Irvington area (it came to naught). On the 25th Bob Hazelwood from International Paper in New York came to see what we had found the day before. Driving around the area Jack Peters, our freight traffic manager for Omaha, was quite helpful in filling in details that Vern and I could not have known. It was obvious that we really did not have much to offer International Paper. I felt that Jack Peters was a diamond in the rough despite his acerbic manner. In his presence I tried to fade into the background and just watch him as he discussed Hazelwood's plans and requirements. Jack was a master at extracting information from the prospect to better serve the potential shipper. It was an education just to listen in on the conversation. My partner, Vern Giegold, however, seemed unable to understand the value of the colloquy happening right in front of him. Over the years, Jack Peters remained a major source of information and a trusted friend. On October 14th when I was in New York City, I called on Bob Hazelwood to ask about the Omaha project, but learned nothing new.

My Nebraska adventures, however, were just beginning. On September 29th an inquiry came in for a plant location near Blair which is about 20 miles north of Omaha on the west bank of the Missouri River. The company was National Alfalfa Company based in Sinking Spring, Pennsylvania. On first blush the inquiry did not seem too important to the office staff; Cermak pooh-poohed it as inconsequential. But, it was my prospect in my territory, and I got onto it at once. The alfalfa company wanted to build a processing plant on the west bank of the river close enough to be served by river barge in the event the railroad's service failed or rail freight rates became unattractive.

This site was an interesting piece of the railroad's history. Here predecessor Sioux City and Pacific Railroad crossed the Missouri by ferry before the construction of the steel bridge in 1888. Opened to Fremont, Nebraska, in February 1869, the SC&P had acquired a vast amount of riverside land to support their ferry operation because long, looping tracks were required to reach the river landings for the

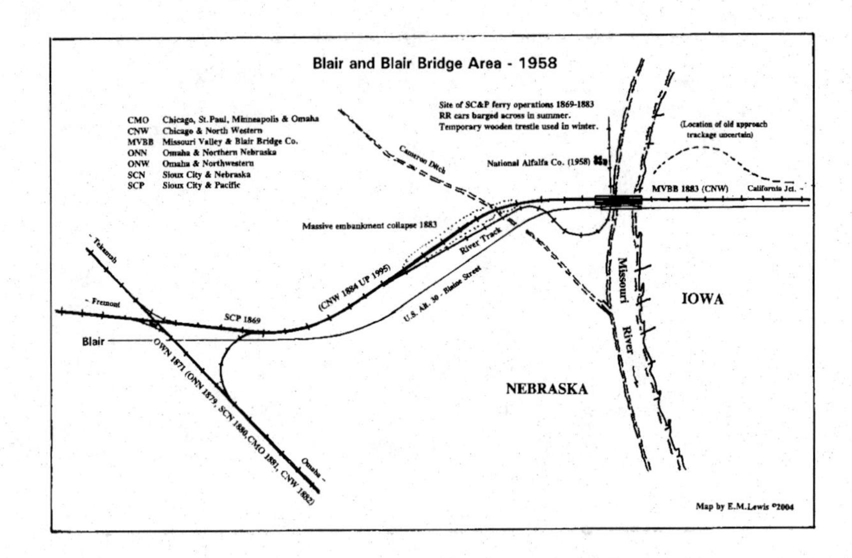

railroad ferry operated in the summers until November of 1883 when the present bridge was opened. The SC&P came under C&NW control in 1884 and was formally merged in 1901. Most of the original land ownership was still held by the C&NW in 1958 along with west side approach now called the river track. This was the site National Alfalfa wanted. The river track was a tattered remnant of the old ferry crossing railroad. Unused for many years, its light rail and questionable tie condition laid there buried in the light brown mud. Before it could be used the entire track would have to be upgraded to handle 100 ton hoppers of alfalfa concentrate. The river track came off the main line about half way between town and the river bridge, crossed highway U. S. Alternate 30 at grade on a long diagonal, dropped down the grade between vertical clay banks, curved 90 degrees around to the north, passed under the overhead highway and railroad bridges to the selected plant site. With its tight curves it looked something like a model railroad design.

By October 6th I had researched soil conditions along the riverbank, discussed upgrading costs with the division engineering staff in Norfolk, Nebraska, and listened to the superintendent's moans about the problems they would have serving the facility. It is a good thing that my youthful inexperience didn't let such negative attitudes tor-

National Alfalfa and Dehydrating Company's new processing facility on the bank of the Missouri River at Blair, Nebraska, April 7, 1962. The track serving the plant was once the Missouri River ferry access track of John I. Blair's Sioux City & Pacific Railroad between 1869 and 1883.

pedo new projects. As a representative of the traffic department, I held firm on this one because my engineering experience told me that the problems could be fixed.. My research brought out much more serious difficulties with this site. The Missouri River valley was shaped in glacial ages by a wind borne clay material called loess, which has some very peculiar properties. The very fine clay particles blown off ancient glaciers do not behave like ordinary soils. A stream cut bank in loess, for example, stands as a vertical wall rather than assuming a gentle slope. After a rain, loess is incredibly sticky which is quickly learned when trying to walk across a wet field in street shoes. The railroad is essentially level from Blair eastwards, but the ground drops steadily away to river level requiring a long embankment to approach the high steel bridge over the navigable river. Grading crews began working in 1882 to build a long earthen embankment using borrow pits adjacent to the proposed track centerline of the new railroad. One morning when they came to work, the new embankment had sunk and the ground on either side was humped up in great ridges. The weight of the new earth works had overcome the resistance of the soft loess subsurface and it boiled up on either side- something like a ship in water. Persevering, the graders finally stabilized the embankments at great cost. On the 7th I rode the CB&Q's *Aksarben Zephyr* to Lincoln and had a fruitful meeting with the State's geologist, a Mr. Reed. Nebraska was the only state I knew of that had a program of systematic drilling. Reed had records of soil borings made in the vicinity of the new alfalfa plant site. After a tour of the Lincoln area with our sales representative, John McClintic, I boarded the Rock Island's eastbound *Rocky Mountain Rocket,* which was right on time. Pulling my reports together, I decided to drive east and report my findings directly to the prospect. On October 13th we arrived in Sinking Spring, Pennsylvania. That Monday morning I sent my wife on the Reading to Philadelphia while I made my presentation to the officers of National Alfalfa. The report was solemnly received and no promises were made, but three weeks later their Bill Evans came to Chicago on November 4th. We negotiated a land sale price with Ken Cady, vice president of real estate, got it okayed by president Fitzpatrick and discussed freight rates with Ed Upland, our grain manager. By the 6th the deal was all wrapped up, and I had my first success as an industrial development agent. On November 20th I introduced four of National Alfalfa's officers to our local people in Omaha. My part of the location project was finished. Holman Braden, the assistant engineer from Norfolk was ready to spit nails because of the expense of rehabbing the old river track. Two trainmasters present at the meeting wondered how they were going to rearrange their train service to accommodate this new customer. But Jack Peters, our savvy Omaha traffic man, took over the meeting and laid down the law on exactly how this new customer was to be accommodated. He was totally in control as this was his customer now. The return to Chicago on the Rock Island sleeper was clouded by rumors of detours, but we arrived on time. On December 4th the C&NW's board of directors approved the land sale.

It turned out later that Vern Giegold was insanely jealous of my success. Remembering him from the vantage of half a century, the image of a stocky little man always dressed in brown, emanating clouds of cigarette smoke comes to mind. Vern was bald, had weak eyes that required thick eyeglasses and seemed to be surrounded by an invisible barrier that did not encourage intimacy. He was a good example of a clerk promoted beyond his capabilities. The successful industrial agent had to develop a comfortable and trusting relationship with the client. The combination of listening and applying imaginative solutions was an art that Vern never mastered. His whole career was marked by missed opportunities and disappointments. As his operating patterns became apparent to me, I found it increasingly difficult to work with him or to even be around him. Besides being a bad golfer, he was a rotten gin player.

While in the East in October 1958, I called on C. A. "Chick" Miller, our Eastern Traffic Manager in his Manhattan offices. He introduced me to Union Carbide on the 15th. While that visit was just a courtesy call, Chick appreciated my spending time with him as Chicago people rarely came to his office to visit. Over the years I made a point of always calling at our New York offices just to let them know I was in town and tried not to make any demands on their time. I knew that in the future when I called to ask for some information or to have an errand done that they might be helpful.

The C&NW had a geologist on the payroll in 1958. E. Bradley Huedepohl worked for Bill Kluender in a small group in Traffic that dealt with mineral, agricultural and forest development. Brad had an imagination and an eye for mineral deposits along the C&NW's lines. On October 21st he called me about a prospective rail client at Rockford, Illinois. All excited, he described the location and requirements of the proposed mining operation. I explained to him that it would be tough and expensive to get trackage to the location in question which calmed him down on that project. On December 1st I noted that he was pushing a limestone mining project for Victor Chemical Company in Minnesota—another prospect that fizzled into nothing. Brad always seemed to be involved in extraction projects of low-rated bulk materials in awkward places for rail service. He did not stay with the railroad long as his boss, Bill Kluender, was not as imaginative enough to match Brad's enthusiams and proved to be a difficult man to work for.

On October 25, 1958, the 110th anniversary of the first train operation on the Galena and Chicago Union Railroad, the drawbridge over the Mississippi River at Clinton, Iowa, caught fire.

November saw a major change in Chicago commuter operations. Authority was granted November 14th by the Illinois Commerce Commission to raise commuter fares, abandon 22 close-in stations and institute a new kind of ticketing.

The changes went into effect December 1st, but public perception was not positive as the Wisconsin Division trains were all gummed-up because someone forgot to fuel the locomotive on one of the first early morning trains. Some of the stations closed on the old Galena Division included Austin Boulevard, Lathrop Avenue, Oak Park Avenue and Ridgeland Avenue. There were many other affected stations on the Wisconsin and Milwaukee subdivisions, but I do not recall all their names. The new passenger ticketing was a break with the past. The new flash passes eliminated the need to punch tickets. Commuters purchased monthly, or weekly, passes and displayed them for visual inspection to the collector. To facilitate display, metal clips were fastened to the seat tops and gallery car foot guards. It was very easy to walk off the train and forget your pass—and many did. Consequently, most regulars just pulled them out when the collector came by and "flashed" it at him. One lady used a long rubberband so that when she got up to detrain, the forgotten pass would snap out of the clip and fly up to her handbag. There were still single ride tickets and employee half rate tickets to punch, but these were few. Fraud followed close on the heels of the new system. The new passes were printed in different styles and colors each month to identify the correct period of validity. It wasn't long before some clever folks used razor blades to split the cardboard and try to use both sides of the pass as two tickets. Our collectors then demanded to see the front surface of passes and soon caught the cheats. Employees received passes *gratis* but were held strictly to the rule that they must forgo sitting on crowded trains. The rule was simple and blunt in definition. In the most extreme case an employee seated alone in an empty car had to give up his seat if a paying passenger came up to him and demanded it. Free transportation for new employees ended in 1966; the current pass holders continued to have their passes until they left the company or retired. So-called "red-X" passes were issued by departments for staff work related travel.

On November 19th the industrial development department

along with the rest of the traffic department moved to new offices in room #1303 of the Daily News Building, and we were authorized to hire two more industrial agents.

Back on the Galena Division, Fred Kruse, my replacement as assistant engineer, abruptly resigned and left on November 21st. Years later, I was to see him again in the Chadron, Nebraska, office.

December came but our activity level did not tail-off. On the 10th Vern and I met with Hugo Hakala, mayor of DeKalb, Illinois, and his friend, Donn Henn. They had heard that Swift and Company was possibly interested in a new plant in DeKalb. Hugo was a cagey Finn who loved to play golf and party. He was to be helpful on several occasions. Vern and I researched potential sites in DeKalb and interviewed a Swift vice president at their plant in the Chicago Union stockyards attempting to get some information on their plans. We were not encouraged nor discouraged, but the plant ended up in Rochelle, Illinois—on the Burlington-Milwaukee joint trackage south of town in September 1959.

December also brought rumors of construction of major missile sites in South Dakota, Nebraska and Wyoming. What rail traffic there might be was not clear, but we pursued all rumors including careful watching construction project announcements in the *Engineering News-Record,* a publication I faithfully read. Many leads from that weekly magazine were fed to our various sales offices for them to follow-up and see if there was any business to be had. While some successes were recorded, the main benefit of these tips was broadening of contacts and experiences by our sales agents, who found new traffic in unexpected places.

At a big staff meeting December 15th for the Traffic Department, Ben Heineman and Gene Cermak spoke about the new emphasis on increasing traffic and revenues. On the 17th I rode the Rock Island's night train to Des Moines (Pullman *Granger*). The next day I drove north to Ames to visit the Iowa State Highway Department and call on the Chamber of Commerce. With Art Knight, our sales agent from the Des Moines office, we made calls on the Iowa Wool Growers Association, Institutional Grocers and Firestone Tire Company. We finished early, and I rode the parlor car on an eastbound *Rocket.* The year ended with a new prospect, General Foods, looming. On December 23rd Gene Cermak and I assembled a packet of information highlighting the C&NW's marvelous freight service and desirable building sites. Propaganda? Of course.

Assessing this first year in a new environment, I was more aware about a number of issues. Number One: I would never play golf. Number Two: tobacco smoking was an abomination. Number Three: I was not alcohol prone. Number Four: I was now associated with a new sort of people—a mix that included a lot of incompetents in positions of power and some very good intelligent people who had important things to say. Number Five: being a good listener was essential to success. Number Six: I would never be a salesman. And, finally, Number Seven: I sensed that my strongest abilities tended toward research, organization and presentation. Despite my initial worry, my engineering training had not faded; I could have returned to engineering at this point without missing a beat. On the plus side, the world of traffic was much wider than the narrower field of engineering at the division level. Despite a premonition that traffic people viewed my technical education with suspicion, opportunities there seemed much more attractive—for now. I was to stay nine more years.

13

AGGRESSIVE INDUSTRIAL DEVELOPMENT, 1959 . . .

The languishing Kroger warehouse project at Northlake came back to life on January 12, 1959 when their engineers furnished a site drawing showing where soil test borings were required. I borrowed rodman Varis Perkalitis from the Terminal Division on the 15th to help me establish a base line for Raymond Soil Testing who finished their field work by the 19th. Next door to the Kroger site, General Foods was interested in establishing a major distribution warehouse project, but they seemed rather slow about making decisions. While these projects were not in my assigned territory, I was often dragged in as a consultant since John Daeschler was hopeless when it came to technical matters and his assistant, Johnny Foxen, was equally inept. Consequently, these new warehouse projects were overseen directly by the head of the department, Gene Cermak. He happened to be ill on March 18th and I was drafted into attending a meeting in assistant vice president, traffic Bob Stubbs' office with Mr. Shearen, president of General Foods, who was just making a social call.

The Kroger site was still unsettled when they decided on April 17th that they wanted to develop another site. A month later, on the 12th, Kroger had reached the point where they had to put up or shut up. At a joint meeting involving Kroger and a General Foods representative, a Mr. Larmon, Cermak and I walked over both properties which lay side by side along the west side of the Proviso North Yard (yard #9). The two food distributors debated the choices at length trying to decide which company would take which parcel. The following day a crew from Wight Engineering and Kaiser, architectural engineering and building contractors, were on site when Larmon of General Foods said they would take the southerly parcel. Kroger, of course, then said that was the piece they wanted. Our traffic vice president Ed Olson and his assistant Bob Stubbs agreed that General Foods had first choice and Kroger could either take the northerly site or politely leave. I went on a brief vacation at that point because I was only a casual consultant in the matter. When I returned on May 27th I found the whole matter was still a horse race with Kroger rapidly fading in the stretch. Site selection was still open on June 11th and remained so until the contract was finally signed on June 25th. The deal was all wrapped up in blue ribbons, signed, sealed and delivered July 15th. I was astonished at the vacillation and indecision surrounding this plant location project. On August 6th I heard that General Foods had let their construction contract and that the City of Northlake had now decided to run a public street through the construction site. The proposed street would also have had to cross the middle of our multi tracked railroad yard at grade. Clearly this was a political stunt to force paying someone off. Kroger finally signed their purchase contract September 3rd. A week later grading for the industrial side tracks to serve both properties was underway, but Kroger did not begin construction of their warehouse until June 22,

1960. I recorded a note that Gene Cermak and some of our high traffic officers made a pilgrimage to Cincinnati October 19, 1959 to settle some loose ends with Kroger. One of our own loose ends was the need to relocate two existing companies operating on leased land in the middle of the sold property. Canceling a lease and telling the tenant to move was one thing, but these lessees were revenue producing operations that we wished to preserve—somewhere else. One was a freight car and locomotive scrapping operation, M. S. Kaplan. We found a nice isolated spot north of Yard 2 east of Mannheim Road for Burt Kaplan's scrap yard. The other business was Patrone's Ready-Mix concrete plant which we arranged to move to a spot north of the old enginehouse along Lake Street.

Returning to January 1959: Gene Cermak and I rode the Milwaukee's *Arrow* to Sioux City, Iowa, on January 20th through a heavy snowstorm. Gene was an avid and intense gin rummy player who flamed bright red when beaten. I won and he glowed as the train of standard heavyweight cars hurtled across Illinois that evening. Perhaps, I should have been more careful about beating the boss, but we were both highly competitive individuals. In Sioux City we met with Wayne Anderson, our DF&PA, and George Wimmer, the Chamber of Commerce's industrial man. The weather plunged to minus 6. At noon Vern Giegold came to town on the noon train from Omaha. Despite the deep snow and penetrating cold we drove out to look at the industrial properties near the airport south of Sioux City. The next day I spent time with the airport manager to learn about FAA height restrictions and the airport's glide path approaches and landing patterns. Industrial usage of the level land south of the airport was somewhat impaired by these restrictions, but careful evaluation showed much valuable property could be used. After some days of work and partying, we returned to Chicago on the 22nd aboard the venerable *Arrow* which was slated to disappear in the near future. During the rest of a very cold January I worked over the maps and notes from Sioux City to establish rail and highway access to the lands south of the airport. We were to be rewarded by new plant locations of Central Soya on a former Mobil Oil Company site and a new coal burning electric generation station of Iowa Power on the east bank of the Missouri River.

February 6th was extremely cold when I caught the first eastbound commuter train at 5:45 a.m. from Elmhurst to Chicago. Quickly running some blueprints in the division's office (I still had my old office key), I made the 6:55 train westbound to Elmhurst and drove quickly to Rockford to meet Cermak and Giegold who were shepherding a group from Brunswick Calde Company around the area. Sometimes we had to go to great efforts to sell prospects on the C&NW's freight service despite the poor condition of the railroad. We lived in the hope and expectation that we could always raise our level of rail service to meet the customer's needs. Sometimes we could; yet, sometimes we failed. The Rockford area was served by a lightweight and weedy branch line which was to be radically upgraded by a major event that lay some four years in the future.

On February 9th I met with chief engineer B. R. Meyers to discuss the Des Moines industrial project at Hull Avenue. Having lost me in the recent past to the traffic department, he was rather cool and correct with me, and I can't say I blamed him. My impression of him, however, remained that he was an aloof, cold and somewhat pompous man impressed with his own importance. I was glad that I had left engineering and his jurisdiction as I was never comfortable when around him. Bernie Meyers never seemed to appreciate the efforts of his staff, but he did play a mean game of bridge, I was told. Ben Heineman surprised him at the American Railway and Engineering Association's annual luncheon on March 10th when he announced to the audience that Bernie had been elected vice president and chief engineer by the C&NW's Board of Directors. Despite his elevation, Meyers had a row with Cermak two days later over a water contract the company was making with the City of Northlake. We previously mentioned the 14″ water line that ran from Austin Boulevard in Chicago to Proviso. That clean Lake Michigan water was a powerful lure to the many cities along its route. Cermak was worried that one of his best selling points for new industrial prospects at Proviso would be piddled away by deals with local communities like Northlake.

The next day I rode train #149 to Milwaukee directly into another snowstorm. Since the man I was to meet, Ed Boerke, a local realtor, was marooned in his driveway by four foot drifts, I went back to Chicago on train #206. Returning on the 13th, Boerke, Harvey Buchholz and I drove around the snow covered landscape in the Carrollville-

Oak Creek section south of Milwaukee. Eventually, I made site drawings of some of this open land for industrial usage, but it was to be a fruitless effort. On the 16th we learned that management had determined to keep our traffic offices open on Saturdays and we all had to take turns just in case a prospect should call or information was needed by some other department. I cannot recall if even one prospect ever turned up on a Saturday. As much as we disliked this half day assignment, we could use it to advantage to catch up on the peripheral piffle that seemed to accumulate during the week.

I took a working vacation on the 18th by riding the GM&O's *Midnight Special* to St. Louis and visiting the St. Louis Museum of Transport at Barretts Station. I was surprised to find C&NW class D, 4-4-2, #1015, moldering away amid the hodgepodge of salvaged railroad equipment. I made a swing to the northeast through Litchfield and Hillsboro, before heading back to St. Louis via Benld, Edwardsville and Madison. The sleeper to Chicago on the GM&O was a comfortable ride.

Another trip to Iowa with Cermak and Giegold on the 25th was on the Rock Island. As mentioned before, Gene loved to play gin rummy on the train which helped the miles fly by. It wasn't hard to best him as he didn't seem to count cards properly as a good player should. Detraining at Des Moines, we drove north to Boone to discuss the proposed Hull Avenue development with the division engineering people. That evening we attended the annual Iowa Development Clinic where Governor Herschel C. Loveless' spoke. Dr. Klemme from Northern Natural Gas in Omaha and Phil Schmidt, the Rock Island's industrial man, made presentations the next day. That night we went back to Chicago on the Rock Island sleeper, an uncomfortable trip as the heat was on full blast and totally uncontrollable.

John Daeschler, our senior industrial agent, was hospitalized March 5th with prostate trouble. They operated on him March 9th, and he was hospitalized for 15 days. The doctors were more concerned about his heart than the prostate. Despite the prognosis, he was back to work April 17th.

On March 16th two C&NW division officers from "my" territory showed up: Al Johnson, superintendent, and Gerry Linn, division engineer, from the Black Hills Division at Chadron, Nebraska. I was pleased to have them make the effort to come to our office and spend some time with me. I promised that I would come to Chadron later in the year.

My wife and I combined a New York side trip with a business meeting to Montreal in April. The annual meeting of the American Industrial Development Council (AIDC) was scheduled for Montreal and I was chosen to accompany Gene Cermak to the affair. Of course, on the day of our departure, April 17th, the Kroger warehouse project broke wide open, but Cermak told me to scram before the top traffic brass learned we were planning to go to Montreal. Barbara and I boarded a Pullman on the New York Central's *Twentieth Century*. The ride was magnificent and was accompanied with a nice dinner and breakfast before we rolled into New York twenty minutes ahead of schedule. We spent the day rattling around New York before taking the Delaware and Hudson north to Montreal (Pullman *Montgomery County*). Still a day early for the AIDC meeting, we took the Canadian Pacific to Quebec City on a chilly April day, returning to Montreal that evening. April 20th was filled with meetings followed by a C&NW hosted cocktail hour. Afterwards, we rode up to the Mont Royal chalet for the buffet. Cermak was elected a director of the AIDC. The next day combined meetings with more sightseeing followed by a noisy dinner enlivened by French Canadian singing and dancing. The final day of the AIDC affair was April 22nd. We left Montreal at 3:30 p.m. on the Canadian National's *International Limited* in the parlor car as far as Toronto where we switched to the Pullman (*Green Brook*). This experience was certainly more entertaining than any golf outing in Des Moines, Iowa.

These were days of a few wins and many losses: the Des Moines grain elevator project collapsed, Kraft Foods optioned land at Marshfield, Wisconsin on the SOO Line, and the Kroger-General Foods situation remained in its usual inconclusive uproar. Through it all, we continued to wave the flag, so to speak. Vern and I went to Springfield, Illinois, on April 29th to attend the first Governor's Industrial Conference at the Orlando Hotel. Speakers included basketball coach Adolph Rupp from the University of Kentucky along with the ubiquitous chicken dinner. On the following day I listened to the barge operators plead for more public funding for their various river projects. This river-railroad struggle was an ancient one that continues to this day. Vern and I drove back to Chicago with E."Smitty" Smith of the Rock Island's industrial department as a passenger as far as Joliet. In early May I returned to St Louis with our chief clerk Don Guenther in tow for his first field experience as he was soon to be promoted to industrial agent. We rode the Wabash (Pullman *Blue Gazelle*) overnight to St. Louis where we rented a car for a swing south to Lutesville, Missouri, where two of my father's brothers operated a wooden pallet manufacturing plant with an attached hardwood sawmill. On May 5th Don and I went east across the Mississippi River to Edwardsville to meet Bill Armstrong, our engineer of buildings. There we went through the recently acquired Litchfield and Madison Railway buildings. Not much to see and the adjacent radiator factory, which had been owned by the former L&M, was a shambles. We put Bill on the GM&O at Springfield and spent the rest of day with Jim Cannon of the state industrial board and Bob Davis of the Springfield Chamber of Commerce. Our tour continued north to Peoria where we inspected the joint TP&W-C&NW land around Mapleton and the river barge terminal at Kingston. On the 7th Don and I met with Mark Townsend of CILCO and marveled at the 2,000 acres of farmland on the west side of the Illinois River opposite Pekin, Illinois. We found it hard to imagine why this beautiful land protected by sturdy levees had not been developed—we would find out later to our cost. That evening we attended a banquet at the Peoria armory on North Adams Street that featured Governor William G. Stratton and U. S. Senator Everett Dirksen. It was a rotten meal not helped by Bob Anderson, our city sales agent, telling us in exhaustive detail about his recent transfer to Peoria from Des Moines. Don and I went back to Chicago on the Rock Island's 7 a.m. *Rocket.*

The poor old Chicago, Aurora and Elgin Railway was to be scrapped. I helped Jim Goinz from our Chicago Sales Office plan a temporary track connection at Wheaton, Illinois, to handle the cars of rail, bridge scrap and miscellaneous track material. Ten miles of their good 10025 pound rail ended up in the C&NW's Spring Valley subdivision replacing the original 60 pound from 1885. On the evening of May 15th I boarded the AT&SF's *San Francisco Chief* with my family for a brief vacation in the Southwest. We went to Gallup, New Mexico, rented a car and drove north to Chinle and the Canyon de Chelly. Later, we crossed the desert to Tuba City and Grand

Canyon before heading south to Prescott, Jerome, Sedona, Flagstaff and so back to Gallup. We returned to Chicago on AT&SF train #2, arriving on the 24th.

The summer of 1959 seemed to be as monotonous as the doldrums in the Sargasso Sea; nothing very important seemed to happen. There was a tentative nibble from General Aniline regarding our Proviso property which developed into a non-event. I made a tour of C&NW lines between Dixon and Peoria, Illinois, on June 9th with Rudy Macen of Warren Petroleum who was looking for fertilizer distribution sites. This was the precursor of the hundreds of rural retail distribution stores which the big fertilizer and petroleum companies were to locate in Midwest farming areas over the next few years. Dropping him off in Peoria, I went on south to Springfield and was invited to a party by Mayor Lester Collins and his cronies. He asked me to meet with his Plan Commission. Why, I cannot say as the C&NW's line was far to the west of Springfield. One could just see the capitol dome from the former station of Bando (which was an acronym for the Baltimore and Ohio which crossed under the C&NW at that point). In this business, however, one never passed up an opportunity to make friends. On June 24th I attended another meeting in Springfield called by the Illinois Central Railroad. Mayor Collins gave the C&NW a nice plug despite our almost invisible presence in his city. I rode the IC's *Green Diamond* on July 9th to Springfield to attend yet another railroad planning meeting.. The city earnestly wanted to get the railroads out of its city streets—they had been there since Abraham Lincoln's day. The principal railroads in the city were the GM&O (former Alton Railroad), Illinois Central, Wabash, Chicago and Illinois Midland and the electric Illinois Terminal over on the east side of town. At this meeting I met Professor William Hay, a renowned railroad authority. After an interesting chat with Ed Grimes, the IC's local traffic man, I headed back to Chicago on the afternoon train.

Our department added a new industrial development agent to the staff, Charles M. Towle (born February 15, 1932). "Chuck" Towle was a pleasant young man whose somewhat withdrawn or guarded manner left one feeling that his attention really was elsewhere. He had a lot to learn about our business before he eventually was promoted to head-up our future Minneapolis office. Over on the Terminal Division my former colleague, Magnus C. Christensen, was appointed assistant engineer and Robert R. Lawton, another former Galena Division man, took the Sperry Car rail inspection engineer's job which Don Arntzen had vacated. That was a bold step on his part as it meant spending many days away from home supervising the operations of the self-propelled testing car. During his tenure of that job Bob got to see most of the C&NW's lines—main and secondary. While the new standard 115 pound main line rail was proving to be relatively trouble free, there was still a lot of the failure prone 112 pound jointed rail still in track not to mention the hundreds of miles of even lighter rail sections. Testing some lines turned up so many defects that there was not enough repair rail available for replacements. Since embargoing service because of rail defects was not acceptable to management, the rail testing programs were carefully scheduled to minimize service disruptions by delaying inspection of lines known to be especially prone to having defect rails until a sufficient stock of repair rail had been laid by. Bob took it all in stride in his usual careful and conscientious way.

During this dull summer period Vern Giegold finally lost the Swift plant to Rochelle. He could never get a clear set of site specifications from them. At his invitation I accompanied him to a couple of meetings with Swift, but by July 29th it was clear they were going to locate in Rochelle, where Vern had encouraged them to visit. Unfortunately for Vern, the site chosen for the new facility was located southeast of the city, served by the Burlington-Milwaukee jointly operated line and not accessible to C&NW service. The final nail was driven in Vern's hopes when they officially announced their choice September 11th. I suspect, in retrospect, that the C&NW's close relationship with Armour and Company had a lot to do with Swift's selection of the site on the Burlington.

July also saw the breakdown of negotiations in the steel industry. The strike began July 14th and dragged on for 116 days until a temporary injunction November 7th broke the impasse. Needless to relate, railroad revenues were adversely affected.

Gene Cermak and I made our first, and often postponed, trip to Wyoming that July. The C&NW's line west from Omaha ran across Nebraska's upper tier of counties through Norfolk, Valentine, Chadron and Crawford to enter Wyoming at the tiny settlement of Van Tassell. The line ran straight west to Glenrock, Douglas, Casper and Riverton with a branch to Lander, the end of track. Like many other expanding railroad empires, the C&NW once had dreams of a Pacific Coast connection. The actual west coast line had been surveyed about 1905 from Riverton, up the Wind River through Dubois (which did not then exist) across 9,600 foot Togwotee Pass to Jackson Lake where the survey turned north along the east face of the Tetons into Yellowstone Park over 8,262 foot Craig Pass and thence into Idaho along the Snake River. (The survey maps and profiles are in the Wyoming State Historical Archives because I salvaged them from the chief engineer's scrap bin and made sure they ended up in safe hands). This was Cermak's first trip to Wyoming; I had ridden Union Pacific trains through the southern part of the state and had traveled through the West before. On this journey Gene and I rode the Milwaukee-Union Pacific's #103 *City of Los Angeles,* playing gin rummy most of the way. I did wish he would read books but his personal engine was turned-up higher than mine requiring more constant action. I can't imagine him sitting quietly under a shady tree reading a book. John D. Boyles, our DF&PA, met us at Cheyenne. This was a magic city for railfans. I found a well groomed 4-8-4 UP #802 parked at the shops, definitely out of service, yet looking like it was ready to roll. In 1983 one of the UP member of the company's steam power program extravagantly told my wife and me that the Union Pacific would never scrap a dirty or broken engine. They would jack up the smokestack and build a new one under it before taking the retired unit to the scrap yard. The UP also had a unique nomenclature for their steam locomotives by abbreviating the numbers in the wheel arrangement, *e.g.* FEF for Four Eight Four (4-8-4).

After a call on the officials at the Wyoming State Offices, we drove north to Glenrock for a tour of the Pacific Power and Light Company's new electric generating station on the north side of the North Platte River. PPL operated a private railroad running 15 miles north to a sub-bituminous coal mine supplying the power plant. Since it was all downhill for the loads, it was a very efficient operation. In 1959 I had no premonition that this was to be my first contact with the famous Powder River Basin coal that was to become such a bonanza for the C&NW, CB&Q and UP in the 1980s. We went on to Casper where we were introduced to some of the local power brokers on the 31st. They included Milt Coffman, a wealthy rancher, a banker (whose name I have forgotten) and the Chamber of Commerce people. The next day John Boyles, Cermak and I drove west to Hudson, Lander and over South Pass followed by a memorable steak dinner at Svilar's extraordinary restaurant in Hudson. We met with Riverton dignitaries on the 23rd including Tom Pickett, a local oil well supplier. Of course Cermak had to play golf. Boyles arranged the game which included four left handed players. Cermak was advised that there were no penalty strokes when killing a rattlesnake. He laughed uncertainly and was seen to anxiously survey the rough. After politicking with Coffman and Jack Perry of the Casper newspaper, Gene and I returned to Cheyenne via Medicine Bow and Laramie. Wyoming's potential for growth was lost on Cermak. He was a city kid and did not understand the West. Wyoming was too far away from the action he reveled in. It was all mine.

I continued to muse about this country, collecting USGS topographic maps and reading the histories. I worked out potential rail crossings of South Pass starting at Riverton or Hudson. Rumors of ancient railroad surveys were a fascination that proved to be elusive. It was tough to cross the mountain crests. Yet, the pioneer wagon trains had found a way. I saved the maps and filed them away. In September I learned that U. S. Steel was developing an iron ore mine near Atlantic City on the south side of South Pass. Their connecting railroad was to run from a Union Pacific mining spur near Kemmerer to handle the ore destined for a Provo, Utah, steel mill. There was always something about to happen in Wyoming.

After a wasted trip August 10th to Crystal Lake, Illinois, in response to a request for rail service, I left that evening for Omaha in a refrigerated roomette on the CB&Q. Decidedly uncomfortable, I got off in Omaha, rented a car and headed for Wahoo, Nebraska, before going on to Lincoln to meet John McClintic, our local sales agent. We took a tour of the streak of rust called the Superior subdivision before I headed north to Fremont where the C&NW and UP maintained a major interchange. The Union Pacific mainline sported a seemingly endless parade of big power, big trains and flashy yellow passenger trains. I made the acquaintance of Howard Shinrock, the local Chamber of Commerce man, and went over potential industrial sites along the C&NW lines to the east and north of town.

Later that day I went north to Norfolk following the branch line from Scribner to Oakdale which ran through what was styled the Bohemian Alps. Another streak of rust through tall grass. As I had plenty of time and had never been there before, I detoured north to the Missouri River and crossed over to see Vermillion and Yankton, South Dakota. On the 13th I drove through rain to Sioux City, Iowa, to see the industrial lands south of the airport without snow on them. George Wimmer, the Industrial Council man and our Wayne Anderson, DF&PA, expressed pleasure that I had come to town and thanked me for the maps I had prepared for them. The next day George and I met with Central Soya's chief engineer to work out a track plan to serve their proposed new facility. Even though this project was in Vern Giegold's territory, I was happy to make the track plans because it was something I liked to do and they were nice people. Following this meeting I drove back to Omaha by way of the Winnebago Indian lands west of the river and north of Tekamah and Blair with time to spare for catching the Rock Island's eastbound *Rocky Mountain Rocket*.

As the hot days of August passed, I could sense change was coming as Cermak wanted to stir things up. On the 17th I worked up a track plan for John Foxen for a Material Service readymix plant at Waukegan. John repeatedly demonstrated that he was totally incapable of reading a blueprint. From his smoke-wreathed desk, Bob Lithgow said he was uneasy about coming changes in the department's organization. Cermak wanted him to move to Minneapolis rather than try and operate that territory from Chicago. Bob's farm at Barrington was too important to him and his comfortable routine of traveling by sleeper Sunday night with a return Thursday evening was just too convenient. Bob could never become part of the business community in the Twin Cities with his travel schedule. Cermak wanted changes. Vern Giegold recommended that I be given the Illinois territory outside of the Chicago Metropolitan district, while Harvey Buchholz wanted me to act as technical staff for all of them. Cermak suggested that I might be transferred to Minneapolis which scared the hell out of me. John Daeschler later reassured me by saying that I shouldn't worry about that move because he had seen what Cermak was recommending to R. C. Stubbs. Combining business with pleasure, I took my family with me to Springfield on the IC's *Green Dia-*

mond. While they toured Lincoln's home, I met with Jim Cannon at the Industrial Commission and later attended a water supply conference. Brad Huedepohl, our geologist, and Howard Azer of the C&NW's testing department were also in attendance. We came back to Chicago on the 27th aboard the *Green Diamond.* August ended when I made a presentation to 152 members of the Carpentersville-Dundee Industrial group. At this remove I cannot recall what I said, but it must have been standard lecture number 17 with plenty of C&NW sauce.

On September 3rd the C&NW's board of directors authorized abandonment of the former Litchfield and Madison's Mount Olive branch from DeCamp northwards. This abandonment did not reduce our industrial site potential in the least.. That evening I rode the Rock Island's sleeper to Peoria to attend a re-zoning hearing the next day (on what property I cannot recall). Don Beste, our local attorney, handled the affair. The sole objector was mollified by concession of a couple of minor points. Later Joe Zelenda, our local sales agent, gave me a tour of the area before depositing me at the Rock Island station for the trip home on the afternoon *Rocket.* After a short meeting September 9th with Chicago's Director of Engineering, our Walter J. Hoffman and Bob Mack from the Chicago Sales Office for the expansion of Chicago's O'Hare Field on the arrangements for rail delivery of construction materials, I went right back to Peoria to collect information for a 300 acre site requested by the Fantus Company.

A short project involving the Milwaukee Gas Company in Harvey Buchholz' territory was a favor deal that contributed no revenue to the railroad. On October 8th I rode train #149 to Milwaukee to meet Joe Waldo of the gas company to advise them on the suitability of a tract of land they were considering for a facility. Since Harvey's image of being helpful was enhanced, I performed as an "expert" to their satisfaction. On train #400 south I rode in one of new doubledeck long distance coaches with I. R. Ballin, the new head of the Real Estate Sales Department. Bob Ballin was an interesting individual. A former U. S. Navy ship commander he was a graduate of the Juilliard School of Music in New York and the University of Chicago. He was the only C&NW official I knew that appreciated classical music—a particular interest we shared. His mission on the railroad was sale of surplus real estate to whomever had the money regardless of freight revenue potential. Our problem with his mission came in defining what was surplus land; there were to be many battles between real estate and industrial development over railroad-owned properties. Hard to warm to, Bob personally was not a bad sort when you got to know him, but he was our implacable enemy in many squabbles over railroad real estate.

The new push-pull commuter fleet was now on the property and open to public inspection in the Chicago Passenger Terminal on October 14th. Other news was not so good. On the 16th we learned that our bonus plan was a gone goose because revenues were down—the C&NW lost $1.3 million in September. Meanwhile, the hot job in the department was a proposed pipeline by Pembina; I was ordered to stay in town and help with maps and brochures. Although this pipeline was constructed, there was no detectable additional revenue to the C&NW. The effort to get something from the pipeline job awakened our sales forces to watch for similar big construction possibilities that lay all about them.

The end of the budget year was coming and I had a lot of unspent funds. Filling out phony expense account reports just to use unspent funds seemed to me like stealing. To use up my surplus, I elected to go on a west coast business trip and take my family along. We rode the Milwaukee-Union Pacific-Southern Pacific's #101 *The City of San Francisco* to Oakland, California. Arriving very late, we were treated to a magical ride on the ferry to San Francisco at midnight. On November 23rd I went to Palo Alto and called on the Stanford Research Institute to gather ideas on how to better identify and qualify industrial prospects. I learned that what science there may have been was pretty crude and that the C&NW was as well organized as any other similar group. I introduced myself to the C&NW's San Francisco office staff and visited with my old acquaintance, Bill Reed, at the Western Pacific Railroad. Reverting to vacation mode, we rode a Greyhound north to Willits, California, where we boarded the Northwestern Pacific's Budd railcar bound for Eureka, California. The NWP once ran passenger service from Sausalito to Eureka, but had recently received authority to drop the service south of Willits. Soon, this isolated run through the redwood forests and the Eel River Valley also would be gone. The ride was memorable especially when the air brakes of the single Budd car failed near Fortuna some ten miles short of the station at Eureka. The conductor was resigned to waiting for relief even though it was Thanksgiving Day and few railroad people were on duty. I introduced myself as a railroad officer and volunteered to pass signals from the motorman to the conductor who could man the handbrake at the rear of the car. We limped into Eureka and had a very late dinner at the Inn. The next day we found the Annie and Mary's (Arcata and Mad River's) old Shay #7 on display in Arcata. Renting a car we drove to Portland where I again introduced myself to the C&NW's Portland sales staff led by Reed Hoover. On December 2nd we rode the GN-UP joint passenger line in UP equipment from Portland to Seattle, Washington. At Seattle we changed trains and rode the GN's *International* to Vancouver, British Columbia. After visiting Victoria we returned to Vancouver via the Esquimalt and Nanaimo Railroad's passenger rail diesel car #9054 to Nanaimo and thence by ferry across the strait to Vancouver. That evening we occupied section 2 of sleeping car *Dunsmuir Manor* on the Canadian Pacific's train #8 bound for Winnipeg. The journey through the snowy Rockies in December was unforgettable. Winter was in full career at Winnipeg when we changed stations to take the GN's train to Minneapolis. Then, on December 7th, after a quick transfer to C&NW train #400, we were homeward bound. Clearing those leftover expense account funds by combining some kind of business and lots of vacation fun was, to me, a practical application of funds. Of course, most of the rail expense was free or on half-rate passes and care was taken to use personal funds for the vacation expenses. From 1948 to 1971 my family and I put in thousands of miles of rail travel in most of the United States and Canada on rail pass privileges. It is hard to describe to non-rail families how important those passes were. Our pay was meager and travel at regular tariff rates was often prohibitive. With the advent of AMTRAK and non-participation by the C&NW in 1971, the free pass system became history and our future rail travel rapidly diminished in favor of air and highway.

On return to the office I found John Daeschler and Chuck Towle preparing a report about the Chicago, South Shore and South Bend Railroad. Looking back it was clear that intelligence gathering on other railroad properties was to become more important than we could guess in 1959. Heineman was interested in many other railroads and would have been better served had a regular group been organized to systematically collect information and monitor elements of the railroad industry. Off and on over the coming years, we often scrambled to hastily assemble data on this property or that responding to management's current interest.

On December 10th we received an inquiry from the California Packing Company which had expressed their intent to build a distribution warehouse at Rochelle, Illinois. I visited with a local land developer, Phil May, and a man from from Rochelle's industrial development group, Bob Hultgren. Our very active local station agent, Russ Fyfe, was included in the booster group. If the C&NW had had a Russ Fyfe at all of its stations, the company's community relations would have been vastly improved The huge warehouse facility was built, as we shall see later in this account. On the 15th Bob Hazelwood of International Paper Company reappeared in our lives with a new project—a paper products warehouse at Proviso. The only piece of land large enough for the proposed building was north of the General Foods site east of the toll road and south of Grand Avenue. The odd shape of the land bothered him, but John Daeschler and I took Bob and another IPC man, Lynn Faulkner, on a tour to examine other parcels at the CENTEX Industrial Park west of O'Hare Field, which convinced them that the irregular Northlake site at Proviso was more suitable to their special needs. We ended up in Kendall Cady's real estate office to discuss land price. After Christmas there was one last forlorn trip to St. Louis on the GM&O's *Midnight Special* when Gene Cermak and I boarded a Texas and Pacific sleeper *Eagle Spirit* on the 27th. We spent the next day in the wreckage of the old radiator plant at Edwardsville trying to lease this piece of the old Litchfield and Madison Railway to an unconvinced prospect. It was hopeless. Afterwards, we visited with our St. Louis sales office before going back to Chicago on the GM&O's #4, the *Alton Limited.*

1959 was a learning time for us all; Cermak was beginning to think about how better to market our services.

14

RAILROAD WARS, 1960 . . . The new decade saw the end of the long-running steel strike. Who would guess that in forty more years this traditional American steel industry would be totally restructured? While we were beginning a new decade, the old one's business was still unfinished. The International Paper Company's interest in the oddly shaped tract at Proviso continued to require our assistance. Within our own organization the care and feeding of this prospect fell to the director of the industrial development department, Gene F. Cermak, because the person nominally responsible, the venerable John Daeschler, retired. His last day on the job was January 8th. Not that John had really been very active with this big prospect anyway because he had started his long anticipated retirement months earlier. His 65th birthday was January 16th and he left the office at the end of the month. Six days later he was dead. His heart quit as he cleaned the snow from his car's windshield preparatory to a long drive to Florida. Cermak was in his element as he vigorously used his skills in selling International Paper on the Proviso site. Bob Hazelwood, IPCO's real estate man, and their engineer, Ian Davies, arrived on January 5th—the vanguard of a larger contingent that soon arrived from their New York headquarters. My role was that of the resident information resource called on from time to time to answer questions of a factual nature concerning the property. The IPCO thing to me was a sideshow and a distraction because I was deeply involved with another big prospect in my own territory that would prove to be the cause of a major legal fight between two supposedly cooperating railroads.

TPW — THE FIRST BATTLE OF PEORIA. On January 8th I received a call from Maurice Fulton of the Fantus Company. Fantus was a site location consultant doing for fee what our industrial development department did for free. Maurie was a cagey guy who knew that dangling the bait of a potentially large freight revenue producer in front of a railroad would elicit large amounts of factual information that he or his staff did not have to dredge up themselves. He was president of the Chicago branch of the New York based Fantus Company founded by Leonard Yaseen. Competent, intelligent and smooth as silk, Maurie had married Yaseen's daughter which didn't hurt his prospects with the company. We fully understood our roles in the game that we were now to play with Mr. Fulton. This was a good solid prospect that wanted a large acreage site in Illinois with rail, utilities and access to navigable water. The C&NW had little to show on its own, but the jointly served lands with the Toledo Peoria and Western Railroad (TPW) around Mapleton southwest of Peoria would be an excellent area to present to the unidentified client of Fantus.

I called Mark Townsend, the industrial development man of CILCO (Central Illinois and Light Company), in Peoria and P. Rice, assistant to the president, of the TP&W, on January 12th to determine their interest in this prospect before going to Fulton's office in the Prudential Building on East Randolph Street in Chicago with a preliminary collection of material. Maurie liked the idea of a Peoria location and urged us to collect data on taxes, power and gas rates and other detailed information. Of course, we knew Maurie was baiting us and that we would certainly do his bidding—while he collected his fees from the client. Still, he was responsible for sifting through the flood of material we produced to make his professional recommendations. On January 20th Roger A. Fischer, freight traffic manager of the TPW, Randy Egbert, their chief engineer, Irving Lawrenz, C&NW rate man, our Vern Giegold (who couldn't refrain from sticking his nose into my party) and I met with Mr. Fulton to formally present the Mapleton area as an industrial site jointly served by two railroads, the C&NW and the TP&W.

Activity subsided for two weeks. On February 8th our J.R (Bob) Kunkel and I caught the Rock Island's *Morning Rocket* to Peoria and happened to run into Fulton in the dining car, Somewhat surprised at our appearance, he graciously invited us to join him. The next day a large party toured the prospective site. The group included Fulton of Fantus; J. Russel Coulter, president, Randy Egbert and Roger Fischer, all TPW; Bob Kunkel, our local DF&PA Fred Jones and myself from the C&NW plus James Gilchrist from the Peoria Industrial Development Council. The tour continued into the next day as we tried, unsuccessfully, to worm the client's identity out of Fulton. On the 12th I presented additional materials to Fantus in their Chicago office. Although Maurie was our host for lunch and introduced his associates, Ken Berliant and Robert Thompson, there were few clues dropped as to the identity of his client.

On March 9th Fulton called to say that he had optioned the site at Mapleton for his mysterious client. On the 24th he invited me to join a party consisting of him and three strangers, who were not introduced, for a trip to Peoria on that ever so handy *Morning Rocket.* We looked at the property while I listened very hard for any clues as to the prospect's identity. Our traffic people could be very helpful if the identity of the prospective shipper could be ferreted out. The next day Mark Townsend from CILCO and Jim Gilchrist from the development council joined the entourage. Several pieces of property were under consideration. One piece north of the new power plant

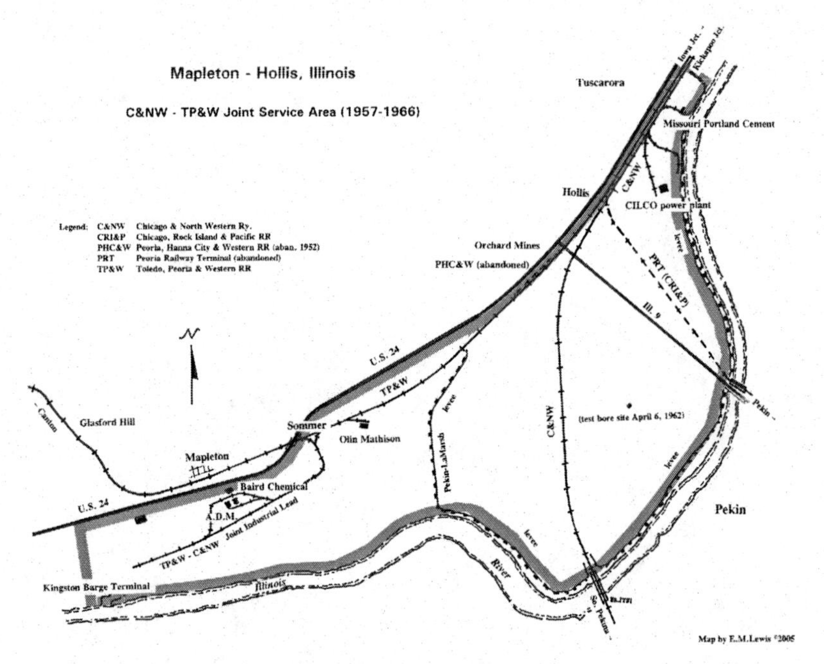

The troublesome joint service area near Mapleton, Illinois. Everything within the solid "red" line was supposed to open to both the C&NW and the TP&W. This area was the center of a long-running series of legal fights in which the C&NW battled TP&W labor unions and local judges to no avail. Here, too, the C&NW's soil investigation resulted in an adventure with the Illinois River which almost changed the landscape.

was quickly scratched, but the land near the village of Mapleton was judged suitable for the new facility—whatever it was. After a purchase deal was struck with Mr. Coulter, president of the TPW, we boarded the afternoon *Rocket* for home. Four days later, we learned that the mysterious client was Archer-Daniels-Midland, a giant food conglomerate.

The next event was introduction of the client in Peoria on the 29th. With Randy Egbert and P. Rice of the TPW, I picked up a Mr. King from a late flight to Peoria. That afternoon was occupied by a long discussion of freight rates and routes. The press release on April 5th. detailed ADM's plans to build a new processing plant in the joint TPW-CNW industrial area at Mapleton. That afternoon Maurie Fulton and Gene Cermak met with C&NW chairman, Ben Heineman, and president, Clyde J. Fitzpatrick, who were invited to attend the announcement ceremony in Peoria May 19th. Everything seem to be working smoothly, but there was trouble brewing. Unaware that there was a snake in the grass, we all basked in the rosy glow of a successful plant location that could bring-in lots of freight revenue. Just as important was the knowledge that our new joint service territory would now have an anchor client which could attract additional shippers to settle there. On February 13, 1961 I had the pleasure of moderating a panel consisting of Mark Townsend of CILCO and Maurie Fulton at an industrial development conference. The C&NW operated a special train for ADM from Chicago to Peoria on May 12, 1961 (I was not invited.) Our department rarely participated in the jolly parts of celebrating a plant location; our job was over, it was traffic's baby now. Actual construction took place in the summer and fall of 1961. In September, 1961 Gene Cermak and I went to Springfield, Illinois on the Illinois Central's *Green Diamond*, to meet with the State Industrial Commission. We ran into Mayor Lester Collins who tried to keep us up all night. Somewhat the worse for wear, we got to Peoria on the 19th for a meeting with ADM and the TPW before taking the *Rocket* for Chicago.

The first hint of trouble came on October 16, 1961 when Randy Egbert, the TPW's chief engineer, and general manager (name lost) came to our offices in Chicago to discuss joint train service to the new plant. As 1962 rolled around there were increasing rumbles coming from our operating department about difficulties in arranging switching service at Sommer and Mapleton in the joint service territory. A problem had arisen about track construction costs for a new Olin Mathiesen warehouse in the joint area. On April 9, 1962 Fitzpatrick took a hand in the growing service dispute. Two days later one of our top attorneys, Robert W. Russell, went to Peoria to force the issue, taking our voluminous file with him. The next day a C&NW switch crew attempted to run their locomotive into the joint service territory, but were stopped when the TPW's special agent arrested our flagman. Bob Russell then filed suit against the TPW claiming that we were denied our contractual rights. The C&NW's president, Clyde Fitzpatrick, was livid as he felt that president J. Russel Coulter of the TPW had reneged on their agreement. The two forces dug in for battle.

On April 16, 1962 I verified that the flagman when arrested was indeed within the designated joint service area. The C&NW's operating unions stirred up the matter further when they submitted time claims covering the past three years for allowing the TPW to switch the new CILCO power plant located in the joint area. Bob Russell drafted me to help gather documents from other departments. The pile of material collected was amazing. By the end of April 1962, our attorneys were crafting a suit to establish our contractual rights to use our own crews to switch the industries now located and to be located in the joint service area. James P. Daley was our lead attorney (Jim was a first cousin of Chicago's mayor, Richard J. Daley). The C&NW often used local attorneys as associates for there is nothing worse in pursuit of a legal action than an outside attorney coming cold into a new jurisdiction—too many toes may be inadvertently stepped on. Our Peoria legal counsels were Tim Swain, a politically connected attorney, and his associate, Mike Gard. On May 1, 1962 a group of us lower types met with Gard in Peoria before Judge Mercer. The hearing opened with Jim Daley presenting the facts of the case in his usual low key and very polite manner. In contrast, the TPW's attorney was a flamboyant flannel mouth who gestured and flailed his arms as he delivered in impassioned and extravagant language a diatribe about how this big outside company (the C&NW) was taking advantage of a poor little local TP&W. In the background the TPW unions sought leave to intervene. R. C. Conley, the Galena Division superintendent, Joe Gallup, our trainmaster, Fred C. Jones, our local DF&PA, and I were merely spectators at this circus. Nothing was decided and we all went home on the afternoon *Rocket* without any idea of how this drama would play out.

On May 29, 1962 a high-level meeting in Chicago was held to discuss the case. Ed Olson, vice president of traffic, R. C. Stubbs, assistant vice president traffic, Gene Cermak, industrial development, Bob Kunkel, coal, Henry Schroeder, rates, Bob Russell, law department, and I were in the group. Everyone present was asked to develop testimony to support our case. Another hearing was held in Peoria June 13th; Russell, Daley and Iver S. Olson (sales) covered that one. Mr. Fitzpatrick again tried to negotiate with his former friend, J. Russel Coulter, but was unsuccessful as Coulter was unwilling to yield on any points. The hearing ended in a continuance at the

TPW's request. At this hearing the TP&W unions managed to become parties to the suit, which in retrospect, may explain Coulter's reluctance to negotiate with the C&NW. A former president of the TP&W (George McNear) had been shot to death in front of his Peoria home in 1947 during a long and bloody strike. In that same conflict there had been instances of striking trainmen driving on U. S. 24 alongside TPW trains being operated by non-union crews and shooting rifles with intent to harm. We did get some good news, however, ADM agreed to appear as a witness for the C&NW. At another hearing in Peoria on June 22, 1962, I was on the stand for two hours, the target for two TP&W union attorneys. My deposition stood up to their assaults and entered the growing record. On March 4, 1963, Judge Mercer ruled against the C&NW. We appealed his decision. In truth we knew we would lose this round because Judge Mercer was a local partisan that believed in that outsider versus local company argument. But this legal battle was not over; it would drag on into 1966.

A brief summary of related events in subsequent years underscores the delay and frustration that can be associated with law suits. On May 29, 1964 I learned that American Cyanamid was looking at TP&W land in the joint service area for a major fertilizer manufacturing plant; I tipped off our attorneys. On June 10, 1964 I discovered that my friend at CILCO, Mark Townsend, was also working on the same prospect. We compared notes. CILCO's top brass confided in me that they were meeting with Cyanamid on June 12th. On the 17th Bob Van Nostrand of Cyanamid called me to discuss the status of rail service in the erstwhile TPW-CNW joint service area. Our attorney, Bob Russell, opined that the large revenues accruing from such a big plant might lever the TPW off their position. On June 19th our president, Clyde Fitzpatrick, said it was all or nothing with Coulter. Gene Cermak and I were authorized to represent Mr. Fitzpatrick in any negotiations with J. Russel Coulter. Too much time was slipping by and nothing was happening. In desperation I called Springfield September 4, 1964 to see if Illinois Governor Otto J. Kerner would take a hand in this dispute between the railroads which, if unresolved, might force Cyanamid to locate in another state. His office must have done something because I heard on the 18th that the Illinois Commerce Commission was getting interested in the matter. It was too late. On October 1st Van Nostrand advised me at a luncheon meeting that while the Peoria site was his company's first choice, the railroad squabble was troubling. On December 1st we heard the facility was going to be built at Hannibal, Missouri, in large part because of the C&NW-TP&W dispute. Mark Townsend of CILCO was heartbroken as this was the biggest user of natural gas that he had ever dealt with.

The law works slowly in contractual disputes. I was not involved with the appeal of Judge Mercer's ruling so had no contact with any matters surrounding the case until it bubbled up again on August 13, 1965. Fred Steadry, then our commerce attorney, grilled me in two sessions. Two days later Mr. Fitz called me in to describe the joint track at Sommer—a track constructed in 1957 when I was assistant engineer on the Galena Division. I wasn't sure what this was all about, but soon learned that another law suit had been filed by the Peoria and Pekin Union Railway who were again trying to "crash the party" at Sommer. On September 23, 1965 I flew to Peoria with Frank Hauff (operating), Bob Kunkel (coal) and Jim Daley (law) to meet with TP&W about construction of additional trackage for coal trains at the CILCO generating plant at Sommer. En route I discovered that I had been elected leader of the meeting, much to my surprise. We outlined our proposal and then toured the "joint" area from Sommer to Mapleton before flying back to Chicago. Mr. Fitz approved our negotiated agreement on October 14th.

More months went by. On March 17, 1966 Jim Daley drafted me to help with the upcoming hearing in Peoria. My job was to lug a heavy suitcase full of legal documents to Peoria for delivery to our local legal associate, Mike Gard. On March 22nd I flew down on Ozark's ancient DC-3 and delivered the heavy load of legal papers. The next day we attended a stipulation conference between the litigants in the morning, and after Daley arrived, developed our testimony in the afternoon. The next day at the hearing before Judge Hunt, I was on first followed by Gene Knecht, a C&NW traveling engineer who described how TP&W special agents (police) stopped him from operating a C&NW locomotive into the joint service area. Entry of stipulated testimony took all afternoon. On the 25th the TP&W put a lot of unrelated material into the record despite our objections that it was immaterial to the matter at hand. In the afternoon blubbery old John Cassidy led president J.Russel Coulter through a prepared litany of leading questions amid our constantly over-ruled objections. Court adjourned over the weekend. On Sunday March 27th I drove Jim Daley to Peoria for a meeting at Gard's downtown office. There was a lot of tough, grinding work to be done to be ready for court in the morning. We finished at 1 a.m. Since the court refused to strike the TP&W's irrelevant material, we decided to throw our "garbage" into the record also. Late in the day Carl Hussey (operating) came to Peoria. At dawn the next day a group us, including Hussey, Knecht, Bill Cook (trainmaster) and me toured the TP&W's mainline track west of the joint service area to look at the "fearsome" Glasford Hill that their trainmen unions kept bringing up. We also refreshed our memories of the track layouts in the area. At noon I served a friendly *sub poena* on ADM's plant manager, Dale Lautenslager, who had been instructed by his superiors to expect, and respond, to the service.

Judge Hunt was not in control of his court as he allowed John Cassidy, the TP&W attorney, far too much latitude in presenting his case. Consequently, time ran out and Lautenslager didn't get to testify after all. That evening Phil Gallaway, president of the P&PU Railway, joined us at dinner which produced some very interesting conversation on the local railroad scene. The last day of the trial was March 30th.. The defense (TP&W) rested their case at 11 a.m. We put Dale Lautenslager (ADM) on as a good rebuttal witness. Then Carl Hussey was called to testify about train operating matters and delivered a concise and cogent report. Rumbling and bumbling John Cassidy then cross examined Hussey to try and show to the court that Carl Hussey was an outsider from Chicago who knew nothing about local conditions especially that dangerous hill at Glasford and that only TP&W men knew how to handle trains in that area. Carl merely replied, " I was there this morning." which took the wind out of Cassidy's sails. Dismissing Hussey, he recalled me to the stand, but soon found that my answers to his sallies were not to his liking and soon dismissed me. That ended the hearing. We could only file briefs and await the judge's ruling. Because of their volume, the briefs took a long time to be filed. It was all in vain; Judge Hunt ruled against the C&NW. The joint service contract of 1957 was canceled. The first Peoria railroad battle (1960–1966) was over and the C&NW had lost. The C&NW never served the ADM, Baird Chemical, Olin Mathiesen,

Kingston Terminal or Caterpillar plants. By the time the final ruling was issued, I was no longer involved in industrial development work. In a few years the TP&W and their truculent operating unions would be swept away. The railroad property went through several owners including the AT&SF. On July 15, 2003 the Keokuk Junction Railway, a short line carrier, sought permission to acquire the old TP&W line from Hollis to LaHarpe, 76 miles, plus the 2.5 mile Mapleton Industrial Spur which connected with this line three miles west of Hollis.

Retracing the chronological thread of this account requires backtracking to January 1960. On the 12th I went to Springfield, Illinois, aboard the Illinois Central's *Green Diamond* to attend another meeting of the Railroad Consolidation planning committee. Two of the railroads, the Wabash and the GM&O, expressed their opposition in no uncertain terms to any scheme that required track rearrangements within the city; the IC remained silent. The C&NW was only a distant and mildly interested bystander because its main line was far to the west of the city and did not connect with any of the affected railroads in the vicinity. Of course nothing useful came from this meeting and I went back to Chicago on the afternoon *Green Diamond.* This whole planning exercise by the City of Springfield to remove the iron grid work from its streets had actually begun in 1924. In 1967 a new effort resulted in formation of a seven-member Capital City Railroad Relocation Authority. As in the past, their recommendations came to naught as Class I railroad mergers, branch line abandonments and shifting rail traffic patterns made their proposals meaningless. In 1994 a new highway bypass around the south side of the city relieved the worst of the highway grade crossing problems.

With John Daeschler's passing and Chuck Towle's arrival, personnel and territorial assignments in the industrial development department were shuffled in February, 1960. In January Bob Lithgow took Chuck Towle on a tour of the Minnesota territory much like my introduction there two years earlier. On February 12th Cermak advised me that I was to take the Illinois territory outside of the Chicago Metropolitan Area as well as Nebraska (except Omaha) and Wyoming as my own responsibility so that Vern Giegold could concentrate his efforts in Iowa. At last, I no longer worked for Vern and could ignore his personal foibles that so annoyed me. (Unlucky in business and in his personal life, Vern was to die suddenly August 14, 1975). Along with this release from Vern's suffocating supervision, my monthly salary went to $675. Charles Towle was the heir apparent to the Minnesota and Dakota region as M. B. (Bob) Lithgow succeeded to the Chicago Area with John Foxen as his clueless assistant. Harvey Buchholz remained in charge of the Wisconsin territory.

On the evening of February 22nd, I headed west to Omaha in a Western Pacific sleeper *Silver Mountain* on the CB&Q night train. Newly promoted Don Guenther went along for another "student" trip. Don was slated to be Vern's assistant in Iowa, but I was asked to give him some direction and monitor his performance. In a driving blizzard we visited Wahoo and Lincoln, the state's capital, with a rental car that had a defective heater which threw a stream of humid air against the cold windshield making vision very difficult. We met our local sales representative in Lincoln, Dick Ryan, as well as becoming acquainted with Fred Berniklau, a local real estate developer. David Osterhout, chief of the Nebraska Development Board, as sober and measured a man as I had ever met, graciously gave us a tour of the Nebraska Unicameral Legislature which was not then in session. Don and I had planned to tour down the Superior Line to the Kansas State Line and the Lincoln line north to Fremont. Since the lines were snowed shut and main highways were barely passable, we struggled north to Fremont with that balky car to see Howard Shinrock, the city's industrial commissioner. After a quick 25 mile trip east to Blair on the Missouri River to view the new National Alfalfa plant, we scooted south to Omaha and caught the CB&Q's eastbound at 11:45 a.m. All in all, not a very useful trip. Earlier, I mentioned that Don Guenther had been our department's chief clerk. Following his promotion to industrial agent, he never lost his clerk's attitudes or mannerisms—always a follower, never a leader. He automatically bowed to authority and was always present in case he was called on for service—just like a little shadow. And that is the nickname he acquired: Shadow. With Don's promotion to industrial agent, a new chief clerk came aboard, Jack L. Sommer. A sober, rather humorless lad of middling stature, Jack's somber demeanor reminded me of a German country parson. He eventually went to Milwaukee to become an industrial agent in Harvey Buchholz' office. Jack and Milwaukee fit together well. The vacated chief clerk's job then was given to Stanley H. Johnston, a former rodman from Galena Division days. I was pleased to have Stan join the department. We were now well positioned to handle the next five year's upsurge in new plant locations—a period the railroad was never to see repeated. Along with these personnel changes, the offices were actually painted in late January. That winter of 1960 was noted for long lasting cold—the high daily temperature did not rise to freezing until March 15th.

The American Industrial Development Council's annual meeting was held in Springfield, Illinois, March 15th, a day of a thawing, but the subsequent heavy snowfall belied the advent of spring. The Illinois Central's *Green Diamond* on the 14th carried a crowd of industrial development types: Gene Cermak and me from the C&NW, the IC's Jack Frost, head of their industrial group and Henry W. Coffman of the New York Central. Fred C. Jones, our DF&PA, and Joe Zelinda, sales agent, came down from Peoria for the meeting along with Mark Townsend from CILCO and Bob Latta, representing the city of Peoria. The day was filled with formal talks and little side caucuses where rumors, lies and misdirections were freely passed around. This development business was a fiercely competitive arena. Chicago's Charlie Willson of the Association of Commerce and Industry, Hack Roth of Northern Illinois Gas Company and Sam Estes from Common-wealth Edison were also present. This group represented a tightly knit group of Chicago boosters that Cermak went out of his way to cultivate—hoping that the names of prospective plant site seekers might slip out in unguarded moments. Gene, himself, addressed the audience for 15 minutes. Sticking to generalities and a couple of good jokes, he cast his charm over the crowd. The evening was filled with conviviality—in excess. We all returned to Chicago on the 17th; the railroad men riding the *Green Diamond*, of course. At the end of March I was supposed to meet Don L. Gunvalson, our sales representative from Rockford, Illinois, in Elburn, but he failed to show up—he had forgotten the meeting. I caught up with "Sarge" in Woodstock, which worked out alright because we then had a chance to met with Bob Rosenthal of Rosenthal Lumber in Crystal Lake, Illinois, to discuss his proposed perlite distribution plans. Perlite was used in insulation of the many residential tracts springing up along the C&NW's northwest commuter line from Des Plaines to Crystal Lake. While the perlite project never got out of the talking stage, Don and I enjoyed the first day of real spring-like weather.

On April 7, 1960 we learned that the C&NW had purchased the Minneapolis and St. Louis Railway effective October 31st. The next day I heard that the Madison portion of the Illinois Terminal Railroad had been acquired by the Alton and Southern Railway in East St. Louis.

The Wyoming gang came to Chicago on April 13th bringing a strong whiff of the far west to our offices. Ed Olson, vice president traffic, hosted a luncheon at the Union League Club. It was quite a crowd, but Ed loved showing off his hospitality. Milt Coffman, a wealthy rancher, and old Buck Buchanan, representing the State of Wyoming, were surrounded by C&NW personnel which included Ed Olson; his assistant, Bob Stubbs; sales manager Iver S. Olsen; geologist Brad Huedepohl; real estate's Bob Ballin; our Gene Cermak and—wonders of wonders—me. To a man they all chattered about the potential and future growth of the great state of Wyoming. Personally, I could not help but think that any industrial development model for Wyoming had to be quite different from the more populous middle west. Wyoming was a raw material state of vast distances and very little water. In my short tenure in this job I knew more about that rickety C&NW line west of Omaha than anyone at the table including the Wyoming visitors. None of us imagined the flood of black diamonds that was to enrich the railroad in another 14 years. Buck Buchanan was a character right out of western movies. With his deeply wrinkled, leathery face, Buck looked like he had been out in the wind and sun all his life. A steady stream of cigarettes kept those wrinkles in his skin deep and solid. He had come to Wyoming in his youth as a cattle buyer for Armour and Company and had just stayed on. He recounted his experiences with the C&NW's passenger train that once ran from Casper to the end of track at Lander. Innocent of local ways, he had boarded the train in Caspar and presented his ticket to the conductor. "Young fella", he said, "Do you plan to ride this train often?" "Why, yes sir", said Buck. The conductor collected the ticket and said, "Next time you ride with us, just buy a cash fare. I throw the money up to the ceiling and whatever sticks there belongs to the company." (This must have been a standard railroad story. I had first heard it in 1947 from a Galena Division conductor on the old Freeport Express.)

I arranged for lease of the old Geneva (Illinois) freight house to the Chambers brothers for a timber roof truss manufacturing plant. The boys were new in the business and could not afford grander quarters. The burgeoning construction of homes in the immediate area was to make them quite well off—eventually. Our real estate man, Bob Ballin, was pretty upset at the cheap rent I had negotiated; he could not appreciate that those carloads of inbound lumber were bringing the railroad a very nice revenue and that the rent was pretty small potatoes. The old brick building located on the north side of the main line east of the depot had been slated for demolition; squeezing some good revenue out of its last years seemed like a good idea. At Rochelle, meanwhile, the warehouse plans for the California Packing Company had gotten into a terrific tangle. I consulted with our commerce attorney, Fred L. Steadry to see if he had some legal insights that would help in a property dispute. There were three railroads competing to serve this big plum. CalPak decided to look at a site on the C&NW's Spring Valley Branch on the south side of DeKalb as an alternative. The project subsided until September, came to life briefly, and then died out again. Though the option on the DeKalb site ran out on September 30th, Don Gunvalson and I were asked to work out a track plan for it on October 4th. The DeKalb site was dropped when Rochelle was chosen as the best place for this huge distribution warehouse. The lure of service by three railroads (CB&Q, MILW and C&NW) was just too important to CalPak. (In 1979 California Packing was acquired by R. J. Reynolds and eventually became Del Monte Foods.)

In early May anticipating acquisition of the M&StL (Maimed and Still Limping or the Midnight and Still Later) Vern Giegold and I explored that railroad's line from the Mississippi River town of Keithsburg, Illinois, east through Monmouth and Farmington to Peoria. We spent the morning with the Davenport (Iowa) office of the Corps of Engineers obtaining information about barge accessible sites at Keithsburg, Fulton (East Clinton) and Clinton, Iowa. That afternoon driving through the flooded countryside along the "Louie", Vern and I realized that we were traveling through a remote, lightly populated agricultural area that was unlikely to have many, if any, industrial prospects. This was not even good grain country which meant new elevators or fertilizer distribution points were unlikely. It was as though that railroad had been built through a kind of desert. In Vern's opinion this must have been where Christ lost his sandals. (This portion of the M&STL was abandoned south of Oskaloosa, Iowa to Monmouth, Illinois on November 9, 1971). Why Vern went with me on this trip—it was out of his territory—became clear when we drove north to Clinton, Iowa, to meet with Robert T. Stapleton, director of Clinton's Industrial Commission. Vern wanted to spend some time in Clinton with Bob who was Cermak's golf and poker buddy. If the boss could always find an excuse to go to Clinton, Vern felt that he could too.

PIONEER, THE SECOND BATTLE OF PERORA (1960–1962)

On May 12, 1960 I learned that Super-Valu, a big grocery chain, was planning a new distribution warehouse in the new Pioneer Industrial Park north of Peoria. Established in 1959 the park was served by a weedy branch of the Rock Island that ran 82 miles northwesterly from Peoria through Dunlap, Princeville and Galva to a connection with the main east-west mainline at Colona, Illinois. By 1960 there was very little train service north of Pioneer Park. The Rock's Peoria switch engine wandered up to nearby Keller station daily to switch a retail lumber yard and a couple of other small businesses scattered along the way. The land comprising the new industrial area was located on a high piece of ground on North University Avenue and west of the old Mount Hawley airport. We were surprised to learn that the developers of the park, who were acquaintances of ours, had acquired this site on the Rock Island. Then, of course, we didn't have anything better in the area to attract their investment money. The C&NW's main track was located about a mile and a half west of Pioneer Park. Our nearest station was Radnor at the top of the infamous Radnor Hill, the ruling grade on the Southern Illinois district. I scouted the site as soon as we heard about Super-Valu's locating there. Harold Austin, president of Pioneer Park, would have loved to have a second railroad serving the property. Given the lack of potential business, I felt that the expense of a C&NW industrial lead track from our main line to the park just did not make economic sense. End of story? Not quite. About a month later on June 22nd, Clyde Fitzpatrick, president of the C&NW, called Gene Cermak to report that International Paper Company was looking for a site served by the C&NW in the Peoria area. Cermak remembered that I had looked into constructing an industrial track to the Pioneer Industrial Park as a possibility and had mentioned it to Mr. Fitzpatrick. Mr. Fitz called me to his office and asked my opinion about a track to the park. Without benefit of an instrument survey, I replied, I thought it was feasible. He then ordered chief engineer B. R. Meyers to conduct a full scale investigation. Meyers instructed the Galena Division to run a survey immediately. The division engineer was Bill Wilbur, the same person to whom I had first reported on July 31, 1947 when I began my railroad career. The two of us reconnoitered the route on June 27th returning to Chicago on the Rocket the next day. On the 28th I told Mr. Fitz that it looked like a pretty straight forward construction job, and he okayed the project. On July 5th Mr. Fitzpatrick added a caveat—no construction unless International Paper actually purchased the site. Mr. Blueweiss, IPCO's agent, made a satisfactory deal with Harold Austin and the plant location seemed to be on solid ground. R. C. Conley the Galena's superintendent growled about having to do the switching service with road power because only through trains operated from Peoria to Nelson and there was little local business along the way to justify a way freight. While this project was hatching, I was trapped into a golf outing affair on July 19th involving the LaSalle Chamber of Commerce and my golf-addled boss, Gene Cermak. With this International Paper thing about to give birth, I made excuses and escaped on the morning Rocket from LaSalle to Chicago. The deal was done on July 22nd.

Harold Austin, Bill Wilbur and I went over the plant site July 25th. Harold was tickled that he was about to get a strong railroad in to serve his industrial park. Wilbur was glum that he had another big job to supervise. I was happy to score a coup on the miserable Rock Island. In my enthusiasm I even tried to interest Super Valu into C&NW service as it was almost next door to the new International Paper plant. Wilbur's assistant engineer (my old job) was John Olson. He arrived at Radnor with his survey crew on a hot and humid July 26th. I ran a track motor car (I still had my permit) from Radnor about half a mile north to the proposed point of switch for the new line and set the boys off through the cornfields. The instrument work soon showed that the new track would have a stiff uphill climb. After a meeting with IPCO's building engineers, the Austin Company, I took Wilbur downtown and put him on the *Rocket* for home before returning to the survey party to help out in getting the first line run. On the 27th the crew had pushed through oat and corn fields, across creeks and a state highway by late afternoon. I bought lunch for them and encouraged the work. But, the party chief, John Olson, prepared to quit for the day at 5 p.m. when they were almost at the end of the job. Johnny was a union stiff and was not about to work any overtime without pay. The instrumentman and rodman tired of the heat and humidity would just as soon finish running the line and not come out again the next day. I encouraged their rebellion and prevailed; Olson grumbled and mumbled until the last notes were taken at dusk. We adjourned to an air-conditioned motel and plotted the notes on profile paper. The next day I was in Chicago with that rough profile to review it with Bill Wilbur. Suddenly he was called to Benld in southern Illinois because of a fatal injury there. I was lucky to have Bill Wimmer, a Galena Division instrument-man, available. A recent acquisition from the chief engineer's drafting room, Wimmer was the neatest and quickest draftsman I had ever seen in operation. Bill, who later became a major engineering officer with the Union Pacific, was a very practical and straight ahead individual who grasped the importance of this project and turned out a finished track plan for the Pioneer Park line in very short order. On the 29th Bill Wimmer and I finished the track design and profile and rushed it to Mr. Meyers' office. Just when we needed him, my superior, Gene Cermak, was away from the office—out on the golf course. Then, the real estate vice president, Kendall Cady, whose approval was needed to secure the right-of-way options, was tied up in a seemingly endless meeting.

At last on August 1, 1960, everyone had signed off on the project, and Fitzpatrick gave us the green light to proceed with construction. Carl "Andy" Anderson, our best land agent, was assigned to acquire the necessary options. On the afternoon of the 1st I rushed to Peoria on that handy *Rocket* train to start the option process. The next day Andy worked up land title descriptions. After a conference with the division engineer, Bill Wilbur, and the staff engineer from Meyers' office, Steve Owens, Andy and I were instructed to seek a wider strip of land to simplify grading. Andy and I contacted two of the three landowners, Mangold and Vetter, and arranged to meet with them that evening. After some posturing and bargaining with Rube Vetter, who in turn was having to negotiate with his wife and sister, the option to purchase was obtained late in the evening. Mangold came around the next day, August 4th. The third owner, however, lived in Florida. Andy declared that we would have to fly down there that evening and get the option signed. Fly? Me? What a revolutionary thought, a railroad man flying. Somewhat bemused by the idea, I drove us quickly to Geneva where Andy picked up his travel gear and then on to Elmhurst to collect mine. At Chicago's O'Hare International Airport we booked a jet DC-8 Delta flight to Atlanta, Georgia, with a connecting DC-7 hop to Orlando. The dam was broken—travel by rail would never be the same for me. The third landowner, Charles Patton, lived in Wauchula, Florida, near Winter Haven. Andy and I had a pleasant meeting with him and his wife, Madge, but there was a catch. Before they would sign the option, we had to do him a favor—address his local Rotary Club that day at noon. I worked up a little talk on industrial development, which went over fine; Patton signed the option. Andy and I then drove over to Titusville to visit his relatives and stay the night. Bedded out on a screened porch, I endured the tropical heat and humidity—worse even than a hot Illinois night. Andy stayed over the weekend while I flew back on a DC-7 through towering cumulus clouds to Atlanta to change to a new Convair 880.

The Rock Island found out about our plans for the track construction as soon as August 10th just a few days after the last option was signed. The Rock's industrial boys were quite upset when the full extent of our planned construction became known, especially since we had already acquired the right-of-way "into their territory". They did not know about International Paper's plans until Harold Austin gave them the details. We shouldn't have been surprised at the ensuing uproar as railroads are intensely protective of their territories. Look at the furor caused by the CB&Q trying to crash into the C&NW's monopoly at Northwestern Steel and Wire in Sterling, Illinois. But in 1960 the C&NW had become aggressive to the point of arrogance. We thought of the Rock Island as some sort of a poor relation who needed our guidance on how to run their business. Of course, and understandably, the Rock's people resented that attitude. In twenty years the Rock Island would be gone and the C&NW's yellow and green diesels would be parading up and down major parts of the Rock's former realm.

The C&NW board of directors was scheduled to meet September 8th. Their agenda was supposed to contain an item authorizing construction of the Pioneer Industrial track, but, I discovered on the 6th that the item had not been included in the agenda. I alerted the president's office through my backdoor contacts. Gene Cermak was called to Mr. Fitz' office the next day. The construction approval item was hastily restored when Mr. Fitz realized that the traffic department had not kept his office fully advised on the matter. Traffic did not gain much luster from this communication breakdown caused by busy social schedules of the top officers laced by the numerous golfing finales of the season. Saved by a whisker, the project's funding was approved by the board. Uneasy at the Rock Island's reported hostile reaction to our plans, I consulted with our commerce attorney, Edgar Vanneman Jr. (who later became mayor of Evanston, Illinois). Four days later Cermak, Steve Owens from the chief engineer's office, Ed Vanneman and I went to Peoria on the morning *Rocket* to Peoria where we met Bob Conley, Galena Division superintendent, Carl Hussey, assistant to the vice president of operations, and Fred Jones the local DF&PA. The next morning September 13th we toured the site of the new International Paper plant and looked at the alignment of the proposed new industrial lead track. At 10:30 a.m. the Rock Island men came to our meeting. We had seen them on the train the day before, but their aloofness projecting hostility had not invited conversation at that time. In today's meeting their bile boiled over. We ventured the idea that the Pioneer Industrial Park could profitably be served by two railroads as a jointly operated facility. It was soon obvious that neither they nor our group were authorized to negotiate such an arrangement—that was management's prerogative. Nothing was accomplished by this meeting save for our appreciation of the depth of the Rock Island's displeasure. We all rode the afternoon *Rocket* back to Chicago—in separate groups. That evening I attended general agent Walter J. Hoffman's retirement dinner at the Palmer House.

After we had all reported to our superiors about the hostile meeting, our top brass decided to approach the Rock Island's management directly. Accordingly, on September 29th, Ben Heineman, Clyde Fitzpatrick, Edward Olson (vice president-traffic), Gene Cermak and Carl McGowan (our attorney) called on the Rock Island to discuss the Pioneer track construction. Their proposals fell on deaf ears; on October 3rd the Rock said they would fight us. I heard on October 7th that Mr. Edwards, president of CILCO in Peoria, had asked Henry Crown, president of Material Services Company in Chicago and a major stockholder of the Rock Island, along with the chairman of Caterpillar Tractor Company in East Peoria, to speak to Rock Island about calming their opposition to construction of the C&NW track. These efforts were in vain. I learned on the 14th that the Rock had filed for an injunction halting our construction and that I was to be a witness at the hearing on the 18th. The day before the hearing, Gene Cermak, Ed Vanneman, a Mr. Eilers (attorney for Muirson Label Company which International Paper Company had purchased in 1960), and I took the afternoon *Rocket* to Peoria to go over our presentation. At the hearing before Judge Mercer the Rock Island opened the session with an exhibit and two witnesses. I was then called, the first of five C&NW witnesses. No decision was announced by the court that day.

There were two legal matters in play at the same time: the defense against the Rock Island injunction and a proceeding with the Illinois Commerce Commission relating to permission to construct a track across a state highway. On October 20th, Dickman (our attorney), Vern Mitchell (signal engineer), Bob Conley (superintendent Galena Division), and I drove to Springfield from Chicago for the Illinois Commerce Commission hearing to be held the next day. Bill Wilbur (division engineer) and Steve Owens from the chief engineer's office, had taken the train down the night before. When the hearing for the construction of a railroad track across state highway 174 opened at 9:30 a.m., the Rock Island's attorney immediately stuck his oar in the water. My testimony was brief and I went back to Chicago with Bob Conley in his hi-rail station wagon—on the highway.

Life went on during the legal lull—until someone issued a ruling everything related to the project was held in suspended animation. The annual traffic department social event was held in Peoria

November 9th and 10th. I took the morning *Rocket* and found vice president Ed Olson and his side-kick, Bob Stubbs, were also on the train. Both studiously ignored my presence; I returned the compliment. In Peoria they went their ways while I made contacts with ADM and my Pioneer friends. I did attend the party and had a tolerably good time as many of my friends also were there.

On November 16th Judge Mercer granted an induction against our construction of an industrial track to the Pioneer Industrial Park. On the 28th Mr. Fitz agreed that we should file an application for construction with the Interstate Commerce Commission as these local Peoria courts were proving to be unfriendly to the C&NW. Just after the first of the year Gene Cermak and I went to Peoria again. Gene had instructions from Mr. Fitzpatrick to personally deliver his letter to Harold Austin, president of the Pioneer Industrial Park, assuring him of our intention to pursue this matter. In Washington the ICC set February 8, 1961 as the date for hearing our application for track construction. Edgar Vanneman worked all through January pulling our case together with me helping where I could. We developed rate schedules, photographs, maps, train schedules and projected traffic flows. Not leaving any stones unturned Ed and I went to Peoria on January 24th despite the bitter cold. He interviewed the principals involved including the investment owners of the industrial park among whom was John Altorfer, a well known local politician. On January 30th Ed Vanneman and I called on Pillsbury in Des Plaines to ascertain the truth of a rumor that they were interested in building a new plant at Pioneer. Their response was neither useful nor encouraging. I learned May 15th that the rumor was correct, but that they were planning a new facility on the TP&W because our Pioneer track had not yet been built.

Although I had made my first flight in August and even flown to Europe in December of 1960, my fellow railroad men still rode the pathways of steel on our trip to Washington in February 1961. Galena Division superintendent Bob Conley and I rode the B&O's *Capitol Limited* in sleeping car *Tygart* on the 6th. It was a rough ride through the mountains. The next day was spent mostly hanging around waiting for the rest of our delegation. In the late afternoon, Mr. Winn, attorney for International Paper Company, and our C&NW people assembled to review procedures for the next day's hearing. The next morning we awoke to find Washington under a blanket of snow. Not an inch or so, a full foot covered everything. Streetcars dependant on underground center track conduits for power were abandoned in mid street. Despite the incredible traffic chaos, the Interstate Commerce Commission's building was open for business. The hearing started promptly at 9:30 and I was the first witness. Since it was our application that was being heard, opponent Rock Island had to wait until we were finished about 2:00 p.m. Of course they needed more time to get their arguments into the record and the hearing was extended to the next day to accommodate them. They finished at noon. After lunch, I was called as a rebuttal witness to testify that the Rock Island could not physically construct a track from their Super-Valu industrial lead to also reach the new Muirson Label (International Paper Company) plant without occupying or encroaching on most of the platted industrial park property. That concluded the hearing; we could only wait for the ICC's ponderous procedures to either reach a decision or order additional hearings. While in Washington, Bob Conley and I took the opportunity of touring the Capitol and happened to meet Illinois Senator Everett Dirksen from Pekin. When he learned we were from Illinois, he shook our hands warmly and turned on his political charm. I was surprised by his short stature. Bob and I went back to Chicago on the old steel rail, this time in sleeping car *Muscatatuck.* At Cumberland, Maryland, there were 30 inches of snow on the ground.

All through February Ed Vanneman worked on his brief because we knew there was going to be a court battle whatever the ICC's decision. My Peoria friends had learned that Moto-Mower Company of Richmond, Indiana, had shown some interest in locating a manufacturing plant in Peoria and asked me to make a sales pitch for a site at the Pioneer Industrial Park. My wife and I took a driving holiday March 24th and met the Peoria crowd at the local airport in Richmond. John Altorfer, heir to the Altorfer Washing Machine Company, had flown them down in his private plane. John was a "blue card" pilot; he would fly only when his blue license card matched the sky's color. Nothing came of this expedition, but we had a pleasant time along the way.

The mills of the ICC grind very slowly indeed. We learned on October 2, 1961 that the ICC had approved construction of the Pioneer lead track. I worked with Hal Lenske of Public Affairs on a press release the next day. The C&NW's stock jumped $3\frac{1}{2}$ points. Of course the Rock Island appealed the ruling on October 10th which put construction on hold again. Their appeal was denied January 19, 1962. But the Rock was not yet finished; they went to United States District Court to try and obtain a restraining order. Ed Vanneman contacted our local Peoria attorney, Mike Gard, to enter an intervenor petition. On October 31st, Ed Vanneman and I went to Peoria to appear before Judge Parsons. The Rock Island obtained a temporary injunction from the judge based on what our attorneys felt was very bad legal grounds. A hearing before a three judge panel was scheduled for February 20, 1962, but two of three judges failed to appear forcing a continuation to March 12th. Another of our C&NW attorneys, Bob Russell, substituted for Ed Vanneman at this hearing. He injected strong arguments for urgency and the loss of revenues throughout the testimony which seemed to catch the Rock Island by surprise. They suddenly realized that their railroad might have to put up a substantial bond of indemnification against fiscal loss should they choose to try and delay this matter longer. Russell reported optimistically to Mr. Fitz, but their elation was premature as the next hearing March 28th was postponed again to April 4, 1962. This time, however, the nickel dropped, and we were in business.

Construction of the industrial lead fell to my old friend, M. C. (Chris) Christensen who had replaced Bill Wilbur as Galena Division engineer; Bill had been promoted to engineer of bridges after Arthur E. Harris retired. Chris organized the field crews and grading contractors so well that dirt was being moved by May 23rd. The survey crew was led by John Olson, the mulish assistant engineer. He had Jim Simons as instrumentman, and Dave Blutt, rodman. The next day I received the Muirson Label Company's purchase order for the private side track coming from the industrial lead under construction. Greedy to be involved in actual track construction again, I spent a good deal of time on the job and got to see first hand what an incompetent head of party we had doing the survey. It was all I could do to refrain from pushing him in a hole somewhere and finishing the job myself. I was convinced he was about as smart as a wooden fence post. After giving him some good advice in an authoritative voice, I left the scene and made sure that Chris knew what he was dealing with. I rarely have had problems with any employee but this guy was a disaster wearing pants. Chris in his quiet Iowa way nodded in agreement and gave me a "what can I do?" look. When Olson transferred to the Twin Cities Division soon thereafter, the entire Galena Division staff breathed a sigh of relief. Now, when I held that job . . .

That summer I made several trips to the construction site to revel in the experience of track construction. I had come into the business at a time when building new track was a rare event. This was to be a sturdy piece of railroad with all new ties and secondhand 112 pound rail. On a June 28th visit to the site with the company photographer, Bob Lindholm, we found the work about 75% complete with a predicted finish to University Avenue by July 10, 1962.

From the vantage point of the present day we may ask if the construction of this long industrial lead was really worth all the effort. The 1.9 mile line meant hauling boxcar loads of paper stock up a stiff grade using road power from passing time freights, an expensive way to provide switching service. While Muirson Label, whose principal business was printing soup can labels, did generate a moderate amount of good revenue freight, it was doubtful that any real profit was made considering the expense of construction and litigation plus the awkward switching service. Quite apart from this local situation, however, was the larger benefit of gaining International Paper's respect for the C&NW's determination, diligence and perseverance which resulted in substantial additional IPCO overhead traffic far from the Peoria battleground. As a sidelight to this affair, R. J. McDonough, who had replaced R. C. Conley as the Galena Division superintendent in 1962, renamed the Radnor station "Pioneer". The old name of Radnor Hill gradually faded from use to become Pioneer Hill as new operating crews replaced the old heads.

Through the lens of history subsequent events are reviewed. In 1984 the City of Peoria acquired 8.2 miles of the former Chicago, Rock Island and Pacific branch line from Keller (Peoria) to Alta from the trustees of the bankrupt railroad's estate. This branch line served a number of small customers located in the northern part of Peoria and in the Village of Peoria Heights. Title was vested in a city-owned entity called the Peoria, Peoria Heights and Western Railroad which had reporting marks of PPHW. The city owned 75% of the company with the remaining 25% interest belonging to Peoria Heights. The

GP9 1758 spots the first revenue car of paper for the Muirson Label Company at the Pioneer Industrial Park north of Peoria, Illinois on July 30, 1962.

property was leased to the Peoria and Pekin Union for operation. On April 7, 1998 the Pioneer Industrial Railway Company (PRY) took over operations from the P&PU. On July 28, 2004 an announcement was made that the Central Illinois Railway would operate the 8.29 mile line.

Up on the hill at the Muirson Label plant rail business gradually dwindled over the years. C&NW service to the plant still required stopping heavy northbound trains and using the road engines to switch the plant—an operation not conducive to efficient train operations. The operating department continued to mutter about providing switching service and were anxious to see this awkward move eliminated. Hoped-for industrial development along the new branch did not occur—Muirson remained the only rail-served customer on the spur. The line eventually went dormant and the automatic crossing protection at Route 174 (now named North Allen Drive) was deactivated. The Union Pacific Railroad sold the unused branch to the City of Peoria in 2001.

In March 2004 the City of Peoria through its Peoria, Peoria Heights and Western Railroad entity filed notice with the Department of Transportation's Surface Transportation Board (successor to the ICC) that they were planning to construct 1,800 feet of railroad. The new railroad would connect the end of the former C&NW

Pioneer Industrial Park
Peoria, Illinois, 1962

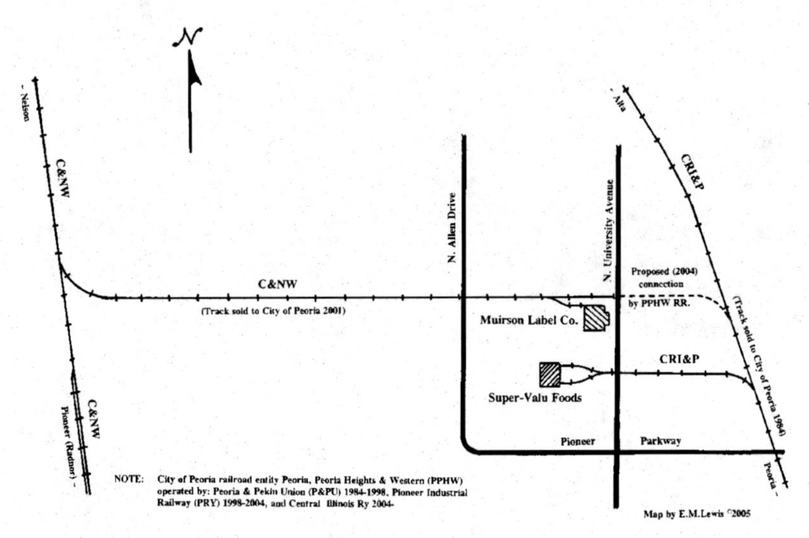

Pioneer Industrial Park north of Peoria where the C&NW and the Rock Island waged lusty legal battles over territory. The tracks were later acquired by the City's Peoria, Peoria Heights & Western RR.

---

spur with the ex-Rock Island branch. This was the very plan that we had proposed to the Rock Island on September 13, 1960 and which they had so vigorously opposed. The 2004 proposal called for abandoning about 7.5 miles of the former Rock Island branch southeast of the Pioneer Industrial Park for conversion into a recreational trail. The construction of the new connection to the former C&NW (now Union Pacific) line would permit continued rail service to rail-served customers from the west. What arrangements the Union Pacific would make to setout and pickup cars from the operator of the branch were not discussed in the application. Perhaps the next forty years will see further development of new rail-users along this star-crossed railroad.

So much for the two battles of Peoria. There was other life and activity in progress during the legal jousting. Returning to May of 1960, we heard a rumor on the 13th that the Chicago Great Western was to be the next acquisition by the C&NW. Four years later that rumor became fact. Another rumor floated by on May 26th that the Iowa Terminal electric line at Mason City, (formerly the Mason City and Clear Lake), was available, but recommendations were against acquisition. We heard a buzz at the end of May that our boss, Gene Cermak, was leaving (not so). July 13th's rumor was about acquiring the Chicago and Eastern Illinois, but no one could figure out why as it was a poor fit for the C&NW.

Drop-in prospects were rare, but I had one on May 19th in the form of Arthur Dixon, vice president of Modine Manufacturing Company. He wanted a new factory site north of McHenry, Illinois, for a manufacturing operation. I learned from him what specific area and what rail services he required. The next day I was off to the McHenry County courthouse in Woodstock to check land ownerships of properties that were contiguous to the C&NW's Williams Bay Line near Ringwood and Richmond, just south of the Illinois-Wisconsin state line. We heard nothing back from them until November 15th when we learned they had acquired property at Ringwood. Eventually, Modine built an automotive radiator manufacturing facility there, but the rail freight traffic never amounted to very much. During May, business activity was slow. On the 31st I rode the Illinois Central's *Green Diamond* to St. Louis and found the city full of conventioneering Junior Chamber of Commerce types. I visited with Kerr at the Chamber of Commerce before catching the GM&O's afternoon train to Chicago at Union Station. Flying was faster but riding the train was still a pleasure because it provided opportunities to relax and get some reading done.

Our fast-driving roadmaster at South Pekin, Bill Baker, died on the job June 23rd from heart failure. He was 45. Bill Wilbur and I attended his funeral services in Peoria on the 27th. A big staff meeting was held in Chicago August 8th. Management singled out our department for its successes in locating new revenue generating plants on our lines. We were urged to develop more outside contacts, rail related or not, as it was good for the C&NW's image. The next day, my friend Carl Anderson from the real estate department, privately advised me that Ken Cady, his vice president, wanted to hire me as an industrial land planner and not to sell myself too cheaply. It never happened because Kendall Cady soon was history, replaced by real estate salesman I. R. (Bob) Ballin whom Cady had recruited. I took another short vacation to New England and found another Fantus Company prospect waiting for me on my return. Described only as a food processor, I pulled together data on several likely locations and gave them to Maurie Fulton of Fantus. The bakery and warehouse prospect turned out to be Sara Lee as we learned October 28th when they announced construction on the Milwaukee Road in Glenview. Never much of a rail user, the plant was torn down and replaced by condominiums in the last years of the century.

In September I was in St. Louis again with Ed Heitz (real estate) and Herman Jacobson (rates) to meet with Commander John Barron at the invitation of Jim Thompson of our St. Louis sales office. Barron was managing director of the Bi-State Agency, a joint operation of the States of Illinois and Missouri for properties along the Mississippi River in the Madison (Illinois) area. John was an old river rat and barge operator on the Upper Mississippi. He had some interesting ideas about rail service to elevators owned by the Agency and found listeners at the C&NW because we were really hard pressed for river accessible sites. We had come to St. Louis on the *Green Diamond* but went back to Chicago on the Wabash in sleeper *Blue Knight*.

October was a busy month. On the 5th we learned of a new electric generating plant for Iowa Public Service at Salix, Iowa, south of Sioux City. This site was one I had mapped for George Wimmer two years ago. Four miles of track would be required to bring coal trains to the plant which was to be located on the east bank of the Missouri River. On the 14th I received an inquiry from Liquid Carbonics who was also interested in being close to the power plant. Since Iowa was not my territory, I turned the inquiry over to Vern Giegold. I do not recall if anything matured from the Liquid Carbonics inquiry. On October 10th Iver S. Olsen, manager of online sales, invited me to lunch with Mr. Fischer, traffic manager of Weyerhaeuser, at the Drake Hotel. On the way there we ran into the City of Chicago's welcome to the King and Queen of Denmark. October 13th was the first day of operation of the new Congress Street Expressway that ran west from the Loop; that first morning was filled with great confusion and traffic jams.

The C&NW's industrial development department under Gene F.

Cermak was expanded in October when Dorvan Skoglund from the former M&StL joined the Chicago office. Another M&StL man, Richard C. Volkert, had joined earlier despite the fact that the official merger of the properties was actually consummated November 1st. We now had twelve people to handle the increasing volume of prospects and attendant correspondence. During this period construction of the International Paper warehouse at Northlake was bait in a sting operation that netted the mayor of Northlake and a couple of other city officials. IPCO had been forced to make payoffs to the city's mayor, Neri, to obtain building permits, water service and related permissions. Complaints to state authorities resulted in an investigation. Cermak was called into the matter November 2nd to provide background information. The mayor was indicted, convicted and served time.

At the request of our Peoria sales office, I rode the *Green Diamond* to Clinton, Illinois, where Joe Zelenda, our traveling agent, picked me up for a call on Edward Aylward of Aylco Chemical in Sullivan, Illinois. Aylward was promoting a $13 million dollar chemical manufacturing plant to be located somewhere in Illinois and the Peoria office thought we should look into it. It seemed to me to be a rather grand project for a man operating a small fertilizer distributorship in Sullivan, but I introduced him to my contacts at CILCO (Central Illinois Light Company) in Peoria on November 7th because the electric and natural gas rates were critical to the success of such an operation. Mark Townsend and I took them on a tour of Peoria area which took so long that I missed the *Rocket* to Chicago. There was always the late afternoon Rock Island motorbus that connected with the eastbound *Rocky Mountain Rocket* at Bureau. Aylward's scheme faded away—another waste of time.

The rest of the year dribbled away in small packets. On November 30th the Rock River Valley Industrial Conference met at Dixon, Illinois. There was a lot of speechifying to the 175 attendees, but not with much visible result. I took another swing south to St. Louis to visit with Jim Thompson and S. A. Keathley in the sales office before heading north by rented car to Edwardsville and Springfield. There was a vague tickle about a Remington Rand project from Fritz Bexten in LaSalle, but the rumor disappeared without a trace. On December 13th Governor Frank Morrison of Nebraska visited our offices and was promptly invited to go upstairs and have a cozy chat with Ben Heineman, Mr. Fitz, Cermak and me. Frank was a great bear of a man and a Democrat which was unusual for such the Republican state. And that was the end of business for me in 1960. Our family vacation was a radical departure from previous years—we flew to Europe to visit Portugal and England with Christmas in Paris.

15

RESTLESS TIMES, 1961. . . Back from Europe with a new perspective I was greeted with the "secret" news that the C&NW and Milwaukee were exploring merger again. Some secret! Just the prospect of combining the two properties disrupted our normal business affairs. The traffic department on the 10th floor of the old Daily News Building was overwhelmed with "secret" demands for information from attorneys and accountants. Estimates of potential traffic shifts, opinions about probable customer reactions and what would the public think (as if the traffic department even recognized the public at all). Detailed estimates and statements were done over many times as new scenarios were presented from day to day. Our industrial development department was located on the 13th floor away from the traffic department so that we were not as involved in the furor. On February 23rd the official line on the proposed combination was released to the press (which had been speculating about rumors for some time).

Business traveling started in January when Gene Cermak, J. S. (Jack) Frost of the Illinois Central and I rode the train together to Milwaukee, Wisconsin, on the 16th to attend the annual Great Lake States Industrial Council meeting. The formal parts of the session were merely interruptions to the gin rummy and poker games which sometimes ran all night. The next week Gene and I were off to Peoria to deliver a letter to Harold Austin, president of the Pioneer Industrial Park, as related before. Returning to Chicago on the *Rocket*, I rushed for home, had supper, packed my bag again and headed downtown on the commuter train. At Union Station I boarded the CB&Q's *Zephyr* for Lincoln, Nebraska, with space in sleeper *Silver Falls*. By chance I ran into my old boss, Harold W. Jensen, on the train. On January 17th while admiring CB&Q motor car #9767 standing at the Lincoln depot, Dick Ryan, our sales agent, appeared to take me to see Fred Berniklau, the developer whom I had met last year. Fred had a piece of property abutting our Fremont-Lincoln line on the north side of town and needed ideas on how to successfully market it for industrial use. After a session with the state's resources division and a guest visit to the Elks Club (which was one of the few places in town serving liquor), I had a bite to eat and boarded the same sleeper for the trip back to Chicago. Ben Heineman was installed as a vice president of the Chicago Association of Commerce and Industry's Industrial Development Commission on January 19th.

This account may be a bit ragged because many of the events surrounding the First Battle of Peoria (the C&NW's struggle with the TP&W) were recorded in the previous chapter and are not repeated in this account. All through this period, this legal action kept bobbing up which tends to leave unexplained holes in the 1961 flow of events. For that matter there are similar holes in all the years to 1966.

Returning to February 10, 1961, I was called to a meeting in the 10th floor traffic department offices to meet with Pete Labagh, traffic manager of the California Packing Company, about their new

Rochelle, Illinois, *circa* 1962

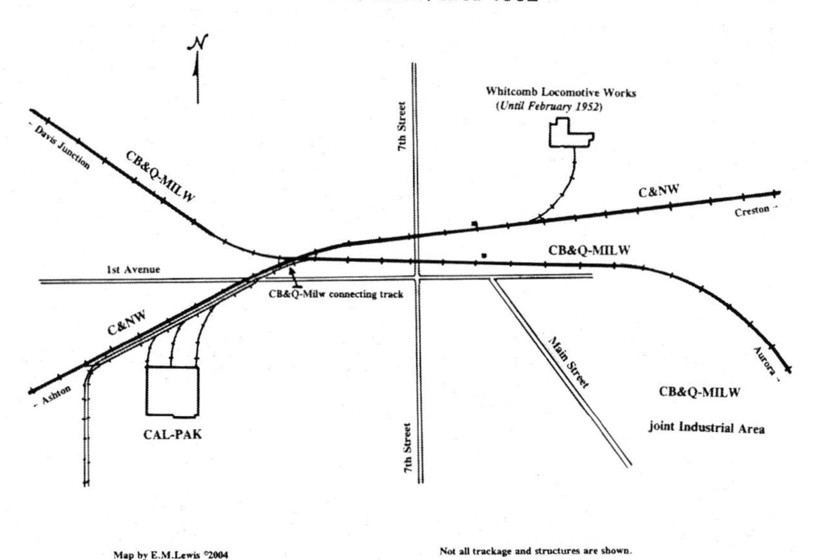

General view of Rochelle about 1962 showing the C&NW and CB&Q/Milwaukee Road mainlines, location of the California Packing Company's distribution center and the former Whitcomb Locomotive plant. The Union Pacific's container terminal constructed in 2003 is to the left toward Ashton and south of the main line west.

warehouse construction in Rochelle, Illinois. I was directed to contact O. O. Waggener, the CB&Q's industrial development chief, and Edward Stoll, who held a similar post with the Milwaukee Road to establish joint switching service. Gene Cermak then appeared to see if the weight of his position would be helpful in my negotiations. The warehouse was to be served by three railroads: C&NW, CB&Q and CMSTP&P (MILW). The latter two were partners because the Milwaukee operated on trackage rights from the CB&Q. Ollie Wagoner and I got hung up on how the costs of track construction were to be allocated. The C&NW and CB&Q track construction policies were quite different with regard to how much of the costs each railroad could legally assume. The cost allocation policies were established in 1918 by the U. S. Railroad Administration when they ran the Nation's railroads. When the roads were handed back to their owners in 1920, each carrier had to either continue using the Federal model or adapt its own individual track construction policy. Once the cost formula was selected, the railroad had to continue its adopted posture or face discrimination suits. The CalPak project subsided from March almost through May. Cermak made a trip to their headquarters in Los Angeles on May 1st to see what needed doing but the delay was an internal company problem over financing that had to be settled. Gene Cermak was a real will-o-the wisp that month. I finally got a chance to talk to him about CalPak and other important matters on May 23rd, 17 days after I had last seen him, although we worked in the same office. Nothing to do but wait and see what transpired. Finally, on June 29th the CalPak principals came to Chicago to seek the C&NW's commitment to serve the new facility and track construction was underway by July 20th. That track cost problem, however, was not yet settled. At this point the CB&Q needed a small piece of C&NW right-of-way to get to the warehouse. If the C&NW had wanted to delay and possibly foil the CB&Q-MILW entry to the warehouse, now was the time to do it. But, CalPak's business was too big and understandings had already been reached. The C&NW sold the tiny sliver of land to the Q to keep the peace. As far as track construction policy was concerned the C&NW could assume the cost of installation of a turnout from a main or industrial lead track and the cost of the new siding from the point of switch to the 12 foot clearance point. Beyond that point the industry assumed all costs. What the CB&Q policy was, I did not record. At this point of complication, our law department took over. I did not, nor did I need to, see the final agreement. The trackage was constructed to serve the huge new warehouse located on the south side of the C&NW mainline just west of the CB&Q diamond crossings. Each railroad in turn switched the warehouse for six month stints. It was a big revenue producer.

The long promised Nebraska visit promised to Governor Morrison last year occurred on Valentine's Day 1961. The day before, Gene Cermak, Vern Giegold, Don Guenther and I traveled west on the CB&Q's *Zephyr* in sleeping car *Silver Bay.* In Lincoln our sales agent Dick Ryan and Lee Rising collected us off the train and took us to see Dave Osterhout at the state resources division before calling on the Governor in his office. We lunched with people from the state utilities and enjoyed an afternoon at the Elks Club playing snooker. The next day, Gene Cermak, was given a commission as Admiral in the

Navy of the Great State of Nebraska. The symbol of their "Navy" was the prairie schooner "sailing" across the great open plains under a billowing canvas wagon top. We also showed the Lincoln chamber of commerce where their new industrial district was (Berniklau's land); they were surprised. There was more food and socializing before climbing back on the *Zephyr's* sleeper *Silver Chasm.* Two weeks later on the 27th Brad Huedepohl, our geologist, announced that he was going to Nigeria as a consultant for the Rockefeller Foundation. Brad's boss, Bill Kluender, was a difficult man to work for—so difficult that Brad decided to quit the railroad. He would bob up again in subsequent years. That evening I was off to Nebraska again in sleeper *Silver Bay* in the *Ak-Sar-Ben Zephyr's* consist. I addressed a meeting held at the courthouse in Lincoln regarding the value of paving of roads in industrial areas. After lunch and a visit to the industrial area, I broke my dependance on rail travel and sleeping cars by flying in a DC-6 (*Mount Vernon*) to Omaha and thence to O'Hare in a 720 jet. Never look back—sleeping cars were becoming nostalgia items.

An important inquiry came from Armour and Company March 10, 1961. This eventually turned out to be a new manufacturing plant for Dial soap. Their site specifications centered on the Fox River valley west of Chicago so I threw together materials about Carpentersville, Elgin and West Chicago for them. They showed mild interest in land along the west bank of the Fox River at Elgin just where the Illinois Tollway crossed it. Our alignment maps showed that a low level timber trestle once crossed the river from the Williams Bay Line near Logan Street, but only the pile stubs now could be seen in the water. The track once served the Elgin Torpedo Sand Company. The proximity of a mobile home park next to this old sand pit doomed the likelihood of it ever being used as an industrial site. The soap factory project cooled off until it suddenly revived on November 8th. The next day I took Chadwell of Armour on a tour that included Proviso, West Chicago, Elgin and Carpentersville. On December 6th he said he liked West Chicago. Six days later he came into the office to look at materials on DeKalb and Elgin. On the 29th I worked out probabilities for 100 year floods for various river sites. The Rock Island's proposed site was wiped out by my hydrological study. The ebb and flow of Chad's investigations gave me a most interesting glimpse of Armour's internal politics when a new plant site was being considered.

The industrial development department rejoined the parent traffic department on the fourth floor of the Daily News Building on March 24, 1961. Happily, I was out of town and missed the fuss and feathers of moving. Unhappily, I did not like my new work space because I was plopped into a large open area far away from any window. In this period Cermak and I had driven to Peoria on the 13th to meet Vic DeGrazia, successor to our friend Jim Cannon of the State Industrial Office. Vic was the cartoon stereotype of a politician—large, unctuous and totally fixated on his image. With Vic in charge, our relationship with the state agency cooled. At least the trip on the newly opened Interstate 80 was a pleasure drive when contrasted to the old two lane roads. While we were in Springfield, I collected soil boring data from the state highway department taken from and near industrial sites in which we were interested. They would be handy for quick reference in coming years. On the 23rd I drove to Urbana to see what information the state water supply people had and brought back another large supply of data. Back in 1959, on August 5th, as a matter of fact, we had received a call from someone asking for information on salt deposits in western Nebraska. At the time we did not have anything relating to the subject. I knew that Bill Kluender, the resources manager, would not share his information, if indeed he had any, so I did not share this inquiry with him. I also knew that Nebraska had been systematically drilling all over the state to better understand what was under their prairie sod. When in Lincoln I made a special effort to learn about this program and was escorted through a large cellar containing rows and rows of soil corings laid out in special trays. In response to inquiry, the director said that they had cored the rock in the Chadron area and there were indeed salt deposits found there. I visited with a Mr. Hopper of Morton Salt Company in Chicago on March 30th. They were on the lookout for both salt and potash deposits in the west. This was before the big potash mining projects in Alberta, Canada had come into the market. Remembering those trays of corings, I arranged to go to the state geologist at the University of Nebraska when I was in Lincoln in August and brought back a 2 inch diameter by 6 inch length of halite (salt) that came from 3,000 feet underneath Chadron. I took the chunk to

Morton on August 15th. Mr. Hopper seemed more amused than excited and sent the piece down to the lab. It was indeed pure salt, but, as with many projects, this one never came up again.

On April 3rd a huge pile of industrial prospect files from the now-absorbed M&StL arrived. We picked through them to separate them territorially. Vern Giegold got most of them as they were largely Iowa based. The few that fell to Illinois did not warrant any follow-up. However, I did a tour with Joe Ferguson from our Peoria sales office on May 17th to look at M&StL properties from Peoria west through Maxwell, Trivoli, Farmington, London Mills, Abingdon, and Monmouth. Joe had just been transferred from the Omaha sales office and was anxious to learn the territory. We called on the superintendent of the strip mines at Rapatee (Middle Grove) before heading back via Berwick and Nemo. The whole west end of the M&StL line was the great Illinois desert as far as we were concerned. The next evening was the 50th anniversary dinner of the Peoria Association of Commerce. Among the dignitaries were Governor Otto Kerner (who later ended up in jail) and John Altorfer, a wealthy local boy getting into politics. He later was to run, unsuccessfully, for governor. Following the dinner the next day, I had a session with Jim Gilchrist of the Association and Mark Townsend of CILCO about how we could get that 2,000 acre dream site just opposite Pekin into shape for industrial development. Little did we know what trouble we were headed for.

Our Jack Sommer moved to Milwaukee and worked directly with Harvey Buchholz there. On April 11th I rode our train to Milwaukee. Jack drove us west to Madison to meet with a land developer subdividing property near that city. Just missing our train back to Chicago, I rode the Milwaukee Road train instead. The next day I happened to run into J. R (Bob) Brennan, passenger traffic manager, C. F. (Charlie) Stewart, his assistant, and Don Gunvalson, our Rockford DF&PA. They were involved in the discontinuance of west line trains #3 and #4. On July 15, 1961 those lightly used passenger runs were gone. I was still determined to make a tour of the C&NW's Nebraska lines. Twice I had been foiled by bad weather, but the third try was successful. Starting out in a Chicago snowstorm on a Sunday night (the 16th) I trudged to Union Station and found my space on the CB&Q's *Ak-Sar-Ben Zephyr* in sleeping car *Silver Palisades.* Was this trip in vain again? In the morning in Omaha I found there was no snow in Nebraska and the day was clear and warm. Dick Ryan, our sales agent met the train in Lincoln. We visited every station on the line down to Superior. An absolutely useless trip. The population was sparse, the stations were mostly names on the map, and the railroad itself was a joke. The sixty pound rail in the line was the original laid down in 1885. We stayed in a crummy country hotel in Superior whose principal reason for existence was the Ideal Cement manufacturing plant (production ended in 1986). The C&NW line ran right up to the Kansas state line, but not across it because that would have meant another set of regulatory agencies to deal with. The Missouri Pacific and AT&SF connections at Superior produced very little traffic. You had to see it to believe it. The next day I made a call on a surprised chamber of commerce manager and dutifully looked at the industrial sites designated around town. That did not take long and Dick and I headed back north to Lincoln. We visited with a section foreman and his two men at Davenport. I asked him what he used for the ties to hold the gauge as the ones I had seen in the track were not very good. He looked at me with a wicked grin and pointed to a pile of new creosoted ties piled on the Union Pacific's land at the diamond crossing. (The 84 mile Seward—Superior line saw its last C&NW service on December 26, 1972 and was sold to the Great Plains Railway which ran its first train June 27, 1974 and its last, April 15, 1975). After a quiet night at a Lincoln motel, I managed a ride to Omaha to attend the Society of Industrial Realtors (SIR) meeting. Governor Frank Morrison was there. John Staley of Quaker Oats Company was the principal speaker. I am sure he did not remember me from my days at Quaker when I was employed as a very junior freight accounting clerk. At the end of his speech, Governor Morrison, tiptoed over to me in his bearish manner and in a hoarse whisper that echoed through the room asked if I would drive him to the airport. Startled, I said yes even though I didn't have a car available. Fortunately, someone else was going to Eppley Field and the Governor went off with them. (Note: Frank Morrison died in McCook in April 2004 at the age of 98.)

After the meeting Dave Osterhout of the state industrial group drove the Governor's car back to Lincoln and left his for my use as I was headed that way for the annual dinner of the Lincoln Traffic Club. Elbert (Smitty) Smith, industrial agent for the Rock Island needed a lift, so he went with me and was fairly amazed (as I was) to be seated at the head table and introduced. By mid-August Smitty was to be pretty sore at us during the second battle of Peoria as related in the previous chapter. The next day hit 85 degrees and it was hard to recall the snow in Chicago just a few days ago. I drove a rental car north from Lincoln to Fremont, visited with Howard Shinrock, and then went to West Point and Norfolk before cutting back to Lincoln. Not much accomplished, but I did get a first look at the Nebraska country and tracks where in another twelve years an immense project was to burst into being and then gradually fade away. The CB&Q train to Chicago was full so I tried the Rock Island which was running two hours late because of a derailment and signal problems. I slept well in sleeper *Granger.*

Freight revenues were down, and we received orders to cut expenses ten percent. The cut was confirmed at a staff meeting April 24th. We were also presented with a book outlining 100 parcels of railroad land that I. R. Ballin was committed to sell. We were invited to review the properties for sale and make comments. Any real objections would have to be substantiated by proving we had a real, live customer actively planning to build before Ballin would eliminate the land from his sale book. Some of our better sites were on the block. The next day I went to Dixon at the request of the local group to look at an old quarry where they thought a fertilizer manufacturer should build a production plant. It was a bad site and I told them so. The next day I attended a luncheon meeting with the industrial group in Elgin, an area just beginning to stir. This little excursion netted a quick sighting of one of the EJ&E's strange double diesels, #915, rambling along with a short local freight. May 5th Alan Shepard went into space and came back safely—a big boost for the space program. Another important railroad event occurred May 10th—my normal commuter train departing Elmhurst at 7:25 a.m. was equipped with the new push-pull double decked gallery cars. This train was always crowded as the first stop was Kedzie Avenue. Since I rode on a pass, I usually had to stand. I divided my train riding between the center vestibule which was mighty cold and filled with tiny icy crystals on snowy days and the front end of the leading car. The lead car was a wonderful way to inspect the track into Chicago. Unfortunately, this car also was the smoker, but I found that by staying close to the front door with its slight leakage of air made the atmosphere tolerable. Years later, crew complaints about the stench resulted in a complete ban on smoking throughout the trains. Financial troubles also afforded management a chance to rid themselves of several unwanted people. Art Mason and H. Coates, longtime junior staff assistants of the old land department, were fired from real estate on May 31st. This was a bit of shock as we had not seen a deliberate firing before from the office staff. Usually undesired people were pressured to resign and quietly go away. The "new" managers ran their departments in a more ruthless management style than we were used to.

Our industrial development department was getting more exposure. I was asked to be the model for a series of pictures to accompany a story about our operations to appear in the International-Stanley Company magazine. Mickey Pallas, a Chicago professional photographer, and I cruised around the Chicago area to find suitable sites for the shots. On June 16th we went to the Kane County courthouse in Geneva for a background in the County Recorder's office, visited a soils testing laboratory, posed with the principal of my old high school in Elmhurst, and stood in front of a firehouse in Franklin Park admiring the equipment. No one was around the firehouse, so I just rolled up the doors, posed in front of the engines, and then closed it all up again. Other chores at this time included helping draft a speech for Ben W. Heineman about the prospects for DuPage County over the next 20 years—as if I had a crystal ball!

Events were stirring—slightly—way out west in Wyoming. I sold the railroad's old Glenrock gravel pit to American Humates Company in June. A start-up company, they had cash problems in establishing a plant and market for their product, a soil conditioner made from the local lignite coal. I finally got a check from them and sent it on to A. G. Johnson, superintendent, in Chadron on October 19th. While this company never really did very much, I got rid of a piece of surplus of property. Curiously, we were to take a look at that pit in 1974 to see if any sizeable ballast deposits remained—a quick glance told us it was mined out. The next event was another inquiry from Armour & Company for a plant site in the Omaha area. Per his instructions good leads were to be brought immediately to Cermak's attention. In this case I had to call him at home because he was never in the office at this time of the year because of his obsession with golf. Then on June 12th I was informed that Northwestern Metals had chosen Fred Berniklau's land at Lincoln, Nebraska, for their new smelter and automobile scrapping operation. This good news, however, set the stage for a major confrontation with my boss. Coming into Chicago on the sleeper *Northern Star* on the morning of June 21st from a short vacation near Ashland, Wisconsin, I just had time to rush across the Loop to catch the Illinois Central's *City of New Orleans* at the 12th Street station. I joined a party on the train consisting of Gene Cermak, Jack Frost, Tom Gage, Keith R. McCullagh and a Mr. Adams, all from the Illinois Central, as well as Carl Ernst from the C&EI, to attend a railroad sponsored industrial development meeting at Southern Illinois University in Carbondale, Illinois. The conference on June 22nd went well despite the fact that many participants had too enthusiastically enjoyed an evening of booze and poker. That evening our planned return on the Illinois Central to Chicago was delayed by a derailment somewhere to the south. At least a three hour delay on all northbound trains was projected. I was supposed to be in Chicago the next morning because the top officers of Northwestern Metals were to be in our office to finalize arrangements for their new sidetrack at Lincoln, Nebraska. An old railroad axiom says that late trains only get later, and I began looking for an alternate way to get back to Chicago to protect my appointment. Jack Frost, as courteous a Southern gentleman as I had ever met, thoughtfully offered an IC driver and a company automobile to take his man, Keith McCullagh, and me to St. Louis so that we could catch a train to Chicago on a line unaffected by the derailment. For some reason that to this day I do not understand, my superior, Gene Cermak, lost his temper in front of the entire group and demanded that I reject Jack's kind offer. In my mind's eye I can still see Cermak's thinning red hair standing on end like fibers in an static electricity demonstration. His prominent freckles stood out in high relief and his light blue eyes figuratively shot sparks. Shocked by this uncharacteristic loss of control before his associates, I considered for a moment that my clients were coming a considerable distance to establish a revenue generating business for our railroad and that I owed them the courtesy of being there regardless of personal inconvenience, I accepted Jack Frost's offer and went to St. Louis. Rolling through the black night with Keith, I was very upset at the uproar. Keith McCullagh, my traveling companion, was also totally mystified by Cermak's outburst. I asked him if I could get a job at the IC because I was surely a dead duck on the C&NW. Keith and I caught the GM&O train at 11:45 p.m. arriving in Chicago rested if not relaxed. Leo Hill, owner of Northwestern Metals, and his son were right on time the next morning. We worked out a track plan for his property in a genial and profitable meeting. Cermak and his crowd on the IC's delayed train had arrived in Chicago at 5:20 a.m. When Gene finally dragged into the office later that morning, he did not say a word to me. This incident marked the beginning of a long and steady decline in our relationship. I was ready to resign, but by now had decided that he was in the wrong and it was his first move. The event faded into the background when on June 26th Heineman publicly congratulated the department for its many successes.

These were volatile times with rumors flying about like pigeons in the park. On July 5th the day before the regular board meeting, we heard that Clyde J. Fitzpatrick, our president, was leaving and that Larry S. Provo was to be the new president with R. David Leach (both former Arthur Anderson accountants) to be vice president. Of course these were rumors, but Provo and Leach did attain those positions in another five years. Provo had served with the M&StL as vice president and comptroller when Ben Heineman controlled that property. Two days later we heard that Mr. Fitz was to be president of the Union Pacific (that never happened). Out at Proviso on the 19th I came across into two GTW locomotives, #3752 (2-8-2) and a big and grimy 4-8-2, #6038, on their way to Northwestern Steel and Wire at Sterling.

Always looking for a chance to extend the C&NW's territory, I paper "engineered" a connection with the C&IM at the point where the C&NW crossed that railroad south of Pekin, Illinois. While the connecting track was feasible, it involved building a long descending track starting at the south end of Illinois River bridge #1731, alongside the embankment on the west side with a sharp right-hand curve at the bottom to avoid the land of an electric generating station. This track was built several years later when unit coal train service was required to the power plant. The whole idea was to cut the Peoria and Pekin Union Railway out of the interchange picture. Not only were there costs involved, but the physical interchange of CNW-PPU-CIM with a 100 car train through congested Peoria was a nightmare operation.

That small town fertilizer distributor, Edward Aylward from Sullivan, Illinois, with his grandiose dream for construction of a major fertilizer manufacturing plant, popped up again. I rode to Peoria on the *Rocket* in the parlor car on a hot July day in a car with a broken air conditioner to meet with Aylward. I introduced him to the TP&W people in good faith because our legal battle was just starting to get serious. The president, J. Russel Coulter, ducked the meeting until someone told him the size of the proposed investment and then he appeared. Of course, the whole scheme was smoke and mirrors, but at that time we just could not be sure. In August I heard that Morton Salt Company had taken a look at investing in Aylward's proposal, but soon dropped it—they saw through the smoke. I took note of their disinterest and put Aylward's project into an inactive file.

I had several inquiries for sites in that TPW-CNW joint service territory. In 1961 the bitter struggle over the so-called joint service was just beginning to surface. The Aylward fertilizer flop was just one of the nibbles. Don Guenther and I were in Peoria July 24th when we ran into Edgar Vanneman and John Danielson, two of our attorneys. They were defending the C&NW in a suit brought by the Peoria and Pekin Union Railway who was trying to become a member of the joint C&NW-TPW service area at Mapleton. The P&PU contended that they owned a right-of-way parallel to and between the C&NW's main track and the Peoria Railway Terminal's (CRIP) defunct line from Iowa Junction south to Hollis and the Illinois 9 grade crossing. If they had a right-of-way, it was not very evident as any old railroad grade was thoroughly obscured by weeds and brush. They admitted it had not been used for 40 years. The next day while Don was busy conducting a market research project involving usage of glass bottle enclosures, I dropped into the courthouse to hear some of the testimony. Not being a witness or involved in the action, I could be an auditor. Later, I met with Mark Townsend of CILCO, Fred Bexten from the state's industrial commission and some people from Baird Chemical, who were interested in establishing a chemical plant at Mapleton. They were not pleased with the price for land quoted by Mr. Coulter and went to Joliet on the 26th to look at property there. On August 14th I heard they had resolved their price problem with the TPW and picked up options on the land at Mapleton with the actual purchase in March 1962.

The end of July was my 14th anniversary with the railroad and there I was on the CB&Q's sleeper *Silver Isle* headed for Omaha. The next day I met three men from Armour Chemical and took them on a tour of industrial sites in Blair, Fremont and Omaha before returning that night to Chicago aboard the Rock Island's *Rocky Mountain Rocket* in sleeper *Plainsman.* Within a week I was back in Nebraska in that same old *Silver Isle* sleeping car but Gene Cermak was with me because there was golf outing in Lincoln he wanted to attend. Dick Ryan met us. I split off the party and went to the University to get that salt sample as related before, rejoining the group to have lunch with our Omaha sales chief, Jack Peters, Lee Rising and Chuck Sayre with the state power group. They all went off to play golf while I found other more interesting things to do like look at the CB&Q's railroad yards and facilities and try to find the Omaha, Lincoln and Beatrice's tiny line. (The OL&B had opened August 25, 1906 as an electric interurban with big ideas. All passenger service ended in May 1928. The Abel interests acquired the line to serve their Ready Mixed Concrete subsidiary and operated it under electric power until 1950). After a big, festive and well-oiled dinner, the party started all night poker games. Fred Berrniklau, newly blessed by sale of his property to Leo Hill for the new Northwestern Metals plant, was happy to be included in the game with Cermak. It was somewhat ironic that Cermak was thrust into close proximity with the people and the plant location that had set the stage for his emotional outburst in Carbondale, Illinois, six weeks earlier. The next day I drove north with Gene to Fremont, introduced him to Howard Shinrock of the Chamber of Commerce, looked at the proposed site for Armour Chemical, and went on west to Norfolk where Gene had never been before. The real gateway to the west lies just outside Norfolk on U. S. Highway 20. The vast horizons, diminution of groves of trees and the undulating land pierced by straight as an arrow highways all contribute to a feeling that this country is different. At Norfolk we visited with Holman (Shorty) Braden in the engineering office of the Nebraska Division and called on Kruger of the Chamber of Commerce. On August 10th we headed back east via Columbus located on the Union Pacific and had a meeting with Chadwell of Armour's real estate department and the folks from Northern Natural Gas Company. And then, we did the unthinkable—we flew back to Chicago on DC-8 despite Cermak's protestations that he was a white-knuckle flyer. After days of traveling in sunlit and dusty Nebraska, the thought of another night on a sleeper was just too much, and I am sure that Cermak really liked the convenience of a quick flight home rather than losing more money to me in a gin game on the sleeper. As an aside: I learned October 18th that Armour and Company had considered offering me a job in their real estate division but decided to promote one of their own employees. The proposed salary of $15,000 would have been substantially more than my railroad pay.

I met three men from the Revlon Company in Woodstock, Illinois, on August 11th. They were interested in establishing a cosmetics manufacturing plant far enough west of Chicago not to have to pay city wages. Woodstock, the county seat for McHenry County, was not very interested in industrial development and there was little property adjacent to the railroad in that area that could be utilized. One possible parcel east of town was tied up in a tangle of ownerships that included (so it was said) Chicago gangsters. I drove the group south to DeKalb for a look at other properties. When in New York August 17th, I called on Clinton Hoch, manager of the Fantus Company which was a location consultant. Hoch was quite open about Revlon being their client and that it was a Fantus recommendation that Woodstock would be the best site for the factory. I wondered how much freight revenue one could expect from cosmetics. I could not imagine, nor could I learn from Revlon, what railroad freight might be generated, yet, the name was prestigious, and we extended our services as best we could. Revlon, however, seemed to prefer doing their own investigations and did not share information with us. In November I heard they were doing an in-depth labor survey at Woodstock. Unexpectedly, Norm Rutkin and Bob Katz of Revlon, came to our office in Chicago on December 1st and said they had to find an alternative to the Woodstock property they had optioned because the land was underlain with peat. We tried to find other good sites in Crystal Lake and around Woodstock without success. On December 8th, Carl Crumrine, a C&NW finance officer, told me his family had optioned 80 acres of land at Woodstock to Revlon. At that point I was pretty certain that there were no benefits for the C&NW to be had from this client and just let it go to a natural conclusion, which it did—nothing came of it. There were many such proposals that would mean little to the railroad; we just had to learn how far we could get involved and still be able to step gracefully away when the prospect proved to be of little value to the railroad. Maurie Fulton of the Chicago Fantus office came up with another Northern Illinois prospect on August 29th. This was a giant electronics manufacturing company which turned out to be the Admiral Corporation of Chicago. They eventually constructed a huge facility on the east end of Harvard, Illinois.

All of our prospects were not large and showy. I helped Bob Seegers, owner of Seegers Grain Company at Crystal Lake, Illinois, find a grain loading facility for multiple car shipments. I showed him the old Terra Cotta pottery works just north of Crystal Lake. It was an ideal location for his operation. Located about two miles north of Crystal Lake on the Williams Bay line, the valley east of the tracks was once the manufacturing plant of the American Terra Cotta and Ceramic Company (incorporated 1887). A substantial hill of clay had been converted over the years into untold quantities of drain and field tiles in coal-fired beehive kilns. As demand for decorative architectural tile and TECO pottery diminished, the business gradually faded and was now totally abandoned leaving the former plant site in brush and weeds hiding acres of terra cotta molds. Some of decorative column capitals at the C&NW's Chicago Passenger Terminal were said to have been manufactured here before 1910. The actual station of Terra Cotta was located at the junction of the long side track and the Williams Bay line. The brick and tile depot which housed a U. S. post office was long gone with only traces of the former cinder platform still to be found. The side track built by the C&NW had even been used by a regular passenger shuttle train service bringing workers from Crystal Lake. The track was still in place even though sunk into cinders and earth. When Seegers filed for zoning permits with McHenry County, the public uproar against his grain loading operation was tremendous. By September 15th he had given up and moved his operation over to Ridgefield, the next station west of Crystal Lake. This was not a big, flashy prospect, but a worthwhile creator of a steady revenue stream. Bob was a persistent fellow; in 1962 he again tried to get the Terra Cotta site despite continuing adverse public reaction. Bob Lithgow, Vince Tumbarello, (C&NW sales agent) and I went back to Terra Cotta on Groundhog Day to take a look at the situation. The little valley still looked pretty desolate and forlorn. Legal matters dragged along without any resolution and included a highly emotional public meeting on June 6th at the Crystal Lake city hall. I was amazed at the ferocity of some of the irate protestors. I had a private talk with our commerce attorney, Fred L. Steadry, on August 28th. He didn't have any new ideas either. The rezoning attempt failed and no grain cars were ever loaded at Terra Cotta which was too bad as it was the perfect place for such activity. Sometime later a steel heat treatment operation located there—I wonder how they got the zoning changed?

This summer season of golf was affecting our abilities to handle good business prospects. Many of our industrial agents emulated the boss and were ready to go at the drop of a tee. Gene Cermak, was a good golfer, but the rest of the boys were duffers. Our top traffic officers were firm believers in the value of golf as an acceptable way to conduct business. I did not win any friends by growling about the long periods when people were out of communication in the far reaches of some distant fairway. There were no pay phones on the greens and cellular telephones would not be invented for another 30 years. On September 7th Gene was off somewhere swinging his left-handed clubs at that little white ball. Ed Olson and Bob Stubbs, our top traffic officers, demanded Cermak's immediate attendance at an important meeting. I said he was out of touch on some golf course. They immediately simmered down as golf was an okay thing with them and directed me to attend the meeting in Gene's place. It was not too exciting a subject—possible C&NW purchase of the defunct Chicago, Aurora and Elgin's former car shops at Wheaton, Illinois. Besides traffic's representation, B. R. Meyers, chief engineer, and I. R. Ballin, real estate, were there. Apparently, Mr. Heineman had suggested that we look at the possibility and report back to him on the merits. We came to the conclusion that purchase of the land would not be a good investment. Bob Ballin, who had a better understanding about property values, disagreed. However, Ballin had not developed enough stature and authority to direct railroad funds into commercial land development. That would come later. The Wheaton property eventually became a block of condominiums. Despite the ripple of frustration surrounding his golfing absences and his absence at today's meeting Gene threw a party the next day for the office staff at his Ruth Lake Golf Club in Hinsdale. Chuck Towle and I begged off to keep the office open.

The real estate sales function was steadily getting more of Heineman's attention. Conversion of vacant railroad land into cash made good sense. Ballin hired Real Estate Research Company, a private consultant, to systematically identify and parcelize the railroad's extra width right-of-way and station grounds. On September 11th I led our Bob Lithgow and Wally Francois from RER on a tour of C&NW properties west along the old Galena Division to specifically identify lands suitable for industrial purposes. I had already warned them we would not find much, and we did not. The fight between industrial development and real estate over the "surplus" real estate continued to escalate. Some positive things happened, however, we got direct long distance dialing in September—no more going through an operator.

Near the end of the golfing season Cermak announced on September 14th that he was to be hospitalized in October for removal of a cyst on his tailbone. While only a minor operation, the recuperation period kept him in bed for several days. Meanwhile, his wife was driving their Jaguar around town. She noticed a little red light glowing on the dashboard and asked Gene about it when he was immobilized in the hospital. Then the car quit running. When Gene finally got on his feet again, he found that the engine had melted into a lump of junk—the red light was an oil pressure failure warning signal. Mechanics found part of a valve lifter had fallen into the oil sump and plugged the system. Irate at everyone, except his darling wife, Gene roared and fumed at the dealership which offered no help and merely wanted $700 to replace the engine. Gene finally conned the C&NW's test department into analyzing the faulty spring and expressing their opinion that the steel was defective. Armed with that, he argued and fussed with Jaguar for months, consuming much energy and time to finally obtain some concession from the motor car company. The frequent absences did not help the vital need to communicate. My records showed that I did not see him in person between September 20th and October 12th despite an often urgent need to talk. Our working relationship over the next five years slid steadily downhill, interrupted by a few level places that soon resumed their inevitable downward spiral. I was doing pretty well in this business, but doing it independently - and that was a problem I had to deal with as best I could.

As a matter of record an Electra crashed into the C&NW's railroad embankment north of Proviso while trying to land at O'Hare Field on September 17th. The site was quite near the high overpass of the Milwaukee Road's Bensenville yards. Luckily, no freight trains were on the track at the time of the accident which involved many fatalities.

In the autumn of 1961 I had an easy location job when the Richardson Company of Melrose Park, Illinois, came to us looking for a site for a new phenol plastic insulation plant in the DeKalb area. I contacted our old friend, the former mayor, Hugo Hakala, who was then working as an industrial realtor. Hugo with his canny Finnish phlegm was glad to be invited to work on this project. To make the project go more quickly, I went to the county seat in Sycamore to research land titles to property along the Spring Valley line south of town. A suitable piece of land was found in the triangle made by the railroad and Illinois state highway #23. Hugo secured the land for Richardson. A week later, on October 17th, I helped their engineers establish a track plan as well as the building's location. The ground breaking was November 10th followed by a welcoming dinner at Northern Illinois University. Would that all location projects would go so smoothly.

In 1961 preparation for leasing two of the four C&NW's west line main tracks located on the elevated line from Kenton Avenue to River Forest was well underway. The Lake Street Elevated Company was the original builder of the double track line that ran at ground level parallel to the C&NW line. The many street crossings required gatemen and slowed travel. The rapid transit company had been ordered to elevate their line as long ago as 1914, but had steadfastly resisted pressure from county, municipal and public sectors until an agreement was made to share the existing C&NW elevated structure—for a price. In addition to the annual lease of the land and payment for the track structure (jointed 112 pound rail) in the two southerly tracks, the Chicago Transit Authority agreed to pay for a new third main line in the vacant center of the elevation. Of course that old 14″

water line from Austin to Proviso was also there, but in former days there had been as many as six tracks through this stretch and there was plenty of room for a third main and the water line. During the latter days when there were five operating tracks on the elevation, the center one was used as a break-in running track for locomotives fresh out of the nearby Chicago Shops. The new third main track was placed in service for westbound trains on October 23, 1961. It was a nice smooth ride because the track was laid with ribbon rail. Eventually, the CTA widened the old westbound track centers at stations to build between-track passenger platforms, installed crossovers here and there and, of course, put in longer track ties to support the electrified third rail. There was hardly enough room left for a stout wire fence separating the transit and railroad lines. The C&NW operation on three tracks instead of four settled down to the new westbound main on the south, the center track (old track #3) was made reversible for the morning and evening fleets and the northerly track was for eastbound trains. At River Forest the four track configuration remained, but the piece of old track #1 was reduced to a passing track where the River Forest local could sit and wait for the westbound through trains to roar past in their evening parade.

I went to Peoria on October 23rd to meet with Mark Townsend of CILCO and the Pekin-LaMarsh Drainage District to discuss development of that enticing piece of vacant land along our railroad just opposite Pekin, Illinois. The parcel was a 2,000 acre expanse of farmland, as flat as table, just sitting there behind a river levee waiting for development as a major industrial area. We were to learn that this attractive land concealed a fatal flaw. Having gotten permission from the land owner, I arranged in November, 1961 to have some soil borings made about a half mile back from the river. The goddess sleeping under the land apparently resented have holes poked into her skin, but waited for her revenge. Based on the soil reports and the lure of the land Mark and I concluded in February 1962 that we should seek a zoning change from agriculture to industrial. On April 4, 1962 when the Illinois River was running high from the meltwater from the very snowy winter season, the goddess opened the test hole from the 3″ soil boring last December with a geyser of Illinois River water that swiftly eroded the earth around it. The water spout which was as high as the river water outside the levee soon threatened to inundate the entire 2,000 acres and render the protecting river levee useless. After some frantic telephoning between Peoria and Chicago, Mark got a local contractor out there with a big load of bentonite clay and rock to seal the crater and hole. It was a near thing and an expensive lesson as the railroad and CILCO shared the cost of the emergency repair. Later we learned that in the postglacial period some 10,000 years ago, a giant icegorge dammed what is now the Illinois River at this point creating a huge lake upstream of which Lake Peoria is a present day vestige. When the ice dam broke, the lake spilled through the narrow opening and scoured the stream bed down to the bedrock some 300 feet deep creating a vast canyon which over time filled with silt and sediment. That fine grained silty fill was dense enough to seal-off the subsurface water table until we came along and punched a hole in it.

On October 24th the Lock-Joint Pipe people came in to discuss location of a concrete pipe manufacturing plant on our extra width right-of-way at South Beloit, Illinois. The next day I was off to Lincoln, Nebraska, with Gene Cermak in the CB&Q's sleeper *Silver Arroyo.* The State of Nebraska sponsored a seminar on industrial development, and we were expected to participate despite the fact that Nebraska was really "owned" by the Union Pacific, Chicago, Burlington and Quincy, Missouri Pacific and even the stumbling Rock Island before the presumptuous C&NW with its spindly tracks could claim any rights. We just made more noise and flash than the others. After a riotous night with an all-night poker game involving Cermak, Fred Berniklau, Don Slonecker and Chuck Sayre, Gene had the honor of conducting a panel discussion the next day. Despite his unavailing labors of the night before, Gene rose to the occasion with just one awkward moment when someone asked him about the future of the Scribner-Oakdale branch. Gene had no idea of where Scribner and Oakdale were much less anything about a railroad. He finessed the question with a bit of stage business and ducked the question. This was an expensive trip for Cermak; he lost at poker with the Nebraska crowd and at gin rummy to me both coming and going on the train. We went back to Chicago on the sleeper *Silver Mountain* on the 27th. The Lincoln crowd came into Chicago November 7th with a hot lead on another Fantus Company client. I introduced them to Maurie Fulton, but didn't learn much as that gentleman was always very careful to conceal his client's identity and never directly reply to questions—always steering the conversation into neutral areas.

I was drafted to substitute for Gene Cermak in a meeting on October 31st. J. R. (Bob) Brennan, former passenger traffic manager, and Bob Christie from traffic, were interested to know what I thought we should ask for in connection with the proposed Northern Lines merger (CB&Q-NP-GN-SPS) then beginning to emerge. If I had any suggestions, they couldn't have been very substantive.

An early and cold winter settled in, and I joined the birds migrating to warmer climes. My wife and I headed south by automobile. The cold and overcast skies stayed with us on most of the journey. Finding SSW #300 (2-6-0) on display in Paragould, Arkansas, wasn't enough reason to linger under the dull skies, and we continued driving southward. In Denison, Texas, the swimming pool at the motel froze over. Cold pursued us across Texas. Southward across the border into Mexico, we followed the old Pan American highway to Mexico City. There were even PCC streetcars operating in the narrow streets of the sprawling Mexican capital. After a few days of exploration, we drove south to the gardens of Cuernavaca and the silver working town of Tasco where, at last, it was warm and peaceful. Hearing that Acapulco had been struck by a hurricane, we stayed in Tasco a few more days before returning north through Toluca and Morelia to Laredo. We got back home just before the snowiest December since 1942 hit the Chicago area.

16

RISING EXPECTATIONS, 1962 . . . My new year began with a public meeting in Wheaton, Illinois. The future of DuPage County as a high growth residential suburban area was made plain to see for all the attendees when the county planners outlined a project for a major sewage collection system affecting every municipality and unit of local government. The impact of constructing many sewer mains under the railroad was trivial compared to the impact on land use—residential, commercial and industrial. The old farms were becoming unprofitable because of increasing tax burdens. Developers recognized gold mines in that open land, if they could just get the timing right. Tieing-up money in raw land is risky as the pace of development is subject to so many uncontrollable events—local and national. The DuPage sewerage plan recognized the inevitable role that the county was to play in the growth of the Chicago Metropolitan Area. We in the C&NW's industrial development department were in favor of the scheme; the tax department was not.

In August 1961 the huge Skelly -Swift fertilizer manufacturing facility had chosen Clinton, Iowa, as its probable construction site. Cermak's friend, Bob Stapleton, the Clinton Industrial Development manager, was as ecstatic as his laconic personality would permit. Early in 1962 the name of the fertilizer company was changed to Hawkeye Chemical and serious investigation work was begun on the selected property south of Clinton near the village of Camanche. Learning on January 3rd that Phase II of Hawkeye's plan involved investigation of phosphate mining in the west, I dusted off my old USGS maps covering the mountains south of Lander, Wyoming, and began researching geologic reports along with material on production and marketing of green phosphoric acid. I wasn't trying to convince anyone at Hawkeye that I was an expert; but, when asked, I was going to be knowledgeable. The project at Camanche collapsed the very next day when the results of soil testing were revealed—bad subsurface and high groundwater. My phosphate research proved to be useful when I was invited to meet with Skelly in Tulsa, Oklahoma in March, to discuss western mining prospects with Dr. Soday of Hawkeye. Our Bill Kluender had hired a new geologist for his agricultural and resource development department. Jim Aase replaced Brad Huedepohl. I could never decide whether to trust Kluender because his furtive nature and secretive behavior put me off. Not only Brad, but his other man in charge of agricultural resources, Dick Hill, had shared stories with me about Bill's odd behavior. I decided not to say anything about my Wyoming map collection and the inquiries about resource development in Wyoming with Mr. Kluender. Looking ahead a year, Phase II of the $100 million dollar project Hawkeye project required another Tulsa trip in March 1963. Cermak, Bob Stapleton and I were joined by Jim Helliker, our Tulsa sales agent, for a meeting on the 12th with Dr. Soday and a Mr. Haslam. The proposed plant site for NITRIN, the again renamed company, had been shifted to a rural site across the Mississippi River about 11 miles south of East Clinton, Illinois, near Albany. Rail service was to be provided by the Milwaukee Road from their freight-only branch line that ran south from Savanna, Illinois, to Davenport, Iowa. At the request of NITRIN, I took a fast look on March 14, 1963 at the possibility of C&NW rail service. First of all the Milwaukee crossed the C&NW twin mains at East Clinton on 58 degree diamonds so that a new connecting track would be required. Obtaining trackage rights from the Milwaukee Road would clearly be a tough sell, while construction of 11.4 miles of new trackage parallel to the their branch along highway 84 would be even tougher because of terrain, public highways, numerous residential properties and regulatory requirements. The new NITRIN facility was designed to convert natural gas delivered by pipeline into chemical products which then would be trucked to Midwestern agricultural areas. This did not leave much potential for rail revenues. I recommended that we just accept the plant's location as an economic stimulus to the Clinton area's economy that could enhance the attractiveness of that area. My position was not universally embraced. The unlucky Vern Giegold with his talent for losing prospects to other railroads was unable to see any good in this latest fiasco and growled privately to Cermak that I was getting too "uppity for my own good". Sometime later the Milwaukee Road, or one of its successors, chopped off the branch about three miles south of the C&NW crossing. Could that have been designed to preclude any future C&NW access or interchange?

Returning to the 1962 thread: on January 8th we learned that a rumor of a joint Milwaukee Road-Rock Island-C&NW consolidation study was really underway. By the 24th the Milwaukee guys were spotted at Proviso observing operation of our freight car humping. This study lasted a long time and ended up as a series of colorful maps and charts. Our department had little involvement with this study. There would be more consolidation studies in the future because Heineman was convinced that there were too many railroad lines in the Midwest for the amount of business available and that consolidation was the only cure for over-capacity.

January, 1962 was very cold and snowy. Many homes developed ice dams on their roofs when radiated heat from inside the home melted snow on the roofs, but not along the lower eaves. The resulting build-up caused meltwater to back-up under the shingles and run down interior walls. Chopping breaks in the ice dams with a hatchet was the only way to relieve the problem even though shingles were often damaged in the process. I tried the hot water method to melt gaps in the ice dams. Railroad track was also being badly affected by ice build-up. Accumulated snow and ice in the track zone tended to thaw when in contact with loose tieplates because kinetic energy from passing trains translated movement into heat. The melted ice water would then run under the tieplate, freeze and expand thus building up a mound of ice that could actually raise running rails and pull track spikes out of the ties. Knocking out the ice lumps and inserting wooden track shims was the only way to restore the safety of the track. Applying salt was tried, but the brine then seeped into the track subgrade creating soft spots in the ballast section. Then, of course, when the thaw finally came, all those wooden shims had to be removed and the ties tamped-up to support the rails. Track forces were hard pressed that winter and spring. On February 10th we still had the same snow on the ground that had fallen in the big storm of December 23, 1961.

In the grip of below zero January weather, our industrial development group went to Peoria on the 8th to attend the annual Great Lakes Industrial Development Council meeting. At the time for our departure, Armour and Company requested a meeting about their proposed Dial Soap plant which was in the final stages of site selection. I arranged for R. P. McDonough, Galena Division superintendent and some of our traffic people to join me in meeting with Mr. Hardcastle of Armour to discuss train service and rates for sites at West Chicago and Elgin. After the meeting, I scooted out to O'Hare Field and caught an Ozark F-127 to Peoria. The 32 minute flight was a nice contrast to the three hour train ride on the increasingly grubby *Rocket.* In Peoria the TP&W laid on an early bird spread that evening. Our court battles hadn't gotten brutal at this point, so I could eat their fare without feeling hostile. On the 10th the council met in the dining hall at Bradley University. This building had not existed when I attended the school in 1944 and 1946–47. Today, however, a frigid blast of air across the floor made listening to the speakers difficult. I had planned to take the day tour on the 11th to the Caterpillar factory in East Peoria, where I had worked nights as a draftsman in 1946, but, Jack Sommer and his Wisconsin pals wanted a tour of the Pioneer Industrial Park north of town. The evening events included the usual booze and card game marathon. On the last day Maurie Fulton of the Fantus Company spoke to the group. Maurie had a great interest in Peoria as ADM was his big client, and our big headache, as related earlier in the First Peoria Railroad War. That afternoon most of us headed home on that worn but handy *Rocket.*

ARMOUR AT WEST POINT. Drawing together the pieces of this episode will drift the story line into 1964—the 1962 thread will be picked up later in this account. Armour and Company was regarded as a special customer by the C&NW's top traffic people. Ed Olson, our vice president, traffic, and his assistant, Robert (Bob) Stubbs, regularly received inside tips from their Armour contacts and then doled out the new plant tips to the industrial department as though we were some kind of fish to be baited. Traffic's moguls seemed to think that our own good contacts with Armour real estate personnel were not important and unworthy of attention. On January 15, 1962 we heard that the Dial Soap plant was going to Montgomery, Illinois, south of Aurora on the CB&Q's main line west. Our site at West Chicago had, supposedly, come in second. An Armour contact said that their meat processing group was interested in a beef plant at West Point, Nebraska. That was my territory, and this tip did not come from our Ed Olson. Obtaining this information put me one up, so to speak. When Ed Olson finally confided the West Point project to us through Bob Stubbs, I was careful not to let on that I was already working on it. I used my own pipeline with Armour and learned the specific details of this Nebraska, and other projects, which were coming fast and furiously because of major changes in the meat processing business. Stubbs wanted me to drop everything and rush out to Nebraska on January 23rd, but the meeting that he had arranged fortunately was canceled.

The traditional production of beef for market involved moving the live animals from ranges and farms to central stockyards in major cities. In earlier days railroads hauled a lot of cows (and pigs and sheep) in huge fleets of specialized stock cars. Drover coaches carried next to the waycars on the end of stock trains were a common sight on these stock trains. The drover cars looked like wooden waycars without cupolas and were specially equipped to carry stock handlers to care for their animals en route. Federal law governed movement of live animals, limiting hours of travel and specifying watering and feeding intervals. The mortality of animals on stock trains could be high as a down animal was often trampled by its fellows. Rough handling by an engine crew could cause great loss. While Federal rules were designed to minimize animal deaths in transit, in reality, the regulations just added cost to rail shipments. As the highway network expanded after World War II, livestock transportation went to the largely unregulated commercial truckers. As live animal rail traffic declined so did the railroad's maintenance of wooden rolling stock and stock yards. The spiral of worsening condition of rail facilities accelerated the shift of livestock traffic to the trucks. Drover coaches disappeared, too, often ending up on blocks alongside switching leads in rail yards to serve out the rest of their existence as shelters for switchmen. Even the mighty Chicago Union Stock Yards imploded as inbound receipts dwindled. A shadow of its former self, the USY eventually survived as a much smaller operation near Joliet, Illinois. Charley Cook, the C&NW's livestock agent, saw his business evaporate year by year. Production costs of processed beef could be slashed when cattle slaughter occurred near the growing areas and meat was shipped as carcasses rather than a s live animals. Soon after capture of the livestock business from the railroads, the truckers' long haul business rapidly declined in turn as modern refrigerator cars equipped with hanging rails and stainless steel hooks came into general use. Shipping quarters of beef by rail turned out to be only a partial solution because the swinging carcasses were damaged by being bumped and jostled on long rail trips. Even though dead, the meat was bruised and degraded by the time final destinations in large cities were reached. Another evolution then occurred in the meat processing business—boxed meat from breaking plants located in the growing areas. At last quality could be maintained by cutting-up the carcasses in local plants and shipping the packaged and inspected meat in waxed cardboard boxes. The local plants also provided employment for rural populations where the company could secure labor at less than urban rates. When a carcass breaking plant was located along a main line railroad with good and frequent service, there was a chance that boxed meat could be shipped in refrigerator cars. In practice the truckers inherited most of that business as well as that of the inbound live animals. The railroad got the non-time sensitive business like lard, tallow, hides and bones. But in 1962 we were focused on the potential for good revenues to be gained from these new meat packing plants.

Tallow and lard went out of what Armour styled an "abattoir" in tank cars labeled as inedible animal products to be processed elsewhere into a variety of consumer products. Bones were shipped to rendering plants for boiling and grinding into bonemeal. Green hides were another byproduct which often proved difficult to handle in rail equipment. They smelled bad and dripped disgusting liquids. Any boxcar once used for hides could never be used for any other kind of traffic. It was the end use for a boxcar, usually an old wooden forty foot car. Shipping hides in open gondolas was not allowed. One story about a hide car may be apocryphal but illustrates the realities of handling this traffic. A car of green hides somehow got lost in transit and, when finally discovered, was rejected by the consignee because it had grown very ripe. The shipper also refused to take it back. The pungent car was bucked from division to division; its papers were lost along the way somewhere. No one had the nerve to open a door on the car as the odor was terrific and a fear that rats as big as small dogs might be inside. The stinking car was a pariah and ended up on the Western Division in Chadron, Nebraska, where the hot summer sun made the car's contents smell even worse. The superintendent and division engineer soon found that they couldn't pass the car off on any other division—they all had been alerted to prevent its return. Slipping it to the CB&Q at Casper or Crawford or the Milwaukee at Rapid City in interchange was considered but the likelihood of its being rejected and returned was quite probable. Something had to be done with this problem at Chadron. The division engineer hired a big Caterpillar front end loader and had a deep hole dug alongside the wye track west of Chadron. The offending car was shoved west with a string of empties between it and the engine to the wye where the big Cat toppled the car into the newly-dug hole, trucks and all. Shoving the excavated dirt into the hole on top of the beast and smoothing it over was quick work. Claims on the car were paid without question as the claim department was officially informed that the car and its lading had been destroyed in a wreck.

I met Wally Beck and Ralph Deemer of Armour's real estate unit

ARMOUR & CO. South of West Point, Nebraska, - 1962

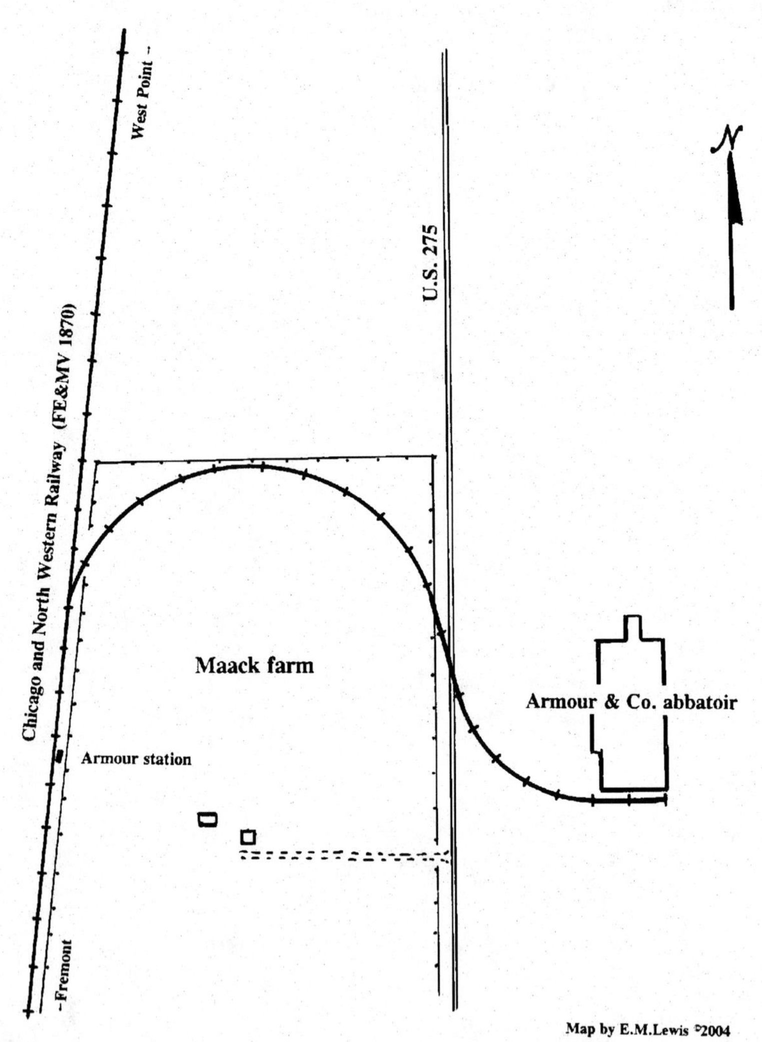

Location of a new Armour & Company meat processing plant south of West Point, Nebraska, in 1962. Construction of the siding required creative engineering to accommodate the intervening landowner.

on April 17, 1962 to discuss the West Point site. While other locations such as Columbus, Nebraska, were also under investigation, the Armour marketing group were sure that West Point was the best place to build an abattoir to harvest that part of the state's cattle population. Located about halfway between Blair and Norfolk, West Point had been established in 1857 in the valley of the Elkhorn River mainly by Pennsylvania Dutch who gave it that name because it was believed to be the furthest west settlement at that time. On May 7th I flew to Omaha on a low flying DC-6B which gave me a lovely view of the C&NW's mainline across Iowa as far as Cedar Rapids. Don Skoglund, our new man inherited from the M&StL, met me at Eppley Field. We toured eastern Nebraska along the C&NW arriving in West Point on the 8th. Using a hand level Don and I ran a line from the mainline to the proposed plant site; it was a pretty stiff uphill grade. Don and I went on west to Norfolk, then northeast to Sioux City, Iowa, and back to Omaha. At Hoskins, Nebraska, we happened to run into Harry Bierma of Real Estate Research, Ballin's consultant, and the Nebraska Division's superintendent, division engineer and roadmaster. This party was reviewing properties of the old Omaha Road (CStPM&O) for potential future rail use or sale. Harry Bierma did not like to see the industrial development department poking around during his property evaluation trips and made it clear that we should not meddle in his affairs. Don and I went on to Omaha and rode the Rock Island sleeper to Chicago. I took my notes on the proposed track grade to Armour's offices in Chicago the next day and went over the topography of their West Point site with Leo Orsi, Armour's engineer. On the 14th of May, Leo called; he was all excited by a new site at Beemer some eight miles to the east. His enthusiasm was not matched by anyone else at Armour and planning the West Point site continued. In August the whole railroad was closed down by a telegraphers' strike which dragged on through September. The railroad shut-down was a good time to be traveling because I avoided the struggle of commuting to the Chicago office. (More of this event will appear later in this account). I flew to Omaha on August 30th in a 720 and met Gene Cermak there. We had a meeting with Dr. Klemme of Northern Natural Gas Company, Dave Osterhout of the Nebraska State Industrial Development group and some of the State's Railroad Commissioners. After putting Cermak on a plane back to Chicago, I drove up north to Fremont for the night. The next day I met with Pliny Moodie, a local West Point attorney, to get some background on the farmer, Don Maack, who owned the strip of land between the railroad and U. S. Highway 275. The proposed Armour plant site was east of the highway and the railroad was west of it with Maack's piece in between. At least the farmhouse wasn't in the way of railroad construction, but Maack's mentally challenged son was. The family feared that a railroad track across the strip of farm would be an attractive nuisance to the lad. When meeting with Maack, I was glad to have had forewarning of the domestic situation from Mr. Moodie. I quickly agreed to provide fencing for the strip of right-of-way and adjusted the alignment slightly to better accommodate Maack's farming operations and to fit the topography better. Maack, his wife and his brother agreed to my option offer. At this time the Nebraska Division's engineering staff were refusing to work and militantly supported the telegraphers' strike. Dusting off my skills as a track engineer, I personally laid out the centerline and grade stakes for the Armour track. Later Shorty Braden, the assistant engineer, had to accept my track layout but ever afterwards complained about the curvature, the grade and everything connected with that sidetrack. In October 1962 Armour obtained title to the plant site. On December 4th B. R. Meyers, chief engineer, called a meeting with S. C. Jones, vice president operations and Bernie F. McDermott, Nebraska Division superintendent, to personally organize construction of the side track while Mr. Jones used his august position to establish future train service to the Armour facility. It seemed to me that this was a lot of high powered brass involved to discuss a modest meat processing plant project in Nebraska, but the Armour name opened a lot of doors on the C&NW. When it came time to cross highway 275 with the new track, Shorty suddenly asked for my assistance to get the necessary permits from the State of Nebraska. Although just a bit irritated by their lack of initiative, I obtained the necessary permit in March 1963. The process was facilitated because Armour paid for automatic flashing light signals. Construction of the track was to begin in mid-March, 1963 but soon ran into a problem with farmer Maack who now would not sign the deed and began harassing our engineers. Charles P. (Chuck) Jevne from our real estate department and I flew to Omaha March 20th on a DC-8, drove to West Point and consulted with Pliny

Moodie, Maack's attorney. Moodie advised Maack to sign the deed and he complied. Grading commenced the next day.

This was not the end of our problems with Armour and farmer Maack. A year later in May, 1964, Armour wanted to construct a refrigerator plant for an icing facility along the new siding and needed three more acres of Maack's land. Ordinarily our department was "finished" with a project when it went into operation, but the Armour name and my involvement with the landowner made me the logical person to fix the situation. I knew that the Armour track and Maack's challenged son was still a problem for Maack and that enlarging operations there could make matters worse. On an earlier visit I had noticed a small oil painting done in primitive style that was a pretty fair copy of a group of medieval burghers sitting at a carpet covered table. The original was copied from the inside lid of a box of Dutch Masters cigars. I asked who was the artist. Maack allowed that he had painted the oil, but that he didn't have a lot of material at hand to work from. When I told him my wife was also a painter; his reserve melted to some extent. Facing this new challenge a year later, I purchased a pretty good book of color art reproductions of famous painters' work such as Velasquez, Goya, Da Vinci, Rembrandt, Breughal and the like at Krochs and Brentanos in Chicago before boarding the CB&Q sleeper to Omaha on May 31st. The next day in company with Mr. Moodie, I went out to Maack's house. When I presented the new book to him, his wary, guarded attitude dissolved like ice under a summer sun. He was absolutely enthralled by the color and richness of the reproductions in the book. Moodie drew up a warranty deed for the additional three acres, and Maack signed it without hesitation—in fact, barely taking his eyes away from his new treasure. Celebrating myself, in a small way, I splurged on a bedroom on the Rock Island *Rocket* that evening.

The West Point plant created substantial freight revenues for the C&NW as well as promoting good working relationships with Armour and Company. Despite Braden's constant harping about sharp curvatures and steep gradient, the road diesels switched there without any problems. Shorty Braden, it seems, was still designing sidetracks to accommodate the long gone class J4, 2-8-4s, which he knew so well when he worked at South Pekin, Illinois, in the 1930s. Even as the West Point Armour sidetrack was being constructed, the clock measuring its short life was ticking. Studies were already in progress to abandon this railroad from Fremont, Nebraska, to Lander, Wyoming. After the coal excitement of the mid-1970s subsided, this line west was indeed dismembered—broken by flood waters, abandoned in part and other pieces sold to new shortline carriers who would eke out a bare existence from the sparse traffic available.

My biological clock was ticking—I was now 35 years old with almost 15 years of railroad service. The past five years had been interesting but professional growth was non-existent. I was uncomfortable with my colleagues whose intellectual interests and motivations were so different from my own. I was not, and would never be, a salesman. In my view of the world a project, service, product, or even an ideal, should be so evidently worthwhile as to make the "selling" of it unnecessary. However, that is not how life is organized so that the salesman must exist. At this point in my life I was a low level railroad officer with professional civil engineering training and experience, stuck on the lower rungs of a promotion ladder which I could never scale. Obtaining recognition and respect would not come through the old boy network of the traffic department whose present leaders could not, or chose not to, see the changes rapidly occurring in the railroad industry. My good old railroad was not leading these changes despite the best efforts of Ben Heineman. Operations were little different from the past—long drag freights crawling from yard to yard fostered by a management style from the 1920s. Solicitation of business was based on personal suasion not service. Early in 1962 I began looking at other job opportunities. Pulling my resume together, something I had to start from scratch, I responded to one job advertised in Sacramento, California. I asked Maurie Fulton at Fantus Company to be a reference—he expressed surprise but agreed.

Our business in this early part of 1962 was as frozen as the below zero February weather. Despite getting new office furniture on Lincoln's birthday and, wonder of wonders, our own copy machine, there was a lot of whining and sniveling among our group. Apparently, my dissatisfaction trickled through to Gene Cermak, who bluntly asked if I was looking for another job. I told him straight out that given my current pay and the slow business, I was looking around. At that time he was away from the Chicago office a lot on very frequent trips to the Twin Cities for purposes that were never explained. Reflecting on the lack of activity our department was experiencing, Cermak now decided to pick a fight with Bob Ballin about real estate commissions and promoted the idea that our department should manage leases of company lands, a function then under real estate's responsibility. I reminded Cermak that the clerical requirement to manage that kind of business was enormous and would distract us from our primary mission which was to locate new revenue producing industry. An aspiring but ill-equipped empire builder, Cermak agreed with my arguments, but soon turned right around on April 20th to push this idea through our staff meeting over my strong objections. Of course, Bob Ballin was never going to give up this important part of his responsibility and the tempest quickly blew out leaving Cermak looking like a loser. Somewhat put off by this unrewarding squabbling, I began collecting materials on the deep water ports accessible to the C&NW in Duluth-Superior, Escanaba, Manitowoc, Milwaukee, Racine, Kenosha and Chicago. The St. Lawrence Seaway was new and foreign trade patterns were beginning to emerge. Although I do not know if any volume of foreign trade ever developed through my efforts, I did identify and promote the various facilities which helped our sales forces show the world we were not totally asleep. My lake port project eventually became an official project as the news slowly came to this Midwestern carrier that commerce was globalizing. The concept of standard containers had not yet been invented. Convinced of the importance of a water-rail connection, I applied for a job with the Port of Portland (Oregon) in May. I was totally bored and quite restive.

In February 1962 I tried a new tack that involved working with rather than against the inland barge operators. My rationale was hinged on that attractive 2,000 acre tract just opposite Pekin, Illinois, on the Illinois River. At the time I had no idea of the hidden danger underlying that land—that would burst upon us on March 30th. On a cold Saturday night in February I rode the GM&O's *Midnight Special* to St. Louis, Missouri, in a dark sleeper *Timothy B. Blackstone.* Despite having no interior lighting, the car was warm and being late into St.

Louis on a Sunday morning did not upset any plans. It was a long cab ride from Union Station out to a hotel on Kings Highway where I joined Jim Gilchrist from Peoria to help man his booth at the annual meeting of the Mississippi Valley Association which was composed mostly of river barge operators. I was the only railroad man in attendance at this affair. With a plentiful supply of my site sheets and other railroad propaganda, I tried to focus the conference visitors' attention on this property which lay right along the Illinois River. Jim and I took turns at the booth, but we both slipped out to attend the St. Louis Railroad luncheon on the 5th. This barge crowd was not interested in property, they were movers of freight not land developers. The conference ended on the 6th, but before I could get on a train headed north, S. A. (Syl) Keathley, our St. Louis traffic manager, asked me to stick around another day as he had a confidential matter to discuss. He was busy all the next day, so I rented a car and prowled the maze of railroad tracks and properties on the east bank. Actually Syl had two things he wanted my help with. On the 8th he took me to National City, Illinois, to look at the properties of the East St. Louis Junction Railway. This property, if controlled by the C&NW, would open up new interchanges to roads like the Southern Railway. The second, and more important property in which he was interested, was the Illinois Terminal Company. He wanted me to survey and render an opinion of the property as a personal favor. I wasn't sure where Mr. Keathley stood in the traffic department's pecking order, but the idea of such a review was well within my training and capabilities. I rode the Illinois Central back to Chicago that afternoon. During March I researched the Illinois Terminal's history using my library of Poor's manuals of railroads. I soon concluded that the ITC would not add much to the C&NW's operations below Edwardsville. There was little local industry, the track was typical lightweight interurban construction and ran down a steep grade from the high lands along the Mississippi bluff line to the flats below. There might be opportunities in the Madison area for direct access to Granite City Steel for iron ore trains, but this could be accomplished without operating on the ITC all the way from Edwardsville. Neither did I see any great potential for industrial development along that line. I wrote a report for Keathley who sent it to his boss, Bob Stubbs. On June 15th Keathley complained to me that Stubbs was sitting on it and could I do something to jar it loose? Just how I could accomplish that was not at all clear. A private word with Bob Russell, our attorney with whom I had developed a working relationship, elicited no solution. Ducking the issue, I wrote an anonymous letter to Ben Heineman as I was leaving for a European vacation July 1st. Of course the C&NW did nothing with the Illinois Terminal. Ultimately, it was purchased by the Norfolk and Western Railway which dismantled most of it.

By March 1962 most people in the Midwest were throughly sick of winter. Our record snow had been 1951–52 with an official 66 inches for the season. This year we had 57 inches of snow and the piles were with us well into spring. On March 9th I was traveling in western Illinois and went to see the remains of a 93 car derailment just west of the village of Franklin Grove. It is hard to totally destroy 93 freight cars in one accident unless conditions are just so. A westbound freight descending the grade on track #1 (the southernmost main) suffered derailment of a car near the head of the train just as it came to the first #20 turnout of the highspeed crossovers at FX. The tracks at that point were on a high fill across a ravine. The derailed car headed for the right-of-way fence taking its mates with it in an elephant chain procession. Normally derailments are halted by the heaping-up effect of wrecked rolling stock. But, at FX the deep drainage alongside the main accepted car after car with no diminution until the ravines on both sides were full of wreckage. Clearing the mains for service was easy—the wrecking crews merely shoved the cars still on top of the grade off the edge to join their companions, pulled whatever was left of the train eastwards into the clear, and patched the mainlines through the former CTC site. FX as a crossover and CTC control point was finished and not replaced. It was a rotten place to put a set of 40 mph crossovers, anyway. The sag in the profile there made it hard to control a heavy freight train at the proper speed to safely negotiate diverging moves. It was to be many months before the salvage contractor had cleared this immense heap of torn steel cars and their spilled contents.

Soon after the Franklin Grove wreck, I shepherded C. T. Smith of International Minerals and Arnold Bockman of Funk Brothers of Indiana around western Illinois to look at sites for distribution centers. They liked Ashton and Peoria, but their project petered out. A lot of folks liked my Illinois tour business; in May it was Jim Feeney, general traffic manager, and John Given from Standard Brands that made the Geneva, DeKalb, Rochelle, Harvard, Woodstock, Carpentersville and Elgin circuit. On May 15th a visitor dropped in unannounced from St. Joe Paper Company. Robert C. Brent, a true Southern gentleman, wanted to inspect sites for a warehouse in northern and western Illinois. Dropping everything, I took him to Cherry Valley, Woodstock, Crystal Lake, Carpentersville and Elgin, but he liked property at Northlake in the same area where General Foods and Kroger had settled. Full of stories about "Mr. Ball" (Edward Ball, the iron dictator of the DuPont's Florida interests which included the Florida East Coast Railway and St. Joe Paper), Mr. Brent insisted on taking my wife and me to an elegant dinner at his expense—contrary to our departmental instructions, but who was I to complain? By August 2nd St. Joe was negotiating with Bob Ballin for the tract they eventually built on, but we held our collective breath because Ballin seemed to delight in scaring our customers away with his idea of real estate prices which were not at all tempered by any thoughts of the potential for future freight revenues. On May 16th I had to rush to Springfield on the Illinois Central's *Green Diamond* to attend a subcommittee meeting of the Illinois State Industrial Commission for the absent Gene Cermak, returning to Chicago that evening with nothing to show for the day's effort. The C&NW acquired some new ALCO diesels (#900 and #901) and proudly parked them in the Chicago Passenger Terminal on April 12th to show them off to the public which seemed to be more interested in the construction of a stairway at Madison Street from plaza level to riverside for the new river taxi service to North Michigan Avenue.

Business was picking up for us. At the end of May I heard from Jefferson Lake Sulphur about a potential new acid plant in the Gas Hills south of Riverton, Wyoming. The plan was to recover sulphur from the sour gas found in the area. Nothing more was heard from them until February 5, 1963 when they urgently requested a site for tank car loading. On June 3rd Gene Cermak and I made our annual pilgrimage to Wyoming just to let our friends know we considered Wyoming to be an active development area—not that very much was happening. Because Continental got us to Denver too late for our connecting Frontier flight to Casper, we were the airline's guests that evening. The next day we boarded a Frontier C-47, a venerable Navy veteran, for the short flight north to Cheyenne and on to Casper. (Yes, it was a C-47 because the civilian DC-3 had the passenger door on the right side while the C-47's was on the left. I remember especially walking up the steep slope of the aisle inside the plane to get to the assigned seat). John D. Boyles, our local DF&PA, met the plane and took us out to look at some property northwest of town that he thought could be a site for a new bentonite processing plant. Access from a brand new freeway, however, was going to present problems. After an evening of socializing, Gene came down with what he styled food poisoning (we all had eaten the same stuff) and stayed in bed the next day. John and I went east to Glenrock to visit Dr. Karcher, the brains behind the new American Humates plant located in our former gravel pit. His concept of manufacturing a soil conditioner from lignite coal may have been innovative, but the product never made an impression in the market. On the 6th John and I collected the recovered Cermak and the two of us headed back to Chicago via Denver on Western and Continental. This short circuited trip, however, apparently had stirred up something because a week later on June 14th, Wyoming State Senator Al Harding appeared in Chicago to talk about a new bentonite processing plant at Casper. Even Ed Olson deigned to descend to our level when he heard a senator was present. The prospect sounded so good that Gene and I were soon on our way back to Casper (on United this time) on the 23rd. We rented a car in Denver and made a tourist trip to Central City, Colorado, before catching Frontier's evening flight to Casper. Milt Coffman, our wealthy rancher friend, threw a party at which we picked up a lot of background material about the bentonite promoter, Al Harding, and his partners. June 24th was a play day for Cermak and Boyles while I took a car and went west to Bucknum, a former station on the joint CBQ-CNW line. There was nothing at Bucknum save the concrete foundation of the depot and a lonely piece of a stockyard. The biggest sign of life was a westbound CB&Q freight with five units (SD9 #330 on the point) storming past. Later, I picked up Chuck Towle at the airport. Chuck's territory included both Minnesota and South Dakota and this Wyoming visit was just a side trip on a jaunt to Rapid City. We planned to drive northeast to Moorcroft, Wyoming, to meet with Senator Harding and his pal, Harry Thorson, when they suddenly appeared in Casper on the 25th. As an opening suggestion for a loading site, I mentioned Bucknum as a possible truck-rail transfer point for bentonite coming south from the rail-less Midwest area. Rail service at Bucknum could be provided by two railroads. This was years before Interstate 25 was constructed and most of the county highways then in existence were surfaced with gravel or native dirt. Harding and Thorson had a third man with them, Red Miller. Red was their operating man and didn't think these dirt roads would hold-up under heavy track traffic., but we visited Bucknum anyway. Chuck Towle had never seen Wyoming before and was dumbstruck at the complete absence of anything—just black sagebrush rolling away to the distant mountains. Cermak and Towle went on to Rapid City, South Dakota, while I went north on the "33 Mile Road" with Harding, Thorson and Miller. The next day I went west to the end of track at Lander while the rest of the group surveyed the back roads from Kaycee to Bucknum as a possible route for the bentonite hauling trucks. I learned a few days later that in their judgement the roads were just too poor and that the C&NW's station ground property at Mills, Wyoming, just west of Casper, seemed be a better choice for a transloading site. Then, the project seemed to cool, and we turned our attention elsewhere. About a year later, in May 1963, the bentonite plant came alive again. John Boyles on his own initiative had obtained options on that property adjoining our railroad northwest of Casper even though the highway access was a problem. Black Hills Bentonite, as Harding had named the operation, was indifferent to the property and John allowed the options to expire. On May 17th Black Hills suddenly wanted John's optioned land. The whole project seemed a bit more serious this time because O. O. (Ollie) Waggoner, head of the CB&Q's industrial development department, called me to discuss a joint C&NW-CB&Q service agreement. The property we had optioned, however, was open only to C&NW service and the CB&Q tried to salvage something. Gene Cermak and I hitched a ride to Huron, South Dakota, in the company's turboprop (N1394Z) on August 5, 1963 along with Charley Cook, our livestock agent, and Ed Upland, our grain man. At Huron we switched to a Western flight to Casper where John Boyles met us. We were supposed to meet with Harding and Thorson but Ollie Waggener and a fellow from their public relations department were entertaining them in the CB&Q's business car parked at the Burlington depot. It didn't matter, of course, because our suggested property was unusable as the Wyoming State Highway Department flatly refused to grant permission for an entrance-exit ramp because Federal money had been used to build the highway and the Bureau of Public Roads would not consider tapping the highway for a commercial entrance/exit. Gene and I flew out of Casper on the 7th on Frontier Sunliner *Zuni* to Laramie, Cheyenne and Denver. Ten days later I learned that Harding had committed to a lease of Burlington property at Casper. Cross that one off the list.

I mentioned above that I had gone off to Lander in June 1962 while the Black Hills Bentonite project was beginning. Flying into Riverton, I rented a car to poke around the Wind River Range and see something of the U. S. Steel project at Atlantic City. The drive was most instructive as to the topography of the area. The lure of an easy railroad grade over the continental divide was just that—something the locating engineers for the Union Pacific had tried in 1867. This was the same place Bernie Meyers and Jack Perrier had explored with their portable barometer in 1956. I had a short visit with Tom Pickett in Riverton to catch up on local news. An oil field supplier in the area, Tom was a source of much valuable information about the country and its people.

The next day I retraced my route to Chicago. Arriving at O'Hare Field, I went directly to an Ozark F-27 flight to Peoria where I met Joe Zelenda from our sales office. He had set up a meeting with a new silica mining company named Blackhawk Silica which was proposing to set up operations at Troy Grove, Illinois, on our Spring Valley branch. The Spring Valley line still had its original 60 pound rail and here was a new customer who would require big, heavy, hopper cars. After the meeting, Joe took me to Bureau where I caught the *Rocket* to Chicago. On September 11th the Blackhawk people came to Chicago. With Steve Owens' help, we worked out a good track layout for the new operation. Three days later M. C. Christensen, now the Galena's division engineer, and I hi-railed the line from Elva to Troy Grove. The railroad was much worse than I had remembered it; Chris was going to have to make some major repairs just to keep the line open. On September 26th I went over the plans on the ground with Blackhawk's engineers. On February 2, 1963, Blackhawk announced a major expansion at the Troy Grove site. They also renamed their company, becoming the Arrowhead Company. Thanks to the good revenues that came from the silica sand traffic, the C&NW was able to buy some good secondhand 10025 pound rail from the defunct Chicago, Aurora and Elgin Railroad and relay it in the worst spots of the Spring Valley branch. Silica sand played an important role in train operations. Dry sand applied to wet rail just ahead of the tread of the drivers often determined whether a locomotive could start its train or not. Not all sands will do in engine service. Silica sand by its very nature is composed of tiny spheres that flow easily through the delivery pipes on the engine. Other sands tend to have irregular shapes that cling together when damp hence the necessity of heating and drying it. The silica sand was a much better type for handling and storage although the tiny round balls tended to bounce off the rail head a bit more than ordinary dry sand. One could always spot the difficult places on a railroad for starting a train by observing the amount of crushed sand heaped along the rails.

EUROPE—1962. I had sent my wife and sixteen year old daughter off to France aboard the Cunard steamship *Saxonia* from Montreal in June. Half rate tickets were available to railroad personnel on Cunard, and we took advantage of them. I could not spend the time to sail across the Atlantic with the girls and arranged to meet them in Paris. To make the travel cost even more reasonable, I combined business with vacation by riding the New York Central's *Twentieth Century Limited* to New York and making business calls with Stanley B. Boardman, now in charge of the C&NW's office in Manhattan. I saw Jim Feeney of Standard Brands (he had been on one of my Illinois tours in May), Tighe and Sunderman of Union Carbide; and, in the afternoon, Rye and Kowalski of International Paper Company. The next day, July 3rd, Gerry Armstrong of our New York office, took me down to Wall Street to see Carl Wheeler of AGRICO to chat about possible rail-river sites served by the C&NW. That afternoon I made a call on Bob Katz of Revlon, and with Tom Hudson, another of our New York men, made a brief stop at Continental Can Company. That finished the business part, and I now went on vacation. That evening I flew the Atlantic in a Pan American 707 named *Endeavor* landing at Heathrow early in the morning of July 4th. Before noon I boarded a parlor car on British Rail's *Golden Arrow (Fleche d'Or)* and crossed the Channel from Dover to Calais on the ferry *Invicta,* a survivor of both the Dieppe Raid in 1942 and the Normandy invasion in 1944. The SNCF train from Calais rolled swiftly to Paris by way of Abbeville, Amiens and Chantilly arriving on time in the Gare du Nord where my family awaited my arrival, just as though we were home in Elmhurst. Four days later we enjoyed a high speed electric powered non-stop Rapide to Strasbourg via Bar-le-Duc. Later, still riding on our sheaf of half rate orders, we went to Basle in Switzerland via Mulhouse. Further rides on the Swiss Railways took us to the capital city, Berne, and on to Visp in the south where the red train of a meter gauge line twisted its way up the long valleys to Zermatt in the shadow of the Matterhorn. The work of Swiss bridge engineers was truly amazing as demonstrated by their soaring viaducts and galleried mountain tunnels. Even at Zermatt there was a further rail journey on a rack railroad to the top of a mountain. Leaving Switzerland for humid Italy through the Simplon tunnel, we came to Genoa, Italy, where the heat of summer made further adventures south into Italy unattractive. We went west back into France. The SNCF line along the Mediterranean shore was then being electrified. We learned to quickly close the coach windows when we passed through the many short tunnels that cut through the rocky noses along the shore. From the train we caught glimpses of the United States Sixth Fleet anchored far out on the blue waters. From Marseilles we rented a little French car and drove north to Paris and Cherbourg where I put the family on the Queen Elizabeth II. I then scooted over to LeHavre and took the night ferry *Normannia* to Southampton where British Rail's 7:03 a.m. made a quick run to London and my homeward flight. In New York on the 27th I rode the New York Central's night train in a sleeper with breakfast in Buffalo where the diner was added to the consist. Of course I was back in Chicago before my family arrived at Union Station on August 1st.

STERLING BOILS OVER — AGAIN. Assistant vice president, traffic, Robert C. Stubbs was meddling in our department's affairs again. His Armour and Company contact had expressed an interest in building a facility in the Sterling, Illinois, area. This was on August 10, 1962. Since the head of the department, Gene F. Cermak, was away—playing golf as usual, Bob Stubbs directed me to get a Sterling site laid out for Armour, pronto. The next week I made a trip to the area and found an excellent site south of the village of Galt on the east side of Illinois highway 2. This was part of the property which the C&NW had purchased in 1956 to keep the CB&Q from sneaking a track into the Northwestern Steel and Wire plant on the west side of Sterling. Checking out a site is not just eye-balling vacant property. Besides describing the topography, information about water supplies, sewerage facilities, highway access, soil conditions, railroad service, zoning, electric and gas utilities, building restrictions caused by nearby airports, zoning, flood plain exposure, ownerships, taxes and nearby land use must be developed. Strangely, price was not important when the "right" piece of land was found. On the 17th I called Ralph Deemer of Armour's real estate department and gave him a preliminary assessment of what I had found so that he could have the facts in hand when he visited the property the next day. At Stubbs' urging, I gave Harry Mathews, Armour's traffic manager, a full report similar to what I given to Deemer—they had interdepartmental rivalry there, too. We were working on the West Point, Nebraska, abattoir with Armour at this time so that we had two good projects going along together. On the 27th I picked up Ed Landgren of Armour in LaGrange and drove him to Sterling where Bruce Meier of Armour in Omaha had come to see the Galt proposal. We tramped through cornfields and bounced over the dry fields in his pickup right to the bank of the Rock River which formed the easternmost property line. A few days later, on October 2nd, Landgren was dead from a sudden heart attack while en route to meet with our Don Guenther to inspect an Iowa location under consideration for yet another plant.

At Galt we planned to extend rail service to the proposed site by constructing about a mile of industrial lead track south from the main line east of Galt, east of and parallel to Illinois highway #2. Since construction involved crossing land that the C&NW did not own, an easement or purchase would be required. And who held ownership of this stretch of property? None other than our old nemesis, the eccentric Paul W. Dillon, chairman of Northwestern Steel and Wire. Since the C&NW continued to block the CB&Q from providing di-

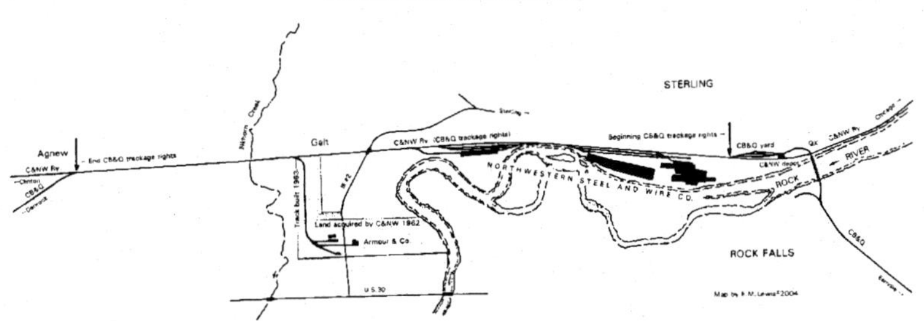

The long-running struggle to retain the Northwestern Steel and Wire Company at a Sterling solely as a C&NW customer required constant vigilance. Repeatedly, the CB&Q sought, and failed, to gain entry to the steel works because of high cost of bridging the Rock River, C&NW purchase of strategic property which blocked access from the west and the strictures of an ancient trackage rights agreement with a Burlington predecessor. Later, changing times caused NS&W to go out of business.

rect service to his mill, he now thought he had a bargaining chip. On October 9th Jimmie Simons in the division office helped work out the alignment and profile for the new lead track east of the highway. A week later Armour requested permission to drill for water on the site. By Halloween they had drilled two dry holes and found a third one the next day. (They eventually hit a good water supply). Harry Bierma, formerly with Real Estate Research, was now working for Bob Ballin and helped draw-up a sales contract with Ralph Deemer of Armour. Then the fly in the ointment commenced to buzz. Paul Dillon claimed that Clyde Fitzpatrick, our president, had promised Dillon first option on the property being sold to Armour when the C&NW had purchased it in 1956. Mr. Fitz denied any such promise. Meanwhile, the C&NW board approved the sale of the disputed land to Armour and Company on November 1, 1962. On November 9th I was the sacrificial goat sent to Sterling to get the easement from Paul Dillon for the Armour lead track to serve the new facility. At the meeting I said we were prepared to pay a reasonable price for an easement, but old man Dillon, his traffic manager, Tom Bellmar, and another steel company man named Laughlin said the price for the easement was letting the CB&Q into the steel mill. I folded up my papers, thanked them for their time and consideration, and left as I was not authorized to negotiate any such agreement. Dillon wrote to Mr. Fitz on the 12th stating his conditions, but our president was not intimidated because we had another ace up our sleeve. We could just as well build the industrial lead **west** of highway 2. W. H. (Heinie) Huffman, assistant chief engineer asked me to accompany him and Chuck Jevne (real estate department) on November 14th on a fast trip to Galt. It was indeed a fast trip until Heinie ran out of gas just short of Rock Falls. After borrowing some gasoline from a nearby farm, we parked Chuck at the Whiteside County courthouse in Morrison to check land titles, while Heinie and I tramped along the alignment of a new track directly south from Galt station and about a quarter mile west of and parallel to highway 2. No difficulties were found; grading would be minimal. Meanwhile Chuck Jevne had determined ownerships of the farms along the west side of highway 2. On the 15th Fitzpatrick ordered Ballin to acquire the farms and easements for the industrial lead track west of the highway as far as the new Armour plant site. By the 28th Ballin was having panic attacks as the land was costing more than he thought it was worth as farmland. Besides, his job was selling railroad land not buying it. A meeting was held on November 30th in the president's office with Mr. Fitz, Bob Stubbs, B. R. Meyers, I. R. Ballin and me on the progress of the acquisition; another meeting on December 7th in Mr. Fitz's office included Meyers, Ballin and me plus Ben Heineman. I was asked to see if Armour could rotate their proposed building ninety degrees to accommodate the side track crossing Highway 2; that was easily accomplished. Another progress meeting was held on December 12th with the addition of Ed Olson and Gene Cermak to the former group—we were really attracting the top brass to this project. On the 19th the official go ahead was given by Mr. Fitz. Heineman was there wearing his trademark bow tie. The first business day of 1963 found Carl Anderson and me at the First National Bank in Chicago to close the land sale to Armour and Company. In March the survey and profiles were all done for the lead track. By April 8th the steel structure of the new abattoir was rising on the site. Early in February 1963, Paul Dillon was still trying to crack our ring of property at Galt. Mr. Heineman had received a private advisory from a Sterling banker alerting us to Dillon's latest maneuver. Our land agents and engineers went to Morrison on February 19th to ascertain what he was trying to pull off. The effort was not seen as too serious and was dealt with promptly. Heineman said he had deposited C&NW funds in Mr. Nice's Sterling bank as a token of appreciation. The industrial lead track was built and in place by May 22nd when permission to cross a county road south of and parallel to our main tracks was suddenly withdrawn by Whiteside County authorities. We suspected that Dillon was still pulling strings in county government, but it was too late, the track was already in place across the road and it stayed there.

At a meeting in the C&NW board room on April 23, 1963 I received permission to promote the newly acquired Sterling-Galt properties for other businesses. The CB&Q was still weaseling around the perimeter of the land ring. Our attorneys opined that the Q might try to claim the right to cross our then undeveloped land, but would have a much weaker legal argument if we could establish a series of additional commercial concerns along highway 2. The old argument or device of keeping ownership of a strip of land fronting a hostile party for "utilities" or a "roadway" had been overthrown by the courts who maintained that retention of such a narrow strip constituted a pointed act which made the action legally ineffective. We needed some bustling commercial activity along that highway to keep our Burlington friends out.

THE ORT STRIKE OF 1962. Ben Heineman really tried to accelerate the modernization of the C&NW through elimination of unnecessary positions and functions. Savings in payroll expense went directly to the bottom line of the financial report. Great advances in communication technology threatened to eliminate the art of telegraphy along with its operators from the railroad's business. The Organization of Railroad Telegraphers (ORT) reacted to this technological threat by demanding in December 1957 that no ORT job be abolished without approval of the union. Heineman did not agree, believing that it was management's right to decide what positions were required for operation of the railroad. Over the next several years some 500 telegrapher jobs were lopped-off. On February 28, 1962 a Presidential commission investigating the controversy found for the company which precipitated the telegraphers' strike against the C&NW on August 30, 1962.

Returning from a European vacation with my family in 1962, I found the ORT strike about to start. The C&NW's chairman, Ben Heineman, and president, Clyde J. Fitzpatrick, were summoned to Washington by President John F. Kennedy's Labor Secretary, Arthur Goldberg, who had also called the ORT for the August 1st meeting. Following the meeting, the strike was postponed on August 2nd for 45 days, but by August 27th it was obvious that all negotiations had broken down. On the 28th the C&NW developed plans to shut down railroad operations. The 29th saw intensive three way negotiations between Goldberg, the railroad's management and the union. No one gave an inch. On August 30th all C&NW train operations ceased. Railroad officers were ordered to report for normal duties despite any personal inconvenience that might be caused by the disruption of train service. The abrupt cessation of commuter service created chaos for thousands of Chicago bound commuters, not only on the C&NW but on other railroads which experienced a surge of our displaced riders. The highways were filled solidly along with most secondary arteries. Even local streets were heavily used by endless streams of vehicles. Parking spaces in Chicago were at a premium. We railroad people used to riding our swift and frequent trains on passes were forced to find alternative ways of getting to the office. My first attempt was September 4th when I tried boarding the CB&Q at Hinsdale very early in the morning. That evening I rode the CTA's Lake Street line to Oak Park where my wife picked me up. Heineman announced it was going to be a long strike. The next day I rode in with M. C. Christensen, Galena division engineer, at 6:40 a.m. in his hi-rail station wagon. This was a long ride on the Congress Street Expressway which seemed more like a slowly moving parking lot. I asked Chris why we didn't try hi-railing on the now rusting mainline, but he was fearful that the Milwaukee crossing at Tower A-2, Western Avenue, might present a problem if the Milwaukee's operator would not line our route through the plant. Also, there would be problems at the many grade crossings through Melrose Park and Maywood because the hi-rail car did not trigger automatic crossing protection signals. The last place eastward we could get off the rail was at the west end of the Chicago Shops at Kilbourn Avenue as there were no other grade crossings east of the Chicago Passenger Terminal on the main line. Since there was no operator in the tower at Noble Street, lining us up for the North Pier line and its several grade crossings where we could get back on rubber, Chris elected to stick with the congested streets. Because Chris often had business elsewhere on his division, I had to improvise other ways to get to the city all of which were a great inconvenience. I decided that this would be a good time to visit my territory and made several trips to western Illinois and Nebraska. On September 6th the Brotherhood of Railroad Trainmen (BRT) tried to force the C&NW to operate the former M&StL lines since the ORT was not represented on that railroad. The C&NW refused that argument, too. On the 14th we heard that President Kennedy was to take a personal hand in the strike issue; on the 17th he asked the union to accept the board's recommendations—all in vain. The ORT demanded on September 28th that all laid-off members be reinstated at once, to no avail. The strike was broken. On September 30th limited operations began and commuter service was fully restored October 1st.

Getting a railroad back into operation after a month of idleness was no easy task. The biggest concern involved safety of operations. The month long build-up of rust interfered with proper operation of signals for train control and grade crossing protection. But, how to get the rust off the rail quickly and cheaply? Wire brushing was hopeless; sandpaper was mentioned and laughed off. Harold W. Jensen had the answer—direct and no nonsense about it—burn it off!

He ordered a special train consisting of an old company flatcar and a diesel locomotive. Tieing down the brakes on the flatcar's lead truck, the engine shoved the sliding wheels at a good clip down the mainline generating a shower of sparks. When the flattened wheels had been ground down so far as to become dangerous, the brakes were unlocked, the wheels rotated a bit, the brakes reset, and off they went again in a cloud of sparks and smoke. The result was a complete scour of the rusted top of rail at the cost of two wheel sets. Train service resumed over the weekend without any signal failures.

Because of the strike and resulting diminished revenues, departmental budgets were cut, and we were admonished to cut back on spending. Our department always had troubles with budget over runs primarily due to Gene Cermak and Chuck Towle's lavish life styles reflected by their expense accounts. Cermak always turned in a monthly expense account of at least $1,200 whether he traveled anywhere or not. We were instructed to overspend our individual allotments or we could be reduced in subsequent years. There were times especially at the end of budget year when we had surplus funds that had to be spent. Creation of fabulous business lunches in December using names drawn from a telephone book was one dodge. After the years of guarding the company's funds in the closely-monitored and sometimes hard scrabble days of engineering, I was sorely disturbed by this attitude of the traffic department's. In my personal view, it was akin to stealing.

After the strike, business gradually resumed. It appeared that there were new players on the railroad messing around in our industrial development business, too. On August 14th Carl Crumrine, one of our financial officers, asked me to look at some property near Cherry Valley, but wouldn't say for what purpose. Russ Fyfe, our Rochelle station agent, collared someone from International Minerals and did such a good sales job that IMC optioned some property there about September 25th. The next month Mr. Knorst of IMC took exception to my offer of the railroad assuming only the cost of the new sidetrack from the point of switch to the 12 foot clearance point, our standard policy which was established under the U. S. Railroad Administration in 1918. Most railroads had their own policies regarding side track construction; some were more favorable to the industry. Knorst did not like our split of the costs. He went directly to vice president Ed Olson complaining that my offer was not good enough. I had assessed the potential from the facility IMC was planning and had determined that there was sufficient prospects for revenue to the C&NW to justify our assumption of the costs of the mainline turnout. I had recently negotiated with Daubert Chemical at Dixon and Globe Union at Geneva with the costs involved for their sidetracks. Neither of these companies was projected to be big revenue producers and the number 10 turnouts in the Galena Division were heavy rail, required electric locks, signal line and circuit changes and involved raising multi-wire communication lines. Jack Reese, our chief lineman, swore that I used his telephone poles as center line stakes for new tracks. Taken together this was a considerable investment by the C&NW. To hedge potential loss to the railroad, I negotiated agreements with these companies that called for their deposit of the estimated costs of the mainline switches and the railroad would refund to them credits for every car of revenue freight they received or forwarded over some specified period of time. The credits were an agreed-upon rate per car which varied depending upon the unique circumstances for each new customer. Accounting hated these track deals I made because special accounts had to be established and monitored using freight revenue figures which sometimes were hard to capture from another accounting stream. Knorst of IMC was convinced, however, that I was not doing his company any favors.

On September 17th I attended the rates and tariff school run by Herman C. Jacobsen of our rate department. The class sessions were held at the LaSalle hotel on Madison Street rather than in a company conference room better to isolate the attendees from the normal pressures of railroad business. Herman did a good job in describing how the rate territories were established, the role of tariff publishers, government regulations surrounding railroad rates, and how rates are abstracted from the tariffs. Irv Lawrenz, also from rates, and a man from the Western Trunk Line bureau gave a further exposition on how the rate process worked. To further complicate matters, the arcane matter of divisions of revenues between railroads was explained. Raymond E. Degnan, foreign freight manager, described the growing field of international rates. This was 1962 and the railroad freight rate business was as complex and convoluted as man could make it. In 1980 the Staggers Act swept away a good deal of this maze through deregulation. In 1962 I had no idea that this week of classes would become an important part of my life twenty-two years later.

The balance of 1962 dribbled away with little trips and contacts that did not mature into anything substantial. In October the Urban Land Institute met in Peoria; Keith McCullagh of the Illinois Central and I attended the tour of the central business district and the industrial growth at Mapleton, the Pioneer Industrial Park and Caterpillar's new Mossville research facility. After listening to the institute's critique of the changes observed since their last visit, Fred Jones, our DF&PA, spirited me away to visit with Henry Zimmerman of American Distillers in Pekin about the possibility of getting C&NW service into their plant. He was generous with maps of the property, but the C&NW passed by his facility on a high fill that formed the south approach to our Illinois River lift span bridge #1731. The New York Central's subsidiary, the Peoria and Eastern Railway, served the distillery.

This tag end of their branch line railroad provided inadequate service and Zimmerman wanted direct access for inbound grain eliminating the P&PU switching charges and delay. Physically and legally, it would be a tough nut to crack. Years later a connecting track on the opposite side of the bridge approach fill was constructed for coal trains to a nearby power station. On August 16, 1963 this area which included Standard Brands and a Quaker Oats Company plant came back to life when Bob Stubbs decided to push construction of a connecting track, but the matter soon faded away again. On the 18th of October I lunched with R. V. Peabody of the Smith-Douglas Company of Norfolk, Virginia. They were interested in establishing fertilizer distribution centers in the Midwest. We had no idea of the number of these little installations that would be constructed all over the middle west. The next day I took a vice president of Kent Feeds, Muscatine, Iowa, on a tour of western Illinois from DeKalb to South Beloit. The poor guy suffered a nose bleed on the drive which certainly inhibited his ability to evaluate industrial property.

Northern Illinois Gas Company sponsored an industrial development council meeting in Joliet on October 25th. Dr. Klemme of Northern Natural Gas in Omaha was the featured speaker. At that meeting I conceived of an idea to form a Chicago club of railroad industrial development people for a monthly meeting. The basic idea was that this group would be the low level industrial agents not the department heads. A boss could be invited, however, providing he bought the first round. I named the group R. I. D. an acronym for Railroad Industrial Development. We met for the first time on December 7th. And thirteen showed up at the Cafe Bohemia located at Adams and Clinton Streets plus one invited boss, Ollie Waggener of the CB&Q.

Rumors were flying again. One said that the traffic department was to undergo major changes (November 2nd) and a second had the C&NW leasing the Gulf, Mobile and Ohio. As usual, they were all smoke and no substance. There were more quick day trips around Northern Illinois in November with a Commonwealth Edison industrial agent, Robert Lippold, and his prospect, Barry MacLean of the MacLean Fogg Locknut Company. Poking around all day in the Huntley area resulted in nothing at all. Armour popped up with a proposed pork processing plant in the Monmouth, Illinois, area. This project was doomed from the start because the M&StL line traversing the area could not have provided the level of service required even if a good site could be found. It just happened that Monmouth was the center of a huge hog production area. Ralph Deemer of Armour's real estate department and I toured through London Mills, Farmington, Monmouth and Little York without finding a really attractive site. Despite the paucity of good industrial sites, I got a lift when we found Midland Electric Coal's #84, a grimy little 0-6-0, still shuffling around Rapatee. I was glad when Armour dropped this pork project. In December Norman Piquet of W. R. Grace Chemicals, Memphis, Tennessee, came into the office to look at properties along the Mississippi River especially at East Clinton, Illinois, and in Minneapolis, Minnesota. Cermak arranged a flying tour the next day using the company's turboprop. Piquet, Cermak, Towle and I boarded at Midway Airport, flew to Clinton, Iowa, to collect Vern Giegold, and flew low at about 3,000 feet north to Shakopee, Minnesota, and return. This was a grand way to see a lot of country in a short time. On December 11th the announcement was made that Charles Towle was to establish a satellite industrial development office in Minneapolis. I made a quick trip to Milwaukee at Bob Tate's (our local DF&PA) request to call on Allis Chalmers. They wanted to look at sites in the Mapleton, Illinois, area. I met McPherren and Walker of AC in Peoria on the 17th, but they liked property in the Morton, Illinois, area better. Missing my train back to Chicago, I took an Ozark Convair to O'Hare—this flying was habit forming.

December turned cold. Suddenly, we were hit by minus 13 degree weather and staying in the office seemed quite sensible. While this had been an interesting year, I sensed that something was still missing. I felt uneasy that our industrial development activities were so dependent on chance. We passively waited for prospects to surface. We acted only when we were contacted. There must be some way to focus our efforts more efficiently and initialize action. We seemed to be like vultures sitting on a perch waiting for something dead to appear. Perhaps I was just somewhat bored with the lack of activity or challenge. Nagging, but unfocused, thoughts relating to this problem were rattling around in my subconscious. Well, just turn the calendar page and maybe something will turn up in the new year.

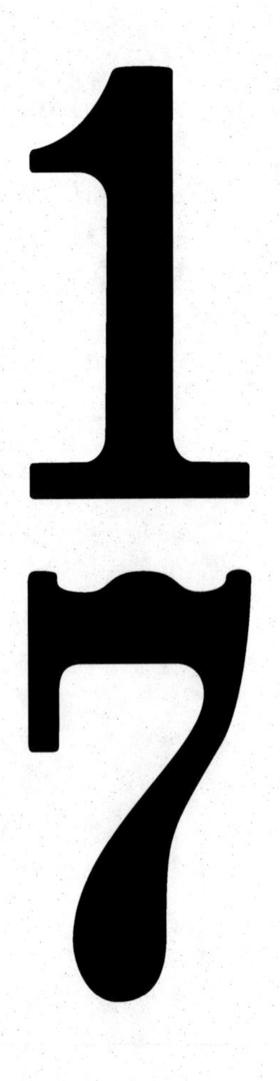

CHRYSLER—THE TRAFFIC DEPARTMENT'S APOGEE, 1963 . . . Our vice president of traffic, Ed Olson, opened the year with an address to an industrial development meeting in Elgin on January 3, 1963. His talk emphasized the importance to the C&NW of the work our department did. Yet, sometimes his actions and policies made us wonder if he really was a supporter or just tolerated us as one of Heineman's public images. Our departmental make-up continued to change. Richard C. Volkert, formerly general industrial agent on the old M&StL, wasn't with us very long. We celebrated his retirement at a luncheon in Chicago on January 7th. His Minneapolis farewell was to be February 4th. I learned that my new annual salary was to be $9,000, if Bob Stubbs approved. He did and sent his assistant, Bob Christie, across the office on the 12th to tell me of the magnificent $70 monthly increase I was to receive. Although the money was meager, I did receive an upgrade in title to General Industrial Development Agent, effective March 1st. I had climbed a rung on the ladder, but where was the ladder leading? A little more money appeared October 7th boosting my pay to $9,300—every little bit helped. As part of the changes in the department, Charles M. Towle and Stanley B. Johnston were to move to Minneapolis to staff a satellite industrial development office there. Stan packed up all the files March 11th, but the transition didn't really happen until April 11th when Chuck moved out of our Chicago office.

For some reason I was invited to participate in a *Duns Review* Round Table discussion in New York on January 10th. About eighteen of us "development experts" sat around a big table and talked about trends in industrial development. I took this opportunity of being in New York to call on Clinton Hoch, head of the Fantus Company, Bob Hazelwood at International Paper, Bob Katz at Revlon as well as make courtesy calls at Penn-Dixie Cement, Union Carbide and Olin Mathiesen. I had come to New York on my favorite train, the NYC's *Twentieth Century Limited* in sleeper *Calumet River* on the evening of the 8th. The ride was comfortable and service outstanding as usual, but returning on the 11th, I was aboard the second section, which curiously left before the first section, in sleeping car *Minnesota Valley*. With no heat in my roomette and the January weather outside, the trip was a very cold ride indeed. Even super trains had their problems from time to time.

In mid January I requested that Bob Ballin, head of real estate, to seek rezoning of our extra width right-of-way at South Beloit, Illinois, for construction of a rail-truck transfer facility for handling and storing cement. Lehigh Portland Cement was positioning themselves for the huge paving contracts coming up for construction of Interstate 90 just a few miles to the east. A month later the petition was heard and was approved. A trip to the Dixon headquarters of the Illinois State Highway department proved to be so much lost motion as Dixon said the territory at South Beloit now belonged to the Elgin office. On March 27th I picked up Al Strauss of Lehigh to meet with the district engineer who just happened to be an old acquaintance, Dick Golterman, brother of my former rodman associate, Ralph, from

Galena Division days. Dick looked at the jurisdictional maps and determined that I had been correct in the first place as the site was in the Dixon district. Al and I drove northwest to the site and then straight south to Dixon to apply for the access permit from U. S. 51 which crossed the C&NW's Harvard-Beloit line at the south end of the property. Our law department approved the sale as did operating although there were rumblings and mutterings about difficulties in switching the facility. Eventually, switching limits were extended south from Beloit to allow a local crew to handle the work without stopping and paying extra for the road crews to switch the new facility. Just when everything seemed to be in order, the CIO Laborers' Union working on construction of the cement storage silos claimed the track construction work. That old labor argument we had faced at Proviso again became a problem again. Railroad crews were also union members and precedents clearly favored track construction by the railroad crews especially on company land. The laborers' union lost this claim, again. The facility was built on time and provided substantial freight revenue for the C&NW. The Harvard-South Beloit line itself lasted several more years as the route for Ed Burkhardt's sand trains. This was a twenty five car set of old ore jimmies that carried sand and gravel from Beloit pits to the Chicago area. The little train terminated at Elmhurst on the tracks serving the Elmhurst-Chicago Stone Quarry on the west end of town, laid over a day and then returned empty via Proviso and the north freight line to Des Plaines and thence back to Beloit via Harvard and Poplar Grove. With the demise of the sand train some years later, the entire line was abandoned west of Harvard.

At dawn on the cold morning of January 31st the Chicago, North Shore and Milwaukee Railway, a heavy interurban electric line, ceased operation. That same day in Rochelle, Illinois, C&NW westbound freight #249 derailed at speed in the track side ditch causing a million dollar wreck. On that same day Armour officially advised that the pork plant proposal at Monmouth had been shelved—to my relief. (The old "Louie" beyond Monmouth did not last long after its abandonment was approved November 9, 1971). January 2, 1963 turned into a busy day for me when I heard that another big chemical plant was looking at the Sterling, Illinois, area. That rumor soon proved to be pure wind in the willows. The next rumor we heard was that Susquehanna Corporation was considering an operation in the mountains 15 miles south of Riverton, Wyoming. This one was serious enough for me to pull out my old set of USGS maps which were covered with my penciled notes from former flurries of interest over the years.

The day of record cold was January 23, 1963 when the thermometer dropped to −18 degrees and the recorded high for the day was −13. Chicago's elevated lines failed to operate and the Chicago, South Shore and South Bend, another electric interurban, closed down because the extreme cold caused the overhead catenary to sag dangerously. Another blast of −18 cold returned on the 28th when I flew to Omaha with Norman Piquet and a Mr. Stewart of W. R. Grace Chemicals in the company plane. The next day we toured Blair, Nebraska, as a possible site for an ammonia production facility. On the return flight, we added J. R. (Bob) Brennan, formerly head of passenger sales and now acting as a consultant to the president, and Roy Erickson, our grain rate manager. We waited in vain for two hours for Charles (Chuck) McGehee, before flying to Sioux City, Iowa to collect J. R. (Bob) Kunkel, manager coal traffic, and someone named Kenny, for the return to Meigs Field and O'Hare. Acquiring a company plane had been Heineman's idea, and it served the company well. Any railroad officer could ride so long as his destination was more or less along the route of the first to sign on. The Grace project heated up in February when a Mr. Siebert came up from Memphis to discuss freight rates. The cold had not abated by the end of January when a minus 14 degrees was recorded. The month ended with 20 inches of snow on the ground and 15 days when the temperature did not rise to zero. The rumor at the time was that the C&NW was buying a 10% interest in the above mentioned CSS&SB. Again the rumor turned out to be just that.

Our little social railroad industrial development group, R. I. D., met for the second time January 25th with 16 members showing up. The third meeting March 1st drew 12 and the fourth on April 15th, 11. May and June meetings attracted 13. After a summer hiatus, R. I. D. met again on September 13th with 13 present plus a head of department, Edward J. Stoll from the Milwaukee Road, who was inaugurated as our third honorary president and was suitably rewarded with a certificate. We met again December 6th and inducted my boss, Gene F. Cermak, as honorary president No.4. Remarkably, this purely social group had actually hung together for a year.

In February Herman Jacobsen, our rate department "teacher" of last Fall, and I were drafted to make a presentation in Sterling on the 20th. Despite the extreme cold where the high for the day was a plus 3, we drove west and made our little talks in the late afternoon and early evening. Two days later Missouri Portland Cement Company came into the office to discuss track construction at their new facility near the new CILCO power station at Sommer. Since this new plant was located inside the infamous red line which marked the troubled joint TPW-C&NW territory, we had to work out a track agreement with not only Missouri Portland but including the TP&W as a participant. Because of the rising tide of woe surrounding that joint service arrangement, Gene Cermak, M. C. Christensen, division engineer, and I went to Peoria on March 7th to meet with Randy Egbert, chief engineer of the TP&W to hammer out an acceptable track agreement. Our attempt was futile because by May 31st the legal thicket surrounding the ADM service at nearby Mapleton had crowded out a mere track agreement with a cement company. Bob Russell, our lead attorney in the case, took over the matter when Bruce Packard of engineering and the contract attorney, Dickman, also failed to approve our work. By July 5th matters were still stalled. In an attempt to resolve the matter, I took Packard with me to East Peoria on the 12th in another attempt to work out an agreement with the TP&W. We succeeded in negotiating a track agreement with just two small adjustments. However, because of the ADM lawsuit, the revised agreement was again stalled. Missouri Portland was caught between two warring railroads.

Cermak and I made another Nebraska visit when we attended the annual governor's conference February 27th. I enjoyed the trip on the CB&Q's *Ak-Sar-Ben Zephyr* in sleeping car *Silver Terrain.* Cermak boarded the train at Omaha for the short run down to Lincoln. We also ran into Bill Spitzenberger from the CB&Q's industrial department on the train. After the conference, which we felt was not as good as in previous years, we attended the annual Lincoln Traffic Club banquet at which Dick Ryan, our local sales agent, was installed as president for the coming year while Gene sat at the head table with Governor Frank Morrison. Of course the dinner was followed by an late-night poker game. The next day while Cermak went back to Omaha, Dick Ryan and I made some courtesy calls in Lincoln, looked at industrial property and had a visit with a Mr. Vihstadt of Brunswick Corporation. I flew to Omaha on one of Frontier's venerable C-47s to catch up with the rest of our group for a DC-8 United flight to O'Hare.

A new man joined our department March 1st, Larry Krause. My first impression was that of an Iowa farm boy who had been poured into a suit two sizes too small topped off by a haircut done by a low ceiling fan. Aggressive and awkward, Larry was going to need a lot of finishing before becoming a successful industrial agent. While first impressions are often wrong, I reserved judgement on this new hire. Over time Larry proved to possess a low cunning supplemented by an absolute lack of couth. I was happy to see him assigned to Vern Giegold and his Iowa territory where he would be right at home with his fellow Iowans. Our inherited M&StL man, Dorval (Don) Skoglund, announced on March 19th that he was resigning. Don was a big fellow with a florid complexion and was notably silent. He rarely started a conversation even when meeting with a client. How he ever got into a customer contact business was indeed a mystery. While a kind and considerate man, he did not fit well in our more aggressive department. His father, Howell P. E. Skoglund, was president of North American Life & Casualty and a director of the former M&StL. On March 6th Cermak and I were invited to attend a ground breaking ceremony for the new Standard Pharmaceutical plant in Elgin, Illinois. This company was located adjacent to the C&NW's old Freeport Line near Big Timber Road. A drug manufacturer, they had no need for rail service, but we went to the affair to show support for our friend, Charles Willson, the new director of the Elgin Industrial Department. Charlie had been with the Chicago Association of Commerce and Industry for some years and was now in his fourth week with the Elgin group. His Chicago friends were convinced that he had taken a backward step in his career by going out to this far suburban area. Charlie was a straight-laced Canadian who didn't think much of the Chicago development group's turbulent glad handing methods of attracting business and was happy to be in a more compatible area where he could better use his cool and honest analytical style of promotion. His sober generosity was illustrated by his gift to me of an small ice crushing machine which he had won in a drawing—he had no use for it in his alcohol-free home.

In March 1963 we made that trip to Tulsa in connection with the Skelly-Swift NITRIN project south of East Clinton, Illinois, which was discussed in the previous chapter. Of greater importance to the C&NW and other Midwestern railroads at this time was the settlement that month of the long running legal battle over division of freight revenues on interline moves on March 28, 1963. Since its earliest days, the Union Pacific Railroad had claimed a larger percentage of transcontinental revenues because its main lines crossed undeveloped and semi-desert territory west of Omaha, Nebraska, to Ogden, Utah. What may have been true in 1869 was certainly not true in more recent years. The "west" had pretty much disappeared by 1900 yet the Midwestern roads were still limited to a 16.5% division on transcontinental business. Settlement of this case gave the C&NW a 22% division which was based on actual mileage operated without any offsetting penalties or biases. The C&NW rejoiced; the Union Pacific sulked.

The American Industrial Development Council's annual meeting was set for San Diego, California in 1963. The C&NW sent a large delegation. My wife and I accompanied Harvey and Ruth Buchholz on a relaxed trip on the AT&SF's *Chief*, arriving in Los Angeles on March 30th. The next day we rode the ATSF along the Pacific shore to San Diego. The Council meetings, tours and activities filled the week. After a quick and chilly visit to Tijuana, my wife and I took a brief vacation in California before flying back to Chicago on April 6th. These meetings were very important in our business, and we attended many of them. Our railroad may not have been among the largest or wealthiest, but through sheer energy was making a name for itself among our peers. These meetings gave us a chance to size up our competitors, brag of successes, trade lies about prospects and establish our presence. We got to know our fellow railroad competitors quite well. On April 10th Cermak and I attended the Illinois State Chamber of Commerce's annual meeting held at Normal, Illinois. Jack Frost of the Illinois Central and Otto Pongrace of the New York

Central moderated one forum. Otto's jokes were better than Jack's. While I drove the car back to Chicago with Phil Schmidt of the Rock Island riding shotgun, Gene Cermak and George Cox of the ATSF played gin rummy in the back seat. Competitors to the bone, but also good friends. Another meeting on April 25th was sponsored by the Northern Illinois Gas Company at the Starved Rock Lodge near Ottawa, Illinois. Howard Roepke, Professor of Geography at the University of Illinois, spoke and very well too, giving me some ideas to mull over. Phil Schmidt of the Rock Island again rode with me so we stopped at Spring Valley to look at our respective company's properties before going on to Chicago.

Implausibility often emerged as an industrial prospect. On April 16th George Bleibtrey, representing something called Motor Wheel Company, cornered me with his plans to build a factory to construct a self-propelled wheel for farm and industrial applications. I do not recall whether his design called for small electric motors or a hydraulic linkage to drive the wheel. Frankly, I was suspicious of the practicality of the product, but agreed to invest some time to take him on the western Illinois tour. We looked at everything. Nothing suited him. I was very happy to hear May 10th that he was interested in properties at Princeton and Mendota, Illinois, which were served by other railroads. Actually, he wasn't looking for a building site; he was looking for some community or organization that would finance his project. I was a bit surprised when he called me June 27th demanding that the C&NW build a sidetrack for him at Dixon, Illinois. I countered with an offer to build the trackage if his company would agree to pay for the initial installation and accept a rebate arrangement based on C&NW freight revenues for a finite period of time. That wasn't good enough for him, and we never heard from him again.

Comparing notes with the my fellow industrial agents, I discovered that our operating divisions on the railroad were using different construction policies for industrial sidetracks. On April 22nd I wrote to B. R. Meyers, chief engineer, asking for a uniform set of track construction specifications as the only ones I could find were issued in 1928. We never heard from him. I developed a handbook for our men that summarized the conditions for side track construction according to the legal constraints we were obliged to follow. Because the various states had differing requirements for clearance requirements at loading platforms, vertical restrictions and horizontal distances to buildings, the book provided a clear guide to the legal distinctions.

United's puddle-jumping local flight to Omaha stopped at Moline, Cedar Rapids and Des Moines. In clear weather it was a terrific way to see the landscape unreel at a relatively low altitude. I rode this flight April 29th to meet with J. M. (Jack) Peters and a Mr. Bachenburg of Westcentral Grain Company who was interested in property at Blair, Nebraska, on the Missouri River bank. I wasn't too excited about a grain shipper having access to barge service, but the Missouri was almost a wild river where commercial river operations were neither easy or common. That same day I met with a Mr. Sorensen of Statex, who was interested in a grain operation at Fremont, Nebraska. I went to Fremont on the 30th and had a meeting with Paul Christensen, owner of a huge grain storage complex on the C&NW's Lincoln line west of town. Christensen's operation was located at the beginning of the Fremont-Lincoln line just before it crossed the Union Pacific mainline at grade. Just south of the river the C&NW line to Superior, Nebraska, began at Platte River. That portion from Platte River to Seward had seen its last train in June 1962 when the C&NW worked out a trackage rights operation from Lincoln to Seward before going on south to Superior. (The Seward-Superior line was sold to a short line operator in 1974 and the Fremont to Lincoln was abandoned March 1, 1981). Before going back to Omaha, I took a quick run northwest to West Point to check on the Armour abattoir then under construction. On May 1st I presented my report to Statex and then flew back to Chicago on N1394Z, the company's plane, with Harry Bierma from real estate. We stopped at Des Moines to pick up Howard Koop, another Real Estate Research consultant who later became an employee of the C&NW. (Koop later left the railroad and became a residential developer in the Winfield area). While waiting for the plane at Omaha, I had an interesting two hour chat with George Behm, a recent immigrant from Germany, who was now working for Ballin's real estate department in the Omaha area. The Statex proposal was not approved by their board. Jack Peters never shared the details of the board's decision.

The Jolly Green Giant wanted to see me at their LeSeuer, Minnesota, headquarters, and I obliged by taking a United Caravelle *Ville de Strasbourg* to Minneapolis on May 3rd. Irvin Gran of Green Giant met the plane and drove me down to the meeting in LeSeuer. They were planning major changes to their old canning plant at Belvidere, Illinois, and wanted my advice on track arrangements. The Belvidere plant was located along the Kishwaukee River at the east end of town. Rail service was provided from the north leg of the former Madison Division wye track which continued north across a steel river bridge to the unused North Yard. I suggested a few minor changes to the track alignment that freed up land for expanding the building. On my return to Chicago I was accompanied by a five foot tall Green Giant doll for my daughter, who at 15 was really too old for that sort of toy, but the gesture was appreciated even though my fellow passengers on the plane had a smile or two.

Somehow, I got involved with a crafty promoter/developer named Sam Lezak. Sam was an attorney and deeply involved in quarter horse racing. He also was acutely aware of the march of suburbia then gathering steam as the new expressways began to reach out from Chicago. He was convinced that farmland in the Ridgefield area beyond Crystal Lake and the Huntley area west of Elgin were prime targets for residential and commercial development. When he discovered that the railroad had an industrial development department with a lot of good information about property, Sam latched onto me to help him develop a new city at Huntley. First he and his sidekick, an old realtor named John Thormahlen, tackled some property at Ridgefield and moved to get the zoning changed to industrial. I suppose that Bob Seegers of Seegers Grain was involved in this action as the site at Terra Cotta had fallen through for his grain handling business. Any rezoning in McHenry County seemed to raise a crowd of protestors and the Ridgefield petition was no exception. On August 16th a mob descended on the hearing. With our railroad attorney's help, we prepared a petition to annex the property to the City of Crystal Lake. I ducked the hearing that evening as feelings were running high and a railroad man might be a convenient target. No decision was reached anyway. Sam, a former Army infantry lieutenant, knew when to press an issue and when to back off. He had an interesting story to tell about his army days in Germany at the end of the war. His squad spotted two Germans burying something in a backyard and had them dig it up. It was a waterproofed parcel about the size of a cigar box. When opened, Sam said it was found to contain German postage stamps. Gee, said I, that would have been interesting to see. As a collector of German stamps for some years, I had a fair collection and was knowledgeable about that country's issues, Did he still have them?, I inquired. Oh, would you like to see them? I was hooked—of course I would. I had visions of the old German States classics and early empire issues running before my eyes. When Sam handed me the box and I opened it, I tried not to let my face betray what I found—all common issues with little value. Why was this worthless collection being buried? Was there something else in the box that Sam had not mentioned? Nevertheless, I was to see Sam again in March of 1964 when ancient John Thormahlen showed up in the office with orders from Sam to purchase 5,000 acres near Huntley, Illinois. That again turned out to be a mere puff of hot air that soon dissipated.

At the end of May, Howard Koop from Real Estate Research became interested in the Peck property west of Geneva for the same reasons that Lezak was looking at undeveloped farmland further north. Howard was convinced the C&NW should buy this desirable acreage on the west edge of the city of Geneva as an investment. I went with him to look at it not realizing that in three years I would be purchasing the very same land for an industrial client. On May 15th I met with the mayor of Geneva and talked to Tony Ware on the Kane County Planning Commission to get some idea as to direction and timing of expected growth in the area. Howard's boss, Harry Bierma, Gene Cermak and I discussed this proposal at some length, but by December other important events had shelved it. Sam Lezak and another pal, Frank Hesser, also got interested in the Peck land in August of 1964, but Sam never seemed to be able to corral the investment money to finance his ambitions.

Life went on. A passenger train smacked a truck at a grade crossing near Milwaukee on May 10th. Oak Manufacturing in Crystal Lake asked their C&NW sales representative, Eddie Carr, to have someone come out and discuss a track change with them. Not knowing about their products or business, I hopped a suburban train to Crystal Lake and called at their plant which was south of town. When I learned that they made electronic components which mostly went to Japan, I was sure that there was not much rail business involved. Since all they wanted was a track extension, I gave them the estimated costs. A few days later a young lady appeared at our office with a purchase order for the track work. I think they were bidding on a government job and had to have a functional sidetrack at their manufacturing site to qualify as bidders. Within days the grading was completed and our crew laid down the couple of rail lengths needed. That was a quick job, indeed. In early June C&NW common stock hit 23½. Who was buying it? On May 31st I met with a Mr. McGill of Elgin Softener Company in Elgin and worked up a little track plan for him for his new production plant. There were a multitude of little jobs like these constantly bobbing up. I rode train #209 north on June 14th to try out the new double-decked gallery cars on a long distance run; it wasn't a very smooth ride—either the cars rode badly at speed or the track condition was poor.

Ben W. Heineman presided at a panel discussion in the Morrison Hotel on May 28th. The subject was development of industrial parks and Heineman's briefing by Gene Cermak made him as aware of the subject as anyone at the affair. He was proving that he was really interested in our department's work and was more supportive than our own traffic department officers. We dreamed up a service program called SITEFAX complete with an advertising campaign which Cermak asked Ray Schaeffer of public affairs to develop. The concept vanished without a trace.

Sometimes projects hang on long after they should have quietly died. Commander John Barron with the Tri-City Port District at Granite City, Illinois, kept promoting the idea that C&NW rail service to his port facility would be in the best interests of everyone. The matter finally arose for discussion at a meeting in Larry Provo's office June 7th with a general consensus that the project did not seem beneficial to the C&NW because direct access was not available. To settle the matter a group went to St. Louis on the GM&O day train on the 19th to take a first hand look at the facility and explore ways of obtaining access. Herman Jacobsen from the rate department, Eugene Anderson, law, Ralph Abbot, finance, and I met three of our operating department people and toured the whole area with Barron herding us about. It was a hot and steamy Illinois June day. Thankfully, Schaller from our St. Louis office, dragged me off to see Monsanto about fertilizer distribution sites before rushing me to the airport to catch N1394Z for the flight back to O'Hare. On the 25th I worked out a Granite City line extension to the port, but it was pure imagination containing few real pearls of possibility. To kill this thing I decided to check out the last loose end—that Illinois Terminal line from Edwardsville to Granite City that Keathley was promoting last year. On June 30th I rode the GM&O's *Midnight Special* in sleeper *Samuel King Tigrett* to St. Louis and drove a rental car to Edwardsville where I hired a taxi to take me to the base of the bluff line where the ITC line entered the Mississippi River flood plain. A cloudless day with temperatures in triple digits, I hiked all the way back to Edwardsville on that weedy, cinder ballasted track. Going through those deep cuts was akin to walking through a working blast furnace. Recovering in my air conditioned car, I cruised north along the old Litchfield and Madison Railway and the C&NW's Southern Illinois district to South Pekin, cut across to see the new Arrowhead silica sand mining facility at Troy Grove and then went straight home.

And now for something completely different. We were well aware that there was rapid growth of population to the west of Chicago and I determined to try using a mathematical technique on available demographic data. I am sure my methodology was known to demographers, who might argue that the basic premise was flawed, but it was cool in the office, hot outdoors, and I wanted to try this out. On June 13th I drew a population distribution map of the Fox River Valley communities and applied a vector analysis to the distributed masses of population. Interesting, but not very useful. It was a first try at organizing data so that they could be usefully employed in industrial development work. Trends over time could be diagramed and displayed. Other matters pressed so I put the exercise away while mulling over the concept. A month later, on July 10th, I was inspired to apply a centroid analysis to the same data and this really produced some useful information. I could now map and analyze population centers, centroids of concentration and develop predictions as to future growth. The germ of the concept was there and next year would see the idea grow into a real plant location study.

Our department personnel changed again in the summer of 1963. Virgil Steinhoff, fresh from the Cedar Rapids sales office, became our chief clerk. Energetic and quick to learn, Virgil would eventually become a successful industrial agent. Our quiet number two secretary, Betty Kauth, retired on July 31st; she was to be 65 years old in September. Donald A. MacBean, a clerk from Bob Stubb's traffic department, was appointed to fill the departed Don Skoglund's position. MacBean had been a railroad person from the day he was born in 1928. At age 16 he began working for the Chicago Rapid Transit—it was wartime and help was hard to find. Over time Don and I made a team in our efforts to bring new concepts into this business. My typist and file clerk, Lena Sbarbaro, was miffed when Cermak did not appoint her to the slot vacated by Betty. But he couldn't because it was a union job open to bidding. June Miller from Stubbs' office successfully bid the job and moved in on September 30th and promptly confiscated Lena's electric typewriter. Lena seethed in her black Italian way and did everything she could to make June feel unwelcome in our little group. Office politics were not confined to the staff—the boss himself was the center of speculation that he was angling for the CB&Q job created by O. O. Waggener's coming retirement. This was only a rumor; we were to have quite a lot to do with Mr. Waggener later in the year. Gene continued to grumble about lack of cooperation from engineering and real estate, the wishy-washy support of his superiors, Ed Olson and Bob Stubbs, and the rotten golf he was playing that summer. His complaining could not have been too loud because he was promoted to assistant vice president at the October 3rd board meeting. About this time I had finagled a private corner in the rear of our open office space and actually acquired the pleasure of having two big windows looking west over the Chicago Passenger Terminal roof. Since there was plenty of room in this odd corner in back of an elevator shaft, Don MacBean also settled in by the second window. The skyline was punctuated by a myriad of water tanks perched atop the many old factory buildings on Chicago's immediate west side. I have forgotten how many tanks were visible—we counted them now and again. This little space was to produce some big and important ideas as Don and I kicked scenarios back and forth in our little private corner.

Another railroad strike threatened July 8, 1963. The basic disagreement involved removing firemen from diesel locomotives. On the 9th we heard that management and the unions had met with President Kennedy and that management had agreed to binding arbitration but the unions had balked. The strike date shifted to the 29th and, again, to August 14th, Congress failed to act on August 22nd and the railroads put up notices that the firemen jobs were abolished the next day. August 27th, all was quiet; no embargoes were announced. On the 28th the U. S. Senate passed a compulsory arbitration bill in which the House concurred. Kennedy signed it and the strike threat was over. Congress had acted with just five hours and a few minutes to spare to avert the work stoppage. The battle over the firemen's jobs was not over, however, just postponed two years.

Not all property searches occur on bright sunny days. On August 2nd while out with Ed Cosnan of Apex and Hack Roth of Northern Illinois Gas Company looking for a smelter site near West Chicago, a violent weather front moved through releasing torrential rain and causing a lot of wind damage in the area. A smelter operation was really not suited to this area anyway and nothing came of this excursion.

Noted in passing: the last Twin Cities 400 and Rochester 400 trains made their final runs July 23, 1963.

At Barr tower on the Southern Illinois district a C&NW southbound freight slammed into the middle of a Chicago and Illinois Midland train on August 5, 1963 killing our engineer. The fireman, who survived, claimed faulty brakes were the cause of the collision, but the evidence pointed to the crew being asleep as they had run the red approach signal and made no move to set any brakes.

The rapid decline of the old M&StL in Illinois was evident when I observed train #20 come into Monmouth on the evening of September 18th. This had been the premier train on the old "Louie" before 1960 and now it looked like a local train with its few cars. I was in that area wasting my time trying to find a suitable place for an ammonia production plant. The railroad's management was not anxious to promote business on this railroad as plans were advancing to lop off the line west of Rapatee Mine, close the Keithsburg bridge and abandon a considerable stretch in southern Iowa. (The line from Oskaloosa, Iowa, through Keithsburg, Illinois to Monmouth was abandoned November 9, 1971). Two days later I went out with a couple of men from Harnischfeger looking for a site for an electrode manufacturing facility. Again DeKalb, Rochelle, Dixon and Belvidere were the cities of interest. On the 26th the annual community development meeting sponsored by the Northern Illinois Gas Company was held at the Pheasant Run resort near St. Charles, Illinois. Both Gene Cermak and Maurie Fulton from Fantus addressed the group which soon adjourned for the main event—golf, of course.

RAILROAD WARS — THE ROCK ISLAND By June 25, 1963 the Union Pacific-Rock Island merger talks had become a serious issue with the C&NW. All of our on-line and off-line general agents were summoned to Chicago. On June 26th the Rock Island's board voted in favor of the Union Pacific's offer to acquire their property. Ben Heineman was confidant that our counter offer would be more appealing to the Rock Island's stockholders. On August 9th John Danielson, one of our law department's brighter lights, pompously (and incorrectly) told me over lunch that the C&NW had veto power over the proposed UP-CRIP merger. Cermak announced on September 23rd that our entire sales department was to be pressed into a proxy solicitation campaign and all our normal work was to be put aside. Mr. Heineman announced that it was a do or die situation for the C&NW—the UP-CRIP merger must be blocked. Rock Island stock price was creeping up; the C&NW's common stock was at 28 on September 30th. The C&NW obtained the Rock Island's stockholder list and convened a meeting of all traffic managers in the board room for assignment of proxy solicitations. I volunteered to help Stan Boardman (New York office) to sort his assigned calls of Rock Island stockholders in New Jersey, Connecticut, Pennsylvania and Rhode Island. The next day we received our marching orders. Both Heineman and Larry Provo gave long presentations complete with slides and were listened to intently. Then we were handed our assignments: I had 36 calls to make in Central Illinois; Bob Lithgow, Harvey Buchholz and Jack Sommer went to Washington, DC; Gene Cermak and Vern Giegold were given Chicago calls; Don Guenther, John Foxen, Stan Johnston and Don MacBean were to stay in the office and deal with current business. Somehow "Little Shadow" Don Guenther managed to wangle an assignment in Boston, Massachusetts, and Vern

Giegold went to Maine. Even though tied to the office, Stan Johnston on his own initiative had secured proxies from three Rock Island stockholders by October 7th.

Funny stories about proxy solicitation began to drift in. The results ran about four to one in the C&NW's favor. Before I could get started on my calls, Cermak and I were directed to take care of a good potential client, Patrick Cudahy and Company, which had approached Bob Stubbs requesting help in finding a plant site in northern Illinois. Cermak and I drove to Milwaukee October 8th to interview Cudahy's principal officers, Richard Cudahy and Charles Watson. They wanted to establish a pork processing plant and knew that we had done a lot of work with Armour and Company which gave Cudahy the idea that we were experts in locating meat processing plants. We agreed to setup an aerial tour using the company plane before winter set in when the weather was deemed suitable. On our drive back to Chicago, I suggested to Gene Cermak that we ought to call the office. He finally agreed and pulled over to a pay phone alongside the highway near Racine. The news was astounding—The Chrysler Corporation, the biggest client that we were ever to deal with, had called the office and wanted an immediate response to their request for a huge site to accommodate a new automobile assembly plant in the Rockford, Illinois, area. Patrick Cudahy's priority tumbled on that phone call—they could wait for good weather. Chrysler was a prize of greater value. We actually shoe horned Cudahy into our frantic schedule November 8th when I conducted an aerial tour of western Illinois with Richard Cudahy, Charles Watson and his two boys. Picking them up in Milwaukee, we flew west to Clinton along the C&NW Galena Division in deteriorating weather. They got a pretty good look at property in DeKalb, Rochelle, Dixon and Sterling with a return loop via Belvidere and Beloit before the weather closed in completely. We set them down in Milwaukee and returned to Meigs Field. I never contacted any of my proxy assignments in this first proxy campaign because we were fully engaged with Chrysler.

CHRYSLER—1963–1964 Bill Young, the head real estate man from Chrysler wanted us in Rockford October 9, 1963. Because Gene

Cermak and I were on the road from Milwaukee when we got his telephoned message the previous day, we did not have time to prepare any promotional material. The day of the meeting I caught the first early morning eastbound commuter train to Chicago from Elmhurst and grabbed maps and some material for Belvidere, Rockford and South Beloit. Of course, the temperamental copy machine wasn't working so I just took the originals from the files and went west on the 8:40 train. Cermak picked me up at Elmhurst for the 75 mile drive northwest to Rockford. We met three Chrysler men at the Faust Hotel, one of whom was John Von Rosen. They explained that they were on a crash program to locate a major manu-facturing site in the area. This project was to be the largest and most important plant location project our department was ever to see. Our vice president, Ed Olson, had taken Gene and me off proxy solicitation as soon as he learned of the inquiry and told us to spare no expense. At the hotel meeting we reviewed our hastily collected materials with the Chrysler men who indicated that their search area was Belvidere-Cherry Valley and Rockford. South Beloit was not in the picture. Because the C&NW did not have much presence in the Rockford area, we concentrated on Belvidere some dozen miles to the east. Gene and I rented the pricey Bernadotte suite at the Faust to demonstrate to these Chrysler men that the C&NW was not a haywire pike and could entertain in style. We also very much enjoyed the upgrade over our more modest normal travel arrangements, the costs of which were always monitored by the critical clerks in the vice president's office. We asked Don Gunvalson, our Rockford district freight and passenger agent (DF&P), to join us for an interesting and lively evening. The next morning the Chrysler bunch showed up with Tom Gage of the Illinois Central in tow along with the local IC sales agent. While they went off to look at Illinois Central offerings, we swung into a furious round of data collection which included tax rates, building codes, zoning, prevailing labor rates, utility availability and anything that could relate to construction of a major plant. At lunch time Gage was still with them. During the afternoon I was lucky to run into the Highway Department's Elgin District engineer and learned about highway access requirements. The intensity of this data collection became infectious as we made every effort to find out everything about everything.

In Chicago on October 11th I pre-empted the traffic department's conference room to spread out the suitcase of material we had collected the day before. With Don MacBean's help we organized the brochures, rate sheets and maps into logical categories. Don and Larry Krause were drafted for copy making and to gather odd bits that were needed to fill in gaps. We worked that evening until 8:15 p.m. The next day everyone in the office worked to pull the Chrysler report together—even Cermak had a hand in preparing the final document. We had ordered aerial photographs of the proposed site southwest of Belvidere. On the 14th the prints arrived completing the data to be presented. While the typing was being done by the three secretaries, Don and I went out to a manufacturer on the west side of Chicago to obtain some fancy presentation binders. The excitement ran around the department and even seeped upstairs as Mr. Fitzpatrick came down to see the production process. It was all finished on the afternoon of October 14th and was ready for presentation to Chrysler the next day. That is, until I remembered at home in the middle of supper, that one exhibit had been left out. I drove into the city to collect the errant exhibit late in the evening. Along with this report preparation, the company treasurer authorized Gene Cermak and me to sign very large drafts on the railroad should we have to obtain land options.

Gene and I flew to Detroit early October 15th on an American Electra II for our appointment with William Young, head of Chrysler Real Estate. On arrival at the Chrysler headquarters at Highland Park we were surprised to see Tom Gage of the Illinois Central, Ed Stoll of the Milwaukee and Ollie Waggener of the CB&Q sitting in the reception area. They had not bothered to get an appointment as we had requested and Young was being polite by seeing them first. A bit nettled by our competition's impolite assumption of our appointment, we had lunch with two men from General Motors realty after rescheduling with Young for an afternoon session. At 2:30 p.m. we finally got in to see Young. Taking a look at the extensive report laid out before him, he congratulated us on the format and content. He said our competition to the Belvidere property was the joint Milw-CB&Q land at Davis Junction, Illinois; the IC had not come up with any specific site of much interest. Relieved that we had made a good impression but realizing that the game was just beginning, Gene and I flew back to Chicago on a DC-6.

The next day was one of nerves as we awaited word from Detroit. Later, we learned that Chrysler's vice presidents had spent all day going over the material we had assembled. At 4:30 p.m. we received instructions from Chrysler to take options on all the land lying north of the Illinois Toll Way and south of the C&NW main track from Stone Quarry Road west, about 1,300 acres in all. Not one technical question was asked. Bill Young and John DiCicco of Chrysler gave us high marks for doing such a complete report with so little advance notice. There was a small cloud on the horizon, however, as Chrysler had also requested that the Milw-CB&Q combine acquire options on land at New Milford, two miles south of Rockford along U. S. 51. They were both hedging their bets and developing a bargaining lever. The IC was out of the running. That same evening Cermak and I drove to Rockford after getting Carl Anderson and Charles Jevne from our Real Estate Department assigned to our project. We needed their expertise in getting those options. While waiting for Carl and Chuck the next day, Gene and I went to New Milford on a foggy morning to see our competition's chosen site. Later our local Rockford attorney, John Holmstrom, came over to the hotel and drew up option forms while Carl and Chuck spent the afternoon going through the land records in the Boone County courthouse in Belvidere. When Carl and Chuck got back to Rockford, I had the hotel bring a large blackboard up to the suite and laid out the land ownerships, acreage and an action column. The key pieces were soon identified, and we divided up the contacts to make four key contacts right after supper that evening. Since Cermak's contact was not at home, he came with me to the Ollmann farm. We offered them $850. an acre for the entire farm. The response was shock and a "we'll think about it". Carl and Chuck had similar tales to tell when we met for a late snack at 11 p.m.

We also evolved a strategy for preserving the secrecy of our client: never mention the name of Chrysler, drive other makes of cars, don't stay at the same place for more than a day or two, and clean the blackboard every night. The next day I arranged to have the option forms printed at a local shop before going to Belvidere to shock some more farmers. One of them was crusty Mr. Shattuck, age 76, who had seen most everything in life. He was a tough nut to crack. Fortunately his property was at the extreme west end of the proposed site and was never needed. Cermak's contacts were the Reynolds and the Willis families. By noon we had contacted nine of the seventeen owners. Carl Anderson was pleased with our progress and predicted that we would probably end up paying $1,200 per acre not the $850 we had started with. Bill Young of Chrysler received daily reports on our progress and was quite pleased at our lightning campaign.

By the 19th Chuck and I had evolved a system for tracking our negotiations. Two days later we were back in Belvidere to make more contacts including the important Lane farm. Although some owners were away, we learned that no one would sign anything until their local attorney, Owen Johnson, approved and he was out of town on a motor trip. Strangely, the farmers all used the same attorney—it was a small town after all. We heard that the Sunday chicken fry social at the Grange was a failure as no one ate chicken—they just stood around talking about this sudden attempt to buy their farms. That afternoon I went to visit an old acquaintance, Fred Shappert, owner of Shappert Engineering. Although I had not seen him for some years, he recalled our work together on bridge #450 near Cherry Valley. He was to be a valuable ally in this campaign. At last the vacationing attorney was back in town on October 22nd. Cermak and I collected Mr. Reynolds and took him to Johnson's office to sign the first option. Our first success. In the afternoon I took Don Gunvalson out to Fred Shappert's home east of Belvidere and introduced them. We let Fred in on some of the big project. He responded with some very important facts about local politicians and activities. One of the pieces of information concerned the MILW-CB&Q competitive site at New Milford. His company had built a highway culvert there some years ago and had a difficult time with groundwater which pointed to some poor subsoil conditions. That evening the Milwaukee Road's Ed Stoll hosted a dinner at the Wagon Wheel restaurant in Rockton and invited Bill Young of Chrysler, Ollie Waggener and our entire C&NW crowd. The evening was a pleasant mixture of good food and liquor spiced with obvious lies and good-natured, but deadly, deception. The next evening, the 23rd of October, Cermak was the host at dinner with the same unlikely group. We chose the Hoffman House on the east side of Rockford. The group was just as jolly, but there was a screaming need to get information lurking just under the surface of our banter. Bill Young enjoyed the whole intricate game hugely. Bill was the kind of guy who was the last to leave a party because he always feared that something would happen just after he had left. That made some late nights for us; we sometimes wished he would stay in Detroit and let us do our work.

We had one absentee landowner who lived in Pasadena, California. Without any discussion, Carl Anderson flew out there and signed them up—option No.2. On the 24th I talked to Gallano and Spencer who were slow to act; On the 26th Lane agreed to sell for $1,200. and would sign the option on the 28th. The break had come when Mrs. Lane said they could take a Florida vacation and quit this miserable farming. A key parcel next to the railroad at the far northeast corner of the area was urgently needed for the lead track to the plant site. The Gustafsons, an elderly Norwegian couple, had been contacted, but I hadn't had much success. I asked Carl Anderson to come with me one evening. As he walked in the door Carl spoke to the couple in fluent Swedish. Gustafson broke into a huge grin, invited us in, and, with tears of pleasure rolling down his cheeks, signed the option form presented by his "landsman".

Although October 27th was a Sunday, Cermak and I went back to Rockford. I set up another blackboard in our motel room and arranged the parcels with a code number in such a manner that a casual observer could make no sense from it. On Monday morning we drove to Belvidere to meet with Owen Johnson, the attorney. A nice old country lawyer, he became very interested in being helpful in our project—whatever it was. I gave him a plat map and a red pencil so that he could keep track of the options. He seemed to be pleased to be part of the team and faithfully colored the map when each option was signed. By the evening of the 29th we had 38% of the property under option. Gene and I had written some very large drafts on the railroad. Bill Young came over from Detroit to check on our progress and was gratified to see how well we were doing. He admitted the MILW-CB&Q people were having trouble with a holdout at their New Milford site. On October 30th a bunch of options came in and we now had 74% of the land under control.

The scent of money being spent in large quantities came to the Rockford newspaper's attention. We soon were aware that reporters were beginning to hover around the area trying to pick up the story. On November 1st they broke the story on page one but did not have any idea as to the identity of our client. I had warned Bill Young three days earlier that we couldn't keep the genie in the bottle much longer. That evening I met Cermak at the Palmer House in Chicago; later we joined Ed Olson, Iver S. Olsen, Bob Stubbs and Mr. Fitzpatrick at the Traffic Club. At the Illinois Chamber of Commerce banquet that evening Lynn Townsend, president of Chrysler, gave an address. Ben W. Heineman was sitting right at Townsend's elbow. Was this a sign that we had won? On November 4th Carl Anderson and I got an option from Woolsey who had been holding out. It was the last important piece; Detroit was very pleased. We continued to talk to the Shattuck, Hawkeye and Bruce families, but their properties were all peripheral to the main block of land. Chrysler sent Bill Young, John DiCicco and Fanning to Rockford to start the next part of the process. I hired a soils testing company to take borings and an engineering firm to make the land surveys. That evening there was another festive, guarded, dinner with the four Chrysler men and the competing railroad men. We enjoyed the show sensing that we had the pig in the poke. On the site November 6th the soil drilling went very quickly as the native limestone bedrock was not very far under the rich black Illinois loam. That evening John DiCicco confirmed to Cermak that the C&NW had indeed won the plant. The next day we learned that Chrysler had accelerated their building plans by one year. There were high level talks between Chicago and Detroit involving Heineman, Fitzpatrick and Ed Olson. What promises, compromises, or guarantees of service were discussed were not communicated to us down the chain of command. Bob Heron from our operating department claimed that he heard a rumor that Ford Motor Company was planning an assembly plant at Geneva. Since he wasn't supposed to know about our project, I soberly noted his rumor for further investigation knowing full well that he had bad information.

After dropping the Cudahy party in Milwaukee November 8th, we had flown back to Meigs Field in Chicago where I took a taxi to the Chicago Passenger Terminal and caught a train to Elmhurst. My wife was waiting on the platform to hand me my suitcase so I could stay on the train to West Chicago where Cermak waited for me for a fast drive to Belvidere. We had dinner at the Shapperts and established a closer relationship with him. The next day Fred Shappert and Cermak went to Belvidere while Gunvalson and I drove to New Milford. Seeing a drilling crew working in the muddy field I asked Don to stop and wait while I hiked over to the man taking notes. I was curious as to what they were finding under this land. The note taker was a Milwaukee Road engineer; I led him to think that I was from the Burlington. Because of the steady rain the Milwaukee engineer invited me to sit in his truck where he proceeded to show me the notes and drilling results accompanied by a litany of woe about drilling in the wrong places and what a screwed-up affair this "joint" railroad project was. Bidding him to take heart, I slogged back across the field and drove off with Don who was so convulsed with laughter that he was a menace on the highway. Later that day we were introduced to Lester "Red" Cunningham, the Mayor of Belvidere at Fred Shappert's. We decided that it was time to bring him into the project as his help would be critical in certain political actions to come. He was both amazed and stunned at our news. On November 10, 1963 Chrysler's Public Affairs released the story to the Rockford paper which emblazoned it across the front page. No need to pussyfoot around anymore.

There was still a political game to be played out far above our level involving the State of Illinois. On Armistice (later Veterans) Day I picked up a big, black Lincoln Continental and drove to Rockford with Carl Anderson as my passenger. At noon the C&NW's engineering staff with a few operating officers arrived in Belvidere. B. R. Meyers, chief engineer, Steve Owens and Maurie Reed, office engineers, M. C. Christensen, the Galena Division engineer, Bill Wilbur, engineer of bridges, Jack McCord, roadmaster and Bob Conley, division superintendent, filled up the fleet of hi-rail cars that had toured the line from West Chicago to Belvidere. Joining this August assemblage, Jack McCord privately told me what a rotten piece of railroad they had just covered. It needed ties, bridge work and rail. The group went on to Rockford in case there was interchange business that might go west from the plant (it never did). That evening our small group met at Fred Shappert's again to plan the visitation of Chrysler brass the next day. The principal Chrysler guy was vice president L. B. Bornhauser. DiCicco prepped us on what to say and when; John wanted this plant site to succeed. Mayor Cunningham was carefully rehearsed for his role. We were all rather nervous about Bornhauser's visit. (Bornhauser was named manager of the year by the Institute for Industrial Engineering in 1966).

November 12th was our D-Day. Eight of the top Chrysler people arrived at noon and everything went well. They had enjoyed their trip, the dinner, and the evening's entertainment. To provide balance and give an air of fairness, the party visited the MILW-CB&Q site at New Milford where soil drilling was still in progress. They were treated to entertainment in the Milwaukee's business car parked on a siding nearby. The Chrysler delegation also made a courtesy call on the mayor of Rockford. The press was waiting with cameras at the ready. At least we had successfully protected Chrysler's identity throughout the entire campaign. There had been a leak somewhere and suspicion pointed to the Burlington. The next morning Mr. Bright, another Chrysler vice president flew in, looked at our site at Belvidere and collected some of the Chrysler people for return to Detroit in his DC-3. By evening we were in DiCicco's term "Ivory Soap" that is 99.4% sure that we indeed had the plant. On November 15th the Boone County Board approved a 30% tax assessment on future real estate taxes. That day the state highway department and our Steve Owens worked over highway requirements while we helped mayor Cunningham compose a welcoming letter to Mr. Bornhauser.

There was still a lot of work to do. Chrysler now decided that they needed a warehouse in Chicago and that it was to be on the C&NW. Cermak and Chrysler's Fanning met to work out a site at Proviso. The next day I went to Belvidere with Chrysler's engineers Silverthorn and Allen and two construction consultants, Brady and Targall. After consulting with our soil tester, DuBose, we walked the site locating buildings and features for the plant. Because the C&NW held title to the property, Carl Anderson prepared a petition for annexation to the City of Belvidere. While I was taking Kami Targall to the surveyor's office, Don Gunvalson took the rest of the engineers to look at the grading equipment of local contractors. After driving the Chrysler crowd to O'Hare in the afternoon for their flight home, I went into the office to meet with the engineering department. Chief Engineer Bernie Meyers with Steve Owens and Maurie Reid got their first look at the proposed plant layout. Consultations with our top engineering brass continued on the 31st.

November 23, 1963—President John Fitzgerald Kennedy was assassinated in Dallas, Texas. And that was the day the C&NW was advised officially that they had won the plant.

BELVIDERE DAILY REPUBLICAN

CHRYSLER CHOOSES BELVIDERE-
TO PRODUCE 960 CARS DAILY

A Day To Remember. . .

Mayor Issues Statement

Robert McClory Congratulates Us

The front page of the Belvidere Daily Republican of November 26, 1963 announcing the good news for the city. Mayor Lester Cunningham, Fred W. Shappert and the new Chrysler plant manager appear in the picture.

Two days later Carl Anderson and I drove to Belvidere in a borrowed Chrysler Imperial while listening to the Kennedy funeral broadcast. The local newspaper of November 26th featured red headlines; I collected a bundle of the papers and distributed them to our farmer friends. That evening the City of Belvidere formally annexed the 500 acres on which Carl Anderson and our local attorney, John Holmstrom, that day had formally exercised options. NBC's Lincoln Ferber and a camera crew filmed an interview on the 27th while standing on the Town Hall Road overpass at the Illinois Toll Way overlooking the future plant site.

Chrysler wanted our railroad people to see an operating example of what they expected at Belvidere. On December 3, 1963 B. R. Meyers, Steve Owens, Bob Conley and I flew to St. Louis, Missouri from Meigs Field in our N1394Z. Jim Thompson, our St. Louis DF&PA, met us at Lindberg Field and drove us south to Fenton where Chrysler's truck plant was located. Bill Allen of Chrysler conducted the tour and explained what railroad service requirements would be expected at Belvidere. After the tour we flew to Peoria, dropped Bob Conley and went on to Meigs.

In December actual construction commenced for the trackage and the plant itself. Theoretically, the industrial development department was finished with the project, but it was hard to let it go into hands that had not been involved with the birth, so to speak. On December 4th Gene Cermak and I drove to Belvidere through slushy snow. Construction of railroad lead track southward to the plant site was already underway. The Willis farmhouse alongside the road was standing right in the path of the bulldozers and earthmovers preparing the track grade. Willis said later that it was very interesting to sit at his breakfast table and see the huge equipment coming right at him. The mainline turnout had been unloaded awaiting tie installation. Communications department linemen were raising their lines. There was hustle and bustle everywhere. We had come to Belvidere because of a promise Cermak made to Mr. Willis that he would address the local Rotary Club. The audience hung on every word while Gene talked about the development of the project.

The next two days were devoted to making arrangements for a special train from Chicago to Rockford for the Chrysler people and their wives. The C&NW really laid on a trip with specially printed menus serving such things as LeBaron beef. By the 9th all was in readiness; Ed Olson and R. C. Stubbs personally inspected the cars and approved the arrangements. Operating vice president S. C. Jones was especially helpful in getting his department behind the operation. Early on the day before the Special's operation, December 11, 1963, Gene Cermak's wife went into labor and was rushed to the hospital so I was to be the host instead of Gene. One of the final arrangements required picking up two new Chrysler Imperials from the Chrysler Center near Evanston. Roy Lawson, one of our Chicago sales officers, and I went up on the train to get the cars. Through a heavy snowfall on slick roads, we slowly crept home.

On December 12th the Chrysler Special Train to Rockford was parked in the Chicago Passenger Terminal ready to go. Cermak was a new father that morning, and I became the official host for this expedition. Roy Lawson and I went to O'Hare with the new Imperials to collect the Chrysler people coming in from Detroit. Their plane was 32 minutes late because of the weather. We packed them into the cars and rolled down the Expressway to the Chicago passenger terminal. With everyone safely tucked on board, the Special left at 11:35

a.m. and ran quickly through snow showers to West Chicago over the Galena Division mainline. The dining car crews outdid themselves in providing our guests with liquid refreshment which was just as well because this was the first passenger train to tread the light rails of the Freeport branch since April 14, 1951, some twelve and a half years before. The 20 mph speed limit was a far cry from the 60 mph old train #703 had reached in the late 1940s. Now, it was a rock and roll ride, but the leisurely luncheon served while we rolled through six inches of new snow masked the poor condition of the track. The train pulled into Rockford at 3:15 p.m. with the passengers in a mellow mood. Don Gunvalson had arranged for buses to convey the party to the Faust Hotel for freshening up for the gala celebration dinner to be held at Belvidere that evening. The Chrysler party arrived by train at the Belvidere station at 6:00 p.m. When the busses bearing the Chrysler party pulled up to the high school gymnasium where the community dinner was being held, we found to our dismay that the place was packed

GP7 1655 at Chicago Passenger Terminal on a cold December 12, 1963 ready to take the Chrysler Special to Belvidere, Illinois.

The C&NW's Real Estate Department played an important part in the successful campaign to locate the Chrysler automobile assembly plant at Belvidere. Charles P. Jevne and wife Myrtle prepare to board the rear car of the Chrysler Special joining Carl A. Anderson.

Aboard the festive Chrysler Special dining car enroute to Belvidere. C&NW engineering's Maurice S. Reid and wife are on the left; vice president-traffic Edward A. Olson is on the extreme right.

The mayor of Belvidere, "Red" Cunningham, visits with Chrysler officials aboard the special train on December 12, 1963.

full of locals and there was no room for the guests of honor. Somehow, more tables were shoehorned in, the guests were seated, dinner was served, the mandatory speakers performed, and the key to the city was handed to the top ranking Chrysler officer present. There had been a rumor that the plant had really been slated for South Beloit. When the Chrysler speaker shot that notion down, the audience glowed with pleasure. Our Chrysler guests were very impressed at the reception for some of them were slated to live in Belvidere. Then it was back to the train and the short trip to Rockford for a night at the hotel. The next day was much calmer. After putting our visitors on a chartered bus to O'Hare Field for a quicker ride, Carl Anderson, Chuck Jevne and I with our wives were the only passengers on the special train returning to Chicago. We enjoyed the attention from the staff and the lovely luncheon on car #401. The ladies especially savored this brief experience of traveling in the luxury of a private train, a rare experience for them. The train stopped at Elmhurst to let them off while we rode on into the city.

The Chrysler plant location at Belvidere had profound effects on the C&NW in terms of an upgrade of track structure, expanded train service and a substantial increase in freight revenues. To me the exceptional performance of all the C&NW people gave me a better appreciation of other people's talents and capabilities. While the project itself was fast moving and intense, each member of the Cermak, Anderson, Jevne, Gunvalson, Lewis team had an important part to play. Unsuspected talents arose as circumstances required. Carl Anderson was a special and key player. His vast experience in land acquisition and profound knowledge of human nature greatly aided our effort. Proud of his Swedish background, Carl fairly bubbled with good humor and a love of practical jokes one of which went further than he intended. At one of our Rockford dinners with the "competition" Carl taught Gene Cermak how to speak a simple Swedish phrase. Gene dutifully repeated the phrase until Carl decided the accent was right. Cermak then surprised Carl by starting to try out his new skill on the buxom Swedish waitress. Carl swiftly clapped his hand across Gene's mouth before the fatal newly learned words could be uttered. There was another story about Carl when he was involved with acquisition of land for expansion of the C&NW's quartzite quarry at Rock Springs, Wisconsin. The owner, who lived on the property the railroad wanted, was not much interested in selling but was friendly enough with these city slickers who had come all the way there to talk to him. Carl and the division engineer, Ottmar W. (Bill) Smith, made a trip to see the obdurate land owner on a very hot and muggy day with a thunderstorm muttering in the distance. As the boys approached the house, the owner invited them to come up on the porch out of the hot sun and have a beer. He pulled a couple of bottles out of an ice tub and handed them over. Carl, trying to think of something to say, looked at the bottle's label and remarked that he had never heard of Potosi beer before, but it tasted good. The landowner took a pull at his bottle and said that it was good beer and cost just 10 cents a bottle. Carl was surprised and remarked that he had never heard of beer that cheap. The man said with a straight face, "Wal, we wondered about that, too, and sent a sample down to Madison for testing. About a week later the report came back and it said—Don't work that horse for a week." While the joke was on Carl and Bill, the communication barrier was broken. Carl got the option and the railroad expanded the quartzite quarry.

One day during the Chrysler blitz Carl and I visited the MILW-CB&Q site south of New Milford. At the time we did not know much about the option status of the land and the other railroads certainly would not have told us, if asked. Carl strolled up to the barn where the farmer was stuffing ears of dry corn into a noisy grinder making animal feed. "Nice day, isn't it?" yelled Carl. "Yep" replied the farmer. "You've got a nice farm here", Carl said above the sound of the feed grinder; "Yep" said our laconic farmer. Carl cupped his hands and said near the man's ear, "I'll bet you'd sell the place to me, wouldn't you?" "Yep" said the farmer. Carl stopped and looked around for a couple of moments, and said "Well, I'll think about it and get back to you." As we walked back to our car, Carl said the man must be holding out on the Milwaukee-Burlington option because he still had the right to consider selling the land to someone else. If he had given an option to the railroads, he wouldn't have talked so positively to us. Carl's philosophy in buying land called for getting a seller in an agreeable frame of mind by getting him to make positive replies to a series of innocuous bits of conversation before mentioning the real question sale of the property. He was a smooth operator with a flair for acting like a country rube who may have seemed to be somewhat guileless but still could be your very best friend. Carl was the right person for his job and was eminently successful as a land agent.

Charles P. (Chuck) Jevne, another member of the team, was a quiet, introverted person whose skill at following chains of title through courthouse records was legendary. While researching land records at the Boone County courthouse in connection with the Chrysler project, Chuck discovered the old 25 foot wide right-of-way of the defunct Elgin and Belvidere and Rockford and Interurban electric lines which adjoined the C&NW's Freeport branch line all the way across the county. The C&NW had to cross this unused and forgotten strip of land to build the new track to the Chrysler plant west of Belvidere. When Chuck learned that the real estate taxes were long delinquent on this old right-of-way, he quickly and quietly wrote a C&NW draft to Boone County to cover the arrearage. The amazed county clerk quickly certified the transfer of the property. The C&NW was now the owner of the old right-of-way which in effect gave the C&NW an added width right-of-way across Boone County. Over succeeding years this extra land was sold off to adjoining owners at a profit except for the stretch we needed at Belvidere to accommodate construction of a new train yard and engine facilities for the Chrysler operation. Privately, Chuck's passion was barbershop singing with a group that was a top competitor in many contests. I suppose the many cigarettes he smoked added timbre to his voice.

Donald L. Gunvalson, known as Old Sarge to his friends, turned out to be an invaluable and handy resource who arranged the ground transportation for our special Chrysler guests as well as providing guidance to the many local services we needed. Don was an innovative salesman and well liked in his territory of northern Illinois. A veteran of the South Pacific war, he was an energetic and most helpful adjunct to our team. I fear he was never rewarded properly for his contributions. I lost track of him when I left the traffic department and often wondered where he ended up.

Our group was deluged with invitations to speak. On December 10th it was the Rockford Chamber of Commerce; a week later on the 18th it was a meeting of Rockford's Winnebago and Belvidere's Boone county planners who asked for details on the Chrysler location project. Keith McCullagh of the Illinois Central, which had lost out early in the process, came to Belvidere and was also invited to address the group. On December 17th the *New York Times* and *Dun's Review* called requesting comments on the "science" of plant location. The C&NW expressed its pleasure by increasing our salaries. Mine went to $10,200. from the former $9,300. It was the last salary increase I was to receive for 39 months. At that golden moment our fame was bright. I couldn't have known that I would reach the halfway point in my railroad career in the first half of January 1964 and that the next three years with the old traffic department was to be all downhill.

One of life's nice little bright sparks flared briefly on September 10th when Governor Frank Morrison of Nebraska invited me to lunch at the Blackstone Hotel (along with a lot of other people) and presented me with an engraved copper plaque proclaiming that I had been commissioned as an Admiral of the Great Navy of the State of Nebraska.

18

DIGITAL GEOGRAPHY FOR MARKETING, 1964 . . .

The big event of 1963 was over as far as the industrial development department was concerned. Chrysler was now the responsibility of the railroad's regular service units. Yes, we had given birth to this lusty infant, but found it hard to stay away from its assimilation into the railroad's family of customers. We grasped at several loose ends that had to be tied up to stay involved with our former prize. On January 5th Cermak and I went to Detroit on the New York Central's *Motor City Special* in a rough riding sleeper *Kings River.* Jim Sullivan and Jim Goinz from the C&NW's Detroit office drove us to Highland Park for a meeting with Bill Young. After lunch with Young, John Von Rosen, Fanning and Stovo, we were given a tour of the Dodge plant then managed by John DiCicco. Gene Cermak took delivery of his pre-ordered new 1964 Chrysler LeBaron (cost $7,549). We spent the evening with John and his wife, Ann. As long as we were in Detroit, we called on a Mr. Mollica at General Motors who was interested in establishing an unloading facility at Council Bluffs, Iowa, for Chevrolet. A Mr. Parrent, also of GM, said he was in the market for an auto parts warehouse in Minneapolis. Just to balance the books, we made a courtesy call on Rudy Powell at Ford. Powell was Cermak's friend from the days when he and Gene had located a major Ford assembly plant on the C&EI in Chicago Heights some years ago. On January 16th Chrysler announced that they wanted another 380 acres of land at Belvidere for a highway truck haulaway staging area. This request would cut a big chunk out of our planned development for the extra land we had acquired last November. On January 22nd Cermak, Bob Ballin, Carl Anderson and I flew to Detroit's Metropolitan airport in N1394Z, the company plane, and were met by Jim Goinz for the auto trip to Highland Park. In a meeting with Young, DiCicco, Von Rosen and Jim Shepheard, we agreed an a division of ownership that suited both groups; Ed Olson and Mr. Fitzpatrick approved our arrangement on the 24th. Meanwhile, our good friend Fred Shappert at Belvidere did not bid low enough on the foundation work and his Rockford rival, Sjostrom, won the work. Construction started immediately; drilled-in caissons were going in by February 4th. The Chrysler project was big news in Northern Illinois that spring. Gene Cermak and I were invited to address the Lions Club of Morrison on February 18th. They did not want to hear about organizing an industrial development group for their town; they just wanted to hear about the Chrysler project. The same thing happened in Freeport the next day at their Lions Club meeting. At least we didn't have to buy our own lunches. By the end of February Chrysler had agreed to let the C&NW retain all the land lying north of Illinois highway 5 at Belvidere for industrial purposes. Over the next few weeks I worked on drainage, subdivision, rail and roadway plans for the former farm land. On March 11th I attended a testimonial dinner for Illinois Governor Otto J. Kerner at Belvidere, photographed the work in progress at the plant site and met with Bethlehem Steel who was preparing to erect the structural steel frame of the main building. Bill Wimmer from the Galena Division engineer's office was in charge of the track construction.

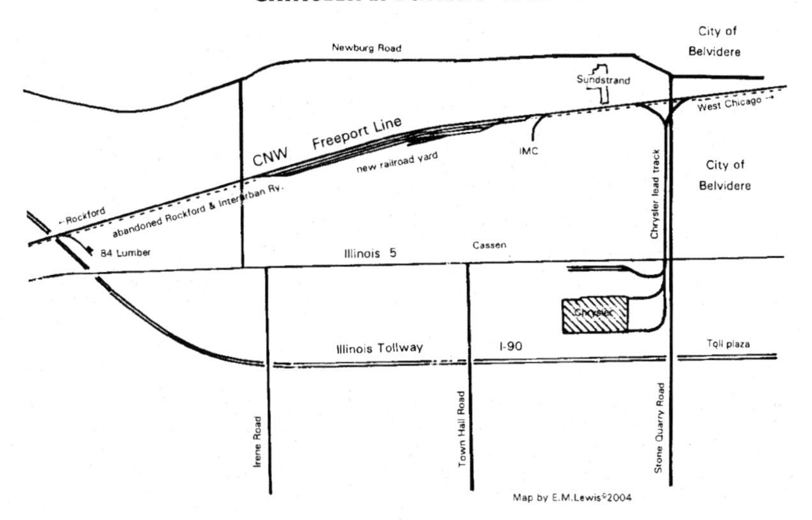

There was a continuing dribble of Chrysler related activities that required our presence during the first part of 1964. There were the farm leases on the purchased land to settle. Since we knew the farmers, our input was of value to the railroad's leasing agent, Tom Lydon. Carl, Tom and I went to Belvidere on March 16th to finalize the lease arrangements. The State of Illinois had agreed with Chrysler to improve highway 5 in the vicinity of the plant site. In May they released plans for a four lane highway to replace the old two lane road. Also in May the railroad received a demand from the City of Belvidere to help pay for the sanitary sewer main required for the new facility. It was not within my authority to do other than make a recommendation. On the 15th I sent the city's engineer upstairs to see Harry Bierma of real estate and Steve Owens of engineering to settle the matter. (The railroad did make a contribution toward the cost.) There were other major expenses facing the railroad as a consequence of Chrysler's selection of this site on its line. The railroad serving the plant consisted of light rail on bad ties in a roadbed innocent of much ballast and full of weeds. This was the original line of C&NW's predecessor, the Galena and Chicago Union Rail Road, constructed 1850–1852. While the rail in the track may not have been the first iron laid, the 72 and 80 pound sections certainly were not adequate for modern traffic. Tie condition was marginal and largely unplated. Bridges consisted of class B timber spans, a few rusty I-beam steel spans, cut stone culverts, and some old wood box culverts. Grade crossings were mostly protected by passive warning signage consisting of standard cross bucks on posts. While some highway crossings did have automatic protection, they were set for the timetable train speed of 20 mph. The direct connection of the Freeport Line to the main line at West Chicago had been relocated some years earlier to accommodate a sale of land to a scrap dealer. The replacement connection north of town to the West Chicago yard involved sharp curves that 89 foot autocarrier cars would find hard to negotiate. The C&NW's management had assured Chrysler that the railroad would be rebuilt from West Chicago to two miles west of Belvidere, a new yard would be built to accommodate Chrysler's traffic and that the new assembly plant would receive all the switching and transportation services necessary for its operation. There were two reasons for not improving the main line west of the bridge over Interstate 90. One was a desire to inhibit or totally bar interchange in Rockford with the Illinois Central, Burlington or Milwaukee Road. The second reason, of course, was the expense of rehabilitation. It was obvious that one of the first activities that had to be done before Chrysler began operating their new plant had to be construction of a better connecting track at West Chicago. The new design cut straight across the old lead track on a diamond crossing into a gentle curve that joined the mainline just west of the JB Tower (EJ&E Ry) crossing. I was surprised that a diamond crossing would be considered as we had always been taught to avoid using this track device because of the expense. Now, it was hang the cost, get the job done. The new connecting track was all ready for service September 29, 1964. In 1965 tie and ballast gangs worked over the track from West Chicago to Belvidere. Budgeted funds ran out much sooner than expected as the poor tie condition required many more new ties than originally estimated. Rail replacement had to wait a few years more so that weird mixture of 72, 80, 90 and 100 pound sections had to perform longer. The track work was substantially completed in time for the first shipments from the new plant. The first auto frame was run through the assembly line August 2, 1965. On August 17th I was invited to drive a new Dodge off an inbound trilevel car. Chrysler surprised us by shipping new cars into Belvidere from other assembly plants so as to optimize delivery to customers within trucking range. The plant was officially dedicated September 13, 1965, just 23 months from the first inquiry. One of the Chrysler officials said that our site was so easy to build upon that a full year of construction work was saved. There was a limestone ledge a short distance under the surface that provided an excellent foundation for the building's caissons, speeding-up their installation enormously.

In September 1964, our automobile traffic manager, Frank Tribbey, warned us that Chrysler's truck haulaway contractor, Cassen of Edwardsville, Illinois, soon would be seeking some of our land near the new plant for their over-the-road operations. Indeed, they did show up on the 24th and asked for a ten acre parcel north of highway 5 at Stone Quarry Road, near the exit from the new assembly line. We were a little uncertain about this request until a call from Bill Young at Chrysler asked for our consideration. We sold them the desired parcel. In October we heard that Chrysler was also considering installation of a light truck assembly plant on the extra land west of the new assembly plant, but that did not occur. Our continued involvement in the Chrysler project gradually tapered off, and we turned our attention to other matters some of which were to drastically affect my future in the traffic department.

Back in January 1964 I had flown to Omaha to meet with John Bullington of Northern Natural Gas Company. Don MacBean, from our office, joined us later. Frontier Chemical Company was interested in a Nebraska manufacturing site and needed a lot of natural gas as feedstock which greatly interested Northern Natural. We drove down to Lincoln through Wahoo, gathered up basic community information, and flew back to Chicago after briefly considering, and rejecting, the CB&Q's night train. Later, John Bullington and John Hahn got rather testy when we would not share our report with them. Those Northern Natural Gas guys always seemed to keep a wall around themselves—ever taking, seldom giving; hence, we felt reluctant to share material with them. Somehow, Ed Olson, our vice president of traffic, got involved through some contact in Birmingham, Alabama, of all places, though the connection with Frontier remains obscure even today. I had a feeling that this prospect was a bit flimsy and it never matured into a credible plant location project.

The railroad's offices were located on several floors of the Daily News Building at 400 West Madison Street. Built in 1928–29 the building was not old but not very modern either. Things sometimes failed like the waterline break at night in the women's toilet just over chief engineer B. R. Meyers' office. The resulting flood pretty well soaked everything in his office. It couldn't have happened to a bigger stuffed shirt. Our monthly R. I. D. luncheon meeting boasted twenty people on January 17th. Our honorary president #5 was soon to retire Henry Coffman of the New York Central's industrial department. On March 6th we had 15 members at lunch plus honorary president #6, Mal Wagner of the Baltimore and Ohio. On April 10th the 12 attendees elected Carl Wilkins of the C&EI as our 7th honorary president. We skipped the summer period as many members were on vacation or deeply involved in their work. Eleven showed up on October 16th. Phil Schmidt of the Rock Island became our 8th "president" November 20th with a lucky 13 present. By December 18th only nine appeared along with Bill Porter of the C&EI. These meetings provided the members a social opportunity to recognize the highly specialized part they were playing in the railroad business. The memories of these gatherings remains warm over the many intervening years.

In January I was asked to assist Eugene Anderson of our law department in conducting the public hearings for abandonment of the Genoa City, Wisconsin, to Hebron, Illinois, branch. This branch line was a vestige of the ancient Kenosha, Rockford and Rock Island that had finally managed to get to Genoa in 1860 and Rockford in 1861. When the C&NW and Galena and Chicago Union merged in 1864, the KRRI, which had been operated by the C&NW, became the Kenosha Division. Always a light traffic line, the west end from Harlem to Caledonia was abandoned October 23, 1937. The segment from Harvard to Hebron saw its last train May 31, 1939 when the ICC gave their permission for abandonment from Harvard to Bain, about four miles west of Kenosha. At the urging of some shippers, service was retained to Hebron and Bassetts, Wisconsin, with the first service effective September 27, 1939. Now the time had come for an end of service to Hebron. Gene Anderson and I attended an low key informational luncheon meeting in Hebron on January 28th. Despite the blow to their civic pride, the Hebronites realized that the line to their tiny city was doomed. The C&NW was also considering abandonment of the Pecatonica-Freeport branch and the McHenry-Williams Bay line at this time. On the 29th I went to Crystal Lake, Illinois, to meet with some of the locals who were all fired up at a prospect of a new electrode manufacturing plant supposedly a project of the Harnischfeger Corporation of Milwaukee, Wisconsin. The site they were promoting would have required a track crossing of an important state highway (U. S. 14) which our Steve Owens thought would surely be a problem with the highway engineers. The prospect actually was Harnischfeger, but they changed their minds February 13th putting the project back into limbo, so they said. Yet, the very next day Ploetz of Harnischfeger came into our office to discuss changes in the track plan which seemed to put the project back on the front burner again. There was a meeting in Crystal Lake on the 19th to discuss the changed layout plan. By March 13th the whole Crystal Lake effort had evaporated and the plant went to Indiana.

In the spring of 1964 we were bedeviled by a plague of small fertilizer-farm store outlets springing up all over the agricultural areas of the Midwest. We had some warning that this new concept would sweep the farming belt in our service area when Smith-Douglas and Warren Petroleum contacted us a year or two ago with the same concept. Now all the big petroleum and fertilizer companies seemed to have adopted programs calling for building hundreds of these retail outlets at major highway crossings of railroads far from towns or cities. Each of these rural outlets usually included a small hardware store, a warehouse for storage of bagged fertilizer, soil conditioner and supplies, a couple of anhydrous ammonia fertilizer tanks, and, sometimes, gasoline/diesel/propane fuel distribution facilities. Aimed at the farmer to make the tools and supplies he most needed so handy that he could avoid that trip to town. While each of these units did not hold much potential for freight revenue, the aggregate business of these big companies commanded respect from the railroads. The positioning of these stores near a paved highway at the railroad track often required a rather expensive track construction cost if there were automatic grade crossing protection circuits in the main track. Insulation of the track was also required to safely isolate the liquid ammonia unloading area and was another cost billable to the industry. The C&NW's policy for construction of new industry tracks required that each installation had to stand on its own merits—there were no wholesale deals cut with the parent companies. The costs of the sidetrack to the 12 foot clearance point included the turnout rail, ties, fastenings and ballast plus whatever labor and hardware circuitry changes required plus relocation or raising of signal and communication lines. The costs were then weighed against the anticipated revenues expected to be received at this point for a ten year period. I negotiated many special arrangements requiring the industry to deposit the full cost of the turnout portion of the siding and then provided for rebates of so much a revenue car over a term of years. This arrangement was beneficial to the C&NW because it did not tie up capital in low return industry tracks. While I am not sure that the fertilizer companies ever fully recovered their side track deposits, I know that they sold a lot of fertilizer from these sites and that the invested track costs probably were just factored into their pricing equations. The bigger firms were easier to convince of our cost-benefit analyses, but the little local co-operatives that soon joined in the merry game, had a tougher time justifying their trackage costs. We were inundated with requests for sites for these farm service units. Locating and presenting potential properties became a routine task, but acquisition of the land was left to the prospect who then had to deal with zoning and permitting with the various county authorities. When the property was finally secured, the C&NW provided engineering for the tracks and usually arranged to construct the industry's part at a standard cost per foot established by the chief engineer. We used up a lot of old suspect heat 112 pound rail on these little stubs. It was strange to see installation of a 100 pound turnout in a 72 pound rail branch line track and then see the rail size step up through compromise joints until the 112 pound rail in the industry's part of the siding could be matched. There were so many of these outlets that I put together a short "how-to" booklet for our sales forces to help them when dealing with new inquiries when out in the country. The flurry of installing these farm stores lasted just a few years before the truckers hauling inbound fertilizer edged out rail tank car delivery leaving many little industry tracks to rust from disuse.

In February it was time for the annual Nebraska governor's conference. On the 9th I went to Lincoln on that old standby, the

CB&Q's *Nebraska Zephyr*, in daylight for a change. I rode with Chuck Sayre and Don Slonecker to Grand Island for the affair which was not really worth the trip. Chuck had a speaking engagement at Kearney, Nebraska, another 55 miles west so I tagged along and learned about the new atomic electric generating station at Hallam, Nebraska. On the 11th I drove to Wahoo and West Point where I looked at the Armour facility still under construction before returning to Norfolk by way of Laurel, Creighton and Verdigre, territory I had not seen before. These spindly branch lines of the old Omaha Road would not long survive. Train service was provided infrequently by low horsepower diesels that crept carefully over the light rail. The weather-beaten little towns were frozen in time. The best maintained facilities in that country were the highways which led to the market towns. The emptiness of northeastern Nebraska was depressing. I went to Omaha and flew in a Caravelle to O'Hare. American Cyanamid, however, liked those fertile lands in Nebraska. Robert Van Nostrand of Cyanamid came to the office March 5th specifically to find fertilizer distributions sites around Laurel. In April their interest shifted to Wisner while International Minerals and Chemicals was poking around West Point, Nebraska. In May I heard a rumor that Cyanamid was seeking a manufacturing site in the disputed TPW-CNW joint area around Mapleton, Illinois, and called Van Nostrand to check out the facts. Indeed they were looking at some of the property there but were fearful that the on-going lawsuit was making railroad service questionable. I had alerted our lawyers to Cyanamid's site search at Mapleton and, on September 18th went to Peoria to meet with CILCO's president Wellington and my friend, Mark Townsend, to discuss the problem. The Illinois Commerce Commission had just interested itself in the case, which, as described before, ended up with dissolution of the joint service agreement and complete defeat for the C&NW. In December 1964 I learned that the Cyanamid facility was going to Hannibal, Missouri.

RAILROAD WARS — THE SECOND ROCK ISLAND BATTLE

Judge Hoffman ruled February 18, 1964 that the proxies collected by the C&NW in October-November 1963 were invalid and that they must be re-solicited. The Union Pacific-Rock Island merger effort was underway again and the C&NW had to try and derail it again. A hopeful rumor circulated on February 20th that a proxy battle would not be necessary because the Union Pacific had quietly withdrawn their offer and the Rock Island's board was debating the matter at length. C&NW stock was at $37\frac{1}{2}$ and on the 25th hit $41\frac{1}{2}$. By the 27th rumors were floating around that the UP had backed down from the Rock Island merger—C&NW stock rose to $43\frac{1}{4}$ the next day. The rumors were a false hope, however, for on March 9th we learned that the Rock Island board had set June 12th as the date for the stockholders' vote on the merger of their company with the Union Pacific. We received notice April 15th that our proxy campaign would begin April 30th as the Rock Island stockholders of record on May 1st were the only ones who could vote. On April 30th some 260 C&NW traffic men from all over the U. S. and Canada met in a big staff meeting held at the Conrad Hilton hotel in Chicago. Ben W. Heineman announced that the Rock Island had shifted the voting date to July 2nd so that stockholders of record June 5th could qualify. He emphasized the importance of our success in this effort. Later, on May 27th, the Union Pacific confused things even more by improving their offer for the Rock Island. In a series of legal maneuvers the proposed proxy solicitation was delayed and rescheduled all through 1964. When it did come, it was a fast call as I was drafted along with Don E. Guenther and Vern Giegold from our department on December 29th for a deadline of January 7, 1965. On December 30th I received my assignment—17 calls in the Maywood-Winfield, Illinois, area. A "no" vote was as good as a vote for the C&NW; I managed to secure a majority in favor of the C&NW. Despite our efforts, the Rock Island stockholders voted for the Union Pacific offer. We had lost, but had we?

A local contact (I have forgotten the company's name) of Don Groves, our Denver traffic manager, displayed an interest in building a chemical plant at Wahoo, Nebraska, with raw materials to come from Riverton, Wyoming. That looked like a nice freight movement with most of the inbound on the C&NW. Our company's response conditioned by the recent Chrysler campaign brought all our involved people to Denver on the run. I rode the Milwaukee Road's *City of Denver* in sleeper *Redondo Beach* with rate department's Ernie Johnson to Denver where Jack Peters from Omaha and Don met us for a meeting with a Mr. Vogenthaler and a Mr. Gleissner. Their principal concern was freight rates from Riverton, Wyoming to Wahoo. Ernie explained what the current tariff covered and agreed to do whatever could be done to set up a favorable schedule for the phosphate and sulphur movements. Scenting a potential "big" one, Iver S. Olsen flew out to Denver in the afternoon, but Ernie, Jack and I had decided to head east on the train. I had a berth in *Pacific Home* on the return. I had a long conversation with Jack Peters on that return journey. He was a gentleman of the old school, endowed with a sharp mind and an abundance of good sense. I went to bed at Grand Island while Jack sat up through the night to get off at Omaha. Later, when Ernie Johnson proposed a schedule of rates to the ICC for the Riverton-Wahoo movements, Swift and Company expressed interest in them. I called Gleissner on April 13th to learn what was happening on the project. Nothing, it seemed—ever.

In March Vulcraft, an electric furnace steel manufacturer, selected a site at Norfolk, Nebraska, supposedly on the Union Pacific, but I later discovered that it was on the C&NW west of town. They manufactured steel truss beams which were widely used throughout the Midwest in commercial applications. That same day (March 2, 1964) a shady real estate dealer from Belvidere called to say he represented 500 acres of land at Belvidere. Although we were already in possession of the land he claimed to represent, he wanted a commission. We went to court, but, to my indignation, settled with him just before the trial began. It was less expensive to buy him off to save legal costs.

On March 24th I watched a "funeral" train roll slowly eastbound through Elmhurst. This was a special move of a good many of the surviving Chicago, Aurora and Elgin interurban cars saved from burning. The cortege was headed by C&NW GP7 #1650 towing old wooden car #317 (Jewett) which had been fitted with a standard coupler. Freshly painted in red and gray wooden car #20 (Niles) was next. Then came Jewett #316 in fresh blue and gray with an old CNS&M line car trailing at the rear. It was a slow trip from Wheaton to Proviso and north to North Chicago where the cars were temporarily stored at the Illinois Tool Company plant until a site for a permanent museum could be established. One such operating museum was founded in South Elgin on the tracks of the former Aurora, Elgin and Fox River Valley Electric.

I rode the AT&SF's *Chief* to Chillicothe, Illinois, on the last day of March. Fred Jones, our DF&PA from Peoria, picked me up for a tour through western Illinois along the old M&StL to Abingdon and Middle Grove. In Galesburg we met with Messrs. Hughes and Schrader of W. R. Grace Chemical Company, about a fertilizer distribution site at Monmouth. Fred and I drove to Keithsburg on the Mississippi River where the only bridge, rail or vehicular, for many miles in either direction spanned the mighty stream. Locating any industry out here would be foolish because it was clear that the days for this line were numbered. The trainmaster at Monmouth, Keith Shreffler, was helpful in reviewing the Grace proposal, but he was somewhat preoccupied with a trainmen's union problem on his territory. On April 2nd we made a call on Peabody Coal and attended an industrial luncheon in East Peoria where Elbert T. (Smitty) Smith of the Rock Island made a strange little speech studded with apologies. I went back to Hughes of Grace with the track plan Shreffler had suggested for serving the Monmouth site. Locations of these little farm retail outlets ate up a lot of time. Fred and I took the approved track plan south to Lincoln where Carl Ward of Grace was located. Fred put me on the Illinois Central's *Green Diamond* at Mount Pulaski. We had a rumor in March that Ore-Ida was interested in the Peoria area. The rumor turned out to be correct and I hustled back to Peoria on tax day, April 15th, on an Ozark Convair. After a brief visit with the Idaho folks at which nothing was accomplished, I rode back to Chicago on the *Rocket*. In April Grace was after another fertilizer distribution site at Earlville, Illinois.

The C&NW's 1963 Annual Report came out in March. Our industrial development department was prominently featured. The March 30th staff meeting at the Union League Club dwelt lovingly on our Chrysler effort and our success was to be the base of a new traffic department program. Little did anyone guess that in two years the venerable, and rotten, traffic department was to be completely made over with many of those present no longer employed.

April arrived and with it the floodgates of new work opened. In addition to the little fertilizer distribution sites, a big Canadian firm came looking for a riverside location at Blair, Nebraska. The Big Muddy with its problematic navigation was somehow thought to be a powerful moderator of railroad rates and a real lure to chemical plants. And the Patrick Cudahy site location which had so awkwardly popped up in the midst of the Chrysler furor in October 1963 came to life again. I went to Milwaukee April 8th to see what new facts Watson might have. I got a sense that there was not much push behind this project. Even so, I gathered Chuck McGehee, our rail/trailer man, and Irv Lawrenz from rates to meet the Cudahy people in Milwaukee. Norm Singer from our Milwaukee office joined us. Watson, Wiley and Jodat of Cudahy insisted that they liked the Harvard area. We explained that the rail rates at Galt, Illinois, (just west of Sterling) were more favorable. Armour's new plant there, however, eliminated that area for Cudahy because they did not like being so close to another meat processor. Following another big C&NW staff meeting on April 30th in connection with the coming proxy solicitation of Rock Island stock, I drove north on May 1st to meet with Harvard's mayor and Cudahy's Watson and Wiley. The city was pleased to be selected as the site for a new manufacturing plant, but, the facility was never constructed.

On April 9th the Nation's railroads prepared to shut down to counter militant union attempts to divide and conquer. By the next day the crisis had been delayed for another lengthy period.

During the Chrysler project late in 1963 the stray thought was tossed out that it would be nice if there were railroad service north out of Belvidere. The old Madison Division once operated from Belvidere's North Yard to Caledonia, Illinois, and Beloit, Wisconsin. This seven mile stretch had been removed in 1942. On a lovely spring day, April 14th, my wife and I hired a taxi in Belvidere and rode to the north end of the abandoned line at Caledonia. A few houses, an old hotel building, a railroad sign and remains of a passenger platform were about all that could be called a settlement where once the Kenosha, Rock Island and Rockford line had crossed the Beloit and Madison Rail Road. We hiked along the old embankment, climbed a couple of fences, and saw at first hand what changes the last 22 years had wrought. Re-establishment of this railroad was theoretically possible but it would be a tough sell. The next day I prepared a map and

Belvidere - Caledonia, Illinois in 1964

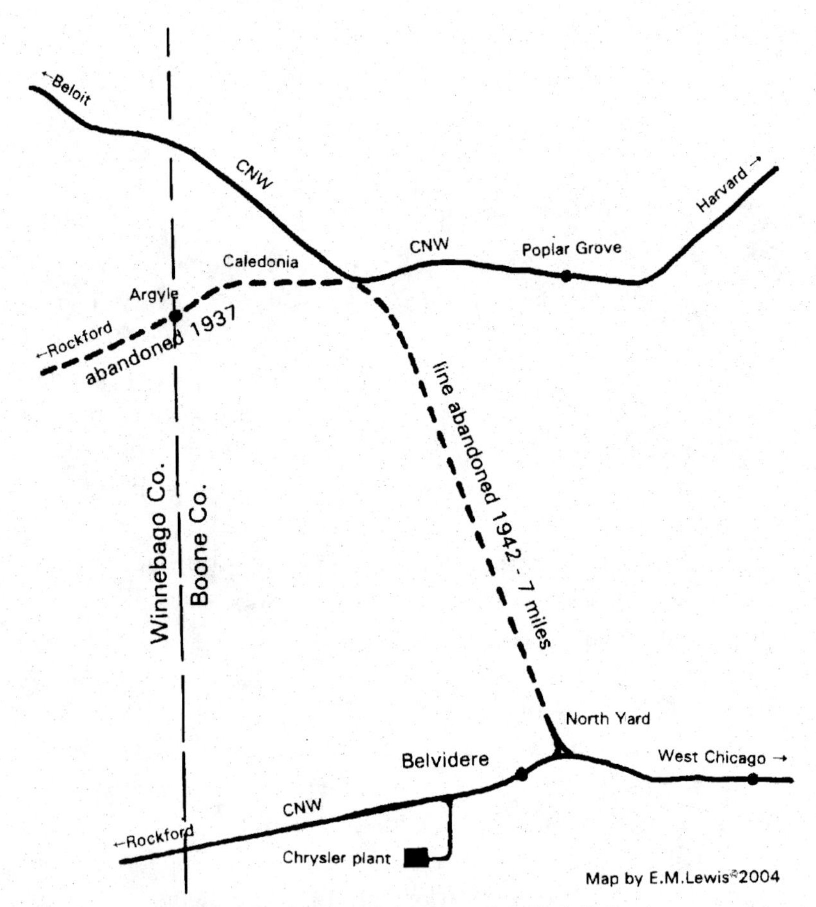

The idea of restoring the abandoned Belvidere to Caledonia link to better serve the new Chrysler plant was briefly explored—and dropped.

written report of our reconnaissance and made a negative recommendation for restoration of this railroad. Mr. Fitzpatrick, however, asked engineering to make a formal survey. Gene Cermak, Steve Owens and I went to the more troublesome spots on June 1st to observe the major problems such as that part of the old embankment now serving as a dam for a recreational lake. To my surprise I learned August 19th that the restoration of this seven miles was budgeted for

1965. Good sense prevailed in the end and no further action was taken to restore this line. I think Mr. Fitz was having personal problems about this time. I was invited to accompany him to a ground breaking ceremony of Standard Packaging's new Elgin (Illinois) plant and observed him close-up. The ceremony was held along our Freeport line near Big Timber Road. The big crowd included Governor Otto J. Kerner who helped by digging with the shiny ceremonial spade and doing a lot of talking. Mr. Fitz headed directly for the bar and was well oiled by lunch time.

IOWA BEEF — (1964–1966) Even as Armour and Company was constructing its new meat processing plants near the cattle raising areas, the beef production industry continued to evolve. The latest innovation was the child of Mr. Anderson, the hard-driving president of Iowa Beef. I rode the Milwaukee Road's Arrow to Sioux City, Iowa April 26, 1964. On the train I ran into John Brugenhemke, one of our smarter sales people, and spent a pleasant and relaxed evening with him as we rolled through the Iowa countryside. As a symbol of the degraded status of this passenger train, we discovered on arrival in Sioux City that the Milwaukee had added freight cars at Manilla, Iowa, thus making us into a mixed train. The switch crews were pretty rough about coupling up the train in the middle of the night. Erv Klover, our Sioux City DF&PA, met the train and accompanied me on a swing through northeastern Nebraska to Laurel, Wisner and West Point where I gathered some hog production statistics for Armour. Why Armour couldn't do this simple task was a mystery to me. The next day Erv and I met John Brugenhemke at the Denison plant of Iowa Beef. Erv and John were convinced that Iowa Beef represented a good freight revenue source. We waited at the C&NW depot for some time until Ray Degnan showed up, late as usual. Ray was such a good hearted fellow that he would listen patiently to anyone's stories long after he should have been somewhere else. We all traipsed into Iowa Beef's headquarters and learned to our concealed horror that they wanted to build a major meat processing facility in sleepy little Wayne, Nebraska. There was barely any train service to Wayne. The railroad there was just two streaks of rust through tall

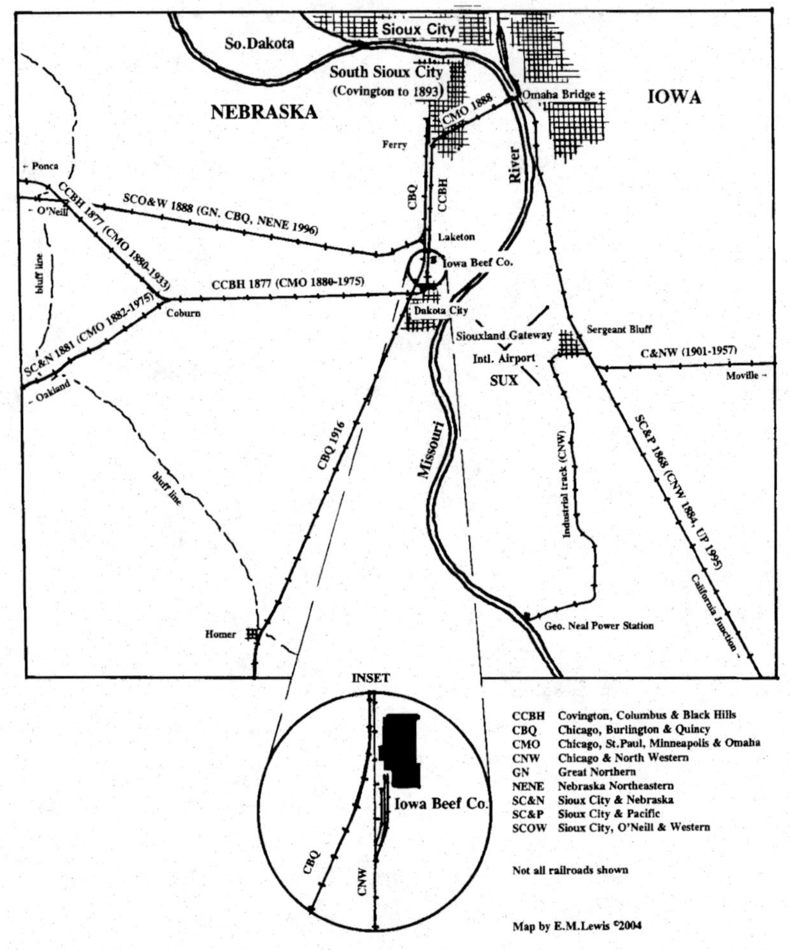

Nebraska across the Missouri river from Sioux City, Iowa, was an active area for industrial activity and railroad contention in 1964.

grass between Norfolk and Sioux City and was little different from the days when the Omaha Road built it. (The last train service at Wayne was to be March 2, 1976). We swallowed our surprise and flew back to Chicago in a DC-6T from Omaha. On the 17th I went back to Omaha on the CB&Q in sleeper NP 365. John Troyke, Degnan's assistant, drove us to Denison on June 18th. While it was a long drive, it was an improvement over the Milwaukee Road's poor excuse for transportation. I had worked up a potential site at Wayne for Iowa Beef's consideration knowing full well that we really did not want them building there. After Erv Klover and Ray Degnan joined us, we again met with the Iowa Beef gang. They seemed interested in the material I presented, but revealed nothing about their plans. We made a quick run to Eppley Field at Omaha and flew back to O'Hare. By the time I returned from a short vacation August 17th, Iowa Beef had turned their gaze northeasterly to South Sioux City, Nebraska, just across the Missouri River from Sioux City, Iowa. My colleague, Bob Lithgow, who formerly had been an instrumentman on the Galena, covered for me during my absence and went to Denison to advise on trackage matters. I think he must have lost some of his track design skills as the proposed rail yard required more land than Iowa Beef had optioned. Iver S. Olsen, on-line sales manager, dropped a lot of the details on me the next day. With Steve Owens of the chief engineer's office helping I worked up a track plan that accommodated Iowa Beef's ideas for rail service at the plant. It was a tight fit to keep the railroad yard tracks within the boundaries of the optioned property. I called Anderson September 2nd to assure him we were working on the plan. He said that they were also considering a site across the river south of Sioux City. Certainly rail service and rates would have been better on the Iowa side, but I soon learned he was just playing off one area against the other to see what concessions he could finagle. Iowa Beef turned out to be a difficult client who bent every rule and took every advantage. Our people soon learned a bitter lesson that this was not to be an easy-to-please customer. Adamantly anti-union, Iowa Beef vigorously opposed all attempts to organize their meat cutting operation and treated their employees like automatons. Injury rates were far higher at Iowa Beef facilities than in other comparable industry operations mainly because of higher production line speeds. Iowa Beef, however, was far ahead of the rest of the beef industry in its methods of handling meat for the retail market. They were the first to break carcasses down into boxed cuts which shipped a lot better than the older method of shipping hanging quarters.

By using some artful engineering, Steve Owens and I crammed

the Iowa Beef yard tracks into the designated property at South Sioux City. We also had to design a similar yard for the Sioux City site as Anderson was still teasing interests on both sides of the river. At the second of two meetings in Fitzpatrick's office on September 29th, it looked like the Sioux City site was to be the choice. I knew that a new Missouri River bridge was being planned by the highway departments and tried to get some information from the Ames, Iowa, office. Both sites at South Sioux City and Sioux City could be affected by the bridge construction. With the decision that Iowa was probably to be selected, I handed off the whole messy project to Don Guenther as his project. Yet, I still had a finger in the pie as I was summoned to a meeting to discuss track construction October 8th—it was getting rather late in the year to be considering track construction. On October 13th I had the whole thing back in my lap—Iowa Beef had settled on the Dakota City, Nebraska, site. Iowa Beef made a mistake, however. They believed that they were to have service from two railroads on the Nebraska side—the C&NW and the CB&Q. Just because the CB&Q had trackage rights from Sioux City, Iowa, to Dakota City in order to reach their branch line that went straight west to O'Neill, Nebraska, Iowa Beef assumed that the Q had rights to serve local industry. That was not our understanding because the old Omaha Road contract did not give the CB&Q such rights. Ollie Waggener of the CB&Q called me to a meeting on October 14th, but no one had told me of any such meeting. I did agree to accompany him to call upon Harold Pike, the traffic manager of Iowa Beef at the Drake Hotel, on the 16th to discuss cost of track construction despite the fact that there was no agreement that both railroads would serve the new facility. Official ground breaking was scheduled for October 22nd and Cermak (who was in New York) was told to attend with me as both Ben Heineman and Mr. Fitzpatrick were unable to be there. We flew into Sioux City in the company plane, N1394Z, and enjoyed the celebratory luncheon laid on by George Wimmer of the Sioux City Chamber of Commerce. Ground breaking was a misnomer as the site grading was finished and some foundation work had been completed. Since the construction was begun, the project was now the responsibility of the sales department and the division engineer; Cermak and I were happy to see the transfer of this troublesome customer to the people who would have to live with it.

It wasn't over—yet. In June 1965 Harold Pike called our office trying to find out what was taking so long about building their yard tracks. The plant was almost finished and about to go into production. The next day, the 9th, Holman (Shorty) Braden, our assistant engineer in Norfolk, Nebraska, called to complain about the track layout created by Steve Owens and me the previous year. He complained, several times, about our tight little package of tracks which was "foisted onto the Nebraska Division by those idiots in Chicago." Steve Owens in the chief's office was not pleased by those comments as he had worked very hard to fit that yard trackage into the space allotted and meet all legal clearance requirements. Shorty was still designing tracks for his long wheel base steam locomotives. The division was too slow in submitting their cost estimates for authority to be approved by the board, which added more delay. On June 15th Iowa Beef flatly declared that they must have two railroad service and that the CB&Q must be allowed to directly serve the new Dakota City facility. Gene Cermak and I predicted that Anderson and Iowa Beef would go directly to Heineman and Fitzpatrick to complain because we would not assume all of the construction costs of their tracks. It was a violation of established company policy and ICC rules. Bob Stubbs, assistant vice president, traffic, advised us to stall Iowa Beef. I met with J. W. (Bill) Alsop of operating and Steve Owens of engineering on the problems at Dakota City on July 14th as again I had been invited to lunch by Ollie Waggener of the CB&Q to discuss the matter. It was a nice lunch with Ollie and his man, Jean Weiland, but I must have been a hard case as he complained of my unyielding attitude to his vice president, G. R. Glover, who in turn complained to Bob Stubbs. By April 1966 the Burlington still had not obtained entry to the Dakota City plant although their local train ran right past it on the C&NW's tracks. On the first of April Ollie and his man George Defiel bluntly demanded that the CB&Q be allowed into the Iowa Beef plant. Of course, we said nothing doing. Iowa Beef, however, had another trick up their sleeves. On July 16, 1966, Bob Wilson, our superintendent of stations, told me that Iowa Beef was trucking their meat across the river to other railroads on the Iowa side. My note for November 14, 1966 shows that our officials were pretty sick of Iowa Beef's shenanigans which were, in some cases, flatly illegal. By then, I had been out of the picture for some time. (In 2001 Iowa Beef was purchased by Tyson Foods.)

Early in May, 1964, the Statex Chemical plant proposal for Fremont, Nebraska, heated up again along with rumor of a Sinclair petrochemical plant somewhere in Illinois. Two men from Stanray Corporation came to the office May 7th looking for plant site material. One of them was a personal friend of Mr. Fitzpatrick. On May 11th I talked to COMINCO in Montreal, AGRICO in New York and worked up a track plan for the Sundstrand Company of Rockford. We were having busy times. On May 12th Cermak and I drove to Peoria for a dinner engagement only to discover that there were two affairs that evening. We split up and I attended the Association of Commerce and Industry. The next day Cermak addressed that same group, but he was not in his top form. Later in the afternoon we drove back to Chicago taking E. T. Smith of the Rock island with us dropping him off at Joliet. Gene was suddenly involved in an intense project in June when he went to Milwaukee to purchase property to keep the Milwaukee Road from invading our shipper Inland Steel Products. We were not the only railroad that aggressively tried to steal another railroad's customers when all around us the trucking industry was taking an increasing share of the longhaul business over the growing Interstate highway system.

In June I met with a prospect at Sterling who wanted 17 acres of our land just across the road from the new Armour plant on Illinois highway 2 for a lumber and home supply outlet. A week later on the 25th a Mr. Erskine representing some Saginaw, Michigan, investors came into the office to finalize the purchase. They were planning five similar operations one of which was to be in Janesville, Wisconsin. By June 30th both contracts had been signed which meant that the two of their five projected Midwestern businesses were located on the C&NW. Establishment of this new business on the "ring" land we had purchased south of Galt to keep the CB&Q out of the Northwestern Steel and Wire Company's mill meant that our objective of filling it up with revenue producing business establishments was happening. By September Wolohan Lumber's trackage was under construction. Don Gunvalson called from Rockford to say Monsanto also wanted a piece of these Galt properties, but that call never materialized as a definite project.

Out west at Fremont, Nebraska, the proposed Fel-Tex fertilizer manufacturing proposal seemed be alive again. I flew out there on June 12th and had lunch with a Mr. Kennard of that company to discuss their specifications for a chemical plant. After eating, we inspected property located north of town at the old West End station site on our line to Norfolk. Kennard ordered soil tests to be made. On June 23rd he announced that he wanted service from both railroads. I called Ollie Waggener of the CB&Q to see what he knew about this location project. Our conversation should have been taped for posterity as we dangled details and generalities but shared no hard facts. Actually, he was already in contact with Fel-Tex which meant our conversation was pure theater. On July 1st a big meeting was held at Fremont to discuss rail service and site questions. Of the sixteen people there, seven were from the Burlington. When I returned from a short vacation August 18th I was greeted with the news that the plant had chosen a site served by the CB&Q north of Fremont. Although Fel-Tex's chosen site bordered the C&NW's right-of-way and construction of a sidetrack would have been very simple, the C&NW was stopped because of the Burlington's retention of a strip of land along our right-of-way fence for a "road". While we could have gone to court and probably prevailed in claiming the CB&Q's *cordon sanitaire* was a pointed act and that crossing a "roadway" with a railroad track would not have impaired the use of the road, there was not enough revenue projected for this new facility to make it worth the trouble and bad feelings that would have been raised. Kennard's wish for two railroads was not enough reason to push the matter further. It was tit for tat: Sterling/Galt, Illinois won by the C&NW; Fremont, Nebraska, for the Burlington. Based on projected revenues produced at each place, the C&NW came out the winner.

On June 17th I worked up a track plan for another lumber and home supply business. This one was 84 Lumber which had acquired property west of the new Chrysler plant at Belvidere, Illinois, just east of the undercrossing of our Freeport line by the Illinois Tollway (Interstate 90). This company came from Eighty Four, Pennsylvania, so named supposedly because there were 84 men from that southwestern Pennsylvania area who fought in the Revolutionary War. 84 Lumber was one of the early entrants in the retail lumber and home supply stores that began to pop up all over the country. Their location at a big curve in I-90 was a beautiful spot for advertising their new establishment.

Unremarked at the time was the demise of sleeping car service on the C&NW on June 22nd.

Locally, the westbound shelter on the south side of the main line at Elmhurst, Illinois, was demolished on June 24th in connection with construction of the new depot. Fred Kruse, my replacement in 1957 on the Galena Division, succeeded Bob Lawton, my former rodman colleague, on the Sperry rail testing car job on June 29th. Jimmy Simons was asked to go to St. Paul as the new assistant engineer, but instead got my old job on the Galena in July. I was proud that "my boys" were rising in the ranks.

Since we didn't have a helicopter available for an aerial reconnaissance, we used our faithful turboprop N1394Z. July 1, 1964 was a beautiful clear summer's day. W. H.(Heinie) Huffman, assistant chief engineer, Maurie S. Reid, office engineer and I flew out of O'Hare. After stopping at Dubuque to drop Reid, Heinie and I flew low over the Rock Island's line from Hampton to Manly, Iowa, counting bridges and noting the lay of the line. We went on to Omaha and had that big meeting with Fel-Tex (mentioned above). After topping the plane's tanks and collecting John Troyke, we went back to Chicago. The next day we flew to Peoria with Huffman, Vern Mitchell (engineer of signals), M. C. Christensen (Galena Division engineer) and Al Handwerker from accounting on board. From the air we inspected the Iowa Junction area south of Peoria where the M&STL, CB&Q, Peoria Terminal (CRIP), P&PU and TPW lines crossed. The Peoria Terminal's line was of particular interest as their parent, the Rock Island, was making noises about reactivating this unused and forlorn track for future service to the Sommer area where the C&NW and TPW were having their legal cat fight. The PTC track had not been operated for years; it was full of small trees west of Iowa Junction. Since the PTC was a subsidiary of the Rock Island which itself was a financial basket case, we didn't give their threats to gain entry to the disputed joint service area much credence. While aloft, we swung over Pekin and took a look at the problems associated with trying to build a connecting track off our high embankment south of our Illi-

Peoria Municipal Airport saw the C&NW's turboprop N1394Z with W. H. Huffman (Engineering), V. S. Mitchell (Signals), M. C. Christensen (Engineering) and Al Handwerker (Finance). The object of this visit was an aerial inspection of the Peoria Terminal Railway facilities.

nois River bridge #1731 to serve the Standard Brands-Quaker Oats plants. On the way back to Chicago we coasted up the Spring Valley line as far as Shabbona Grove before heading straight to lakefront Meigs Field. (Meigs Field was partially demolished by the City of Chicago in March 2003 preparatory to creation of a new park).

Our department was often tapped for special duties that had little to do with our primary mission. On July 9th I rode the electric Chicago, South Shore and South Bend to Michigan City, Indiana, to help Rhodes Blandford, our local sales agent, collect an overdue freight bill from the Arrowhead Company based in Chesterton, Indiana. After collecting the check, I rode back to Chicago from the Tremont stop. I did not mind these trips which were like little vacations to someone who loved railroads. On July 10th John Hahn offered me a job in Omaha with Northern Natural Gas Company. He must have gotten over his January pique when he had taken exception to our reluctance to share a report with him. I took the offer as a compliment, but said I was still a railroad man.

It was time to go out west again. On July 20th I flew to Casper, Wyoming, by way of Denver. Arthur E. Knight had succeeded John D. Boyles as our local sales representative at Casper. We looked at land along the C&NW east of town for future industrial use. I was very dubious that this gravelly, windswept property would ever be used for more than poor pasture. We went west the next day to Shoshone with a side trip to the Wind River Canyon where the Burlington threaded its way through deep and rocky gorges. At Riverton we called on Teton Timber who was harvesting lodgepole pine for 2×4 inch stud lumber near DuBois and hauling it to Riverton to load railroad cars. I drove to Burris near DuBois to talk to a Mr. Ellerby of Vipont regarding their gypsum mining activities. Coming back to Riverton through Ethete, I was struck with how empty this country really was. On the 22nd I inspected the vacant station grounds at Arapahoe which had at one time been a real station, visited with Ed Axe of the Susquehanna Corporation which was processing yellowcake uranium ore at their Riverton plant, and then headed back to Chicago. Arapahoe became a problem in November. An Indian sheriff from the Wind River Reservation argued that the Arapahoe tribe had a claim on the station grounds where Jefferson Lake was loading liquid sulphur into tank cars. I turned that matter over to the real estate department. Chuck Jevne went through the original land records from the Wyoming and North Western Railway's construction in 1906. He could find no support for the sheriff's argument. He did find some very amusing Indian names on the old land records.

The real estate department was grabbing more and more power during this period. By late August its leader, I. Robert Ballin, appeared to be trying to eradicate the industrial development department. Cermak's combative spirit rose when sales of company land at Norfolk, Nebraska, and a study of land at Des Moines, Iowa, were pushed through without our approval. Ballin's new salesman, Richard (Dick) Taylor, spent a lot of time picking our brains about our plans for development of the "extra" acreage next to the Chrysler plant at Belvidere. Cermak and I sensed that Ballin was itching to get his hands on that prime property to fatten his land sales figures.

On August 27th Don Guenther and I flew in Ozark's puddle-jumper to Fort Dodge, Iowa, with stops at Rockford, Dubuque, Waterloo and Mason City. Ed Jeffries, our local sales representative, took us on a tour of the gypsum quarries and through the wallboard manufacturing plant of U. S. Gypsum. I also took a good look at the track layout at the Iowa Beef plant located there on the Illinois Central. We stopped at Duncan, Iowa, to call on Container Corporation of America before flying back through some dangerous thunderstorms that extended east as far as Waterloo where the weather cleared. Not all the activity in our department was in my territory. Bob Lithgow and Cermak were successful with a warehouse for Anchor Hocking at Gurnee, Illinois, in early September. Terra Chemicals first surfaced in September when Cermak met with Allen Grey in South Carolina. By December they had picked their plant site at Sioux City, Iowa basing their location choice on a study made by our department. Gene Cermak, however, made a political mistake when he told Frank Koval of public affairs about the Terra location first; Frank ran right to Heineman with the news before Gene's boss, Ed Olson, could convey the news to Mr. Fitzpatrick. Another rocket landed in Cermak's lap. (As information the Terra Industries manufacturing facility actually located at Salix, Iowa. On December 13, 1994 the 29 year old plant was leveled by an early morning explosion that injured 18 and killed 4.)

My career in the traffic department began to slide more rapidly during the last half of 1964. Not just one incident marked the decline—there were a series of unrelated events that seemed to tip the scales against my continuation in this career. One of the more interesting tasks turned into a political mess that tarred everyone associated with it. On April 29, 1964, just after I had returned from a call at Patrick Cudahy Company in Milwaukee, Mr. Fitzpatrick strolled into our office and personally asked me to do some research for him into C&NW-UP relationships in the 1904–1906 period. He had a hunch that there might be some old agreements, now forgotten, that could embarrass the Union Pacific in the coming fight over the Rock Island. Tom Ross, the C&NW's Secretary, gave me permission to browse through the corporate archives. There was a lot of material to review and some interesting items turned up. One was an oil field development proposal by the Belgo-American Company in central Wyoming that came to naught because of under capitalization and the opposition of Standard Oil. I worked through the minutes of the C&NW's executive board and the records of the Wyoming and North Western Railway, which was the C&NW's construction company west of Casper. (I wish that I still had all my notes). On June 22nd Jay Hillman, one of our top legal people, asked if I had found anything yet. I shared my early findings with him, an action that caused a real dustup.

This was summertime. If my superior, Gene Cermak, was not on a golf course, he was playing handball at his club. He was almost impossible to contact during regular business hours. I fear that my disapproving attitude of his constant absence and inattention to business had been communicated to him by some of his sycophantic colleagues. Jay Hillman reported my early findings to Heineman and Fitzpatrick. On June 25th they called for an immediate conference with Gene Cermak, who happened to be in Elgin playing golf. When Gene finally responded to a frantic summons to return vice president Ed Olson's telephone call, he received a 45 minute tongue-lashing from his outraged superior whose principal complaint revolved around the breakdown in communications that allowed my preliminary report to go to the top without being first reviewed by his staff. Ed Olson was no dummy—he knew very well why my preliminary report had not gone through normal channels. The normal channels were not functioning because Cermak was not paying attention to business. Smarting from that sandpapering, Cermak sent me a preemptory note to see him first thing on Monday morning, July 6th. I was there at 8 a.m., but he didn't arrive in the office until 8:40. By then I was on my way to a 9 a.m. Illinois Commerce Commission hearing for the abandonment of the Genoa City-Hebron line. At the hearing I was called to testify as a rebuttal witness and had to stay all day. Cermak never did say what he wanted to see me about; I suspect he had ordered me "front and center" to pass along the verbal hiding El Olson had given him.

Heineman and Fitzpatrick were still keen to find something useful and urged me to continue delving into the records whenever normal business permitted. It was clear that this project had a high priority, and I devoted as much effort to it as I could. By July 12, 1964 I had gone through as much material as I could uncover and wrote-up my findings. It took days to get the typing completed and the maps prepared, but by July 23rd the report was as ready as it would ever be. The report was entitled *Chicago and North Western Ry. and the Projected West Coast Extension 1904–1906*. The key element was a very interesting October 22, 1902 agreement with the Union Pacific on how joint traffic would be handled. My next problem was one of protocol. The report had been commissioned by Mr. Fitzpatrick, but he couldn't have it until it had been reviewed by my superior, Gene Cermak, his superior, R. C. Stubbs, and the top traffic man, Ed Olson, in that precise order. I dumped the report into Cermak's overflowing in-basket (he was off golfing somewhere), let Hillman know that it was finished and left town for a driving vacation with my family in eastern Canada.

In early September our vice president of traffic, Edward Olson, was stricken with rheumatoid arthritis. Unable to function in the extremes of Illinois weather, he was soon put on the shelf in a low-key job as sales representative in Phoenix, Arizona. Robert C. Stubbs, was promoted to the vice presidency. He was not industrial development's biggest fan, either. The first part of September was spent in shepherding a Mr. Galbreath of Sutherland Lumber Company on a tour of the Peoria area and attending a Great Lakes Industrial Council meeting held there on the 18th. While there, I drove out to Kickapoo to see the new M&StL-CNW connection. The former M&StL line at the foot of its Maxwell hill was rerouted across a new diamond crossing of the CB&Q to a new switch in the C&NW main. The rest of the M&StL line to Iowa Junction including their old engine servicing facilities was soon to be abandoned. On September 23rd came the announcement of the proposed merger of the C&NW and the CMStP&P into a new company to be called the Chicago, Milwaukee and North Western Railway. Rumors were also flying about that Mr. Fitzpatrick wanted to buy the New York Central's Peoria and Eastern Railway from Pekin, Illinois, to Indianapolis, Indiana. That would have solved our plan to serve American Distillers, Standard Brands and Quaker Oats at Pekin, but, that rumor was just another emanation from the third stool on the right. Quite by chance I had acquired a copy of Nelson Trottman's *History of the Union Pacific* last May. In September Bob Stubbs heard I had a copy and requested that I loan it to him. This strange book had been written by our law department's leading senior when he was yet a callow youth, he later told me. Mr. Trottman was not proud of the book and said it ought to be burned. I understand that he had made serious efforts to corral all the copies and destroy them. (I donated my copy to the C&NW Historical Society in 1992).

APPLIED GEOGRAPHY September 1964 saw the beginning of a career change that would extend beyond my railroad days. I had always been interested in geography, had traveled extensively, been interested in why and how things are transported, read history as a hobby, received an engineering education that dealt with physical matters and had a penchant for organizing data. These interests and curiosities coalesced in September 1964, eventually becoming a commercial product widely used in the coming age of computers. Earlier in June of 1963 I had tried the idea of calculating population centroids for DuPage and Kane Counties in Illinois. (Later I learned that the Bureau of the Census performs a similar calculation every ten years to determine the population center for the United States). While my results were mildly entertaining, they were of little practical application. On September 24, 1964 we heard a rumor that Jones and Laughlin Steel Company was looking at the Midwest for a steel production and distribution center, but could not settle an internal dispute as to whether Chicago was the best location or if someplace further west might be better. By October 6th Don MacBean and I were considering how we could develop information about steel consumption in the C&NW's service area. Profiting from our recent explorations in locating data sources, Don and I decided to try using available business data from Dun and Bradstreet organized in geographic terms. For a modest expenditure we purchased business information data from D&B. We specified that we wanted sales volumes and employment for selected businesses engaged primarily in the fabrication of steel products in a seven state area west of Chicago. The data arrived in the form of several boxes of the ubiquitous IBM 80 column cards. Don and I had to plow through these boxes to determine if the data appeared to be valid. We ended up with 13,250 steel consumers located in 515 of the 643 counties in seven states of our service area. While tabulation of the information by political units was a major achievement and was quite informative, somehow, the results seemed inadequate. Then the figurative light bulb came on. Vector analysis of these data might provide some insightful meaning to these geographic summaries. After all, the quantities of steel and numbers of employees were finite numbers and places had physical attributes of location in the form of X-Y coordinates. By October 12th Don and I had worked our way through the cards for Illinois and Iowa and finished the raw data on October 15th. I set up a map of our seven state study area and drew a horizontal base line (X) and a vertical (Y) axis which crossed in Cherokee County, Iowa. From a county outline map I measured the X-Y coordinates of 515 of the 643 counties in our study area and sent the coding sheets to the Ravenswood computing center to be keypunched. After we had done a first rough shot at the calculation on October 15th, I made an educated guess that the point of interest was going to be west of Chicago all right, but not conveniently on the banks of a navigable river as had been specified by J&L On October 20th I took a break from data bashing and drove around in the Ladd-Seatonville area as well as across the Illinois River to Hennepin, the county seat of Putnam County. I had a hunch about this place and collected plat books and other community material even though the C&NW did not go south of the river. Don and I sent our cards to Ravenswood for tabulation and calculation on October 31st. The first results obtained were full of strange numbers that required review and some adjustment. By chaining Don to his desk, so to speak, we tediously waded through the printouts fixing obvious problems so that we were in fair shape by early November.

I had an interesting telephone chat with a Mr. Longini with the Mellon Bank in Pittsburgh as they were acting as agents for their unidentified client (whom we all knew was Jones and Laughlin, but no one would admit to it). Despite many interruptions and delays while waiting for translation of our basic grid data into machine readable form, a start was made on the steel consumption report November 13th. Four days later we had an arguable draft in hand. On November 20th the first output from the computer came in. Calculation of the centroids was next. We got an answer on November 24th, but a rerun was necessary when a significant data error was discovered. The next day, November 25, 1964, we had a final solution. The center of steel consumption for our seven state study area fell at a point nine miles northwest of Dixon, Illinois, near the village of Polo. The center of employment was calculated to be just one mile away. At the time I had little concept of how basic the center of gravity concept was in the science of market analysis. All we knew is that we had managed to organize a lot of data into meaningful tables and to successfully calculate valuable geographic information by using emerging computer technology. Even our boss, Gene Cermak, got interested in what his two strange birds had created. Longini was extremely anxious to learn what we had wrought and urged us to come to Pittsburgh as soon as possible for a presentation. Production of 52 copies of the report and preparation of the color slides took some time. When we were to go to Pittsburgh, December 4th, O'Hare field was closed by an all day snowstorm. Two days later, Cermak, Don MacBean and I were off to Pittsburgh on Baltimore and Ohio's train #8 in sleeper *Monocacy*. Ellis W. Ernst and Tom Hunt from our Pittsburgh offline office met us at the station. We called first on Mr. Tietjen, a vice president of Jones and Laughlin, and gave him a preview of the report. He was rather patronizing until he saw that our study confirmed his belief that the steel consumption center was indeed west of Chicago. That afternoon we met Longini at the bank and made our formal presentation to him and a tape recorder. We were rewarded with his statement that it was the best presentation he had ever heard. He had only one question, "how many adjustments to the Dun and Bradstreet data had we made?" Don and I explained that restatement of quantities were made to achieve uniformity but no items were altered in scale. We said that we would be happy to furnish a detailed report on our methodology as well as samples of the editing. That evening my reward was a ride on the famous Fairview streetcar line that ran high above the city along the edge of the surrounding mountains. We flew back to Chicago on the 8th. Two days later Longini had our supplementary material, but he had already passed the report on to his client (J&L) the same day we were flying back to Chicago. Tietjin of Jones and Laughlin told Ed Olson December 11th that our study was a valuable item and much appreciated. The next day he called me from Pittsburgh and wanted information on 1,000 acre tracts of land located along waterways in Illinois. I already had that feeling in October about where this data bashing was going. Very quickly, rural Putnam County became the proud, and surprised, home of Jones and Laughlin Company's new steel plant complex.

The methodology used in the Jones and Laughlin study evolved into creation of a geographic database that utilized internationally recognized grid coordinates expressed in degrees of latitude and longitude. Here was the new thing I had been searching for these past two years. There was a potential for a radical new approach to site location. In the early days of the computer revolution, this geographic database would provide a basic requirement for positional logic. It could be thought of as sending a computer to geography classes. Now, it was possible to do electronic spatial analysis involving real places of the world. My life was to take a drastic change in direction. Metaphorically, my working career could be likened to riding a horse named RAILROAD across a vast and rolling plain. Along the way we would come across a new-born foal, GEOGRAPHY, lying along the path. Gathering its tiny legs, it wobbled along after us. Sixteen years later my main horse, RAILROAD, was dead and GEOGRAPHY had become my main steed. But growth and training of this new concept was not to be easy. There were many problems to be solved.

There were some left over items in 1964 to be mentioned for the record. The Northern Illinois Gas Company held an open house on September 25th to celebrate moving into their new headquarters building on Illinois highway 59 and the Illinois Tollway near Warrenville. On the 27th the CB&Q's *Ak-Sar-Ben Zephyr* smacked into a detouring CRIP *Golden State Rocket* at Montgomery, Illinois; I was glad not to be aboard. C&NW stock hit 59⅞ on September 30th. I flew to Omaha on October 8th (no more riding those dangerous trains!) and connected to a Frontier 580 prop-jet to Scottsbluff, Nebraska. There I boarded an old C-47 for the short flight to Chadron. The Black Hills division officers, Al Johnson, superintendent, and Gerry Linn, division engineer and Bill Cook, our Denver sales representative, joined me in a hi-rail trip to Gordon, Nebraska, to see the Shuster brothers who were proposing construction of a new rendering plant. This was my first close-up look at a rail line that I was to get to know very well ten years later. We interviewed the Shuster boys and, after stopping to visit with a banker in Rushville, drove back to

Chadron where Bill Cook and I flew south to Sydney, Nebraska, before going on to Denver, Colorado, where I caught my Chicago bound 727. That was a lot of flying for no visible result, but it was important to respond to local requests like this to demonstrate our department's commitment and to prepare the ground for future requests for assistance that we might someday need.

The Chicago Great Western became a target. On October 13th Ed Olson hosted a big dinner reception for CGW people in Chicago. In December a group of citizens from Oelwein, Iowa, came to Chicago to express their worries that a CGW merger would seriously injure their city's economy. The Soo Line also felt they would be damaged by the proposed merger and filed suit to stop it in December. On October 31st the South Beloit property came up again. Lehigh Portland had established a cement unloading and storage operation and now Wickes Lumber Company thought they would like to have a lumber distribution site there, too. But October ended on a sober note when two of our traffic people were discharged for alcoholism; one from the West Coast office and one in Chicago. It was no wonder that the whole department wasn't fired as the entertainment of clients usually entailed consumption of a lot of booze. Fortunately for some of the high traffic officers, the spotlight did not shine upwards.

In early November office alterations corralled two of our heaviest smokers to my great relief. Don MacBean and I had already figured out how to use that odd space behind a stairwell for our office but getting Lithgow and his tobacco fumes behind partitions was a big plus. About this time I made a framed display of railroad stamps for Ruth Buchholz who was presenting it to her husband as a birthday present. I took it to Milwaukee on the 19th to see Harvey's new offices and make the presentation of this beautiful array of postage stamps. In November we heard that Caterpillar Company was planning a new foundry complex to be located at Mapleton on the TP&W property. The rumor was true and the plant was announced December 22nd. Since we were deep into the lawsuit over access to the "joint" area, the event was not considered important to our railroad. And the track agreement with Missouri Portland Cement for the track at Sommer was still hung-up in the legal mill. On November 20th another rail strike loomed for the 23rd, but it didn't happen. On December 3rd John Hahn of Northern Natural Gas and two people from Knox Gelatin arrived at Meigs Field in Chicago. After a rather brief meeting in which they expressed interest in our property at Fulton, Illinois, I took them to O'Hare for their flight home. Nothing evolved from this contact. The end of the year wound down with the assignment of proxy solicitation from Rock Island stockholders as recounted earlier. This year had been interesting. I could hardly wait to begin development of the new computerized geographic concept.

19

THE GODS OF TRAFFIC HAVE FEET OF CLAY, 1965

. . . The Rock Island proxy chase continued over the New Year's holiday to January 6, 1965. Of my assigned 2,900 votes, 1,100 went to the Union Pacific and 1,800 for either C&NW or not voted, which was equivalent to a C&NW vote. But, the C&NW had started too late on this campaign. The Rock Island meeting was delayed to January 8th indicating something was up. Then on that day, the meeting was postponed again until Saturday morning, January 9th. On Sunday we heard that the Rock Island board had voted 4 to 1 in favor of the Union Pacific takeover of the Rock Island. It was not the final act of this drama, however. The C&NW made a new offer to the Rock Island shareholders on October 7th. The State of Wyoming came out in opposition to the C&NW's offer for the Rock Island. After all, Wyoming was a Union Pacific state. John Danielson, our fiery young attorney, was assigned to respond to the Wyoming position. In November I helped assemble maps and informational materials for him. Iver S. Olsen, manager of on-line sales, organized a big sales staff meeting November 29th to plot a strategy to calm the Wyoming problem. The proposed Rock Island and Union Pacific merger was so long a running saga that it eventually failed from legal exhaustion. Why did the C&NW oppose that merger so vigorously? Because the Union Pacific would have gained direct access to Chicago with great negative impact on the C&NW's share of transcontinental traffic. The C&NW had always been the Union Pacific's natural partner with the best profile and alignment, but it would be April 1995 before the obvious union was consummated.

The Jones and Laughlin study of 1964 was translated quickly into actual plant construction. J&L acquired property near Hennepin, Putnam County, along the south bank of the Illinois River just east of the point where the river made a big swing south toward Peoria and its connection with the Mississippi River below Alton, Illinois. On January 7th Maurie Fulton, president of Fantus Company, complimented us on our interesting and professional report. Maurie had been retained by J&L to investigate the railroad service at Hennepin. Jim Ronayne, our manager of steel traffic, agreed to work out schedules of service. On the 22nd, Cermak, Ronayne and I gave the schedules to Fulton and Bob Thompson of Fantus. They expressed surprise and pleasure at our quick response. The New York Central was the lucky recipient of this big traffic producer. We approached them for trackage rights from our connection at Ladd, north of the river, to Moronts, south of the river, using their Illinois River bridge at DePue, Illinois, a distance of about nine miles. They would have none of it. On the 29th Fulton called to say that J&L would help the C&NW obtain access south of the river, but we were not at all sanguine about a positive result. I dug into the history of the little-known DePue, Ladd and Eastern Railroad, a 3¼ mile subsidiary of the C&NW, to see if there were any possible justifications that it could be used to gain access to the NYC's line from Zearing, near Seatonville. By March 3rd I was fairly certain that the C&NW had no rights. In retrospect the C&NW would have had a major line reconstruction facing it had the trackage rights on the NYC materialized. The railroad line from Ladd

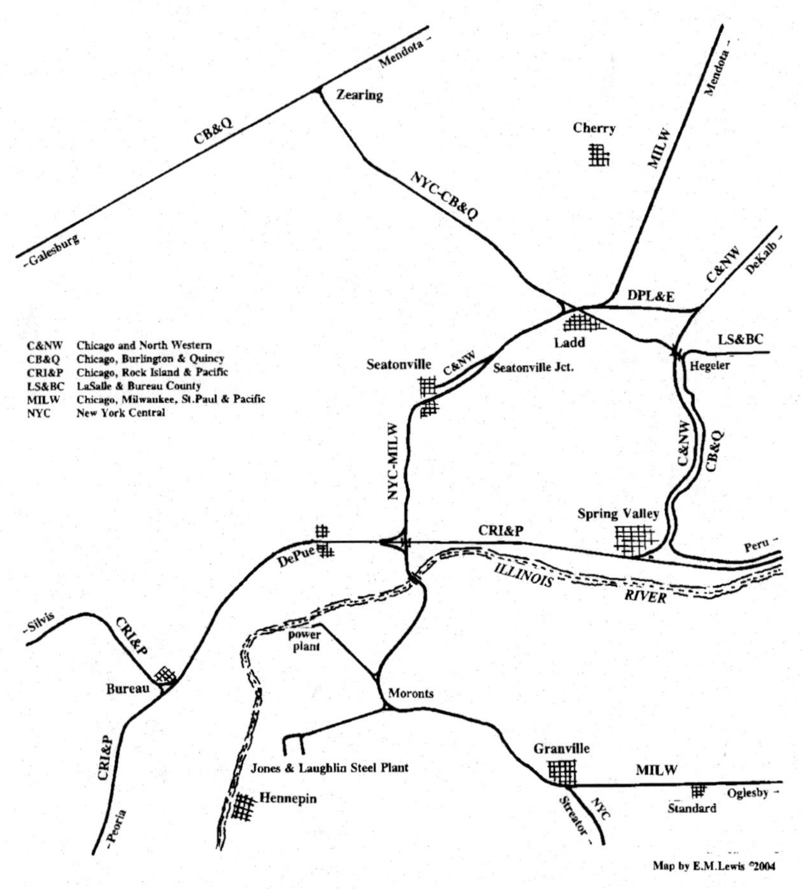

Geneva, Illinois, where R. R. Donnelley's plans for a future printing plant triggered a huge land acquisition of prime real estate by the C&NW only to see it metamorphose into a golf course.

to the main line at DeKalb was 50 miles of light rail, bad ties and weak bridges—just another broken down branch line. The three mile piece west of Churchill extending to Ladd was a 60 pound rail relic of a coal line which had served a mine at Seatonville. The mine had closed before 1930. A dribble of interchange traffic, mostly slow-routed lumber, kept the line alive. It was also a key link in the shortline mileage tariff—a dubious claim for continuation, but our traffic people were sure it was beneficial to the company to maintain that connection and the little-used one with the Rock Island at Spring Valley, Illinois.

From review of the old agreements and the assessment of the fact that the Milwaukee Road had trackage rights on the NYC and included Moronts as a local station in its timetable, the effort by the C&NW to crash the party was not worth the potential expense. J&L realized that their precipitous land acquisition without really considering the rail service available, had not been their smartest move. On April 29th they formally announced the new steel mill to be built at Hennepin. The State of Illinois agreed to construct a new highway bridge across the Illinois River and a highway extension, Interstate 180, north to a connection with Interstate 80. Meanwhile, a river barge collided with and severely damaged the NYC lift bridge at DePue on April 8th. With the weak rail service at the western tail of the NYC and the spindly Milwaukee Road service on this remote extension, it was apparent that most of the J&L traffic would be outbound product carried by steel hauling trucks on the expanding Interstate system. What did we gain by this plant location project? J&L favored the C&NW by increasing our share of their other overhead business. In our shop, we went on to other business. The last echo heard was on May 5th—a stockholder of J&L asked their management why the C&NW did not serve the new plant. Nice of him to ask, but no thank you.

Commander John Barron of the Tri-State Agency at Granite City, Illinois, continued to scheme at getting C&NW direct service to his grain elevator facility. I rode the GM&O's *Midnight Special* to St. Louis on January 24th to attend John's meeting the next day. He was now promoting a new rail yard complex at Mitchell, Illinois, to consolidate the numerous tiny rail yards scattered around the East St. Louis area. The man had vision all right, but he was used to getting funding from governmental bodies. Railroads just did not have piles of loose cash around to build railroad yards for future industrial growth. Jim Thompson and S. A. Keathley also attended the meeting just to keep tabs on John Barron in case one of his proposals should actually catch fire. On the 26th I visited with the chief engineer of the Illinois Terminal about their new connection with the Norfolk and Western at Edwardsville. The N&W had recently merged the old Wabash Railway. Later, I went with Jim Thompson to see Bob Hall at Monsanto and got a preview of their plans for the coming year. The next day, the 27th, I added a new piece of railroad to my travel log—the New York Central from St. Louis to Indianapolis, Indiana. I bought a sleeper space on this day train to use as an office to write my reports on the Granite City non-events. The Great Lakes Industrial Development Council was meeting at Butler University. The speeches were about the importance of research and the developing high tech industry beginning to emerge. Catering to high tech business did not seem to have much potential for generating rail revenue so I went to the concert hall nearby to listen to the Indianapolis symphony rehearse. That evening the C&NW sponsored an open house for the council delegates—our food was better than at the official dinner served later. Unfortunately, strong drink overcame our Chuck Towle who ended up with his shirt ripped to tatters in a brawl with a bartender. Our return to Chicago on the NYC the next day was uneventful.

Now John Barron pulled another arrow from his bulging quiver—convert the Chain-Of-Rocks bridge to a railroad span. That idea quickly died. But, John was persistent—he managed a meeting with Mr. Fitzpatrick, Gene Cermak and Bob Stubbs on July 27th to promote getting direct C&NW service to his port. A committee was appointed to investigate the feasibility of the project. The committee sort of faded away after a few inconclusive talks with our grain people in August. I did have a chat with Mr. Brown of the Dreyfuss Corporation (a big grain merchandiser). On December 9th, John Barron was back in Chicago to discuss rates and service with W. E. Braun, assistant vice present traffic, and Roy Erickson, our grain rate specialist. A week later, Braun, Erickson and I flew to St. Louis so that Bud Braun could have a first hand look at Barron's facility—I think he wanted to see if it really existed. Jim Thompson and Syl Keathley from the St. Louis office joined us, and we all went to see Wil Edmonds, traffic manager of Granite City Steel. A nice trip, but no results.

Returning to January 1965, the 29th saw changes in the top people of the traffic department. As mentioned earlier, Edward A. Olson, vice president, was put out to pasture in Phoenix, Arizona, as the C&NW traffic representative in deference to his arthritis problems. Robert C. Stubbs took his place, Walter E. (Bud) Braun was promoted to assistant vice president-rates, and Iver S. Olsen took Stubbs' avp-sales and service title. Ed Olson suffered another disabling attack January 8th and had already gone to Arizona by the 13th. As feared, Stubbs was to prove to be no friend of the industrial development department.

The first two months of 1965 were spent dealing with the bits and pieces left from former projects. Site location work seemed to die in the cold winter weather. Our principal activities were giving speeches, attending conferences and waiting for spring. On January 30th I drove west in cold icy weather to DeKalb, Illinois, where Mr. Bird of Chrysler addressed the Genoa Chamber of Commerce annual dinner held at Northern Illinois University. Genoa was just 11 miles southeast of the new automobile assembly plant and supplied a fair number of workers. On February 3rd I rode train #149 to Milwaukee because Harvey Buchholz asked me to join him for a trip to Madison, Wisconsin. We made a few calls including one on the railroad's lobbyist, a Mr. Ostby, who was working on the repeal of Wisconsin's full crew law. The next morning the thermometer stood at minus 23°F. I met Governor Warren P. Knowles and participated in a panel discussion that afternoon. The subject? Chrysler, of course. The rumors were rampant that Chrysler had considered many locations before settling on Belvidere and why was their town left out? I tried to make it very clear that the Rockford area had been selected long before their site acquisition team had been activated. After the meeting, I rode the Milwaukee Road's afternoon train to Chicago and chanced to meet our J. R. (Bob) Kunkel also heading home. The CB&Q would not give up on the Northwestern Steel and Wire Company at Sterling. I prepared some maps for Stubbs to show to Mr. Fitzpatrick what the latest Burlington ploy involved. Whatever was afoot did not materialize in any overt action. In the middle of summer on July 28th, K. Julian of Commonwealth Edison confided that he had heard that Northwestern Steel and Wire was looking at river sites elsewhere in Illinois. Their efforts soon faded, if, indeed, there had been any at all. Paul Dillon was not about to move his immense steel works away from Sterling where he was the *de facto* king.

Our little social group, Railroad Industrial Development (R. I. D.), attracted only six members January 22nd. February 26th and March 26th were a little better, but April 30th's four industrial agents present made me think that this group had run its course. May 28th, however, a healthy 13 showed up. June 25th's drop to 10 present could be attributed to the summer season. Autumn's first meeting September 24th at the Illinois Athletic Club on Michigan Avenue attracted 15. Gene Cermak, liked to attend, but as an honorary president, the first round bar bill was all his. The last meeting of 1965 was December 10th with a dozen present. It looked like R. I. D. was good for a another year. An interesting note about the attendees—it was always a different mix as new people came into the railroad industrial field replacing transferees and drop outs.

We did have a little industrial development activity during the cold part of the year. Bob Lithgow, who was responsible for the Chicago metropolitan area which mostly included the Chicago switching limits out to the line of the EJ&E Railway, invited me to lunch February 7th with John Given of Standard Brands and Roy Newman of Curtiss Candies which was a subsidiary of Standard Brands. They were interested in railroad land at Proviso which was outside of my territorial assignment, but I enjoyed the free lunch anyway. Maurice Fulton of Fantus Company was retained by Ben Heineman on January 14th to do a confidential study of Oelwein, Iowa, to better understand what impacts merger of the CGW might have on that community. The new depot at Elmhurst was opened February 8, 1965; the old depot was demolished on April 1st. Henry Schroeder, the head of the rate department, retired February 5th and was feted on the 9th. On February 11th I introduced the slightly shady Sam Lezak to Frank Merrill of Barber Greene Company. Merrill operated a manufacturing plant on the C&NW's Sycamore branch about half way between DeKalb and Sycamore. He was concerned about the lack of low cost housing in the area which severely impacted his ability to attract and retain a good labor force. Sam, who had been contemplating a new town in the Huntley area, shifted his direction further west and got very interested in the area north of DeKalb near the crossing of the Milwaukee Road branch south from Davis Junction and the CGW at a long forgotten place named Wilkinson on the old maps. By August 20th the project was stalled by rezoning problems roiled by sturdy local opposition. In the spring of 1965 I met with Sam, John Thormahlen, (Sam's real estate partner), Frank Merrill and Warren Osenberg of Barber Greene to discuss formation of a new town to be called Five Points. By October 11, 1965, Sam claimed he had secured Federal Housing Authority (FHA) approval. This turned into another leaky balloon that soon ran out of gas. The land was to remain in agriculture for many more decades.

On Valentine's Day 1965 I went to Omaha on the CB&Q's *Ak-Sar-Ben Zephyr.* The next day amid drifting snow I carefully drove up to Fremont to see a couple of prospects (Devol and Johnson) for some reasons now lost before going south to Lincoln for the State's annual industrial development conference. Ron Reifler of Fantus was a principal speaker and made an interesting presentation. I saw many people taking careful notes. On the second day it was my turn to perform for just over an hour. At least, no one got up and walked out. I drove Reifler back to Omaha and had a session with Jim Heidkamp, our local C&NW real estate representative. The next day I flew on Braniff north to Sioux City, Iowa, and on to Sioux Falls, South Dakota. Charley FitzGibbon, our local sales representative, drove us 20 miles to Canton, South Dakota, to see a prospect who had expressed interest in a Norfolk, Nebraska, location. This was a pure waste of time, as it turned out. Charley took me back to Sioux Falls for a flight to Minneapolis where Stan Johnston (my old rodman from years ago) gave me a tour of their new downtown offices. The next day I contacted General Mills about a rumor of a new facility and learned they had a project slated for Indiana or Ohio. Chuck Towle then took me on tour of the hot industrial sites around the area before I left for Chicago. One of his friends named R. Scott was quite disdainful of our J&L Steel Report. How he had seen it was a mystery, but I detected more sour grapes than genuine disagreement with the methodology. Tuloma Gas was rumored to be again looking at the Peoria area (February 23rd). On the 26th the chimera of a C&NW line from Wyoming to Ogden, Utah, surfaced briefly and again swiftly disappeared. Despite terrible weather on March 1st I addressed the Lions Club in Marengo, Illinois, on the hottest topic in the area—Chrysler. Marengo, just nine miles east of Belvidere, was also positioned to furnish labor for this new plant. The C&NW-CGW merger hearings were still under way on March 3rd. (That merger was finally consummated July 1, 1968).

INTERNATIONAL MINERALS AND CHEMICALS — COLONY, WYOMING — 1965–1966 International Minerals and Chemicals of Skokie, Illinois, had plans for a new bentonite processing plant near the end of the C&NW's Belle Fourche-Colony branch line. On

March 5th Gene Cermak, John Brugenhemke, Ray M. Roth and I trekked north to their headquarters to learn what exactly they had in mind and how we might help them. IMC was a big customer of the railroad which translated into our close attention when they wanted service. The new processing site being considered was near the end of the so-called Bentonite Spur right next to an existing plant belonging to Baroid (National Lead). IMC wanted to know what could be done about extending the Bentonite Spur further northwest quite close to and, perhaps, even into the extreme southeastern corner of Montana. All the raw bentonite was now being trucked from beds along the Belle Fourche River in northeastern Wyoming and the truck hauls were getting longer and longer as the nearer beds were depleted. John Foxen and I had driven into that very terrain in 1958. My memory of that eyeball survey suggested a line extension might be possible, but a more careful look was necessary. On March 15th I flew to Rapid City by way of Denver to meet R. C. (Dick) Christenson, the Black Hills Division engineer, and Bill Carland, our division freight agent at Rapid City. Using Dick's hi-rail station wagon we rolled north on the railroad from Rapid City to Belle Fourche and then another 19½ miles on the Bentonite Sub to Colony, Wyoming. Switching to highway U. S. 212 we went northwest to the tiny village of Alzada, Montana. It was obvious that there would be a lot of heavy earthwork to extend the railroad. The Bentonite Sub was surveyed in 1948, and I wondered if the survey crew had made any notes beyond Colony. The next day we drove from Spearfish, South Dakota, into the back country on the clay haul roads and sheep trails. With dry conditions and extreme cold there was little danger of getting stuck; spring would be a different story. After Bill Carland went back to Rapid City, we contacted a Mr. Arthur at the local IMC office to borrow some aerial photographs of the area. Dick and I returned to Alzada through a driving snowstorm. The fine particle dry snow blew like dust and drifted in the fence corners and leeside of the highway. Using the photos was not much help in this weather. We retreated to the IMC office and spent some time plotting the bentonite reserves on a territorial map that Dick had brought from his Chadron office. As we went south Dick decided to show me the famous Deadwood tunnel that had been carved out by convict labor about 1890. We hi-railed up the branch from Whitewood. Three miles up the line, we hit a snowdrift and derailed the station wagon. Digging out in that cold wind was a chore. Luckily, we were near a small grade crossing, which made the re-railing job easier or I would have frozen more than just my right ear. Returning to Chicago on March 18th by way of Pierre, Sioux Falls and Minneapolis, I wrote to the chief engineer requesting an instrument survey for extension of the Bentonite Sub, but Meyers came back on the 26th and said he didn't have any spare engineering forces available. With no substantive facts available about topography, land ownership or good revenue projections, I let the line extension subject lie dormant. The whole project went to sleep for a year.

IMC's Colony proposal came to life again March 11, 1966 when Ed Harney and Ray Roth from the Chicago sales office asked me to attend a meeting in Skokie with them. Another three months slipped away. On June 13, 1966 I was again summoned to Skokie to meet with IMC's engineering staff, Heidle and Wilbens, to try and fit their proposed track layout on their property at Colony. It didn't fit. Just like Iowa Beef last year, some creative track engineering was required. The problem was a corner of land common with National Lead. All that empty land in Wyoming and we were stuck trying to give legal clearance at a small piece of land about the size of a dining room table. By June 22nd there was general agreement that we had to avoid that corner at any cost as Baroid was not anxious to have a competitor next door. July 6th found me in the co-pilot's seat on IMC's Piper Aztec from O'Hare Field in Chicago to Belle Fourche, South Dakota. Menze, Wilbens, two other IMC engineers, and the pilot filled every seat in that small plane. We didn't fly very high and followed the iron trail of the Milwaukee Road in South Dakota with a refueling stop at Sioux Falls. The pilot allowed me to fly the plane awhile, but kept his hands firmly on the controls. I could feel his gentle corrections through the wheel. At last, I was "flying" a plane. Some 23 years after my enlistment in the United States Army Air Force in 1943, I at last had my hands on the controls of an aircraft. And, it was only five years ago that I had taken my first commercial flight. In Belle Fourche Bill Carland, our new division engineer from Chadron, Dallas B. Carlisle, and his instrumentman Peterson met the plane. At the plant site we staked out our careful track design and it fit.

We weren't out of the woods yet. Baroid (National Lead) now came up with the fact that they had never granted the C&NW a right-of-way easement across their property when the line was constructed in 1948. This put a severe crimp in IMC's plans. I immediately called Edward Warden, our attorney in Chicago, Bob Christie, assistant to Bob Stubbs, and the real estate department. On September 14th I was on a bright banana-yellow Braniff BAC-111 to Houston, Texas, for a meeting with Loucks, Ham and Greene at IMC's Drilling Mud division whose project at Colony was imperiled. Lou Verdin, our local sales agent, was very helpful in getting me around the area, which was totally new to me. After a short session with the IMC people who were extremely anxious to have this problem resolved, Lou and I went to see Baroid (National Lead) top management, Rowand and Eisely. To my surprise, they said they were planning to build a new plant and that the IMC construction wouldn't bother them at all. It was their local people at Colony that had taken such a stiff posture. Between planes at Dallas that day I called Menze in Skokie, but he was not quite satisfied. I called Ed Warden in Chicago and then went on to Kansas City as I had to meet with Howard Thompson in Blair, Nebraska, and make a short trip to Norfolk before going back to Chicago. Ed Warden's suggested resolution of the right-of-way problem did not satisfy either Menze or Heidle of IMC. Under Ed's guidance I wrote to Baroid (National Lead) requesting a formal release of rights. Then Gene Cermak, without reading his mail, jumped into the middle of the affair and thrashed around which created more delay to IMC's and my distress. On September 29th Menze called and asked me to come to Skokie quickly. After a fast trip north, Wilbens, Menze and I redesigned the south end of the track plan to conform with a site drawing Baroid had provided that morning. The next day I sent copies to all concerned and called Dallas Carlisle so that he could assemble and organize the necessary track material. By October 10th IMC was unhappy again—why wasn't the track under construction? Because Carlisle was having problems finding the material. Then the Public Service Commission of Wyoming stuck its official nose into the project on October 25th causing still more delay. Yes, the track was eventually constructed in 1967. By then, I was on a new career and had lost track of many of my old associates and the life I had led in the traffic department.

Picking up the thread of 1965 again, I flew to Sioux City March 22nd with Gene Cermak, Vern Giegold and Larry Krause from the DuPage County airport near West Chicago in good old N1394Z. After dropping Vern and Larry in Sioux City, Gene and I flew on to Omaha where we picked up a car to look at various places like Blair and the new Iowa Beef plant site at Dakota City. We talked to the South Sioux City Chamber of Commerce about the possibility of locating a hide processing plant proposed by John Laurie. If it meant employment, they were for it despite the poor reputation tanneries possessed. Meanwhile, the Iowa Beef plant at Fort Dodge, Iowa, was on strike—a portent of what was to come in the yet-to-be built plant at Dakota City. For some reason now forgotten, Gene Cermak and I were feted in Fremont, Nebraska, and treated to a city tour of 35 miles that seemed to last forever. There were calls to be made on all the important people in town. Nine of our C&NW sales people from Omaha were also present. On the 25th of March we attended J. M. Peters' retirement party in Omaha along with about thirty of Jack's men including off-line offices in Denver and Kansas City. Jack was the very model of an old-time, sensible railroad traffic officer and had all of my respect.

At the invitation of the Sterling-Rock Falls Co-Op I went to Sterling April 2nd. They wanted to relocate their old lumber yard and retail outlet from the CB&Q at Rock Falls to our new land south of Galt. On the 13th I rode train #1 to Sterling (just like old times) and met with the Co-Op's board that evening. They outlined their proposal; I laid out what the railroad could do for sale of the land and what track construction would cost. They retired to discuss the matter and voted to accept the deal. I caught train #2 eastbound the next morning. Claim agent Bill Cottrell was on board; we had a pleasant chat about division affairs before he got off at DeKalb.

The company plane with J. L. (Jack) Perrier (engineering), J. W. (Bill) Alsop (operating), R. P. McDonough (Galena Division superintendent) and J. R. (Bob) Kunkel (coal) and me flew to Peoria April 5th. Kunkel got off and the rest of us went on to St. Louis where Cermak met us with a huge Cadillac he had rented. Gene decided not to accompany us as I drove the group around East St. Louis, Madison and Granite City looking at railroad yards and port facilities. I put them on a plane in St. Louis during an intense downpour—so intense they were grounded for some hours. I left them at the airport and went on to attend the C&NW's hospitality hour at the annual American Industrial Development Council (AIDC) meeting. The next day Don MacBean and I took Cermak on a repeat of the previous day's tour. After meeting with two Pillsbury men about possible industrial projects in the St. Louis, Springfield and Peoria areas, we rode the Illinois Central's *Green Diamond* back to Chicago. I met Bob Wilson, superintendent of stations, on a commuter train April 7th. He was just back from a freight train wreck at Malta the previous night. He said the wreck was a jumble of pianos and baby food caused by a burned-off journal. High water in early April damaged many C&NW lines in Iowa and Minnesota. As the meltwater moved south it put the Galena Division main lines under water at East Clinton on the 26th. In the office Cermak and Ballin were locked in a fierce jurisdictional struggle over Standard Brands' desire to purchase some of the railroad's Proviso property. The purchase was finally accomplished and construction began on their new distribution warehouse north of Grand Avenue and east of the Illinois Tollway (June 3rd). American Cyanamid elected to build a new fertilizer distribution facility on the old Litchfield and Madison line at Staunton, Illinois, despite my warning that switching services would be meager as there was no wayfreight service south of Benld all the way to Madison Yard. Stopping a time freight to switch a car of ammonia in, or an empty out, was just not cost effective. (On May 5, 1966 I photographed the new installation and its rusty side track). On April 30th, we heard that W. R. Grace Chemicals had selected a site at Blair, Nebraska, for a new facility.

The new traffic department regime now gave us a real indication of what they thought about our department. On April 8th I heard that all traffic department personnel, except us, were to get a 7% increase. Cermak asked me if I would be willing to relocate. My reply was neither yes or no—I avoided a direct answer while reflecting on why I would want to live somewhere else. That relocation threat in the engineering department was one of my reasons for taking this job back in 1957. Our family had deep roots here, we had aging parents to consider and our home was mortgage-free. Cermak never pressed the issue, and I did not bring it up again.

John Foxen, Bob Lithgow's assistant in the Chicago area, was the best example I had ever seen of someone promoted to a job above his abilities. Watching him read a simple blueprint was almost like a Chaplin routine. He hardly knew which way to hold it much less be able to glean information from the document. In later life he was elected to serve on the West Chicago city council which may have been indicative of his political skills, but being an industrial development agent required more ability than poor Johnny was ever able to summon. On May 3rd he suddenly asked me for help with one of his prospects, Guardian Packing Company. I met their president, Floyd Carley, and executive vice president, Lloyd Rock, at O'Hare and took them on a site tour from Berkeley to Gurnee including the fast growing CENTEX Industrial Park at Elk Grove Village. We provided them with a lot of descriptive material although I felt zoning restrictions for their kind of business would be tough to overcome in these suburban areas. It was hard to find sites for meat processors close to populated areas. They eventually chose a site for their processing facility at Genoa, Illinois, on the Illinois Central. By June 23rd Guardian had given up that location because there were no municipal utilities available. I suggested that a site on the CGW at St. Charles might suit them. On June 24th I met with Mr. Root of Guardian and gave him the information sent to me by the CGW at Kansas City. I reported back to B. R. Harris, the CGW's industrial officer, as a matter of courtesy, but never heard anything from him. The meat processing business was then in such rapid transition that I believe most of the product from these smaller operations would have moved over the highway rather by rail.

The C&NW traded for a new airplane. On May 4th Ralph Gauen (rate department) and I flew in the new Kingaire N403NW to Peoria to talk rates and service with Harold Austin, president of the Pioneer Industrial Park, and a Mr. Laidlaw whose affiliation was not identified (or I have forgotten). The airplane's registration of 403 was as close to the desired 400 that could be obtained. The radio call for N403NW was "403 November Whiskey" which often brought smirks from more than a few of its passengers. A public relations firm hired by the C&NW to help in the UP-Rock Island proxy battle visited our offices on May 5th to see how we operated and to determine, I sensed, what level of sophistication we possessed. While their representative, Mr. Hammerstrom, seemed to appreciate our operation, we were not very impressed with them. On May 10th Bill Kluender was dismayed by the news that Jones and Laughlin was developing an iron ore mine at Tracy, Michigan, which could not be directly served by the C&NW. Strange how those ore deposits cropped up in places other than alongside our tracks. Also in early May I heard that Spencer Packing Company was interested in establishing a meat packing plant at Wayne, Nebraska. A few days later on the 14th we heard that their interest had shifted to Wakefield—an even stranger place to locate given that rail service there was almost non-existent. (The last train from Wakefield to Sioux City operated March 15, 1977). By the time I had worked up a map showing the operation of our eastbound train #256 which catered to movement of meat products, Spencer had selected Schuyler, Nebraska, on the Union Pacific mainline for its new plant. Bill Cook in Omaha called on May 11th for help with an AGRICO riverside plant site inquiry. I hastily flew to Omaha on an old DC-6 and went directly to the Corps of Engineers to get updated maps of the Blair area, but the office staff had all gone home by 4:30 p.m. The next day I met with a Mr. Martin, manager of engineering for AGRICO. Collecting a party of C&NW people which included Jim Heidkamp from real estate and Boomer Smith, our city agent, we went on a tour of the Blair properties along the Missouri River near the National Alfalfa plant I had located there in 1958. After finally collecting the maps from the Corps of Engineers, I joined Max Jauch (rates), Charley Cook (livestock) and Chuck McGehee (sales) in the new Kingaire N403NW for the flight to O'Hare. On May 13th AGRICO formally requested a 50 acre site at Blair.

GEOFAX Looking back with the advantage of years, I am now more aware of the unease that permeated my notes from the early 1960s. I was bored with the routine of the job, had little respect for many of the people managing the traffic department, and was vaguely aware that there was something important lying just out of my sight that involved computers and data organization. While I tried to avoid criticizing my colleagues on their handling of customers, it seemed to me that they were locked into old style methods of promoting industrial development without any recognition that the climate was changing.

Factory smokestacks belching decorative black smoke as depicted on the official letterhead of the City of West Chicago was a good example of an image that was no longer acceptable. Zoning of land for specific purposes began to appear in our territories in the late 1950s and now environmental concerns about the impact of industrial operations were beginning to be voiced. Industrial development had lost its luster in many communities. Even the once welcomed railroads running through the many small towns were becoming targets of dissatisfaction as the highway system provided greater mobility and access while the railroads acted as walls separating and isolating communities. Oblivious to the changing atmosphere, our traffic and industrial agents just waited for customers to walk in the door. This kind of passivity was not going to work in the future—we had to invent better reasons for prospects to seek our expertise. And, we industrial development agents needed to get into the plant planning cycle earlier. A new attitude oriented to customer service had to be developed. We needed smarter people armed with new tools.

An early attempt to develop one of those new tools was a big data collection project which I named GEOFAX. Essentially it was a geographic data base for computer applications. At the time I had no clear idea of how such a data base could be used, but I had a gut feeling that it must be a worthwhile because I had been unable to find anything like it in the big world outside of the railroad. Building this data base was no easy task; it involved a lot of grinding drudgery over a long period of time. Mark Twain stated the case quite well when he said, "The secret of getting ahead is to get started. The secret of getting started is breaking your complex, overwhelming tasks into small manageable tasks, and then starting on the first one." And that is the course I took. I have always been attracted to solving puzzles. GEOFAX was as convoluted and complex a puzzle as I had ever tackled. Mulling over the astonishing reception we had to our Jones and Laughlin study which used centroids of market and population, I was concerned that the concept was too crude. Using X-Y coordinates acting at the center of counties had a potential for skewing data because an oddly shaped county, for example, could have its center fall outside of its boundaries. What if actual city coordinates could be used? Logically and pratically much better, indeed, but development of thousands of coordinates would be a monumental task. Collecting geographic data for a specific study was such tedious work that there must be a way to retain the information for future re-use. I hated the idea of doing anything twice. What base lines would apply to any designated study area? On a trip to Omaha I spent the evening of May 11th in the library mulling over these unfocused ideas. I decided to have a chat with Dan Fliss, manager of computer operations at Ravenswood, about number systems—he didn't have any idea of what I was talking about. Collecting the pieces of my concept into a rough format, I sent it in to Gene Cermak and followed up on May 20th with a proposal to build a database and call it GEOFAX. He called me at home late that evening and said he liked the idea. After Don MacBean spent an entire day telling me it wouldn't work, he agreed to help me assemble it.

The concept was there, but the methodology had to be invented and implemented. I visited Rand McNally's map library in Skokie June 2nd and had a fruitful talk with the librarian, Luis Freidle, about how the Bureau of the Census handled large masses of data. He showed me Rand's collection of geographic material including the vault where he handed me a secret Soviet NKVD atlas of the USSR which used 23 different colors to portray Mother Russia. The railroads were depicted, of course, to a level of detail that showed which side of tracks the station buildings were on. It had been smuggled out in a diplomatic pouch. This was exciting stuff; much better then trying to find a site for a fertilizer distribution store in southern Illinois. Sudden inspiration came about adopting a standard base grid when I determined to use international standard latitude and longitude coordinates for named locations. Every place has to be somewhere, populated or not. Dun and Bradstreet in their eagerness to sell us more data now presented us with a computer generated printout of their United States city list. A two inch thick pile of paper, it not only included city names presented in alphabetic order by state but also included the county and the D&B numeric codes for state and city. Small places or former named locations, were referenced to larger post offices in the D&B printout. Don and I agreed that we had to try and identify all of these named places because we could not determine the potential value of any particular item. We tackled Illinois first. While a lot of the names were familiar to us, many were strange, in fact so strange that we could find no record of them anywhere.

A new science edged into our lives—toponomy or the science of how places are named. First, however, we had to figure out where to get latitude/longitude coordinate information. Surprisingly, the answer was found on the state highway departments' county maps. Every county in the United States was required to produce maps drawn to a set of standards specified by the Federal Highway Administration. The scale was usually a half inch to the mile. Most importantly to us were the latitude and longitude tics shown along the map borders. It was quick work to draw the horizontal and vertical coordinate lines on each map. *Voila*! We had a grid. I developed two scales on strips of cardboard to measure the exact coordinates to one hundredth of a degree—one scale for latitude and a set for variable longitude. I learned early in this process of the value of expressing degrees of latitude/longitude in decimal format rather than in the traditional degree, minute, second values. In those early days of computer science it was a lot easier for machine calculation for data calculation. Picking up a scale, measuring and recording information got to be a very tedious procedure. I fixed-up a set of transparencies with grid lines. The clear acetate could be rapidly slid over the face of the map making coordinate determination a lot faster.

As we gathered the coordinates Don and I recorded them in the Dun and Bradstreet book alongside the place names. All through June and July we plowed through the Illinois listing. The locations of some of the illusive ones we found by writing to postmasters of nearby communities. The answers received were often marked with neat red "x"s on the accompanying maps and contained bits of Illinois history. Many of the strangest ones were former names or ghost towns. It seems that Dun and Bradstreet Company, which had collected and rated credit for merchants since 1847, always added names to their city list but rarely, if ever, eliminated any. We determined the coordinates of every point in the Dun and Bradstreet book. On August 19th Maurie Fulton at Fantus pooh- poohed our project saying it was worthless. Why, he had a book in his library that had all that latitude/longitude stuff in it. Later I found a copy of that book; it was a book on astrology and the coordinates were pretty bad. Maurie was to regret his supercilious remark. Larry Provo heard about GEOFAX from Gene Cermak and issued instructions to our computer managers (who reported to him) to be cooperative when running our material. Don and I plugged away at our research for the rest of 1965. We amassed a great collection of state-county highway maps which proved invaluable for this project as well as for our industrial development work. By December we had completed the states served by the C&NW. I then started on Indiana because I was now sure that we had a winner. Eventually, the geographic database included some 44,000 records.

The work on GEOFAX was done on a time available basis. While the spring and summer of 1965 were fairly busy, there were no large projects requiring concentrated attention. Most of these inquiries never developed into serious projects but just kind of hung around like flies buzzing on a hot summer day. There was Mr. Brown of Dreyfuss Grain, who was mostly a source of information rather than a serious prospect; Webb of the Blair Cattle Company; Wil Edmonds of Granite City Steel who was always seeking a direct C&NW connection; and the American Colloid Company in Skokie who were sort-of interested in the Alzada, Montana, area. Packing plants kept bobbing up in response to Armour and Company's construction of abattoir's in the cattle raising areas. On May 24th I flew in 34 MIKE, a leased railroad company plane, to Des Moines with Jim Feddick, John Brugenhemke, two of our sales people, and Ray Janer of Penn Dixie Cement. We dropped them off and flew on to Fort Dodge to meet our local man Ed Jeffries for a trip to see Spencer Packing Company in Spencer, Iowa. We drove back to Fort Dodge by way of Algona, Iowa, in plenty of time for my flight to Sioux City. The plane stayed on the ground a long time because of the threat of tornadoes in the area. On May 25th Raskin Packing of Sioux City said they would build a plant at Wakefield, Nebraska, if the city would finance it. I drove to Wakefield to see just what this proposal was all about and met with the city fathers who said it wasn't likely. I went back to Sioux City for a flight to Omaha where I was met by Mr. and Mrs. H. F. "Boomer" Smith for the short drive north to Blair, Nebraska. I was to be the main speaker that evening to a crowd of 191 people. They seemed to enjoy my industrial development story which had a lot of Chrysler and J&L material thrown in. We then adjourned to the Legion Club for more talk and visiting before we could escape back to Omaha. The next day was a throw-back to the old west. Bill Cook, Boomer and I called on Webb and Thompson of the Blair Cattle Company. They were creating an immense feedlot operation in the steep hills west of Blair. Since the C&NW's Fremont-Missouri Valley mainline passed through the area, the operators were sure we would want to build a stockyards and siding there to haul their critters to market. I politely wished them well in their endeavor and said I would look at the economics. In response to their persistent heckling I went out there again in September. Heavy rains had made the feedlot a sea of sticky mud and manure. That soil was the infamous loess clay that plagued the Missouri Valley. One step in a wet puddle and your foot was encased in a ball of brown goo. Mixed with the cattle manure, that feedlot was a prime place to not visit. The rail service concept was a charade—those cows were going to market in trucks on a paved highway. The C&NW was unofficially killing-off its livestock traffic by letting the stockyards and cattle cars quietly deteriorate to the point of uselessness. Livestock's principal advocate on the C&NW, Charley Cook, retired May 27th.

May 28th, C&NW common stock hit 75; August 2nd it was 90⅝; on October 7th, 118.

Cermak and I called on Joseph Hepburn of National Can Company in Chicago on June 8th. The headquarters of the company was located on Cicero Avenue just across from Midway Airport. Don MacBean and I had done a study of tin can usage in our service area using some of the methodology of the Jones and Laughlin steel consumption report of last year. By July 16th we had developed a report for National Can. Four days later, Hepburn and I went on a tour of Elgin, Marengo, Belvidere, Rockford, Huntley and South Beloit looking for a potential can production plant site. After that tour, Don and I refined the report using our new coordinate system and drew up maps showing the centroid results. That report was finished August 31st. On September 13th we heard that National Can had selected a new plant site at Loves Park, Illinois, to be served off the C&NW's KD branch that wandered through east Rockford. Our report had pointed to the Rockford area as the theoretical center of distribution for a can producer. By November the plant was under construction. Score one for research using a new approach to industrial development.

In June there were rumors of C&NW acquiring the Velsicol Chemical Company and the Fort Dodge, Des Moines and Southern

Railroad. The Velsicol rumor was correct; an announcement was made on June 14th.

At Chuck Towle's invitation I went to Minneapolis June 16th to address a convention of Indian agents. Although it was probably a waste of time, I gave them my Chrysler slide show during which my voice gave out. The next day I called on ADM to croak about flour terminals and bentonite mining in Wyoming. I gave up on making any more calls and went on a leisurely tour of the Twin Cities area with Chuck and the company photographer, Bob Lindholm.

On June 23rd an ore train destined to Granite City went through one of those old class B timber spans on the Southern Illinois at Hubly. Don Gunvalson, our sales manager in Rockford, asked me to go with him on a call at Northwestern Steel and Wire Company at Sterling, July 1st. He said he needed a back-up because they were always beating him up about something or other. Given our long-running problems with the steel company, I was hesitant to go, but Don assured me that I would not be lynched. The call on Mr. Hasselman (an appropriate name) proved to be calm enough. Don and I then went over to Fulton where the recent floods had destroyed the old C&NW trackage serving a feed mill. Amazed at the degree of destruction caused by the high water, we walked over to the Milwaukee-C&NW interchange nearby that was still intact. We suggested to the mill operators that given the amount of business they gave the railroad and the heavy expense of restoring the washout, they would be well advised to contact the Milwaukee and see what they could do for them in handling their rail business over the piece of C&NW track still operable. On July 2nd I heard that Hunt Foods was interested in acquiring the Green Giant Company which operated that cannery on our line in Belvidere, Illinois.

The C&NW certainly was looking at its connections these days. On July 12th Stubbs called a meeting to consider the benefits of acquiring the Chicago and Illinois Midland Railway. Owner, Commonwealth Edison Company of Chicago, had allegedly offered it to the C&NW for one dollar. John Brugenhemke and I volunteered to help frame questions for the negotiators. I went to Peoria on the CRIP's *Rocket* and was surprised to find the diner had been replaced by a sandwich bar service. This handy little train was on its way to extinction. My riding on a pass didn't help their revenues either. On the

13th John and I along with Joe Zelenda from the Peoria office drove southwest along the C&IM to Havana and then southeast to Springfield. We inspected the rail to river coal dumper at Havana, the yards, track condition and local industries (not many). That afternoon we went on to Pawnee and Taylorsville. We telephoned our report to Chicago, collected county maps and returned home from Springfield on the Illinois Central's *Green Diamond.* The C&IM was one of the most over-maintained railroads I had ever had the pleasure to inspect. Heavy rail, good tie and track condition, neat structures and good motive power. The railroad, however, just didn't add anything to the C&NW at that time because its principal traffic was coal from Taylorsville going north to Havana from whence it was barged to Commonwealth Edison's riverside Fisk electric generating station on the southside of Chicago. C&NW ownership would not change that operation as the power plant was not accessible by rail. Any potential for industrial development on the C&IM in this rural area appeared to be remote and in the congested area of Springfield, unlikely. Nevertheless, we continued to collect C&IM data. For some reason I was thought to be our resident expert on the C&IM, a title which really meant very little.

That forlorn M&StL line from Peoria to Keithsburg was still in operation. Maurie Fulton of Fantus had a prospect interested in a rail-river site and asked about Keithsburg. I steered him away from it. Keithsburg's future as a rail point was very dim at the moment. After the traffic dried up, the C&NW applied on June 12, 1970 for abandonment of the Keithsburg, Illinois, to Oskaloosa, Iowa, 95 miles including the Mississippi River bridge. Approval was granted August 4, 1971 and was effective on September 8 1971. The final day of operation was November 9, 1971. The rest of the line from Monmouth to Abingdon, Illinois, saw its last service November 17, 1975. The last piece of the former M&StL beyond Middle Grove died July 30, 1976. Mid State Coal's strip mining at Rapatee ceased in 1996.

On July 23, 1965 Gene F. Cermak was promoted to "assistant vice president sales and service on line and industrial development" which must be the record for sheer length. The July 26th letter announcing the new position was hastily retrieved within the hour as Ben Heineman had forgotten to get board approval first. On September 9th Robert C. Stubbs was made vice president-traffic. He had given Cermak permission July 30th to establish an industrial development office in Des Moines. I wondered if this new office was what Cermak had in mind when he asked me if I would move. I was glad to have Vern Giegold take that job and get out of my immediate vicinity. In October I. Robert Ballin finally was made vice president-real estate.

August was a slow month despite brief flurries involving AGRICO, Sinclair and R. R. Donnelly who were all reported to be nosing around the territory. Gene Cermak and Harvey Buchholz were quite busy with a paper making group in the Appleton, Wisconsin, area. Gene continued his past summer schedules that revolved around golf and poker making him absent from the office for long periods. His in-basket stacked up as he did not bother to have anyone summarize the material and give him, at least, a telephone report. Places like off-line Dubuque were great for golfing. His secretaries, Naomi Scharenberg and June Miller, were underutilized and chafed at the boss's work habits. Gene, however, claimed golfing was an important business activity as well as being fun. On August 19th he hosted a golfing outing for Maurie Fulton of Fantus and Larry Provo, then our vice president-accounting, and drafted a willing Don Guenther to tag along as his "little shadow". On August 9th Stubbs asked me for a background report on the Baltimore and Ohio branch to Beardstown and some history of the B&O-IC trackage in the City of Springfield. The next day I met with the industrial commissioners of Geneva, Illinois, and agreed to make a presentation to the city council on the 24th. On August 12, 1965 the C&NW took over the switching rotation at the Cal-Pak warehouse in Rochelle, Illinois. The CB&Q, the Milwaukee and the C&NW took six month turns switching this facility. After a visit to Rochelle, I drove to Dixon and Galt, to check on developments and happened to see the new trailer depot at Earlville before heading home. Mid-South Chemical expressed interest in a fertilizer outlet at Benld, and American Cyanamid wanted a site at South Pekin. On August 23rd I drove to Peoria with Herb Steutz, a new Galena Division instrumentman fresh from Austria, to see the Cyanamid people and then flew back to Chicago from Peoria. Two days late Bob Kionka from the Galena went with me to Belvidere to look into the track changes desired by Green Giant, check on the possible restoration of the Caledonia-Belvidere line and make a quick run to Winnebago to see work in progress for California Chemical's new facility. This was just like my old days as assistant engineer on the division. We heard rumors August 30th that the C&NW was negotiating trackage rights on the Alton and Southern in East St. Louis. R. P. (Mac) McDonough, the Galena Division superintendent and I flew to St. Louis on September 2nd with J. R. (Bob) Kunkel of the coal department joining us. Jim Thompson, our St. Louis guy, took us over to meet with Wil Edmonds of Granite City Steel and the Alton and Southern people. It was apparent from the conversation that the C&NW and A&S intended to set up a direct service arrangement where C&NW crews would run a train (or trains) into A&S's modern hump yard for switching and forwarding. The Alton and Southern was to make up a train (or trains) for the C&NW which our crews could take directly out of the A&S departure yard. That traffic diversion would take the pressure off that poor excuse of a yard at Madison that we had inherited along with the Litchfield and Madison Railway. After lunch with the vice president-sales of Peabody Coal Company, I flew back to Chicago on a Delta DC-7. In September rumors of a C&NW-Norfolk and Western merger floated around; someone really had to stretch his imagination to fabricate that scenario.

Bill Kluender, our director of resource development, suddenly appeared in our office September 1st and, humbly, for him, begged the loan of any Wyoming USGS topographical maps in the Lander area I might have. I had developed a very complete set over the past seven years. Two days later at a meeting including B. R. Meyers, chief engineer, Bob Stubbs, vice president traffic, Gene Cermak and me, an expedition to Wyoming was organized to meet a party from Pittsburgh Plate Glass (PPG) who were interested in a project south and east of Lander. Kluender dashed off to Pittsburgh to see PPG even though the project was, in our mind, properly the responsibility of the industrial development department. Meyers and his assistant, Jack Perrier, pored over my topo maps that Kluender had given to them—I had to reclaim them later. (This is when I learned that they had prowled the area in 1956). The project site was south of Riverton and, from the maps, would require about 11 miles of new railroad to reach it. On September 8th Kluender and I flew to Riverton by way of Denver. Art Knight (traffic) from Casper, R. C. Christenson (engineering), from Chadron and Larry H. Jacobson, Kluender's forestry/minerals man from Rapid City met us. The next day Jack Perrier, Chris, Bill Kluender, Larry Jacobson and I packed ourselves snugly into Chris' hi-rail station wagon and rolled west toward Lander on the rail. The proposed branch extension up the Lyons Valley west and south of Hudson began about milepost 741 (miles west of Fremont, Nebraska). We dropped Perrier at the motel to examine the maps while we drove south to Atlantic City on the highway before returning to meet with Sid Forbes, chief planning engineer of PPG. He announced that the proposed plant site had been moved to a place called Onion Flats on the map. Very early the next morning we returned to milepost 741 to eyeball the terrain to Onion Flats. Railroad construction was not as simple to Onion Flats as the earlier proposal. Almost a month slipped by before the next chapter unfolded. I was in Minneapolis when word came to go to Pittsburgh. On October 6th I flew to Chicago on United's Caravelle *Ville de Lille* where I connected to a 720 for Pittsburgh. Kluender, Ellis W. Ernst, our Pittsburgh traffic representative, and I met with PPG to discuss the proposed new mine (I seem to remember that it was phosphate ore) and rail service. The big problem was lack of electric power in that area. Since Pacific Power and Light (PPL) was the main utility in central Wyoming and based in Portland, Oregon, I gave Reed Hoover in Portland a call to determine if he could obtain any cost data from PPL. Larry Provo received our joint report on the project October 12th and assigned Ron Schardt, his top financial man, to develop a cost analysis. Three days later the project was dead—PPG had allowed their land options to expire. Wyoming continued to be the land where something was going to happen but never did. We had to revise that statement in seven years.

At the same time the PPG Wyoming project came to our notice, another potentially huge revenue producer in Illinois emerged. A large rail user was poking around the Fox River Valley from Elgin southwards to Geneva. Phil Schmidt, the recently retired chief of industrial development for the Rock Island, told us that J. J. Harrington, a prominent Chicago industrial realtor, had been retained to find a big site for an unnamed client. Phil also said that I was high on the list as a possible candidate for his old job on the Rock Island, but nothing came of that bit of gratuitous information. Despite our recent series of Rock Island-C&NW spats, Phil was a likeable fellow and friendly to us. On September 15th he called again and reported that Harrington's client was very interested in the farmland southwest of Geneva, Illinois. I swept together some general information about the area and marched over to John Harrington's office that same day. I never quite understood how Phil Schmidt fit into this puzzle except he may have been used as an intermediary by Harrington to get our attention. Phil now dropped out of the play as I was now in direct contact with Harrington. (Phil died suddenly September 30, 1969). Harrington, the sly old geezer, dropped hints about his mysterious client (that we already suspected to be R. R. Donnelly) including a clanger that it was Volkswagen. Time eased along to late October when Harrington called to say that we were number 2 in a list of 29 sites under consideration. Number 1 was a property on the CB&Q at Naperville, Illinois. Don MacBean and I launched a crash program October 29th to collect detailed community data for Geneva and Kane County, learning in the process that rezoning lands in Kane County from agricultural use to industrial would be a tough proposition. I presented our collected data to Harrington on November 1st.

Harrington kept referring to me as "Mr. Avis" because we were number two. National advertising for the Avis Company used that theme in its competition with the giant Hertz Corporation which had the largest car rental business. Harrington now wanted a letter from our president Clyde J. Fitzpatrick describing our railroad service at Geneva. Gene Cermak quickly said nothing doing as such a letter could be construed as a guarantee for future service. I agreed with that position and pointed out to Harrington that this was the C&NW's mainline railroad not some rickety branch line. On November 10th Harrington was back again with another request—truck driving times from Geneva to Chicago. I knew I was being used as a go-fer, but in industrial development work, prompt and accurate responses were marks of professionalism. Up to now the precise amount of property of interest had not yet been specified.

On a hunch I called Tom Peck, owner of 160 acres of Geneva property southwest of the city on November 3rd and inquired if the land could be acquired. To my utter surprise he was all for selling it and quickly too. He was tired of farming. Ten days later I contacted other adjoining owners and wrote a detailed report. On November 18th Lindholm, Mayor of Geneva and a Mr. Brannon of the Planning Commission came to our offices to discuss future development of this property which was right next to their city. They were concerned about uncontrolled growth and wanted to be sure that the city would have a voice in the future use of the property. The next day Gene Cermak came up with his concern that John Harrington didn't have all his marbles and that this whole thing was a hoax put into his head by Charley Willson of Elgin. Charley and John Harrington had known each other when Charley was with the Chicago Association of Commerce and Industry and Charley was not sure John was firing on all eight cylinders. On December 6th Howard Koop of real estate came up with the Volkswagen story. It was nice to be able to let smarty Howie know we had heard that phony tale before. That day when I called Harrington; he wanted me to obtain "preliminary options" on 616 acres of land. Just what a preliminary option was I have never learned. On the 10th he came into our office saying that he had an appointment with Ben W. Heineman to disclose the identity of the mysterious client interested in the Geneva property. I drafted a letter for Harrington to sign authorizing the C&NW to take options on the Geneva property and checked it through Fred Steadry and Ed Warden in our law department. I then alerted Carl Anderson in real estate that we had another big land acquisition project and could he be of help? It was literally right in his own Geneva backyard and he already knew some of the owners. Harrington was all excited the next day as he had been instructed to arrange a meeting with the land owners. We still did not have the option forms and John had not yet signed the agreement that I had drafted. The signed letter and option forms arrived on the 16th and the owners all proved to be avid to sell. On the 17th I presented the options to the three sets of owners. At that point Harold Smith, our local attorney, bowed out as he had a conflict of interest. I found Joe Radivich, another local attorney, who took on the matter. Harrington okayed a $2,000 per acre price. On December 22nd Tom Peck signed for his 160 acres; two sisters (whose names I have forgotten) signed for sale of 370 acres and an old bachelor, Mr. Walsh, okayed his option for 86 acres subject to the Catholic Church's Joliet bishop's approval which he eventually gave. A slight modification in the option on Walsh gave him a life lease on the property in the event of sale. This wrapped up all the vacant land from Western Avenue in Geneva, west to Randall Road, all of it lying south of the C&NW's Galena Division mainline. On December 23rd I consulted with our counsel Fred Steadry and got Mr. Heineman's approval for a letter to Harrington detailing the land transactions.

We heard that the CB&Q was going crazy over at Naperville. Just after Christmas, the CB&Q abruptly extended the Chicago Switching District out along their mainline to Naperville. This would radically change freight rate bases by extending the commercial zone. Ben Heineman on vacation in Aspen, Colorado, judged that the CB&Q's move had serious implications and called Cermak to discuss the matter. Acquisition of that much property so close to the Chicago area was bound to attract attention as much as we tried to control information. On January 4, 1966 Carl Anderson and I met with our local attorneys, Joe Radivich and Dick Cooper, about developing a strategy to start the rezoning process. Ben Heineman took a hand in this game, too; he had a friend who was a Kane County judge in Aurora and with whom he consulted as to how this process could be handled. The next day Harrington called and alerted me that a soil testing program would soon be needed. A local realtor named Herron, now tried to horn in on the option process; I turned him over to our law department. On the 6th Carl Anderson and I took the signed options to a Chicago attorney, Harry Baumann of Sidley Austin, who represented the unnamed client (that we already knew was Donnelly). Afterwards, we called on John Harrington who assured us that the buyer would be making good use of the property and would be beneficial to the city of Geneva. On January 7th Cermak suddenly and coldly ordered me to stay away from Harrington. A few days later I heard that Joe Radivich in Geneva was complaining about some perceived slight on our part. These Croats must have thin skins because I could not think of any instance where he might have been insulted. I called him and had a friendly chat. I also arranged with my friend Dr. DuBose of Testing Services in Wheaton, Illinois, to start a test drilling program on the property. Despite Cermak's peremptory order I could not very well avoid John Harrington who invited Carl and me

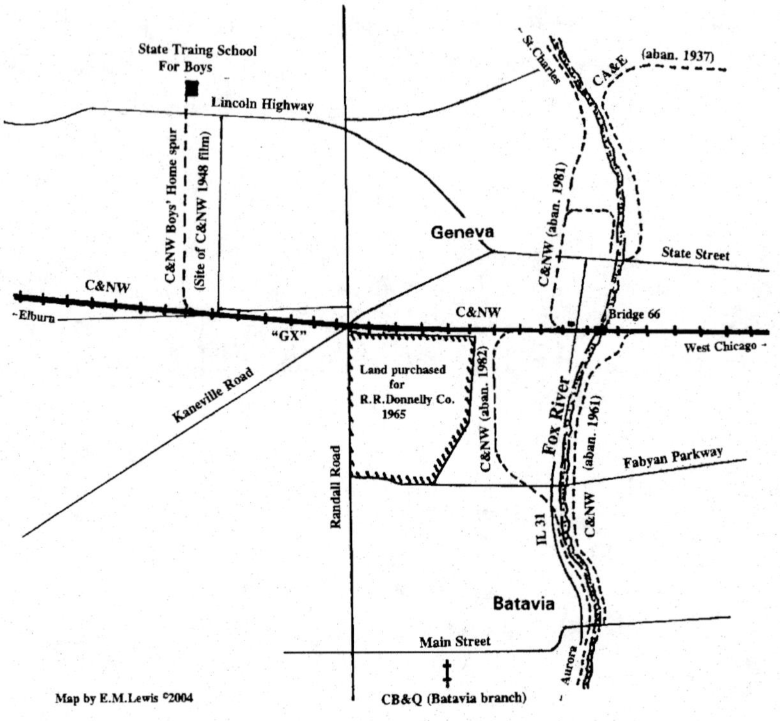

The area in northern Illinois near the big bend of the Illinois River where Jones and Laughlin built a new steel facility in 1965 using guidance from a C&NW market study. There was no chance for C&NW rail service for this site because of the old and tangled railroad ownerships in this region.

to lunch at the Union League Club. Our competition was at the next table which made a strange tableau, all in all. Gene Cermak now decided to get involved and went to Geneva with me on the 17th to see Radivich and then Dr. DuBose in Wheaton. By the 19th word began to seep out about our options at Geneva. A man from Naperville called but didn't learn very much from us. On the 20th I hired a local land surveyor, Donohoe, from Geneva to begin the field work and map preparation. Soil testing started January 26th. In extremely cold weather on January 28th Cermak and I met with Mayor Lindholm of Geneva and five other city officials to discuss the future of this 616 acre site. The meeting was cordial and cooperation of the city was offered. Our work with Chrysler at Belvidere in 1963 was still fresh in their minds. On February 14th John Harrington now wanted presentations from Commonwealth Edison and Northern Illinois Gas Company. The utilities quickly responded. Edison, especially, because the City of Geneva had their own power distribution grid and Edison wanted to keep this new customer off the city's system. The next day while Gene Cermak and I were flying to St. Louis, he confirmed that the mysterious client was R. R. Donnelly's Lake Side Press—my guess had been correct. Donnelly's main business was printing huge volume runs of things like magazines and the Yellow Pages. We had a small crisis in March when we almost let the second option period expire. Checks were sent quickly written and sent out by registered mail. On April 2nd Cermak said that Chuck Towle, our man in Minneapolis, had full details on the Donnelly project from some private source. Harry Baumann, Donnelly's attorney, repaid all our option money and expenses on April 12th.

I set up a luncheon meeting with Harry Baumann, Mayor Lindholm, Gene Cermak and me. This was the first face to face contact of Donelley's representative with city officials. On May 11th Gene and I attended a dinner meeting of the St. Charles Chamber of Commerce and, later, a meeting of the industrial commission in Geneva. Mayor Lindholm reported June 14th that Donnelly was to disclose their plans in two weeks. The next day Harry Baumann returned from a European vacation and asked us to extend the options. I spent another $22,000 which he repaid to the C&NW June 20th. Two days later he called and said that Donnelly had selected Geneva and was contacting Mayor Lindholm with the news.

At this point there was no further information about anything associated with this project until September 29th when Donnelly asked the C&NW to quit claim any interests it might have in the property. Ed Warden saw no problem with that. The next day attorney Radivich called to complain that he was being by-passed in the rezoning process. I called Baumann suggesting that he might do well to give Joe a call. The rezoning hearing was held October 19, 1966 and passed unanimously with no objections voiced. We weren't finished yet with that local realtor, Herron, who was claiming a 5% commission of Tom Peck's 160 acres. I got a copy of Tom's letter dated in 1962 to Mr. Herron in which he had terminated Herron's exclusive right to sell the property for Peck. That letter put Herron in a bad position, and he scuttled out of sight. The final official act was rezoning of the Peck farm by the Kane County Board of Supervisors on November 15th. Now all the options could be exercised. I wrote a history of this project for Stubbs and Heineman on November 17th and was commended by my boss, Gene Cermak for a professional job. On December 12, 1966 Donnelly acquired the 616 acres which included a famous octagonal sheep barn. Later, we learned that Donnelly as a corporate strategy acquired such tracts of land for future development as a sort of land bank program. The next facility they were to build, however, was in Mattoon, Illinois. The Geneva property was never used by Donnelly as an industrial site. It became the Eagle Brook Country Club with a championship 18 hole golf course and 92 acres of protected wetlands.

Turning back to October, 1965, our department went to Minneapolis on the CB&Q's *Twin Cities Zephyr* to attend the annual industrial convention. After entertaining a large crowd in our suite at the Radison that evening, we joined a tour the next day to look at industrial lands near Shakopee on the C&NW and at Pine Bend on the Chicago Great Western. Gene had to catch the 6 a.m. flight to Chicago to meet with Heineman in connection with the Iowa Beef problem described earlier, but the meeting was postponed and he flew back for the rest of the convention. We learned from Ed Upland, our grain man, about the new concept of building huge new grain elevators on designated branch lines to replace the many old and inefficient country elevators. This program would require co-operation of the growers, grain companies and the railroad to make the system work. On October 7th I made a presentation to the Illinois Association of Real Estate Boards at the Sherman Hotel in Chicago. The next day I went to Rockford to see a new prospect who planned to establish a plant to lithograph tin-plate. Don Gunvalson, our Rockford sales agent, now had an assistant, Harold Gardella, to help cover his ever more active territory. The battle for the Rock Island was not yet over. On October 12th I wrote a report on the Chrysler plant for Glore Forgan who had been retained to develop C&NW propaganda for the campaign. Later, in December, we heard that we were to go on a lecture tour in January to tout C&NW's successes. Luckily, that idea did not see action.

At Cermak's urging, Don MacBean tackled the problem of reallocating sales territories. We discussed the matter at some length as we were about to seriously upset the comfortable old territories by aligning the new assignments along county boundaries. In the future it would be possible to do comparisons of C&NW sales and traffic statistics with other data collected by county units. The Battelle Institute of Columbus, Ohio, sent a Mr. Biedermann to our office October 13th to see the clever system we were creating. On November 29th all sales department heads and industrial development people met at the O'Hare Inn in Des Plaines, Illinois. It was Gene Cermak's big staff meeting where he revealed his plans for conducting on-line sales. With Don McBean's help I had created a large map detailing the new sales areas which rigidly conformed to county boundaries. The tense air crackled with good ideas and brave new plans. Greater use of information from the car movement system (CARFAX) was stressed. Ray MacCarthy, our old line traffic manager from Green Bay, could not stomach all this radical restructuring and boycotted the meeting. The next day was the rate department's turn. Herman (Jake) Jacobsen and Irv Lawrenz discussed rate making and a new methodology using a structured approach for analysis of commodities. The old way of just counting the number of loads without caring about contents was about to pass away. Walter (Bud) Braun had made the classic remark sometime earlier when he said with an absolutely straight face, "We lose a little bit on every car handled, but make it up on the volume." Some of the new concepts sailed right over the heads of many in that audience, but it was clear to all that the old online sales empire of Iver S. Olsen and his predecessors was to be ripped to shreds. One of the big changes was elimination of much of the onerous and generally meaningless reporting which had been branded as a major curse to good salesmanship. Better and more timely information was available from the CARFAX reporting. Other data based on county boundaries (labor, government payments, health, demographics, transportation, for example) could now be applied for comparative purposes. Some of our old hacks were visibly shaken by the new program. As part of this realignment the Iowa industrial development office finally became a reality. On October

20th we shipped Vern Giegold's and Larry Krause's furniture and files to them. That was not the only change as a power struggle was underway between Cermak and his on-line sales assignment and two of the off-line powers: C. R. (Clem) Bair in San Francisco and S. B. (Stan) Boardman in New York. Gene's aggressive style of sales was encroaching on Bair and Boardman. They felt threatened. Gene relocated from our department across the floor into R. C. Stubbs' old office. The move was accompanied by an announcement that a new director of industrial development would be sought. The sales side of Gene's dual responsibilities made access to him even more difficult leaving us to make daily business decisions on our own where we felt we could or else impose onerous delays on our projects, which in our line of work could be fatal to their success. Cermak's play days were over—now he had to run hard just to keep up with his new responsibilities.

Industrial prospects and plant sites continued to appear—with, or without, our encouragement. The Omaha Public Power District (OPPD) announced on October 15th the planned construction of a new electric generation station at Blair, Nebraska. That area had proven to be very attractive since National Alfalfa put up their plant there in 1959. The lure of the Missouri River as a navigable stream continued to an illusion over the years; yet, industry still held to the belief that availability of water transport would keep rail rates down. I flew to Omaha on the 19th in an old DC-6 to see Howard Thompson of the Blair chamber of commerce and Howard C. (Howdie) Hanson Jr., vice president of the Blair Bank. We tried to look into the future by planning an industrial area for the farm fields that stretched east of town to the river. That was a quick one day trip that yielded large dividends over the next 30 years as industry after industry located major production facilities there. On October 20th we heard that Carnation Company was looking for a Rochelle, Illinois, location. We were a bit late on this one, and they ended up on the Milwaukee Road in an area open to reciprocal switching. Ellis W. Ernst, our Pittsburgh off-line sales manager, asked for a copy of the Jones and Laughlin study to show to Wheeling Steel. The next day McGuffin Lumber Company came up with a project for a planing mill at Benld, Illinois. This request died in infancy, however. There seemed to be a curse on Benld as no new industry located there dur-

The author as General Industrial Development Agent in Chicago in 1965.

ing my career in this business. Perhaps their reputation as a wild town in coal mining days was hard to shake. Leath Furniture Company asked for help in finding a site for a new furniture warehouse on the extra land we had purchased at Belvidere. With our sales representative Jim Tobin I called on their Mr. Hartman in Rogers Park on the north side of Chicago and soon determined that there was little potential for rail business with this firm. I was glad they eventually dropped the matter as furniture was one of our worst kinds of freight for railroads because of the high damage claims associated with its movement. Another company that showed interest in 40 acres of this Belvidere property was Goss Printing Company of Rockford, Illinois. On my trip to see them November 9th I saw that Green Giant on the east side of Belvidere had finally started on their planned plant expansion. On Veterans Day I rode the dingy old Rock Island *Rocket* to

Peoria to attend the annual Peoria Traffic Club banquet with Fred Jones, John Brugenhemke and some of our other sales people. Maurie Fulton of Fantus Company came up with another prospect for either the Peoria or Clinton, Iowa, areas—Marbon Chemical. In December there was another trip to Omaha for another Blair and Tekamah visit. On the 8th I met with Bill Cook about Midwestern Beef's track problem at Norfolk, Nebraska, attended a luncheon meeting of the Nebraska Division of Resources and then flew back to Chicago with Ed Wozniak, the Rock Island's new industrial agent.

Personnel changes continued to occur. William H. (Heinie) Huffman was appointed prime engineer for the Rock Island-Milwaukee-C&NW merger studies then underway. Our Stan Johnston was reassigned from the Minneapolis office back to Chicago. The industrial development department was cast adrift, our momentum was gone. The news on November 18th came that our president, Clyde J. Fitzpatrick, would be away for five months for unexplained reasons. This was a surprise to many and a relief to others. By early December the grapevine was saying that he would not be returning.

The last event of the year was on December 29th when the Interstate Commerce Commission approved the merger of the New York Central Railroad and the Pennsylvania Railroad. Almost unnoticed, the ICC also approved the C&NW's bid for the Rock Island.

20

THE FALL OF THE HOUSE OF TRAFFIC, 1966 . . .

Carnation Company had an aggressive traffic manager named Al Davis. He knew how to make the railroads work for him as he controlled a lot of lucrative traffic. On this January 5, 1966 Al was in Bud Braun's office with a bag of new ideas about his new distribution warehouse at Rochelle. Located on the Milwaukee Road, the facility was open to reciprocal switching, but Al wanted direct C&NW access. I took his ideas over to the Galena Division engineer's office in the Chicago Passenger Terminal where Bill Wimmer and I worked out some alternatives. Unfortunately, all of the plans required at least one mile of new track construction over privately owned properties. Davis insisted that we pursue the idea. Three days later I contacted the farmer whose land would have to be crossed; he was adamantly against any railroad track on his land. On the 12th I called

Al Davis in California and reported my failure. I sensed he was already off on a new tack and really didn't care about the Rochelle track plan. North of Rochelle the extra property we had at Belvidere was a continuing attraction, but its proximity to the Chrysler plant was also a curse. Word on the street held that the labor pool in the Rockford-Belvidere area wasn't large enough for another big employer. The State of Illinois held its annual economic development program in Springfield in January; I went down on the GM&O's day train. Governor Otto J. Kerner addressed the 75 attendees at an otherwise unmemorable meeting.

Relations with my boss, Gene F. Cermak, now took another downward turn in an incident orchestrated strictly by chance. R. P. (Mac) McDonough, the Galena Divison superintendent, M. C. (Chris) Christensen, the division engineer and I arrived at O'Hare Field January 18th for a flight in the company's turbo prop to St. Louis. Cermak and his sales gang arrived at the same time and found that the company's jet was in Rapid City, South Dakota, and that a Beechcraft had been leased for their flight. Cermak turned his nose up at the substituted aircraft and, pulling rank, took our plane. The Beechcraft took a long time to get to St. Louis. Bill Newgent and Jim Drinkard met us at Lindberg Field to take us to East St. Louis for the meeting with the Alton and Southern. The A&S people included E. B. Hardwig, superintendent and a Mr. Hartenberger who informed us that the St. Louis Southwestern (SSW) had made an offer for their property and that the direct service with the C&NW was on hold. The Cotton Belt's bid was not welcome news to Wil Edmonds who said that Granite City Steel would oppose their action. The episode of the "hijacked" airplane was reported to Heineman whose office controlled the air fleet. Cermak received a severe reprimand for commandeering our airplane. He then blamed me for whining about the plane swap not realizing that the chairman's office had been informed directly by the pilots. Heineman also dumped the Alton and Southern matter into Cermak's lap on February 7th so that he had to depend on me to help him sort out the problem as I had been involved with that area for five and a half years. Gene and I flew to St. Louis on the 15th. After lunch with Smith Reed of the Missouri Pacific Railroad, Gene and I went to see Commander John Barron and a Mr. Baebler of Union Electric Company. The next day I met with Jim Reed of the

Southwest Port development before heading for Minneapolis and on to South Dakota. Not long after these events I heard that Heineman was interested in acquiring the A&S; Stubbs asked me on May 23rd to fill him in on the details.

January also saw the retirement of Hack Roth, venerable leader of the Northern Illinois Gas Company's industrial department. His big farewell party was held January 20th at the Elmhurst Country Club—about 175 showed up. A few days later Sam Lezak and his eternal land promotion schemes turned up again. Sam's current project was a revival of last year's bright idea for a new city near the old CGW-MILW crossing in DeKalb County. The site was west of Sycamore on Illinois highway 64. Forty years later it was still farmland. Sam's other "hot" area was still the dream he had for the village of Huntley, Illinois. Forty years later there were substantial residential settlements northeast and southwest of that small place. Whether Sam Lezak had a hand in their development is unknown. Other projects dribbled in, were examined for viability and mostly ended up in cold storage being deemed unlikely or inconsequential. Ballin picked my brains on the Milwaukee Road's industrial department and contributed that information to the merger committee (February 2nd). Actually, that should have been Cermak's responsibility. The next day I drove south to Peoria to give a luncheon talk to a small group, visited Green Valley (just a railroad station sign at the site of a former passing track) and South Pekin to check on sites before going home. February 4th, Ben Heineman was appointed to the Civil Rights Commission by President Lyndon B. Johnson. He was also named Chicagoan of the Year.

The South Dakota IDEA banquet was held in Pierre on the 17th of February. The temperature hovering at −5°F seemed to be concentrated right in my bedroom. Despite borrowing blankets from the second bed, I was still cold. We convened the next day in the chambers of the South Dakota House of Representatives which was not in session at the moment. Ron Reifler of Fantus Company gave another of his fine talks; I showed slides of the Chrysler development. We didn't hang around to socialize—it was too cold. J. R. (Bob) Kunkel, our coal traffic man, and I called on Commonwealth Edison about their proposed new electric generating station at Cordova, Illinois. The site was on the Illinois side of the Mississippi River and served by that Milwaukee Road branch line that ran south from East Clinton. It was to be a nuclear plant, however, so that rail revenues would probably be limited to inbound construction materials and machinery. When Bob Stubbs received our report, he insisted that we go back to Edison and push a site at Keithsburg on the moribund M&StL. Edison gave us a flat "no" on February 10th. That was a bad day for railroad operations: there were two rear end collisions on the C&NW, one at the truss bridge over the Milwaukee Road's Bensenville Yard and the other at Franklin Grove on the Galena Division.

Stubbs sent me out on an emergency job to Aurora, Illinois, to perform a Saturday survey. I drafted my wife to hold the measuring tape—it cost me a lunch. Montgomery Ward and Company plans for a new retail store on the west bank of the Fox River had not accounted for presence of the C&NW-CB&Q interchange track which ran through the building site and an adjacent street. Monday morning I presented my findings and a suggested track rearrangement to a group meeting in the chief engineer's office. The proposed relocation was approved when Stubbs said the Burlington would pay the costs. The Galena Division was completely by-passed in this matter by these corporate officers who normally were so conscious of protocol. I took the plan to Wards on February 18th and met with a Mr. Wilson and a Mr. Kiernan. They were very pleased with the quick response and indicated that the C&NW would be considered favorably in future routings of Ward's rail shipments. The same day a rumor was running about the office involving purchase of a large block of AT&SF stock by Mississippi River Fuel, a subsidiary of the Missouri Pacific Railroad. Armour announced a big expansion of their new Sterling/Galt operation February 11th. I reviewed their plans with them on the 14th. Our department also heard that the Sinclair site for a major chemical plant at Muscatine, Iowa, had not proven to be usable because of soil conditions. We immediately launched a campaign to find an alternative site—on the C&NW, of course. Since this problem was in Vern's Iowa territory and I had no suitable sites available in Illinois, I just let him struggle along in his own inimitable way.

Despite the growing turmoil in our organization, plant location work continued apace. Early in February we learned that Gulf Oil's interest in the Blair, Nebraska, site had rekindled. I flew to Omaha on United's Caravelle *Ville de Coutances* for a meeting on site with them. This was a big prospect that attracted a lot of attention so Cermak invited himself to the party. Gulf identified some problems with the property some of which we could immediately address. They needed an easement from National Alfalfa to obtain access to land north of that installation. On March 7th I flew to Kansas City on an old Electra. Because this was my first visit, affable George White, our local sales representative, gave me a tour of the city. This visit proved to be just the first of many in the years to come. I went out to Shawnee Mission, Kansas, where National Alfalfa's operating headquarters were located, to try and negotiate a lease revision at their Blair operation to afford Gulf access to their desired site. To my relief, they were reasonable and said they would refer the matter to national headquarters at Sinking Spring, Pennsylvania, for corporate approval. I ran into George Falconer and George Dean of Gulf at the Kansas City airport; we flew back to Chicago together on a tired old TWA Constellation. On April 26th Gulf asked me to set up a meeting with Grace Chemicals to explore the idea of setting up a joint river dock at Blair. The Big Muddy continued to be a lure for barge service, but the river was too wild and unpredictable to ever be of much use for transportation purposes. On May 20th National Alfalfa released the land for Gulf and by June 2nd the Gulf project was ready to go. It took another trip to Kansas City on June 23rd with our real estate's Jim Heidkamp to adjust the Alfalfa lease and clear the way for Gulf Oil. They were driving piling on their site October 24th. This proved to be a big revenue producer for the C&NW, but, after the initial surge of enthusiasm in February, no one seemed to care by October.

There was a brief flurry with an Austrian company, Plasser International, manufacturers of very fine track maintenance equipment. Herr Newhofer was looking for some isolated place where they could test their equipment on sorry railroad track. I suggested the North Yard at Belvidere which was pretty sorry, indeed. There was still a lot of old, light weight rail in the rarely used yard tracks which were sunk in cinder ballast. I offered them a lease on the whole yard, but their interest went elsewhere.

The Railroad Industrial Development (R. I. D.) luncheon group was still in operation. We met April 10th and the 14 attendees were introduced to the new director of industrial development for the

Rock Island, a Mr. Orrico. That was a job that someone had mentioned me for last year, but I probably would have passed it anyway as the Rock was a pretty feeble operation by this time. During the rest of the year R. I. D. had only 7 attendees at the May 27th gathering; 19 from 9 railroads on September 30th; 12 on October 28th. We honored George Cox of the ATSF as our honorary president on December 2nd.

On March 1st I went to Rolling Meadows northwest of Chicago with Harry Bierma and Keith Feurer of the real estate department to meet with a consultant about development of our Belvidere property. Keith was an energetic new boy who was to have his moments of glory on the railroad in 1983 and 84. In March I visited with Russ Fyfe, the C&NW station agent at Rochelle, Illinois. As an unofficial member of the industrial development department, Russ was always on the lookout for new business to bolster his job at Rochelle. The railroad would have been fortunate to have had every station agent with his enthusiasm and good sense. He had picked up a rumor that Caron Spinning Company was looking for a plant site. The rumor didn't stand up, but Russ was unfazed and continued to work hard at growing his accounts. I drove north to Rockford to address the Delta Nu Alpha traffic fraternity meeting. On March 15th there was a luncheon meeting of Illinois banks where Cermak, Jack Whittle and Charley Willson from Elgin, led a discussion on trends in industrial development. On the last day of March the firemen-off issue boiled over and some railroads including the Illinois Central, Missouri Pacific, Union Pacific, Grand Trunk Western, Pennsylvania, Seaboard Air Line, Central of Georgia and the Boston and Maine were struck.

I left Sioux City, Iowa, April 4, 1966 to go to Blair to see my local contacts as this area was becoming a hot property, and we needed all the local cooperation we could muster. Howard and Ruth Thompson (Chamber of Commerce) and Howdie and Mary Maud Hanson (Blair Bank) were especially helpful people. We took a good look at the farmed land sloping down to the river and tried to envisage the changes that were to happen there over the next 35 years. (Cargill, among others, established a huge corn-based chemical plant there.) Later, I had a visit with Bill Spitzenberg and Walter Gephart of Northern Natural Gas in Omaha before flying home on a United Caravelle *Ville d'Rouen*. In mid April Dun and Bradstreet's Chicago representative, Al Smith, setup a meeting for me at their headquarters in New York. Ed Wozniak of the Rock Island came to the office to learn something about our organization. He must have been discouraged with the Rock because he had switched over to the Milwaukee Road by October 19th. Bob Kunkel had a new project and asked me to go with him to Litchfield on the Illinois Central's *Green Diamond* on April 14th. Bill Newgent from St. Louis met us. The proposal was construction of a new connection from the old station of Lemmon on the C&NW's Southern Illinois district to the Burlington and GM&O lines to the east to serve a new coal mine. We went south to Virden and then to Compro which once had been a connection with the Chicago and Illinois Midland Railway. After poking around the territory, we went to Springfield to catch the afternoon *Green Diamond.* Since Bob lived in Olympia Fields and commuted to Chicago on the Illinois Central, he preferred riding the IC train from Springfield to the Gulf, Mobile and Ohio which went to the Union Station making a long ride home for him. While the IC train was more inconvenient for me, Bob was the senior officer, and we did it his way. About this time another big project arose when W. R. Grace Chemicals became active in establishing a major ammonia fertilizer production plant along with a phosphate mining venture. Gulf Oil was also pushing a fertilizer distribution facility at Franklin Grove. There was a lot going on in my territory this spring and the other industrial department people were equally busy in their areas.

Late in January I heard that Cermak wanted to establish a research department and that Don MacBean and I were to be interviewed for the proposed new unit. Cermak was in no position to establish such a function, however. On February 10th Cermak, his aide Don Guenther, Ralph Gerrard of Northern Illinois Gas and I discussed over lunch the importance of research. There was a growing awareness that traditional industrial development activities were fast becoming obsolete; data collection and research were growing in importance. I wrote up a proposal for a modest research function and gave it to Cermak on March 2nd. I was aware that my friend Herman Jacobsen, a strong advocate of rate research, had fallen from favor in recent weeks. If his bent for research was the reason he was being treated so badly, then top traffic management must not think much of the concept. My proposal seemingly dropped into a black hole. My relations with Gene Cermak also were gradually growing stiffer and more formal. No more scribbled personal notes or quick phone calls—everything was formalized in memos or letters with replies in kind. By March Cermak was said to be upset that the industrial department did not tell him everything that was going on. How could we? Our business often required fast responses to client inquiries with no time to wait for his decision on matters we could easily handle. My journal note for March 9, 1966 indicated that communication with the titular head of the department was non-existent and that our organization structure was broken. Abruptly, I was ordered to meet with him March 10th at 8:30 a.m. I was there and waiting, but he had something else to do and kept me waiting—all day. Of course I did not wait around his office door until his royal personage was available. After a decent interval, I went on about my business. Two days later he finally remembered our appointment and came striding across the office with his thinning red hair standing on end, electric sparks shooting out his blue eyes and every freckle standing out in high relief. I was in a meeting with a client from Barton Aschman, the consulting firm. He reined in his horses when he saw me with a visitor and went back to his office without saying a word. By now our relationship was stone cold. When we finally did meet face to face, I told him directly that he was doing a rotten job of managing. Of course, this was not the way a mere lackey speaks to his master and my stock continued to sink. Despite this dismal episode, friendly contacts informed me quietly that my research proposal had attracted a lot of attention and that Bob Stubbs was professing support for it (March 24th).

My ego had gotten a lift February 23rd when Maurie Fulton of Fantus offered me a job with their company. Since I was very depressed about my railroad career in that wintry season, I pursued the offer. I floated $15,000 as a desired salary which would be a hefty jump from my $10,200 railroad pay. They countered with my need to take a Wunderlich test which I did well on. On March 31st Maurie offered $11,500 to start with a rise to $12,000 in six months. After checking through the fringe benefits, I judged the offer was far too low. Knowing that I had some value in the marketplace stimulated me to pull my resume together. Access to Cermak remained difficult these days. It seemed that a crisis was necessary to get his attention.

Finally, on April 1st I had a chance to meet with him and told him flat out that I had job offers. He called me at home on Saturday saying he was concerned about my possible departure from the C&NW. He had mentioned my problem to Bob Stubbs who said he would try to talk to Heineman before the 11th when the chairman was going to Italy on vacation. Meanwhile, Maurie Fulton wanted an answer by May 8th. I made a career decision and turned him down. Cermak later told me that Heineman was aware of the situation but I must be patient for a few more weeks.

Don MacBean and I went to New York in late April. G. A. (Gerry) Armstrong of our New York sales office and I called on a Mr. Doyle of W. R. Grace Chemicals to discuss mining of phosphate ore in Wyoming. He said they would attend any meeting I could set up with local interests. I knew a Mr. Schenk with Susquehanna Western who had operated the yellowcake processing plant at Riverton and was knowledgeable about resources in Wyoming. On May 2nd I introduced Mr. Merryman of Grace to Schenk whose office was in the Civic Opera building just across the Chicago River from our own offices. The key to mining phosphate in Wyoming turned on being able to move sulphur from Alberta to Wyoming in large quantities and at low costs. This was a job for our rate research people. The project never developed.

Stan Johnston was now back in Chicago and went with me to familiarize himself with the Illinois territory where we had worked together as young engineers so many years ago. We flew to St. Louis May 5th to collect Ken Hessler and Don Rostenstoy of International Minerals and Chemical north along the Southern Illinois line. IMC was interested in finding a good place to install a potash distribution depot. It was a gritty trip as we ran into dust storms between Springfield and Peoria. Unsuccessful in finding a suitable site, our biggest entertainment was watching a wreck crew pick up a boxcar and waycar derailment at the south end of the Nelson Yard. Later in the month Cermak and I drove to Peoria for the annual Illinois State Chamber of Commerce meeting where Bill Quinn, president of the Milwaukee Road was the featured speaker. Gene sat at the head table while I was out with the regular folks. Two days later I went to Nebraska where Bill Cook, Jack Peters' successor in Omaha, had lined up a meeting with Nebraska Consolidated Mills. I went on to West

Point and Norfolk to review the prospects for a feed mill at each of those places. Just missing the Chicago flight, I rode CB&Q's train #3 in sleeping car *Silver Crag* that evening.

On May 18th Ben Heineman was elected chairman and chief executive officer by the C&NW board of directors. C. J. Fitzpatrick, the former president, was gone—forever. Larry S. Provo, from the accounting firm of Arthur Anderson and the C&NW's current chief accounting officer was elected executive vice president.

International Minerals and Chemical now came forward with another plant location in my Illinois territory. The potash terminal project of early May had now been supplanted by a request for a foundry supply facility site. My past acquaintance from the Wyoming bentonite project, Menze, took the northern Illinois tour with me on May 26th. He liked Belvidere and DeKalb. We stopped to chat with Frank Merrill at the Barber Greene plant between DeKalb and Sycamore because Menze wanted to learn about labor availability. His project must have had some urgency because he was avidly pushing both DeKalb and Belvidere just a week later. On June 3rd I took him and Heidle to DeKalb to have a session with my old friend, Hugo Hakala, the former mayor, who had real estate interests south of the city along our Spring Valley branch. Since Belvidere was also high on their list, I arranged to have Keith Feurer from our real estate department join us to look at our land near the Chrysler plant. IMC liked what they saw at Belvidere and wanted to proceed quickly on acquisition. Ballin's real estate department then commenced to drag their feet on the sale. To break the impasse I drew a track plan for the foundry supply building and dragged Harry Bierma, Howard Koop and Keith out to Belvidere to finalize the sale. I ordered soil testing on June 29th and was gratified to hit solid limestone at a 5 to 7 foot depth—just right for foundation work. On July 5th Keith and I took the final sale contract to IMC's Skokie headquarters. Ground breaking for the new building started before August 15th although the formal ceremony didn't occur until September 1st. I had the distinct impression that real estate wouldn't know a freight revenue producer if it bit them on the bum.

Apparently the idea of a research function was still very much alive and I was a leading candidate to head it up. Gene said on April 26th that my new title was to be director of research and that he was head of the selection committee. Three days later, however, everything had changed. On May 2nd, Gene and I first met with the brand new corporate development department. Staffed by three young, highly educated, focused and very hard-nosed men: Maury H. Decoster, Vic Preisser and Ted Struve, Heineman had charged them with examining, studying, recommending and reorganizing several vital railroad functions. Maury Decoster's assignment was to restructure and modernize the traffic department. He was to become the Archangel Michael complete with flaming sword. Vic Preisser got into the freight claims morass and I have forgotten what area Struve tackled. At that first encounter Gene and I were interrogated at length about our operations. The trio thought that the GEOFAX geographic data file had merit which made me feel good, but Cermak came away from the meeting somewhat subdued. On May 13th Decoster, Bud Braun and Cermak reviewed an IBM proposal for programming our coordinate file and I was told to make it happen. Two days later Decoster and I met to size each other up. The hour long meeting was hard, quick, intense and bare bones ending with Maury loaning me a text on statistics which he said I needed to master. On May 27th Cermak showed me Decoster's approval of my having the research job, but Bob Stubbs would have none of it. It was about this date that corporate development hired an advertising agency to develop an image survey of the railroad.

In early June I went to Wyoming to address the Wyoming Bankers' Association at Jackson Lake lodge. The invitation had been tendered in March before the internal turmoil which was to convulse the traffic department had begun. I had created a slide show describing the development of railroads in the State of Wyoming and had really hung my presentation on the accompanying slides. On the morning of the presentation, June 10th, there was no electric power in the lodge when I arose. Scrambling frantically, I recast my presentation to compensate for not being able to show the slides. To my great relief the power came back at 9 a.m. and I went back to my original script scheduled for 10:30. The half hour presentation went well. Wyoming Governor Clifford P. Hanson was in the audience. Not knowing that, I had gently twitted the state highway department for omitting our Belle Fourche to Colony bentonite spur from the official highway map. Hanson came up afterwards, introduced himself and said he would personally make sure that the next maps would include that line. Art Knight from Casper who had accompanied me to Jackson Lake, was rather impressed that the governor had attended the session much less coming forward to speak with me. I enjoyed my 15 seconds of fame.

In early July I was again out in Wyoming when Erv Klover in Sioux City sent me an urgent message. He wanted me to meet with Vose and Schuller, consulting engineers, about a proposed municipal dock at South Sioux City, Nebraska. Construction of a dock at this site was a dubious proposition for several reasons: it would have been on the inside of a Missouri River bend subject to heavy siltation, there was little prospect for barge operations on this capricious river and a connecting rail line would require construction of track down a right-of-way in a city street, which was politically unacceptable. Tangled into the background of this dock and access track was a little known predecessor company of the Omaha Road. The Covington, Columbus and Black Hills was chartered in 1875 to construct a 42 inch gauge line from Covington (now part of South Sioux City) near the site of the proposed municipal dock mentioned above. This company actually built 26 miles west to Ponca City by 1876 before running out of capital. Acquired by the CStPM&O from the bankruptcy court, the line was standard gauged and in 1907 extended another 19 miles to the new community of Wynot, Nebraska. The displaced narrow gauge Mason bogie engines of the CCBH were said to have been used to move cars on and off the Missouri River ferries before the construction of the railroad bridge at Sioux City. (The Wynot branch was abandoned April 19, 1933). Meanwhile, back in Chicago, the Rock Island case dragged on. We had 52% of the stock proxies. I was asked to preview the testimony of Ben Heineman and Iver S. Olsen, manager of on-line sales. Heineman began his testimony on November 14th. Back in my Illinois territory, the Downing Box people were looking for a plant site but had ruled Belvidere out because of Chrysler's big impact on the labor pool and the resulting high union wages being paid. When I took them to Belvidere June 23rd, however, they liked the site a lot (but never did anything more about it).

It was obvious by late May that polarization of the traffic department had commenced: either you were for Decoster and helped him with his work or you were against him and all he stood for. There were no neutrals. Cermak lined up on the positive side and helped pull the old traffic department structure down but, like Samson, he was to be destroyed in the collapse. Don MacBean and I also took the pro-Decoster path. I escaped before the ruins crushed me; Don barely escaped only to land in another frying pan before leaving the company for good. In late June I began to prepare written reports for Decoster on C&NW markets. Maury was beginning to learn that the department's complacent and some what arrogant attitude toward our shippers was in marked contrast to other railroads' more customer oriented attitudes. Decoster continued to make direct requests to me for opinions on our present methods for managing commodities in the sales department. Separating commodities and territorial responsibilities was almost impossible as the two functions overlapped and intermingled in many aspects. It was obvious that the sensible thing to do was establish a marketing function geared to a specific commodity regardless of territorial assignments. Given my negative attitude toward the present traffic officers, I had no compunction about working for Decoster and did not share my reports with Stubbs and his adherents. Stubbs seemed to be in a boozy state much of the time anyway, a fact that Decoster noted. A massive reorganization plan that Maury was building would completely remake the sales function. Because I was about to leave for a three week European vacation, I asked Don MacBean to do some of the commodity analysis work in my absence. Don understood that he and I were already "dead meat" in the Traffic Department so working with Decoster now would not make any difference to our future careers.

To correct any impression that all traffic department personnel were tarred with the same brush, I really liked and respected the abilities of a great many of them; it was their leaders and their bad management practices that earned my disrespect. At this remove I can well recall the dependable work of a number of some very good Chicago office men. Bob Cumbey, Herman Jacobsen, Ralph Gauen, Art Reno, Kurt Ramlet, Ray Degnan, John Troyke, Jim Ronayne, John Brugenhemke, Jim Goinz, Dick Ryan and Walt Conard are some of the faces that are still fresh in my memory. Each of them had a positive role to play in the railroad's development. It was Ray

Degnan and Jim Ronanyne that dreamed up the concept of the famous Falcon train service which was to begin operation in early 1973 and flourished vividly for a number of years. Overpowered trailer-on-flatcar trains flew out of the Falcon "nest" at Wood Street westbound to Council Bluffs on advertised schedules with rights over all trains. One carman at Wood Street was seen running alongside a Falcon getting ready to fly and was asked what he was doing. He replied, "checking the air on MY train." These trains even had their airline joints taped together to prevent unwanted emergency stops that sometimes occurred at high speed on our less than perfect track. The corresponding eastbound Falcons were similarly expedited. These trains were among the first C&NW train operations aimed at providing superior, dependable and fast service and the shipping public responded. The Falcon concept expanded with trains added to the Twin Cities. Whether we made any money on them is beside the point. Another story concerned the quietly efficient Art Reno. Someone said of Art that if it was necessary to know the number of mail boxes along highway 52 between St. Paul and Rochester (Minnesota), he would ask Art to count them. Art's report would be an accurate count of the boxes, how many were on each side of the highway and what color they were painted. It was a shame that Art's assignments were not at a decision level commensurate with his experience and capabilities.

Our European excursion was possible through the courtesy of my artist wife who had a rather successful year. Because of the U. S. airlines' strike which began July 10th, Air France laid on a special flight that left O'Hare at 7 in the morning on July 18th for Paris with a stop at Montreal. After several days in Paris, we traveled by car east to the Vosges Mountains and then southeasterly to the old cathedral cities of Fontenay, Vezelay, Autun, Conques and Albi. A visit to the Lascaux caves was memorable. At Foix we turned northwards to Pau, St. Emilion, La Rochelle and Niort with a loop through Brittany to the city of Quiberon before going easterly to Caen. 1966 was the 900th year celebration of William the Conqueror's invasion of England and many museum displays were devoted to that event. After staying overnight in an ancient hotel on the mystical Mont St.-Michel, we walked Omaha Beach and visited the sobering cemetery behind its border hedge of red rugosa roses. We sailed to England from Le Havre on British Railway's night ferry, struggled with left hand traffic as far as Stonehenge on the Salisbury plain, before returning to London. We had been assured in the French newspapers that TWA was again flying. They weren't. By sheer luck we got the last three seats on a Sabena jet to New York. Since we were in London and Sabena was leaving from Brussels, we had to make a predawn flight on British European Airways (BEA). The flight to New York was uneventful—the trouble began after we landed. Since nothing commercial was flying, I called Vince Holland in our New York office and told him of our plight. He managed to get three slumber coach spaces on the NYC. We dashed uptown to Grand Central where, in a guarded voice to the ticket clerk, I asked for the Pullman space held in my name. Although the crowd around the wicket was in an agitated mood, I was able to get the coveted tickets without any difficulty. Despite the crowded and somewhat untidy condition of the train, we were delighted to be aboard and moving west. I woke up in Rochester but let the girls sleep longer. Breakfast was in the half diner after the Pullmans were cutoff at Cleveland. Rolling on westward in the coach, the burly conductor came through asking for transportation; I told him that ours had been collected last night by the sleeping car conductor. He said he would check and passed on. An hour later, somewhere near Toledo, Ohio, he came through again. He announced that we had only paid fares as far as Cleveland, and that we would have to pay cash fares to Chicago. As I reached for my very thin wallet, he spotted my yellow C&NW annual pass. "Oh, are you a railroad man?" I admitted I was. He said, "Shove over and let me tell you about this god-damned outfit." The conductor was highly critical of the New York Central's current management (Alfred J. Perlman, of course) and its "crazy" innovations like that jet airplane engine mounted on a rail car that couldn't be stopped until they opened the front windows. He vented his frustrations in a rising crescendo for half an hour. Then he stood up, straightened his cap and made a half bow. "Mr. Lewis, it is my pleasure to have you and your family ride with me to Chicago as my guests. Enjoy your trip." And off he went. (As a note: two General Electric J-47 jet engines from a vintage 1948 B-36 bomber had been mounted atop a 13 year old stainless steel Budd RDC-3 combination baggage, mail and coach railcar. The experimental run near Bryan, Ohio, on 40 miles of tangent track with the so-called *Pride of the New York Central* set a speed record of 183.85 mph in just four minutes. Not repeated, the test yielded valuable information for future development of the 120 mph Metroliners).

On August 10th I was back in the office to find a new chief clerk, Keith Peterson from Sioux City, had replaced Virgil Steinhoff who had gone to Minneapolis as a new industrial development agent to work for Chuck Towle. That new research position remained a chimera and the directorship of the industrial department also was closed to me. In July Gene Cermak had recommended Charles Towle, our Minneapolis manager, or Jack Shaffer from the New York Central as candidates for his old job. Time slipped by. In November Don Guenther let slip that the search for the directorship had gone outside the company. The department's discontent and disarray was evident to the most casual observer. On August 22nd Decoster had unveiled his radical plan to Bob Stubbs, Gene Cermak, Stan Boardman and Roy Erickson for restructuring the traffic department. The polarization within the department deepened. By September details of the new marketing function beginning to emerge. Decoster held a big meeting on October 3rd and Bob Stubbs dug in his heels. Rumors and counter rumors bubbled up all around as each party strove to advance or derail the inevitable changes. Maury was said to have accused some traffic people of incompetence; I couldn't believe he actually had said that in public because Decoster was an intelligent person. Maury said to me on October 12th that the whole package was now up to Heineman. Matters stood in limbo until November 25th when the announcement was made that a new marketing department was to be established on December 2, 1966. A week later Maury asked Frank Jenko, Don MacBean and me to design a sales coverage reporting system. We put together an interview process that could be contained on a punched card so that summaries could be quickly prepared. At that point Decoster put a hold on our group as we were getting ahead of his thought processes. In the new structure Cermak's title was assistant vice president-sales planning. To many that appeared to be a demotion and his reward for cooperating with the reformers. The other "traitors" were said to be MacBean, Stan Johnston and myself.

It was a relief to be away from the office. On August 24th Don Pacey, the new director of industrial development at Elgin (replacing Charley Willson) introduced me to Mr. Funk the president of Elgin Paper Company. The company occupied an old building in East Elgin that was to lose its rail service as the railroad was planning to vacate the old industrial section there. I helped him locate a nice piece of land along the Freeport Line near Big Timber Road for a new warehouse. It was not a big revenue producer, but it was a steady one. The next day I heard Swift and Company was planning 300 new fertilizer distribution facilities in the mid-west.

On September 15th the Northeastern Illinois Planning Commission set up a map and document display in two C&NW parlor cars parked in the Chicago Passenger Terminal. A steady stream of visitors demonstrated the public's interest in this graphic depiction of what the metropolitan district was today and where it might be headed. The next day I went to Nebraska where the State's Resources Division sponsored a meeting of railroad industrial agents to create site evaluation teams. After a lively discussion the matter was put to a vote: Union Pacific opposed, everyone else was in favor of the concept. Later in the month Fred C. Jones, DF&PA at Peoria, retired on September 26th and a large delegation of traffic people flew in an old Martin 404 from Chicago to attend his party. Fred was a gentle old fellow who realized that the new structure of the sales department was going beyond his abilities and that retirement offered an easy solution for gracefully bowing out. Further south, the never solved problem involving Commander John Barron and his Bi-State Authority at Granite City ended abruptly September 28th when he suddenly died from heart failure.

In October I attended a reception hosted by Governor Nils Boe of South Dakota; J. R. (Bob) Brennan attended for Mr. Heineman. The Governor gave everyone a present of a Black Hills gold tie pin in the form of the state bird, the pheasant. On October 13th Quinn became president of the CB&Q and Crippen became head of the Milwaukee Road. We heard on October 17th that eight diesel units and 34 cars had derailed in Minnesota. The wrecking train arriving at the accident scene then smashed into the surviving waycar and damaged two more diesels.

Determined to salvage something positive as my career in the traffic department imploded, I vigorously pushed ahead with the GEOFAX coding work during the first months of 1966. National coverage was my aim, but I was a long way from that goal. Correspondence with postmasters all over the country was yielding some fascinating information on local history. Dun and Bradstreet's cities list contained a lot of defunct places which I gradually identified and either scrapped or included in the growing file. I made a design decision that if I could correctly identify the site of a named location, it went into the file. I also discovered there were books on toponomy (naming of geographic places) and began to accumulate a collection for my private library. There was an inquiry from Walter Gephart of Northern Natural Gas Company in Omaha about using the coordinates, but we just did not have enough coverage yet to get involved in a joint project. In May I purchased a book on FORTRAN programming. The language was cumbersome and designed for use in the scientific community. The business oriented programming language, COBOL, which was still under development at this time would be a better software for the applications I had in mind. As the year progressed I expanded our map collection to include the USGS 1:250,000 scale maps which were the only government maps that covered the entire country using a single scale size. If spread out and pieced together, the maps from Maine to California would have covered a basketball court. As I plodded through each state, county by county, capturing the coordinates for city names, I found the work acted as a palliative to the stress, turmoil and tension crackling around me.

On April 17th I had flown to Washington, DC on a DC-6 to pursue inquiries into the use of coordinates. After a courtesy call on our Washington office run by J. F. (Jim) Williamson, I visited the Potash Institute, which was composed mostly of lobbyists who were not very knowledgeable about the business they were supposed to be promoting, and met with Huss and Koehler of the United States Post Office's Customer Relations office to learn about naming conventions and postal logistics. Don MacBean arrived that evening. The next day we met with Don Church of the Bureau of the Census in Suitland, Maryland. He was most helpful and showed us the PICADAD system they had used ten years earlier for structuring data. At the Association of American Railroads (AAR) director Carl Byham expressed interest in what we were doing but offered no new ideas. On the 20th I called on R. L. Banks, railroad consultants, but the visit turned into requests for our resumes and references. Declining their overtures, Don and I drove a rental car north to Wayne, New Jersey, where we called on American Cyanamid. The next day Don and I met with Dun and Bradstreet in their New York offices. They liked the concept of our geographic data base and offered full cooperation in its development. These contacts were to become the foundation for my new career path. My little horse, Geography, was getting frisky, but was not yet strong enough to carry me. Don and I came back to Chicago on the Pennsylvania's *Broadway Limited* which, we both felt, had a worn and dingy air about it. Less than a week later, on April 27th, the merger of the NYC and Pennsylvania was announced. But the Interstate Commerce Commission turned down the merger of the Northern Lines consisting of the CB&Q-NP-GN and SP&S.

Don MacBean and I were very proud of the Jones and Laughlin Steel Study of 1964 even though management had not shown any interest in it nor had even given us so much as a pat on the back much less an increase in salary. Although I had successfully acquired 616 acres of land at Geneva, Illinois, for R. R. Donnelly, industrial growth at Blair, Nebraska was generating impressive freight revenues and National Can was building a new plant at Loves Park, Illinois, based on our research, my salary had been stuck for two years at $10,200. Don and I were not the only ones held in their pay grade—the whole industrial development department had not benefited from any raises. Yes, there was a sour taste in our mouths. Don and I decided to send one of the extra copies of the J&L study to a national contest sponsored by *Industrial Marketing Magazine* in Atlanta, Georgia. The contest was aimed at rewarding good marketing ideas. To make sure that I was not violating any company policies, I asked Ed Warden of the law department to review the entry form. He said it looked okay to him on January 20th; we filled it out and sent it out the next day. In the tumult of business we forgot about the contest entry. On July 7th a very excited Don MacBean called me to say that we had won second prize. The next day Don advised our superior, Gene Cermak, that we had won an award. Ordinarily, our messages to Cermak would go through "Little Shadow" Don Guenther who was acting as Cermak's screener. MacBean knew that there was little love lost between Guenther and us and such "good news" might never get to Cermak. The official notification arrived July 11th. Late in October the magazine called to make an appointment to present our award. Don MacBean advised Cermak's office that the ceremony was to be November 4th in our offices. Len Ernst of the magazine arrived at 11 a.m. for the official presentation. No one showed up except Don and I. Cermak was unavailable; the company photographer, Bob Lindholm, was out of town; no one in the traffic department evidenced any interest in the event. After an embarrassing moment or two in the deserted reception area, Mr. Ernst handed us our framed medal and bronze plaque, shook hands and left. That afternoon Maury Decoster learned about the sorry event and told Ben Heineman. The chairman was exceedingly distressed by the callousness of the traffic department. That afternoon Cermak rushed into our office, collected copies of the magazine award article and ran upstairs to Heineman's office to explain why there had been such a monumental foul-up. Don and I at that point in time couldn't have cared less; we were both so disgusted with the organization that a personal apology would have been meaningless.

IBM had spotted the magazine article and called us August 15th. Their local man, Bob Donaldson, came to the office in October. Don and I showed him how the GEOFAX work was going, but seemingly the slow but steady progress we were making only seemed to excite this neatly dressed young man. Apparently, there were potential users of geographic coordinate systems out there in IBM land. In September I stuck my toe into those waters by sending out letters to some of the people that we had met recently describing the concept of computerized geography. If the railroad didn't want to promote this thing, I would try and stir up some interest on my own. No one responded to this first sally.

One of those eventful days, not recognized at the time, was November 3, 1966 when I first met Dr. William P. Allman. Bill was one of those over-educated, rather pompous young men, who knew they had answers to every question especially with regard to railroad prac-

Winners of a silver medal for industrial marketing research: Donald A. MacBean and Eugene M. Lewis on April 7, 1967. Their railroad careers were not advanced by this national award.

tices. Although Allman had never held a job in private business, Ben Heineman hired him on the strength of the doctorate. Allman was temporarily assigned to the corporate development department with Decoster, Preisser and Struve until he could get his own operation in computer sciences underway later next year. On the 10th he invited me to lunch with him and a vice president of the Illinois Central, Richard P. DeCamara, one of the more important individuals with whom I was to interact in subsequent years. DeCamara was chair of an AAR *adhoc* committee developing a Standard Point Location Code (SPLC) for the transportation industry. Now there was a subject I could really appreciate as it closely paralleled my thought processes in my developing geographic coordinate system. December 6th was another seminal date. That was the day I determined to market the coordinate system as a private venture. Avon Products had called in response to the magazine article and asked if it was for sale. Concerned that marketing the database by an employee might be construed by the railroad as a conflict of interest, I sought and received permission on December 12th from Gene Cermak and Maury Decoster to privately market the system. The "approval" by Cermak and Decoster, however, was later challenged leading to a rather unpleasant series of events which will be related in the next chapter. At this time Howard Roepke, professor of geography at the University of Illinois and editor of the annual bulletin of the American Industrial Development Council (AIDC), asked me to write an article about the use of coordinates in industrial development work. Then just before Christmas MacBean heard that the proposed market research unit was still under consideration and that we were the prime candidates. At that point neither of us really cared anymore. I was now bound to shake loose from this stultifying atmosphere because I could see there was a whole new world out there and it was calling to me. Market research for the railroad was beginning to look like another dead end. It was almost anticlimactic to fly to St. Louis to call on Ralston Purina and W. R. Grace to talk about industrial sites. My mind was elsewhere.

21

SYSTEMS IMMERSION, 1967 . . . This January began with an intense and prolonged cold spell. The excuse to stay indoors seemed to stimulate my creative juices. This month I invented a name for my coordinate database—MAGIC for Market Analysis by Geographic Index Code. MAGIC was to become the focus of a lot of pain, anguish and personal satisfaction. Gene Cermak took the trouble of now telling me why I had not been considered for the director of industrial development position. I was being saved for the market research slot. In this conversation on January 5th he seemed worried because Ben Heineman was taking a direct hand in finding a new face for Gene's old job just as he had hired the new boys in corporate development. Heineman obviously was not pleased with the talent available within our company. I was sure that I really didn't want promotion to Gene's old job. I was getting bored with this whole business. But, then again, not being considered was a blow to my self esteem. Both Cermak and I sensed that industrial development as we had known it was rapidly changing. It was time to set a new course.

I received an unexpected invitation from Dick DeCamara of the Illinois Central to attend a Data Systems Division (AAR) meeting in Jacksonville, Florida. The invitation came January 11th and Cermak approved my attendance the next day. I had begun to see Maury Decoster in the new corporate development group more frequently and often asked his advice on matters relating to research. On January 20th I showed him the advertisement for the director's job. Knowing I was becoming restive, he asked if I would be interested in joining Bill Allman's new group since the market research project was taking so long to mature. I thought that might be a good answer to the current situation and threw my name in the hat, so to speak. Gene and I had another civil chat on January 30th. Cermak was having a lot of problems with his new job in the clash of new versus old and agreed that going to Allman's new endeavor might be a good move for me. Professor Roepke at the University of Illinois, called about my article that I written for the AIDC bulletin—it was too short, could I make it longer? Certainly, how much more do you want? On January 5, 1966 the C&NW's acquisition of the Fort Dodge, Des Moines and Southern Railroad was announced at a Chicago Association of Commerce and Industry luncheon. My old friend and colleague of 20 years, Marion Boyd (Bob) Lithgow retired and was feted at a luncheon February 3rd. I planned the party which was held in Chicago and attracted 32 people. Bob seemed to enjoy the proceedings. He still owned a lot of valuable farmland near Barrington, Illinois, an upscale and rapidly expanding bedroom community. Bob's land was situated in a choice location northeast of the city. He gradually sold it off, becoming very wealthy. His love of flying was satisfied when he purchased his own small plane. In August 1969 Bob's beloved wife, Mary Lithgow, died suddenly leaving him alone. I attended the service on the 23rd; he expressed his gratitude for my solicitude as they had no children and few relatives. Bob himself died in 1977.

Geocodes now became a larger part of my life. DeCamara sent me copies of the newly developed Standard Point Location Code (SPLC). This six digit code was developed to uniquely identify all transportation points in North America. Don and I compared the coded points in the SPLC to our Dun and Bradstreet city list and found many differences which we could not immediately explain—that would come later as we became more involved with the coding process. I had also gotten approval for Don MacBean to go with me to Jacksonville to attend the Data Systems Forum. On arrival, I found Bill Allman was also there. At DeCamara's invitation we attended the *ad hoc* Committee on SPLC. I gave a brief account of the MAGIC coordinate system. The committee members' enthusiastic responses and questions indicated that we were on the right track. Allman latched onto us like a leech. He asked in his unsophisticated direct manner if I would mind being stolen from the industrial development department. While in Jacksonville I also attended meetings discussing development of a standard route code. This subject was a thorny one and would never be universally accepted. Sixteen years later the industry still would be wrestling with adoption of a route code. The Data Systems Division also touched on UMLER (Universal Machine Language Equipment Register) and ACI (Automatic Car Identification). This was a whole new world to me. The ACI committee's chair was an old friend, Arthur E. Pew from the CB&Q. Art had been a rodman on the Galena Division many years ago. I made a point of visiting the Seaboard Coast Line's industrial development department as a courtesy call. Then DeCamara announced that I was scheduled to speak for 45 minutes on the subject of geographic coordinate systems at the annual Data Systems meeting in Houston, Texas, in September. This new life was already rolling along faster than I could imagine.

Back in Chicago January 26, 1967 began as just a cloudy morning. By noon the snowfall was so heavy that many downtown offices closed early sending hordes of commuters to the railroad stations. Normal midday three car train sets were overwhelmed. The railroad scrambled to assemble more equipment in deteriorating weather conditions. Schedules were thrown out—it was load and go on all C&NW commuter lines. I trudged home from the Elmhurst station through deep drifts blocking the streets. The total snowfall in one day was 23 inches, a record. Chicago closed down. Skiers appeared on major highways. My daughter driving home from college in Madison, Wisconsin, almost managed to get the car home, but had to abandon it in a schoolyard a block away. It sat there for three days. By February 22nd the season's snowfall totaled 53½ inches.

At the annual meeting of the Great Lakes Industrial Development Council held at the Drake Hotel January 30th, I was elected a director. I never attended a board meeting because already I was beginning to separate from industrial development in anticipation of a career change. Our department's nemesis, I. R. Ballin, the vice president of real estate, also took over the materials department on February 2nd. His new title was a mouthful: vice president materials procurement and real estate. The Chrysler project echoed one more time when the United Auto Workers (UAW) contacted me February 1st to purchase a site for their union headquarters across the road from the plant. I referred that request to real estate.

Events moved swiftly in February. Gene Cermak announced February 3rd that Jack Shaffer from the former New York Central had been appointed director of industrial development. Our Chuck Towle in Minneapolis was so incensed that he vowed to resign. Both Stan Johnston and I expressed our dismay to Cermak who said that the decision to hire Shaffer was Heineman's. Jack stuck his nose into the office February 24th and tried to see his new boss, R. C. Stubbs. He had to wait a long time which was a foretaste of what the future was going to be like. The end of February was Cermak's last involvement with industrial development on the C&NW—a run of almost ten years. Shaffer arrived on March 1st bringing his own secretary to our staff's consternation. He also purchased new furniture which again set the wrong note with his new staff. Three weeks before Shaffer's arrival I had attended a Transportation Research Forum (TRF) meeting. Bill Allman was also there. Later in the day he called me to set up a meeting for the next day at which he offered me a job in his new advanced systems development group at a salary of $11,000. He had already cleared the offer with Stubbs, Cermak and Decoster. I checked with Decoster who assured me that working for Allman would not preclude me from consideration for the pending market research position. I called Bill Allman from Peoria on the 10th and asked for $12,000. He did not seem shocked by the amount; we eventually compromised on $11,500 effective April 1, 1967. That was my first salary increase in 39 months. Keeping quiet about the new job, I began to wind-down my industrial development affairs. All during this period I steadily continued recording coordinates from my growing map collection. Knowing that my job change was now a certainty, I gradually began removing the map collection, correspondence, coding ledger and all materials relating to the coordinate project from company property. After all, I had permission to pursue this project on my own.

The work of industrial development did not slacken despite our awareness that a change in leadership was coming. "Customers first" was the motto graven on our minds. Jim Heidkamp, C&NW real estate in Omaha, called on February 23rd to report that Olin Mathiesen was interested in a site at Blair, Nebraska. The location of Gulf Oil last year was attracting attention of other chemical firms. The day Shaffer came on board as the new director found me in Nebraska meeting with Howdie Hanson, my Blair banker friend, and our Jim Heidkamp. We spent a lot of time on the undeveloped lands east of town and established the location of a new county road to provide better access. The next day one of Bill Cook's sales reps, John Nelson, introduced me to Paul Newsome of Nebraska Consolidated Blenders who was also interested in a Blair location. After a long discussion on a complicated land swap deal, I went to the courthouse and checked on the land records. The news that I was leaving his department finally trickled up to Jack Shaffer's ears. As soon as I appeared in the office, he began dangling a new assistant directorship position and extra money as inducements to stay. Of course the proffered position did not exist and Stubbs would never have approved such an arrangement. To protect the many projects in progress, Jack assigned Stan Johnston to my Nebraska territory whereupon Stan promptly announced he was taking a job in the real estate department, leaving me still responsible for the Nebraska activities. On March 9th I was in Lincoln, Nebraska, attending the state's annual industrial development seminar. Both Harold Thompson and Howdie Hanson from Blair attended the meeting. They had picked up the rumor at the meeting that I was leaving the business and expressed great dismay. We huddled to work out the Olin Mathiesen and Consolidated Blenders proposals because I knew that Larry Krause, my successor, was absolutely clueless about the properties, the people involved much less any idea about negotiations in progress. The next day I collected Larry in Norfolk and took him to Fremont and Blair for introductions. It was hard to hand off these live projects into the tender care of someone whom I did not regard very highly. I shuddered to think how Larry would behave with my Nebraska friends. Three days later I was back in Blair to meet with Howdie Hanson, Omaha Public Power District, the Northern Natural Gas people and the Washington County surveyor to plan the future layout of the lands at Blair. The Board of Supervisors had approved the new county road into the riverfront area. I suspect that Howdie Hanson and Harold Thompson had something to do with that swift approval. On March 17th I flew to Memphis, Tennessee, where Bob Seibert, our Little Rock traffic representative, met the plane and drove me to Little Rock for a meeting with Bob Trusheim of Olin Mathiesen to talk about the Blair location which they said they liked. I had also helped Harvey Buchholz in this interim period when he, Jack Sommer and I had gone to Milwaukee to meet Walter Wiese about a paper distribution warehouse for the Madison area. I rode the Milwaukee's *Pioneer Limited* back to Chicago. I was a lame duck and eager to get out of this work.

Shaffer reorganized industrial development agent territories: Johnny Foxen was assigned to temporarily take over my Illinois territory, Vern Giegold and Larry Krause in Des Moines added Nebraska to their Iowa territory; Chuck Towle in Minneapolis already had the Dakotas and added Wyoming to his area. My bridges were now on fire, and I was running across them. Outside the C&NW the word began to spread through our railroad grapevine. At our Railroad Industrial Development (R. I. D.) luncheon meeting February 24th when we elected Joe Arado from the EJ&E as our honorary president #12, there were a few cautious inquiries which I tried to brush off. While I knew I was changing careers, I didn't want to say anything at this meeting. Keith McCullagh of the Illinois Central called on March 21st to verify what he had heard. By March 31st at my final meeting of R. I. D. the genie was out of the bottle. After we honored Jack Shaffer as our 13th honorary president, I resigned from the organization which I had founded because I was to be an ex-industrial development agent the next day. Ed Wozniak of the Milwaukee Road took my place as head of R. I. D. By October 3rd Ed had quit railroading and had gone into the insurance business. R. I. D. quietly died.

Things did not go smoothly for Mr. Shaffer in his first month. I had firmly rejected Jack's offer of a phantom job and increased salary. Stan Johnston had slipped away into the real estate department. As a final fillip the department secretary, June Miller, and her pal, Annie Holden, Stubbs' file clerk and special good friend, took Jack's secretary, Grace, out to lunch and got her swozzled. Welcome to the C&NW, Jack. Returning from Little Rock on the 17th, I got a late night message from my friend Don MacBean that Shaffer was asking about the coordinate program. He had quizzed Don at length and said that he was going to base a national advertising campaign on the concept. He did not know that MAGIC and all the supporting material were gone from the railroad. On the 28th of March I flew to Des Moines and turned over my Nebraska files to Vern and Larry. After a very brief meeting, I bid them adieu and flew back to Chicago on United's Caravelle *Ville de Rouen.* I wasn't the only person changing jobs; I heard that Norb Kraegel, vice president-accounting, had resigned to take a new job with HOMCO in Kansas City, Missouri. He left because the railroad was so slow in adopting computer applications. My last day as General Industrial Development Agent was March 31, 1967.

THE NEW BOY — AGAIN April 3, 1967 was my first day in Advanced Systems. I took two steps down in grade and got a 12% increase in pay. I also had my pride intact and was heartened by new opportunities beckoning. This was my 20th year of service with the C&NW. My railroad experience was one reason Allman wanted me on board. Another, was growing national recognition of the value of the coordinate system which suggested to him that I might have a few more bright ideas he could use. His new department was a formless blob in April 1967. Everyone was new. My boss, Frank I. Stern, had just arrived from the former New York Central. Edward A. Garvey, a bright Irish kid from Baltimore, had a recent brief experience with the Rock Island, Viswanath Khaitan, an Indian national related to a wealthy Calcutta family, Warren W. Davis, a computer genius, Jack Rezek, another computer type, Dominick D. Violante, a practical systems man from the Illinois Central and Ron Cassen from Canton, Illinois rounded out the new group. Ron was a stocky, fair complexioned, very slow-spoken person who used unexpected fifty cent words in ordinary conversation. Solemn to an extreme, his slow motion pace reminded one of a three-toed sloth. We were all crammed together in a chaos of temporary cubicles until July when we were supposed to get our new offices. The atmosphere in our temporary quarters was something like freshman week at college. Since Norb Kraegel had resigned to go to Kansas City, Provo needed a new vice president. He chose an old friend and fellow Arthur Anderson alumnus, R. David Leach. While Leach's father had been a bridge engineer on the Illinois Central, his son had no railroad experience at all. He also lacked good management skills especially when handling the strange and volatile computer folks. A consummate bean counter, Dave Leach proved to be an unapproachable boss who was never to be a friend but often seemed to be an enemy. He became the black cloud on my horizon, but at the moment the sun was still shining.

This new crew needed training and familiarization. I was supposed to be the guide into the intricate and mysterious railroad operations on the C&NW. On April 4th we all attended a lecture on the CARFAX system given by its creator, Fred Baldauf. CARFAX was the punch card solution to moving freight trains across the C&NW, finding freight cars, locomotives and waycars (cabooses), and keeping IBM rich through leases on a lot of ancient unit record equipment. Fred Baldauf was the guy who had gotten me into hot water years ago when CARFAX was new when he overheard my comments about the obsolete technology he was using. Part of Fred's program to our group included viewing the 1957 film featuring Ed Olson, our now retired vice president of traffic. The film gave our new kids very little insight into real railroad operations. Fred droned on all day and into the next, building the entire C&NW, tie by tie, rail by rail. Decoster, now running the new sales department with Stubbs up front as titular head, asked me to develop a model for running a census of his sales territories, but Allman thought I should be doing something else. Frank Stern was also beginning to think of things for me to do; most of them involving creation of ways to describe the railroad's business. We decided to build a commodity grouping using the Standard Transportation Commodity Code (STCC) for C&NW business. That turned out to be something useful. Already, Bill Allman was beginning to grate on people's sensibilities and to evolve into a figure ripe for ridicule. For instance, he insisted on being addressed as Doctor Allman, not realizing how arrogant that sounded. Poor Bill, he had never had a job in the real world and did not know how to behave in a corporate environment. Academia was his home. Even his marriage kept him in the educational world; his wife, Alice, was the daughter of a professor at Ohio State. On April 17th he was awarded the title of assistant vice president which was to be the highest position he was to attain on the C&NW. By the end of April the principal subject for discussion among his wondering crew was the strange way Allman ran his department. I was becoming aware that these system guys were such a disparate group that they would never fit into the railroad's normal society. Railroading was not a way of life or even of much interest to these people. Systems people solved operational and data collection problems using rapidly evolving computer technology. They did not hesitate to change jobs to follow the newest computer applications. A feeling of isolation from the C&NW's operations began to drag at my spirits even at this early point in my new career. I found that working at further development of my coordinate database at home helped offset the increasingly strange atmosphere of the peculiar advanced systems group.

April saw the end of the winter season's record of 68.4 inches of snow. April also brought tornado season. One ripped through Belvidere leaving 20 dead. Bill Wimmer, assistant engineer of the Galena Division, lost the roof and one room of his house there. The Chrysler plant was undamaged. I took Ed Garvey to Proviso on April 24th as he was interested in train operations. Ed decided for some reason that a railroad tunnel under the hump would make operations at Proviso easier. Tunneling here was not an option and his reasons for making such an expensive effort did not stand up to close scrutiny. Ed had seen such a tunnel under another railroad's hump lead in a trade magazine and thought it would be the solution to our undefined operating problems. Dropping that idea, he then went to the operating department and began reading dispatcher's train sheets trying to get a feel for train movements. Ed had a short attention span in those days.

Toward the end of April I began to sense that Jack Shaffer, now in his second month as director of industrial development, was going to try to grab my coordinate database. I was being kept informed by Don MacBean. I studiously avoided the old department and Mr.

Shaffer. My MAGIC coordinate file now included 44,200 records covering most of the United States. The file existed only in handwritten format. I had not yet decided how to translate the data into machine readable form, but by working with programmers in the advanced systems area, I was learning how to accomplish that vital next step. Decoster wanted to use MAGIC for sales territory analysis. Jack Shaffer now suggested that Don MacBean and I go to Los Angeles to meet with Litton Industries to see if they had an application for MAGIC. It was easy to get out of that trip. Prudently, I duplicated the work book and stored it away.

Frank Stern had worked as an analyst for the New Haven Railroad and forValu-Line before going to the NYC. His *forte* was costing, something the C&NW had little ability to develop given the scattered nature of the pertinent information. Selling me on Frank's methodology was impossible, life was too short. I had worked for many years with accountants and had observed first hand that all many of them had to show for their years of being buried in ledgers were green eyeshades, gaitered sleeves and bunions on their butts. At least Frank taught me some new terms and concepts like matrix mathematics. I began developing a commodity flow analysis. Back in the industrial department Shaffer hired Jack Wauterlek from the NYC and Bill Wimmer off the Galena Division. On May 1st Bill Allman acquired a new secretary, Phyllis Nyboer, a naive young blonde girl from Woodstock, Illinois. During the interview Phyllis asked just what she was supposed to do. Allman fatuously replied "fulfill my every wish and anticipate my every desire." While Bill in his innocent-arrogant way did not mean what might be construed from that silly answer, in a different age it could have been grounds for a sexual harassment suit. Bill employed two and sometimes three secretaries because he was a one man paper mill. He wrote lengthy reports to Mr. Heineman and Dave Leach which kept the girls typing endlessly. It was difficult for the rest of us to get any of our own meager correspondence done. It would have been a blessing if Bill had a modern-day personal computer and done his own keyboarding instead of producing those pages of scrawl that consumed so much secretarial time.

On May 3rd Herbert Landow of the Illinois Central gave a presentation to our group on simulation modeling. Now, here was something practical. Seven years later I was to use the technique in a project. Inspired by Herb's approach, I began to try a node and arc network concept for tracing commodity flows. Frank was bogged down in his costing morass where I could not help him. Gently ignoring him, I went ahead with my railroad segmenting scheme.

One of the perks which most officers of the railroad received was the annual physical. I wished that I had skipped the 1967 appointment at Passavant Hospital in Chicago. To this point in my life my hospital career had consisted of a tonsillectomy in 1931, as an outpatient. My examining doctor was a very nice lady from Nova Scotia. In the X-Ray procedure the technician failed to fasten the safety catch on the heavy film plate allowing it to fall. The film canister landed squarely on the great toe of my right foot. That smarted. Luckily, I had my street shoes on. I hobbled down the hall for the rest of exam and told madam doctor what had occurred. My foot throbbed but she brushed it off after a cursory glance. Finishing the exam, she was about to tell me to go away when I again complained that I had a very sore foot. Oh, all right, go back to X-Ray and have a picture made. I hobbled down the long hallways to X-Ray and took off my shoe. When brought to the doctor, the exposed film was so fresh that it was still dripping. She took one look and ordered me into a wheelchair. Yes, indeed, the toe was broken—and it was totally their fault. Now nothing was too good for me. The hospital sent me all the way home in the suburbs by taxi at their expense. Luckily, the break was clean and really didn't hurt after the outraged flesh had calmed down. I whittled a cane out of a tree branch, bark and all. A few days later as I tapped my way down the marble hallway to the elevator lobby, I ran into Mr. Heineman and his executive secretary, Bernice Boehm. Surprised at my three-legged gait, he politely asked what had happened to me. When I related the circumstances, he broke into laughter so intense that he had to lean against the wall. Recovering, he begged my pardon for his insensitivity for my injury, but that story was the funniest thing he had heard all day. I replied that I was not offended and was glad to share the moment with him as it really didn't hurt that much.

Being semi-immobilized, and enjoying it immensely, I began researching the history of latitude-longitude determination, the development of cartography (map making) and the science of spatial distribution. This was an interesting subject that opened many lines of investigation and proved to be rewarding. On May 16th I was back in the office, with my crazy cane, wondering what I should work on next since my director, Frank Stern, was not being very directive. The idea of simulation was still rattling around in my mind. I decided to try and recreate the movement of a freight train across the railroad from whatever records could be found, filling in the gaps from personal observation. Being cooped-up in the office made me feel I was losing touch with the railroad. Perhaps this train tracking scheme could help re-establish my connection with the real railroad. Allman's crew was feeling much the same way. They were becoming more vocal about his apparent lack of goals, bad communication and general ineptness for managing people. Some very strange ideas were being hatched by this group. One was construction of a hump yard at Nelson to take the load off Proviso. As these people were all "foreigners" to the C&NW, they really had little concept of how the railroad operated. Before big operational changes can be made, it is absolutely essential to understand how the system really works. I suggested that we concentrate on current operations, determine where and why the system fails and work out how it could be made better. I was ignored.

To test my car load flow through a network idea, I selected a minor commodity and tried tracking its movement through the traffic lanes across the railroad using existing CARFAX reports. Apple juice may have been a strange shipment to trace but the exercise was designed only to illustrate and highlight the reports available from CARFAX, such as they were, in an effort to better understand what we had to work with. Decoster thought it was a creative approach; Frank Stern was unimpressed. Frank did tell Allman about my idea to document operation of a freight train's journey across the railroad, but Bill was not interested in actual events, just theoretical models.

There was a rumor mill working in advanced systems, too. On May 26th the buzz was that C&NW was merging with Essex Wire Company which sent the stock up 19½ points to 126. On May 29th the stock hit 134. I escorted Frank Stern to that black hole, Ravenswood, on my 41st birthday, June 5th. He was looking for information and data about car movements and freight revenues because the CARFAX reporting did not contain any dollar elements. Every request he made at Ravenswood was instantly shot down in flames by the gnomes running the place. The next day I thought we could try and sort 1966 data by revenue/ton miles which occupied us for some time. Decoster thought that exercise was a waste of time and that we should try something else. On June 12th a new man appeared in our group, John Roser. A rough-cut young man from Carmi, Illinois, John was quick to grasp operating situations, but lacked polish and style in presenting his findings which greatly hindered his abilities of communicating good ideas to management. On the 14th I drove Roser and Stern to Belvidere to observe yard operations there. This was the new Chrysler plant set-up and was Frank and John's first view of the C&NW at work. Thinking nothing of it, I filed my expense account report for mileage incurred which gave Bill Allman a heart attack. The result was a notice to all staff that they had to get pre-approval on all expenses. Just how one did that was a mystery. His pompous directive was treated with hilarity by this fractious crew. On June 16th Frank and I interviewed Walter E. (Bud) Braun, assistant vice president traffic, and E. J. (Ernie) Mueller, rate department, about their responsibilities and inquired about the type of reports they received and how they used them. The sales department didn't get many reports and did very little with what they did receive. The rest of the gang went out to Proviso to observe operations there. In the office Warren Davis was doing a "secret" job for Gene Cermak involving tracing of cars. Another Arthur Anderson man now appeared, Dick(?) DeCaire—we wondered if he was a plant.

By the end of June I was beginning to wonder if I had made a career mistake. My railroad experience was untapped and I was not really learning much new. Cabin fever, I suppose. On July 1st we moved the entire department south two blocks to 120 S. Canal Street into the fifth floor of a brand new steel and glass structure that straddled the northside tracks of the Union Station. The entire floor was filled with fabric covered cubicles. The smallest cubes were 6 × 6 feet, but some of us had 6 × 9 foot spaces. Telephones were shared by two people through an opening in the common wall. My first thought was PRISON; my second, EGGCRATES. Managers and directors had offices with doors—and windows. Bill Allman occupied the corner office overlooking Canal and Monroe Streets. Frank Stern had a lovely view of the Rowley's Artificial Limbs building across Canal Street with its graphic depictions of their products. There was Muzak, too.

The speaker directly above my space quietly had some surgery performed on it after hours.

Personnel changes came fast. R. C. Stubbs was retired June 30th to be replaced by W. E. Braun. Larry S. Provo was elected president by the board of directors July 6, 1967. A new man, James Woodruff, was also elected vice president corporate development the same day. This new position was a Decoster-Cermak plan. There were also rumors of some big project in the operating department involving the use of 50 people borrowed from the sales department.

On July 7, 1967 I heard about a new computer simulation program developed by the Canadian National Railway called Train Performance Calculator (TPC). This computer application was to become an extremely important element in the development of the Western Coal project 1972–1980 and the locomotive training program in 1977–78. Another important personal event on July 12th was a luncheon with Dr. W. Edwards Deming, famous for developing the science of quality control. Not particularly successful in the American business world, in July 1950 Deming had found a willing audience in Japan where he was revered as a god. His methodology played a key part in the economic recovery of that nation's industry. I don't recall what we talked about although we sat together. DeCaire gave a presentation on statistical sampling which was Deming's speciality.

I reminded Bill Allman that he had agreed I could go to Houston for that September presentation to the Data Systems meeting. His reaction was surprise— because he had forgotten about that January commitment. On August 22nd he whined again about the expense, but finally okayed the trip. To chafe Shaffer a bit I showed him the program for the Houston affair and told him I was preparing a presentation. Things had not been going well for Jack Shaffer; I had been told privately by my former colleagues that the department was a disorganized and dispirited mess. Jack reacted visibly to my announcement of the Houston talk; I could barely suppress a smile while I rubbed it in. Later Don MacBean said that Shaffer asked him if he could take my place and give the presentation. The concept of a geographic database had now crystallized in my mind and Don was not privy to my new concept. Shaffer was beginning to realize his big plan to promote a new sort of service to attract industrial prospects using my innovative MAGIC program was in ruins.

That big program involving sales department people was a good move—assignment of some of the junior sales force to service with the operating department as coordinators of customer service. At last, oil and water were brought together and the program actually proved to be beneficial. On July 14th I ran into Jim Drinkard, late of the St. Louis office, in his new role as a coordinator. He was very enthusiastic about the assignment. As part of personnel shifting August M. (Gus) Malecha became the new Galena Division superintendent. Gus was a soft spoken person, a bit rumpled in appearance, but with a good head on his shoulders. We were to work together on many projects over the coming years. Gus had been an Air Force intelligence type who had spent his service years eavesdropping on the Russians from Turkey. Gus slowly made his way up the operating department ladder with his ever ready bottle of Maalox (an antiacid medicine) in his upper right hand desk drawer. A shop union machinists strike on July 16th against the nation's railroads lasted one day; Congress acted the next causing the union to back off. Trains ran normally on the 18th.

Dick DeCamara of the Illinois Central invited me July 19th to join the Data Systems' *ad hoc* committee for the Standard Point Location Code (SPLC). Frank Stern okayed the appointment as he really had little for me to do just then. Bill Allman was impressed and passed it on to his boss, R. David Leach, with a recommendation for approval. To my great surprise Leach approved despite the obvious need for expense money for travel to meetings. By October, however, Allman was convinced that the *ad hoc* committee's work was completed, which was not at all the case. I was offered the chair of the whole SPLC committee on October 19th. Dave Leach turned that down because of my "corporate commitments"—whatever they were. He must have felt a twinge of remorse because he then gave me a nice salary increase of $1,000, which raised my annual pay to $12,500.

I had another old commitment to give an address to the Great Lakes Industrial Development Council. I had been a director a short time from January 30 to March 31. I rode train #209 to Oshkosh, Wisconsin, July 25th. Harvey Buchholz and Jack Sommer hosted a reception at the Pioneer, a brand new hotel complex, partially financed by the real estate department of the C&NW. The Pioneer was constructed on the site of a former freight house. I suspect that part of the railroad's investment was the value of the site. My talk was keyed to the coming revolution in industrial development work made possible by the burgeoning computer technology. There was a lively question and answer period following my remarks, but I sensed that the main thrust of the talk had sailed over everyone's heads. I knew I had to do better in Houston next September. Returning to Chicago on train #216, I was bemused at the lack of enthusiasm in that group for considering new ways to meet customer expectations.

Dave Leach became a managerial nightmare. He issued office rules that our bristly group felt were designed for the guidance of children. He acted like a General Douglas McArthur running post-war Japan. Despite his overbearing style our group continued to grow and change. Don Fryer joined us from the operating department. I believe that operating wanted to get rid of him and gladly gave him to us. Don tended to be an autocratic old line manager. He was to have some unfortunate experiences when he eventually returned to operations as a division manager. While he added a sense of railroad reality in associating with these peculiar systems people, Don did not really contribute very much when he was with us. This job was more like a vacation for him; he kept a neat desk and regular hours.

By late July Maury Decoster was becoming discouraged by the way the marketing arm in the traffic department reorganization was being de-emphasized by new president Larry Provo. We noted the shift in our own work when Ron Cassen's monumental (expensive) data collection project called TDB (Transportation Data Base) was first delayed and then completely sidetracked. No one had the guts to just kill it because TDB was a basic need for any sensible management control system. The staff meeting held to start TDB was a disaster. Ron Cassen, its originator and director, should never have been allowed to speak in public. He just could not articulate what this grand plan was supposed to do and what benefits would be gained. The audience sat there in leaden silence amid the falling incomprehensible words like "heuristic" and "paradigm" hoping that the meeting would soon end. Ron's instincts were correct—you cannot manage what you cannot measure.

Despite the ominous start Allman took his directors, and me, to a meeting with Walter (Bud) Braun, then in charge of rate services to traffic. He thought that we should interview each official function in traffic and determine what information was important to it and how the data could be obtained. We went beyond traffic and interviewed Tom Harvey, superintendent of car service, Jim Ronayne, iron and steel, George Maybee in operating, Bob Macomber of freight claims, Don Guenther, Cermak's assistant, and Bob Wilson and his stations department. We discovered they didn't know what information they needed to run their jobs. We opened up the rules of the interviews by asking what pie-in-the-sky reports would be useful despite the fact the data did not then exist. Many of our interviewees still were completely lost when asked to think about non-existent reports. Of course we were barging in on their normal hectic work schedules and asking "stupid" questions—in a word, please go away, we don't have time for this B. S. Nevertheless, our three interview teams pressed on. Stan Boardman (off-line sales), Jim Feddick (grain), Chuck McGehee (TOFC), Bob Kunkel (coal), Frank Tribbey (intermodal), Ray Degnan (foreign freight). Frank and I learned that most of these managers were running their jobs in the style of their predecessors with little appetite for innovation because there just wasn't time, or desire, to experiment with new and untried methods. To make sure that we were not just reflecting the Chicago milieu, Frank and I began a series of interviews in Green Bay (Ray MacCarthy, Omer Torme, Jay Orr, Paul Hintz), Cedar Rapids (Ed Jeffries) and Sioux City (Erv Klover, R. N. Kennedy and trainmaster Rosenbaum). Back in Chicago we tried Jim Feddick again and John McClintic, formerly sales agent in Lincoln, Nebraska. The final interview was with W. B. (Bill) Selman at Ravenswood to talk about his freight claims department.

Frank was right at home in Ravenswood. It was a scene from the 1920s. Huge rooms were filled with lines of desks, heaped with paper, the atmosphere a blue haze of cigarette smoke. Frank Stern was a bean counter at heart and was proud to be a member of the New York Central's Green Eyeshade Gang. (After he left the C&NW, Frank worked as controller in a plastics manufacturing plant and for a sausage factory). We continued our meaningless interviews in New York on August 7th. There we saw Tom Hudson, Gerry Armstrong and Jim Wignott. Their problems while different from the on-line sales force were still centered around lack of information. The next day Frank I and rode the Pennsylvania Railroad to Philadelphia in a dingy coach and a hot and dirty diner. After a chat with our man John White, we flew to Pittsburgh where Jerry Hunt, the local sales department man, met us in United's Red Carpet Room for an interview between planes (I had invested in a lifetime membership from United). We caught a United Caravelle *Ville de Nice* to O'Hare. At least these little flights got me out of the sterile atmosphere of my cubicle and gave me chance to add to my collection of ZIPcode maps. I was surprised that even the U. S. Post Office in Washington, D. C. did not have a central collection of ZIPcode maps. These handy maps were locally produced and sometimes included in local telephone directories. It was the work of seconds when visiting a new city to carefully remove the map from a public directory—after all, the books were renewed frequently. I had a brainstorm that adding the ZIPcode to my geographic database would be a smart idea.

Interviewing people who had no idea of what information they needed was a farce. We tried to introduce ideas such as what commodities were shipped or received by their customers (they had only a general idea), what rates were moving the business (no one knew), comparing changes in traffic patterns over time (nice, but what do we do with such information?), car types required for each season (good to know), forecasting (an alien concept to every one). Ben Heineman had often said that he wanted a black box on his desk that would display, when asked, such information as the freight revenues today, number of cars on line, claims paid, cash on hand, obligations coming due, and the like. Heineman's dreams of data access plus a good deal more would take twenty years or more to become fact.

Frank liked this interviewing so much that he kept extending the process although we had quite enough information to know that we were wasting our time. We again tackled Ray Degnan (foreign freight) before boarding train #153 to Milwaukee to see Bob Tate and his chief clerk, Tom Minard. Back in Chicago we spent time with Roy Erickson (grain rates). We finally ended up the sequence with Bob Cumbey (rates in general), Ron Schardt (finance) and, the last one, Gene Cermak (sales-planning) He was hard to catch in August when the golf courses were open. I sensed from my interview with Gene, my former boss, that he was not as keen on his job as he had been. Frank Stern shared his summarized findings on September 15th with

Stan Boardman, Gene Cermak, Bud Braun, Maury Decoster and Jim Feddick. He did very well with the presentation leaving us all with the feeling that there was many a mile to go before anything useful could be made available.

Someone didn't like one of Allman's secretaries and fired her when he was out of town. His second girl, Phyllis, was in a panic because of the endless stream of dictation emanating from Bill's portable recorder and midnight scribbling. Despite her fears that she could not handle the flood of paper, she soon discovered her innate basic strengths enabled her to keep the first secretary's job under control. During this period the company instituted a testing program to identify hidden talent. I took my test August 15th and found that it included logic, math, word association and a personality profile. (Three years later I learned, unofficially, that I had tested "high"). One direct result of the screening program was identification of potential computer programmers from the railroad's lower ranks such as brakemen, clerks, and others. The systems department organized training classes in basic computer language and logic for these recruits. This approach saved a lot of money as hiring through an agency required payment of high fees. Allman was still adding to his advanced systems group: Richard Lewis, a systems manager and S. R. (Bob) Tamura, a programmer formerly with the U. S. Air Force on Guam. Of course Bill lost people too: Dick DeCaire, the ex-Arthur Anderson man, departed August 31st after a scathing exit interview in which he blasted Bill Allman as an immature and clueless manager whose arrogance antagonized everyone he encountered. He had a lot more to say—all of which ran off Bill like water off a duck's back. We speculated that Bill probably had not even listened to DeCaire's verbal diatribe. Bill's insensitivity was clearly demonstrated by his thoughtless habit of convening meetings during the lunch hour. Dom Violante, Ron Cassen and John Roser were horse collared into one of these sessions and had to sit there while Bill ordered his secretary to fetch him a sandwich and a carton of milk which he then had the temerity to consume in front of his hungry audience. That story, with embellishments, was repeated far and wide. Yes, DeCaire's comments were on the mark.

After Labor Day I took my family on vacation. We flew to Houston and had a pleasant time on the beaches of Galveston before going up to Houston for the Data Systems annual meeting. Jack Shaffer had asked Bill Allman to get a copy of my address, but I had already left town. At the committee meeting on September 11th I was again asked to take the chair and again declined as instructed by Dave Leach which was just as well because I really did not know the members of the group very well and they did not know me. Bill Allman attended the main meeting and even acted like a normal human being when we were with outsiders. On September 12th I delivered a 45 minute talk to about 250 people on the subject of using geographic coordinates in solving marketing problems. I used the Jones and Laughlin Steel study as an example. It seemed to be well received. Years later, an attendee recalled some of the slides I had used which I felt was a pretty good indicator of the talk's reception. Reporter Nancy Ford of *Modern Railroads* magazine came up after the presentation and said she wanted to do a story. Allman also congratulated me. I told him I had again turned down the chairmanship of the SPLC committee just to let him know that someone appreciated my abilities and that I was sticking to Mr. Leach's orders. Don MacBean and Dave Thelaner (systems department) also attended the meeting. After getting clearance from R. David Leach to grant Nancy Ford an interview, we worked out a story line September 18th which was an extension of the Houston talk. I was concerned that there might be unresolved questions surrounding the ownership of MAGIC. Nancy said, "Let's get the snake out of the box and kill it." On September 26, 1967 in a meeting with Nancy Ford, Bill Allman approved the text of her article and said, to Frank Koval's (public affairs) amazement, that advanced systems was far too involved in other projects to be concerned with this MAGIC thing. Shaffer, who was not present at this meeting, would not have agreed. Two days later I repeated the Houston show for 50 of our people. Jack Shaffer did not come but sent his chief clerk, Keith Peterson, with a tape recorder. Neither Cermak or Decoster showed up.

On September 25th I began to study a geography-based tariff called *Leland's Open and Prepay Station List*. This publication of a privately owned company based in St. Louis listed all of the actual active stations on all railroads in line order with their assigned station numbers. The tariff further contained all the notes and conditions surrounding each station such as whether it was an open (manned) station or if shipments had to be prepaid. Ron Cassen's Transportation Data Base (TDB) was going to need a lot of geographic material to make it work. This started a chain of thought. I knew there was a freight accounting station number and now here was another series for the rate department. Since there were more than one set of station codes in existence, shouldn't they be pulled together to provide a method of relating freight revenues with car movements? A station number bridging file could be a valuable asset. I was desperately bored with the non-sensical assignments Frank had given me and needed a good data collection job to keep from going stir-crazy. Allman decided that the TDB approach needed re-doing and came up with a modification renamed Transportation Data System (TDS). The authors of the original TDB, Frank Stern and Ron Cassen, were in despair at this rewrite of their project. On October 25th Allman ripped Ron's work apart and rewrote the proposal for TDS—again. The TDS proposal (called "tedious"by us) was in final typing in November. Allman in his characteristic juggernaut approach to practical matters insisted that the typing of the proposal be done directly on mimeograph masters (yes, we still used mimeographs). Correction fluid was banned so that a single typing error could not be repaired. Consequently, each page of typing had to be letter perfect, something achieved only after repeated attempts. The work took at least five times as much effort as just making a simple photocopy master. Bill, however, bruskly refused to heed our suggestions—he knew best.

C&NW purchased the Philadelphia & Reading Corporation in September. That brought in Fruit of the Loom underwear, Lone Star Steel and Acme Boot into the corporate family. September 18th was the first day for our new vice president of marketing, John (Jack) Kane, formerly with an aluminum pan manufacturer. Rumor said that the C&NW was looking at the Ahnapee and Western Railway in Wisconsin. Our railroad was turning into a conglomerate. By late November speculation was that Heineman's interest was shifting away from the railroad. In October Provo was beginning to worry about the physical condition of the railroad as W. H. (Heinie) Huffman's work with the merger committee was finding that reports on bad track conditions were alarming. While track condition later proved to be the key element to the railroad's operations, no one in management in 1970 really understood how important it was. Much of the C&NW's woes over the next decade could be traced to its deteriorated physical plant. If the problems caused by bad track had been recognized sooner and corrective programs begun, the C&NW could have been a stronger property through the 1970s.

Jack Shaffer was not done with me and MAGIC. In October he shoved Don MacBean aside and hired an outsider to handle industrial development research. On the 19th he tried once more to have me come back. Knowing that he had no position from which to negotiate, on October 30th I demanded $16,000 and all rights to MAGIC. After a chat with Decoster and Cermak the next day, I decided that swimming with that shark would not be healthy. The industrial development department had fallen from management's favor and was now utterly at sea. I was more and more certain that MAGIC had commercial value and was developing some clear ideas about how to further enhance the database. I told Allman November 2nd that I was staying in advanced systems; Shaffer was miffed. My inside informant reported that Jack was combing the department's files to find the MAGIC material. Of course, he couldn't find anything as it had been all quietly removed a long time ago. By the end of November Shaffer was ready to seize MAGIC claiming that he had outside counsel's opinion that the material belonged to the railroad, that Decoster and Cermak's approval of my ownership was beyond their authority, and that any attempts to market MAGIC would be met by injunctions, *subpoenas*, and other legal actions. In a meeting with Dave Leach and Bill Allman, I adopted the posture of being an unjustly treated party and that all the hostility was on Shaffer's part. I also said that I would be happy to cooperate for the company's best interests, knowing full well that I had a second copy of the handwritten master carefully hidden away. R. W. (Bob) Russell, my old friend from the law department, worked with Decoster to try and attach a value to MAGIC. The concept was so alien to their level of experience that they were unable to ascribe any value to the data collection. On December 8th I agreed to hand over the Dun and Bradstreet city list with the attached coordinates to Jack Shaffer. On December 12th Don MacBean called to report that the *Modern Railroads* magazine containing the Ford article was all over the railroad. What great timing! I carefully "packaged" the 3" thick book and turned it over to Bill Allman on December 18th and requested a signed receipt. Because I was so "helpful" in resolving this controversy, I was rewarded with another $1,000 salary increase effective January 1, 1968, bringing me up to $13,500. Of course all of the data may not have been quite accurate, but that was a risk Shaffer had to take when he tried using strong-arm tactics on me. Years later, a friend in real estate (which took over the industrial development department about 1980) and who was on the verge of leaving the company, quietly returned the infamous book—by now quite valueless. I carefully burned it.

There was a great deal of interest generated by that magazine article. Fantus Company in New York now thought the use of coordinates was a terrific idea forgetting that their Chicago president, Maurie Fulton, had already pooh-poohed the concept. Art Johnson of the Baltimore and Ohio requested a copy of my Houston presentation for his boss. I gave him the text, but not copies of the slides. A highly respected railroad economist, Herb Whitten, invited me on November 29th to make a presentation in March in Washington, D. C. I asked permission to do so. Also in November Herb Landow, formerly with the Illinois Central and now with the accounting firm of Peat Marwick and Mitchell, proposed a profit measuring system to advanced systems. We were so far from having any reliable data available that the idea of measuring profit was ludicrous. On November 22nd we heard that Ed Upland, the traffic department's grain man, had been booted out, but had landed a job in the operating department where he coordinated movement of grain cars to elevators during the harvest season. His replacement was James R. Feddick who later became a vice president with Archer-Daniels-Midland. Pretty good for a kid who started out in the freight house in Green Bay, Wisconsin. Elsewhere, I heard of complaints about the disorder in computer operations at Ravenswood; even the unflappable Ed Lillig, a manager in the systems department, was quite upset at the inability to get simple jobs done in the computer facility. It was no wonder because the union shop rated computer machine operators as an entry level job which often placed inept people in charge of the operation of some very technical—and expensive—equipment.

As the year ended I was feeling pretty good about the way things had turned out. I had resolved the MAGIC ownership problem and was rewarded for being a good boy. I also had set a course in a new direction that was to run parallel to and quite separate from my railroad career. My geography "horse" was almost strong enough to ride making the railroad into an iron "hobby horse" where I might yet have some fun in a business I truly loved. This account, however, will deal only with the railroad side of my career; the geography aspect is quite another story.

22

ADVANCED SYSTEMS STUMBLES, 1968 . . . I began to live a dual life this year. My public face was the career with the C&NW while the private self was deeply involved with computerized geography and spatial analysis. Frank Stern lobbied hard for the manager of market research position but Maury Decoster of corporate development chose an outsider, Bill Stevens from Celanese Corporation. One of my sources reported that Decoster was then getting $32,000 in annual salary, more than twice my $13,500. As much as I respected Maury, I felt that I was worth a lot more than what I was earning and resolved to do better. Chuck Towle was so depressed by his prospects in industrial development that he carried through with his threat to resign as the manager in Minneapolis. As his star fell, mine was slowly rising. Dick DeCamara and Art McKechnie of the Illinois Central tried to recruit me. I was flattered, but the prospect of developing my private business and making some real money while keeping my invested time going with the C&NW seemed a smarter course to follow. Undeterred by my decision to stick with the C&NW, DeCamara and Tom Desnoyers tried to establish a code manager position at the Data Systems division for either Don MacBean or me, despite my repeated protestations that I did not want to relocate. January also brought me a new title in advanced systems: Systems Planner, a step-up from Senior Systems Analyst. Others in our group were showing signs of moving on. Both Jack Rezek and Warren Davis were getting their fill of Bill Allman's absurd behavior. It was more and more obvious to me any attempts to gain company loyalty from these computer people was a waste of time. They were dedicated to their craft not to the company. A new manager appeared in January, Robert (Bob) Kase. Born in Philadelphia, a graduate of the U. S. Naval Academy, and married to a beauty queen, Bob had the good looks, people skills and calm, decisive mind that marked him as a good manager. He came to the C&NW innocent of the turmoil within the organization, became embroiled in the chaotic computer systems struggle and was spat out during a downsizing.

Now that Shaffer had his hands on the MAGIC code book, I distanced myself from any contact with the program. At Shaffer's frantic urging, his long delayed circus band began to play. Mostly they had clowns instead of skilled artists. Unbeknownst to Shaffer, Bob Donaldson, the bright young IBM salesman who had contacted Don MacBean and me last year, had a customer that wanted to purchase a copy of the still unformed MAGIC System. Not yet ready to begin my own geographic business, I referred him to the new vice president of marketing, Jack Kane. Kane was thoroughly hooked on the concept of MAGIC and pushed Shaffer to show some results. Since Jack Shaffer had failed to lure me back to his department, he now had to find someone else willing to tackle the MAGIC project. He hired young Edward L. Nordan to manage and promote the database. "Fast Eddie, as he was soon labeled, was a bit strange. Wary and suspicious of everyone's motives, he was a product of a New England military school and a stint in the Army where he had developed all the earmarks of a low level intelligence officer afflicted with the paranoia that went with that assignment. Years later I looked back at Ed Nordan and was reminded of the notorious G. Gordon Liddy, the chief Watergate burglar in 1975. It didn't take a rocket scientist to realize that the manuscript of a database was useless in that format, but Ed's computer knowledge was so limited that he could not fathom just what he should do to make the data useful. He asked me for help January 17th; I side-stepped him by telling him that he must get Dave Leach's approval first as I had agreed not to get involved with that subject. I was then very much occupied with coding new sales and marketing territories. Ed must have gotten clearance because he and Don MacBean showed up the next day with an armload of MAGIC material. First he wanted to know what I thought about a sale to IBM's customer; I replied "sale of what? That pile of paper?" Besides, vice president Dave Leach would have to approve of any sale of software to an outside party which was really a bogus answer because MAGIC was only data not software. Nordan tried to weasel me into admitting that I had a copy of the database. Of course I did, but didn't admit it. His blandishments were as transparent as a glass of tap water. They went away—Nordan mystified and irritated; MacBean with a secret smile on his face. Eventually, Ed figured out the next logical step and asked for help from the systems department. Leach told Allman to assign someone to the project; Bob Kase was it. Poor Bob, who had no knowledge of the history of MAGIC, innocently showed me the correspondence from vice president Kane and said he would keep me informed about progress being made to get the database into a useable form. Another new player came into this brew. Deepak Bammi from New Delhi was assigned to do the programming for Nordan. Since my cubicle was adjacent to Bammi's and we shared a telephone through an open space in the common wall, I was able to keep a covert eye on developments. Now, Fast Eddie began to worry about expanding the file and performing maintenance. Kase arranged to get the data keypunched at Ravenswood, amid a lot of growling from the machine room manager. Nordan now decided that my name MAGIC for the data base was tainted and came up with GAMIC as the working acronym.

Kase now was faced with the ancient mariner's problem—how to calculate circular arc distance between two points whose coordinates are expressed in decimal latitude-longitude degrees. He fell back on that venerable Navy reference work, Nathaniel Bowditch's *Practical Navigator*, first published in 1797 and still in print from the Government Printing Office. The formulae in Bowditch involved use of haversines and other mathematical abstractions that had been discarded by the United States Geological Survey many years ago. The calculation routines were impossible using the railroad's adopted standard machine language, COBOL, simply because that computer language did not include the pi (π) radical.

By July GAMIC still was not operational. Shaffer pressed for hiring an outside counsel to patent the GAMIC system. This effort must have failed because I later obtained my own copyright on my new CENTRE-US series and was advised that it was the first copyright ever issued on a machine language database. My source in traffic told me in August that the GAMIC materials were locked up in a limited access repository so that the company could maintain first rights should a similar program appear on the market. What a charade! A comparison of the old MAGIC (GAMIC) material with my new format would have shown little in common; the contents, coverage and formats were completely different. GAMIC languished in its locked box. If it was ever used in a practical application, I never heard about it.

That MAGIC-GAMIC sideshow was, to me, irrelevant. I was now off in a new direction that sprang me out of my cubicle "jail", allowed me to meet new people and become involved in some very interesting activities. On January 19th I had been asked to participate in the planning committee for the Data Systems Division's annual summer convention to be held in Chicago. Frank Dillon of the CB&Q was the chair and asked Leach and Allman for their approval in my participation. The 5th annual winter forum meeting of the DSD was held in Philadelphia. Bill Allman, Louis Haberbeck from systems, Bob Lehnertz, a new recruit to advanced systems, Don MacBean and I flew to Philadelphia on January 21st in the company plane, N403W. The next day I participated in my first SPLC committee meeting. Remembering my old traffic department manners, I called on John White, the local C&NW salesman. He was distraught because at age 61 he was being replaced in the coming realignment of sales and marketing. The planning committee for the summer convention met on February 12th. I was assigned to entertain the ladies, an odd thing to ask me to do. Vic Rymarowicz, the DSD secretary, came to our office from Washington, DC, to help me make the proper arrangements for the meeting next fall.

Back at the funny farm, as I began to style Allman's operation, we caught him violating his own instructions when he snuck away from the Philadelphia DSD meeting to visit his parents in New York. Why that wasn't OK is now obscure, but he had issued a written directive against mixing personal and company business. On January 24th he came up with another hastily issued order stating that the expense account reports in the future would allow only 25¢ as the maximum tip for a hotel porter. Soon afterwards, Dave Leach whined about the high cost of the "extravagant" meals we were claiming on our expense reports. He reminded me of our old bridge engineer, Art Harris, who used to dole out carfare to his bridge inspectors. Coming from the free and easy old traffic department, I found these petty attempts to control expenses ludicrous. Another new recruit to advanced systems at this time, was Richard Ogle, fresh from college and very, very reserved. Assigned to Frank Stern and me, Richard was sure that we could change the world by restructuring the railroad's information systems. Soon he began to see that any positive changes were going to take a long time. In late January Frank and I met with Al Handwerker of finance and Maury Decoster of corporate development. They were quite annoyed to hear that the time horizons for getting something useful from systems were so far out. I had developed an Area-Region-Branch (ARB) code structure for Decoster's new sales and marketing areas. It was a cascading numbering scheme which provided a quick and easy way to aggregate statistics at various levels of detail. This simple building block scheme along with other similar data collection subsystems had to be implemented before any grand system could be designed.

I introduced John Roser to the Galena division dispatcher and gave him a tour of the 1948 CTC machine installation. I also wasted a lot of time discussing future machine-generated reports with Cermak, Handwerker and Decoster—reports they would never get. More importantly for the railroad's future, Ed Garvey and I learned to code railroad track, assemble "trains" and establish speed limits for the Train Performance Calculator (TPC); I got quite good at it. This powerful computer application along with our ability to quickly translate track code for the program was to pay enormous dividends to the company in the future. On February 29th I tried to interest J. R. (Bob) Kunkel into a trial TPC run for his coal trains, but he was not ready to use something so strange to his staid and static world of coal rates. I think that we could have helped him better manage rates on coal by determining the relative costs of movements from different sources. In five more years he would be ready to listen.

On February 1, 1968 the New York Central and the Pennsylvania Railroad merged into Penn Central. The two cultures were now forced together and began their disastrous downward spiral. Someone in our shop called the new railroad PANYC, pronounced "panic". The term "green" team and "red" team was then heard for the NYC and PRR.

February 7th saw the demise of the old traffic department. A huge staff meeting was held that day at the Ambassador West Hotel in Chicago. Jack Kane, vice president marketing, addressed the group the next day. He dwelt on the great MAGIC system that was going to revolutionize marketing and introduced Jack Shaffer as the magician. Of course the fact that no system existed, no software had been written, and the data were just handwritten entries in a book of printouts was not mentioned. While Bill Allman and Frank Stern were in attendance, I was excluded as too lowly a person even though I was the inventor of that marvelous, earth-shaking scheme. After the meeting, the newly assigned marketing people moved into their assigned quarters on our floor at 120 S. Canal. My old pal in Rockford, Donald L. Gunvalson, was by-passed in the shuffle. Don MacBean escaped from Shaffer's industrial development department by joining Vic Preiser of corporate development on February 16th. By May, however, Don was as malcontent as ever. Gene Cermak was stuffed into Frank Tribbey's little windowless 2 × 4 office—a big comedown for him. Gene told me March 22nd that the "pots and pans guy", Jack Kane, had betrayed him. Gene's " little shadow" Don Guenther ended up in a dead-end job checking customer lists. There were rumors floating around February 19th that Ben Heineman was slated for a federal government post at HEW (Health, Education and Welfare). We also speculated (hoped) that Bill Allman might be drafted into the military as the Viet Nam war was raging, but felt the military services would suffer unduly should that happen. We now had a new vice president of operations: Harold Gastler from the Toledo, Peoria and Western. Somewhat ironic, given the nasty court battles we had with that small railroad in Peoria. George Paul, a group vice president in charge of personnel resigned and returned to his old job on the Southern Railway. My friend from the law department, Robert W. Russell, was appointed vice president personnel. Bob had been a navigator in the Eighth Air Force during the war. He was an intelligent and very bright person who had been a good C&NW attorney and who was now to manage an important, and delicate, function for many years.

A video terminal displaying tracing of freight cars was demonstrated in our office on March 4th. Now here was something that had real potential. While cathode ray tubes were commonly used in television sets, their use for data display was new to us. At this time Dick Ogle and I were immersed in designing mock report formats for Frank Stern. The coding pages were 11″ × 17″ pages lined to represent IBM printer format driven by the 80 column punch card. Dick and I spent endless days putting crosses in little boxes trying to design readable reports. I absolutely hated every moment of that job. Tom Rathenau at Ravenswood was responsible for getting the cards punched and then printing these stupid report mock-ups. He said to me that he was ready to use Frank's head as the ball in a soccer game. We also had a demonstration of Automatic Car Identification (ACI) by a committee of the Data Systems Division chaired by Arthur E. Pew III. The March 8th presentation looked good on paper, but optical reading of a barcoded plate on the side of a moving freight car (or locomotive) proved to be too prone to misreads because of dirty plates, poor lighting, train speed and incidental damage. The railroads had installed ACI code plates on a great percentage of the car fleet before giving up on them. There was some effort to use passive transponders mounted on the cars. This latter technique did not get very far either; the whole concept of identifying each piece of equipment flying past on a train was rendered obsolete by the rapid improvement in data communications.

C&NW stock was now off the market and replaced by Northwest Industries @$116 per share (March 27th).

I flew to Dulles Airport March 18th to deliver a talk the next day to the Physical Distribution Institute. This was a commitment I had made to Herb Whitten before I had changed jobs. I met Harry Bruce, then with Spector, a trucking firm, Dr. Church of the Bureau of the Census, Warren Stockdale from the Baltimore and Ohio Railroad,

Jim Wahl of General Mills, a professor named Robinson from American University, and, of course, Herb Whitten, my host. The concept of a geographic database really made an impression. I skirted the issue of who actually had such a database and stuck with concepts. I went to New York on the shuttle in a BAC 400 and then back to Chicago—the fare was the same.

Ron Cassen, our advanced systems manager whose efforts to create a transportation data base had been so badly mismanaged by Bill Allman, refused to sign his annual evaluation report. Ron, normally a placid individual whose slow, thoughtful speech tended to lull his listeners to sleep, worked himself into a smoldering rage that finally erupted in a bitter confrontation with Allman. As before when Dick DeCaire left, Allman shrugged off the vitriolic accusations of lacking management skills, inconsiderateness of others and general incompetence. Cassen should have saved his breath. Allman seemed to be unconscious of his surroundings and business situations. Bill would leave a meeting with his boss, R. David Leach, and Bill Zimmerman, head of the systems development group, not realizing that the thin red line circling his neck meant that his head had been figuratively sliced off.

The so-called Northern Lines merger was approved about April 22, 1968. The Great Northern, Northern Pacific, Chicago, Burlington and Quincy and the Spokane, Portland and Seattle were to become one big railroad. On May 6th Art Pew stopped by for a visit. He was job hunting because the former CB&Q officers and staff in Chicago were being moved to St. Paul, Minnesota. Arthur, an heir to the Pew millions, really wanted to stay in Chicago. Railroading was his life, however, and he did move his family and private railroad car to the Twin Cities.

TRANSPORTATION GEOGRAPHY Everything is somewhere. "Everything" includes physical objects, concepts, activities, or data which can be described by words, colors, shapes, sizes, densities, or weights. "Somewhere" is a locus which also may be described by words or by other symbols—reference to a known place, description of the place, perhaps by its attributes such as atmosphere, ambiance, or historic significance. None of the above is intuitive to the "stupid" computer—instructions must be conveyed to it in machine readable language. Management in the railroad industry was similarly crippled by being unable to relate anything to anywhere unless specialized reporting was organized. From the very first days of the railroad industry, revenues were of vital importance to management. Establishment of freight and passenger accounting functions followed soon after a company opened for business. The accountant quickly realized that simple ledger entries recording income by stations were difficult to tabulate especially as railroad operations expanded. Similarly early railroads often named their shiny new steam locomotives for company officials, stockholders, politicians, war heros, local places, animals, birds or atmospheric conditions. Keeping track of individual engines identified by name became burdensome—a situation eased by simply assigning numbers to each locomotive. With the advent of telegraph dispatching, engine numbers became a necessity in order to safely operate the road. Establishment of station numbers represented one of the earliest coding systems instituted by the railroads. In the beginning little forethought was given to assigning numbers to stations. Every railroad had its own ideas and, even within the company, stations were often known by several different numbers depending on what function was involved. Most often station #1 was the company's starting terminal with ascending numbers along the route. If the next station was identified as #2, a problem could arise should an intermediate place be established at a later date between #1 and #2. Some roads used multiples of five digits to provide for future expansion. The C&NW's operating department used an internal station number based on the miles from Chicago, *e. g.* DeKalb, Illinois was station #57. That worked until branch lines were constructed and other companies were merged forcing use of new series of numbers to avoid duplications. The freight accounting officer could not (or would not) utilize either the car accountant's or the operating department's number schemes much less the number used in the tariff (heaven forbid)! On North American railroads a Freight Station Accounting Code (FSAC) was applied to every point where freight revenues were generated which, of course, left out many passenger only stations. An FSAC number series was created by each railroad which then filed its list with the Accounting Division of the Association of American Railroads (AAR) in Washington, DC. The AAR annually published a one inch thick book which contained the freight station numbers for all of the railroads in the country. These accounting station numbers were the basis for settling interline accounts between the railroads. The FSAC number was the station number used on freight bills. On the C&NW the FSAC number for Chicago was station #5—there was no number 1.

As mentioned above, many railroad "stations" do not generate freight revenues; some only see passenger revenues or may be freight yards, interchanges with other railroads, or shop facilities. These "stations" were omitted from the FSAC coding. The car accountant, who was responsible for tabulating car miles and related statistics, developed his own set of station numbers which, on the C&NW at least, did not correlate with the freight, operating or tariff set. Similarly, the baggage and express function used a number set that suited their special needs. The C&NW's passenger department also used a station number set which was mostly an adaptation of the operating department's numbering scheme. In the rate department station numbers were published as a tariff filed with the Interstate Commerce Commission as a legal publication. Physically, this well-thumbed book with gray covers was two inches thick. Published by a private tariff printing company in St. Louis, *Leland's Open and Prepay Station Number List* contained station numbers for all railroads. It was said that the originator of this compilation was Peter Leland, who had come across two old fellows who were rate clerks on the Missouri Pacific. These men had made lists of railroad station numbers for all the railroads with which the MP interchanged freight. Leland hired them to obtain their handwritten lists, provided them with the booze they were so fond of, and created this tariff used by all railroads in rate matters. This publication was the only one that lists stations in number order as they occur along the railroad lines. Station numbers are maintained in station order but may not be a closed set like 101, 102, 103, 104 . . . To accommodate new entries the number series are usually in an open sequence like 101, 106, 110, 115 . . . Should there be a new point to be added it was easy to drop it into a vacant number slot. Should the slots all be filled, the publisher then used decimals, *e. g.* 115, 115.15, 116 . . .

On April 25, 1968 I began investigating these disparate station numbering schemes used on the C&NW because it was obvious that there was a terrific discontinuity of coverage between them. After talks with accounting, rates and operating, I found FIVE different station number lists in daily use and none of them contained common numbers for the same point. No wonder we could not design computer systems to integrate rate, revenue, and operating statistics. Despite Frank Stern's opinion that this was a useless project, I gathered the three major station code lists (FSAC, Car, Tariff) into a single file using the name of the station as the base and hanging the appropriate numbers on it. Some stations had three station numbers, some only two while some reflected their uniqueness with just a single number. I called this file CLASP for Consolidated Location and Station Project. Further investigation of the CLASP file showed 16% of the named places were no longer active. That was just so much wasted effort to keep churning these old names through the system. I split the CLASP file into three sections: Active, Hospital, and Cemetery. Active stations were important freight revenue, car movement and tariff points. The hospital list contained places that had little freight revenue activity, or car movement data or just appeared in the Open and Prepay tariff. The cemetery list detailed stations that were on abandoned lines and had never been purged from the station lists, or were no longer required for any company business. Transferred to punched cards and printed the CLASP file vividly showed what a ragged mess it was. Even producing the first printouts seemed to reflect the aura of chaos surrounding the project. Jack Rezak, our laconic programmer in advanced systems, went with me to the computer center at Ravenswood June 27th to run the station program. First the card reader failed and then the console blew a fuse. Jack was philosophic about the wasted time—he was planning to leave for another job anyway. By July 26th I had determined that we had to clear out the "hospital" and the "cemetery" portions of CLASP. I sorted the file by operating divisions and eventually got Ravenswood to print them (August 7th). The superintendent of stations agreed to send them out to his people in the field. Nothing happened until I gave Pete Rasmussen in the stations department a firm push to get the program underway early in October. By the 25th returns began to trickle in from the divisions. There were some amazing stories attached to some of these so-called "stations". One station had been inactive since 1917. Tabulating these cripples by November 18th we had enough information to obliterate 10% of the active station list. Places supposedly active like Blue Mound, Wisconsin, for example, were found to have no sidings and no facilities for handling freight business. The divisions were delighted to clear away this dead wood. To fill in this ragged table and make cross referencing possible, I requested that the freight and car accountants assign station numbers to the blanks in their sequences. There was a little grumbling at first, but the numbers were plugged in. Likewise, Kurt Ramlet, our tariff officer, obliged by filling the blanks in his Open and Prepay list. To improve CLASP even further, I tucked in the new Standard Point Location Code (SPLC). The resulting database gave us our first opportunity to relate car movement with revenue. While Bill Allman gave me good marks for this job, my supposed leader, Frank Stern, seemed embarrassed because he had not realized what positive results could come from this simple approach. No one wanted to "do" data—just design systems that depended upon data that did not exist.

This station code confusion was one of the principal reasons the AAR had established the Standard Point Location Code (SPLC) committee. In 1965 the railroads and the trucking industry's tariff association had agreed to jointly create standard codings for commodities and named places. The railroads took on the commodity coding which resulted in the Standard Transportation Commodity Code (STCC) while the truckers did the SPLC. That there was basic philosophic difference in defining geography by the highway boys and the railroads was not immediately recognized. The problem arose because none of the people involved in those early formative meetings were geographers or had any experience developing coding systems. I was told that one of the key railroad representatives thought more of going for his morning run on the beach than attending those crucial first meetings at Swampscott, Massachusetts, where the SPLC coding rules were established. The primal concept—what defines a "point"—was not addressed. A point or place name to a trucker was some undefined area around a named city, town, village, borough, or crossroads community. The railroads, however, viewed the SPLC as representing a finite point called a station. Mashing the two concepts together created a lot of friction, muddled definitions—and, sometimes, outright hostility. I got some curious answers when asking railroad men where, exactly, was Minneapolis or Kansas City or Chicago or Philadelphia? When asked one of our officers said "Chicago" was located about 400 feet north of the bumping post on track 4 of the C&NW's Chicago Passenger Terminal because that was the equivalent milepost zero as established in 1848 on the Galena and Chicago Union Railroad. That arbitrary point had been translated from Canal and Kinzie Streets through the reconstruction of 1908–1911 to the newly constructed station at Canal and Madison Streets. For exact geographic positioning that definition presented real problems.

Personnel changes increased during that summer of 1968. John Roser resigned May 10th for a new job at the consulting firm of Peat, Marwick and Mitchell, and left a week later. Allman hired a new office administrator May 3rd—Larry Creighton. I wondered where he found these people! Larry seemed to have little education and few managerial skills. He was a pleasant young man, somewhat crudely assembled, who consistently demonstrated his inability to deal with these peculiar systems people. Jack Rezak left for IBM on July 29th. Max Davis in marketing quit without notice September 11th. Bob Macomber of freight claims resigned and was replaced by that jolly Irishman Jim Flynn August 28th. George F. Maybee escaped from advanced systems August 29th to joyfully return to the operating department. (George was to be the last division manager at Boone, Iowa, in 1995). An old associate in engineering, Gene Bangs, once an instrumentman on the Wisconsin Division, died July 25th. Vic Khaitan, one of our advanced systems programmers now half a world away from his native land, liked life in this country very much. Currently, he was working on a survey of commuter motivation. To ask a native of Calcutta to determine how Chicago commuters regarded their transportation service seemed a bit odd. Vic labored on this abstruse subject all summer. He was becoming thoroughly Americanized which worked out well for me. I liked tea while Vic preferred coffee. His uncle in the Indian parliament regularly sent shipments of the choicest tea—golden flowery orange pekoe (GFOP)—from the family tea farms. Vic passed the parcels directly on to me. Not only did I obtain some very fine tea, but the high value postage stamps went into my collection. At lunch time in those days our gang often played a game of hearts with as many as seven players and just one deck of cards. There was no science to these games, it was just a collective effort to stick someone with the black queen. We all knew when Vic had the Queen of Spades as he would always break out in uncontrollable giggles.

I lunched on May 10th with our new research guy, Bill Stevens, and Douglas E. Christensen, an acquaintance from industrial development days. Doug was with the C&EI and was soon to become a C&NW employee, as we shall see.

Because of my familiarity with the railroad's geography and stations, Ray Schaffer of public affairs and Pete Rasmussen of stations asked me for help in adding the newly-acquired Chicago Great Western Railway to the C&NW system map. That merger occurred July 1st. The railroad map in Iowa was a tangle of intersecting branch lines whose linkages were quite unknown—rationalization was urgently required. Isolated in the systems area, I then had little concept of the unexpected woes soon to face the operating department caused by the CGW merger. Iowa became the center of increasing problems of not only on the old C&NW and CMO lines but also with the additional routes added by the M&StL merger in 1960. Now the problem was to be compounded by adding the CGW, FDDM&S and DM&CI. The operating department's Iowa indigestion was not to be cured until hundreds of miles of redundant branch lines had been abandoned and a strategic piece of the Rock Island acquired in 1980. In 1968 we were barely aware that there was a problem.

On May 29th we had a big session on freight claims with Vic Preisser and recently appointed manager Jim Flynn, a Mr. Ducret, Bill Allman, Keith Waldron from systems development, with Frank Stern and me bringing up the rear. Or so I thought, until they all pointed to me to be the leader of the meeting. Later that day Ted Struve, the third member of the corporate development group, Stern and I wrestled with development of a branch line reporting system. We were also involved with an accident and loss prevention study. While the freight claims problem was interesting, it was really not my kind of project. I was relieved when the project stalled in July; Preisser and his group had serious problems in defining their terms and goals. In June Susan Freleng appeared on the scene as another outside consultant with a lot to say. Her cynical manner and tone of voice set my teeth on edge. Sarcasm coupled with a sour view of the world was her style. At least Dorothy Parker had clever turns of wit that Sue never displayed. (When she eventually left, she opened a kitchen wares store in the Hyde Park neighborhood in Chicago). She was part of the freight claims team that expected systems to deliver a useful product in a few weeks; actually, they should have planned for years because the data did not exist in any form that could be given location, reliably collected and meaningfully organized.

On June 9th Don MacBean, Bill Allman, Bob Kase, Bob Lehnertz and I flew to Montreal in the company plane, N403W. On the 10th the SPLC committee met and welcomed two new members, Charles S. Holden and his "scientist" George Chabot from the Canadian National Railway. The Data Systems meetings held the next day were interesting, but I told Allman I was leaving early on my scheduled vacation to attend EXPO then underway in Montreal. He was annoyed until I pointed out that Kase had already started back to Chicago after lunch and Bob Lehnertz had gone home yesterday. Herb Landow, Clark Hungerford Jr. and I went out to the fair. The next day I rode a CN train to Ottawa to make a call on Mr. Emmerson at Canada Post. I also met with Frank Pope of Dun and Bradstreet Canada. I was learning that market analysis did not stop at international boundaries. My geographic database should be expanded to include Canada. On the evening of the 12th I rode south on a D&H train from Windsor Station, Montreal, to New York where I made a courtesy call on our New York office the next morning. I had a meeting with Bob Baechtel and Dick Looney of Dun and Bradstreet in their Wall Street headquarters where I explained the transportation industry's effort to establish a geographic standard. Baechtel wanted a cross-reference table of SPLC and ZIPcodes. Dick Looney said my coordinate file was five years ahead of the marketing industry. On a rainy afternoon, I boarded the southbound *Silver Meteor* on track 12 at Penn Station for a 2:40 p.m. departure. The ride south to Washington and beyond to Fort Lauderdale, Florida was not particularly memorable. The return from Florida to Chicago with my family was, however. We rode Seaboard chair car #6262 north on acceptable track and moderate adherence to timetables until we came to Indiana and ambled along at a discouraging 30 mph over PennCentral's rotten track all the way to Chicago.

The Joint Tariff Computerization Committee met in Chicago. Its chair, Bob Walker from the Grand Trunk Western, invited me over on June 26th to discuss the SPLC and how it could be incorporated into railroad tariffs. Early in July Allman's reworked Transportation Data System (TDS) monster seemed to have come back to life after everyone had read its obituary. The Freight Claims project needed data that TDS would have produced. Because the railroad had some new management since the first attempt at a transportation data base was proposed (and died), we pulled the remains together and laid the resuscitated corpse on Jack Kane, vice president marketing, and Harold Gastler, vice president operating on August 1st. The result was a deafening silence—TDS was still a non-starter. Al Handwerker and Ed Kreiling of finance gave the rejected project another poke a week later to no avail. In September TDS quivered and emerged phoenix-like from the ashes as TDS II, a much modified and cut back project. Frank and I were forced to start the interview process with the sales department again. TDS II gasped but did not die, and soon Frank and I were off to Kansas City to see George Lund, Ed Ruche and John Reimers. On the 12th of September we were in Minneapolis interviewing John Boyles (my friend from Casper days), Harold Schonning, Joe Mullen, Wayne Anderson and Bill Dougherty (from Duluth). I visited with Don Guenther who had taken Chuck Towle's place in the Minneapolis industrial development department and George Nick, who had joined Don from the division engineer's office. Frank Stern and I flew to Detroit from Minneapolis and stayed at Greenfield Village. In Detroit we interviewed Jim Sullivan, who remembered me from the Chrysler project in 1963, Ray Clelland and Jack Sheeds. I called John DiCicco at Chrysler just to say hello. He was now in charge of Latin American operations. Thankfully, Larry Provo put a stop to pursuing TDS II saying that we needed a better revenue system, a pronouncement that panicked Allman (August 19th).

Another old friend from industrial development days, Jack Frost of the Illinois Central, retired. I applied for his job directly to Wayne Johnston, president of the railroad, on July 22nd. Four days later Dick DeCamara called and said he had put in a good word for me. An appointment for an interview with vice president Howard S. Powell was set up but on July 31st I heard that Charlie Catlett, Frost's assistant and one of the good ole' boys, had been appointed. Charlie, a native Mississippian, was firmly rooted in doing business the good old southern way. They had a lot of chemical plant activity in the Avondale-Baton Rouge stretch of the Mississippi, and Charlie did not have to do much development work in dealing with these walk-in-the-door customers. That brief episode marked the end of all my ambitions for a return to industrial development.

As mentioned above, the merger of the Chicago Great Western on July 1, 1968, was barely noticed in the rarified air of the systems department. Management's pleasure with having its own entrance to the Kansas City gateway would soon develop into unforseen and monumental headaches over the next few years. Our department would be called upon to help sort it out. The acquisition of the Fort Dodge, Des Moines and Southern and the Des Moines and Central Iowa properties on July 28, 1968 was painless and did not present as many serious problems as the CGW merger.

Warren Davis in our group gave a lecture on network modeling August 7th. I was quite taken by the concept. Later that month I worked on an idea for mapping economic areas using that computer technique. A geographic database would have been a key element in constructing this model. Since Shaffer's primitive version (GAM-IC) did not contain the requisite data elements, it was not considered at the C&NW for modeling purposes. Privately, I could see that my more comprehensive geographic data base, CENTRE-US, (now in commercial production) could provide a solid base for such modeling projects. Because a modeling project never evolved at the C&NW, the need for a geographic base did not arise.

The infamous 1968 Democratic National convention riots in Chicago occurred during the Data Systems Division's annual meeting at the Pick-Congress Hotel on Michigan Avenue. Registration for the meeting was way down because of the political turbulence. I took my 65 lady guests on buses to show them around the Loop, Lake Shore Drive, a visit to the Elks Memorial, and, finally, lunch at Stouffer's restaurant at the Old Orchard Shopping Center. I had never been a tour leader before. The whole episode was a bit strange. I had gifts for every lady—in fact, I had too many left-over decorative perfume bottles and had to sell them off to anyone I could to cover the original investment. The banquet that evening was festive but the

Chicago police battling with demonstrators outside the hotel added a tang to the occasion. At the SPLC committee meeting on the 24th I presented the group with my exploded drawings depicting the SPLC code structure. The neat thing about that six digit code was its ability to aggregate data by county or state or region depending on how much of the code was utilized. Pete Conway, one of the AAR's administrative people, was quite impressed and became an active supporter.

The vice president of corporate development, Jim Woodruff, wrote to Provo May 1, 1968 outlining possibilities for his department interacting with the railroad's traditional departments. He felt that corporate development could continue to be a trouble shooting group or accept assignment of specific responsibilities within the company. Jim had been wallowing about for several months and desperately needed to define just what his department was supposed to be doing. Ten days later he wrote to Provo again suggesting a management rationale for the C&NW. He defined three types of departments: Independent (Operating, Sales/Marketing and Finance); Directly Dependant (Accounting, Purchasing/Materials, Systems, Labor Relations and Law); and Administrative (Personnel, Real Estate, Corporate Development and Public Affairs). In this memo he suggested an executive committee approach for decision making. Provo filed it. On September 20th Jim tried again by setting forth a topical outline for developing corporate objectives. This one called for assessing the C&NW's environment, inventorying the railroad's strengths and weaknesses, and developing objectives. This seven page memo detailed four development teams drawn from a wide array of departments. This proposal didn't go anywhere either. Provo liked his centralized control and did not want to share it with any "teams". Woodruff quietly left having accomplished exactly nothing.

Larry S. Provo was now chief executive officer as Ben Heineman was becoming more distant from railroad affairs. Advanced systems was doomed. Its leader, Dr. William P. Allman, had failed to demonstrate to Mr. Provo why we were needed. The exodus began. Warren Davis went to Europe September 30th and did not intend to return to the C&NW; Dominick Violante gave notice on October 3rd and left on the 15th. The various projects sputtered and mis-fired leaving the rest of us listening to rumors and feeling abandoned. On October 25th a rumor arose that advanced systems was to be taken away from the heavy hand of R. David Leach—not true as it turned out. I continued to work at a segmentation or network scheme for the railroad to provide capability for isolating costs and revenues on individual lines. Given our uncertain future I was surprised when Bill Allman said I was being promoted to Senior Planner and was to receive a 9% raise in salary to $14,700 effective January 1st. John Danielson, one of our more aggressive and ambitious lawyers, asked me to present a talk on the MAGIC system to the Chicago group of Interstate Commerce Commission (ICC) Practitioners at a luncheon meeting November 8th. I carefully passed him on to Jack Shaffer. Danielson later told me that Shaffer had ordered him not to talk to me and that Ed Nordan would give the talk. John felt like he had stepped into some sort of hidden bear trap. The day before the talk Nordan appeared in my office cubicle and demanded that I hand over my set of colored slides for him to use. Not anxious to ease his pain, I dragged the matter out for some time before I grudgingly loaned them to him. John invited me to attend the luncheon as his guest. Ed did a fair job and even introduced me as the inventor of the concept. On November 12th Frank Coyne of the Data Systems Division invited me to join the parent Standards and Coding Committee of which my SPLC group was an *ad hoc* committee. I dutifully passed the invitation along to Allman for official approval. He was reluctant for me to take on the job as he was having another brain wave called TRACS which was awaiting official blessing and would fully occupy his staff, if approved. The chair of SPLC committee, Lou Kenzil, called and asked me to chair the next meeting. The next day, November 20th, Tom Desnoyers asked me to be vice-chair of the geographic code committee. To my utter amazement Leach approved my vice chairmanship (December 6th). This was a major breakthrough in my involvement with geographic coding at several levels and in several organizations. Being involved on a wider national scale made the current railroad situation bearable.

I was surprised, but not shocked, November 1st when I heard that my old boss, Gene F. Cermak, had resigned from his dead-end job in the reorganized sales department to go into the real estate business. I ran into him on the street a few days later. He said he had talked to Ben Heineman about the turmoil in the new sales organization, but Heineman was now so far from the railroad scene that he could only commiserate. Iver S. Olsen, the other half of the sales organization and Gene's counterpart, was a master of internal politics and had gradually squeezed Cermak out of the sales department. Gene had a run of 11 years with the C&NW and left some memorable accomplishments. His rapid decline, however, was sobering. His last day was November 15th.

I went to a reception hosted by the Great Lakes Industrial Development Research Council in Chicago on November 12th and ran into a lot of my former associates: Chuck Towle, Mark Townsend from CILCO in Peoria, Bob Stapleton from Clinton, Iowa, Charlie Catlett and Tom Gage of the IC. My pleasure at seeing the old gang was tempered by sadness at seeing them still playing the same old development game using out-of-date rules. Transportation and customer service supported by market knowledge enhanced by rapidly advancing computer technology was completely lost on them. I was happy not be a part of this crowd anymore. Two days later Bill Allman was full of himself with a big secret about a major shift of railroad operating divisions. His refusal to share knowledge seriously interfered with my network segmentation scheme. A little *sub rosa* investigation in the operating department soon developed full details of the "secret". On December 2nd Dick DeCamara at the IC wanted to know if I was still available. He also asked about Vic Preiser in corporate development, but didn't say why. Later in the month Frank Stern said that Jack Shaffer was putting out the word that I was in the job market and that Allman had picked up the rumor. Although Allman had already given me a fat raise and a promotion, he expressed concern that I might elect to go elsewhere. On December 4th I flew to Washington with Bob Walker of the Grand Trunk Western who chaired the Joint Tariff Computerization Committee (JTCC). This was my first appearance at the Standards and Coding group which included the SPLC committee, Route Coding, Packaging and Customer Codes.

Another highly regarded associate from my engineering days, Al Johnson, the practical roadmaster at Sterling, Illinois, retired. His party was held in Chicago on December 19th, and I made a point of attending the celebration of his working life. Al seemed to be relieved that he now could lay down his responsibilities for keeping the Galena Division mainlines across Illinois safe. Almost at the end of the year Carl R. Hussey was assigned to the C&NW-Milwaukee merger study group and said they could use me in their work. So the prospects for 1969 were looking up: I had a promotion, pay increases, my services were in demand, the work I did in 1968 was proving to be useful, and I was getting more deeply involved with interesting geographic matters. The poor old C&NW was not doing that well. The reorganizations of old line departments were exposing a lot of bad situations. The deteriorating condition of the railroad's structure and operations was becoming more apparent. Incompetence of some managers was also there for all to see. While some operating officers seemed to have forgotten how to run the railroad, the mergers of the CGW, FDDMS and DM&CI had added so much new track and created so many new traffic lanes with their built-in peculiarities that many unforseen problems were arising to challenge the best of our people. Because most of the merged railroads' management people elected not to join the C&NW, our own line managers trying to integrate the new lines were overwhelmed with operating problems. The C&NW Railway had grown too large for its managers to successfully run this bloated tangle of trackage. Some radical pruning was needed along with an infusion of more good operating managers.

23

ABOARD A SINKING DEPARTMENT, 1969 . . . The C&NW Railway hit a low spot in 1969. The newly reorganized sales and marketing departments were not functioning very well. The distractions of being part of the conglomerate Northwest Industries did little to modernize the railroad company with improved track and motive power. Many of the old heads left, or were pushed-out, leaving their replacements and a swarm of outside consultants without the critical knowledge of how to operate the railroad. The company was bloated with thousands of miles of non-productive and redundant branch lines. Specialized departments like our advanced systems group were so far removed from reality and practical purpose that they scared president Larry Provo with their costly projects that produced so little benefit toward solving basic structural problems. The railroad did not need "advanced systems"; it needed immediate and practical systems. In my sterile steel and fabric cubicle I tried to concentrate on what humble programs would be beneficial to the company. One of our major managerial problems was the lack of meaningful and timely data on such simple things as freight revenues by line, by commodity or as measures of interline business. It seemed that every department had erected barriers which blocked the free flow of data. Such data as were available were jealously held until the numbers could be reviewed and "sanitized". Cars on line, freight revenues, freight car demurrage and car hire, ton miles, real estate sales, payroll data, new industrial plant locations, capital and operating expenses, interline settlements—all were carefully massaged before general release. There was no "real time" reporting. Rationalizing different types of expenses or relating cost to revenue was impossible. There were no cross reference computer files by which one event could be related to another. In my humble opinion the greatest need for this railroad was creation of simple cross referenced computer files to bring all this disjointed data together in meaningful arrays. Since I was very much into geography, the creation of the cross referenced station codes master list (CLASP) seemed to be a primary step. It led directly into my next project—development of a network model. Again, this was a lot of definition and data collection work. All of the railroad's links and junctions had to be identified and described in computer sensible formats. I named it the Segmented Railroad File (SRF). Here, again, my "leader", Bill Allman and my immediate superior Frank Stern, disapproved of the SRF project and its building block, the CLASP station master, saying that there were more important things I could be working on. Of course, they never identified what those more important things were. I did not wait for them to come up with meaningful projects; I calmly, and quietly, went ahead working on my two unofficial projects. I could not sit around twiddling my thumbs when these potentially powerful tools cried out for assembly. I believe that neither Allman nor Stern understood that patiently assembling data was the fundamental building block that had to be accomplished before a system could be properly designed. I was substituting my judgement for theirs which again garnered neither praise nor encouragement while I steadily pushed the CLASP and SRF projects forward. By the end of January Bill Allman's stillborn TRACS system continued to dominate and drain the life from our dwindling department. Despite the poor prospects for any success with his fatally flawed big system, Bill made arrangements to borrow Tom Martin from finance on a half time basis to help put flesh on the cadaver. Tom was soon joined by another loaner, Bill Goldammer, from the costing section. Many years later I happened upon Dr. John Gall's book, *The Systems Bible*, which was a somewhat tongue-in-cheek guide to systems development. Gall postulated an "Inaccessibility Theorem" that succinctly described the situation we faced in 1969. The theorem states: **"The information you have is not the information you want. The information you want is not the information you need. The information you need is not the information you can obtain".** By mid-February TRACS (or TRASH as we styled it) was in such disarray that a month later on March 18th, when a presentation of it was given to Larry Provo, the monster came away without legs, arms, eyes or a brain. All the work I had done for Frank Stern over the past two months to dummy-up reports was totally scrapped by the severe surgery inflicted on the TRACS proposal. By April 22nd we determined that of the 75 proposed reports for the sales department, only 16 were possible to produce because of a complete lack of data. Allman took the bleeding TRACS program and tried to salvage something from it. Frank and I were assigned to write parts of the patchwork creature with a deadline of May 1st· We finished our pieces on time, but no one else had theirs ready. Frank then had the temerity to show his wild-eyed reports to various sales managers; they were aghast. Most of them were totally perplexed as to just how these cockamamie concoctions would help them run their departments. The abbreviated TRACS proposal was finally completed May 22nd. At which point Allman in his usual creative frenzy tore it all apart to reassemble into a new structure which, of course, forced a complete retyping. This caper earned him the undying enmity of the two long-suffering secretaries, Phyllis Nyboer and Kathy Gordon. The deadline was missed—again, but no one cared. The TRACS proposal died on arrival.

I had continued work on the station list (CLASP) in January-February without telling Allman because I had been encouraged by the favorable reception other departments gave the concept. There was a growing appreciation that this was an important tool. I deliberately hid office copies of correspondence relating to the subject so that Bill would not tumble to the fact that I was doing forbidden work. Frank, my immediate superior, seemed content to let me run by myself as it was easier for him to leave me alone than devise some official project for my occupation and entertainment. Someone must have said something about my subterranean work because suddenly in March Bill Allman unfroze his injunction on my CLASP work and asked me to bring him up to date. Bill always reacted when someone outside of the department said something good about our work because he so rarely heard anything positive from anybody. In response to his request for a briefing and by fudging the timing of each event, I showed him that we had advanced a long way toward accomplishing the development of these two (now critical) data files. He was left wondering how so much had been accomplished without his immediate supervision. Gaining official sanctity, I was named to head the station code clean-up committee on May 29th—about a year after the work had begun. In the summer Lloyd Erickson, the freight accounting officer, finally woke up to the fact that his station numbers were not the only station code on the railroad and that with all these mergers and additions of the newly acquired stations he had a massive headache in identifying these new (to him) stations. I helped him weed out the duplicates and cripples which made a pretty clean list for the AAR to publish. Also in July I helped Ed Evert in accounting at Ravenswood get the CARFAX station number list corrected. It took so long to get these old-line departments to change anything. Just when we thought that we were about finished, we hit a bump in the road when the rate department tried to axe some obsolete stations in Wyoming and Illinois. Since the Open and Prepay station numbers were published in a tariff filed with the Interstate Commerce Commission (ICC), the publisher, Leland in St. Louis, would not publish the changes without state regulatory commission approvals. The law department filed notices with all eleven states in which the C&NW operated for blanket approval of an attached list of obsolete stations. Nine of the eleven state railroad commissions agreed; Wyoming and Illinois did not. Since there were few items in the Wyoming list, the law department was able to settle them one by one and get approval. The Illinois Commerce Commission, however, decreed that each station could not be officially abandoned until after advertisement of the notice of discontinuance for three weeks in the closest local newspaper to the site and an official hearing was conducted. One of these so-called stations was a railroad crossing of the

C&NW's southern Illinois district with a New York Central (Big Four) line near the city of Benld, Illinois. The Gillespie paper printed the notices, we nailed a legal notice poster on a telephone pole near the crossing and went to the hearing—no one from the commission showed up. That was the end of that one at any rate. This kind of detail was absolutely necessary to get the good stations properly cross-referenced. Eventually, we had freight station, car accounting, O&P tariff numbers and the Standard Point Location Code (SPLC) all together in an electronic file. Increasing use of CLASP in many software applications proved the project to be a real benefit to the company. Now we could relate freight revenues to car movements at specific geographic locations.

The Segmented Railroad File (SRF) was the second important tool in building a network model. This was a node and branch model which established station locations by mileage along the segments and defined connectivity at the nodes. Using an algorithm derived by a very smart young man (whom I never met), the network problem of tracking data through a complicated web was simplified for the IBM main frame computers then in use at Ravenswood. The station list database (CLASP) was the critical element to make this model work. The first time we looked at freight revenues and car movements together, we were truly amazed to find so many zero business links. Startled, we did a rigorous review of the logic in the model, tweaked a bit of code here and there, reran the model and got essentially the same results. There was a lot of redundant trackage out there—especially in Iowa. A big benefit derived from this network model was development of annual density statistics which were in demand for branch line abandonment studies, engineering justifications for track improvements and traffic flow studies.

The first half of the year saw many personnel changes. M. C. Christensen, my mentor in engineering 20 years ago, became assistant engineer of bridges. Harold Keeler, another former colleague, became Galena Division engineer (January 6th). On January 20th we heard that Northwest Industries had made a bid for B. F. Goodrich —that was in the morning. In the afternoon we learned that Larry Provo had fired Jack Kane, vice president-marketing. Later Kane said he didn't know why he was bounced. (Six months later Jack Kane became vice president-marketing at RCA). Walter E. (Bud) Braun, paunch and all, became Kane's successor February 1st. After complaining a lot after Kane's dismissal, Maury Decoster of corporate development, quit without notice April 7th. His restructure of the old traffic department remained, but the survivors of the old traffic department had taken control back and quickly reverted to their former ways. Lew Marshall and another personnel officer left in April. Another of Heineman's whiz kids, Vic Preisser, had taken over personnel in the reorganization, but left the railroad on February 15th to become an officer at Litton Industries. No one was sorry to see him leave. Arrogant and devious, he was a danger to anyone associated with him. That infamous hidden tape recorder went with him. Deepak Bammi, our crack programmer, departed January 31st. Later, he taught computer science at the University of Illinois in Chicago. His brother worked for DuPage County as a respected urban planner. Not everyone left, however. J. R. (Bob) Kunkel was promoted to assistant vice president-rates and divisions February 13th. Harvey B. Buchholz, my former industrial development colleague in Milwaukee, retired and was honored by a formal dinner February 27th. I attended; Gene Cermak, his former boss, did not. In less than a year Harvey was dead—January 2, 1970. Joe Verona, former head of the dining car department, who had done such a nice job for the Chrysler special train in January 1963, went with me to Harvey's funeral on January 6, 1970. Also in January Alan Boyd was elected president of the Illinois Central.

Marketing moved out of our office space to their own newly constructed quarters in the Daily News Building on February 21, 1969. Advanced systems and systems design and installation (SDI), Bill Allman and Bill Zimmerman's groups, were the only occupants left on the 5th floor at 120 S. Canal Street to run out the lease. The place not only had a haunted air about it, but it was so far remote that we felt we really were not part of the railroad at all. I made an appointment with Bob Russell, now head of personnel, on March 18th to see if he had any ideas about my future on the railroad. Although he said he would look around for opportunities, he never came up with any. I was now firmly convinced that my fate was my own to shape and that I must continue to steer an independent course.

That study group to merge the C&NW and Milwaukee continued to work at their task even though there was a strong feeling that the merger would never occur. Seeing what good results we got from the C&NW station list merger, I worked with our programmers Warren Davis and Bob Tamura in April to produce a master station list for the merged companies. Luckily, that was exactly what the merger committee soon called for, and we were happy to be able to provide it in a timely manner. The merger committee also settled on using the six digit standard point location code (SPLC) as the master number for the merger (April 18th). In June the merger committee shifted their focus to motive power. Data are data. Pulling together a simple listing of C&NW and Milwaukee diesel power by class, horsepower, age and manufacturer did not take much effort. Our motive power people were very pleased with the new computerized roster and eventually took over its maintenance until a better, more comprehensive database was developed. Milt Crandall, a good fellow from South Dakota and a practical hands-on manager, was especially helpful in assembling the proper information. In April the Illinois Central unions struck over local issues; Boyd, the new president, had a fight on his hands.

January 1969 had been unusually bitter cold. I flew into National airport across the Potomac from Washington, DC, on the 9th for an SPLC meeting at the American Trucking Associations' (ATA) building. Former Canadian, Bob Hennell, represented the National Motor Freight Tariff Association (NMFTA) which was one of the numerous components of the ATA. Bob was one of the founders of the Standard Point Location Code structure and felt that it was his personal creation. He had been responsible for applying the SPLC code to actual place names. When I demonstrated to the committee that he had egregiously violated the logic structure in some congested areas, especially in Pennsylvania, his hostility toward me was palpable. Granted that Pennsylvania is the most difficult state to properly identify in any code scheme because of a disproportionate large number of named places, Bob adopted the posture that his code assignments were cast in bronze and that my proposals for fixing some of the irregularities were just not going to happen. Bob's problem was his lack of knowledge of the geopolitical structure of the United States and, as became apparent, his home country of Canada. The AAR Data Systems Division's representative on the committee was a young fellow named Roy Amburgey who muddied the waters further by taking some arbitrary actions on his own. I was increasingly uneasy about the vast difference in code definition: the truckers' area versus the railroads' point philosophy for place identification. Another problem looming was the lack of sufficient code numbers for Canada and Mexico. The code designers had allocated 80% of the available code numbers for points in the United States because they had no idea about the total number of place names that had to be encoded. For just a temporary committee there was still a lot of work to be done. Later in January Frank Stern and I flew to Kansas City in the company's DH-125 for the Data Systems meeting. After a long day of committee meetings on the 27th, Frank, Don MacBean and I missed our scheduled flight from Kansas City's Municipal Airport which was located just north of the business district. We were forced to take a 17 mile ride north over an ice-coated highway to the sprawling and brand new Metropolitan airport designed to supersede the close-in, and convenient, Municipal field. Our flight was delayed at the new airport because the crew was stuck in Kansas City. The weather closed in with a freezing fog coating everything in a film of clear ice. Our plane, a TWA 727, was parked inside a hangar at Metropolitan, fully loaded, awaiting the crew which belatedly appeared. Just before midnight the hangar doors were opened and the 727 roared down the icy runway headed for Chicago. While we were in the air, Chicago was closed because of ice. We were diverted to Dayton, Ohio, which was still clear of the approaching ice storm. Landing at 3 a.m. the plane unloaded at the terminal in the midst of the icy downpour that had caught up with us. There were no taxis, airport buses or live airport people in sight. At least the lights were on. I spotted a car rental counter with a lone attendant and quickly rented a car. Loading as many as would fit, we slithered our way along the icy roads to a motel for a few hours sleep—at airline expense, of course. We finally left Dayton the next day at 11:45 a.m. for the short hop to Chicago but still had to endure another 45 minutes of circling while waiting our turn to land.

In March the SPLC committee met in the AAR's Chicago office at 59 E. Van Buren. Toward the end of March the committee chairman, Lew Kenzil, suddenly died and was buried on the 26th leaving the chairmanship open again. This time Dave Leach approved my taking the appointment April 28th. Now, I could tackle these thorny coding problems from a position of authority. There were a growing number of complaints about duplicated code assignments and differences between truck and rail versions of SPLC. Bob Petrash, director of the Data Systems Division (DSD), now became involved. We discovered that his clerk, Roy Amburgey, to placate member railroads requesting code number assignments had applied SPLC numbers on his own initiative that often either did not conform to code structure logic or duplicated numbers already assigned and published by the NMFTA in their tariff. It took years of argument and negotiation to extract and correct those bad codes Roy had slipped so carelessly into the system. Meanwhile, he had resigned and gone to work for a local school district.

Jack Shaffer and his NYC pal, John Wauterlek, presided over the crumbling ruins of the industrial development department. They were not at all happy in their work. Silverman in personnel said on January 24th that he doubted Shaffer and Wauterlek would be around very long. Shaffer had hired Ed Nordan to manage and promote my old MAGIC coordinate system. Ed had failed abysmally. The original old database was never used in any useful application by the ID department and was now moribund. The program had two fatal flaws: there was no application software program, and its application had to be done on the client's computers because the input to be analyzed was likely to be proprietary business data. Swift and Company had showed some interest in the concept but had refused to furnish their private data to the railroad for processing. None of the Shaffer crowd understood that the MAGIC file was just a specialized collection of interrelated data. To make any money from this geographic database, electronic copies of it had to be sold as a product for use within the buyer's own private environment. Because they did not understand what they had, Shaffer and friends had no concept about its marketability or application. Blinkered, they adopted the posture of clasping the database close to their chests in fear that someone might steal it. While industrial development did not see this opportunity to make money and locked up their aging database in a vault, I was busily marketing a much more comprehensive geographic database under the name of CENTRE-US. This database was sold as a software product on magnetic tape. Realizing that the data were alive and ever changing, I advertised an updating service for a rather low fee. While the initial sale brought a nice solid amount into my bank account, the continuing subscriptions for periodic updates really added to the income stream. I kept that update service simple by supplying a complete updated version of the database twice a year which avoided the problems associated with creating software, in many formats, to insert new data or to change existing records. Making a tape copy of the entire current version and shipping it to subscribers in either the rapidly declining 7 track or the more common 9 track versions, was purely a mechanical copying process that I had done by an outside service bureau. In his final days in the industrial development department Ed Nordan tried to recruit Ed Garvey from our advanced systems group and again failed. On February 1st Nordan was transferred against his will to schedule planning in the operating department.

A face from the past appeared in April bringing with it a short and crazy adventure. Norman Hillegass, once a rodman and instrumentman on the Galena Division, was now a salesman for Pettibone-Mullikin, manufacturers of railroad work equipment. Norm's dream was founding a company to construct and repair railroad freight cars. He wanted me to quit the C&NW and join him and his associates in the venture called the Tuscarora Corporation, or TUSCAR. (Norm was a Pennsylvania boy, born and raised at Mechanicsburg near Carlisle and Harrisburg. Tuscarora Mountain and the Tuscarora State Forest lay just to the west—hence the name). He dangled a vice president job before me, but it was not at all clear what the job function was to be. I had no knowledge of the freight car construction business or possessed any mechanical skills that would be helpful. I listened politely but remained dubious about this scheme. All through the summer Norm waltzed with, plotted, cajoled, romanced or otherwise tried to con about every sizeable community in Nebraska and Wyoming trying to find a local group that would finance his carshop. Pettibone got wind of the project and fired Norman's friend, Pat Rogers, who was to have been the shop manager. After a lengthy period of not hearing from him, I got a telephone call on October 19th from Omaha—Norm was in a hospital following a heart attack. TUSCAR never got started.

Personnel continued to depart in May and June. Stanley B. Johnston, another associate both in engineering and industrial development, went to work for Ziegler at the Beardsley Company, a Chicago realtor. Three of Bill Zimmerman's systems design group left including Bill Knight. Tony Rathenau resigned June 26th. Kreidler and Stusek were dismissed July 8th with 60 days of severance pay. Richard Lewis from our group resigned August 15th and ended up at Canteen Corporation (April 15, 1970). Don MacBean now worked for Bob Russell in personnel. By June 5th he was beginning to complain about that job. Don never found his niche at the railroad even though he dearly loved the business. There were some new faces in advanced systems in early 1969: Julian S. Eberhardt, an industrial engineer, came from the CB&Q bringing some operating experience from his days as a diesel locomotive fireman and Jerald B. Groner, a cynical, hard working grump of 28 who had been with the Elgin, Joliet and Eastern. Smart, dedicated and skeptical, he was the perfect person to deflate foolish projects and concentrate on the important ones. Jerry went on to work for the Union Pacific Railroad in 1995 after the C&NW merger. Jules went on to an illustrous career at the C&NW as a vice president of planning and systems, retiring in 1995 when the UP merger occurred.

A new wrinkle in geographic coding emerged May 9th when G. William (Bill) Wright, traffic manager at GATX called and wanted to know about the SPLC. Bill and I were to work together many years on implementation of SPLC. He was instrumental in opening contacts with the Mexican railroads for developing an SPLC for that country. Despite the chaos within the C&NW, I continued to attend the SPLC committee meetings and add more hard data to that body's efforts to gain acceptance within the rail industry. We were being resisted by the freight accounting officers and systems people because our 6 digit code did not fit in their current 5 digit code standard for existing computer programs. Because settlement of interline freight revenues was based on the FSAC (freight station accounting code, maximum 5 digits) introduction of a six digit code gave systems managers heartburn when they contemplated all the software that would require changes. Most systems people in those days were working in the COBOL language which was pretty rigid in field definition—that unit record mentality was still strong all over the country. Getting SPLC accepted was a tough sell. By November I was having a hard time with our own John Butler, vice president finance, and Bill Zimmerman, director of systems design, over that very issue. How was I to sell the railroad industry on using the SPLC when my own company was resisting change? The answer, it turned out, was to be patient and continue urging companies to plan for the future. When new systems and programs were written, open the field for the station number. As programmers finally got away from rigidly defined fields, making the change to a larger station number became easier. Eventually, the old hard coded accounting systems were phased out as newer more comprehensive designs were implemented. There were some instances of railroads seeing other values of the code structure and implementing it for other than freight accounting.

On May 12th I was in New York and went to Washington on the new Metroliner. I was invited to the cab in New Jersey and got a photograph of the speedometer at 115. We flew over the Susquehanna River bridge at 70 mph. The committee meeting in Washington the next day included Tom Desnoyers (N&W), Jim Woods (ATSF), Bob Petrash (now AAR formerly PRR) and Otto Hermann (SP). Desnoyers had been fired by the N&W because he spent too much time on AAR standards work, but was given a half year's severance pay. Don and I flew back to Chicago by way of Cleveland just for the novelty of it. Ed Lillig from systems design and installation, Frank Stern, Don MacBean and I flew to Detroit on June 9th in the company plane, N403W, for the Data Systems Forum. We came back on an American BAC 400 landing at Midway Airport. During July the south train shed at the Chicago Union Station was demolished preparatory to construction of another steel and glass box by Tishman Realty and Construction Corporation.

In June I worked with Edward A. Burkhardt, an officer in the operating department, and our Bob Kase on a new locomotive data base system. As mentioned above this was the program that was expanded to include the Milwaukee Road power for the merger committee. Jerry Groner was supposed to be developing a way to assign costs to the different arms of the segmented railroad file (SRF), but the locomotive project had a higher priority.

BLOCKING BOOKS On July 24th Bill Allman assigned me to work with the operating department to develop a blocking scheme for directing freight cars across the C&NW. Now we had something related to the real world to work on. I was glad to escape the sterile atmosphere at 120 S. Canal. I met with Ed Burkhardt, Bill Allman, Ed Nordan and our programmer, Warren Davis, on August 4th to discuss how a network model could be constructed to address the operating problems caused by too many small switching yards that consumed transit time. Frank Stern, Jerry Groner, Ed Nordan and I settled on using the existing segmented railroad (SRF) program as the best structure for building the blocking book scheme (August 22nd). Given the technology available to us, we were forced to use hand sorted decks of the familiar 80 column punched cards to print "books" for each switching yard. The blocking books consisted of cardboard binders with thin metal binding tabs containing about 50 pages of computer generated listings that specified the blocks in which cars destined for a particular destination were to be placed for outbound movements. A yard clerk preparing the switch list at Sioux City, Iowa, for example, could refer to the book produced for his yard when confronted with a load destined for Wales, Wisconsin, and send the freight car to a "Proviso" block being made for the next eastbound train. Given the complexity of the C&NW's system at that time because of all the M&StL and CGW lines that crisscrossed in Iowa, Minnesota and Illinois, the transportation planners in Chicago were hard pressed to impose an efficient traffic flow through the system. The blocking books delineated management's desired train service policy and brought some direction to the sometimes capricious forwarding being done by field forces far from Chicago. To indicate his instructions for assigning cars to desired blocks Ed Burkhardt marked a set of division segment maps for each yard using colored pencils. Nordan then used the maps for guidance when assembling the decks to fit Burkhardt's colored lines. The theory worked, but producing the blocking books was extremely labor intensive.

The concept of blocking freight cars was not a new idea. The Rock Island faced with growing traffic and inadequate yards for sorting freight cars in 1923, developed a scheme for grouping cars headed for the same destination. The Rock learned that the key to a successful scheme was establishment of a system wide car classification system and a detailed plan of each switching yard. Coupled with proper train schedules, the blocking plan produced dramatic reductions of the time trains spent in the yards, reduced engine hours and accelerated delivery times, but there was a hidden downside. Freight trains in the late 1920s were becoming heavier and longer as improvements were made in braking systems and draft gears. To maximize tonnage trains, blocking schemes were stretched out over an entire day rather than the eight hour periods used when trains were shorter and more frequently operated. By the 1960s cash starved railroads tried to save money by running heavier and less frequent trains. Pre-blocking schemes were intensified in attempts to offset yard delays, but the lack of strictly scheduled and operated trains caused cars to miss their connections which exacerbated yard congestion. To make matters worse railroads were disinvesting in yard facilities. Blocking worked well in a disciplined environment. Without strictly controlled train schedules, blocking schemes collapsed and were discarded in favor of just running heavy trains when there were enough cars on hand. Here we were in 1969 with no knowledge of what had worked in the 1920s nor any idea of why that concept had failed. Yes, we could specify what blocks should be made at each yard and which trains were to handle them. What we didn't understand was the need for a disciplined operating environment where trains ran on schedule regardless of tonnage available coupled with eliminating some of those time-eating yards. And with the ragged condition of the main lines, regularity of operation was only a dream.

Fast Eddie Nordan was difficult to work with. He was confrontational about everything and had the temerity to argue with me about the correct spelling of station names. I was the unofficial master of geographic station files on this railroad. He would not settle on a course of action and move along to a conclusion. On October 15th I told him that he had to freeze the blocking book project design soon or support for the program would disappear. While Burkhardt was somewhat tolerant of the slow progress, he was getting antsy about getting these books into production. Don Fryer of operating laid down the law to Nordan to get the project going or get out. To be fair, Ed had a sick wife and baby that may have been distracting him. (His wife died early in 1973). But his nit-picking and inability to decide on details seriously delayed the project. Ed Nordan did not believe in the 90–10 rule where a project can be pushed into operation when it is at least 90% OK and the remaining 10% can be cleaned up later. The first book's card decks eventually went to Ravenswood for processing in late October but computer services failed to do our work. Computer accessibility grew worse in November when the computer developed internal problems resulting in no progress for the blocking book project and seriously compromising normal accounting processing.

Continuing this tale into 1970: Nordan's project stumbled along in January with no production largely due to transportation department changes in train service. Every book printed so far had to be redone. On January 15th Warren Davis, Bob Tamura, Nordan and I reviewed the logic in the print program and decided to rewrite the code to make it work better and faster. Bill Allman earned a plus mark for advocating a magnetic tape based scheme. Of course, Ed Nordan opposed any "hard" coding—for what reason is lost to history. By January 23rd it was obvious that Allman's directive to use mag tape was impeding progress on this project as the railroad's computer center's ability to handle tape was in its infancy. In a meeting February 5th with Allman, Nordan and me, Allman came out on the short end of that argument and gave up pushing mag tape usage. Ed Burkhardt wisely stayed out of the gunfight. The whole Ravenswood computer center was a bit primitive, to say the least. The machines were serviced by computer operators who fed card-based jobs into card readers or mounted tapes on the few drives we had. The operators were entry level types as it was believed that almost anyone could be trained to place card decks into the readers and mount a tape in a drive. Unit record equipment still dominated data entry—key punch machines were scattered all over the railroad; we even had one in our office. Warren Davis, however, did not allow the monkeys in the computer center to run his jobs. When word came that he was going to make a run, the operators would assemble to watch. Warren's first instruction to them was " Don't touch anything." He programmed the run sequences on magnetic tape which then ran every operation in the proper sequence without any human interaction. I wonder if those computer operators realized they were seeing a vision of what the very near future held and that their jobs were very much at risk? Warren W. Davis was an unconventional person as well as a computer genius who never received the respect at the C&NW he so richly deserved. I was glad to have worked with him as I learned a lot about the rapidly evolving technology.

The Northern Lines merger was approved by the Supreme Court February 2, 1970. On February 25th Nordan was instructed to produce a special blocking book for the Northern Lines in Minneapolis by that evening. Ed and I borrowed vice president Gastler's company car and drove to Ravenswood to push our card decks through the computer center. We bound up 50 copies and made the deadline with time to spare.

On March 19, 1970 Ed Nordan was forced to join our advanced systems department. In preparation for a big staff presentation slated for May 27, 1970, Bill Allman insisted that Ed Nordan rehearse his portion of the program before the meeting. Ed flatly refused. Allman fired him on the morning of the big event. Ed took himself off to his office in the transportation department where the blocking book programs and card decks were kept. He systematically destroyed master card decks, control cards and documentation before calmly leaving on vacation to Florida. Someone in the operating department sounded the alarm. I ordered Nordan's office locked and sealed and that no trash containers were to be emptied. As soon as I could get there, I went through the trash baskets and retrieved all the pieces of ripped-up cards and papers. After reporting the situation to both Allman and Harold Gastler, vice president-transportation, Gastler ordered the room to remain locked until Nordan returned, if indeed he would. On the 28th Nordan called me and warned that I had better not mess with his office as everything was "coded", whatever that meant. I suspected he was trying to set himself up as the only savior of the blocking book project. The Nordan affair was not resolved by June 3, 1970 and Frank Stern was annoyed that I was unavailable to help him with a car management scheme which was, by decree, not to require any computer involvement. The next day the decision was made that Fast Eddie was not to be rehired and I was delegated to put the project back together. On my 44th birthday I was the new occupant of the blocking book office, a neat little private inside room in the heart of the operating department where I once more could be part of real world railroading.

It was apparent that Nordan had scrambled the master files but not fatally—I carefully reconstituted them. The torn cards were repunched and placed in their proper places. To really frustrate the production he should have distributed the destroyed material in trash baskets outside of his office which would have made reconstruction much more difficult. I had a lovely summer away from Allman and Frank Stern in my little office where I could work without interruption on other things besides producing blocking books. Later, Nordan had an interview with Burkhardt but pointedly avoided inquiring about what I was up to. Nordan's systematic method of removing critical control cards had been easily deciphered and reconstitutd—just like finding the key in code breaking. As a saboteur he was a failure.

In late summer of 1969 the SPLC committee met on August 24th in Penn Center, New York. I was now the chair and ran my first meeting. In the afternoon Tom Desnoyers, Herb Landow and I rode the old Myrtle Avenue El in the last wooden cars still in operation—it was due to be closed the next week. After more meetings on the 26th I flew to Montreal for meetings the next day with CN, CPR, the Canadian Freight Association (CFA) and Canada Post. On the 28th I presented a proposal to the Data Systems Division's General Committee for an SPLC education program; they approved. Three weeks later I went to San Francisco for the Data Systems' annual meeting and gave my SPLC presentation twice. In San Francisco we were given a tour of the BART (Bay Area Rapid Transit) system then under construction. After the meeting was over, I took my wife on a short vacation to the Big Sur region before coming back to Chicago.

More personnel changes occurred that autumn of 1969. The vice-president-finance, Truman Brandt, was fired August 4th. In advanced systems Bob Lehnertz resigned on August 22nd to return to the Denver and Rio Grande Western. (He later became their chief communications officer). Vic Khaitan, our Calcutta Indian, privately told me he was going to American Can Company before the end of September. Bob Tamura was negotiating for a job in Colorado. Warren Davis, our most competent systems man, was looking at a job in Rome, Italy, with the United Nations. Out in Iowa Del Perrin, superintendent at Mason City, was fired. Known from his bullet shaped physique as "Junebug" from his football days, he had been one of the Illinois Central crowd brought in by C. J. Fitzpatrick in 1957. A major shake-up in the operating department created four new assistant vice presidents which were really just changes in functions not people. Ed Ellis worked there then and was very unhappy with the changes. (He later went to AMTRAK, becoming vice president mail and express in 1999). The axe also fell on advanced systems. On October 23rd Bill Allman announced to what was left of us that there would be no more long range planning and that our staff was to be reduced. All of us wondered who would survive. We soon found out. Al David (a Christian East Indian national working as a programmer), Richard Ogle (analyst), Larry Creighton (administrator), and Ron Cassen (manager) were fired. The survivors were Frank Stern, Jerry Groner, Ed Garvey, Bob Tamura and myself. I was 23 days off in my prediction that the department was headed for a crash; I had selected October 1st as doomsday. Our new mission according to Bill Allman was concentration on quick flashy projects like my locomotive database to help restore his credibility. Warren Davis and I told him that none of our current projects had the potential for saving big dollar amounts quickly. November 7th was the last day for Al David and Dick Ogle. (Dick went to the Great Northern in St. Paul). The third typist position was abolished. The incumbent, Cecilia Smith, had an abysmal attendance record in any case. On December 17th Bob Kase resigned. (He found a job with a suburban computer services company before moving back east to work for the Norfolk and Western). After his exit interview with Bob Russell, he told me later that my situation was discussed and that Russell said that I was underutilized but that would change. I did not hold my breath as I had heard that same song before from the same singer.

In the reassignment of projects in our reduced department, I ended up with responsibility for operating CADFAC, a rather simpleminded software program for calculation of site grading quantities. Nordan had brought it with him when he joined the industrial development department some years earlier. (I never had a call for it.) I also was made responsible for the consolidated station list project (CLASP), STRESS, a bridge department calculation model, SPLC committee work, something called TRAIN-OFFf, traffic related items for the merger committee, and, of all things, GAMIC (my old MAGIC invention). I strongly objected to the last item because I had signed an agreement to have nothing to do with it; Allman agreed and removed it from my plate. On November 6, 1969 Allman suddenly reorganized his group again. I no longer reported to Frank Stern (a blessing) but directly to Allman. Frank was visibly annoyed as he had been using my production to shield his inactivity from close scrutiny, but he still retained Jerry Groner as his donkey. In addition to the projects listed above, I also gained something called car transit network. I was now number three in this smaller group something like being third officer on the *Titanic.* On December 4th there was another strange shift in the operating department when all AVPs were directed to report directly to Larry Provo. Costs were out of control and the railroad was in terrible shape. After a big staff meeting December 9th to discuss corporate objectives, we heard that Ben Heineman was refocusing his attention on the failing railroad part of his Northwest Industries. There were more rumors that we were to lose another position. Systems design and installation under Bill Zimmerman was the other part of Leach's empire that still occupied the 5th floor at 120 S. Canal Street. SDI's top employment figure had reached 105 at their peak. Now; they were at 78 and sinking fast. Despite all this turmoil Allman cranked out reports, memoranda, and proposals for grand schemes that kept the two secretaries busy and consumed reams of paper. No one paid any attention to his flood of paper.

President Richard M. Nixon signed an emergency order to stop a railroad strike threat October 2nd. By December 1st another strike was looming. (It was postponed until January 31, 1970 when the unions struck and the railroads locked them out. Luckily, it was a Saturday. A hastily obtained injunction forced the unions back to work on Monday morning). There was good news, however; I heard on December 3rd that my nemesis Jack Shaffer, director of industrial development, had finally had enough and was leaving the company on December 5th. He had lasted just two years and nine months. In that period the industrial development department was utterly destroyed. I really should have thanked him for accelerating my departure from that line of work. While he was in command when the ship sank under him, he was not totally to blame. The old guard torpedoed this outsider, and the techniques we had used so successfully were rendered obsolete by changing attitudes and requirements that he could not satisfy. His furious determination to seize my geographic material seemed to cloud his abilities to make the adjustments in service to the industrial client that had to be made to remain successful. His vendetta against me had the effect of focusing my energies to create a new and vastly superior series of geographic databases covering the U. S. and Canada. By the end of 1969 these databases existed as electronic data files, maintenance software was complete and tested, expansion was underway, and commercial applications were in production. All of this activity was performed away from the railroad and involved no railroad employees or equipment. My reputation as a geographer was growing outside of the C&NW. I wondered how long it would be before someone on the railroad would stumble over my outside life—apparently, no one ever did. In late December 1969 Dick DeCamara, vice president of the Illinois Central, asked me to take the new job of leading the standard coding effort for the newly organized Transportation Data Coordinating Committee in Washington, DC. I replied that the C&NW was too poor to consider loaning me out for that project. As things evolved, TDCC turned into a big operation that required hiring a full-time director. In this, my 22nd year at the railroad, I was achieving my personal goals, making money and had little incentive to pull-up roots for a new and unknown job in Washington. I looked forward to the new decade because I felt that I had gotten back into the pulsing heart of the railroad and my private development was beginning to yield substantial rewards.

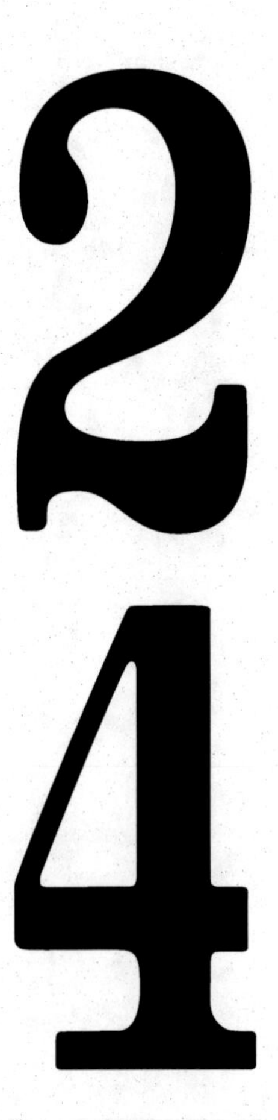

THE INTERNAL CONSULTANTS, 1970 . . . These Januaries in Chicago were brutally cold during this period. The cold seemed to affect the C&NW's internal functioning and many more people left its employ. Among them was Don Guenther, a former industrial development colleague, who resigned from running the Minneapolis satellite office and joined Swift and Company. John Danielson, our over-active attorney, said he was watching a power struggle in the operating department between Bill Alsop, Ed Burkhardt and Harold Gastler. Bruce Stewart, the number two man in personnel, resigned January 15th. Bill Allman announced that we had to cut expenses as the railroad was almost broke. Dave Leach canceled my planned SPLC committee meeting in St. Louis because of the expense freeze. Then I learned that I was getting a $1,200 increase which boosted my annual pay to $15,900. When Bill Allman advised me of the raise, he also told me not to tell the new lads the whole truth about the railroad's bad situation, an instruction that was as obscure as it was unnecessary.

Ed Marquardt in the corporate secretary's office called on me to do some historical research in connection with a line abandonment. After all, the big set of *Poor's Manual of Railroads* containing the needed information had been "loaned" to me many years ago on the condition that I would be called upon to look-up items from time to time. This was the first time in 22 years that I had a request. (These volumes were donated to the Railway and Locomotive Historical Society in Chicago in 1992. Eleven years later they turned-up in a member's home in Hastings-on-Hudson, New York).

As related in the last chapter, the blocking books project was a very hot project during this very cold period. There was an urgency for completion of the segmented railroad project (SRF) because I knew that our programmer, Bob Tamura, was planning to move to Colorado and wanted to finish the coding before he bailed out. I was also very deeply involved trying to get railroad acceptance of the Standard Point Location Code (SPLC). A report and review designed to help railroad financial officers understand the benefits of adopting this six digit code to replace the current Freight Station Accounting Code (FSAC) was prepared. Our SPLC committee recognized that there would be a ripple effect through all existing railroad freight accounting systems by expanding the station number field, but that unless there was a goal established, future systems yet to be written would not include the proposed standard. I had included a description of the SPLC's code structure and its ability to define geography by state or part of a state with just the first two digits and county with the first four digits. The last two digits defined a specific point. It was a tough sell and the C&NW's vice president of finance, John Butler, was especially adamant that they would not accept the standard. Larry Provo requested a show-and-tell session on January 27th. The hour and a half allocated stretched to three and a half. Later I learned that the meeting was triggered by my report which had been obtained by Bob Russell who gave it to Provo. Bill Allman and Bill Zimmerman never discovered why this meeting had been called.

On February 2nd the United States Supreme Court approved the merger of the Northern Lines. Our C&NW-Milwaukee merger studies dragged on amid strange statements to the press. William J. Quinn was the Milwaukee's chairman. By March 9th rumors circulating that the MILW-C&NW merger was dead were confirmed on April 13th when the Milwaukee said they did not want the C&NW. Heineman's plan to sell the C&NW to the Milwaukee had failed, which in retrospect was bound to occur because the destitute Milwaukee was already reeling toward the bankruptcy which would occur in seven more years.

In the late afternoon of February 6th the infamous personnel policy #19 was presented to our group. The policy stated that all intellectual property created by employees belonged to the company. Warren Davis, our ace programmer, immediately went to his cubicle and packed his possessions. I was ordered by Bill Allman to sign the policy by Monday February 9th. On that date I flatly refused to sign the policy much to his consternation. I later learned that our own attorneys believed the policy to be unenforceable and that many other employees had refused to sign it. By February 13th Allman had collected just two signatures with modifying codicils; seven of our group had flatly refused to sign the hated policy. Allman asked Dave Leach what to do. By February 20th Allman was telling us that the policy was never meant to be a condition of employment, but the damage was done. Warren Davis was heavily courted by the Illinois Central and went to work for them. Fast Eddie Nordan was also a non-signer and adopted a very belligerent attitude. I also refused to sign the policy for very good private reasons and continued to resist efforts to force me to sign for many years. I was the last C&NW officer to sign that one-sided surrender document.

Gene Cermak, my former boss, called on February 18th with congratulations as he had heard from the grapevine that I was to be the next director of industrial development. That grapevine must have had a bad stalk in it because Douglas E. Christensen from the C&EI was appointed to that post March 13th. Larry Krause, our Iowa farm boy galvanized into an industrial development agent, was very upset because he did not get the top job. I congratulated Doug and wished him success while breathing a silent prayer of thanks that I had not been tagged with that position—I had grown far away from that profession. Doug was an interesting person and proved over time to be a smooth and consummate politician. With interests in such disparate activities as theater pipe organs and magic effects, he was willing to try new approaches to old problems. At this point I lost track of the industrial development department which ended up about 1980, I believe, as a small operation in the real estate department while Doug went on to other more responsible positions with the railroad.

I requested approval of a trip to Washington in connection with my chairmanship of the SPLC committee. Leach declined because of the expense. I offered to pay my own expenses. A shocked Allman relayed the offer to Leach who again said "no", but this time on the basis that I was needed in the office. Leach and I were definitely going to have difficulties as I was becoming extremely annoyed at his arbitrary management style.

More people left. Dave South and two more marketing people departed February 27th. Bob Russell, vice president of personnel, was reported to be in a sweat over this erosion. That seemed to me to be an overstatement because I had never seen the cool and laid-back former trial attorney ever exhibit any agitation or anxiety. Ed Garvey from our group was in Philadelphia on the 27th to interview at PennCentral for a position. He said he wanted to be closer to his old home in Baltimore, Maryland, and actually did move his family to Annapolis when he left Chicago. Al Duntun resigned March 4th to go to Colorado and left on the 13th. In March the winds of change blew more strongly. Advanced Systems died and was officially replaced by Staff Services on April 1, 1970. Allman retained his title of assistant vice president reporting directly to president Larry S. Provo. Bill thought that arrangement was pretty special, but it would prove to be his ruin. Staff Services (*Der SS* according to some) was to be an industrial engineering based operation. I was not sure I liked that bend in the road as industrial engineers, in my opinion, were not really engineers. I had always felt that the IE discipline belonged in business college. Jules Eberhardt, recently arrived from the CB&Q, was an IE. Allman's undergraduate degree was in IE, Bill McKenna, newly acquired from the mechanical department was an IE, and Ed Nordan, who was drafted into staff services, was also, I believe, an IE. My civil engineering education was in danger of being buried in work/time measurement exercises. Bill Allman asked me about my commitment to the SPLC committee. I said that I was deeply involved, but, given the C&NW's opposition to adopting the code, perhaps another chairperson should be found. My friends in operating told me to hang loose. Bill Wilbur, engineer of bridges, said that engineering wanted me back. The best news about this reorganization was we were out from under R. David Leach as of March 31st. There were other railroad people besides us that didn't like this Arthur Anderson refugee. Bud Braun scored Dave and his "garbage producing computers" at the quarterly vice president's meeting. I could see that the new staff services group had a lot of fence mending to do, and our biggest problem was its leader, Bill Allman.

On March 10th systems design and installation, marketing and engineering all agreed that they needed a railroad segmentation scheme to effectively capture and analyze freight revenues especially on branch lines. They were lucky, we had already created SRF (segmented railroad file) and the accompanying station master file which made the task possible. The tangle of outmoded branch lines from the mergers of 1960 and 1968 had be rationalized. Determination of revenues coming off these lines was the first big step in separating the quick and the dead.

The big news, however, came on March 19th with the announcement that Northwest Industries had agreed to sell the C&NW to its employees. Heineman, failing to peddle the railroad to the Milwaukee Road, was now willing to sell it to anyone with cash, including us. Taking a spring break my wife and I went to Washington on March 23rd on the Baltimore and Ohio's *Capitol Limited.* I went to meetings of the TDCC (Transportation Data Coordinating Committee) task force while Barbara viewed the art scene in Washington. She went to New York on the 26th but her train was badly delayed by a broken wheel. I made a quick trip on a *Metroliner* to Wilmington, Delaware, for a meeting with DuPont, returning to Washington on the *Senator* and my flight back to Chicago. Barbara was stuck at LaGuardia airport all night by the combination of a massive 16 inch snowstorm in Chicago, the Easter Holiday rush and an air controllers' strike.

My new title of project manager-staff services was effective April 1st in the midst of another record spring snowfall—12 inches was down by ten that evening. Our first task was moving back to the Daily News Building where we settled in room 712 which had been used by the long gone corporate development experiment. Remodeling and painting was still in progress on April 2nd. That evening the survivors of advanced systems, Jules Eberhardt, Ed Garvey, Warren Davis (he was soon to depart) and I celebrated the demise of our old department. The accident and loss prevention crowd took over our old space at 120 S. Canal. We physically moved April 8th. I got Vic Preisser's old quarters complete with floor to ceiling walnut bookcases. Frank I. Stern and Jerry B. Groner shared a long and narrow office overlooking the Chicago River. Edward L. Nordan never did have a desk in our new set-up as he was reluctant to leave his office in the operating department on the first floor of the Chicago Passenger Terminal. Jules Eberhardt and William R. McKenna shared an office as did Edward A. Garvey and George E. Ingram. Provo dropped by on the 9th to see how his new bunch was settling in. Bill Allman, occupying a big office on the south side of the space, was slow to get us into useful motion, but he kept the two secretaries, Phyllis Nyboer and Bonnie Michalek, furiously typing out his golden thoughts. Thanks to my wife's generous loans many of our blank walls were decorated by her large and colorful paintings.

April 10, 1970 should have been marked in big red letters in my journal because I officially returned to the real world of railroading by attending the morning meeting in the operating department. This daily meeting, sometimes called "Chapel" or "Prayer Meeting", began at 8:15 a.m. when the communication links to all the divisions were activated. The meeting convened around a long conference table set up in a glass-walled enclosed room in the operating department. The same chairs around the table were claimed by senior operating officers every day. On occasion heads of departments who were scheduled to make some presentation or had been invited for discussion of some particular subject also were invited to sit at the table. The perimeter of the room was lined with chairs for the rest of us lower types. Provo sat at the north end. Microphones lined the center of the table. Under the table at J. W. (Bill) Alsop's place a concealed master switch allowed him to shut off transmissions if he judged the world should not hear certain conversations. Tom Evans, communications engineer, often gritted his teeth when Alsop flipped the cut-off switch and blamed the loss of communications on poor lines. The meeting always started on time with the divisions taking their turns making reports on the previous day's train operations. The irregularities and failures from the previous day already had been tabulated by the power desk and distributed before the meeting so that Provo and the other operating officers could asked specific questions and then sit back to listen to the sound of spears being thrown as each division sought to shift blame to someone else. The failures recorded were many and varied. They included such mundane things as late and delayed trains, too many cars on line, derailments, bad track, outlawed crews, repairs to track and bridges, bad weather, accidents, shortage of crews, signal systems out, train control circuits inoperative, diesel engine failures, terminal congestion, communication lines down, late connections, IBM card reader out of order, high winds, deep snow, high water, ice in the track, grade crossing accidents, trains damaged by vandals (in urban areas), no pool waycars available, work slowdowns, switching out bad orders, or wet rail . . . the list was imaginative and endless. A good division manager had to invent a variety of excuses to avoid repetition because using the same one all the time meant that he didn't know how to fix the problem. Shower Chicago with new stories; baffle them with bullshit. Provo was usually glum at these meetings, but sometimes a new and innovative reason for failure brought a grin to his face. Over time, however, the litany of woe became ritualized as the division managers wearied of creating new excuses. To me this was contact with the real world where I could hear first hand what was going on over the entire railroad. I made a habit of attending the morning meeting as often as I could. Some meetings were grim like the one on April 24th when the air was full of complaints about bad train service. At the meeting May 15th Provo asked many probing questions but got no answers. His irritation was visible by the way he pulled vigorously on a steady parade of cigarettes.

A wildcat strike April 11th by one of the smaller unions, Shops and Sheet Metal workers, was swiftly halted by injunction. Rumors circulated in late April that the Santa Fe or the Norfolk and Western or the Missouri Pacific were looking at the C&NW as a merger candidate. Whoever started these rumors must have been hoping they were true.

Back in our brand new staff services group, Allman assigned me my first job: investigate company communications. I prepared a report, Allman rewrote it. I straightened it out. While the report was pretty skimpy and did not really shed much light on the communications aspect, we had a good man in charge, Tom Evans, who could have done a lot more given proper funding which is what the report said in essence. On April 21st Bill dropped Car Fleet Management on me. Frank Stern was also involved and went around questioning car distributors about how they made decisions in their work which made the car "disturbers" both irritated at him and our fledgling group.

The recommendation I had made to the AAR last year for an education program to promote use of the Standard Point Location Code (SPLC) resulted in plans for an SPLC Congress to be held May 7th at the Twin Bridges Marriott in Alexandria, Virginia. I flew down to Washington on the 5th to attend TDCC and AAR committee meetings held before the congress. The turnout for the congress was large and enthusiastic as advances in standards in geography were discussed. I participated in a workshop panel chaired by Herb T. Landow (avp planning, Illinois Central Railroad) and included Robert G. Hennell (chief, tariff research section of the National Motor Freight Tariff Association), Littleton H. (Larry) Fitch (manager of traffic and planning analysis, Mobil Oil), F. W. Gibson (manager statistical analysis, Flying Tigers), Robert A. Lawhorn, (systems analyst, Bethlehem Steel), and Ken A. Bluett (traffic analyst, Bethlehem Steel). I was grateful that Bill Allman appreciated the value of this type of involvement and gave me as much latitude as he could. I also was glad that the C&NW's consolidated station number file was almost ready to hand over to a permanent maintainer. Bill Allman was irritated that no user would step up and continue the maintenance on the file which meant that I had to do it. I really didn't mind as it kept me in direct touch with changes being made in the field. Bill was not a great detail man; his sight was always fixed on the far horizon. The CLASP file was in heavy demand now that its versatility had been demonstrated. This tool wouldn't stay bright and shiny very long, however, if unceasing and careful maintenance was not performed. I finished the maintenance documentation and was informed that systems was to take over the file. I held a meeting with Dick Zogg on July 7th to discuss the turnover process. Zogg was Dave Leach's administrative assistant and demonstrated his antipathy toward maintenance of CLASP by adopting a rather supercilious attitude that said without words that this job was beneath his dignity. He made little effort to understand how these station codes were woven together, their individual purposes and that breaking their intricate relationships would kill the file. The file maintainer had to monitor code changes from four different sources to make sure old stations were removed and new ones were properly fitted into the file. I had severe doubts about Dick's capabilities, but for better or worse, Zogg took maintenance of CLASP as his responsibility. We needed that database because on May 14th I had begun working with Maurie Reid on branch line assessment. Pruning the railroad of redundant and unprofitable branch lines was one of Provo's big objectives and he was absolutely correct in pushing these abandonments. The branch line group became an important and long-running operation.

Jules Eberhardt and I went to Butler, Wisconsin, on May 19th to meet with R. P. (Mac) McDonough, the Wisconsin division's AVP and his car distributors, Robb and Burkee. I dropped in on the engineering staff who were mostly former Galena division graduates, Jim Simons and Ray Snyder. Even Joe Gallup, a former Galena trainmaster, was now working in the Milwaukee area.

Orders were issued May 22nd that the Galena division had to be repaired by July 1st to eliminate the crippling slow orders then in effect. The work was pursued vigorously until August 17th when all track maintenance programs were pared back. The railroad was broke again. We heard that an ICC inspector was on the property looking for illegal links between Northwest Industries and the railroad. As related before, Ed Nordan was fired and tried to mess-up the blocking book project on May 27th. Recovery of the destroyed material took awhile as I gratefully eased into the operating department atmosphere. On June 8th I was called on to make a short presentation at the morning meeting on standard codes for Provo and the department heads after which I began the task of reassembling the scrambled blocking book master files. The next day my wife and I flew to New York where I caught a Metroliner to Washington to attend a TDCC meeting while she prowled around the art galleries. Later we took a brief vacation driving along the Delaware and Maryland coasts to Virginia's Chincoteague Island and Williamsburg before circling back to Washington via Fredericksburg, Manassas and Falls Church. The return to Chicago on the B&O's *Capitol Limited* wasn't all that much fun as they were short a Pullman and we had to sit up all night.

On June 22nd the PennCentral collapsed into bankruptcy. July 1st saw the opening of the new Northwest Passage which connected the Chicago Passenger Terminal directly to the Lake Street Elevated station at Clinton Street. On that day I was introduced to a newly hired operating officer from the old Wabash Railroad, Robert M. (Bob) Milcik. Bob was a close personal friend and college chum of Ed Burkhardt. Ed was from St. Louis, Missouri, and Bob came from upstate Moberly. Bob had kept his good ole country boy mannerisms along with his enthusiasm for hunting and fishing, traits that bothered Ed's mother who looked down her nose at Ed's gross and crude friend. This translated into Bob's nickname, "Gross" while Ed, in sympathy, became "Crude".

The Interstate Commerce Commission (ICC) examiner ruled on July 9th that the Rock Island should go to the Union Pacific and Southern Pacific. Of course, the C&NW protested the ruling. The next day at the morning meeting Jim Feddick gave a very interesting account on gathering grain from the elevators in Iowa demonstrating that he had a good grasp of the process. ADM must have thought so too, because Jim later went to work for them and prospered. Allman sought authority to fill Nordan's vacated job and was turned down because of the financial woes the railroad was experiencing. To make matters worse for Bill, his lead secretary, Phyllis, with her new husband, Dennis Lind, abruptly left for Texas. Bob Russell, vice president-personnel, cornered me after the morning meeting on the 17th and asked about Allman and the staff services department. There appeared to be a lot of dissatisfaction with Allman's group which seemingly was not producing anything useful to the company. Sensing trouble, Allman leapt into the blocking book project to develop a PLAN to move the project along. I was doing just fine and did not need him roiling the waters. Any plan that Bill produced was doomed if it involved Ed Burkhardt—the two were like oil and water. Ed would not cooperate in any proposal that Allman had a hand in. In this period I managed to get some of my work through the Ravenswood computer morass and actually begin producing the long awaited blocking books. By the end of July we had 13 of the 69 books completed. But Allman would not stay out of the way and called a meeting to discuss ways to distribute the completed books to the yards. Jim Ronanye (TOFC), Bob Tamura (my programmer on the project) and Ron Brezinski of systems sat there in stunned amazement while Allman's wild ideas flew around the room like demented sparrows. Ronanye shook his head in disbelief as he walked out of the room.

On July 29th I was privileged to deliver a presentation on geographic codes and coding to about 65 attendees of the Federal Statistic Users Conference at the Pick-Congress hotel in Chicago. While the transportation industry's SPLC was included, the primary focus was on the newly evolving FIPS (Federal Information Processing Standards) series of codes for countries, oceans, continents, places, states, counties and congressional districts.

Burkhardt asked me if I would like a job in the operating department August 5th. I demurely said that whatever was best for the railroad. Nothing concrete developed although he asked me again August 28th. Actually, I liked my present situation very well indeed. I maintained an office in staff services but was seldom there because I had a lot more privacy in the heart of the operating department arena. Allman fumed that I wasn't nearer (to be harassed) but Burkhardt seemed happy to have me around. During this time Jules Eberhardt in staff services was working on the possibility of a new freight yard at Merriam, Minnesota, where the old CStPM&O (Omaha) crossed the M&StL. Bill McKenna and Ed Garvey were looking at a new freight yard at Des Moines, Iowa. Neither of these projects ever materialized which was just as well as they would have contributed little to increasing traffic mobility. In fact even considering building another freight yard was off the mark. Little local yards needed to be closed and the bigger yards modernized.

I went east to Washington again on August 12th and visited the AAR headquarters where I found them in the process of laying off 160 employees because bankrupt PennCentral had failed to pay their dues. The new president of the PennCentral was William H. Moore formerly the operating vice president of the Southern Railway. After the SPLC meeting, I flew to Boston and met with Professor Fisher's seminar on computer graphics at Harvard University before returning to Chicago. Their use of the computer for mapping data arrays was dazzling and opened up a whole new world for thought. On my return home Maurie Reid asked me to contact Robert Palmer, city manager of Elmhurst (where I lived), and determine what Bob thought about abandoning the CGW mainline through town. Palmer didn't seem to think it made any difference except that city streets could be improved when the crossings were removed. Another rumor had William H. (Bill) Thompson, vice president-operations of the Illinois Central, coming to the C&NW on September 15th and that this was very confidential information. Bill did not come to the C&NW. Up north in Duluth the Burlington Northern dropped a train through a swing bridge on August 29th.

I marvel in re-reading my journals that I was able to travel so much with the C&NW being in such rotten financial shape. On September 1st I was off to New York again and went by limo to General Foods' headquarters in White Plains where the TDCC met the next day. We discussed the importance of a master geographic code file containing all geographic code elements. This comprehensive database was exactly the same concept that I had used on the C&NW to match the five different station code lists. Attendees included Herb Whitten, Nevil Black (Southern Railway), and John Emery. On my return to Chicago I found that I had a new typist, a good, diligent girl named Barbara Sidor. In my absence she had taken over collating and binding Ed Burkhardt's train service book on her own initiative. By September 11th the last book was sent to the field and the project was finished—I thought. Bob Milcik was now in charge of the blocking book production and sent another seven out September 21st. What we didn't realize was the volatility of change in blocking policy due to seasonal traffic flows and variability of railroad operations. Looking back it is easy to see how easily a computer model could have been constructed that would have adjusted blocking instructions at each switching yard and displayed management's policies online. Such technology did not yet exist. In 1970 we were still in the days of unit record equipment fed by punched cards with magnetic tape drives just beginning to come into use. Bob and I managed to squeeze more books through Ravenswood until October when they went through a massive computer conversion that stopped all production. Provo, in exasperation, demanded daily progress reports from the systems department. While creation of blocking books would never end, the function had to be taken over by the operating department.

I managed to stay with the project until December 17th when I finished training Ralph Johnson to manage their production. Ralph had been station agent at Geneva, Illinois, and still looked like he had just come off work out in the back 40. Ralph may have had straw sticking out of his collar, but he was industrious and careful of detail taking only a month of training to be qualified to create and distribute those books.

The diesel firemen were still trying to keep their unneeded jobs and called another strike on September 15th. A Presidential order was expected to stop their job action, but the midnight deadline approached without any news. Being associated with the operating department made me one of the emergency crews in the event of a strike. I got a call at 15 minutes to midnight to appear at Lake Geneva, Wisconsin, at 5 a.m. After a few hours of uneasy rest, my wife and I drove 80 miles north through a thunderous rainy night to protect my assignment. On arrival we found that the expected injunction had been issued. Back home we went. That afternoon I met Tom Desnoyers of TDCC at Rand McNally & Company's headquarters at Skokie, Illinois, to discuss geographic location codes. I was invited to attend an American National Standards Institute meeting. Allman, a bit peevishly, refused to let me serve on the ANSI committee or any other non-rail industry group because of the travel expenses involved and would not hear of my paying my own. I did not press that issue because he might get suspicious of how I could afford to pay my own way. I was supposed to be a wage slave like the rest of the department. Another strike alert came September 23rd ; it was delayed two weeks. Harold Tell in operating was in charge of the emergency crew calling and these strike threats kept him busy positioning people to protect train service. December 7th brought word that another fireman's strike was called for the 10th. Congress was expected to act, but Harold Tell had to put his plan into operation and give orders to his crews. I was ordered to Lake Geneva again. I arrived there at 4:45 a.m. to find no pickets in sight. The conductor said he didn't see any pickets and he was taking his train to Chicago. One of his crew refused to go, and I was pressed into service as flagman on train #624. The C&NW was the only major railroad in the country operating that day. No settlement occurred during the day so I went to Elmhurst as a ticket collector on the 5:17 p.m. westbound. The train was crammed with Burlington and Milwaukee riders stranded in the city. The next day, December 11th, all the trains were back to normal. I asked Harold Tell why he had sent me to Lake Geneva, Wisconsin, from Elmhurst. His reply, "Isn't Lake Geneva near Elmhurst? It isn't? Oh, you should have been sent to Geneva, Illinois."

Ed Burkhardt and some friends operating as the Pea Vine Corporation had purchased a tiny short line railroad, the San Luis Central, in Colorado in 1969. Ed was the official president. When this became general knowledge there were mutterings in some quarters that this involvement in another railroad was a conflict of interest. However, Ed maintained that since he was not an elected C&NW official and that the C&NW did not connect directly with the SLC, his "hobby" did not conflict with anything. Ed had a lot of interesting experiences with this operation including acquisition of a fleet of retired Railway Express Agency refrigerator cars. The first batch of these wooden bodied cars were refurbished in Green Bay, Wisconsin, by a private contractor before being sent out into the railroad world to haul such products as potatoes. Ed hoped that the SLC's cars would never see home rails at Monte Vista because he doubted that there was enough trackage to park them all. These free running cars were money makers for Ed and Pea Vine. At the end of their lives, some of the high speed trucks ended up under AMTRAK express cars. (The SLC was still alive in 2004 with a fleet of 843 cars. If they all wound up on home rails, their combined length of some 8 miles would have occupied more than half of the railroad's 13 mile main line.)

In late September I flew to Atlanta, Georgia, with Herb Landow to attend the Data Systems committee meetings. A few days later Harold Gastler, vice president-operations, asked me to accompany him to a high level presentation by General Mills on September 30th as I had been referred to them in connection with the SPLC. We learned October 5th that the Employee Railroad plan was awaiting ICC approval. Someone styled the name of the new company " the People's Agrarian Railroad". Allman's problems with his department continued to increase. He called a staff meeting September 25th and one of his key people, Bill McKenna, forgot to attend, later saying he didn't think the meeting was important. At my annual evaluation on October 16th Allman criticized the paucity of my production and failed to mention the good quality of my work. I pointed out to him that much of my recent work was now in daily use in many computer applications and that I didn't see that his floods of paper had produced much of anything. When he called a special staff meeting December 3rd to discuss Christmas decorations, the response he got from his unruly crew was pretty rude. To add to his burdens, he hadn't found anyone willing to work on freight car scheduling, a project that had been and would continue to be, a graveyard for many ambitious reformers. To be fair, car scheduling was beyond the technology of 1970; communications and computers had to grow in scale and flexibility to handle this enormously complex network problem.

In mid December I attended an AAR class on networks even though the C&NW was in no position to utilize the techniques in 1971. Dick Ogle, formerly an advanced systems employee and now with the Burlington Northern, attended the class along with a lot of Illinois Central and Santa Fe people. Bob Tamura, my main programmer, finally got his job with Gates Rubber in Denver and departed January 9, 1971. My wife finished her masters degree work at the Illinois Institute of Technology and went with me to Portugal for a well deserved vacation over the holidays.

On November 9th my well regarded friend, Tom Hudson, suddenly died in his New York office. Ten days later, in a change of attitude that really surprised me, John Butler of finance, withdrew his objections to future use of the Standard Point Location Code (SPLC). John Butler was one of the better heads on the C&NW and remained a respected friend for many years. His acumen and practicality were to be major assets to the company in the years ahead. It had been a productive and interesting year, all in all.

25

THE LURE OF KANSAS CITY, 1971 . . . Our new project for the year was to be Service Definition. Jim Ronanye from the TOFC marketing group was tagged to help us on this one. Trouble commenced immediately in trying to define the scope of the project because there was little common ground between the providers and the recipients. We needed a gimmick to make this project viable. Bill Allman's goading did not help; long conversations with Bob Milcik and the other operating people underscored the many difficulties at arriving at a consensus. Even a simple definition of what constituted "service" proved to be a stumbling block. By the 18th Allman had begun to back away from some of the messier components of the project. A few days later it was obvious to me that this entire subject was being viewed as a hair triggered menace by many of the operating officers whose help would be critical if Service Definition was to be successful. We were antagonizing the very people we should be cultivating. I covertly met with some of my friends in operating and concluded that their department should have time to get its internal political war settled before embarking on this controversial project. Allman refused to heed my advice because pleasing Provo was his main object. By January 28th I had finished my report on the subject. Bill ripped it to pieces and reassembled it the way he liked it. Then he found his clerical staff had vanished—all were out looking for new jobs. Bonnie Michalek, our second secretary, was the first to bail out followed by my very capable secretary, Barbara Sidor, who went to Shell Oil across the street on February 12th. In my report I had outlined a recommendation that there should be an audit of train service established and again told Allman that this was the wrong time to press the matter. A week later Larry Provo told Allman to do an investigation of locomotive usage—no ifs, ands or buts. Allman sent the service definition report to Provo on February 16th and was immediately told that the locomotive project was far more important and that we should forget service definition. Allman complained bitterly that Provo was wrong; Bill did not seem to understand that the president possessed a much bigger cannon and, possibly, had a better idea of what would best serve the railroad's needs. Eventually, Provo approved the Service Definition project in March. Concentrating on the locomotive project, I spent February looking through overtime slips for yard engines at Proviso while the rest of our gang searched dispatchers' train sheets trying to establish what diesels had been where and when. The frightful fact soon emerged that no one, not even the power desk in the operating department, knew where all the diesel units were at any given moment. While our crew was trying, unsuccessfully, to identify operating patterns from existing reporting, I felt we were viewing the problem from the wrong end of the telescope. A better approach would have been determination of power requirements to move the traffic, but that required a more complex and bigger effort than Allman was willing to tackle. He was not interested in locomotives even though his pet service definition project depended heavily on having the proper power in place to move the trains he was evaluating. Yes, I got a raise in January; my annual salary soared to $17,100.

On January 7th I entertained the twelve member SPLC committee in our offices. Nevil Black (Southern Ry.), Tom Desnoyers (TDCC), Larry Fitch (Mobil Oil), Ken Bluett (Bethlehem Steel), F. W. Gibson (Flying Tigers) and Guerin (affiliation forgotten) were among the dozen. Despite his prior approval for the meeting, Bill Allman seemed irritated by having the group use our conference area as he kept barging into our meeting to ask me trivial questions. On February 24th I flew to Washington National for a TDCC meeting at the Aviation Club. Herb Landow, the chair, resigned as he was changing jobs again. Now that old devil lurking in the Standard Point Location Code (SPLC) surfaced—the difference of place definition between the railroads and the trucking industry sparking a furious debate. The railroads were determined to have the point concept paramount, but the trucker's tariffs were based on an area concept using rate basing points. Bob Hennell, the trucking association member, who had done most of the early coding of the SPLC, now refused to add new points to the code that were railroad specific. The ensuing chaos threatened to scuttle the entire code. Luray McHargue, the code manager from the AAR, knew so little about the code and its problems that for many years she mostly ignored attempts to repair the damage. Asking her why Duluth, Minnesota, was miscoded always elicited a tongue lashing until, one day, she quietly fixed it.

Although Nevil Black was fiercely proud of his Southern Railway, he certainly had some tales to tell about the railroad and its people. There were a lot of stories involving their colorful president, D. William Brosnan Jr., whose authoritarian style of running the property was legendery. Nevil said that Mr. Brosnan became president in 1962 and liked to prowl around the railroad by himself to see firsthand how things were going. One fine day he visited the new Chattanooga hump. The yardmaster's office was located atop an elevated tower by the hump crest which gave a commanding view of operations in the yard. Mr. Brosnan trudging up the stairs inside the tower to the office passed a young man who was standing at a stair landing watching from a window as freight cars glided over the hump. Without making any comment Mr. Brosnan proceeded upwards to the top. On his way back some time later, he saw the same man standing in the same spot watching the hump in operation. Pausing, Brosnan addressed the man without formality, "How much do you make a week?" The fellow replied, "about 100 dollars, sir." Brosnan reached into his pocket, pulled out $200 and handed it to the man with the admonition that he did not want to see him on the property again. "Thank you, sir" the young man said as he took the money, went down the stairs, climbed into his Coca-Cola truck and drove off.

Our staff services personnel continued to change. Ed Garvey went to Washington February 26th to interview at the AAR. He left on March 31st, but we were to see him again in September when he came back to work for Ed Burkhardt in the operating department. (Garvey did very well in later years. In 2004 he was senior vice president of Helm Financial Corporation, a privately-owned rail equipment financial firm based in San Francisco, California). Ron Rudolph from United Airlines joined us March 15th. George Ingram also came to staff services in March. George, who was about my age, had been hired to teach Construction Planning Management (CPM) to railroad officers who seldom, if ever, used logic to plan anything. The CPM technique was a standard industrial engineering tool, which George patiently explained to his classes using practical applications such as describing how getting dressed in the morning could be viewed in logical steps. Possessed of a lively curiosity, George looked at problems from angles we often overlooked. Many of his "solutions" were too far out on the rim of the bell curve of probability to be practical, but, once in a while, he latched on to a good idea. Since I was his nominal boss, it was my job to pull him out of the hot water he so often got into. Always on the move, he suggested using a video camera to record car numbers in switching yards and went ahead with a demonstration without ever asking either mine or Allman's permission. While this was not at all a crazy idea, it was too visionary for immediate acceptance. Eventually, George enrolled at Roosevelt University in Chicago to complete his degree work. While he worked for me, I made every effort to keep him in school for the late afternoon and evening classes until he had earned his degree. Not only did school benefit George, it also kept him out of trouble on the railroad. Allman finally got a new secretary in April. Arlene Cabai, a pleasantly rounded Polish girl with freckled skin and a mop of auburn hair. She came from the Northwest side of Chicago and was a good worker despite being a bit of a flirt. Bill McKenna had been drafted into staff services from the mechanical department and didn't like the move, or Allman, very much. Bill, a stocky little fellow with a good head on his shoulders and no neck to speak of because of a birth defect, was feisty and made no bones about his lack of respect for Allman which bordered on open hostility. McKenna left for Anaconda Cable in New York April 16th. I had a long meeting on April 7th with Bob Russell, vice president-personnel who knew of McKenna's departure before Allman was ever informed. Russell said Allman could be the next to go.

Another strike threat surfaced March 6th. The morning meeting of March 12th was extremely interesting. Alsop had to use his cut-off switch because of the loud and undignified shouting and scurrilous recriminations about lousy C&NW train service and attendant loss of business. Stone-faced Bill felt that all that discontent need not be spread over the whole railroad. Sales and operating continued to be at each others throats. On March 18th an unidentified person (or persons) stole a diesel engine (GP7 #1647) idling at Crystal Lake and started it up the line toward McHenry. The resulting crash at 35 mph into a standing set of doubledeck gallery cars headed by E8 #5032B at Lake Geneva, Wisconsin, about 4 am. caused a lot of damage. The guilty party, or parties, was never identified. The next morning's meeting included the details of a head-on collision in Minnesota. When I heard that a wreck had occurred on the Freeport line near Gilberts, Illinois, on March 22nd, I drove out to see the damage because this was the Chrysler plant's "lifeline" connection with the mainline. These incidents were typical of the ragged operations we were experiencing in early 1971. Chrysler was treated with great care by the C&NW's operating department. There were instances of special trains being operated to rush so-called shut-down cars of auto parts from Chicago connections to Belvidere. These one car trains had right-of-way over everything. Bill Alsop, himself, skippered one of these emergency trains.

In the spring of 1971 I worked on a Quality Control concept that would pinpoint service failures so that corrective actions could be taken. Of course, Allman always had to "improve" my methodology. By March 16th he had so "improved" the proposal that the prospective users refused to cooperate any longer because the convoluted Allman plan was too complex for practical use. On April 2, 1971 a Quality Control Center (QCC) was actually established in the operating department—without any staff, at first. The idea was to set standards for train operations, measure deviations, ascertain causes for the failures and recommend changes to avoid future problems.

QCC was slow to begin as they did not even have telephones installed ten days after the official start. Tom Minard was part of this operation; I had last seen him working as a clerk in the traffic office in Milwaukee some years ago. The first QCC report was made by Tom with my assistance on April 16th.

For some unknown reason Bill Allman assigned me to help Jules Eberhardt with his study for a major new switching yard to be located east of Marshalltown, Iowa. Jules also was puzzled by the assignment because he didn't need any help and was at a loss to know what I could do to help him. We agreed to continue on our own projects and ignore Allman. Construction of a big yard facility at Marshalltown yard was another example of wasted effort by a lot of people.

Before the M&StL merger in November 1960, Marshalltown was just another important Iowa city along the C&NW's Iowa division. The Louie crossed the C&NW here and maintained a yard and engine servicing facilities. After the 1960 merger, Marshalltown suddenly became an important junction as export grain coming from Northern Iowa and Southern Minnesota on the Misery and Short Life (as some called it) was yarded there and sent east on the C&NW rather than continuing to Peoria on the M&StL as in the past. The C&NW incorporated the cars off the Louie into its existing trains which turned south to Madison, Illinois, at Nelson. The diverted traffic had the immediate effect of drying-up the old M&StL because there was so little local business on the line between Oskaloosa, Iowa, and Peoria, Illinois. In 1968 the Chicago Great Western (also known as the Cinders, Grass and Weeds) came into the C&NW family. The CGW crossed both the C&NW and the M&StL in Marshalltown at a closely packed bunch of diamond crossings just west of the downtown section of the city. An unexpected result of the merger was the sudden creation of Kansas City as a major C&NW gateway for export grain which caused another massive redirection of export grain. Southbound trains off the Gut Wagon (CGW) now could roll across the C&NW and head straight for the Bell Avenue Yard at Des Moines before proceeding to Kansas City. Eastbound C&NW and southward M&StL traffic, however, required reverse moves at Marshalltown to swing west onto the CGW for the Kansas City trip. The CGW could yard trains in the C&NW and M&StL yards which lay side by side along the mainlines east of the interlocking. Marshalltown now had

C&NW in IOWA - 1967

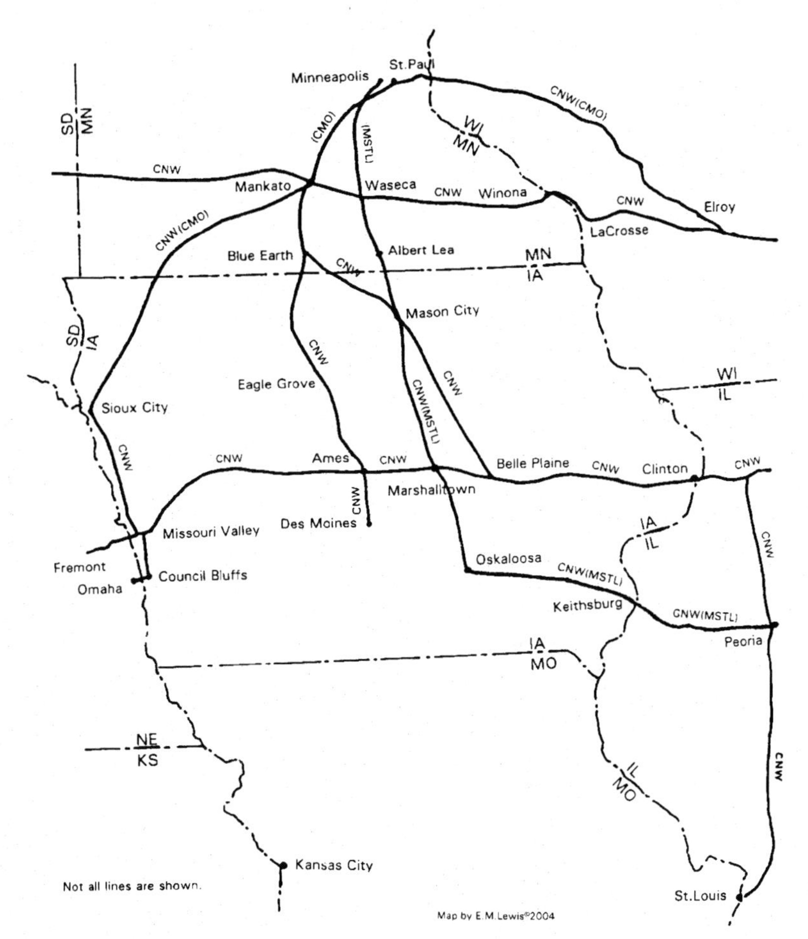

The C&NW's major Iowa lines in 1967 included the former M&StL north-south line to Peoria—soon to be dismantled.

three rail routes all under the same ownership. Train operations regularly blocked automobile and pedestrian traffic in the city to a great degree eliciting many municipal howls. To solve these problems and deal with the sudden concentration of traffic, the idea of building a new yard east of town was proposed. The new facility was to be about two miles long with the east end anchored at Timber Creek, a substantial stream. The terrain was not level, but earthwork to grade the area was not impossible. The major problem, as it turned out, lay in

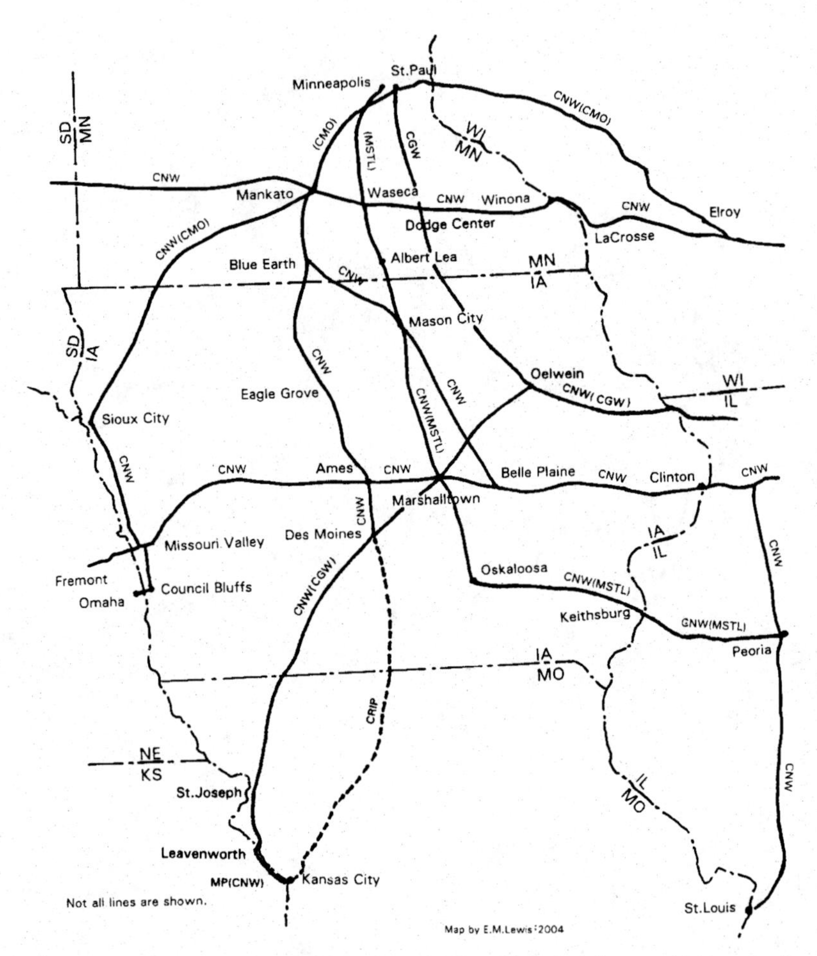

By 1971 the Iowa map of the C&NW's major lines included the recently acquired CGW and its access to Kansas City.

the character of the subsoil which tested poorly for compactability. Little thought was devoted to raising line capacity by improving the mainlines for increased train speeds. Events involving the Rock Island in 1980 outran the planners. Fortunately, a new Marshalltown yard was never constructed and most of the CGW line was abandoned in 1984.

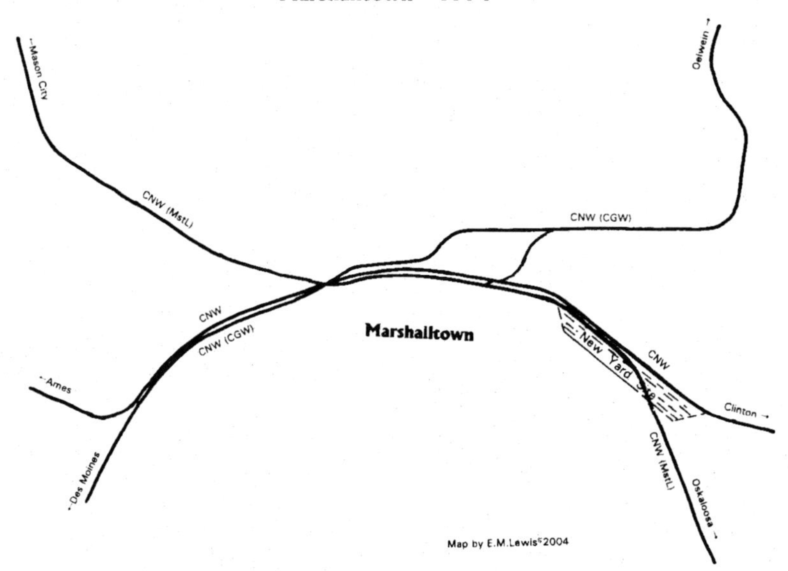

Marshalltown became more important to the C&NW point after 1971. This crossing of three major routes confused the railroad with dreams of a new switching yard when abandonments and upgrades to the main lines was the better answer to improving traffic flow.

I rode to Champaign on the Illinois Central's *City of New Orleans* March 24th to meet with Professor Howard Roepke at the University of Illinois and to visit their library map collection. The next day Dick DeCamara called with a proposal for me to take over the maintenance of the SPLC for the Transportation Data Coordinating Committee. While flattering, the job would have been a full time task and my present dual occupations were consuming all my time. Two days later I ran into Ron Reifler of Fantus Company whom I had not seen for several years. I was glad I had not taken that job with them years ago.

DES MOINES TO KANSAS CITY—1971–72 At last our department had an opportunity to deal with a real physical operating problem bedeviling the C&NW. On April 21, 1971 I was assigned to locate new passing sidings on the former Chicago Great Western line south of Des Moines. The first step was finding maps and profiles as this was new railroad to us following the 1968 merger. I flew to Waterloo, Iowa, on April 26th and drove over to Missouri division headquarters in Oelwein, the CGW's old main headquarters, for a meeting with the superintendent George R. Hanson, division engineer Del Swenumson, Ray Snyder (engineering) and Mr. Gifford, the chief train dispatcher, to discuss the problems that were disrupting train operations.. Building new passing sidings seemed to be the immediate solution to the congestion problem, but the real solution later turned out to be something quite different.

The CGW's railroad was a single track, unsignaled line marked by numerous and abrupt changes of grade through the hilly country between Des Moines, Iowa, and Leavenworth, Kansas. Acquisition of the CGW by the C&NW July 1, 1968 may have been viewed by Mr. Heineman as just getting rid of another redundant railroad in the Middle West and, ultimately, most of the CGW was indeed abandoned. But, the one leg of the CGW going to Kansas City coupled with an extraordinary jump in export grain traffic moving to Gulf of Mexico ports created massive changes to the C&NW's traditional east-west traffic flow across Iowa. Now the grain production regions of northern Iowa and Southern Minnesota had another route to New Orleans and Houston through the old CGW's Kansas City line south of Des Moines, Iowa. The traffic surge overwhelmed the modest CGW line. As mentioned before, one solution to the traffic jam was thought to be a new hump yard east of Marshalltown where three of the C&NW's routes crossed.

Train operations on the line south of Des Moines were so erratic that discerning and categorizing operating patterns was very difficult if not impossible. Since we had to define the operation before we could devise solutions, we fell back on the old string line diagram technique. The cats cradle technique was old fashioned but proved to be helpful. We put the first one together in Oelwein by scrounging a piece of corrugated board about 30 inches square from a local grocery to support a scaled distance-time graph of the railroad south of Des Moines. After fastening the graph paper to the backing, we were then able to define train movements over the line by sticking pins into the station locations and the times reported on the train

Example of String Diagram - Time versus Distance

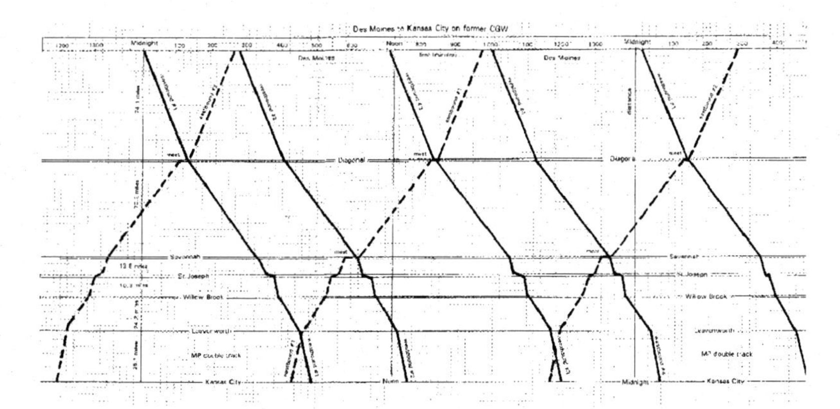

The classic "string-diagram" for analysis of train operations quickly jointed out the choke points on the Des Moines-Kansas City line of the old single-tracked CGW. Passing siding were few and widely spaced.

dispatcher's actual OS (on station) sheets. Stretching black thread along the pins for each train movement gave an instant picture of operations over a 24 hour period. To improve the picture we extended the time chart a half day before and after our planning day because some trains were still moving to their terminals in those periods. The very first try at stringing train operations on the Missouri Division below Bell Avenue, Des Moines, revealed a picture of the chaos caused by outlawed trains (12 hour maximum for crews on duty) and delays resulting from waiting for the dogcatching (relief) crews. George Hanson wanted to achieve some sort of stable operating plan that would guide the train dispatchers. Working with him we estimated reasonable train running times and assumed that desired train schedules were attainable despite the railroad's demonstrated utter lack of success in achieving any sort of regularity. George also decreed that westbound (south) trains had the right over eastbound (north) trains. We developed a pattern that seemed to work for the single track line. The principal difficulty seemed to be the long stretches without passing sidings. Most of the sidings were on the south end with 70 mile intervals in the northern part. Laying out the new schedules I noted where the strings crossed and made a list of potential sites for field examination. Those assumptions were soon to be disproved, but, as of the moment, we had nothing else to base our planning upon and, like Candide, hoped for the best of all possible worlds. George Hanson, Del Swenumson and I headed south in a hi-rail station wagon from the CGW's Bell Avenue yard in Des Moines on April 28th.

Passing sidings a mile or more long are hard to properly locate along an existing railroad. First, the placement of the entrance and exit turnouts has to be done carefully to avoid starting heavy trains from a siding into a steep grade. It made a lot more sense to drape a passing track over a crest. Second: public and private road crossings must be accommodated. Including them as grade crossings within a siding location makes a farce of train operations when a standing train has to be cut to permit free vehicular traffic across the railroad. Third: major bridges, tunnels or overhead rail or highway structures have to be avoided, if possible, because of heavy construction expense. These restrictions didn't leave a lot of obvious opportunities, but we found possibilities at Lorimor and Diagonal, Iowa. We got back on the highway at Sheridan, Missouri, and went on to St. Joseph where we again went down the railroad through the maze of railroad crossings, joint track, and trackage rights that characterized the CGW's route south. It was easy to follow the CGW's route through the St. Joseph maze. So much grain was being shipped in 40 foot boxcars with paper grain doors that leakage of corn left a neat double line of dribbles all along the tracks. At the BN crossing at Beverly, Missouri, we again went on the highway to Leavenworth, Kansas, where we marveled at the sharp curvature at the west end of the Missouri River bridge. Because the trackage rights on the Missouri Pacific required their dispatcher's OK for our vehicle, we stayed on the highway into Kansas City. I had seen enough for this trip and flew back to Chicago.

I had been impressed by this strange CGW line which traversed such hilly terrain on high cinder fills, its remote, lonesome territory, the tortuous, meandering route through St. Joseph, the ancient (1893) swing bridge at Leavenworth, and the weird collection of worn vari-colored F units used on their long, long trains. I wondered how a modern railroad could be operated reliably through this corridor with its bottlenecks and the many chances for failure of both track and structures. Those cinder fills, for example, were time bombs. When this waste product of steam locomotion was dumped for use as embankment material, the hard, crunchy nature of cinders and clinkers concealed its true nature as an active chemical agent. It was just waiting for moisture to continue its slow decomposition and compression. Over the years the fill structure crumbled and settled with resulting loss of volume and shrinking of embankments. Tracks on top of the fills sank, more ballast was added which increased the load on the crumbling cinder base which accelerated the compression of the cinders which required addition of more ballast which increased the load *ad infinitum*. In retrospect, I wonder if the cinder fill failures we had experienced at Radnor, Illinois, on the Galena Division in the 1950s were not the same phenomenon. I wrote my report on the condition of the line and Allman cautioned me (May 3rd) to keep it secret. The next day I waded through the contract for the trackage rights agreement with the Missouri Pacific for the Kansas City-Leavenworth line. Carl Hussey hinted darkly that he would take care of an old trackage rights agreement between the CB&Q and the former CGW for the Q's access to Des Moines from a connection at Talmage some 53 miles south of Bell Avenue. Although the Burlington included this line in their system map in the *Official Railway Guide,* to my knowledge, they did not operate any trains over it after the C&NW took over in 1968. Meanwhile, R. C. (Bob) Conley, chief of the power desk in the operating department, announced his intention to show those clowns on the Missouri Division how to run a railroad. He went to Des Moines and applied his expertise to the constipated railroad with no visible signs of improvement. Bob Conley was one of the Illinois Central "mafia" who had come to the C&NW with Clyde Fitzpatrick in 1957. Called "Number One" by his colleagues (because every statement he made was preceded by those words leaving the listener waiting for the "number two" that never came), he had served as Galena Division superintendent for a number of years before graduating to the power desk where he acted like the king dispatcher of the entire railroad amid ever present clouds of blue cigarette smoke. No one in operating dared comment on his Des Moines exploits.

That Missouri Division string diagram continued to prod our imaginations. We experimented with different combinations in an effort to stuff more trains into that narrow corridor. The more we crammed trains into the system the worse the delays became because of the few meeting points. The most comfortable solution turned out to be 3 trains south and 2 trains north in a 24 hour period—an imbalanced scenario that would fail to move the available traffic. Fleeting, or bunching, of trains was considered. By sending trains south at 30 minute intervals as many as seven trains could sail down the line in a long elephant chain. Northbound trains, however, would be forced to stand idle for many long hours at remote sidings to accommodate this type of train operation which was not practical because of crew and locomotive availability even if the loads were all lined up and ready to roll. Bell Avenue was not that big a yard for staging a fleet operation; Kansas City's tiny yard made that plan ridiculous. Another "solution" was alternate day operation: westbound on even dates; eastward on odds. That might have worked if the traffic, crews, yard capacity and locomotive power had been available. And they weren't. The other imponderables included equipment failure, weather, different levels of locomotive engineer skills, track or structure conditions and other quirks and difficulties that tended to arise when the operation was already under stress. Other than recommending extension of sidings to accommodate two trains and improving the speed over the line by making massive track repairs, we did not see a quick, clever or inexpensive way to provide a solution to running that congested traffic artery.

Another assignment involving the use (or mis-use) of waycars (cabooses on most other railroads) sidetracked me until June 21st when I got back into the Missouri Division's problems. By the next day I had come to the conclusion that the C&NW just had too much traffic to try and cram it all down the decrepit CGW line. In my view it was a serious mistake to think that the line could be upgraded to a viable operation. Consulting with Ed Burkhardt, we concluded that a better move would be securing trackage rights on the parallel Rock Island. To demonstrate with hard facts, pure logic and attached dollar signs that the Rock Island was a better route to Kansas City and that the old CGW should be abandoned, I applied Train Performance Calculator (TPC) to the problem to see where the trains were most prone to operating difficulties, how the siding locations benefited train operations and calculated running times and fuel consumption. The program calculated drawbar forces in tension or compression at regular and close intervals along the line. Reviewing the gateway reports at Kansas City I found many inconsistencies in car counts when compared to the notations on the actual train sheets. I read the operating department's files on the Leavenworth bridge across the Missouri River and found out what a rattletrap structure it was. Built in 1893 by the Leavenworth Terminal Railway and Bridge Company to supplant an even more ancient structure constructed 1868–1872, it was used initially by three railroads: the CGW, CRIP and CBQ. The former "Fort" bridge, despite being only 18 feet wide, served as the highway access to Kansas from Missouri as U. S. 92 and was reputed to be the oldest bridge spanning the Missouri River.

On July 1st I began writing my findings which included a tabulation of all the ups and downs in elevations and the sum of central angle for the curves. Traveling to Oelwein again on June 29th on Ozark's little Fairchild aircraft via Cedar Rapids, I collected more train information and met with a harassed George Hanson and the voluble Ray Snyder, who just had to tell me about "my" George Ingram. George had come to Oelwein to teach CPM to the staff. He visited a derailment at McIntire, Iowa, involving some of those old F units. When no one else would, or could, George climbed into the rerailed unit and started it up although he knew nothing about diesel locomotives. The mechanics on site had refused to mess with the engine and were furious at George's success.

One warm evening in June I experienced a moment that seemed to epitomize train operations on the Missouri Division. Standing alongside the main line, I watched a grain extra southbound leave Oelwein behind its usual collection of old F units. The train did not roll along smoothly and silently—it heaved and bucked, each loaded grain car ponderously rocking from side to side out of synch with its neighbors amid a chorus of creaks and groans from track and cars. In the half light of the early night that long and heavy train resembled a caravan of laden camels plodding along in file toward far-off Kansas City. On another railroad with better track these grain trains would have had a steady monolithic unity as they rolled along effortlessly and silently with that silky smooth motion characteristic of heavy loads on good track. Not on the Missouri Division where the rail was too light for the heavy loads and the track surface too uneven for any speed above 30 mph. It was sobering to reflect that this departing train probably would never go any faster in its entire southward journey. No wonder the division was constipated.

By August 5th I had completed a cost analysis that said if the C&NW spent $7 million on the upgrading of the CGW the same money would have paid for trackage rights on the Rock Island through 1974. The whole DM-KC project became a hot potato because of the politics involved. People began to choose sides. To reassure and protect my understanding of the problem I decided to get into the heart of train operations because I knew that direct, personal knowledge carried weight in a controversy. On August 15th I flew to Kansas City with Bill Allman. The next day Bill and I went around with Jeff Koch, the trainmaster, touring the Missouri Produce Terminal; the Union Pacific's Armourdale Yard; and the MP-MKT joint yard, all destinations for a lot of C&NW's Kansas City traffic. Leaving Allman to his own devices, I rode #162 north with 80 cars in tow. We got out of Kansas City's postage stamp sized yard at 2:45 p.m. That evening, 7½ hours later, we had only gone 59 miles. Jeff rescued me at Willow Brook, south of St. Joseph. On the 17th I sat in on two disciplinary hearings conducted by Jeff Koch before going out to the ATSF's Argentine hump yard. A discipline hearing simply meant that a rule infraction had been detected. The miscreant had the riot act read to him for the failure (real or perceived) and was given an opportunity for rebuttals, excuses and pleas for mercy. Punishment was assessed either as outright dismissal or being taken out of service for some period of time that varied according to seriousness of the violation. The hearings reminded me of a military court martial. After a look at the Santa Fe's Argentine yard with Allman, I flew to Washington for a TDCC meeting while he went to Chicago.

With Ed Burkhardt's help copies of the Rock Island's track chart, profile and operating timetable were acquired. Their line from Des Moines to Kansas City, constructed in 1914, was a much better engineered (and financed) railroad than the haywire CGW line. The profile data from the Rock Island material was swiftly translated to TPC code. We ran identical test trains on both the Rock and the old CGW to compare the results. No surprises there; the Rock was far superior. This was our first translation of a foreign line railroad to our TPC program; it was not to be the last. The TPC program became one of the C&NW's most important intelligence tools when dealing with other railroads and was used in pricing coal moves, training enginemen, and understanding power to weight ratios for the development of the fast TOFC trains in 1973, the Falcons.

On August 3, 1971 staff services was shifted away from direct access to Provo and moved to the supervision of James R. Wolfe, vice president labor relations. (Jim became president of the C&NW after Provos' death). After some effort I finally got the Des Moines-Kansas City report out of Allman's hands on September 2nd and into Wolfe's. Jim forwarded it to Provo the next day. In the report I recommended that the C&NW should negotiate trackage rights on the CRIP and junk the CGW line. Reaction from the operating department was swift. Harold Gastler suggested bringing the Burlington Northern into the picture by securing trackage rights on their line south from Beverly, Missouri, to Kansas City along the east bank of the Missouri River. He completely missed the point of the entire report—the worst part of the C&NW line was north of Beverly. Harold was not the sharpest pencil in the drawer. A committee was formed to consider the report's findings. I called a meeting October 25th and two people showed up. This underscored the politicized nature of this report because no one knew where Provo stood on the matter and being on the winning side was to be greatly desired. There were a couple of lingering questions that could be answered only by going out on the line. I flew to Des Moines, rented a car and headed south along the Rock Island on November 10th to sample the track structure. Tie condition was pretty good for the 112 pound jointed rail. Line and alignment were good for at least 40 mph. Bridges looked OK. I was pleasantly surprised at the track's generally good condition considering the Rock Island's shabby financial condition. I went south as far as Polo, Missouri, where the joint Milwaukee-Rock Island line into Kansas City commenced.

That evening I caught up with division manager George Hanson, division engineer Del Swenumsen and Len Clemmons (operating department, Chicago) in St. Joseph. On the 11th George, Len and I walked through St. Joseph following the CNW (former CGW) route through a maze of tracks. Len had maps and contracts so that he could tell which railroad owned what trackage. There were several stop boards where rail lines crossed which raised havoc with automobile traffic in St. Joseph when a long freight inching through town had to stop and proceed at each board. On the way we met some AT&SF officials and asked about the future of their line that ran from Henrietta to St. Joe as the C&NW used a portion of their railroad from St. Joe to BC Junction (Bee Creek) on trackage rights. Of course they had no idea of their corporate plans, but the traffic was so light that it probably would be a candidate for abandonment. Retrieving our hi-rail car we retraced the route again and went on south to Leavenworth before transferring to the highway for the run into Kansas City.

The next day I met with superintendent Bill Apple of the Kansas City Terminal before riding Rock Island freight train #68 north. Ed Burkhardt had fixed me up with the necessary permits because he supported use of the Rock Island and did everything he could to make it happen. The Rock's train was called at 2:30 p.m. but didn't leave Kansas City until 4 p.m. It was an all night trip with a set of old diesels that labored to get over the road. One unit kept going down leaving the rest to struggle. We finally rolled into Trenton, Missouri, the crew change stop, before midnight. I grabbed some food at the local beanery before going on north with a new crew. During our stop at Trenton, the mechanics had fixed the ground relay problem on a trailing unit so that the last leg into Des Moines went quickly. The Rock even provided a driver to take me to Bell Avenue where I checked into the nearby Holiday Inn. (I was told by some of our trainmen that calling down to front desk and asking for an extra pillow would bring a warm and willing lady to your door. I didn't try to verify the truth of that remark). Riding this Rock Island train had another impact: everyone in the general offices who knew of my trip now believed that I knew more about the Kansas City situation than they did because I had been there. Years ago in Peoria I had learned from Carl Hussey the lesson of the importance of going out to see an operation in person.

I returned to the Des Moines-Kansas City study with my new Rock Island information. After reviewing the roster of the merger protected trainmen at Kansas City on November 18th, I began assembling a presentation on my updated findings. Jim Wolfe was overseeing our work because our department now had no director—just three of us managers each working on separate projects. I think Jim was letting us run along unsupervised to see which of the three could rise to the director's job. On November 30th he approved the development of an educational show on the DM-KC study. All through December I developed charts depicting the physical differences in the two lines, comparison of fuel usage, train speeds, line capacity car and traffic flows. I had a long session with Don Yeager, late of the CGW, getting details of how traffic had changed at Kansas City since the C&NW take-over. Ed Burkhardt was especially helpful by bringing up discussion points that I had not thought of. During the winter the whole subject took a back seat because of management's preoccupation with immediate problems such as the December 27th wreck at Conception, Missouri. The Missouri Division was the center of attention at the next morning's meeting.

On January 17, 1972 Jeffrey S. Otto joined our group and, luckily, was assigned to me. A diligent, rather intense young man, his insight and willingness to plunge into the task at hand was much appreciated. While his education as an industrial engineer was not a plus in my prejudiced view, his energy and knowledge about the railroad business quickly eliminated any reservations I may have had. Within days Jeff discovered a flaw in our TPC program—the simulated trains ran flat out at the speed limit without any acceleration or deceleration. The program modification was not difficult, but we had to redo all our TPC runs to get corrected running times and fuel consumptions. We also had to recheck the critical spots in the runs where drawbar tensions exceeded normal limits. The new results further confirmed our original recommendation that we dump the CGW line and run everything on the Rock Island. We gladly redrew the exhibits before photographing them. In January we fleshed-out the study with additional detailed information. At the end of January Jeff and I flew to Waterloo and drove over to Oelwein to gather more crew cost information. George Hanson, the superintendent, was a very uneasy man as the Missouri Division was overwhelmed with traffic and plagued by bad track, old power, and a shortage of crews. We heard sad stories about miserable track in some of the branch lines. One section foreman responsible for the line through Dayton, Iowa, was said to have felled trees growing on the right-of-way to manufacture track ties because he could not obtain any usable ones through normal requisitions. At least he was able to hold the gauge of the rails. Another story told of the resourceful section foreman who used salvaged dunnage lumber from cars of farm machinery unloaded on his territory. It was certainly uneconomic to spend labor on installing six inch timber of dubious species just to keep the rails from spreading. The old practice of reusing salvaged ties by plugging spike holes in them before turning them over was officially frowned upon but was often done. In the mainline production gangs the new hydraulic tie cutters chopped old track ties into four pieces for ease of removal. Consequently, the supply of whole usable ties was greatly diminished. When improvisation by using non-standard material was the only way to avoid embargoing a line, official notice seems to have been easily deflected—providing nothing bad happened, of course. Then the whole blame dropped on the foreman's shoulders for using unsafe materials.

While we were there, Jeff and I toured the Oelwein shop where both F and GP diesel units were being rebuilt. On February 1st we drove west to Mason City, Iowa, to meet with Dallas Carlisle, Central Division superintendent. On the way we took a look at the Hampton, Iowa, area where the CRIP and the old M&StL lines ran parallel about a mile apart. How can rail fans go through Charles City and not look-up the electric Charles City Western? In the late afternoon we found the carbarn and peeked in. It was obvious that this railroad had not operated for some time.

Harold Gastler, vice president-operations, pitched another curve at us by demanding a new connection with the Norfolk and Western's former Wabash line at Conception, Missouri. Our report was almost ready when I ran it past George Hanson in Oelwein, John Danielson (an attorney now assigned to work in the operating department) and Jim Wolfe. Two days later I previewed Ed Burkhardt and Al Myles (of labor relations). Wolfe requested February 18th that I give Provo a copy. He came back five days later requesting a ten year projection of costs and savings. On Leap Year's Day I asked Lee Fox of engineering for a ten year projection of maintenance costs for the former CGW Des Moines to Leavenworth trackage. Lee let out a loud snort—the railroad wouldn't last that long! Calming down, he gave us a horseback estimate to our hypothetical question. Marketing's Jack Toren did a ten year car forecast for us which completed our very broad review for answering Provo's question. We went to press.

Ed Burkhardt was already rerouting trains over the Rock Island using various emergency reasons that did not include the fact that the old CGW, now C&NW, trackage was getting pretty shaky. He overdid it though, because the Interstate Commerce Commission inquired on March 21st why so many C&NW trains were going over the Rock Island. We couldn't figure out who alerted the ICC. The Rock Island was quite happy to have the C&NW trains as the detours generated a lot of cash for the poor, starving Rock which had surplus capacity to handle the extra moves.

The big day arrived for Jeff and me to make the formal presentation of our Des Moines-Kansas City report. On March 27, 1972 about 20 of the railroad's top managers got the whole show—slides, charts and all. Using color slides Jeff and I unveiled our charts and graphs to an attentive audience. First came maps showing the railroads and routes to Kansas City for Plan I (current C&NW access by way of the former CGW railroad) and Plan II (using the Rock Island south of Des Moines). There were detail maps of Kansas City showing how a 150 car train tied up the city's railroad grid under Plans I and II. A chart portrayed CRIP and MILW daily train movements over the Plan II route. Another graph showed TPC simulation results for both routes compared to speed limits and the before-and-after repair track condition of the CGW. The CGW's minimum vertical clearance of 17 feet above top of rail was compared to the 19 feet available on the Plan II route. We even added up the total central angle for both plans: the twisty CGW route (Plan I) contained 6,458 degrees versus Plan II's 4,669 degrees. Driving in that point on the CGW's inferior alignment a chart compared grade changes (348 on CGW versus 191 on the Rock Island), grade differentials exceeding 1% (344 on the CGW, 13 on the Rock), and total rise and fall (6,458 feet on the CGW versus 4,669 on the Rock Island). Clearly, the Rock Island possessed a far superior railroad in this corridor. We wound up the show by showing the significant differences of train operation under slow orders versus no slow orders on both routes. When the lights in the room came on after the slide show, Jeff and I waited expectantly for the only reasonable response to our proposal. Engineering and the irrepressible John Danielson both had negative opinions and did not hesitate to express them. The rest of the crowd silently adjourned. On March 28th when Jules Eberhardt presented a report to Provo on the Marshalltown yard project, Jeff and I were asked to repeat the Des

Moines-Kansas City show again for him and many of the same people who had been present the day before. At the conclusion all eyes were fixed on Provo to see what his reaction would be. He muttered those fatal words, "Gee, it's nice to have our own line to Kansas City." The audience at that point all figuratively moved over to Provo's side of the room leaving Jeff and me standing alone.

There were believers in our recommendation who were politically astute enough to know that publicly standing against the president was not a healthy posture. Reaction to our presentation was slow to come. By April 3rd to Jeff's and my disappointment, the decision was made to repair the Missouri Division. Plan I (fix the present railroad) had won despite all the arguments against it. Harold Gastler began the process by throwing money at engineering to eliminate the slow orders on the Missouri division. Provo believed that the proposed new hump yard at Marshalltown, Iowa, necessitated use of the more direct CGW line to Des Moines and Kansas City. Later, after the Marshalltown bubble had burst, it was obvious that pouring money into the Missouri Division had been a bad decision. Ed Burkhardt continued to stuff as many C&NW southbound trains down the Rock Island as he could. It was cost effective in transit time and eliminated a lot of crew changes. Trains did not run out of legal time on the Rock Island; on the old CGW, "dogcatcher" crews to relieve outlawed trainmen were a daily necessity. The big increase in north-south traffic which gradually changed the C&NW's historic east-west orientation seemed to catch many of the older managers by surprise. They tended to view the new traffic flows as temporary things and continued plans for superstructure improvements and services along the old east-west axis. Acquisition of the M&StL and CGW permanently altered traffic flows even though most of those two railroads were eventually scrapped. In April 1980 the north-south line of the Rock Island, and some other pieces, came under control of the C&NW. Eventually, full ownership brought the newly christened Spine Line into the C&NW's network along with considerable upgrading and new connections to the east-west main-line at Nevada, Iowa. Emphasis was placed on making direct deliveries to connecting railroads in Kansas City to obviate the need for much of a yard there. The CGW's minuscule yard and antique engine servicing facilities were downgraded and the former CGW office building was sold to commercial interests. The rest of the CGW line north to Des Moines was dismantled in 1985. The former CGW's Kansas City yard was finally abandoned February 25, 1987.

In retrospect the good work that went into the Des Moines-Kansas City study was not entirely wasted in spite of the bad management decision made to continue use of the old CGW line. The basic concept underlying the study must have had merit because it refused to die and was revived from time to time until 1980 when full implementation began. Jeff and I were to become deeply involved with the western coal project late in 1972. The analytic tools developed for the DM-KC project would prove to be extremely important for the future coal line development and corridor studies.

In April, 1971 I picked up a strange story from a relative who worked for General Motors Electro-Motive Division, manufacturers of a good percentage of the diesel locomotives in the United States. He said that the Union Pacific had arranged with EMD to pay $1,000 more for each C&NW order of the popular F units which we needed to handle our growing traffic base. I reported this to Bill Wetherall, director of motive power, who was then scrambling to acquire more power. On April 29th the railroad was faced with another wave of cost cutting; the C&NW was slowly collapsing. But, we weren't the only railroad having hard times on April 29th—AMTRAK, created to consolidate national passenger train traffic, was having trouble being born. Even the name was uncertain then—some called it "RAILPAX" but that was discarded when "RAILPOX" soon appeared. Of course, the epithet "ANTHRAX" also was quickly applied to AMTRAK. C&NW management's attitude to the elimination of long distance passenger trains was a feeling of great relief as it meant the end of costly hearings and seemingly endless regulatory processes that tied up so many resources for long periods.

On April 30th long distance passenger service died at the Chicago Passenger Terminal. I went to the CPT to witness C&NW's train #2 and the Baltimore and Ohio's *Capitol Limited* arrive in the morning for the last time. I went back in the afternoon to see C&NW's train #209 and the B&O's #2, *Capitol Limited* depart. Threatened injunctions made the last day event questionable until the writs were denied and AMTRAK was born. Because the C&NW chose not to join AMTRAK, our annual foreign line free transportation vanished forever. The end of an era.

WAYCARS (CABOOSES) On May 4, 1971 Bill Allman dropped two new jobs on me. One was to take responsibility for becoming the Kansas City expert, something I was already becoming; the other, to find out why the waycars were creating operating problems. Other names applied to these end of train cars included "cabins" on some eastern roads, "doghouses", "crummies", "shacks", and even worse appellations on others. Those special cars tacked on to the end of all C&NW freight trains were either in short supply, or were in such a state of maintenance that crews refused to ride in them. There had been comments from time to time in the morning meeting about delays to trains because suitable waycars were not available, but experiences were different all over the railroad. I talked to the Federal Railroad Administration (FRA) on May 26th about caboose safety and was appalled to learn just how dangerous riding on the end of a train could be. I was just beginning to get into this project when I was directed to other higher priority jobs. Allman was great at shifting priorities without notice which often left us spinning around in futility. When I got back to the waycars, I learned that there were three basic types: assigned, pool and transfer cars. To complicate matters some of the assigned cars and all of the transfer cars were not usable for high speed mainline train service. It soon became apparent that servicing was the main culprit in the shortage side of the equation. I went to Proviso with Allman and the devious Mike Caputo to see the waycar servicing operation which was located north of the original mainlines of pre-Proviso near a place called "the Middle". Consisting of two tracks full of waycars, the cleaning operation was primitive, unsupervised and very labor intensive. The insides of the pool cars could be pretty grim. I saw the inside of one that was slathered with beef stew because a brakeman was sore at the world.

The caboose cupola was supposedly invented by a C&NW conductor, T. J. Watson, in 1863 when he had a hole cut in the roof of a boxcar and sat a pile of boxes to see over the top of his train as it went over the road to Clinton, Iowa. In those early days a closed car was needed to protect the brakemen and conductor from the elements. Over time the waycar became an assigned car to a regular conductor and his crew. It became their home away from home complete with cooking utensils, wardrobe and larder, sleeping accommodations, curtains, and convenient niches for personal items such as clothes, wet weather gear, fishing rods and guns. The conductor did not take a run if his car was unavailable because of repair or some other reason. Likewise, the waycar did not travel without its own conductor. The same was true of enginemen and their locomotives in the old days. That all changed over time as railroad operations expanded to meet the demands of service. As the years passed waycars on the branch lines tended to remain assigned to specific conductors or crews, while mainline trains were served by pool cars. The unassigned pool waycars were generic steel cars, painted red or yellow on the C&NW, that freely floated between terminals as needed. These steel bodied waycars were fitted with high speed trucks, electric light generators and were supposed to be used on the time freights running between Proviso and the Twin Cities, Proviso and Omaha and St. Paul to Kansas City. The cars were treated poorly by the crews who regarded them as "loaners" in which they had little vested interest. The interiors grew grimy, fixtures were broken and the tool kit gradually disappeared. Sometimes, the pool cars were deliberately trashed by rowdy crews. Tools were lost because it was easier to throw a wrench into the weeds along the right-of-way after fixing a broken airhose than carry the tool back to the waycar. Of course, when that second broken airhose couldn't be fixed, big delays resulted while waiting for a car mechanic. Added to the cleaning and outfitting expenses was the recurring cost of switching the cars in and out of the cleaning track as well as the extra moves needed to put them on and remove them from their trains.

I tried tracking individual cars around the railroad to see what had happened to them. The CARFAX reporting system was helpful but some people knew how to fritz the reporting. On further probing I discovered that the divisions were playing games with the cars by hiding the good ones and sending their junk cars out on mainline trains. The Central Division was especially adept at keeping good mainline cars on their local wayfreights. When confronted with specific examples, superintendent Dallas Carlisle said they were short of assigned waycars and had to grab a pool car to maintain service. Dal tried to run trains without any waycars and generated a lot of heat from the trainmen's union. He was on the right track, but far too soon to make it stick. I suggested that painting a red stripe on the mainline yellow cars would make it more obvious when a good car had been swiped and was being used in local service. At least the transfer crummies weren't useable on the time freights and tended to stay near their terminals. I developed some cleaning and outfitting standards for the guidance of the cleaning track contractors, but that was not solving the problem. I even suggested that division officers should make unannounced inspections of their waycar cleaning tracks. Another suggestion was a quality control program but that would have required assigned personnel and drew no support. I wrote a report, Allman ripped it apart and reassembled it, gave it to the operating department who accepted it and sent it on to Provo June 17th. It disappeared without further comment.

The problem was that no one understood the real function of the waycar. The only reason a live person had to ride in that dangerous car in modern railroading was to read the air pressure gauge. When the air had reached the end of the train from the locomotive, it meant the brake lines were charged and the train could proceed. The Florida East Coast got it right when they invented the airline pressure detector for the end of the train. Called FRED (Flashing Red End Detector) the device hung on the last coupler and was coupled to the air line. Not only did it provide a flashing red light, the most important function was the radio signal to the locomotive when the air pressure had reached the proper level. So, the solution to the waycar problem was to get rid of them. When Jim Wolfe became president, one of his first acts was just that. Like Alexander slashing the Gordian knot, he got to the heart of the problem quickly.

The ICC hearings for the employee-owned company, then called NETCO, were held May 4, 1971. Both the ICC and the Union Pacific Railroad attacked the proposal. Despite the opposition the ICC approved the NETCO proposal September 2nd. In the dimness of the wings the Union Pacific-Rock Island merger still lurked. The first morning meeting held under the new name was September 3, 1971.

Back in May Jules Eberhardt, Van Schwartz and I had traveled down the Nelson-Peoria line to review possible sites for passing tracks. We looked at an area south of Nelson and one at Manlius before cutting across the country to the Burlington Northern's line between Buda, Elmwood, Yates City and Farmington to investigate this railroad as a possible by-pass route. We stopped at Elm on the old M&StL and at Pioneer (formerly Radnor) on the C&NW before heading back to Chicago. The increased export grain traffic was putting big strains on the Southern Illinois district as well as on the Kansas City line.

The subject that was to overshadow the rest of my C&NW career first appeared on May 10, 1971 when Ray Gotshall, marketing, asked for help on the coal traffic in the Peoria area. We agreed to run some Train Performance Calculator simulated coal trains for him. Ed Garvey had been our resident expert on TPC, but he had since taken a job in Washington. Before leaving us Ed left instructions on how the program was operated. Over the next few days we puzzled through his instructions and finally got it to run. As related before, Jeff Otto and I used the TPC program to great advantage in our analysis of the C&NW (CGW) versus CRIP routes between Des Moines and Kansas City.

On May 17, 1971 the nation's railroads were struck by the signalmen's union with resulting chaos that day and the next. The C&NW came back to life on the 19th. As a member of the emergency crew I was assigned to the yard office at West Chicago and acted as a switchman, bill clerk, train inspector and telephone answerer until 8:30 p.m. that evening. The first day of the strike also brought the news that my vice-chairmanship of the Transportation Data Coordinating Committee had been approved. My first trip on AMTRAK was May 23rd when I rode #3 to St. Louis over the GM&O. The fast run to St. Louis was marred by the hour and a half wait just outside the Union Station for two long freight drags to clear. The next day I attended the Data Systems Division's forum and watched Ralph Dunn of the Reading deliver my SPLC report to the audience. That evening I dined with Bob McKnight of *Railway Age* magazine. Carl Rogers of the Milwaukee's computer group and I flew back to Chicago on the

25th. On June 14th I attended a TDCC task force meeting in Washington chaired by Ken Bluett of Bethlehem Steel.

An event affecting most Americans occurred on July 1, 1971 when the United States Post Office became the U. S. Postal Service.

Frank Stern became deeply involved with a very peculiar assignment in that May of 1971. He was supposed to project a picture of the railroad as it might appear in July. This involved collecting all manner of data and displaying them in some meaningful way. As usual, the data he needed were elusive, hard to collect or did not exist. Drafted to help him, I finally suggested that he creatively invent some of the missing figures or he would have nothing completed before July arrived. Rejecting my tongue-in-cheek advice, he was at the edge of hysteria by May 26th. He was consumed with worry over variations in fractions of percents when the data were so bad as to be almost unusable. Frank as a forecaster needed a wizard's crystal ball because the railroad data did not contain anything smacking of predictability. On May 28th the entire staff had to pitch-in and slap some sort of a report together because Provo was going to be in Washington all the following week presenting rebuttal testimony on the NETCO application. I managed to slip out of this stupid project because my waycar study suddenly jumped in priority thanks to the capricious Allman. By June 23rd the C&NW seemed to be at death's door—train service was awful, costs were being severely controlled and all activities were cut back to the quick. There just was no money. At this point I went on vacation in Canada but returned to work July 16th because of the probability of a national railroad strike. The Union Pacific and Southern Railway were hit first. At the morning meeting Provo, just back from Washington, filled us in on the situation. The C&NW settled with the unions July 22nd, but the UP and Southern were still out. Two days later the strike began to spread; the other lines were mad at the C&NW for settling. By the 26th the Norfolk and Western and Southern Pacific were added to lines shut down with five more scheduled for the last day of July. There was a lot of labor discontent at this time—telephone installers were also on strike. (I did not record how or when the strike was settled other than Jim Wolfe spoke at a Transportation Research Forum (TRF) luncheon meeting September 22nd about the recent strike settlement).

Personnel changes noted included O. D. Campbell returning to the Illinois Central and, my old colleague, Stanley Johnston was now a vice president in a two man real estate operation at Oak Brook. The 1972 budget meetings were now underway, and we were invited to attend to see how our proposals faired. This was Allman's first exposure to the budget item selection process; he was appalled at the speed with which decisions involving vast sums were handled. The next day I flew to Escanaba, Michigan, with J. W. Alsop, Harold Gastler, W. H. Huffman, and James J. Johnson for budget meetings with Leo Tieman (a long ago Galena instrumentman) and Don Schwarz, superintendent at Green Bay. We inspected the ore dock and related operations which turned out was the prelude to assigning me to a project there to improve the flow of traffic through the facility.

The City of Escanaba had lodged some serious complaints about freight operations through the heart of their community. Don Schwarz came to Chicago from Green Bay September 1st to give me some insights into the ore traffic. A week later I flew to Green Bay and drove to Escanaba to interview Ozzie Brookes, ore dock superintendent. I prowled that area for several days looking at the ancient engine servicing facilities, car repair tracks, the "new" yard where

Escanaba, Michigan in 1971

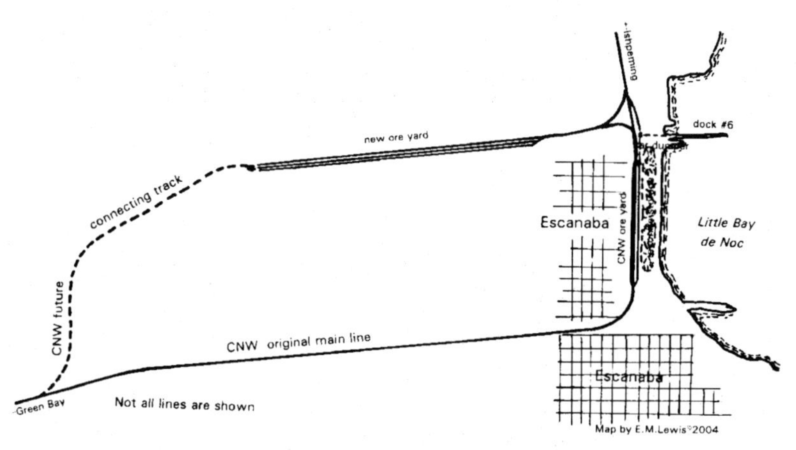

The taconite ore traffic of the 1970s created problems at Escanaba with dust and slow train movements through the city. A new operating pattern was required to better facilitate all-rail shipments and improve air quality.

loads were sorted for dumping into the lake carriers, the conveyor lines to the ship loaders and the mountains of taconite stored awaiting lake transport. A few miles north was the major interchange with the LS&I at Little Lake. Recalling the railroad north of Escanaba, I remember the entire track was coated with taconite pellets—little hard balls of sintered clay and iron ore about three quarters of an inch in diameter. Looking, and acting, for all the world like rusty ball bearings, these pellets were a very real hazard to pedestrians and to vehicles at road crossings. Since they were so hard, they held up pretty well in the ballast, but their presence was not really beneficial for stabilizing track. Signal circuits were badly affected, too. I stayed up most of one night watching the car dumper in action, timing some of the operations, and observing the switcher's movements. Part of the problem with Escanaba was staging all-rail movements of ore trains headed for steel production centers around Lake Michigan and to Granite City, Illinois. Loaded and empty trains handled at low speeds dictated by the short wheelbase ore "jimmies" used up a lot of track capacity. The slow moving trains headed south into the city of Escanaba and made a 90 degree right hand turn to the west before easing down the railroad to Green Bay—leaking taconite pellets all the way. All through this investigation safe operations was the key subject stressed by management. Given the congested traffic pattern and the heavy tonnages involved, I could not have agreed more.

Progression of this project was constantly interrupted by the Des Moines-Kansas City study and other "rush" jobs, but ore car flows at Escanaba were eventually charted showing the traffic movement in the ore dock yard and through the connecting lines. Convinced that coping with the present all-rail ore movements, which were expected to increase substantially, required a more flexible track layout, I "pioneered" a new railroad connection west of the city from the existing main line to the stub ended new ore yard. After researching land ownerships and finding that the low sandy land covered with scrub pine was held by just a few owners, I wrote a report and recommendation September 24th for a new line connection and left on a short vacation to California. For once Allman did not tinker with my proposal and it passed out of our department. I heard no more about the matter until quite by accident some years later I was surprised to learn that the new track connection had actually been constructed. We never did solve the environmental problem of iron dust coating everything near the dumper. The neighbors had a legitimate complaint as their houses, cars and yards were all mono-colored by a coating of windborne red iron ore dust which may explain why the Union Pacific withdrew from that business and sold the iron ore lines and lake transfer facilities after the 1995 merger.

At Green Bay on September 17th I was presented with a new but very old problem. Dick Ryan, my old friend from Lincoln, Nebraska, and Don Schwarz, superintendent, took me down to Fond du Lac to sort out the grain car cleaning operation. We were facing a huge environmental problem there. Empty grain cars had to be cleaned out before reloading at the elevators. The easternmost track of the yard was used for this work. The west rail of the track was slightly elevated so that the entire car canted toward the low boggy land east of the cleaning track. Cars were manually broomed and shoveled with the sweepings going directly into the track side swamp. While the smell of fermenting grain was very strong, the bigger problem was the leachant that was polluting the surrounding groundwater. I had no immediate answer to this situation other than suggesting outrageously expensive special buildings and mechanical disposal units as possibly the only solutions. After explaining the problem at the morning meeting on October 12th, I coopered-up a temporary palliative for Fond du Lac and later was amazed that my estimated costs were actually budgeted for 1972. In November I again visited the site to see a demonstration of a mechanical vacuum system that sucked the grain and dust into a covered truck which took the material to a dump—somewhere else. It was on this trip that the division's assistant engineer acted as my chauffeur from Green Bay to Fond du Lac. Otis Pritchard had an interesting story to tell about being a crewman on a B-17 bomber until he was shot down and spent the rest of the war as a POW in Germany.

Deeply involved in geographic code activities, I flew to New Orleans with Herb Landow to attend the Data Systems Division 7th annual meeting. At this meeting I presented my SPLC report which clearly outlined the serious controversy surrounding adoption of an "official" Canada SPLC as an industry standard. This report ignited a struggle between the railroads and the truckers as to whose version of the Canada SPLC was to be adopted. The ATA (truckers) version had been hastily cobbled together using the U. S. geographic structure as its model. They had not enlisted any geographic experts or even any Canadians in developing their version of the SPLC. Politically, Canada is not structured like the United States. Its "counties" are not equivalent to United States counties and have no geographic significance which was the key element designed for digits 3 and 4 of the 6 digit SPLC. Consequently, the trucker's version of the code was gibberish. The railroad version of the Canada SPLC was developed by members from the CN and CPR plus the Canadian Freight Association, publishers of Canadian tariffs. This problem of the two Canada SPLC codes and the struggle to determine which was legitimate would go on for years. There was no way that a cross reference bridge between the two versions could be made. The stage was set for another bitter struggle.

Returning from a brief vacation I attended the quarterly management meeting at which the vice presidents explained why they had not met their goals. Revenues were off $12 million, yet the railroad still showed a profit because of creative accounting. In October staff services was asked to organize an inventory of the 40th Street yards. Out on the railroad the October 19th commute on the west line was badly screwed-up when a Milwaukee Road train ran a red board at the Western Avenue interlocking and locked its brakes in an emergency stop. On the 21st I reverted to one of my past lives by attending the retirement luncheon for Larry Trimble of Commonwealth Edison where I saw many of my old associates from industrial development days. Gene Cermak, my former boss, and Don Guenther were both there. On October 26th staff services were asked to evaluate fast, short trains versus slow drag freights—Frank Stern and Jules Eberhardt undertook that project. During October the railroad experienced a power shortage as many diesel units were plagued with electric wiring failures. In November I ran some TPC runs for W. H. Huffman for the Tilden Mine ore train operation. I was glad to provide this service for him and urged other departments to take advantage of this very useful computer program.

Bob Russell of personnel had asked me on November 3rd if Allman had improved. Apparently, Russell received negative answers from other people because Bill Allman suddenly announced to us November 8th that he was leaving the railroad. That afternoon, however, Jim Wolfe and Bob Russell clarified Allman's statement by making it quite clear that he had been fired. Since George Ingram now reported to me, I told him that he was now marching to a new set of orders (November 9th). My first assignment for him was to help Dan Fliss train new computer operators at Ravenswood. That way I got him out of engineering's hair and gave him an opportunity to make flexible work hours so that he could continue his university studies.

Jim Wolfe, vice president of labor relations, had been in charge of our group since August and was still trying to assess the individual values and the leadership qualities of our three principal managers: Frank Stern, Jules Eberhardt and me. Jim played us against each other to see how we reacted to the various twists and turns of policy and what useful work we could produce. One of his ploys was asking each of us to write a job description for the open assistant vice president position. Instead of the three different versions he expected, we put our heads together and created a unified one for his perusal. The assistant vice president job was still unfilled as the year ended. At the end of December I received another 5.5% increase, the maximum allowed by President Richard Nixon's freeze of ages and prices, which put me at $18,040.

There was more activity, however, before the calendar could be turned. On November 15th we heard that the Winona drawspan across the Mississippi River had been knocked into the river by an errant barge tow. On the 17th Jim Wolfe handed out two more projects: (1) review all of the charts in the morning meeting room and (2) review all reports for content and utility. A week later Wolfe dropped an industrial engineering consultant's package called WOFAC on me for evaluation and monitoring. That proved to be a bad move because I neither liked, nor understood, industrial engineering. Consequently, that assignment suffered from profound neglect. It turned out that departments afflicted with WOFAC programs did not like the methodology either and dealt with it as I had by ignoring it as much as possible. At his invitation, I met with company psychologist, Barry Lewis, on the 24th and had a pleasant, if unmemorable, conversation. On November 29th I received a call from my future employer, Thomas Hodgkins, president of Distribution Sciences of Rosemont, Illinois. Tom's company was a computer services company involved in rating freight bills for industrial clients from computerized motor tariffs. Tom was interested in the Standard Point Location Code (SPLC) for his business. In nine years I would become his geography expert. Away from the railroad the geography part of my life was maturing nicely. My collection of North American maps had expanded and were now encoded into my data files. I had developed a good client base of major companies including several government agencies. My financial dependence on the C&NW was becoming marginal. Through lucky circumstances, I did not wear the iron collar of a house mortgage or any other major debt. In a word, the railroad was becoming a hobby and geography, the breadwinner.

On November 30th I flew to Washington and stayed at the Watergate Hotel soon to become notorious in our Nation's history. A Geo-Coding Conference sponsored by TDCC, DOT (Department of Transportation), CDC (Canadian Data Commission) and MIT (Massachusetts Institute of Technology) was held December 1st through 3rd. Some 70 code experts representing government, private industry and universities met to discuss the progress and value of a standardized geographic code. The group decided that latitude/longitude was the best common denominator for the many geographic code systems then extant. That decision underscored the correctness of my selection of that system in 1965 as the base for my geographic files. TDCC adopted the definition of a geographic standard point as being a specific point rather than the area concept espoused by the trucking industry which again supported my original stance. That point definition was the basic foundation concept that should have been rigorously applied at the very inception of the creation of a standard **point** location code for the transportation industry. Failure to stick to this primary definition when assigning the actual code numbers led to years of argument and dissention. Because I was a year late in getting involved in development of the geographic code, I spent most of my time battling to get the basic structure both repaired and expanded to include Canada and Mexico. The NMFTA claimed they had put a lot of money in the code and, by God, they were going to recover their investment through sales of the printed tariff—bad code structure or not. Obvious errors brought to their attention were shrugged off and railroad industry needs were not deemed important. Gradually, over time, some of the most egregious errors were fixed. (In the year 2004 the SPLC code was still being published by the NMFTA as a tariff according to their Internet website).

Frank Stern and I met with Larry Provo on December 9th as part of the assignment to evaluate existing reports and their utility. Larry confessed that he was a detail junkie and liked receiving a lot of computer generated paper covering levels of detail that he should not have bothered with. He suffered through these heaps of computer fan folds because he feared the loss of information as it made its way up the pyramid of management. His CPA days conditioned him to reading a lot of fine detail reports that he really should not have received. We did not succeed in eliminating a single page of his daily intake. It was also at this meeting that Provo confessed that big computer system projects scared him because they always cost more than projected, took longer to install and produced fewer benefits than advertised. If Allman had been more perceptive and understood Provo's attitude about big computer applications, perhaps he would have survived longer.

26

THE EMPLOYEE-OWNED RAILROAD IS BORN, 1972

. . . J. William (Bill) Alsop, the tough, laconic, top operating officer, was pulled out of service in January. I had first met Bill when he was running Proviso. The Proviso yards were a big part of Bill's life. His office in Chicago was wired right into the yard's radio circuits and always turned on. Even at home Bill had a yard radio which carried only his favorite music—the sounds of switching operations at Proviso. Sometimes, he would trigger the send button and ask "what's that move you're making?" to the consternation of the switch crews, yardmasters and the current superintendent who all instantly recognized that gravelly voice. Now the doctor said Bill had to cut out booze and quit smoking if he were to come back to his operating department job because his liver was severely damaged.

Early in the year I knew that my shot at the director's job of staff services was hopeless because our group was rapidly evolving into an industrial engineering operation, not my dish of beans at all. I had been newly saddled with managing a program of value engineering and had rushed to do some quick research to at least learn the modern buzz words. The cost analysis class I had in college some 17 years earlier covered only part of the subject. Those departmental charts in the morning meeting room were also part of our responsibility. I convinced personnel that theirs was hopeless and suggested a better design. That miserable WOFAC program the company was considering was another industrial engineering exercise that burdened my soul. WOFAC was being touted by an outside consultant for examining operations in several railroad departments to attain greater efficiency. On January 13th a WOFAC vice president and I met to take each other's measure. The next day I met with W. H. (Heinie) Huffman and the signal department to review the WOFAC proposal for their area and which we concluded had to be restructured and resubmitted to include desired changes. On the 25th I sent Jeff Otto, my assistant, to the WOFAC class as he was better equipped to deal with these industrial "engineers" and their efficiency studies. Jeff had been with us just a week having started on the 17th. He was to have a long and valuable career with the C&NW right up to merger in 1995 and then went on to work for the Union Pacific until 2002.

On January 21st we learned that president Larry S. Provo wanted our space for his office and a company board room as he was anxious to remove himself from proximity to Ben Heineman and the Northwest Industries climate. In our office space Frank Stern and Jerry Groner shared a rather long and narrow room overlooking the Chicago River. My wife created a special painting for them to cover the long blank wall in front of their desks. The work, a three part canvas of large abstract black and white paintings linked together with strings, was named by Frank *Ein B'rera* (Hebrew for "There is no choice"). For two years the boys endured this art work hanging before their eyes not really knowing if they should complain about or enjoy this abstract work. Their complaints were muted by favorable comments from many of their visitors. Other examples of Barbara's abstract expressionism works were scattered around the offices enlivening otherwise bare walls. Provo's need for the space forced us to relocate next door and across the hall. Our new main office was moved to the west side of the building. Here we set up the department head's office, a reception area, the department files, a conference room and Frank Stern's office. Across the corridor the rest of us were given offices in the space once occupied by the defunct pass bureau operated by the venerable O. H. Woods. Today, OSHA would have declared these work places unsuitable as they were all windowless boxes. Work began May 1st and we moved into our new quarters on May 18th. I greedily kept my three ceiling height walnut bookcases inherited from Vic Preisser and the former corporate development group. This was to be my last railroad office. In this 9 × 12 foot space I arranged my desk facing a wall with a large work table alongside leaving the center of the room as open area. I even had my own private doorway opening directly into the corridor so that I could enter and leave without being observed. The vacant walls were decorated by my wife's works and a big C&NW system map which I kept current as the abandonment advices flooded in. I also had a remarkable large color photograph made from a low flying military reconnaissance aircraft depicting a destroyed oil refinery complex which may have been in North Korea. This picture of desolation was merely a nice abstract of browns and grays when viewed from any distance. Over the next few years I was fortunate to have some very bright young people assigned to me for training, guidance and assistance. To encourage their questions I maintained an open door policy and urged them to interrupt me as often as they needed. The system worked well as my assistants freely shared their progress and problems with me.

Geographic code activities continued. On January 27th I gave my Standard Point Location Code (SPLC) report at the committee meeting held at the Illinois Central's offices in Chicago. On February 9th I flew to Washington to attend an American National Standards Institute (ANSI) meeting chaired by Walter L. Schlenker of General Electric. I was appointed as an alternate to Tom Desnoyers of the Transportation Data Coordinating Committee which made me an official member of the ANSI group. At the TDCC meeting the next day, the AAR agreed to creation of a central registry for the Standard Point Location Code despite NMFTA (truckers) opposition.

By February it was becoming obvious that staff services could not go much longer without a designated leader. The trend was definitely tending toward industrial engineering applications although I felt that railroad operations needed more serious and practical applications rather than just trying to improve current practices. Jim Wolfe, however, came down on the IE side and named Jules Eberhardt as the director. Jules was the right choice for this role. Frank Stern was put-off by the promotion, but his skills were in accounting and economics—areas where our department was not deeply involved. I did not mind Jules's elevation to department head of an IE operation because I was a civil engineer and did not enjoy the tedium of performing efficiency measurements and analyzing work processes. Besides, I was living a quite separate existence in my special interest of geography. A positive benefit with Jules' promotion meant that George Ingram, an industrial engineer type, now came under Jules' supervision along with that boring WOFAC project. George didn't last long under Eberhardt; his last day on the railroad was March 30th. Jeff Otto and I remained a team. We went to Milwaukee on AMTRAK February 24th to inspect the Butler diesel shop and the one-car spot RIP (Repair In Place) track. Jeff and I were at the morning meeting on the 28th when Nancy Ford of *Modern Railroads* attended as an invited guest. It was her article about my MAGIC coordinate system in 1968 that had triggered that career threatening uproar with Allman and Shaffer. I took Jeff to the American Railway Engineering Association's (AREA) annual meeting in Chicago March 7th, (I had maintained my membership all these years). Later, we did the suppliers' show at the Amphitheater which was good exposure for Jeff whose railroad experience was limited to a brief tour with the Chesapeake and Ohio. Jeff was a quick study, however, and a great model railroader which greatly increased his knowledge about the details of railroading.

On March 8th we heard that NETCO was about to be activated. The ICC approved creation of the company March 20th just after Larry Provo went to Hawaii on vacation. On April 18th NETCO was given a deadline of May 20th to get its money together for the purchase of the C&NW from Northwest Industries. At first the target date for acquisition was July 1st, but the timetable was accelerated in mid-May to June 1st. The last day of the stock offering was May 24th. Two days later we learned that 950 employees had purchased 72,000 shares of the NETCO stock—somewhat less than expected but enough to swing the deal. Jim Wolfe had acquired 2,000 shares. On June 1st the employee owned Chicago and North Western Transportation Company was born. I calculated that with our purchase Barbara and I owned about three miles of track—somewhere. The first sales of the CNWT's stock were rumored to be $115 to $200 per share—depending on the information source. Not publicly traded, the stock's valuation was hard to establish. By the end of October it was easy to sell the stock for $115 per share. I quietly acquired 20 shares for one of our finance officers at $200 a share on November 15th. The buzz was the stock was worth $300 and could go to $900 by January 1, 1973. In 1990 an original share actually sold for $1,990, considering the stock splits that had occurred along the way. On November 10th the Dow Jones stock index hit 1,000 for the first time in history.

By July it seemed clear that Provo's heir apparent was to be James R. Wolfe, if the responsibilities and assignments he was getting were any clue.

While Jeff and I were deeply involved with the Des Moines-Kansas City fiasco, Norm Hillegass popped into my life again. He was still promoting his TUSCAR freight car project. On March 14th over lunch he glowed with prospects of locating the new car shop at Plattsmouth, Nebraska. He was baiting Union Carbide with tales of the huge amounts of welding gas and supplies he was going to need. America may have been built by dreamers, but not by such folks as Norm Hillegass. Other events in March included the ICC examiner's second report on the Rock Island on the 22nd that said the railroad should be split three ways and the C&NW should be merged into the Union Pacific (it was in 1995). In Wisconsin the state government's Full Crew Law for railroad operations in the state came into effect. The new ruling caused a lot of operating problems and some great inconveniences for the union trainmen. Stopping trains at the state line to board or unload the extra men was a nuisance to the railroad and to the men, but many of the railroads did just that to drive home the point that the Wisconsin law was a regressive and a counterproductive political statement. We gained a new recruit to staff services March 27th—Jack Ripple. A tall thin young man, Jack brought a lot of energy and good ideas to the group. An innovator and not bashful about pushing projects, he soon got under Bill Wetherall's skin in the mechanical department, and I had to step in and smooth things over. On November 3rd Jack was assigned to the materials department.

After the decision was made to fix the old CGW line between Des Moines and Kansas City, Jeff Otto and I went to the south end on

April 5th for a familiarization trip. While poking around the area, we found ATSF #811, a neat little 2-8-0 Consolidation type on display behind a sturdy fence at Atchison, Kansas. We stayed in St. Joseph, Missouri for the night. That afternoon we had been invited to ride a maroon CGW switcher #28 north to rescue train #143 stalled on a hill north of town. Train #161 and a local wayfreight were stuck behind the first train. We got around #161 and the wayfreight, and shoved #141 over the hump before running back light to St. Joe. On the 6th Jeff and I went to Savannah, Missouri, where a section crew was struggling with cinder fill subsidence at the south switch of the passing track. Later that day we watched a train of welded rails snake through the switches at BC Junction. Jeff then got involved in analyzing yard operations in the Kansas City yard.

Back in Chicago I wrote to the vice presidents about their departmental charts in the morning meeting room. While some of the charts were neither accurate nor useful, we offered to work with their staffs to develop displays that would better portray their department's activities. On April 17th I learned that the feisty John Danielson, formerly an attorney in the law department and lately assigned to the operating department, had gotten into an argument with the vice president, Harold Gastler, and abruptly resigned. Continuing my familiarization trips, I took Jeff Otto and Jack Ripple west along the Galena mainline April 19th to visit the car shop at Clinton, Iowa, and returned via Belvidere. That Segmented Railroad File (SRF) was needed again; I helped Ralph Johnson of operating correct the data file as a favor because I was so familiar with its internal structural details.

The establishment of a central registry for the Standard Point Location Code (SPLC) was not at all settled when I arrived in Washington April 25th for the TDCC meeting. Jim Harkins and Bob Hennell, both from the National Motor Freight Tariff Association (NMFTA), a member of the American Trucking Associations (ATA) waged a bitter battle over the concept claiming that their copyright would be violated by a centralized registry. After more acrimonious debate, the TDCC committee voted to give control of the SPLC to the Association of American Railroads (AAR) which led to another lively shouting match. That evening I had dinner with Jack Jones of the Southern Railway (at heart a good old Iowa boy), Jack Carter of General Foods, Nevil Black (Southern Railway.) and a Mr. McCullough, also of General Foods. The next day I had a planning session with Ed Guilbert and Tom Desnoyers of TDCC before flying back to Chicago. This "standard" geographic code was getting to be a nasty battle over turf and was sweeping away any sense of cooperation between the rail and road carriers. The argument was forcing our customers to choose which version of the geographic code to use in their computer systems—a divisive choice not likely to promote standardization.

I went to Proviso May 4th with Richard (Dick) Hoffman of train accident reporting to consider what to do about the state of the yards. I suggested that we might try a five year plan for rebuilding the key elements. Bob Milcik from operating toured the yards with me before I rode to Elmhurst on diesel switcher #1641. Three days later Bob Lawton, from engineering and I discussed how such a plan could work to catch up on the deferred track maintenance and do some desperately needed upgrading. Jim Zito, then superintendent at Proviso, wanted to talk about this plan on the 23rd. A week later our little group of Zito, Lawton, Milcik, Otto and I took a good look at the County Line switches at the west entrance to the yards in connection with work needed for the developing plan. Jeff Otto threw himself into the Proviso plan with enthusiasm. After the morning meeting which we attended at Zito's office at Proviso on July 31st, Jeff presented his map of Proviso track conditions. Dave Boger, the division engineer, was a bit sour about the method used, but I was quite proud of the quality and methodology of Jeff's work. Getting a plan in place took more months of consultations with Zito and Boger, but on October 13th Jeff presented the 5 year proposal to Larry Provo with Zito and Boger's approbation. The reception was positive, studded with comments on the excellence of the work. Privately, I was pleased to see approval of such a massive project that did not depend upon a single time and motion study. To my mind the Proviso plan was a good example of the kind of projects the railroad most needed; projects that could be accomplished fairly quickly by careful, methodical analysis. Industrial engineers, stand aside.

Frank Stern and I flew to Minneapolis-St. Paul May 11th for the 11th annual Data Systems Division forum. While most of the affair was devoted to Dick DeCamara's new freight car control system for the Illinois Central, I did have an SPLC report to make. On return Jim Wolfe handed us another project—a manual simulation of the proposed Marshalltown yard. Jules having recovered from his bout with chicken pox flew with me to Des Moines June 22nd to visit the Marshalltown site. The new yard was to lay east of town on gently rolling terrain requiring considerable grading. The next day in the office we worked out connections with the former CGW and M&StL lines in the northeast part of the city. When contacted, however, the owners of the property refused to sell which meant condemnation proceedings—with the prospect of lengthy legal confrontations with doubtful outcomes. On July 13th I obtained maps for the territory east of Ames, Iowa, as a possible alternative to the Marshalltown location. By the end of the month the projected Marshalltown yard began to fade as Provo said that our cash position was so poor that we would have to sell scrap and take depreciation on debt service (whatever that meant), leaving no funds for any capital improvements. When Wolfe got back from vacation August 9th, he resuscitated the project telling us to first try to get the preferred site south of the mainline east of town and, secondly, try to obtain land north of the mainline. Harold Gastler, vice president -operations, vacillated; he was not convinced that the C&NW needed a new yard at Marshalltown (and he was right). Harold was voted a golden spike award at the morning prayer session August 16th. He would need it. When I returned to the office on September 18th from a three week trip to Alaska and Hawaii, I found Harold had been demoted after a run-in with Larry Provo and W. H. Huffman. Jim Wolfe was now slated to become vice president of operations.

Our old project to review computer generated reports and recommend elimination of the unnecessary ones lumbered along. As we found ones that seemed candidates for scrap, we went to the recipients and asked how the report was utilized, whether it helped them to manage, and what were the consequences if it were no longer available. As a test we arranged to stop production of a particular monthly report to see what reaction there might be. There was only one inquiry—from a chief clerk who was used to filing the document and had missed the last one. On May 25th I interviewed a wary Roy Miller in real estate about their reports. I came away with a profound feeling that the quiet atmosphere of that department revealed a low level of activity that was in marked contrast to the frenetic air evident in many of the railroad's other departments. I couldn't believe that so much calm was merely efficiency. In June I was deep into the data processing department's wonderland of reports and at the end of the month, the mechanical department's. The idea of storing records on microfilm made a lot of sense, but that was quickly shot down after a July 14th presentation by a microfilming company, Many people just had to have their paper in front of them on their desks. The subject of paper records also came up on June 9th when I was asked to help clear out the old corporate development files that had been dumped in the Secretary's office. While sorting what should be kept and what was scrap, I found some very interesting items including some sharply critical memoranda from Ben Heineman to his four "hot-shot" consultants. In our own office the paper flood of Allman's days was past, leaving the third stenographic position under-utilized. Penny, the most junior, was let go on July 1st. Reduction of forces did not always mean pay cuts as I received another salary boost June 2nd which upped my salary to $18,250. (Allman's abrupt dismissal from the C&NW failed to staunch his flow of abstruse theory. He was recorded as the author of a work on Linear Programming in the Railroad Industry which appeared as a chapter in his book entitled *An Optimization Approach to Freight Car Allocation Under Time-Mileage Per Diem Rental Rates*, Management Science, June 1972. This work typified his theoretical and impossible-to-apply solutions to pressing and very real problems facing the railroad industry. He went back to academe as a teacher.)

On June 12th I flew to Washington for another TDCC meeting and ran into Dick Shelgren of Trailer Train. The truckers (NMFTA) were supposed to have presented a plan for their maintenance of the SPLC, but they did not offer anything. The other committee members spent the rest of the meeting tossing verbal rocks at the NMFTA. These meetings were getting vicious.

On the railroad another round of management reorganization was underway. The concept now was to be decentralization. The venerable Galena Division was replaced by the Illinois Division. George Maybee, briefly with our group, went west to run Oelwein. Dick Hoffman assigned to Boone, Iowa, as administration officer, took a job in

Washington with the AAR before going to Iowa. (Later he told me that our Bob Russell was too cheap with salary levels). There was a suggestion made to move Dave Leach and his systems design and installation gang to the third floor of the Chicago Passenger Terminal from their plush quarters at 120 S. Canal Street. He successfully fought off the move. Now Jim Wolfe wanted to rename our staff services group and democratically asked us to vote on it. We all said "no, thanks", but he changed the name anyway to Corporate Industrial Engineering and Staff Projects—a real mouthful. Next door, Provo's new offices were coming along famously complete with a fake fireplace and decorating by Marshall Field and Company. In late July a rumor surfaced that Lee J. Fox was to become our assistant vice president. It seemed scarcely credible because Lee's action-now style was in marked contrast to our more deliberate pace for project analysis. On July 21st Lee J. Fox was indeed announced as our new director. So, now we had a zoo—a Wolfe and a Fox. Lee J. Fox, born in 1934, had served as a U. S. Marine in Korea. Built like a halfback football player, he sported the marine hair style atop a square stocky frame. He had a solid, smoldering, deliberate aspect that commanded respect. His *modus operandi* was full speed and straight ahead, to hell with the consequences. Yes, he got things done, but often left a lot of wreckage behind him along the way that had to be picked up or smoothed over. An artful politician, he was not.

Jeff Otto and I flew to Mason City in July on an old Ozark Fairchild. We again visited the Hampton, Iowa, area to look at the land between the Rock Island and the old M&StL lines that ran parallel in this area. A connection there would make a better traffic corridor for southbound grain trains. (In October 2002 the M&StL line was dismantled from Hampton to Sheffield. Railworks, a Lakeview, Minnesota, contractor constructed a half mile long connection to serve an elevator at Chapin using secondhand 112 pound rail). Jeff and I continued south to Iowa Falls, Jewell, Story City, Roland and Nevada to see what other connections with the Rock Island were feasible. At Nevada we tramped around the soybean fields in the northeast and northwest quadrants of the intersection of the Rock Island and the C&NW examining the terrain for a possible connection of the two railroads. This was the more sensible place to connect the Rock and the C&NW. At Nevada a connecting link south of the

C&NW main line was not practical because U. S. 30 closely paralleled the right-of-way line. North of the C&NW was the logical place for a connecting trackage. We determined that the topography in the northwest quadrant would be the best for a long track that would reach north from the C&NW to a point on the Rock Island where that railroad began a sharp descent to pass under the C&NW east-west main. (Nine years later a much grander connection involving several looping tracks affording universal train movements between the two main lines was built when this part of the Rock Island was acquired). We also went over to Marshalltown to look over that troubled yard site. Returning to Des Moines we boarded #143 southbound on July 26th departing Bell Avenue at 10:01 a.m. After a long day we rolled into Kansas City at 7:30 p.m. We were disappointed at the small amount of track repair so far accomplished—there was a lot of 25 mph territory still out there. That day saw operation of the first 50 car grain train destined for export to the USSR.

I flew to Montreal on Air France July 31st. Transferring to an Air Canada plane for a short hop to Ottawa was easy for me, but my suitcase went on to Paris. I spent some vacation time at a UNESCO-IGU (International Geographic Union) convention which was held in Ottawa on August 1st. There were some very interesting presentations about solving problems of geographic display. The science of geographic portrayal was beginning to break away from purely mapping techniques in use since the Middle Ages. Ideas involving the use of satellites in space were just then being proposed—ideas that would soon make the geographic science of the present totally obsolete. I was greatly stimulated by this brief exposure to this evolving technology. A few days later I rode a CPR train to Dorval where Dick Braley of the Canadian Freight Association picked me up. He was so proud of his computerized tariff publishing operation that he had to show it off to me. He was years ahead of anyone in the United States and had done an "impossible" task with limited funds. Dick, a big hockey fan, took me to a Canadiens game that evening before I went on to Toronto for my connection to Chicago.

Personnel changes continued—Jack Battel of marketing was asked to leave. We were told that there would be 2% shrinkage of officer ranks each successive year. I decided to make some comparisons of other railroads' capital spending, revenues, expenses, mileage, traffic mix, *etc.* from publicly available data found in trade magazines and annual reports. Collecting the data and getting the apples and oranges all together, Jeff Otto molded them into a package that one could argue made some economic sense. Lee Fox bought the report and passed it up to Wolfe who was supposed to whisper in Larry Provo's ear. The report contained a couple of hot potatoes, one of which was that the C&NW had 170 more supervisory officers than other comparable companies. Meanwhile, out on the railroad, the rivers were full and train service was slow.

A service request system called PACT was inaugurated by the systems department to regulate requests for computer programming services. The system soon deteriorated into a bureaucratic mess because priorities established became meaningless as changing events could, and did, alter the relevance of requests. Systems staff demonstrated time after time their total inability to switch to new approaches and kept doggedly plugging away at requests that were no longer needed. Progress was often measured in years for writing and testing programs. When Jules Eberhardt was appointed to head the department (sometime after 1980), he took one look at the bloated, convoluted PACT system and killed it with a stroke of his pen. That was in the future, however. In 1972 we had to fight to get our requests through the morass. Not having a Warren Davis or any other programmer in our CIESP group, we were totally dependant on systems. They in turn were utterly opposed to any local "cowboy" programming done outside the systems department's standards. When purchasing and stores department went to an outside vendor to purchase a Honeywell computer system (that couldn't "talk" to the main IBM equipment), Dave Leach and his directors had a fit that went all the way to Provo's desk. Purchasing, however, was so desperate for the computer support they could not get in-house for their inventory system, that they successfully stood their ground and got their Honeywell equipment and vendor support.

The report review project continued to rumble along without much enthusiasm on our part. I consulted with Carey Pierce, real estate sales, on September 19th about their display without making any changes. Carey was a practical guy who had a lot of common sense behind that leathery wrinkled skin heavy smokers so often have. One story about him followed him throughout his railroad career—the day he had his physical exam at Passavant hospital in Chicago. This episode occurred when he was undergoing the lower GI (gastrointestinal) exam in which a barium solution is circulated through the colon while the doctor watches the fluoroscope images as he rolls the patient this way and that on the table. At the end of the exam the doctor directed that the barium supply be unplugged. Carey twice warned him that the staff might be in danger. The doctor, thinking Carey was just being funny, pulled the plug. The duty nurse reported later that the ceiling of the examining room had to be washed saying, "Oh, dat Mr. Pierce, he sho made a mess!"

We never ceased to be surprised by new bits of railroad lore. On September 22nd I learned about the "slaughter" file in the claims department. Bill Selman said they often just paid small claims as it was too expensive to investigate all of them. While he could justify this wholesale trashing of claims, there was no way of knowing whether some smart individuals or companies were deliberately cooking the system by filing a multitude of small claims that they knew would not be noticed in the periodic payment system. If the computer could have been harnessed to match these claims, we could then be sure that we were not being cheated. In just a few years, this type of audit would be routine as computer technology, storage and ease of programming rapidly evolved.

Sensing that there was something to value analysis, Lee Fox laid the project on me but soon discovered he had selected the wrong spear carrier for that project and turned it over to Ron Rudolph on September 28th. Lee now jumped into his favorite project— track gang organization and track maintenance manuals. He created a big "book" of diagrams and descriptions that he urged me to edit. By October 23rd we had a huge production track maintenance manual assembled and sent it out to be printed on October 27th. Armed with this material, he then attempted to sell the concept to Provo and Wolfe. We worked up a presentation for him and practiced and practiced. Finally, we tried a dry run with Jim Wolfe and Al Myles (labor relations). Their acute and pungent criticism of basic assumptions caused massive changes to the data and forced Lee to seriously rethink his methodology. Lee was all for charging right ahead despite the obvious changes required. Jerry Groner from finance combing through the data found many errors in calculations which he lovingly showed to Lee who tried to ignore the ribbing he was getting. Jerry didn't win many friends with his forthright manner of advertising their errors, but Jerry was just Jerry doing his job. At last on November 22nd Lee Fox made the presentation to Wolfe, Myles and most of our department. The session ran from 9 a.m. to 5 p.m. with a short break for lunch. It was obvious that Lee's transparencies needed changes—Fox was learning that we could teach him a thing or two about selling an idea. The big day for Lee's project was Pearl Harbor Day, December 7th. My part of the program was the introduction. As I started the program W. H. (Heinie) Huffman, a key engineering figure, got up and left the meeting to go to Proviso instead of staying for our presentation. Wolfe, visibly annoyed by Heinie's abrupt departure, ordered him found and told to return—pronto. We took a break for 45 minutes until Heinie reappeared, somewhat abashed. Following the presentation all division engineers were ordered to be in Chicago the next week. Engineering reacted to Lee's massive track maint-enance project as though someone had kicked a hill of fire ants in their backyard. On the 12th the same presentation was laid on the division people who, quite honestly, wanted to believe that the program would really come true, but feared to hope that it would. Lee Fox and his crew repeated the show for Harold Gastler on the 14th. Sensing general acceptance Heinie Huffman now claimed that the concept was his all along. But, Lee Fox was not yet home free on this one as the Iowa Division on December 19th threw a monkey wrench into some of his basic assumptions of track production rates which lay at the core of the program.

Back on September 27th I had attended the 8th annual Data Systems Division annual meeting held at the Palmer House in Chicago. I ran into Bob Lehnertz and Bob Kase both former managers in Allman's old advanced systems group. Lehnertz was back with the DRGW and Bob Kase was working for Computer Sciences. I also saw Ed Callihan from the Alaska Railroad and invited him to attend our morning meeting as my guest on the 28th. He expressed pleasure with the opportunity and enjoyed a tour of some of our commuter equipment parked in the station. It was small recompense for the courtesies extended by Ed and his wife on our recent Alaskan visit.

My former life in engineering seemed to be coming back. On October 9th I inspected a new track geometry car with Bob Lawton and Bob Kionka. The self-propelled Budd car had been equipped with special computer sensing devices that measured freight car rock and roll. The C&NW was acquiring more and more center flow covered hoppers for grain. The cars with a higher center of gravity than older hopper cars had a tendency to rock off poor track at certain speeds because of a harmonic resonance that was a function of the car length, length of rails (position of rail joints) and state of maintenance of the track as well as alignment and grade. On the 26th the C&NW operated a test train from West Chicago to Belvidere with the AAR and Department of Transportation track measurement cars. Since each of the cars operated on a different principal, comparison of the results gave important data for calibrating our test car. I enjoyed the ride because I believed that looking at the test cars and their results would be the extent of my involvement in track testing, but I was wrong. As a result of the rock and roll tests speed limits on branch lines were set at either 10 mph or 30 mph because the worst speed of operation was found to be 18 mph. Instructions were issued to locomotive engineers that when the big grain hoppers were in their train, it was imperative that they pass through the red zone to 30 mph quickly or maintain train speed at 10 mph. Derailments on the light rail branch lines dropped measurably.

Jim Wolfe announced that the C&NW had resigned from the Association of American Railroads (AAR). A scheduled vote on October 12th was not held because so many railroads were affected by the C&NW withdrawal. I wondered how my involvement in standard coding would be affected if the C&NW were not a member. It turned out to be a tempest in a teapot. On the 17th I made a quick trip to Washington to attend a DOT conference on construction of a geographic convertor file. I knew something about that technology.

In October I made a comparison chart showing the disastrous rate of track tie replacement on the C&NW since 1958. There were some big and obvious differences between what the divisions reported as being installed versus what was reported to the ICC in 1971. Exposure of the faked reporting created a good deal of internal consternation, but the parties responsible for the bad numbers were never discovered.

WESTERN COAL At this point the long running and very complicated development of the C&NW's entrance to the Powder River Coal fields in eastern Wyoming begins. Since the events were sometimes infrequent and intermixed with normal activities, the items specifically oriented to coal will be prefaced by the symbol ▲.

▲ The C&NW's Western Coal story began with a late afternoon meeting June 16, 1972 when the possibility of moving coal from Wyoming was discussed. I was not present, but Keith Feurer, who had been designated as the real estate department's representative, believed that this was the first event in what proved to be the long running western coal project. At this meeting a coal committee consisting of Lou Duerinck (law department), Bob Sharp (marketing) and Keith Feurer (real estate) was formed. On September 25, 1972 Raymond E. Gotshall of coal marketing contacted our corporate industrial engineering department about the possibility of moving Wyoming coal from Gillette, Wyoming, to the Bentonite Spur at Colony, Wyoming, and thence south to Rapid City, South Dakota, and east to Pierre. I recalled that many years earlier I had read that a railroad had been projected to run east from Belle Fourche to Newell and thence southeasterly to a connection with the Pierre-Rapid City line at Philip. I found an old system map showing that proposal as a dotted line. There must be survey notes filed away somewhere, and I began to look for them. In the Secretary's office I found some information about the Belle Fourche Valley Railway. A few clues were found in the real estate department's old land maps, but the survey notes were lost and forgotten.

▲ I decided to recreate the old survey using modern maps and

▲ As noted before, this symbol will be used to denote material relating to the Western Coal Project.

ordered a bunch from the USGS. The proposed line would run east of Belle Fourche southeasterly across the Cheyenne River to the Pierre-Rapid City (PRC&NW) mainline. We also had to pioneer a line across northeastern Wyoming westward to the Gillette area from the end of the C&NW's track at Colony. We tested the suitability of the Belle Fourche-Newell branch by running a 100 car coal train grossing 13,000 tons in the Train Performance Calculator (TPC) and found that the hills and sags of that branch created drawbar stresses far in excess of safe standards. If that railroad were to be used for coal traffic, a complete rebuilding would have been necessary. On November 1st I was appointed as the manager of the Western Coal project. I did not know at the time that Lou Duerinck already had been designated as its leader last June. It seems that Kerr McGee planned a mining development in the Powder River basin to supply coal for a gas manufacturing plant to be located on the upper Missouri River north of Pierre, South Dakota. Ray knew about our TPC work and asked for some practice runs on the proposed line. I said that we needed some basic information before we could even say such a line was feasible. I still hoped to find some old records and actually did locate a few tantalizing scraps from a survey west of Pierre in 1880. On the 14th Ray and Brad Huedepohl (who was once our C&NW geologist and was now a private consultant) discussed the known coal reserves in eastern Wyoming. We considered hiring Brad for a special investigation, but the project seemed to have cooled off. I wrote a memo to Wolfe, Gastler, Owens (engineering) and Burkhardt entitled *Railroad Survey—Pierre to Belle Fourche (1880–1909)* summarizing what we knew so far. On December 21st in a basement on the east side of the Chicago Passenger Terminal I found the old survey maps of the Philip-Newell segment. Uncared for all those years, these drawings on linen cloth were indecipherable because of water and mildew damage.

On November 2nd I attended a planning meeting of the vice presidents and listened to their plans for 1973. Provo now wanted to establish a central reference library, and I was appointed to investigate what was required. Bill Merritt and Lee Castelvechi of materials were helpful, but, somehow, we could never find a suitable space nor determine what reference items people wanted. My next assignment, November 20th, was to a task force to investigate the safety of suburban equipment. Steve Adik from mechanical and Ron Ehlert met November 24th to discuss collision worthiness of the bi-level gallery cars and reported our findings to our leader, James A. Daley, law department. The car specifications were known and deemed to be adequate so all we could do was suggest the end cars should carry a brighter paint job.

The standard coding efforts continued. I met with Bob Petrash of the AAR in Chicago on November 14th about the need for a standard patron code. Jim Woods of the ATSF, Ralph Dunn (ex Reading now PennCentral) and Otto Hermann (SP) were among those tackling this difficult subject. At a meeting in Chicago on November 29th they voted 14–1 in favor of a central SPLC registry.

As related above, December 7th was Lee Fox's track productivity presentation. Mixed in were quick activities such as assisting Bill Jacques in identifying locations for Automatic Car Identification (ACI) scanners at Proviso for a test. I attended Frank Koval's retirement party at the Tower Club in the Opera House on December 28th. Our Tim Killinger, another industrial engineer, quit on his own December 4th. Jeff Otto was promoted December 8th, and he had earned it. Barry Lewis, the company's psychologist, was killed in the crash of United flight #553 approaching Midway Airport on December 8th. Thirty of the forty-three passengers were fatalities. Among the dead was Dorothy Hunt (Mrs. E. Howard Hunt), whose purse contained $10,000 in 100 dollar bills. The next year would see the unfolding of the Watergate conspiracy and reveal the depth of E. Howard Hunt's involvement. But Barry Lewis' death also brought a big surprise the next year when I received a telephone call from a woman in New York. Our telephone operator connected her to me as I was the only Lewis she had on her officer's list. I said that I was not Barry, but did not tell her he was dead. I reported the call to Bob Russell's personnel department where Barry had been assigned. The next thing I knew, special agent chief Charles Miles was on the phone advising me not to talk to this woman again under any circumstances and to direct any future calls to Mr. Russell. It seems that the caller was not the bereaved Mrs. Lewis who knew nothing about her dead husband's friend in New York. And chief Miles wanted it kept that way.

The game rules were now different with the C&NW; we were masters of our own fate with our new employee-owned company. If we failed, we had only ourselves to blame.

27

THE FUTURE IS BLACK, 1973 . . . On the first business day of 1973 I was appointed to the Standard Point Location Code Council. This was a 17 member organization composed of transportation, business and government interests. I was one of the three railroad members. At the first meeting on January 10th in Washington, I presented many agenda items for resolution of code irregularities that I had discovered. The American Trucking Associations' tariff group, NMFTA, had created these problems and now was called upon to explain the errors. Their poor record of maintenance lay there on the table for all to see. The next meeting was February 14th at the Mayflower hotel in Washington. The council discussed their role in depth. Bob Hennell of the National Motor Freight Tariff Association, the principal culprit in creating the problems, just sat and silently glowered. This was going to be an acrimonious period in the otherwise calm, logical world of geographic coding. With the advent of the year my railroad salary went to a nice round $20,000. In five years I had almost doubled my old traffic department pay. Within our group Jeff Otto by virtue of his promotion was now working with Jules Eberhardt and

Rich Howard, another recently hired industrial engineer. Howard was to prove to be the bane of my existence. I interviewed Mark A. Johnson for Jeff's old position on January 24th. Mark then was working for Ed Burkhardt and transferred to our gang February 1st. Mark, a tall, thin bespectacled young man with a tendency to be shy, developed into a good, dependable hand. He eventually went on to get his master's degree in business by attending evening classes, the same route I had followed for my undergraduate degree many years earlier.

▲ The South Dakota maps I had ordered from the United States Geological Survey arrived just after the New Year. On January 4th I laid them out to begin plotting an imaginary railroad from Newell to Pierre. The aim of this exercise was finding if a railroad could be built to haul coal from Wyoming to a Panhandle Eastern gasification plant proposed for a Missouri River location along the Oahe reservoir north of Pierre. The original Kerr McGee inquiry had cooled off, but Ray Gotshall was still interested in information about coal train capabilities. The basic premise that low-sulphur Wyoming coal could be used to generate electricity in the Midwest and reduce air pollution had promising economic prospects. By January 23rd interest in the gasification plant project had revived. I gave Mark a quick course in map reading, railroad location, profile construction and operation of the Train Performance Calculator (TPC) program. Northern States Power now appeared to be interested in using western coal in a new electric generating facility at Henderson, Minnesota. Steve Owens from engineering was working with the NSP engineers to solve a local trackage problem at Henderson (February 22nd). Since Steve was supposed to be investigating access to the Wyoming coal fields, but was preoccupied with the NSP project, I took on the task of reviewing maps of the country west of Belle Fourche, South Dakota, because that was the first point that seemed to be far enough north of the Black Hills to offer the chance of extending a railroad into the coal fields in Wyoming. When I traced the route of a long abandoned short line (the Wyoming and Missouri River Railroad) along meandering

▲ As noted before, this symbol will be used to denote material relating to the Western Coal Project.

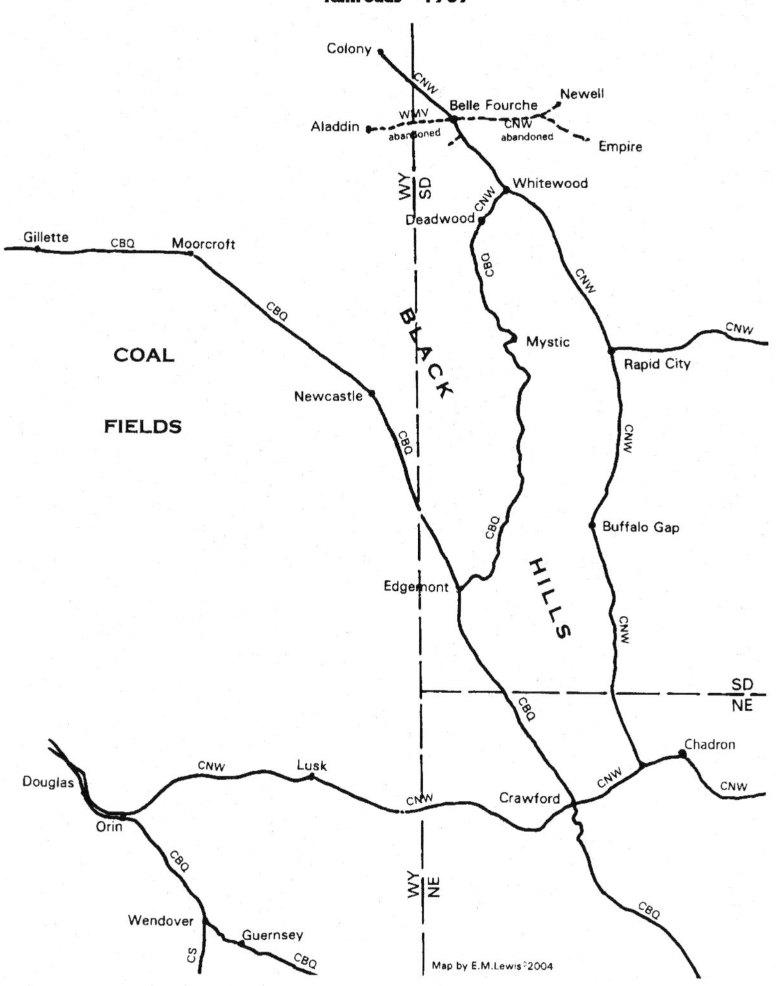

In 1969 the Powder River coal fields of central Wyoming were known but unexploited. The decade of growing rivalry between the C&NW and CB&Q in the coal hauling business was about to commence.

Hay Creek west to Aladdin, Wyoming, it soon became obvious that the country to the west and south of Belle Fourche was far too rough for easy railroad construction. The only possible route appeared to lay along the twisting Belle Fourche River which ran northwesterly from Belle and then curved around southwest past the Devil's Tower National Monument in Wyoming, continuing its course to its headwaters

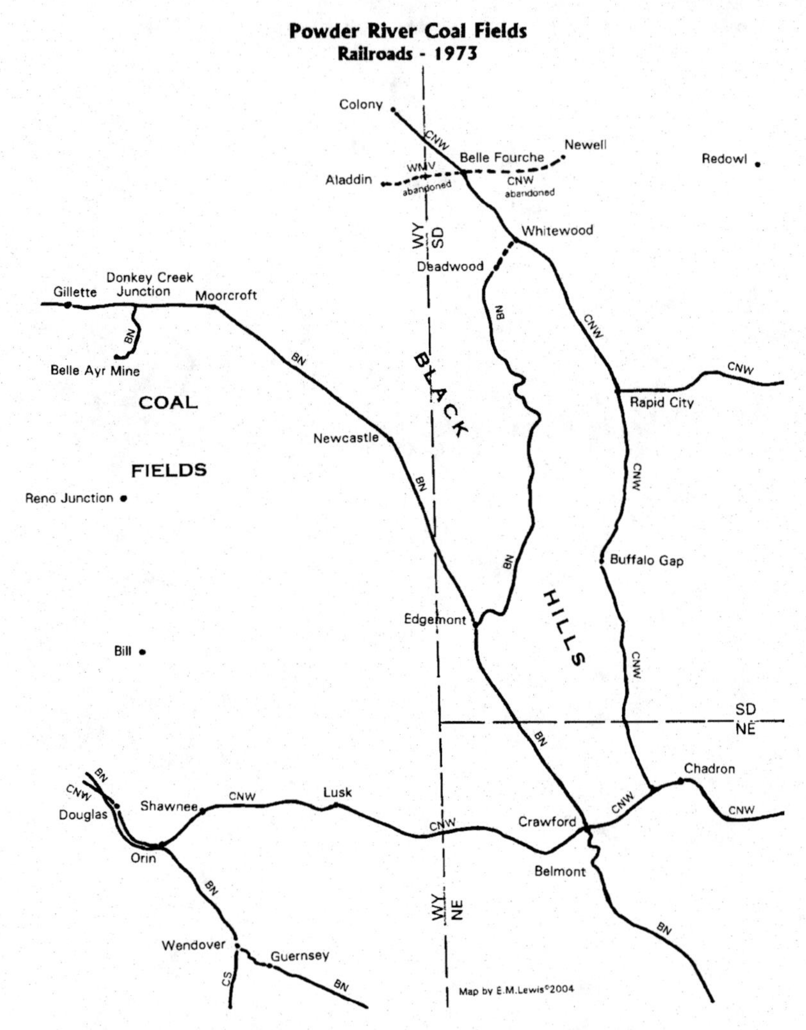

By 1973 the first of the surface mines was opened south of Gillette and the coal boom was on in the great lonesome expanse south to Douglas.

in the coal fields. The BN's Gillette-Alliance mainline track crossed the river west of Moorcroft, Wyoming. The C&NW track charts for Belle Fourche south to Rapid City revealed the sawtooth profile of our existing railroad which would have made heavy train operations difficult if not impossible. Rugged terrain, crossing a hostile railroad with probable legal consequences, the potential for severe environmental problems and the obvious need for much more detailed surveys, made the northern approach to the Powder River coal fields very unlikely. We had to consider another access.

▲ Given all the negatives with this northern approach to the coal fields, I shifted my attention to a southern access route coming off our Chadron-Casper railroad. Coal mining and the C&NW were not strangers in Wyoming. The first coal mine opened near the newly constructed Fremont, Elkhorn and Missouri Valley Railroad line about 1½ miles west of Fetterman in Converse County. Called the Inez Mine, it produced 12,986 tons in 1888. Further to the west mines at Hudson, a boom town of 1,500 people in Fremont County on the Lander branch, were in operation when the railroad arrived in 1906. By 1910 the railroad had at least two locomotives converted to burn the low btu lignite coal. Back in eastern Converse County the Lost Springs Coal Company at Shawnee and the so-called Wyoming Northern spur serving the Rosin Coal Mine north of Lost Springs shipped loads of coal on the C&NW from 1910 to about 1919. The last of the Hudson mines in Fremont County closed in the early 1940s. In 1973 this past history was unknown to us—only traces of abandoned spur tracks remained on valuation maps as clues to the mining activities of sixty or more years before.

▲ I asked Ray Gotshall what the country was like north of Douglas, Wyoming, toward the Powder River coal producing area. He said that he and George Lindquist of Foley Brothers of St. Paul, Minnesota, had flown low over the route in November 1972 in the company plane. Lindquist said there were no physical problems that he could see to building a rail line in that country. That casual statement bothered me a lot because a proper reconnaissance had not been made on the ground—just that one overflight. Although I had figuratively planted a tombstone on the northern access route from Belle Fourche, I decided to give Mark some useful training by continuing the Cheyenne River survey. On March 13th we began developing an imaginary railroad from Newell, South Dakota, the end of a now abandoned short line 23 miles east of Belle Fourche, to a spot on the map named Red Owl. After some more training sessions in the art of drawing a profile plotted from elevations taken from the topographic maps, Mark got into the concept and really seemed to enjoy the process. We "crossed" the Cheyenne River on March 21st and finished the line to a connection with the PRC&NW (Pierre, Rapid City and Northwestern also known as the "Pretty Rough Country and No Water") near Philip on March 22nd. Translation of the line and profile data into TPC coding was accomplished quickly. The keypunchers at Ravenswood cooperated promptly for a change so that we were able to quickly set up the train decks for the first trial runs. Because it was mostly a downhill railroad, no surprises emerged from the TPC runs. Although this was just a practice run, the exercise proved that a better railroad for coal trains could be built east of Belle Fourche to a connection with the mainline east to Minnesota and avoid the dips and sharp crests of the existing line to Rapid City and eastwards. Satisfied with the result yet knowing it all to be in fun, the creation of this paper railroad to the Cheyenne River proved to be a great preamble to the emerging coal project.

▲ A month earlier Jeff Otto and Mark Johnson had set up some TPC runs from the proposed Antelope Mine in the Powder River coal fields to Henderson, Minnesota, for Ray Gotshall using that Foley Brothers' eyeball survey which was deemed to be a nice, flat grade from the mine to our existing railroad at Douglas. I was still very uneasy with that assumption because my Wyoming maps showed drainage areas which belied the flat land theory. But, how extensive and deep valleyed were these streams? In 1973 there were no USGS topographic maps of that area available except at a scale of 1:250,000 scale which was useless for the level of detail we needed. While the TPC runs were pretty sketchy at best, they gave Ray some basic ideas as to the practicality of such train movements. We shared this information with marketing March 15th. Engineering howled about the expense of rebuilding existing lines to accommodate these heavy trains, but we calmed them by reassuring them that this was just a hypothetical exercise. Although the northern access route around the north end of the Black Hills was pretty well discarded by early April, I was about to start preparing maps showing the mythical line to Philip and Pierre when the project's urgency abruptly faded. This would not be the only time that interest in the coal project would swing from hot to stone cold—that was to be a hallmark of the Western Coal development. Steve Owens never did get around to connecting the coal fields with Belle Fourche as it became obvious that the need for it had disappeared. Given the difficult terrain and management's lack of interest, there was no longer any point in pursuing a railroad location around the north end of the Black Hills. After a meeting with Robert A. (Bob) Sharp, director of marketing, about western coal, I was surprised to find him to be quite unknowledgeable about actual coal mining activities in Wyoming. Bob may have started his railroad career as a fireman on the Union Pacific's big engines at Rawlins, Wyoming, but he did not know anything about the Powder River area. From working on the various maps at hand I knew that my knowledge of the area was too slim to confidently project costs for rail transportation. I resolved to go see for myself.

▲ On April 4th I flew to Denver and on to Cheyenne where I picked up a rental car. The drive east to Egbert, Northport, Yoder, Torrington and on to Lusk was my first real exposure to the area that would claim my attention for the next seven years. I first toured north along the Union Pacific branch line running north 60 miles from the mainline connection at Egbert to Yoder, Wyoming. Inspection of its rail and alignment, made it obvious that this was not suited to heavy coal train movements. (Years later, this line was used upon occasion as a relief to move empty trains north and to bring a few loaded trains south during the annual track maintenance program. In 1997 27 loaded trains were handled in a two week period.) At Yoder the branch connected with a Union Pacific secondary main that left the mainline at O'Fallons, Nebraska, 182 miles to the east. This UP line continued west and north of Yoder another 19 miles to South Torrington, Wyoming, where the rails disappeared under roadside dust and gravel short of the North Platte River. On the north bank of the river through Torrington, Wyoming, the Burlington Northern's Casper (Wyoming)-Northport (Nebraska) mainline operated on a more substantial railroad. Turning northwest I drove up U. S. 85 across an empty landscape to Lusk on the C&NW. Early the next morning, April 5th, I set out to trace a possible route for a C&NW coal route into the Powder River basin.

▲ Eight miles west of Lusk I stopped to take a close look at the existing railroad. The 72 pound rail rolled in 1920 looked pretty feeble in the weedy track but the gentle grade on the line was generally downhill to the east, an admirable advantage for heavy trains. At Keeline, Wyoming, 16 miles west of Lusk, the track laid straight and level. Just over the Niobrara County line in Converse County at Lost Springs a big bridge across a wide ravine gave notice that any loaded eastbound trains would have to apply a lot of power beyond the third throttle notch to make the grade out of the draw. Five miles west I passed through, or more accurately, by, Shawnee and went on down the hill to Orin where the CB&Q and C&NW had interchanged cars since 1900—infrequently in recent years. At Douglas I took highway 59 north. The coal line railroad alignment that was eventually built connected with the C&NW at Shawnee some 25 miles east of Douglas, but in April 1973 I did not have enough information about the terrain to place a railroad that far east and assumed the southern connection to the existing railroad would be near Douglas. That is where Lindquist and Ray Gotshall had projected the line on their aerial trip last November. Twelve miles north of Douglas I turned off the highway and drove several miles east on a gravel side road. Parking the car, I got out and walked some distance along the road. The great sweep of this immense country was impressive. The bare and open hills studded with low cedar and sagebrush rolled away in all directions. To the far west the snow-capped peaks of the Medicine Bow Range were visible. Given this terrain my engineering sense could find no physical reasons why a railroad could not be built through this open rangeland. However, it wasn't level. While there were few drainages and public roads for a railroad to cross, there was this tremendous open space which could accommodate many alternate routings to achieve an optimum alignment. I could not have know then that I was looking at the future railroad's highest point. The elevation of 5,200 plus feet at what became known as Flat Top Road was the top of the southward grade later known as Walker Hill. Ten years later this spot would be crossed by a heavy haul railroad.

▲ Driving another thirty miles north through the same open landscape brought me to civilization at Bill, Wyoming. This was the only "town" shown on the state highway map between Douglas and Gillette. On a gentle rise at the Dull Center crossroads a tall Conoco sign proclaimed that gas was available from the two pumps. This was Bill, population three. (When the owners's wife and children were in residence, the population rose to six.) A one-story building on the west side of the road with a house to the rear, a small schoolhouse and a couple of trees made up the entire town. I pulled up to the store building which was made of corrugated iron painted white and was amused to see a parking meter out in front. A big sign

Bill, Wyoming, on August 7, 1973. The only inhabited place in the 112-mile stretch between Douglas and Gillette had a fluctuating population of 3 to 6 people. By 1984 this lonesome spot had become a major railroad operating yard along a heavy duty main line filled with coal trains.

A 1973 postmark from Bill, Wyoming, ZIPcode 82631.

over the door proclaimed that Bill was a United States Post Office, ZIPcode 82631. I had to go in to see this set-up. Inside was the prescribed fenced area for the postal facility on one side and a substantial bar complete with a jar of pickled eggs on the other. A table or two, shelves of food supplies and hardware items and a big cooler full of beer filled the rest of the room. I asked the proprietor, Dean Munkres, about the parking meter. He said that he had picked that up for the cowboys who got a big kick out of feeding it nickels when they parked their dusty pickups in front of the store. I asked him why he lived way out here in this wide open country. He said he had gotten fed up with the hustle and bustle of city life and liked living here. Well, where had he lived? He replied "Douglas"—a town of 4,000.

▲ North of Bill I went out to a wagon mine located on Antelope Creek in northern Converse County. About sixty miles north of Douglas, the road to the mine site was rough gravel crossed by many faint trails that the county highway department optimistically labeled as roads. Through this area ran the famous Bozeman Trail established in 1863 from its junction with the Texas Trail near Douglas. This historic and troubled trail began at a tiny settlement called Sage Creek (later named Douglas for Illinois senator Stephen A. Douglas) and ran across the dry alkaline plains in the shadow of the Wind River and Big Horn ranges passing through the site of present day Sheridan into southern Montana. On the banks of Antelope Creek I found a wagon mine. No one was around. There was a truck dump, an elevating conveyor for truck loading and a black open pit not more than an acre or so in size. The coal stood in a 20 foot vertical face along the bank of a dry creek. The rough working showed steep walls of solid coal. A couple of old work trucks were parked near a rough wooden building that served both as mine office and workshop. This part of Wyoming was peppered with these wagon mines which may be described as small operations where soft lignite coal was extracted in small amounts for local use in domestic coal stoves and furnaces and hauled out in wagons—later trucks. Finding no one around to ask permission of, I just helped myself in collecting samples of the coal out of the creek bank. (The site later was purchased by NERCO and was producing trainloads of coal by 1986).

▲ I continued to scout the surrounding empty country looking for major impediments to railroad construction. Other than a few drainage structures, I did not see any major problems. The empty black sagebrush flats were all gentle slopes. There was no human habitation in sight save for a couple of ranches tucked down into draws out of the ever-present plain winds. On this trip I did not go far enough north and east to find the one spot that would present some construction difficulties—Logan Draw. I made many notes and took color slides before driving south to Cheyenne to visit the Bureau of Land Management (BLM) office where I obtained as many maps as I could. Returning to Cheyenne I spotted a hulking Union Pacific Bigboy, 4-8-8-4, #4004 brooding behind a cyclone fence in a park east of downtown. Exhilarated by this most productive trip, I boarded that Frontier flight to Denver with my coal samples wrapped in paper deep inside my suitcase. And I knew that construction of a coal railroad into the Powder River Basin was practical but that the grades were not as flat as our marketing people had led us to believe.

▲ On April 9th I assembled a slide presentation, but Jim Wolfe had gone on a three week vacation. The railroad was being severely battered by storms in Missouri and Iowa leaving freight trains stuck in many unlikely locations. Despite the fact that our boss, Lee Fox, showed little interest in the possibilities of the coal business, I stuck to developing my information. On April 12th I drew a mass diagram of potential coal traffic for the 1977–1981 period based on Ray Gotshall's projections. It was so unbelievable that when Mark Johnson scaled it down by dropping off some of the more "iffy" prospects, it still was awesome. I gave the chart to Gotshall for presentation to Larry Provo on the 13th, but Provo was then busy with testimony in the Chicago Metropolitan Area Transportation Study. At last on the 16th the big presentation in the board room was made to Provo and most of the vice presidents using flip charts prepared by marketing. The outcome of the meeting was a decision to offer Panhandle Eastern a definite rate on coal from Wyoming to Pierre, South Dakota, and to file a construction petition with the Interstate Commerce Commission (ICC) for a new railroad north from Douglas to the Powder River fields. I was designated as the engineering representative to work with the law department on the construction application as Steve Owens was planning to retire in December. W. H. Huffman muttered under his breath that he had better leave too. On April 18th I completed a 1981 coal density projection based on the best estimates then available from marketing. It was sobering.

The C&NW faced another cash crunch in early 1973. All its officers were mobilized into teams to search-out any company property that could be sold. On January 4th I rode north to Kenosha on a commuter train led by F unit #415 to scan the right-of-way for disused sidings and passing tracks. Riding the cab car of the southbound commuter train was another excellent spot to verify possibilities seen on the northward trip. There were many rusty old sidings lurking in the weeds at stations along the way, but the costs of yanking them out and loading for scrap would probably have exceeded the value of the metal recovered. Ed Burkhardt was of the opinion that 25 miles of third main on the Wisconsin Division's Milwaukee sub could probably be removed without affecting commuter service. Lee Fox got very interested in this north line situation and went so far as to recommend single tracking the line north of Waukegan which was far in excess of Ed's proposal. We also had the locomotive reporting system as a project, but Frank Stern was not making much headway with his assignment. Lee Fox then dropped another task on me—develop standards for track inspections (this was before the Federal Railway Administration had established their standards). About this time Larry Provo began to think long range planning did have merit after all and laid the job of organizing a suitable program on us. James A. Zito, superintendent at Proviso, now wanted to automate the hump yard which required another meeting in his office on February 7th to find out what he had in mind. On February 15th purchasing announced a program of systematically reclaiming secondhand track materials. Jack Ripple, now assigned to purchasing, and Robert R. Lawton of engineering were instructed to establish a track material reclaim center at Council Bluffs. This project received wide publicity as it became a laboratory for development of some very innovative material handling devices. Rail, fastenings and some car department parts were centrally graded and sorted into usable and scrap. The rapid dismantling of branch lines was producing a wild array of different weights of rail, angle bars, tie plates and turnouts much of which might be reusable elsewhere. The center operated successfully for some years before the inflow of material slackened to the point it was no longer profitable. During the latter part of February most of our group were sent north to ride switch engines in Green Bay to measure utilization—a giant industrial engineering project. I have no notes about whether anything constructive was learned or any better operating practices were identified.

Our Des Moines-Kansas City study of 1971–72 had not quite died following that disastrous meeting on March 27, 1972. Our old study was revived January 12, 1973 with a request from Larry Provo that our data be reviewed because the operating situation on the old CGW line had gone from bad to worse. Mark Johnson went through our material with a fine-toothed comb and finished his review February 8th. Jules Eberhardt needed the results for Provo who was reassessing the Marshalltown new yard conundrum. Provo reluctantly approved renewal of options on the Marshalltown site even though there was considerable doubt that the land was suitable for a major railroad yard construction. Jack Perrier of engineering, despite his deteriorating health, struggled with yardage calculations.

The previous October Jerry Groner, now in the finance department, and Ralph Johnson of stations, had begun adapting my old segmented railroad file model (SRF) to automatically capture ton mile statistics. Generating the annual density map had been an onerous and imprecise task. The figures were sometimes fictitious and unreliable at best. It was important to know what part of the railroad was carrying tonnage so that scarce maintenance dollars could be carefully channeled to the most needful areas. Early in January I brought the SRF numbering scheme up to date. In my off moments in January and February I also prepared division maps for SRF—all hand work and rather tedious as each junction had to be described as to how traffic could flow through the node.

I ran into Bill Allman at a Transportation Research Forum luncheon meeting January 17th; he was still unemployed. The speaker was Worthington Smith, president of the Milwaukee Road, who talked about reorganization of their traffic department. I hoped that they had done a better job than we had. The C&NW had a bad January with 116 locomotives out of service, income down and train service terrible to awful. The excuse given at the morning meeting on the 19th was too much business. Yet, at the end of February I heard that the railroad had made a lot of money in the last two months because of the lack of snow and the absence of interfering ICC inspections. *Forbes Magazine* gave the C&NW bad press but seemed more interested in berating Ben Heineman than panning the railroad. On February 2nd rumor said that the C&NWT stock was to be split 33:1 and sell for $50 per share. That meant my investment of $1,000 would be worth $33,000. Six weeks later on March 16th the stock price was said to be in the $500-$800 range. The Union Pacific was said to be going to offer $1,300 per share in connection with the ICC examiner's report of February 15th which awarded the Rock Island to the UP only if the UP also took the C&NW, if we would agree to a merger. It looked like another ten years of litigation. On May 24th the C&NWT board approved a stock split of 60:1 which raised my holdings to 1,200 shares. The estimated stock value was $5.00 per share—a twelve-fold increase in one year.

On a March 5th flight to Washington I was involved in an emergency landing at Indianapolis, Indiana, when the pilot thought he had lost the hydraulics. A successful landing was made on a foam covered runway. The problem was a faulty instrument panel light. Soon, I was on my way in another plane. The SPLC Policy Council met in the AAR building on G Street, elected officers and set some rules and procedures. In the afternoon our quorum vanished and things spiraled down into another spirited debate between the rail and truck interests. On March 7th we met again at the AAR building; again, there was no quorum. I presented the documentation I had developed on the SPLC logic breaks, a proposed code for Mexico, another for Puerto Rico and an SPLC Canada code developed by the Canadians. The truckers held fast to their version. The railroad position on Canada was simple- the truckers should scrap their version. This was going to be a tough fight and I wished I had more backing from my own company which really wasn't interested in standards and coding given they were in dire straits just trying to survive.

On March 8th John Kenefick, president of the Union Pacific, attended the C&NW's morning meeting. I started my week of reading the morning report on March 21st. Every department took a turn which was an excellent way to bring the reality of railroading to those not in contact with operations. Provo was on vacation during my reading week until the last day, March 27th, when he appeared with a deep tan from his Hawaiian trip. That morning the railroad was in a real mess. Because Harold Gastler was on vacation, he did not hear all the barbed comments. Keith Feurer from real estate had the next week's duty. Since he usually was late, I stood by ready to read the next morning when he came rushing in at the last second. A little incident of interest occurred on March 27th as we passed the Chicago Shops on the train, I spotted the first homemade diesel #504 manufactured by Milt Crandall from an old Union Pacific E9B unit. Milt was full of ideas on how to do more with less. The cab was mostly flat steel plates welded with a sloping front so that it looked something like a snowplow. The newly designed cab was smaller than the one on a regular E9; a standing person, unless very short, had to stoop to see forward.

My first boss, William Wilbur, retired as engineer of bridges March 1, 1973 and promptly went to work as a bridge designer for a consulting firm in Chicago. Herb Landow left the ICG March 2nd and set up as a private consultant from his home in Tarrytown, New York. Ron Rudolph of our group resigned March 19th to take a job with Haskins and Sells, one of the big eight accounting firms; his last day was March 30th. The next day I heard that former advanced systems member Dom Violante was moving to Mobile, Alabama, and that there many more changes in store at the Illinois Central Gulf. On May 1, 1973 Joe Mesa, a Cuban, joined our group, reporting to Jules Eberhardt. With the advent of spring, our department was scattered all over the railroad. On April 23rd Rich Howard was looking at the tunnel problem at Camp McCoy, Wisconsin, Lee Fox and Jules Eberhardt were in Rapid City, South Dakota, Mark Johnson and Jeff Otto were on their way to Escanaba, Michigan and Frank Stern was in Minneapolis. Lee wanted to reorganize our department with Jules Eberhardt as director of planning; I was to be the public works and special projects guy. To help Fox understand where our requests came from, we developed a brief report on each—Frank Stern invented most of his. Lee Fox became interested in concrete track ties in late February and asked me to develop information about this relatively (in the U. S.) new technology. This project slowly evolved as knowledge was acquired. Lee also agreed that we needed to assemble a track data base (March 27th) and was all ready to leap into the project without bothering to establish definitions, structure, methodology and sources of data. He was like a bull in a china shop. On March 23rd we received a lecture on job enrichment delivered by Bill Clark of personnel with the aid of his lissome blonde assistant, Adele.

The problem of too many branch lines had to be addressed. Abandonment was not an easy process as each state had its own regulations and the Federal Government through the Interstate Commerce Commission prescribed a regular maze of hurdles that must be addressed—one by one. My long-ago boss, Maurice S. Reid, was appointed to chair a branch line committee to remove these unneeded lines. He organized a group composed of all departments that had some interest in the lines under study—law, finance, accounting, real estate, operating, stations, marketing, and sometimes—us. One of the big problems was assembling the data pertinent to the branches. Maurie traveled each line, photographed the track and structures, assembled the statistics to build as complete a dossier as possible for the law department's use in penetrating the regulatory thickets. Abandonments often took years to accomplish. We actually welcomed a natural disaster such as a washout, serious fire, a bridge collapse or other calamity which took a line out of service because the obvious was there for all to see. The resulting drop in freight revenues also added to our claims of decline of business on the line knocked out of service. Of course there was the counterclaim that the railroad did not move briskly enough to restore service and that was why revenues were so low. The railroad could counter that it couldn't afford to spend money to restore service on a branch earmarked for abandonment. Eventually, the ICC required periodic filing of maps that identified branch lines being considered for abandonment. This often worked as a two edged sword—the maps alerted possible opponents of the coming action on one hand, while on the other, letting the shipping public know where not to invest money in new facilities.

On April 24, 1973 I met with Tom Hodgkins and David Grumhaus of Distribution Sciences to discuss incorporating the SPLC into their freight rating software (in seven years I was to join their company). The next day Ron Boesin of marketing, two Commonwealth Edison men, and I drove to Dixon, Illinois, to look at Edison's Dixon electric generating plant. The plant was an old coal fired plant located on the Rock River. The coal handling equipment was not very efficient and was designed to handle a few cars at a time. The Rock river was in flood on that date; the Mississippi was then at its highest stage in history.

▲ Louis T. Duerinck (ex New York Central) assembled the ICC petition for the coal line project. When we met April 9th, I showed Lou pictures of the area and what we had developed regarding the terrain and related matters. W. H. Huffman had been assigned to organize a survey and was not making much, if any, progress. Heinie had his own priorities which often did not coincide with the company's. Steve Owens was pressed into the survey and cost estimating by May 2nd. Early in May I had new projections from Ray Gotshall and again produced 1981 tonnages—they were outrageously massive. I was acting as a sort of gadfly at this point by staying in constant touch with engineering, law and marketing—poking, urging, encouraging, pushing and activating every element I could imagine might be involved in this project. After meeting with Bill Wetherall and Ray Gotshall on May 7th to develop car requirements for 1981, I produced an exhibit to accompany the ICC petition. Two days later I worked with finance to develop costs for construction and operation. On May 10th Ray had Cabot Corporation in town to talk about development of their Wyoming lands which were completely surrounded by Pacific Power and Light's holdings. The petition to build 76 miles of new railroad into the South Powder River Basin was completed but could not be presented to the ICC until the board blessed it May 24th. It was mailed to Washington the next day.

▲ I was supposed to go to St. Louis with Bob Sharp and Bill Kluender, but they left without me. Ray Gottshall, who worked for Sharp, said that he was glad I was still in town as he had a luncheon lined up with a consultant studying feasibility for a new coalfired electrical generating station in the Great Basin area. Ray and I flew to St. Louis May 11th where we picked up Steve Owens for a call on Peabody Coal. We had a very interesting meeting about coal property development in Wyoming with their top officers. On May 15th Larry Provo met with Peabody and got a good response. Engineering, the designated coordinator of this project (I thought that I had been appointed as such, but something changed along the way), was not making much progress. Jim Wolfe stepped in and bearded Harold Gastler about that arrangement. There was not much love lost between those two. Meanwhile, I kept the fire going as best I could by providing more information and new maps. There was some peripheral piffle with the coal project—an eager young man gave us a demonstration May 22nd on his process of spraying latex on loads of coal to prevent flying dust and diminished loads. (After years of moving millions of tons in thousands of open top hopper cars, the coal dust mixed with abnormal rain and snow caused massive track failure in 2005.)

On February 20th I had gone to Jacksonville, Florida, for the Standards and Coding Structures (S&CS) meeting. My first trip here since 1967, I was glad to see the old ACL #1504, 4-6-2, still on display in front of the SCL headquarters and in pretty good shape. (It was removed September 27, 1986). In May Frank Stern and I went to Roanoke, Virginia where I conducted my first meeting as the chair of the Standards and Coding Structures (S&CS) committee of the AAR's Data Systems Division. Bill Greenberg, (B&LE) was my vice chairman. It was a lively session. Frank and I stayed at the N&W's Hotel Roanoke and enjoyed the fine Southern ambience including the hotel's famous peanut soup. On the 16th I met with N&W's Bob Kase (formerly with our advanced systems group), Leon Atkinson and Louie Newton about possible acquisition of the old Nickel Plate (NKP) yard at Madison, Illinois. The yard was immediately adjacent to the C&NW's former Litchfield and Madison Railway yard. We also talked about the practicality of moving coal trains over the old Wabash line from Council Bluffs to St. Louis. That May visit to Roanoke was a real eye-opener. This was the heart of the N&W's terminal operations complete with immense shop facilities, a hump yard and a lot of train activity with good vantage points to see the action. The company also made every effort to show the AAR group around to see the operations close-up. The old steam power was gone save for a freshly painted N&W "Y" class and the C&O's #1604, a 2-6-6-6, tucked away under a highway viaduct. These locomotives belonged to a museum group organized to preserve some of the region's big power. The next day I flew to Washington with Dick Braley of the Canadian Freight Association for the SPLC Policy Council meeting which turned into another tumultuous affair. Braley would have none of the NMFTA's version of a Canadian SPLC code and told them so in no uncertain terms.

Jim Wolfe stole our secretary, Arlene Cabai, effective September 1st. Her replacement was a young man named Darrell Treptow who was given the title of chief clerk rather than secretary although the male secretary had been a common fixture in the railroad industry many years ago.

A UP-C&NW special inspection train was scheduled to go west on the Illinois division May 21st but a freight wreck at Geneva, Illinois, blocked both mains. The next day's destination was North Platte, Nebraska. Larry Provo was back in Chicago on the 23rd expressing surprise at the rough talk of the UP's top brass.

▲ I spent May 25th analyzing land ownerships in eastern Wyoming. After calling the USGS in Denver to order additional topographical maps, I began plotting a new railroad line north from Douglas on the C&NW's line on 20 foot interval maps. Five days later I reviewed the proposed alignment and profile with Steve Owens of engineering before setting off to the west coast with Ray Gotshall. We flew to Portland, Oregon, and met with Garth Duell of Pacific Power and Light (PPL) to discuss their land holdings north of Glenrock, Wyoming. I mentioned that I had visited their coal mine and its rail line in 1959. In exchange for information on our proposed railroad they provided maps of their property locations. After Ray and I made a courtesy call on our Seattle off-lines sales representative, Jerry Elliott, we flew south to Burbank, California. The next day was the first birthday of C&NWT. We met that day with Ken Canfield and Bob Steele of Atlantic Richfield Company (ARCO) in their Los Angeles headquarters for most of the morning. They had been quite close to the Burlington Northern who were planning a line extension of their Donkey Creek branch southward to Orin, Wyoming, to connect with their existing railroad. At Atlantic Richfield Canfield and Steele were visibly affected by the news we had filed a petition with the ICC to tap this area as our proposed railroad ran right to their new coal mine. They realized that they now had a chance for service by two railroads, a highly desirable circumstance. Ray flew back to Chicago while I went to Oakland to visit my daughter. California was becoming more of interest to us now that our daughter had settled there. Why, there were even steam engines on display. In Oakland the SP's #2467, a 4–6-2, recently painted, was parked, unfenced on a grassy lawn. That wouldn't last long.

▲ It was much later when we understood why the BN had to build southward to serve the Powder River coal fields. The massive coal tonnage projected could not be handled from the mines northwards on the branch to the main line and thence east to Alliance via Newcastle and Edgemont. The BN's Donkey Creek line was constructed in 1972 essentially as a 15 mile branch to serve AMAX's Belle Ayr mine. The lead track was built the shortest way over the hills with 1.4% grades eastbound and 1.25% westbound. Another easier route south from Rozet would have required another seven miles of railroad, but the BN had no idea that 100 million tons a year were to move over this line and opted to build the branch with steep grades, sharp curves and no passing tracks. The Donkey Creek branch line was good for only one train at a time which translated to about one a day each way. But, more serious problems lay under the mainline subgrade east of Moorcroft. The native soil contained a lot of that slippery water adsorbent clay known as bentonite. As coal traffic tonnage increased over this railroad, Burlington Northern faced severe problems in maintaining line and surface during the high plains' brief wet spells. A surfacing crew had to follow almost every coal train to restore line and surface. Adding to the transportation problem the ruling grade from Crawford to Belmont (Nebraska) was a helper district for southbound tonnage. Coupled with sharp curvature (10 degrees), a long tunnel and a long 1.55% grade, Crawford Hill's operating difficulties were ultimately eased in later years by a massive grade and curve reduction project which included daylighting the tunnel.

▲ On June 7th I attended a Railway Systems Management Association (RSMA) seminar at the Drake Hotel in Chicago. The subject was coal transportation. On my return to the office I found Lee Fox in a state—he had been transferred to the engineering department effective July 1st. That was the consequence of developing that grand scheme for production track maintenance—now he had to make it work. I spent some time on June 11th with Steve Owens refining the paper reconnaissance for the new coal rail line north of Douglas. The next day after a conference with attorney Lou Duerinck, Steve

Owens, and Ray Gotshall, Ray and I flew to Oklahoma City for a meeting with Kerr McGee's traffic manager, Joe Dewey, and their Messrs. Hall and Zitter on the 13th. Kerr Magee was well advanced with development of a coal mine at Black Thunder Creek in the Powder River Basin even to the point that the loading tipple site had been established. This was a critical piece of information—now we had a definite destination for our rail line. That afternoon Ray and I called on Oklahoma Gas and Electric to find out about their new coal-fired Muscogee electrical generating plant. From Oklahoma City Ray and I flew to Denver via Liberal, Kansas, on a very bumpy flight. Our Denver people, Don Groves and Gene Gillespie, met with us before we called on ARCO's local representative, Mr. Shearer, to talk about environmental matters.

I got back in Chicago June 15th just in time to hear Larry Provo express his dissatisfaction with the operating department not only for their inability to run trains on time but also for their lame excuses. Tom Minard quit that day to take a position with a new start-up short line called the Oklahoma Western. Ed Burkhardt struggled with Gastler's draconian orders to cure the train deficiency problem. The morning meeting of the 21st was another litany of screwed-up trains. The operating department was under great pressure and was threatened with massive changes. Jim Wolfe promoted Jules Eberhardt to take Lee Fox's vacated position and gave him the title of assistant vice president. To my surprise and with a considerable shock to my sensibilities, Wolfe promoted Rich Howard to be the director of our department leaving me at the manager level. Jim Wolfe couldn't have realized that he had set the stage for my certain departure from the company. Frank Stern was warned on July 3rd that he was close to being separated. While I was piqued at being passed over for promotion, I decided to stick it out until September to see how the coal project would evolve—if it did. Bob Russell, vice president personnel, asked me on July 5th how I felt about the recent promotions. I diplomatically masked my disappointment by saying that an industrial engineering department should have an industrial engineer in charge. Russell and Wolfe never understood how much I disliked industrial engineering and its demonstrated failure to provide any substantive benefits to the railroad. Neither of these officers nor any of my colleagues knew about my second life and my long range plans for an early retirement.

▲ On June 18th I consulted with Lou Duerinck about the progress of the coal line construction application to understand what the next steps would be. A call to Joe Dewey at Kerr McGee revealed that he had talked to ARCO who expressed the opinion that the Burlington Northern was going to oppose the C&NW's application. Indeed that was true for the next day the Burlington reacted by claiming invasion of their territory. I briefed Jim Wolfe on the situation. Joe Dewey called in the afternoon and wanted an audience with Larry Provo on June 27th. On June 20th ARCO said they were pleased with the C&NW's quick action and would support us against the Burlington, if need be. Fortuitously, I heard from Joe Schmidt at the Chesapeake and Ohio Railway that they had a new and improved version of TPC for sale (we bought it). A special meeting followed the regular morning session on June 25th. A core group of people involved in the coal line was established to coordinate the project which by now was being called the Western Coal Project. Steve Owens worked at calculating cut and fill yardages for the new railroad. To speed up that work I loaned Mark Johnson to him both as a helper and to give young Mark experience with the engineering process. The next day, June 26th, Bob Sharp of strategic planning, joined the coal group. Logically, Bob should have been the leader of this whole project, but he really did not seem to grasp its enormous potential to change the entire character of the railroad. The others, of course, thought I was a single-minded fanatic. On the 27th Joe Dewey of Kerr Magee came to town as planned and met with Provo and ARCO. The C&NW committed to proceed with the new railroad despite the Burlington Northern's opposition. Provo said he would have a chat with Lou Menk, chairman of the BN, at an AAR board meeting on June 29th. For some reason the conversation didn't happen and Provo said he would try again on July 13th. (Louis Wilson Menk retired from the BN in August 1981 and died November 23, 1999).

I had gone to Washington on June 27th for an SPLC meeting with Bob Petrash (AAR), Miles Manchester (Federal government), a Mr. Moses of Goodyear, Pete Conway of the AAR and Chuck Holden

(CNR) The fight over the two Canadian versions of SPLC was in full cry. Knowing that this war was not going to end soon, we scheduled meetings in September and October with the AAR's secretary, Vic Rye.

A half year had slipped by. C&NW train service was no better in July than it had been in January. In fact, the situation now was described as a crisis. There were a few grace notes to report. I ran into Ben Heineman July 20th just as he was leaving his office with his private secretary, Miss Bernice Boehm, who was carrying a bouquet of flowers. I remarked about the attractive bunch of flowers and Heineman said she looked like a well decorated grave. In the general laughter that followed, he admitted the quip was Dorothy Parker's. On July 25th the president of the Rock Island was observed calling on Larry Provo. The next day Bob Russell, vice president-personnel, dropped in—his visits were always to be regarded with suspicion—to chat about the Watergate scandal which was just then in the news. On August 6th Provo came back from vacation and attended the morning meeting. For a change the railroad was actually in pretty good shape except for the Missouri Division which was swamped, as usual, from too much business going south to Kansas City.

▲ On July 5th I calculated velocity curves for the new coal railroad. When I was "overlooked" for a coal meeting convened by Larry Provo on the 6th, it was apparent that this coal project was going down a long and bumpy road. Wolfe intervened and got me back in the group, but this sort of marginalization was to occur time and again. There is something in the human makeup that causes society to shun those who are too focused on an objective. Recognizing that and allowing for people's inertia, enabled me to survive several detours on this road to the coalfields. Actually, there were only a few "true believers" on the C&NW that had the foresight and determination to see this project through.

▲ From a source that will remain nameless we obtained a copy of the Burlington Northern's alignment and profile for the Donkey Creek branch to the Belle Ayr mine south of Gillette, Wyoming, and their proposed extension south through the coal fields to Orin Junction. We quickly translated them into our TPC program, assembled the card decks and rushed them to Ravenswood. On July 9th we received the results and were shocked to see what a terrible railroad the

Burlington was proposing to build. Operating heavy coal trains on their new alignment and profile would have been a nightmare. The new BN line was projected through the Rochelle Hills several miles east of our line rather than over the gentle slopes where we had located our line. I suspected that this BN location was a slapdash effort done in St. Paul just to get something on paper and was not much better than the Tsar of Russia's location of the St. Petersburg-Moscow railroad using a ruler and pencil on a map—including the little curve caused by his thumb overhanging the ruler. Since the C&NW already had filed a petition with the ICC for this construction, it seemed to me to be a good idea to get a press release out, but Jim McDonald, Frank Koval's successor in public affairs, and I. R. Ballin of real estate vetoed the idea. The next day Provo called in the whole coal group to work out a strategy for dealing with the Burlington Northern. Lou Duerinck (law) Steve Owens (engineering), Ed Burkhardt (transportation), John Butler and Ron Schardt (finance) and I (CIE&SP) suggested various approaches including a jointly owned construction company—an idea which proved to be unworkable because of the terms of the NETCO ruling by the ICC that required the C&NWT to own 51% of any joint companies.

▲ Provo went to St. Paul July 13th and laid out the idea of a joint operation in the Powder River coal fields to the Burlington Northern. Steve Owens was present at the St. Paul meeting and said later that the BN was taken aback at the degree of our prepardness. Our Train Performance Calculator (TPC) runs were a sensation to the BN who had not thought to use this analytic tool, if, indeed, they had it in-house. When Provo produced our TPC chart, Robert W. Downing, the BN president and chief operating officer, asked acidly why BN staff had not done that. Provo turned the diagram so that the Burlington men across the table could read the results but at the same time kept his finger firmly on the paper so that they had to stand up and huddle together to peer at it. Larry Provo hugely enjoyed the moment. Bob Downing shrewdly asked Provo how he intended to pay for this new construction. Provo said, "I guess we'll use cash." As our people left the conference room Provo quietly asked them "Just where in hell is the Powder River?" Three days later Larry Provo, Ray Gotshall and Bob Sharp were in Houston, Texas, meeting with Panhandle Eastern who were now aware of the C&NW's determination to build into the South Powder River Basin. On July 19th the C&NW completed and filed the ICC's Return to Questionnaire. That action seemed to be a signal to the coal group that the project could now go dormant; everyone reverted to their normal duties. The coal committee was supposed to meet every Monday, but few regularly showed up. The 1974 budget preparation was the current crisis. At least the railroad seemed to be doing better and business was reported as good.

▲ While it seemed that everyone had gone to sleep on the coal project while waiting for the ICC's next pronouncement, I was very restless because I was quite certain that there was a lot more planning to be done. For example, there had been no thought given to the condition of coal routes beyond the C&NW. The obvious gateway points would be Fremont, Nebraska, Kansas City, Missouri, St. Paul, Minnesota, and Chicago. Taking advantage of the lull, Mark Johnson and I went to Omaha on July 6th to check on the old Wabash Railroad route that cut across Iowa and Missouri from Council Bluffs, Iowa, to Brunswick, Missouri. Why had Harold Gastler been so eager to have a new interchange with the N&W at Conception? After a tour of the Union Pacific facilities in South Omaha and Council Bluffs, Mark and I headed south to follow the N&W line after a pause at a coal train derailment at 2nd Avenue in Council Bluffs. We worked our way down the line through Malvern and Imogene to Shenandoah, sampling the track condition, alignment and grades. Now owned by the Norfolk and Western Railway this light traffic line laid with 90 pound jointed rail went up and down the rolling hills following the terrain with a minimum of grading. The railroad was like a museum piece from 1925. We had a track plan dated 1910 and found it little different in 1973. The next day we went on southeast to Blanchard, Iowa, Burlington Junction, Missouri, Maryville and Conception where the line crossed over the C&NW's Missouri Division. We decided that this N&W route could be crossed off the list of possible viable coal train connections. Later, we obtained a more detailed track chart and converted the alignment data into a TPC run which quickly confirmed our initial assessment. Mark and I also inspected the Union Pacific line from St. Joseph, Missouri, to Hiawatha, Kansas, and the Missouri Pacific line between Nebraska City and Omaha. By August 3rd I had completed an analysis of the Omaha-

Kansas City corridor routes. Our Missouri Division was not a viable alternative to the other railroads between Omaha and Council Bluffs; the Burlington Northern and Missouri Pacific routes were operationally far superior for loaded coal trains.

▲ We now had a new face in the coal project. James F. (Jim) Brower was appointed as our new assistant chief engineer-construction as of August 1st. Freshly hired from the PennCentral and a veteran of the New York Central, Jim and his wife, Lee, had just moved to Prospect Heights, Illinois, from Delaware. Jim previously had done a lot of work in southern Michigan when the NYC rationalized its tangle of lines there. A big man, brusk and direct in manner, he was an energetic force embodied in a somewhat rough exterior. We worked together on the coal project for the next six years.

▲ Monday, August 6th, Lou Duerinck reported that no protests to our construction application had been received by the ICC—not even from the Burlington Northern. Jim Brower and I flew to Casper, Wyoming, where Earl Root, the roadmaster, met us with his hi-rail pickup truck. We rolled over the highway east to Douglas and north on Wyoming highway 59 into Campbell County. This was Brower's first look at Wyoming, in fact, his first trip west. The rangelands were greener than when I had been there in early April. When we got to Bill, I was surprised to see the parking meter was gone from in front of the store. I went in and asked what had happened to it. A drunken cowboy had taken exception to it, roped it and yanked it out with his pickup truck. Jim, Earl and I drove around the dirt county and ranch roads trying to get as close as possible to what we thought was the alignment of our new railroad which we had developed from maps and aerial photographs. It fit the land pretty well, too. At the Converse-Campbell County line we spotted the Antelope mine near the Best Coal Mine that I had visited in April. While cruising the country we kept an eye out for a rock source for ballast. In that land of black sagebrush and scoria bluffs, there was very little, if any, hard rock suitable for the huge quantity of ballast we were going to need. Underground it was all coal. Jim seemed somewhat stunned at the immensity of that open country—I think he missed seeing trees.

▲ The next day we looked at Orin (Orin Junction on the BN) and then headed east on the railroad. We stopped often to sample the embankment and marvel at the worn rail and terrible tie condi-

Lusk, Wyoming, on a sunny March 11, 1974. The inspection crew at the depot included Jim Brower in the foreground, M. C. Christensen talking to Jim Simons in his hi-rail car while the city's water tank loomed on the hill over its owner's well-patronized "yellow house."

tion. To imagine this railroad carrying a unit coal train was a real stretch. Shawnee, Lost Springs, Keeline, Manville, and Lusk were names that would become very familiar to us. At Lusk Earl introduced us to the station agent, Karen Groves. She had an interesting tale to tell about the madam whose establishment was in the yellow house just across the street from the station. It seems she had purchased ownership of the city's water works some years before and that consequently the municipal authorities did not interfere with her business for fear she would cut off their water. Karen said she was a nice old bird that often came over to the depot to visit. We rolled on eastwards stopping to check on some section men installing ties at milepost 488. Wyoming changed to Nebraska just beyond Van Tassell. The downhill run through the White River canyon east of Harrison was a beautiful ride on old 110 pound rail. (This part of the line washed out in May 1991 and was abandoned May 27, 1993 from Van Tassell to Crawford). We rolled past historic Fort Robinson, now inactive. Here Chief Crazy Horse was killed while being held as a prisoner in the stockade. At Crawford we crossed the Burlington Northern's heavy traffic railroad and took a good look at the interchange. Passing Dakota Junction with its wide sweeping wye tracks, we tied up in Chadron that evening. For want of any entertainment we went down to the depot and watched switching operations that evening. I got up early the next morning to attend the morning meeting in superintendent J. C. "Pete" McIntyre's office. (Pete later became president of the Dakota, Minnesota and Eastern Railroad).

▲ On our return to Chicago Jim and Steve Owens made some slight revisions in the proposed coal line alignment which I then factored into the TPC decks. We were able to fine tune the grades with this excellent software program. On August 17th R. A. Sharp was appointed head of the Western Coal Committee with Jim Brower as the vice chair. An aerial survey was authorized and a contract for a complete engineering survey was in the works.

The national effort to establish the standard geographic code continued; I managed to fit that work in with my other activities. On August 16th I went to Washington for the SPLC Policy Council meeting. The group voted 9 to 4 to adopt the Canadian Freight Association's version as the official Canada SPLC code. Back on the railroad a management inspection trip to Kansas City was delayed by locomotive failure. Eventually, they got back to Oelwein and traveled on the old CGW north to St. Paul on the 22nd. Jules Eberhardt asked me to review the soil tests from the Marshalltown yard project which was now dragging badly (August 21st).

Frank Stern was fired on August 31st. A few days later on September 4th Harold Gastler, vice president-operations was summarily dismissed and replaced by James R Wolfe. That last shift promotion cost our department dearly—we were once more put under the thumb of the obnoxious R. David Leach. The next day our department name was changed again. We were now to be Systems and Corporate Industrial Engineering. Jeff Otto left our group to go to work for Lee Fox in engineering as a production control engineer (September 17th). My aide, Mark Johnson, was promoted to Jeff's job and assigned to Rich Howard which left me short again. Another new manager was hired on the industrial engineering side, Barry Ptashkin, who would work for Rich Howard for a rather short period. That IE part of our group was growing while the consulting function shrank. I concentrated on the coal project as the main event.

▲ Toward the end of August I tried to find a computer software program that would help us determine where passing siding should be installed to maximize line capacity. The only one I could find was so ancient that it needed an obsolete computer to run it. Even the vaunted Southern Railway did not have such a model and the AAR was completely oblivious of the existence of any such a program. In those days all simulation models were custom coded for specific applications; there was no off-the-shelf general package. There was no help for it, I would have to run a manual simulation to understand how train operations dictated passing siding locations. And that would take a lot of work.

▲ In a presentation to the American Railway Development Association in May 1984 at Dearborn, Michigan, Keith Feurer then director of land acquisition and leasing of C&NW reported that the BN had declared in a letter dated August 28, 1973 that the South Powder River Basin was "exclusively Burlington Northern territory." On August 31st I heard the Burlington Northern had scrapped their original construction alignment which had been projected east of our proposed line and had their surveyors on the ground running lines west of highway 59. Our proposed line was east of the highway. It was beginning to look like the railroad wars of the 1880s with rival track crews battling each other through a narrow corridor. In the end the BN finally adopted most of our proposed alignment. On August 27th Lou Duerinck and I set off to Colorado to interview environmental consultants. One of the firms interviewed was owned by astronaut Wally Schirra; another company had worked on installation artist Christo Javacheff's famous Valley Curtain at Rifle, Colorado; a third was an academic non-profit group at Boulder; and a fourth company in Cheyenne proved to be more of an engineering firm. Later on August 29th we met with Bob Rose, our local attorney in Casper to discuss the political situation in the state. Keith Feurer of real estate and Bob Sharp of marketing joined us to plan strategy. Calls on the Bureau of Land Management revealed their rising concerns about construction activities projected within their jurisdiction. Lou, a long time associate of Jim Wolfe, was a good man to have on this project and adeptly handled the environmental aspects of the project for the ICC petition. By September 6th the BN had turned down our offer of a joint operation and reaffirmed their posture that we were invading their territory. Lou Duerinck and Bob Sharp met with the Bureau of Land Management who controlled much of the land in that part of Wyoming. By September 18th I was definitely out of the line of action in the coal project—again. Well, this had happened before and would probably happen again, so I concentrated on thinking about what elements of the project were being ignored, overlooked or just plain hadn't been thought of. I was somewhat distressed that the complexity of this massive undertaking was requiring a lot more effort and dedication than the company seemed willing to invest. The train simulation model concept had now expanded beyond just the new coal railroad; we would have to consider the entire Shawnee-Fremont line, all 519 miles of it. To prepare for a model for the big simulation "game", I pushed marketing to provide an updated projection on coal traffic.

▲ Keith Feurer told of the strange episode of September 25, 1973 when Provo and five of the railroad's top officials flew into the tiny Douglas, Wyoming, airport. The airport (later converted into a drag strip) was a tight fit for our jet plane, but local officials were impressed by the delegation of C&NW officials visiting their community. After the luncheon meeting, Provo and his gang had to be driven to Casper for a commercial flight because a windshield on the company jet had popped out while standing at the Douglas airport. Because of the group's late arrival at the Casper airport, there was a scramble for remaining seats on the departing flight which resulted in Provo and his vice presidents being jammed into center seats in the coach section—a highly unusual circumstance for these gentlemen accustomed to first class accommodations.

▲ Returning from a New York trip, I found a bunch of newspaper clippings had arrived and got them up to Lou Duerinck as he was preparing another petition to the ICC. That simple action seemed to get me back into the coal project stream again for I was invited to attend a meeting with Panhandle Eastern in Jim Wolfe's office (October 8th). The next day I looked into the remote possibility of electrification for the new railroad and quickly determined that capital costs for structure and locomotives could not be amortized by fuel savings in a reasonable number of years.

The first Chicago and North Western Transportation Company stockholders' meeting was held September 6th. in Chicago. Four days later the split stock was pegged at $11.00 per share so that my initial investment of $1,000 was now worth $13,200. Not bad for one year. I flew to Montreal on September 12th for a Standards and Coding Structures meeting. On October 2nd I went to New York to meet with the Traffic Committee on Computer Services (TCDS). This was one of the most useless groups I had encountered so far. A creature established by the railroads' chief traffic officers, the group had little idea of direction and should have been put out of its futile existence. The General Committee of the Data Systems Division voted to recommend that my *ad hoc* SPLC group be abolished. On October 5th I was elected to head the Standards and Coding Structures Committee, a step up. On October 10th the SPLC Policy Council met in Boston shadowed again by the muted rumble of the rail versus truckers controversy.

INTELLIGENCE AND DATA BASES ▲ The rapid development of the coal line project required acquisition of a huge amount of information not usually found lying around in the streets or even in libraries. I became a bottomless pit for other railroads' proprietary information that included such mundane items as timetables, special instructions, track charts and profiles, yard capacities, physical condition of tracks, and any written or published material that had appeared in the public press. As mentioned previously, C&NW's upper operating management had a deep-rooted aversion to rail fans. Employees tainted with that label were held at arms length by the old line officers. But, it was the covert FRNs that were the greatest contributors of vital information to my growing stock of material. Recent hires from other roads sometimes had timetables and profiles from their previous employer or knew someone who could get them. Retired employees often had a box of old material in their homes. I spread my want list widely through the C&NW and my outside contacts with other railroads from industrial development days. Bob Milcik came up with Wabash track charts, Ed Burkhardt obtained Rock Island material, Bob Sharp was great with CB&Q profiles, Doug Christensen cast a wide net and brought in some essential items—the contributors were varied, many and, often, anonymous. The effort at one point assumed a real cloak and dagger aspect when some especially choice BN material was available to me only on a timed turnaround basis. These we quickly copied. I can only presume that some word leaked out about our intense data collection activity because Jules Eberhardt, a former CB&Q employee, said he had heard from one of his friends that all BN track charts were now to be locked up. (It was too late; by February 11, 1974 we had "borrowed" the most needed items, copied them and returned the originals). My shelves and file cabinets filled rapidly. We gradually reduced the timetables, profiles and special instructions (usually speed limits) into the Train Performance Code (TPC) format creating digitized railroads. In a few months we had converted enough material to TPC that we could "run" trains over all the possible routes east from Fremont, Nebraska, as well as look over the BN's shoulder to see how well their Gillette-Alliance trains ran or where problems were encountered on their Alliance to Denver run—like that helper district at Crawford Hill for example. We also collected Union Pacific and Missouri Pacific items that gave us the secrets on their lines south through Topeka and into Arkansas. Some material relating to railroads east of Chicago we merely filed for future reference. Coding was laborious as each change in gradient, every curve's beginning and end, all stations and landmarks (bridges, tunnels, crossing of other railroads) had to be recorded. The CB&Q line along the east bank of the Mississippi River north from Savanna, Illinois, to LaCrosse, Wisconsin, was the crookedest line I had to deal with. TPC coding was the kind of detail work that proved to be an excellent odd moment job. I took a set of charts with me to jury call and filled in the entire day of waiting for selection by coding track—it beat playing pinochle with strangers.

▲ To bring establish fair comparisons we adopted a standard coal train to run over all these digitized railroads. The power selected was the SD-40-2. After some experimentation, we found that four units for 100 loaded cars grossing 13,226 trailing tons (the odd 26 tons was for the then required waycar) made a good test train. After the coal trains began to really move in earnest, additional cars were tacked on to our basic 100. (Aluminum cars of greater capacity and higher horsepower locomotives later appeared making our "test" train look rather puny). We "ran" the test train over all the segments and combinations that could conceivably handle a real coal train. Each run was identified by railroad, origin/destination, mileage traversed and efficiency as expressed by gallons per gross ton mile. The results were tabulated in descending efficiency order with the most efficient occurring at the top. This data index was to prove to be a valuable resource over the next five years. We did not need to guess at a competitor's ability to handle trains—we knew—exactly.

The October 18th morning meeting was interesting—we had a visitor, Fox of the Union Pacific. The interchange traffic increase at Fremont had become tremendous as more and more trains were stuffed through the C&NW's single track from Fremont, Nebraska, to Missouri Valley, Iowa. The line was "dark" in that it was unsignaled. The decision to install CTC on this section was made at the meeting. The next day I "ran" some test trains on the UP from Fremont to Kansas City and was quickly convinced that the Union Pacific would be our most important partner in coal traffic going to southern power plants. In October a petition was filed with the ICC by the C&NW suggesting that the BN and C&NW's competing railroad construction petitions be consolidated. The ICC agreed and suggested that the C&NW and BN develop a jointly operated facility, which is exactly what happened after another six years of fussin' and feudin'.

Nationally, on October 24th, President Richard M. Nixon fired the special prosecutor investigating the Watergate break-in. That same day I flew to Philadelphia to attend the American Railway Engineering Association (AREA) meeting. I was still pursuing that concrete tie and concrete bridge idea because I was very much aware that providing treated timber ties for repairing over 500 miles of railroad in a very short time would strain national timber tie supplies. The concrete tie manufacturers were aware of the C&NW and BN construction proposals and were willing to bring their demonstrations to Chicago. Here I was in a peculiar position as I had no official standing as a representative of the C&NW. Yet, I could see that our engineering and purchasing people's attention was not being directed to solving this very basic requirement. If they reacted in their regular true to form manner, the dust and commotion of their last minute rushing about would darken the skies. Perhaps our people felt that reaction was better than planning ahead. To be fair, however, there was no official direction to them from the top to plan anything. I fed as much information as I thought appropriate to the concrete tie people to keep their interest and tried to provide technical material about this new track structure for C&NW engineering people, whether they wanted it now or not. They were going to need it eventually.

Jeff Otto had spent the past few months studying the traffic flows and general confusion at Duluth, Minnesota, and Superior, Wisconsin, also known on the C&NW as the Head of Lakes. Thorough and compelling, he presented the results of his study October 26th. The railroad was definitely beginning to have a strong north-south traffic component. The work that Jeff and I did on the Des Moines-Kansas City fitted into that overall picture quite neatly.

▲ Jim Brower and I went to Wyoming again October 31st into the teeth of a threatening snowstorm. Jim wanted to check the topography north of Douglas. After following the proposed alignment north to Coal Creek, we came south to Irvine where the C&NW once operated a gravel pit for ballast. The rock in this abandoned operation was river cobbles mixed with fines—of no use as railroad ballast. We woke up on November 1st to find 2 inches of snow down. We went back north on highway 59 to Gillette in dim weather. The Belle Ayr mine south of Donkey Creek was marked by a tall concrete silo where BN unit trains were loaded. A conveyor ran to the top of the silo at a 30 degree slope from the crush house which in turn was fed by conveyors from the mine dump. In that cold and dusky light the railroad signal system was the only color visible. The scene reminded one of a sci-fi film. Despite the height of the towering silo we discovered that the whole mining complex disappeared from view when we looked back after only a few miles. In the years to come there would be many similar installations, and they too would be swallowed up in the vastness of the country. At the end of the Amax mine spur we found the Burlington engineers' stakes for the first proposed railroad extension south to Orin Junction. We followed the line stakes using what roads we could find. Some of these "roads" marked on the county's official map were little better than cattle paths. Traversing this lonesome back country contained an edge of danger heightened by our lack of experience and by the accumulating snow. As we cautiously edged through the Rochelle Hills along the BN's first proposed line, Jim and I could not imagine why their engineers had even suggested a railroad through this broken terrain. On the Red Hills Pass road we saw eagles, deer, antelope and buffalo, but no good route for a high tonnage railroad. Crossing the Cheyenne River through a shallow ford, we came across herds of sheep almost invisible in the snowy landscape. The next day we poked around Douglas looking for a railroad by-pass and not finding a good alternative. There really was no need for a Douglas by-pass as the coal line connection to our existing railroad was later established at Shawnee some 26 miles further east. North of Glenrock on a distant ridge we saw the little diesel powered 11 car Pacific Power and Light Company train bringing coal from their shallow mine fifteen miles north of the Dave Johnston power plant at the river.

Larry Provo went to Spain and Portugal in November. That month brought a rash of derailments which seriously messed-up operations (November 7th). I interviewed Robert S. Fried for my assistant and got him started December 3rd. His first assignment was helping Jim Brower calculate earthwork yardages on the coal line. On November 26th Lee Fox tried to entice me back to engineering; I said "thanks but no thanks". Again, on December 13th, he tried again by dangling the title of engineer of planning. It seems that many division engineers were now under fire for perceived managerial failures, and the department was desperate to find promotable talent. While the opportunity was flattering, it was actually an invitation to personal disaster. On the 17th I told Fox that becoming involved in the engineering turmoil was not on my career path—whatever that meant. Actually, I had no idea where my career was going but the pull of the coal project was too strong and my independent (stubborn) attitude made me a poor candidate for planning track maintenance for a Lee Fox whose capricious and untested approaches to the huge problem of track upkeep would soon have driven me crazy. The engineering job also would have meant the end of my involvement in the national coding effort where I was getting a great deal of personal satisfaction. I got a call from Jack Carter of General Foods on November 27th saying that the NMFTA (truckers) was up to some new SPLC mischief. I flew to Washington December 3rd for the annual TDCC meeting at which I was supposed to deliver a report. On arrival I discovered I had forgotten to pack my slides which were a key part of the presentation. Through a lucky break I remembered that Lou Duerinck lived in Lombard just five miles from my home. He collected the slides from my wife and took them to Larry Provo who was coming to Washington on other business. That was really up-market to have the president of the company acting as a delivery boy. Larry kind of enjoyed this little job which saved my presentation. The strained relations between the truck and rail elements on the SPLC committee were evident when the council voted the Canadian version as the official SPLC for Canada. On December 5th I sat on a panel discussion with Dick Hinchcliff of NMFTA and a Mr. Dunn of U. S. Steel. Hinchcliff was affable enough, but his background as an attorney and affiliation with the troublesome trucking tariff bureau gave me pause.

▲ Bob Steele, ARCO's traffic manager, came to Chicago November 8th to discuss their company's role in our application to the ICC and to warn us about the BN's intentions to obstruct our plans. Lou Duerinck, Ray Gotshall, Jim Brower, Jim McDonald (public affairs), Ron Schardt and I attended that November 9th meeting. Jim McDonald, formerly with the *Chicago Tribune,* advised against making any public news release. (That seemed to be his policy in general as he rarely did anything about publicity). Marketing finally came forth with a coal traffic projection that seemed probable even though the large tonnage figures were awesome. Mark and I translated the tonnages to what would be required in terms of cars, locomotives and trains to move the tonnage from mine to connections for our simulation exercise. (In July 2003 the Union Pacific averaged 44 trains daily from the South Powder River Basin). Mark and I ran a series of TPC trains to develop the actual train running times across each segment. The hardest and most thoughtful part of development of the simulation game involved writing the rules of play. The actual trial runs would be done next year.

▲ A coal related trip involved Gene Cunningham (car department), Milt Crandall (motive power), Ray Gotshall and me. On November 14th we flew in the Merlin II (NW401) to Zanesville, Ohio, where we rented a car and drove to Noble County to observe the Muskingum Electric Railroad which operated 15 miles of line between some open pit mines and a coal preparation plant. A conveyer system took the coal to an electric generating station at Beverly. The railroad was completely automatic. We were permitted to ride the GE E50C electric locomotive provided we did not touch the controls. At 25,000 volts overhead, I was not touching anything! Loading 15 cars took 50 minutes; unloading, just six. (The railroad was built in 1968 to the northernmost mining area. As the mining operation moved south, the railroad shrank until by 2001 only 2½ miles remained. Because of the high sulphur content of the coal and dwindling reserves, the railroad ceased operation in January 2002).

▲ On November 15th I made arrangements with John White of the Costain Concrete Tie Company to make a presentation to C&NW engineering staff on December 12th. Our top engineering people attended and actually took notes. On November 19th I had heard Larry Provo give a talk to the Western Railroad Club. He touched on the timber tie situation as hardwood was scarce and its treatment cycle took too long. I had a bright idea (that quickly dimmed) to utilize the immense forest of dead elm trees standing in the cities and towns of Illinois. The elms had died suddenly from Dutch elm disease whose vector was the bark beetle. The main wood of the elm was hardwood and not affected by the beetle which carried the virus that killed the tree's circulation. The fatal flaw in my idea was urban metal. One overgrown clothes hook or brace rod, could instantly destroy a $500 saw blade. In those days metal detectors were not sensitive enough to pick-up metal bits in the tree trunks. I interviewed suburban foresters about this elm log project in January 1974 and got Dave Bernier and Jack Ripple of materials interested. The forestry department of Cook County came up with a lot of oak logs, but Jack Ripple couldn't find a sawmill to take on the work. Actually, the elm log proposal could not have been carried out because later environmental rulings banned transportation of the logs to reduce the spread of the bark beetles and the fatal virus.

▲ By the end of November Ed Burkhardt was expressing displeasure with marketing who was proposing rate scales based on too many loaded cars in unit coal trains. Ed demanded that the simulation runs be limited to 100 car trains as there were too many variables introduced when more cars were added. Maurie Reid and his branch line committee now wanted TPC runs to determine fuel savings in abandonment cases. This was the period when OPEC had shut off oil destined to the U. S. from the Middle East because of the Israel-Palestine conflict. On December 5th Larry Provo met with the Burlington Northern under ICC auspices. An agreement was reached to build and operate a joint facility railroad into the coal fields. As the year ended Jim Brower and two of our boys (Mark and Bob) were calculating excavation and fill yardages. Lou Duerinck remarked that I was respected for my persistence in keeping the coal project moving. Lou's comment was welcome as I had felt that I was becoming a nag and a scold. The author William James wrote "The deepest principle in human nature is the craving to be appreciated." Perhaps I did feel unappreciated but my deep belief in the absolute positive benefits that would come to the railroad from that limitless coal reservoir in Wyoming kept me going along a sometimes lonely course until the project either succeeded or had utterly failed.

I was invited to attend Larry Provo's strategy session December 11th. Carl R. Hussey was a key player in this meeting. No role was assigned to Dave Leach and his systems people, nor, disappointingly, to our department. On the 18th R. A. Sharp, Ray Gotshall and I flew to Los Angeles on a 747. When traveling with the brass, one flew first class, and I liked it. We met with Steele, Evans and Zamba of ARCO to get the location of their tipple site in the Powder River country. Despite a big winter storm in the Midwest, Ray and I then flew to Oklahoma City to see Joe Dewey, Hall and Kemp of Kerr McGee. They were very cagey about their coal country activities. Dewey really pushed sales of their treated timber ties as bait. (They went out of that business in 2004). The next day we saw Gibbons and Taylor of Oklahoma Gas and Electric and got some details of their new power station in northern Oklahoma. Lou Duerinck and Jim Brower had some news about the BN's ideas for the joint railroad and felt that some of the proposals for construction and operation were a bit strange (December 21st).

It was time for a holiday break; we all needed one as this had been a very intense run of events. But, we had made considerable progress. We didn't know enough to be discouraged. If we had known what was ahead, we may all have quit pushing this dream.

28

THE CHIMERA OF COAL, 1974 . . . ▲ The first business day, January 2, 1974, found me totally immersed in the coal project. There were several rail routes for coal movement to the proposed Oklahoma Gas and Electric coal-fired generating station at Perry, Oklahoma. But which was the best? Using our intelligence archive and TPC results, it was quickly obvious that CNW-Fremont-UP-Topeka-ATSF was the most efficient route for unit coal trains. I advised Ray Gotshall that this route was the best one to use in his rate negotiations with OGE. Ed Burkhardt agreed on January 9th that the UP at Kansas City was the best southward route and agreed that the C&NW had no viable direct way to get to Kansas City. Out in the real world, however, January was a snowy month accompanied by extreme cold. By January 15th every day had been below freezing.

January also abounded in rumors. On the 3rd we heard again that the Missouri Division was to be wiped out and divided between the Iowa and Central Divisions. On the 8th the buzz was Provo was going to

▲ As noted before, this symbol will be used to denote material relating to the Western Coal Project.

move up to chairman, but who then would be president? A new man joined our CIE&SP department, Lee Eschelman. Computer savvy, his first task was to refine the TPC program and make it easier to use—those card decks were quite cumbersome.

SIMULATION — WESTERN COAL ▲ Our train simulation model of the coal route between the Powder River coal fields and Fremont, Nebraska, was "played"on a big layout that resembled a large family board game. The board diagram was a geographically correct line drawing at 1:250,000 scale portraying the entire railroad from Fremont, Nebraska, to the Wyoming coal fields. The heavy stock paper was 14 feet long by 2 feet wide. Along the thick black line that portrayed the 580 mile railroad line were noted the stations, passing tracks and junctions. Mileposts were marked as well as the mileages between stations. "Trains" were lead weighted markers bearing paper signs that indicated loaded and empty trains and their direction. Tally forms provided a record of the train movements which were made in one hour increments. We developed tables of TPC running times between stations at the design maximum speed of 40 mph for loaded and 50 mph for empty trains. Hazard cards for unexpected events or changes in local conditions were used to bring a flavor of real railroad situations into the model. The starting rules were simple: the railroad was assumed to be in good operating condition without slow orders, communications were reliable, all crews reported on time, there were no locomotive shortages and every unit worked to specification, there were always empty cars available (to start, anyway), the supply of eastbound coal never varied and all movements would be at one hour intervals. The object of this exercise was determination of line capacity, where choke points would develop and where passing tracks should be placed. To get meaningful results the simulation run had to be a continuous all day process requiring at least three people to move and monitor the activity. The more people involved made the simulation go that much faster.

▲ All through January we worked at the simulation model refining its methodology. Rich Howard brought his rowdy gang into the process one day but their light-hearted approach to this "silly" endeavor soon had the model so screwed up that we had to abandon that day's results. On January 29th George Gibson, the company photographer, came to our conference room to film a simulation in progress. His film was later used in a company orientation movie. Bob Fried was a big help in getting this simulation to work as well as it did. He was convinced that a mathematical computer model could be developed to determine our model's objectives if only we used the correct parameters. He worked very hard to develop the all important algorithm that would correctly relate the logic streams. Despite the obvious fact that there were just too many independent variables, he persisted in his efforts until I put my foot down and said "enough". He had prepared a detailed thesis of his work which he gave to our newly promoted department head, Jules Eberhardt, who had a minor fit when he saw the size and complexity of his effort. I pointed out that Bob's work was not something that should be released from our department as it was only an academic research effort, but Jules was under the gun to show that our department was producing useful material because the industrial engineering group was having a miserable time showing any positive results in any of their projects. Between the three of us, we hacked out a summary that said, in essence, that the problem of siding location contained so many inter-dependant variables that a good engineering solution could not be obtained with the computer technology then available.

▲ By February 4th we had determined from the model that we could handle, with luck, about 45 million tons a year over the Shawnee-Fremont single tracked line with additional passing sidings, assuming a complete upgrade of the railroad. As a frame of reference there were 326 million tons moved out of the Powder River basin in 1998 on both the BN and UP. Selection of passing sidings turned out to be harder than I had thought. Jim Brower, Bob Fried and I spent all day on February 5th trying to crack that riddle. The next day a giant snowstorm hit Chicago. I managed to get to the office and began to ponder about where we would find all the people we would need for operating this coal railroad. When talking to Pete McIntyre, the division manager, at Chadron about some of these problems, he suggested we bring the simulation out there for a field test. We welcomed the opportunity and were pleased to have the division people interested in participating in this process. Hopefully, we would learn some practical lessons from them because it was dangerous to allow headquarters staff to incorporate only their opinions and ideas in a project of this magnitude.

▲ Bob and I boarded AMTRAK's *San Francisco Zephyr* to Denver, Colorado, on February 17, 1974. We had all of the simulation materials with us. Unfortunately, I was coming down with something nasty. Arriving in Denver the next morning we were surprised to see the Union Pacific's famous steam locomotive, 8444 (4-8-4), on the point for the run to Cheyenne. Bob was almost out of his head at the exciting thought of riding behind steam and shot a lot of film while leaning out of the Dutch door at the end of the sleeping car. We rented a car and drove to Chadron by way of Sydney and Alliance. I had galloping influenza and felt absolutely rotten. Trying to keep myself together I managed to make the morning meeting on the 19th in Pete's office with Jimmie Simons, Fuhs, Boeslager, and other division officials. Afterwards, we set up the simulation downtown in a side room at the Chuck Wagon restaurant. Being away from the railroad's distractions while working through a simulation cycle was important in developing a good cohesive run. The first cycle was run by division manager Pete McIntyre and his division engineer Jimmie Simons. Later we got the chief train dispatcher into the spirit of the "game" He had a lot of practical experience with this old railroad, and we much appreciated his good natured involvement. He was vastly amused, however, to think that he would ever live to see coal trains running on that poor old rickety railroad at 40 miles per hour. That evening Bob and I drove east to Hay Springs returning to Chadron in time to see train #320 arrive from the west. On the 20th McIntyre and Simons were summoned to Chicago for a staff meeting. They left at noon for Rapid City and a flight east. Bob and I continued the simulation with motive power superintendent Fuhs and other division people. Afterwards, we drove to Orin, Wyoming, following the mainline before turning south to Cheyenne. By the time we got to the Little America motel just west of town I was in a state of collapse and went straight to bed. Feeling somewhat better the next morning, I was the passenger as Bob drove west over Sherman summit to Laramie before returning to Cheyenne to catch our train to Chicago. I gratefully collapsed in my berth and slept through Denver. Later, I revived a bit and made it to the dome car before crawling back to bed.

I awoke the next morning in Burlington, Iowa, amid a violent snowstorm. Western Illinois was a mess—cars stuck in drifts, wires down. Our train crawled through the storm to Mendota at 20 mph. In Chicago the short walk from Union Station to my office through the snow almost did me in. I went home early, struggling through 6 inches of drifted snow. Despite my physical distress, we had gained important insights into operation of the model and demonstrated to the division people that this coal project was not just a Chicago staff smoke and mirrors idea.

In January track inspection became a hot topic. Recovered, Jules and I drove to Clyman Junction, Wisconsin on January 17th to sample a piece of the Milwaukee-Wyeville mainline. The weather was raw and windy with no snow on the ground—just the brooding threat in the gray skies. The 112 pound jointed rail in the main tracks was clearly not in the best condition; we found many loose track bolts—some of them had fallen completely out of the joint bars. I placed a couple of the missing bolts on the center of a tie as a test. Jules and I observed the track inspector sail right past them on his motor scooter. Under this new hot subject all officers were supposed to go out periodically to inspect track. The problem was many of the "inspectors" didn't know what they were supposed to look for. (The Federal Railroad Administration had not yet written their rules for track inspection). At the end of January I took Bob Fried on an inspection at Glen Ellyn, Illinois, on the west suburban line. As we hiked west to College Avenue, Wheaton, I gave him an onsite course in spotting track deficiencies. He learned quickly. Fed up with the office, I headed south on March 27th on an AMTRAK French-built *Turboliner.* Renting a car at Springfield, I drove south to Benld where I inspected the yard area and two places on the mainline. One spot was at Nilwood where the track looked pretty good. That was the location for the first hotbox detector I had seen in action. Further north at Virden I found a dangerous defect requiring a call to the dispatcher.

My committee work continued; on February 26th I flew to Washington for an SPLC Policy Council meeting where we elected Harvey to the chair for 1974. I was saddened to hear my very good friend Littleton (Larry) Fitch of Mobil Oil had suffered a stroke early in the month. At the morning meeting of March 7th the C&NWT share price was announced to be $13.75 which made my $1,000 original investment now worth $20,000. The Illinois Division was currently in a state of turmoil. Don Fryer, who once had been associated with our advanced systems group, was now in charge. His inept handling of people coupled with his combative attitude was sparking a lot of resentment.

▲ The coal mining companies in the Powder River basin were now sharing their closely guarded tipple sites to us as they at last believed we were going to build a railroad. Bob Steele of ARCO and Joe Dewey of Kerr McGee called on January 23rd to inquire about the progress of negotiations between C&NW and BN. The next day they called again complaining about BN's poor attitude toward their companies. On January 25th the BN sent their people to Chicago to obtain a copy of our proposed alignment north of Shawnee so they could design their part of the joint railroad. During March I worked up equipment and manpower requirements while Jim Brower met with Northern States Power. We re-ran the TPC program developing fuel consumptions under varying conditions. Leaving Bob Fried and Lee Eschelman to plow through heaps of TPC coding, I went to Wyoming on March 10th with Jim Brower and M. C. Christensen (engineer of bridges). Chris and I almost missed our flight because of a coal train derailment at HM tower in Elmhurst that halted commuter service. In Casper we added Richard W. Bailey (engineer work equipment) and Jim Simons (division engineer) to our party. This was to be a long hi-rail trip over that lonesome railroad we were planning to make into a major coal hauling line. On March 11th we started at Douglas, Wyoming, and headed east on the light rail stopping at the many timber spans for Chris' inspection of condition. The timber spans were all class B bridges which meant that everyone of them would require additional pilings in the bents and more stringers under the rails. The steel bridges were mostly adequate for the proposed traffic. We met westbound #319 at Lusk. In the White River canyon east of Harrison, Nebraska, the hi-rail station wagon died. We had to push it down the track a fair distance until we reached Glen, a "station" where a trackside telephone was found. Jimmie called the dispatcher who sent a track inspector up on his speeder to jump start our car's battery.

▲ While waiting we had a good look at the track through the canyon. The rail was secondhand 110 pound suspect rail removed from mainlines. This section type and design had been a failure and was especially prone to cracking in the joint area, hence its relay on this light traffic line. We stopped in Chadron that night. Snow fell during the night coating the track we were to pass over the next day. The track east of Chadron was the worst track that I had ever seen in an operating railroad. Broken ties were more numerous than intact ones. The running rails were odd lengths with many extra joint bars spanning breaks or fissures. Ballast was a thin layer of gravel, cinders and dirt. About twelve miles west of Gordon, Nebraska, we came on a broken 112 pound rail at milepost 371.8. Jim carried a red flag in the car and jammed it into ballast next to the defect as well as setting torpedoes on the approaches. The rail had been broken for some time as the separate parts showed signs of wheel batter. At the tin shack that served as a depot at Gordon, Jim left instructions for the section foreman to make repairs. Since we had to clear for an eastbound extra #1765, we went to lunch. As we went east the snow cover disappeared revealing the awful track condition which seemed to get worse the further we went. The track was so poor as to be barely safe for very low speed operation. To add insult to injury there were stretches of spilled bentonite clay that covered the broken ties with a slippery gooey mess (milepost 326.5). At the booming city of Eli, Nebraska, population hopefully five, we met the section gang with their orange hi-rail pickup at the wooden shack that served as a depot. Looking beyond the present track condition, however, the bright spot was that the railroad was all downhill from Bordeaux just east of Chadron. We tied up that evening at Valentine and enjoyed some very tasty thick T-bone steaks. Jimmie Simons was very depressed by the condition of "his" railroad—this was his first real trip over the line and Jim Brower was also very glum at what he was seeing.

▲ On March 13th we crossed the high bridge (#478) over the Niobrara River east of Valentine. This steel structure constructed in 1909 opened May 2, 1910. Realignment of the mainline required 5.73 miles of new track to eliminate the original tortuous low-level river crossing built by the Fremont, Elkhorn and Missouri Valley in 1883. The new trestle, 140 feet above the river, was carried on 12 steel towers whose foundations were pinned to the bedrock by old rail sections driven into the rock and encased in concrete. The bridge was formally opened to passenger train service on May 22, 1910. And that was not the only high steel bridge on this railroad. Fifty one miles to the east, the deep valley of Pine Creek was spanned by bridge #410, a steel trestle 120 feet high constructed in 1905. (When the bridge was officially opened January 18, 1906, one mile of curving mainline was eliminated). The station of Long Pine east of Pine Creek was the crew change point for this section of line. A well maintained structure housed the agent and dispatcher. Nearby was a white painted dormitory for crews. East of Long Pine the track continued to be unbelievably bad. Mile after mile of "happy" track greeted us. "Happy" was the sardonic term applied to the broken tie ends standing up at 45 degree angles and "waving" to us. There was a stretch at Bassett where the broken tie centers were even with the top of the running rails which meant that the base of rail was riding on the ballast—what there was of it. It really was quite hopeless to patch this track. Installing a single new tie to hold gauge was a waste of time and material because the new tie would soon break if not supported on either side by competent fellows. This railroad could only be repaired by wholesale installation of sound ties in clusters of at least ten, or preferably, twelve ties per 39 foot rail. A new track requires 19 ties per rail so this repair job would require replacement of half of all the track ties in the line.

▲ We visited the agent at O'Neill, John Beck, who gave me some antique Railway Express stationery as souvenirs. The depot building was a substantial brick building (built 1910) with a hard-fired brick passenger platform. An old Pullman car painted oxide red with a silver roof and an ancient outside framed wooden box car was the communication foreman's home and tool car. East of O'Neill the track was marginally better. We met westbound #355 (diesel #4307 on the point) at Ewing (milepost 155) and spotted a tamper that had been used to pound the sandy dirt ballast under the track ties in this stretch. Raising track on sand didn't seem like a very productive activity, but you used what you had, not what you wanted. As we went further east the track seemed to have more mud in the tie cribs. The country changed, too; it was a wetter part of the state. We crossed the Elkhorn River on a substantial through girder bridge(#234). Five miles later we had lunch at Neligh and were somewhat heartened by the improved railroad we had just come over. And the condition was much better on this piece into Norfolk. We crossed the Elkhorn River again at milepost 84.9 on a lattice truss so beloved by the C&NW's bridge department in years gone by (#147). Along the Elkhorn's outer bank were a number of automobile bodies. They made good riprap to control stream bank erosion and were inexpensive to obtain. In those days the Environmental Protection Agency (EPA) had yet to develop its standards on such use of old car bodies which had the potential of residual petroleum leakage. Norfolk with its big switching yard was a weedy shadow of its former glory. There was substantial business from inbound iron and steel scrap generated by Nucor Steel north of town on the old CStPM&O line. A new Vulcraft Steel truss manufacturing plant west of town at milepost 82.4 shipped some completed product by rail but more by truck. (The ICC granted permission to abandon 319.7 miles of the line west of Norfolk on March 31, 1992. The last C&NW train service was December 1, 1992. The section from Merriman to Chadron became the Nebkota Railway in 1994. The line from Chadron to Crawford was sold to the Dakota, Minnesota and Eastern by the Union Pacific in 1996. The remaining 247 miles from Norfolk to Merriman was acquired by the Games and Parks Commission of the State of Nebraska for conversion to a recreational trail on December 5, 1994.)

▲ Friday, March 15th, was the final day of our long trip. The track was in much better condition but still not good enough for coal trains. I was surprised to see twin circus-loading style truck ramps at the Armour plant east of West Point. Meat wasn't going east in railroad refrigerator cars; it was going in refrigerated trucks either over the highway or on a flatcar. Also new was a passing track and a white-painted waycar body established as a station with a proper sign on it saying "ARMOUR". The track also was newly surfaced with crushed limestone ballast. Five miles east we crossed the Elkhorn River again at bridge 46, another steel truss like #147 west of Norfolk. Another surprise—there was a rail gang at Scribner installing some second-hand rail. We went to within four miles of Fremont where we got back on the highway and headed for Omaha. Jim Brower commented that the railroad laid well, but all that track had to be plowed off the embankment and rebuilt with heavy rail and sound ties. All through this trip I made notes on the possible passing siding sites as there was no better way to inspect the right-of-way than from a slow moving hi-rail vehicle. The good news from this trip was the growing realization that the job was do-able; the bad news was it would be horribly expensive. (This line was broken by washouts between Pilger and Stanton in July 1982. Local service from each end was ended in July 1984 when further flooding occurred.)

▲ At this time the infamous slurry pipeline proposal surfaced. On March 26th marketing asked our department to help frame a response to the idea that a pipeline from the Wyoming coalfields to Texas for the transport of finely ground coal suspended in a water carrier could be more economic than handling coal by rail. The railroads, of course, reacted to this threat with a great deal of energy. The entire concept in my view was flawed because there was little or no water in Wyoming to make the slurry for the coal. And then what happened to all that "black" water at the Texas end? This was an environmental nightmare that was never fully explained by the pipeline advocates. This pipeline issue had a long life in the courts as the promoters sought recovery of their costs by claiming that the railroads illegally conspired to thwart their plans. Of course, it was plain economics that killed the slurry pipeline concept.

In connection with a study to install Centralized Traffic Control (CTC) on the southern Illinois district, I invited the big electronics firm of TRW to make a presentation for a state of the art electronic installation. On April 3rd they brought their team to Chicago to give their spiel to Bill Zimmerman (systems), Tom Evans (communications) and Vern Mitchell (signals). Afterwards, I gave the TRW people a tour of the power desk and the Illinois division's 1948 CTC machine. I was able to introduce them to Jim Wolfe which made me feel better about asking them to come to show us their wares. Later in the year on October 30th Bob Thompson of TRW presented a computer driven CTC system concept to another group of our officials; it was well received. They offered to do a study for a CTC-CTD (Centralized Traffic Control—Computer Train Dispatching) on the Southern Illinois. Gus Malecha, Vern Mitchell and I discussed the matter at length on November 19th and came to no conclusion.

After the morning meeting on April 1st, the operating department was rearranged. J. W. (Bill) Alsop (returned from his drying-out) was made responsible for train operations and Edward A.

Burkhardt became the operations planner. I had a private meeting with Ed to warn him that Bob Russell and chief special agent Chuck Miles were planning to make an issue of Ed's owning the shortline San Luis Central Railroad through his involvement with a freight car deal. Ed had purchased a fleet of old Railway Express Agency refrigerator cars which were being rehabbed in Green Bay, Wisconsin, and there were questions being raised about Ed's use of C&NW yard trackage to store these cars. Burkhardt had been careful to be in the clear on this matter; Russell and Miles backed off. Three days later Provo was pressing Alsop to improve train performance and Burkhardt to improve service. As for those idiotic departmental vanity charts in the morning meeting room, our group was still responsible for their clarity and timeliness. We would work with any department that asked for help, but we did not go out of our way to solicit business. On April 19th I heard through the grapevine that both Ed Burkhardt and Gus Malecha wanted me to come to work for them in the operating department. No firm offers appeared, however. I did spend some time with Malecha on April 25th showing him how the TPC program could help them predict train operations.

▲ On April 2nd I put on a slide show from the recent trip over the Shawnee-Fremont railroad. Jim Brower, Dick Bailey, Lee Fox, Jerry Conlon (political affairs), Chris Christensen (bridges) and Jerry Groner (finance) were in attendance. The color pictures of ties with ends in the air, the rippling rail wobbling into the distance, the shoulder less, narrow embankments, tired looking timber bridges and general air of an abandoned line evoked lots of comments and loud remarks. Afterwards Bailey and Brower got together to estimate cost of repairing the line to a safe condition as it was obviously unsafe the way it was. Jim and I worked on scheduling track repairs April 4th—not that anything would be done. John Butler, vice president of finance, asked for a repeat of the show on June 7th.. We, of course, obliged.

▲ In our intelligence gathering effort we found that TPC runs on the BN from Orin Junction, Wyoming, to Kansas City were needed. We already had much of the data but this new project, called XRAY, had some gaps. Jules had some special instruction folders from his days at the Burlington. A retired CB&Q B&B supervisor had some old track charts in his basement and Bob Sharp managed to secure a great piece of the missing material from his contacts. Bit by bit, we obtained the track data needed to "run" coal trains in TPC on the BN (May 1974). By then we also had obtained the Missouri Pacific charts we required from Omaha south to Kansas City and beyond.

▲ Our immediate problem for the coal line, however, was an urgent need for a ballast source. The C&NW was one of the leaders in advocating and using hard rock ballast. It was said that a good ballast section makes up for light rail and marginal tie conditions. In my early engineering years the ballast installed in the main lines was limestone quarried and crushed in the Chicago area. One of the big suppliers was the Elmhurst-Chicago Stone Company located right in my home town. The rock was available in quantity and was relatively cheap—but was much too soft when used in heavy traffic mainlines. Track ties under a loaded train move up and down abrading the rock ballast pieces until the original two inch rocks are reduced to a fine dust. Rain and surface water first turn the pulverized rock into mush which then dries-out into a concrete-like material that does not provide good tie support which in turn accelerates the degradation of the wooden track ties. Engineering recognized the weakness of limestone ballast and selected crushed slag as its replacement. This waste product of the steel mills, became a favored main line track ballast in the 1950s. Slag had its own problems as it was not chemically stable and its metallic content sometimes caused problems with the signal circuits. Faced with finding something suitable, the C&NW found a good source of quartzite on line at Rock Springs, Wisconsin, and determined that this material was very satisfactory after they figured out how to reduce the hard material to ballast size. The C&NW opened its pit there and hired Foley Brothers of St. Paul to operate the facility. The product was called "pink lady" from the rosy hue of the broken rock. But, Rock Springs, Wisconsin, was a long way from Nebraska. We needed a local source. There were few quarries in Nebraska. Most of them were located in the southeastern quarter of the state and were all limestone—too soft to use in heavy haul trackage. River outwash gravel pits in Wyoming were too limited in capacity and contained such a conglomeration of rock types that the material could not be used. The Union Pacific used a granite ballast from a quarry half way up Sherman Hill west of Cheyenne, Wyoming. We needed a quartzite outcrop handy to the work. There were undeveloped quartzite deposits in southwestern Minnesota near, but not close enough, to the old Omaha mainline. I scanned geologic maps in the Lusk area for known quartzite deposits and called on the Wyoming highway department in Cheyenne where John D'Amico and Bill Sherwood were somewhat helpful (April 17th). Their records showed hard rock of the Guernsey formation to be found at Elkhorn (near Orin) and some in the Lusk area. I drove to the sites and collected some samples before heading back to Chicago. On May 2nd James A. Barnes, now chief engineer, asked me to collect some more rock samples at Lusk. On June 12th Jim Brower and I flew to Rapid City, South Dakota. On the flight I reviewed the draft report of the Environmental Impact Statement (EIS). At Rapid City Doug Oakleaf and Jim Simons (from Chadron) joined us for a meeting with a quarry operator in the Black Hills area who was sure he could supply the quantity of rock we needed. Afterwards we drove to Belle Fourche on the highway because the railroad was blocked by a derailment. Simons had his hands full with this light rail railroad that was handling a lot of very heavy bentonite clay traffic. Because of the heavy traffic the railroad embankment itself was failing. To cure the problem the railroad hired a contractor to inject a lime slurry into the road bed to stiffen it up—a giant chemical reaction. The injection process was interesting to watch as the probes were hydraulically forced down into the soft roadbed about twelve to fifteen feet before the slurry was pumped into the ground. Doug managed to be where he shouldn't have been and got a good shot of the white lime all over him. That evening Jim was called away from dinner by news of another derailment east of Rapid City at Wasta, South Dakota. He came back in the morning and drove us out to the site. Ten cars were derailed just short of where the tie gang had been working. The railroad out here looked almost as bad as the section east of Chadron. We put Doug on a plane; the rest of us went to Chadron with a side trip to Crawford, Nebraska, to look at the Burlington Northern's line and their ruling grade south of the town. They were using pusher crews to boost heavy coal trains over the hill and through a tunnel. (In later years the track was realigned, the grade moderated and the tunnel eliminated, but it still remained the ruling grade).

▲ On June 14th we looked at a site south of the depot at Lusk as a possible quarry—it didn't appear to be adequate. Jim Simons went back to Chadron and Earl Root, the roadmaster, ran us into Casper with a side trip to the Elkhorn rockface south of Orin. We had not solved the ballast problem. In December Lee Fox was all for hiring our former geologist, Brad Huedepohl, to find a rock source for us. (In 1983 when track reconstruction began, a quartzite mine was opened a few miles south of Lusk in an outcrop west off highway 285. The enormous tonnage of crushed material was hauled to the railroad site for loading on ballast cars by truck which did not improve the unpaved county roads very much). At Lou Duerinck's request I worked up a density map for probable coal movements beyond the C&NW and got a surprise. A large percentage of the traffic was headed to Arkansas and Texas which meant that Kansas City would be a key gateway. On May 6th I had to rework the projection because marketing had revised their estimates—upward. Kansas City became an even larger gateway in the new forecast. Engineering finally recognized the western coal project as they included it in their strategic plan submitted April 23rd. The next day Ray Gotshall and his right hand man, Ron Boesen, got me involved in Northern States Power's new facility south of Minneapolis. They had some severe right-of-way problems complicated by the NIMBY (not-in-my-back-yard) factor. I was of little help in this matter. I learned September 25th that the NSP plant at Henderson, Minnesota, was a dead issue. In late April I had begun laying out the passing siding sites on the Fremont-Shawnee line based on what we had learned in our simulation runs. Even in that vacant north Nebraska country getting the right combination of location, topography, roads and drainage proved to be a challenge. I set up a contest for naming the new sidings and got a lot of suggestions from our group—few that could be decently used.

I went to Washington April 15th for a meeting of the TDCC coding group. Ken Bluett of Bethlehem Steel was the chair, but could not attend as his company was moving him to a new facility at Burns Harbor, Indiana. I was there again on April 30th for the SPLC Policy Council which actually addressed and fixed some structural problems in the code. Back in Chicago Bob Drengler, the trainmaster in charge of crews operating from the Chicago Passenger Terminal, asked for help in laying out a new trainmen's layover room (May 3rd). The same day personnel's Warren Bruce, Bob Russell's go-fer, came to see me about new space for Russell's department. Suddenly, I had become an architect. On June 3rd I went over space on the 18th floor of the old Daily News Building (now the Riverside Plaza Building) with Russell and Al Myles from labor relations. This space had been occupied by Northwest Industries. When they moved out they left everything behind even the pictures on the walls. (On December 19th my artist wife, Barbara, was asked by Tom Lydon of real estate to look at these pictures and make some estimate of value). Remodeling work started June 10th and Russell moved in on the 17th while the work was still in progress. Engineering moved into the first floor of the Chicago Passenger Terminal August 3rd.

Larry Provo instituted one of his most brilliant innovations of office life—the quiet hour. All officers were ordered to refrain from holding or attending meetings; making or receiving phone calls; or engaging in any activity that would disturb the calm that settled on the railroad between 3 and 4 p.m. each day. Secretaries patrolled the phone lines and shooed visitors away. This time for a quiet hour fit my biological clock very nicely, and we all got a lot of things accomplished during this period. Some, I am sure, took a short nap which was not a bad thing to do.

Track inspection became a fad. Everyone wanted to be a track inspector. If only they knew what to look for! Bob Fried and I paired-up because I wanted to make sure that he knew bad track when he saw it. On May 11th we walked the Illinois Division main from Geneva to West Chicago getting thoroughly soaked in a sudden downpour when we were far from shelter. Don McCracken, a freight conductor on the Illinois division, was a great source of first hand information about rough track. A former high-school acquaintance who discovered that I was a railroad officer, Don would send me a list of the spots he had noticed when he encountered bad track on his trips over the old Galena Division,. Using his list was a lot better than randomly selecting a section of track for inspection. On the July 6th weekend everyone was sent out to look at track. Bob Fried found and reported an especially bad defect and was commended for his work—my instruction had paid off. On July 9th all engineering and operating people were out on the railroad between Chicago and Fremont, Nebraska, inspecting track. This intense concentration may have been a good idea, but the disruption to normal and program work was tremendous. McCracken slipped me another list of bad spots on the Illinois Division on July 11th. I was a little surprised when J. R. (Bob) Brennan, formerly the chief passenger agent and now a presidential consultant, asked me to inspect track with him. I took him to Glen Ellyn, where he lived, and gave him a short course in track construction and inspection. Bob was the most gentlemanly railroad official I had ever met. He was firmly convinced that the C&NW's salvation lay in Omaha in the Union Pacific's hands. (Bob died in the autumn of 1999). On July 25th I took Lee Eschelman to Ashton (one of McCracken's bad spots) and inspected the CTC crossovers at AW, Malta, DeKalb and GX (west of Geneva). The #20 turnouts that I had helped install in 1948–49 were now in a sorry state of maintenance which explained the rough rides conductor McCracken was getting.

Our corporate industrial engineering group (oh, how I hated that industrial engineering title) lost its chief clerk/secretary when Darrell Treptow took a job in the labor relations department on May 15th. On June 10th our new secretary arrived—Joann M. Tsoodle, a full blooded Kiowa Indian. Joann was smart, quick to learn and eager to expand her expertise to jobs outside her secretarial role. We taught her how to run the TPC program and put her to work coding track. She became a valued extra hand for our coal group. Other personnel losses included Barry Ptashkin and Joe Mesa, both part of Rich Howard's gang, on July 30th. Joe Mesa, a native born Cuban, wanted to work in the operating department and became trainmaster at Mankato, Minnesota. But, Barry was just fed up with working for Howard.

Grain traffic in May was substantially down because of falling commodity prices and wet fields. Our C&NW stock hovered around $10 per share. All through May rains kept falling so that we began to worry if we had enough stores of gopherwood for building arks. By May 21st declining revenues could be attributed directly to softened roadbeds and disarrayed train service. One of the criteria used to measure the railroad's health was the cars-on-line figure. The number fluctuated wildly from day to day as the divisions reported the count on their territory as of midnight. Why the numbers gyrated all over the place was a problem I wrestled with for two months and never did get a conclusive answer. The problem seemed to lie with the method of counting and the divisions' bad habit of shoving cars onto the next division late at night which resulted in a lot of duplicate counting. Until a good car accounting system was established, the figure was to remain a fiction at best. After all the wet weather, we heard on July 12th that the winter wheat crop in South Dakota was burnt out by drought.

I was supposed to attend the Data System Division's meeting being held in Vancouver, British Columbia, because I was chair of the Standards and Coding Structures Committee. Dave Leach vetoed all trips because of the expense. Those of us with roles in the Data Systems Division were understandably annoyed at being grounded. Then Leach had the temerity to send a recently hired programmer, Sharon Wolf, to the meeting in direct contravention of his own instructions (May 28th). Needless to say, my opinion of him, which was already at zero, sank to a very negative value. Although I reported to him through Jules Eberhardt, I directly felt his cold, clammy bean counter's grip on several occasions. Leach did not like my involvement in outside affairs especially at the national level. I think he was jealous. Yet, on June 25th I was allowed to go to Washington for the TDCC geocode meeting. The next day at that meeting in the American Trucking Association (ATA) building, I figuratively dropped a bomb on the truckers by seeking postponement of a motion and requested that this committee adopt by-laws citing *Roberts Rules of Order,* a copy of which I displayed. The committee had operated like some social club up until now. As an effective working group it had to have rules to function properly. My view prevailed. When I got back to Chicago, I found Leach had issued instructions that no one was to go out of town without his express approval. As he intruded more and more into our operations, my resentment grew apace. Now the fun began. Ray Gotshall asked me to go to Beaumont, Texas, on a coal related call, but did not want to fight with Leach whose reputation for being very tight-fisted was well known. In September, however, I took a couple of days of vacation and flew to Madison, Wisconsin, to appear in Dr. Edward J. Marien's seminar which included a discussion and presentation on the use of standard codes. I had to step carefully, because Dick Hinchcliff of the ATA was present and I did not want him to spot my *alter persona.* By September 18th when I went to Washington for the SPLC Policy Council meeting, Leach had eased up on the travel ban. Progress had been made on a set of bylaws. Miles Manchester, the chairman, had provided himself with a copy of *Roberts*, which I took as a positive step. Dick Braley from the Canadian Freight Association was as obstinate as ever and refused to concede any point. He declared that he would get an official Canadian government ruling on the SPLC, if necessary. These were stormy meetings. From Washington I flew to Atlanta for the Data System's annual meeting. I ran into John Roser, our former advanced systems man, who drew me to one side and confided to me that Provo had a drinking problem. I did not believe that to be true and told him so. My committee included Bob Petrash (AAR), Otto Hermann (SP), Bob Ingermansen (Rock Island), Bill Greenberg (B&LE) and George Leilich, (Western Maryland). On September 20th I was awarded an appreciation plaque for my services as chair of the Standards and Coding Structures Committee in 1973–74. My final act as chairman was to preside at a meeting that would radically reshape the Data Systems Division. In November I again went to Washington for the Policy Committee which adopted those vital by-laws. The furor over the Canadian code continued unabated, however.

Other interesting things happened in the summer of 1974 including management's delayed realization that the second main removed from West Denison to Missouri Valley a decade ago had been a major mistake. Jim Brower was assigned to look into restoring that double main and the double tracking the Fremont-Missouri Valley segment as well because most of the Union Pacific eastbound traffic was going through that single-tracked bottleneck (July 10th). George Hanson, the former superintendent of the Missouri Division, was now assigned to the operating department in Chicago and was showing interest in having TPC runs made (July 18th). Ed Burkhardt also was a convert and asked for a variety of runs some of which were for C&NW trackage we had not yet coded. I redid the division segment maps to accommodate the disappearance of the Missouri division. To my annoyance I was invited (ordered) to attend a class in negotiation July 1st. On July 5th an order was issued that all track work was suspended unless it was on the Chicago-Fremont mainline.

▲ One of the dead horses in the road I was dealing with in July was the notion stuck in top management's brain that our railroad had a viable all-C&NW route from Fremont, Nebraska, to Kansas City, Missouri, by way of Marshall-town and the former CGW line south. Not only was its circuitous mileage a major problem, but the condition of the railroad would not support heavy coal train movements. Our collection of TPC run results in the UCTSIGMA database gave us factual information for upsetting the dead-wrong idea a lot of people had about the value of that old CGW railroad. On July 19th I had a quiet conversation with Lou Duerinck (law) about my dilemma and showed him the huge competitive advantage the BN had over an all-C&NW move to Kansas City. We decided to consolidate our findings and ask Bob Sharp to look at it first. Ray Gotshall, Sharp's coal man, was appalled at the comparisons. I then released the report to a few more people, Ed Burkhardt among them. Ed requested that runs be made over the entire Fremont-Kansas City route on June 7th. Interest in the coal project began to rise again. I put a big planning chart together for the Western Coal Project on August 9th; Bob Sharp approved it three days later. At a big meeting on the 12th Larry Provo helped kill that unworkable round-the-horn route (all C&NW Fremont-Kansas City) by saying he liked the idea of interchanging coal trains at Fremont. That put the skids under any plans to run those heavy trains east to Marshalltown and around the corner south to Kansas City through Des Moines. When both Bob Sharp (strategic planning) and Ron Schardt (finance) heard those words, they realized that this definite change of policy required radical adjustments to their costing of coal train moves through Kansas City. Two days later Sue Frehleng, that caustic tongued consultant to the finance department and a leftover from corporate development days, let her cynicism show in her abrasive and negative remarks about the entire idea of moving western coal. But, Provo was enchanted with the idea of the Union Pacific becoming more involved with these future coal trains (August 15th). I went to California on vacation for three weeks.

▲ On the national scene another potential problem arose when a proposal was made in the House of Representatives to ban coal mining in the National Grasslands (July 31st). The principal coal mining area in eastern Wyoming was within a designated National Grasslands although the land was covered mostly with black sagebrush and darned little grass. In fact the Bureau of Land Management had graded this land as "40" which meant that a single unit of livestock (a cow and a calf) required 40 acres of land to subsist there. In September there was another alarm when the Environmental Protection Agency (EPA) was said to be preparing a ban on strip mining in the West.

President Richard Milhous Nixon resigned August 8, 1974.

On return from a California vacation and a visit to what was to become our future home, I discovered that little had changed in the office. Even the coal project had gone dormant—again. There was a distinct air of unease in the place and some growling about Jules Eberhardt's perceived lack of leadership. At the September 26th morning meeting a rather interesting study of crew costs west of Winona, Minnesota, revealed extraordinarily high expense with operating the railroad west to Rapid City, South Dakota, The rotten track and resulting slow orders combined to drive crew costs skyward. Jules called a staff meeting October 1st with Rich Howard and me to discuss how things were going under Dave Leach's supervision; we agreed it couldn't get much worse. Lee Eschelman took a job with A. T. Kearney and left on October 31st. His acceptance of the job offer on the 13th seemed to awaken Jules to the internal problems we were experiencing and privately consulted with me after Bob Fried, Eschelman and Mark Johnson had all unloaded on him. It was my opinion that a major part of the blame rested on director Rich Howard's shoulders but I could not say that to Jules who seemed to be enchanted with his fellow industrial engineer. Perhaps I was unfairly prejudiced and should have adopted a more tolerant attitude toward Howard who had a lot of inferiority issues complicated by a failing marriage. Rich offended me by his devious nature and uncouth social behavior. I would not trust him as far as I could throw a chimney by its smoke. While Rich Howard at 31 was very bright, he failed to generate much respect from his employees. As an industrial engineer he knew the techniques but never succeeded as a manager to establish efficient work plans for any of the problem areas he and his crew were asked to study. The rapid turnover of a parade of intelligent young engineers assigned to Rich bespoke his inability to build an efficient staff. As a manager he consistently failed to motivate his subordinates, a fact underscored by the rapid turnover of bright young people assigned to him. Some of my lads promoted into his care, soon sought other opportunities within or outside of the railroad. At the North Fond du Lac shops, where repairs to work equipment was centralized, Rich's assignment was a pure industrial engineering work analysis exercise. It turned into just an excuse to be absent from the Chicago office and enjoy a dandy lifestyle at the company's expense. Howard was my personal *bete noire* and was the proximate cause for my planned departure from the railroad in 1980. But in 1974 he was a director that seemingly could do no wrong in his superior's opinion. Soon after I left the railroad, Rich Howard was fired.

On October 2nd I had a surprise visit from Harold W. Jenson who had been my first division engineer in 1947. Now 75 he was living in retirement in Mountain Home, Arkansas. Harold had served in both World Wars, acted as a forceful leader on the C&NW and had been a big factor in shaping my early railroad career.

Bill McGovern, a former junior naval officer, came to work for Howard and occupied the office next to mine. Bill was a slow, methodical person whose productive output was minuscule. I was reminded of a three toed sloth as each movement seemed to be thought through before taken. He had trouble from the start with Rich Howard and Jules Eberhardt. Why he was ever hired remains a mystery. The best thing that happened to him was being fired. He got a more suitable job with the Association of American Railroads' lab in Chicago and did a good job editing their news bulletin. Bill was not dumb—just slow. I took him out to Rochelle and Ashton October 7th for track inspection. Some good things were happening to the track. New 136 pound rail was being installed at RX, the CTC crossovers. At Malta we saw a gang installing more 136 pound rail. We happened to run into Lee Fox under whose leadership this track work was being done and inspected more track at Cortland.

Jeff Ottos's Five Year Plan for Proviso now lay in shambles—no work was done in 1974. Bob Milcik, now the superintendent there, was as livid as such a placid individual could be and blamed engineering for the abject failure of the entire project (October 15th). At a meeting on the 25th systems was told to get involved in helping run the railroad not just recording past events. Something must have occurred because Fred Yocum said December 13th that systems was much better about responding to operating department requests. Dave Leach kept tight control over the "his" computers but his accounting background stifled any attempts to expand the machines much beyond mere record keeping. On October 29th I happened to see Dick DeCamara on a commuter train; he said that he had been at a meeting with Inland Steel and that my name had come up. He didn't enlarge on that cryptic statement. Meanwhile, Provo was again hammering the operating department about unsatisfactory train operations (October 31st).

At the end of October I had updated the railroad's density statistics which involved grinding through a vast amount of data. Mainline traffic was up 31% in 1973 over 1972—no wonder the rail was wearing out. This turned into a three week effort, and I vowed (in vain) that next year the computer would do this work. I made out a PACT request for systems assistance to mechanize generation of the ton mile density statistics.

▲ By October track building consultants began to appear—drawn by the scent of new construction. On the 2nd Jerry Neben, then with DeLeuw, Cather of Canada, came into town to get the details of the reconstruction job from Jim Brower, Bob Fried and me. At this point the coal project went into back into the freezer—again. In a sense I welcomed the pause as we now had a chance to do some essential planning. The downside was everyone else went back to their normal duties and the coal project was further chilled by the breeze of their departures. On December 4th Jules and I reviewed the draft of a joint operating agreement prepared by the BN. The next day's meeting in the law department on this draft was a lesson on how unfair contracts can be artfully constructed and the matter was again addressed on the 9th. While the railroads elbowed and postured over their legal agreement, Doug Oakleaf and Jim Brower in engineering were developing the rehabilitation costs for the 519 miles from Shawnee to Fremont. Filing the return to questionnaire to the ICC was a big event that was done on time. Meanwhile Bob Sharp had acquired from sources unknown the Burlington's annual density statistics which revealed a great deal about some problems they were going to have with their existing railroad serving the coal fields. I prepared a comparison of the Burlington and C&NW coal projections for Sharp on December 19th. That ended coal project activity in 1974.

There were other things going on that autumn and early winter. On Halloween I was handed the Sears Roebuck project. Sears wanted to build a transfer warehouse at Proviso—shades of the long departed

LCL Freight house. I gave the job to Bob Fried who got involved in several meetings at Proviso to evaluate the Sears project. The idea was just what Federal Express developed at their Memphis, Tennessee, base—consolidate shipments going to similar destinations. This proposal was all boxcar traffic. The last meeting on the subject was November 27th with Jules Eberhardt and Bill McGovern. No one was in charge; the meeting was a chaotic affair. Doug Christensen of industrial development should have been the leader according to the rules that I once operated under, but he was opposed to the concept and wanted to save the property in question for new industrial prospects. I argued that this Sears thing was really a new industrial prospect, but he did not view it that way.

The ICC finally concluded November 12th that the Union Pacific should have the Rock Island. The news was greeted with yawns as the eleven years it had taken to come forth with this opinion had seen some major changes in relationships between the C&NW, UP and CRIP. Now, the merger was obviously unworkable. The UP now had a better eastern connection partner in the C&NW than the Rock could ever become. And there was real doubt that the Rock even could be rescued from the scrap heap. The Rock Island was to close down late in 1979.

On December 5th I rode AMTRAK to St. Louis on the *Turboliner*. The grand old Union Station was now a vast ruin. I took a taxi over to our Madison, Illinois, yard where Bob Lawton waited with business car #401 which had formerly been the private car of the CGW's president. The car now had been equipped with track analyzing equipment for finding out of standard levels. The machine also measured distances and profiles. Bob was there to teach me to operate the machine. I enjoyed the master bedroom that night and the full breakfast provided in the morning at 8 a.m. We departed Madison at 6 a.m. on the tail end of train #380 headed for Proviso. We ambled up the southern Illinois district, meeting two southbound extras, and stopping twice to release balky brakes on the 401. At South Pekin where full propane tanks were installed, we picked up an assistant roadmaster for the run to Nelson. After a stop at West Chicago to set out a block of cars, we went on to Proviso arriving about 9:30 p.m. Three days later 401 was on the rear of freight train #253 when it left Proviso that evening with Lee Fox and me aboard. The lamb chops served were delicious. Because of air brake problems at Clinton, Iowa, we had not made much progress over night. Eventually, we arrived at Marshalltown where Dallas B. Carlisle and Richard C. Christensen joined us for the trip north on the rear of train #702. I was successful in operating the track analyzer on this my solo voyage and was glad to know that I would not be doing this as a full time job. It was boring work with occasional bursts of energy when a bad spot was hit. The run went quickly north of Mason City, Iowa, and Albert Lea, Minnesota. We dropped Dallas and Dick at Merriam, the end of their territory and rolled on into Cedar Lake yard, Minneapolis, about ten hours ahead of schedule. After a quiet night aboard 401, we were attached to a transfer run and moved through the Twin Cities on the former Great Northern. I left the measuring equipment turned on even though it was a foreign line move. The worst spot we found on the entire trip was on the Burlington Northern just before arriving at our Westminster Yard in St. Paul. Lee Fox and R. P. (Dick) McDonald had driven ahead to East St. Paul. Collecting Lee, we rolled north on the very rough New Richmond sub. After picking-up 20 bad spots, the machine went insane requiring an instant shut-down which transformed me into a tourist for the rest of the trip. At Itasca, Wisconsin, I called Bob Lawton to get his instructions on what to do with his idiot machine, but he was away from home that evening. I remained a tourist. That evening I hung around the Itasca yard office waiting for Bob Lawton to return my call. Meanwhile, Lee had been summoned to Chicago and went off to find a flight, leaving me to bring the 401 back to Chicago. On December 12th the 401 was attached to the rear of train #408 which had a very large consist that day. Charlie Hellem, Dick McDonald, a roadmaster (whose name I did not record), and I were the entire complement when we started south. At Spooner, Wisconsin, we picked up Ray Burks, the roadmaster for that territory. Ray had been a Galena Division rodman years ago. The train broke apart twice on Sarona Hill south of Spooner, losing three hours. We dropped a couple of our men at Altoona, Wisconsin, and Ray got off at Wyeville, Wisconsin, where he was replaced by Glenn Kerbs, assistant division manager-engineering (the new style title), who boarded for the ride to Butler, Wisconsin. Glenn, an ex-NYC employee, was a good steady type who would be a player in the coal project. I went to bed at Adams, Wisconsin, when they stopped to change crews, and awoke the next morning just as we passed St. Francis on the south side of Milwaukee. The trip was slow because of some bad brakes in the train. We slowly inched past Blodgett, Illinois, where a bad wreck had not yet been fully cleared—there was still one diesel to be picked-up. Kerbs dropped off as we passed Valley. I detrained at Mayfair to catch a CTA train into the city leaving the 401 to rumble on its own to the coach yard.

It had been an interesting year, all in all, but there were a lot of unresolved issues to be settled in the next. In Japan Larry Provo was invited to ride the head end of the *Shinkasen* (Bullet) train.

29

THE DREAM OF COAL TURNS YELLOW, 1975 . . .

▲ A fresh new year and a fresh new idea for the east end of the projected coal line near Fremont. Restoring the recently abandoned Arlington-Omaha line for direct service seemed like a good idea. Instead of trying to run heavy coal trains through the hills to Blair and around the connection south to Florence or across the Missouri River to Missouri Valley and thence south to Council Bluffs in order to open up the Omaha gateway for C&NW trains was awkward. Of course the Union Pacific had a perfectly lovely direct line from Fremont to Omaha. Yet, we C&NW people believed that by keeping control of our coal trains as long as possible, we could squeeze just a little more revenue out of the moves. Our independent attitude sometimes got in the way of practicality. Luckily, these were just abstract studies of possibilities.

▲ Our department had acquired a remote terminal connected to the Ravenswood main frame computer. We could now enter TPC data and set-up runs without using those clumsy punched card decks which

▲ As noted before, this symbol will be used to denote material relating to the Western Coal Project.

required transport back and forth to Ravenswood. Getting that terminal installed in Frank Stern's old office on the west side of the our building required stringing a coaxial cable between floors which presented a lot of problems including drilling holes in concrete floors. I suggested using the elevator shaft and the cable was installed over a weekend. Still exploring the alternate routes east of Fremont and despite the plain fact that the C&NW already had sold the Arlington-Bennington right-of-way and the awkward interchange in Omaha would have made coal train movements an operating nightmare, we cranked the Arlington segment through the TPC machine and got satisfactory running results. This endeavor proved to be a complete waste of time because it was now obvious to everyone that Fremont, Nebraska, at the end of the 519 miles from Shawnee had to be our major interchange with the Union Pacific and the Burlington Northern. Jim Mann of personnel came into our offices on January 9th to begin planning the people requirements of the coal project. On the 15th I got Gus Malecha in operating to start considering where the crew change points were going to be. Jim Brower was already developing rehabilitation costs for the railroad when the wheels came off the project again. Other than reviewing a redraft of the C&NW and Burlington joint line contract on February 28th, there was little or no activity on the coal line development until April 11th. Everyone involved seemed to have fallen asleep. We won't disturb them for awhile as we pursue this tale.

Early in January our department received two more requests for assistance—help in the design of new division office buildings in Boone and Mason City, Iowa. The next day heavy cuts in capital and operating spending were announced because of a dramatic drop in business. Much of our cash squeeze was coming from recent purchases of locomotives. On the 14th bad weather struck Iowa, Nebraska and Minnesota. Revenue continued to slide as the national depression deepened. President Gerald Ford was blamed for not solving the financial decline. The cyclical nature of the market's movements were not well understood so sitting Presidents usually were handy targets. By the 17th plans were made to scrap more obsolete freight cars at Clinton, and, worst of all, to sell the stockpile of new 136 pound rail at the welding plant in Tama, Iowa. Jim Zito said in later years that he cried when he heard they had to sell that rail. Jim Zito was one of the few C&NW officers that had correctly identified our major problem as worn-out track. The Sears Roebuck transloading project at Proviso appeared to be dead January 9th when I went to see Bob Milcik at Proviso—operating thought they needed the site for future rail operations and industrial development thought they needed it for new industry. Another of Rich Howard's young industrial engineering men left; January 10th was Frank Bebjl's last day. On January 13th we heard that high winds had blown open Dave Leach's office window and the cold blast of arctic air had frozen his toilet. A water pipe on the 14th floor then broke and flooded his office. It couldn't have happened to a nicer guy. On January 16th James A. Barnes, chief engineer, warned his people that the Federal Railroad Administration (FRA) inspectors were on the prowl and that there had better not be any phony inspection reports filed. Provo added, "It's your job if you are caught filing a false report." Fleeing the winter, I took my wife to Florida going by way of Washington and then south on AMTRAK's *Silver Meteor* in sleeper *Mobile River* to Fort Lauderdale. My salary had been increased to $22,500 which was comfortable but now secondary to my other income.

My former colleague, John J. Foxen, left the railroad February 1st under a company program offering early retirement to those people from 60 to 65 years old. Al Simandl, chief clerk when I worked on the Galena division in the 1950s, died suddenly February 3rd. He had worked at his regular job all day and was struck down at home that evening. On February 13th I learned that Gene Cermak, my former boss in industrial development, had moved to Greenville, South Carolina. He certainly could play golf longer in the year that far south. That early retirement program also attracted Charlie Stewart, Roy Lawson, Howard DePierre, Ray MacCarthy and Ellis W. Ernst—all people I had worked with in traffic. Fred O. Steadry, our commerce attorney, retired March 31st. Bob Russell got up at Fred's party and told a long story about how I had saved him in a court appearance. I still do not know what I had inadvertently done that had saved his bacon.

Richard J. Howard and Mark A. Johnson became deeply involved in a purely industrial engineering study at North Fond du Lac. They were running a detailed time measurement analysis in the maintenance of way repair shop. This effort was, as stated previously, a monumental waste of time and energy. It served as a refuge for Rich for months and reduced his availability to work on really important projects. He milked it to the point that his boss, Jules Eberhardt, finally said enough—come back to earth. In all probability I had made a number of caustic comments about this boondoggle that irritated both of these men, but this industrial engineering baloney really was getting under my skin.

Track inspection was still a hot subject. On January 31st I attended a track inspection class being run by Raymond E. Snyder, who was once my junior rodman on the Galena Division. Jocularly nicknamed Whiplash Snively, Ray had steadily grown from that very green young smarty-pants into a respected track engineer. He had also grown from a skinny kid into a well-padded adult. He did a good job with his track inspection class. We will have more to do with Mr. Snyder later in this account.

The economic news got worse in February. A 4% cut in personnel was in the works with special attention to engineering personnel. I was happy that I had not listened to Lee Fox's overtures last year to join the department. Compounding all railroads' difficulties, the ICC turned down a requested 7% rate increase. Twenty assistant trainmasters were axed on February 3rd and pressure was increased on those over 60 to take early retirement. Most railroads were retrenching leaving deep wounds in their organizations. Jobs were abolished and expenses were pared to the bone. Business remained flat through the cold and heavy snows that shut down our western lines in February. On February 13th Provo announced at the morning meeting that revenues were down 27% from the previous year. He also wanted to know why the operating department could not get the car count down and improve service. There were no answers offered. On February 26th the C&NW offered the entire suburban fleet to the Regional Transportation Agency for $51million dollars—we needed that money.

During this depressed time the coal line project was relegated to the furthest back burner. Taking advantage of the lull and to fill in gaps on our railroad where TPC had not yet been run, I coded track incessantly. We would be ready to run ballast trains over just about any piece of the railroad on our new TPC III software. Bob Fried had a new task to perform for Jim Wolfe—develop a fuel consumption study using our TPC program. Having that terminal in the office was a great help although we sometimes overloaded the computer at Ravenswood with too many jobs. Requests began to flow in from many users. Provo wanted certain runs made on February 27th. Then disaster struck—our terminal was taken away because it was only a "loaner". At the heart of this problem was Dave Leach. Always on the lookout for saving money he had struck a deal with RCA for new main frame computers at Ravenswood replacing IBM equipment. RCA was so tickled at ousting IBM from an installation that they agreed to Leach's demand that the new machines would be rent free until they met specifications. Somehow, they never seemed to work just right according to our systems people until, at last, RCA demanded payment or they were yanking their equipment. Just when the railroad was so desperate for cash, Leach had to pay up. I don't think he saved the company any money with that "smart" deal. On March 6th Provo asked what happened to his request for a TPC run. We told him our terminal had been taken and that we were back to using card decks again. Provo quickly reversed Dave Leach's decision to yank our terminal. Following the big snowstorm April 2nd, a meeting was called for the next day to develop a program for fuel conservation. Larry Provo was now a strong advocate and customer for use of our TPC routines and made several new requests for runs. We were really struggling without our terminal and the unstable Ravenswood mainframe. At the morning meeting of April 10th there were several questions relating to the TPC program; I invited Gus Malecha of operating to come to our office for a demonstration. The next day I took Bob Fried to the morning "prayer session" at which he successfully fielded several questions. On the 14th a delegation from operating arrived for the demonstration and, of course, the Ravenswood computer went down. On the 17th Provo walked across the hall from his office with Ed Burkhardt to get a closer view of the operation. That same day Ray Gotshall asked for Union Pacific runs. Then, to our surprise, the Union Pacific called the same day, and requested TPC information—our fame was spreading. Ravenswood was completely down April 21–23 and sporadically through the 28th. Even though the main frame was very wobbly into May, we continued to try and run requests using our newly replaced cathode ray tube monitor and keyboard.

A rumor had arisen February 11th that Ed Burkhardt was to become vice president operations of the Kansas City Southern. That didn't happen. But, Steve Adik did become vice president operations of the Lehigh Valley, a bankrupt carrier in the east. Another rumor on the 14th was confirmed four days later: Lee J. Fox was leaving the C&NW engineering department for a position with the Illinois Central Gulf. At the morning meeting on the 19th James A. Zito was introduced as the new assistant vice president-chief engineer of maintenance. Rumor also had Bob Sharp, strategic planning, going to the ICG, and that didn't happen either.

On March 10th Bob Fried and I attended a meeting of the Wisconsin Department of Transportation and the Federal Environmental Protection Agency (EPA). They were trying to establish a method to measure fuel consumption of diesel locomotives on branch line operations. Maurie Reid's branch line committee had organized this effort to demonstrate the C&NW's care and diligence in presenting abandonment applications. We prepared a demonstration TPC deck to show how easily answers to questions about fuel consumption could be obtained. On March 12th Bob Fried and I finished a preliminary fuel study for Reid. (Bob Fried later became a fuel conservation officer at CONRAIL).

In late February the Rock Island collapsed into bankruptcy. I was drafted on the 28th to help assess the value of certain of their properties to the C&NW. Furthering the fall of that railroad was a goal of the C&NW. Larry Provo attended the hearings on the Rock's bankruptcy at the ICC on March 4th. By the 13th the ICC was making noises that the C&NW and other connecting railroads might have to operate the Rock Island. The bankruptcy was confirmed March 17th. We got ready to take over various parts on the 21st, but the freight embargo was set back two weeks. One interesting effect stemming from the Rock's troubles was the sudden revival of interest in our discarded Des Moines-Kansas City study of 1972. I still had a strong feeling that the basic premise of that work was sound and would become a reality in the future..

The Marshalltown yard project officially died March 19th. To save face a smaller yard was proposed between Nevada and Ames,

Iowa. Later, a small holding yard was built east of Ames, but it was never used to any great extent. The whole Marshalltown episode had eaten up a lot of valuable time and resources. On March 27th the morning meeting revealed the railroad was still a tangle of good intentions and bad train operations. The Milwaukee Road had troubles too; they had derailments at Tama, Iowa, and Medary, Wisconsin.

It was March 27th when Bob Wilson of personnel and Doc Davison of the medical department came to us for help in designing a testing facility for new train service candidates. The medical facilities were located under the former Galena division tracks off the Suburban concourse on the west side of the Chicago Passenger Terminal between Washington and Randolph Streets. The area had been used for immigrant waiting rooms when the facility was built in 1911. Wilson and Davison wanted some sort of wooden mock-up that replicated the dimensions of a freight car with a ladder whose first step was so many feet above the ground, an air brake hook-up that required the candidate to reach the requisite distance without placing a foot inside the running rail, and an 80 pound weight that had to be lifted waist high. When Ira Kulbersh, chief mechanical engineer, came up to our office to talk about constructing such a mock-up, I volunteered that creating any test replica was going to generate arguments that the artificial device did not match reality. Ira said "Why not install a real piece of rail equipment in the testing area?" Pure genius and a wonderful idea. With a real car in place, candidates could demonstrate their agility in mounting the end ladders, lifting a real knuckle up to the coupler, avoid stepping on a real rail while demonstrating their skill in joining air hoses by using actual equipment. The best part was that the candidate could not complain that the test rig was not real. Ira arranged to have three feet sliced off the end of a condemned gondola car being scrapped at Clinton, Iowa, complete with ladders, brake wheel and all components intact. Two pieces of rail were also furnished. With some muttered complaints and a lot of heavy lifting the B&B crew struggled to maneuver the heavy metal into place and bolt it to the wall (June 20th). When they finished, the end of the car on the wall looked like the famous surreal painting by Rene Magritte showing a steam locomotive emerging from the back wall of a fireplace. The real reason behind all this emphasis on physical testing came out sometime later. Management believed that women seeking railroad jobs as brakepersons would bring big problems to train operations. Bad times and a growing awareness of their equality were bringing more women into the work force seeking jobs in areas traditionally held by men. Never mind that the women had performed well in some of the dirty railroad jobs during the war. Now the war was over and the women were supposed to stay at home tending their children. Railroad management was not yet ready for the female revolution and were dead set against hiring women for road service and yet could not overtly forbid hiring them. This rigorous physical testing was supposed to eliminate women from road service on the premise that they were not strong enough to handle the work. That was an incorrect assumption, of course. To the surprise of many officers, a lot of women passed the test and became good brakepersons as well as other train service employees. (Note: Ira Kulbersh died July 6, 2003 after a long and distinguished career).

The railroad began to recover in early April despite trackage badly affected by the thaw and a very wet spring. By the end of the month revenues were still rotten with some signs of gradual improvement. During this low period I quietly increased my holdings of C&NWT stock. Over time I acquired a substantial block at an average of $2.25 per share by offering cash to disappointed holders who wished to unload their stock. All of this was by word of mouth. I was surprised by some of the people who chose to dump their investment. My wife and I held that stock for 17 years and sold it for $50.00 per share. Also in April I was having some success at mechanizing those miserable annual ton mile statistics and made a trip to Ravenswood with Sy Berman from systems to interview the car accountant about his records. On the 14th I was drafted to help with a head count of passengers on west line train #61. Since there were just two cars on this train that departed Chicago at 6:40 p.m., the task of counting ins and outs at each station was not too difficult.

▲ The coal project revived briefly April 11th when Oklahoma Gas and Electric came to Chicago. There were some minor tasks that were done involving TPC runs for the new ARCO mine and Bob Fried's comparison of General Electric new U30 diesels and EMD's SD-40s for coal train service (no big differences were found). After contacting the Missouri Pacific to obtain more of their track charts for lines in Arkansas, I began re-engineering Bordeaux Hill east of Chadron, Nebraska, as the TPC runs pointed to this hill as one of the big problem spots on the Shawnee-Fremont line. Each reworking of the grade and alignment was tested by using TPC. Eventually, I got the best result possible, and wrote it up for future consideration when we got down to doing the real job (July 10th).

In May the company's cash was so bad that a system wide scrap drive was mounted. I was drafted to go to Mason City, Iowa, and look for anything that could be sold for hard cash. Ray Snyder and I drove to Mason City by way of Oelwein. After we arrived, I was shifted to Tracy, Minnesota, as my base station. Tracy was located in the southwestern corner of the state in an area that I had never seen before. On May 19th four us—roadmasters Erickson, Clayton Cockrell, Royce Dyslin and I—drove north to Tracy from Mason City via Sherburn. At Tracy we had lunch before Clayton and I set off looking for scrap. First we set out to do the 58 mile line to Gary, South Dakota. One mile out of Tracy the hi-rail truck derailed because of wide gauge in the track. Re-railing our wagon we had to proceed slowly because it was hard to see the rails through all the high grasses. I quickly learned not to go wading through those weeds as they were so high that the journal boxes of passing freight cars had left an oily black coating on them. Just past the state line near Gary we ran into a strong summer thunderstorm that cooled down the 92° heat. (This piece of railroad was abandoned January 26, 1981). On May 20th Clayton and I did 23 miles on the Wabasso line and 12 miles up to Vesta, Minnesota. Again, the rails were invisible under the waving fields of grass. I was reminded of Ed Upland's exploits when trying to get empty grain box cars moved for loading on the infrequently served branch lines in South Dakota where the high grass and weeds of late summer seriously impeded diesel locomotive operation during the annual grain rush because crushed stems mashed on the rail heads caused the units to slip and stall. Ed went to a local Sears Roebuck store, bought two 24 inch rotary lawn mowers and had the mechanical department rig them on the front of the locomotive. The mowers did a good job of whacking the high grass a foot on either side of the running rails. Ed said they sailed right along until the mowers ran out of gas which required a short stop for refueling then

The weathered depot and spindly track through the daisies at Lucan, Minnesota, were too typical of the faded branch lines burdening the C&NW in the 1970s. Lucan first saw rail service in 1902 when the Minnesota and Wisconsin Railroad (an Omaha Road subsidiary) built the line to Marshall. A post office opened in the small community in 1908. Now it is May 20, 1975 and rail service would end early in 1979.

off they went again. The Vesta line was laid with 50 pound rail, a light section I had never seen before. (The Wabasso line and the Vesta line were abandoned February 9, 1979). From Vesta we cut across to Redwood Falls by highway in time to see one of the ALCO units, #1665, at work and then rolled 25 miles down that branch to Sleepy Eye, Minnesota. (This line was abandoned January 19, 1981). After lunch, we returned to Tracy on the main line. The most iron or steel materials we saw were the rails in the track we were riding on. Clayton had all of his section crews together to install a turnout at Lamberton, Minnesota, to serve a new grain elevator. After westbound train #167 cleared, we hi-railed to Mankato where we met Butch Drews, track supervisor, for lunch. Bidding Clayton goodbye I went off with Butch to Merriam, Minnesota, where we got on the rail to Waseca before returning to Albert Lea, Butch's home base. On May 22nd Butch and I hi-railed 42 miles north from Albert Lea to Waterville and then 6 miles up a former CGW branch to Morristown, Minnesota. This was another branchline that was slated for early retirement (October 1,

1976). There was an old CGW switchstand there that I would have liked to have collected—an ancient type rarely seen— but there was no way to earmark that one for preservation. Again, we did not find much scrap lying about. Retracing our route to Albert Lea we went south another 49 miles to the Milwaukee Road diamond at Britt, Iowa, (also known as the Hobo Capital of America) before getting back on the highway to Mason City. (The line from Albert Lea to Lake Mills was abandoned November 16, 1976; the next segment south to Luverne, Iowa, went on April 28, 1981). On the 23rd a debriefing meeting was held in Mason City at which it was obvious that there was not much cashable scrap in this territory. Ray Snyder and I drove back to Oelwein where we got on the old CGW mainline and hirailed 74 miles east to Dubuque, Iowa—in a heavy rain. (That railroad was abandoned April 22, 1981).

The railroad's revenues continued to decline. On May 27th new cuts were announced—15% below April 1974 levels. On May 30th we learned that 125 people had been cut from the payroll including my friend, Don MacBean. Bill Weatherall, chief mechanical officer, also retired. Revenues were running 30% below 1974 levels. The mood around the company was not very good. A big freight wreck at Chelsea, Iowa on the 28th merely underscored the mood of depression. We were told that our department would remain at seven positions. On June 17th business was still bad and there was a threat of a clerks' union strike. Anything that would generate cash was considered. One idea involved invoking heavy usage of AAR Car Service Rule #70 which required the re-weighing and stenciling of freight cars either when major repairs were done or every five years. A look at the weigh dates on the cars of passing freight trains revealed a large percentage of the cars were past their five year limits. A railroad could generate a lot of cash by simply reweighing expired date cars and billing the owners. To verify our estimates regarding the possible number of reweighers, I spent June 6th at the Proviso hump to sample how many cars would qualify—there were a lot of out dated cars in service. A salesman from Toledo Scale came in with some information about a portable electronic model that would be useful for such a Rule 70 program. The idea gained some support and the rip tracks around the railroad were instructed to raise their AAR billings by including reweighing expired date cars.

▲ I was sure the coal project was dead. The C&NW did not have the money to fix the decrepit Shawnee-Fremont line much less pay our half share of the new coal line north. While the agreement for sharing the costs of the joint coal line between the BN and the C&NW had been signed in May, the C&NW had not addressed the question of developing an energy policy—there were too many distractions around the place dragging at our attention. On June 19th I completed a report on BN train operations north of the Powder River which showed the utter lack of capacity for that line. The Sierra Club won a big delay in any mining operations by a lawsuit they had filed against this project which further dampened any enthusiasm we had for the coal line's prospects.

Train Performance Calculator (TPC) was getting a lot of use and a bit of a reputation. Lee Fox even sent one of his ICG men over to see the program run (June 2nd). The next day a man from the Milwaukee Road came in to see it operate. We did not tell them that we had acquired all their track data required to run their railroad in the program. We did continue to broaden our coverage during these slow times. Coding for Butler to Wyeville on the C&NW was finished on June 20th. In late July John Sward of the ATSF supplied a huge collection of their track charts for our use. I did the Colorado Southern line south of Guernsey into Denver, Colorado, in early August, before tackling the ICG line across Iowa. Bob Lehnertz on the DRGW in Denver managed to find an ardently desired set of Special Instructions for the C&S from his contacts in Denver. George Hanson in operating requested a coal train TPC run from the Twin Cities to Fremont via the old CStPM&O's Omaha Line, Sioux City, Iowa, to California Junction, and over the hills west to Fremont. This was not a good move as we pointed out the bad operating locations to George on the printout. Joann Tsoodle proved to be a capable coder and operator. Her secretarial duties were not enough to keep this bright girl busy. She said that she felt her efforts were helping our department enhance its reputation by doing this important work. I mentioned to Joe Marren of public affairs that he might want to do a story about TPC (July 18th), but the suggestion fell on deaf ears. Perhaps, that was just as well to let it remain a secret weapon.

In early June our tariff people came to us with a very important project that was proving impossible for them to solve without computer assistance. As the C&NW absorbed the M&StL, CGW and FDDM&S the resulting spider web of lines had rendered our mileage tariff obsolete. The mileage tariff was the governing authority for rating shipments that required a mileage factor. Any two points on the C&NW had a theoretical least distance mileage that was supposed to be shown in the mileage tariff. This meant that any rusty, weedy, branch line could be the key link in calculating the shortest distance even though there was no train service on the lines used in the calculation. Every time an abandonment was approved, the mileage tariff was supposed to be updated. It was an impossible manual task—the network was too complex. The answer was a computer generated matrix that used our reliable Segmented Railroad File (SRF). I knew that thing would be useful someday.

The new training center at West Chicago opened June 10, 1975. The railroad was making a major commitment to train personnel to properly operate locomotives, move trains, switch cars as well as learn management skills. On June 13th Jeff Otto and I went to Monee, Illinois, to see the ICG's new ballast undercutter at work. The ICG was systematically reducing a long grade by repeated passes of the machine which clawed-out old ballast and subgrade from under the ties. While standing out there on their railroad ballasted with white limestone rock, we were struck by the vivid color contrast with our C&NW pink quartzite roadbed. It was easy to identify a C&NW line by the color of the ballast—when it was ballasted, of course. Jeff had a chat with Provo June 25th about the vital need for a ten year plan for replacing rail; Provo agreed that such a program was necessary. When things got better, Jeff would get his chance to develop a rail plan.

▲ The coal project was still in a swoon in July. Gotshall had an inquiry from Brown Brothers Harriman (the Union Pacific's major stockholder) about our 1974 coal traffic projections. I talked to Bob Lehnertz about the D&RGW's burgeoning coal traffic out of Colorado (July 9th).

Lock and Dam 26 came up at this time. The Corps of Engineers were planning reconstruction and enlargement of this facility located across the Mississippi River near Alton, Illinois. Barge interests had been pressing for a larger lock for years while the railroads had vigorously opposed the project. The Western Railroad Association

(WRA) solicited our help in opposing this project. I met with Paul Gleim of WRA on July 11th to get background information.

Gus Malecha and George Hansen of operating must have thought Bob Fried and I needed a break because they invited us to join them on an expedition to Wisconsin Rapids, Wisconsin, July 14th. There was a proposal on the table to use the Milwaukee Road from Necedah to Port Edwards to serve the paper mills and get rid of our worn-out trackage. On the 15th we met some Milwaukee Road men and hi-railed on their railroad from the undercrossing at Necedah to Wisconsin Rapids on pretty good track. Later we drove to Adams and inspected Adams yard under a very hot sun. The yard trackage was describable in one word—bad. A lot of benefit could have been achieved by simply replacing the many missing track bolts which would have helped extend the life of the worn-out rail. On the 16th Bob and I drove over to Marshfield while the others went back to Adams. We inspected track at Granton, Merrillan, Warren and Necedah before returning to Adams for long discussions about train service—or lack thereof. The next day we all went back to Chicago with stops at Oxford, Dalton, South Randolph and Clyman to see if existing sidings could be extended. I noted that the crossing diamonds at Clyman had been removed. After meeting Gus and George at Jefferson Junction, George headed for Milwaukee while Bob and I went on to Janesville.

On July 21st Bob Fried interviewed at the ICG for a job. He was fearful that the C&NW either would die under our feet or his job would be abolished. Freight revenues were still poor with no improvement in sight. On July 28th the big 4-8-4, Reading #2124 operating as America's Freedom Train #1 derailed on the sharp curve at Canal Street on the Milwaukee Road right under the C&NW's approach to its Chicago Passenger Terminal. The locomotive was carefully rerailed without damage even though a wrecking crane could not lift the big engine underneath the C&NW viaduct. Freedom Train #2 represented by Southern Pacific's #4449, a 4-8-4, operated on the C&NW from Proviso to Crystal Lake and, on August 14th to Green Bay, Wisconsin.

The morning prayer session on July 30th was more than interesting—it was an event. Last April personnel had conducted an attitude survey during the depths of the railroad's worst financial times and reported the results at this meeting. The lowest ratings went to R. David Leach's area of responsibility—systems and our CIE group. Bill Zimmerman's programmers were the worst of the worst. We could have told management that without a survey. A meeting was called August 7th to figure out how to better integrate Zimmerman's malcontents into the life of railroad. The key factor in this whole problem was simple—programmers were not railroaders first; they were more oriented to their craft than to the company. It was true all through the rapidly developing computer industry. Programmers possessed little company loyalty; challenges presented and solutions achieved were the important things. Leach could have managed a happier group if he had lightened-up his suffocating managerial style, improved the working environment, utilized free choice of hours, provided better access to the computer and stop micro-meddling in their work at every turn. Life in Stalag 520 did not materially change after the attitude revelation. Dave Leach presented an attitude improvement plan to Provo August 20th. Five weeks later Jules Eberhardt, our leader, was called into conference with Leach and Provo and was told that our small group was too passive. If so, it must have been Howard's industrial engineering bunch because I had been accused of being too pushy. Personnel brought up that policy #19 thing again. I had consistently refused to sign my intellectual inventions away. Dave Leach again demanded, amid dire threats, that I must sign the policy. I again refused and said I had the advice of counsel in this matter. At my annual evaluation October 17th, I received a just average rating which, I was firmly convinced, was directly related to not signing that hateful policy. The fact that I had launched a huge project that would radically change and enrich the railroad did not alter the unfriendly atmosphere of the moment. My 28th anniversary with the company was July 31, 1975 and I was 49 years of age. It was obvious to me that my career under Dave Leach, Rich Howard and Jules Eberhardt was about at an end. I needed two more years to nail down the 30 years needed for full retirement, with an additional 16 years to reach the traditional age 65 threshold. A careful review of the C&NW pension plan then in effect revealed a weak spot—there was no age requirement, just the 30 years of service. My outside income was holding up very well, if I could hang on for two more years at the C&NW, perhaps this coal project might come back to life and make an interesting finale to my career. As for that odious policy #19, my attorney had said that it probably was unenforceable even if I did sign the thing. I was now fairly certain that the uproar of 1967–68 had been largely forgotten by now. After due reflection, I decided, firstly, that the risks to my private business were small and, secondly, to make my long term plan for "retiring" in two years more workable, I signed the hated policy on October 21st. I was told that I was the last officer to sign. There were never any repercussions from that signature and my profitable geographic sidelines continued until I sold that business in 1980.

Freight revenues in August looked a little better. My wife and I took a three week vacation in Alaska, British Columbia and the Yukon. On September 10th we rode the narrow gauge White Pass and Yukon Railway—all 111miles to Whitehorse. Soon after this trip the railway closed its line north of Lake Bennett to passenger service when the ore traffic disappeared. When I returned to the office a week later, nothing had changed. Some of my old jobs resurfaced. As feared, the station number consolidation list (CLASP) had fallen into disrepair and systems was now aware that they were churning obsolete information because there were increasing numbers of reports without any data attached making some of the reports look like Swiss cheese. I was asked September 19th to again take-on responsibility for the consolidated station file that Dick Zogg, Leach's assistant, had failed to maintain despite direct instructions to do so. Dick never got the message that maintaining that file was important; he just ignored the whole thing because it was beneath his dignity. Another past project was that waycar study. On October 22nd an update was requested, but all my old work papers had been lost. Over the next few days I put together a plan for tracking the errant cars as they wandered around the system. Knowing where a particular car had been in the previous 24 hours did not solve the problem of having a suitable car ready to go tomorrow. On November 4th I took a quick trip to Proviso and looked at the cleaning track. It was as bad as ever. Those pool cars took an unbelievable amount of abuse from the crews. The study dragged on through November and December as collecting movement information proved to be a time consuming chore especially since some divisions did not properly report movements of these cars. Gradually, we could see a pattern evolve that allowed us to track the strays. The best waycars seemed to gravitate to the Central Division where they ended up on branch line locals while the worst ones ended up in pools where they remained a cause of complaints by the crews that had to use them. There was supposed to be a big discussion about the waycar situation at the morning meeting January 6th, but the huge snowstorms in the west pre-empted the conversation. I finished my draft report on the 7th; Jules Eberhardt approved in on the 9th, but Leach was still sitting on it on January 21st. Leach, it seems, wanted to add an alpha character to the waycar number to make it easier to track in the CARFAX reporting system, but couldn't afford to redo the programming to accommodate the larger number.

By September 22nd my projects all seemed to be dormant if not dead. I fell back on my time filler—coding track for TPC and pushed ahead on the Union Pacific as I had a fresh supply of track charts come into my "library". I was now firmly convinced that there would never be a single coal train come east from the Powder River over the decrepit Shawnee- Fremont railroad. Last May 7th Jim Brower had estimated a cost of $21 million dollars for rebuilding tracks and bridges just to run the first train. Using that piece of junk railroad was a crackpot idea, and I was very depressed to think about it.

▲ THE LIGHT BULB SHONE BRIGHTLY ▲ My depression of September 22, 1975 vanished the next day in a burst of new energy. I am not referring to the deranged Sarah Moore's assassination attempt on President Gerald R. Ford in San Francisco on that date (my daughter's married name was Sara Moore), but to a flash of inspiration that turned the entire coal project around in a new, and ultimately, successful direction.

If we could not use the Shawnee-Fremont railroad, let us consider connecting the C&NW and the Union Pacific along the Nebraska-Wyoming state line.

▲ The next day I was the first customer at the Rand McNally store in Chicago to get the USGS topographical maps of that area from Torrington, Wyoming, north to our existing railroad at Lusk. I turned all my railroad skills into engineering that 55 miles of paper railroad connecting the two companies. It looked feasible on this scale of map (1:62,500) with no major problems that I could see. I had a very private discussion with attorney Lou Duerinck who thought it a good idea. But would it work? The UP's North Platte subdivision had not yet been coded in TPC, so I worked at that the next day and drew a line and profile drawing of the new connecting link. I showed Duerinck my rough draft describing the link on the 26th; he stressed that I must keep this very quiet. To protect the concept I named the new railroad PROJECT YELLOW. The name came to me from a history of World War I that I had recently read wherein the German high command before 1914 had developed the Schlieffen plan to invade France through Belgium and code named it Gelbe (yellow in German). Since this new rail link was also an end-around move and both the C&NW and UP locomotives were painted yellow, the name seemed appropriate. Lou said he would meet with our top law officer, Richard (Dick) Freeman, and then let me know what steps I should then take. On September 29th I glued 1″=2 mile scale county road maps together for use as a base map. The proof of the feasibility of the new route was demonstrated when I ran a TPC run over the paper railroad at the newly established maximum speed of 40 mph for a loaded coal train. This project consumed every minute of my time over the next few days. By October 6th I had finished the last exhibit for Project Yellow. Duerinck said Freeman was on vacation, but he thought so much of the concept that he would talk to Provo the next day. To protect my departmental status I showed the material to Dave Leach on the 6th; he actually seemed impressed. (Ten days later he would give me that awful review. Something did not compute, as they say.) Provo invited Jules Eberhardt, Lou Duerinck and me to meet with him on October 7th at 10 a.m. He grasped the significance of the new connection at once and seemed to be somewhat gleeful at being one up on Union Pacific's president John Kenefick and Burlington Northern's Bob Downing and Lou Menk. Keep it quiet was his order.

▲ The new connection shifted emphasis for reviewing other future coal routes. I switched my TPC coding to the BN's lines south from Northport, Nebraska, October 8th. On the 10th I coded the Union Pacific west of North Platte, Nebraska, and finished the BN south to Brush, Colorado. All of these lines were to see heavy coal trains in the future. The UP lines were done October 15th. Project Yellow remained a deep, dark secret for the time being. At the morning meeting October 10th Provo had just returned from a trip with John Kenefick of the UP. He said we could expect an increase of traffic through the Council Bluffs and Fremont interchanges.

▲ The whole coal project and Project Yellow slid into the background for the rest of 1975. Fred Gerber of ARCO stopped by October 16th to be updated on events. Project Yellow remained a secret. At the end of October Provo was in Washington explaining to the ICC where we intended to get the money for our half of the joint line project.

At the management conference November 25th, the company was characterized as being in a holding posture with no prospects for attracting capital and little chance for improving the physical plant. There was some comment that the company might not continue to be a railroad in coming years. I was one of the few that believed that we could change that forecast if only everything came together in the right order. Other little incidents included a wreck at Elgin, Illinois, when children threw a switch under a train on November 6th—they were caught. There was another strike threat on December 3rd. I was assigned to protect a job at 40th Street yard at 4:30 a.m. Luckily, the strike was settled that evening. I set up a PACT request for Rollins W. Coakley of public affairs on December 11th for computerization of commuter fare schedules. Bob Million of roadway machine maintenance was fired December 18th. It was just as well that Bob Milcik's good ole pal was dismissed as he did not have much to contribute to the railroad except endless stories about hunting and fishing. Ed Christie of labor relations had his company car stolen in the nightclub district of North Rush Street and had a lot of explaining to do. December floated past without much activity. Bill McGovern and I boiled down a Bob Fried 50 page treatise on the Butler-Wyeville train operating problems to a 9 page summary (December 23rd). The important point of Bob's study was that the passing sidings on that stretch of railroad were too short for the length of trains operated. When the report was presented to division management, the dispatchers were visibly pleased because it clearly demonstrated that the poor train operations were not entirely their fault. Fried was learning, however, that writing up a railroad study was different from a school assignment where one had to show all one's work in arriving at a conclusion.

Project Yellow was ready to spring into life when the human actors had learned their parts. There was a lot of engineering, planning and negotiating to be done. I was primed and ready to roll.

30

YELLOW IS A PRIMARY COLOR, 1976 . . . ▲ Ray Gotshall came to me on January 2, 1976 and asked for a private briefing on Project Yellow. While I was not sure where he had picked up the information about this secret project, Ray was one of the insiders so I gave him an overview of the new connector railroad in eastern Wyoming. On January 12th the U. S. Supreme Court agreed to hear a Wyoming anti-mining suit and the next day dissolved an injunction against ARCO and Kerr McGee. The coal project was on again. Provo issued a public statement that we may yet build the railroad.

Given that I had a big shiny project to work on, all the little distractions in January seemed unimportant. As a manager, however, I had to deal with the smaller problems too. There was that on-going task of screening-out obsolete stations. After seven years one would have thought that someone else would be doing that detailed task, but it ended up in my lap again. Jerry Iwinski, now the engineer of bridges, wanted help with bridges—just what he thought our group could provide was not clear. In Chicago the Rockwell Line's gang problems came up for

---

▲ As noted before, this symbol will be used to denote material relating to the Western Coal Project.

discussion. It seems that young hoodlums from disadvantaged minorities were attacking freight trains stopped on this west side line and robbing from the cars. The railroad traffic to and from Wood Street was increasingly trailers-on-flatcars (TOFC) and was growing by leaps and bounds. Congestion at Wood Street caused trains to be held out on the approaches and the street gangs were taking advantage of the situation. I asked for police reports on the incidents to determine specific locations and frequencies. I wrote a report February 4th with recommendations after reviewing them with Bob Milcik and his people. (I do not recall if I advocated razor wire and submachine guns or more extreme measures). The Chicago and Illinois Midland Railway was again the subject of a meeting between Provo and the president of Commonwealth Edison on January 26th. Lou Duerinck suggested that C&NW should acquire the Green Bay and Western—for reasons now obscure. I began coding the Milwaukee Road's tracks from Chicago to Council Bluffs as it possibly could become a competitor for eastward coal movements. The old track charts and books left over from the merger talks years ago proved to be extremely useful for this new application. My library stock continued to expand as more material was secured. Larry Provo gave our small group a vote of confidence on January 15th and urged us to do more. I lost another good assistant when Bob Fried elected to attend trainmasters' school in April. Jules and I asked Bill Clark of personnel to find some candidates for us (March 25th). Bob graduated from the class and was posted to Mason City, Iowa in 1976. Early in 1977 he was to relocate to Oelwein, Iowa, but instead took a job with CONRAIL in Philadelphia.

The Chicago Transit Authority had a bad accident in which hundreds were injured at Jefferson Park January 9th. On the 29th negotiations resumed between the C&NW and the Regional Transit Authority (RTA) for purchase of our suburban fleet.

▲ On January 27th we learned of a new coal train movement destined for Texas. This was a sign that the coal project was slowly reviving. Our library did not have much available for the Texas region so I called Harold Gastler, now vice president of operations for the Missouri-Kansas-Texas at Denison, Texas, January 30th, and asked him for a set of their track charts on the route of the proposed coal movement. I agreed to furnish a copy of the TPC run in return (and did so April 8th). His material arrived February 5th. From another source I obtained the Fort Worth and Denver (FWD) and Colorado and Southern (C&S) track charts from Denver to Fort Worth which opened our eyes to the competition this Burlington Northern line might give us on Texas movements. The C&S-FWD lines crossed the eastern outwash from the Rocky Mountains which gave this railroad a profile that resembled a washboard. Coal could be moved but the horsepower to tonnage ratios and fuel consumption were much higher than anything we faced. On January 18th a new proposal involving Burlington Northern and Illinois Terminal service to the Union Electric Company at St. Louis arose. I ran a rush TPC job on this combination for Ray Gotshall and met with him and Fred Yocum on the subject (January 23rd).

On January 28, 1976 I was offered a job by Thomas Hodgkins of Distribution Sciences, Inc. Thanks, said I, but I am into a big, important project that I wish to see either completed or dead. Tom was understanding. Over the next few years, he continued to renew his offer until, at last, I accepted and went to work for him July 1, 1980. I was happy to know that I had value in the job market. At this point I still needed another 18 months of C&NW service to reach my goal of 30 years to establish my planned early retirement.

The railroad's big effort to decentralize decision making by establishing division managers was not working as planned. The divisions were becoming independent little kingdoms. On February 6th there was considerable discussion that the concept was failing. The month slid along with its usual storms resulting in soft track and snarled train operations. On February 24th a major structural change to our office culture was proposed—elimination of our secretaries and stenographers to be replaced by a centralized word processing center. There was a lot of resistance to this practical proposal. Confronted with this radical change the attitude of the old line managers could be expressed in one phrase: no one can handle MY work as well as MY secretary. At the morning meeting of February 19th we had two guests from the Department of Transportation (DOT). They announced that they had $600 million to spend on railroads—we could have used all of it. Jeff Otto gave them his presentation on the rehabilitation plans we had for rail and tie replacement—if we could get some funding. And that is what eventually happened; the C&NW became very successful at tapping the Government's till.

A big nasty project involving Car Fleet Management was not going very well. Estimates that a year's programming was required to get anything useful from the program discouraged the operating department. I asked Bill Zimmerman of systems if the traffic data could be put up on line using the Virtual Memory Operating System (VMOS) terminals giving access to the data from many scattered locations. He was shocked—give live unfiltered information to users? It was so frustrating to know the data were there but could not be accessed in real time. Bill said that such an approach was not within the capability of our computer systems. We hammered away at the car problem again in a meeting March 1st with Bill Zimmerman (systems), Tom Harvey (car service), Jim Thompson and Gus Malecha (operating) with emphasis on using current reports which were incomplete and far too late to be of much good for planning car distribution. I proposed the VMOS terminals again and Zimmerman said nothing doing. I am not sure whether the technology really was not available or that Bill Zimmerman just did not want to get involved. Bill did demonstrate a VMOS terminal on March 19th to Tom Harvey who wanted it right away. As an aside: VMOS was one of the first applications stemming from the new open architecture of the 3600 series of IBM computers. In those early days when transistors were replacing the old vacuum tube technology, VMOS was a jump into the future. On February 27th CNWT stock was quoted at $6\frac{1}{4}$–$5\frac{3}{4}$, the highest in two years. My investment in stock still sat at $2.25 per share so I was well ahead at this point. On May 14th the stock was essentially trading in the same range. Bob Conley, my crusty old friend from the Galena Division days, retired at the end of February. Shortly after retiring to his country place near Necedah, Wisconsin, he suddenly passed away.

▲ On March 1st Tom Brooker, a director of the railroad, was named to head the coal committee. The coal project of 1972 was quite different from the one we were trying to assemble in 1976. There had been a series of project leaders to suit the program in its several incarnations. I had been one of the first designated leaders, Bob Sharp of marketing had a short reign and now Brooker was to be the man. I never met him and he never directly asked for details from us. Ray Gotshall of marketing came to us on March 2nd for materials relating to the project. I furnished Wyoming maps even though I was slightly miffed at being set apart from the project—again. I concentrated on my TPC analyses of the MKT and other southern railroad routes to order to prepare route maps and fuel comparison tables which I intuitively felt, would be urgently required in the near future.

▲ Among the new maps was one I prepared showing the Project Yellow connecting line to a point on the C&NW called Node. Node in reality was just a small wooden store building with a tiny post office on the south side of U. S. highway 20 about 8 miles east of Lusk, Wyoming. No longer a railroad station, Node was just a convenient name on the map in that empty country to use for the junction where the new line would turn south. Later, when the design was refined, the junction shifted slightly to the east. (On January 17, 1979 I named that point Crandall for my friend Milt Crandall, chief of motive power, who had died of heart failure aboard a stalled diesel on a snowy night that month. Jim Zito had tears in his eyes when I showed him the map with Crandall on it). The new connector railroad was projected to run 56 miles south through country uninhabited save for a few scattered ranches. There were no utility lines and the few roads were of native dirt. To the west were the broken remnants of foothills. To the east at the Nebraska state line the land again became broken and rough. The corridor we were to use was an open, relatively level expanse several miles wide that had once been a migration route for long-gone buffalo herds and the gateway for the historic Texas Trail on which long horned cattle were driven north. A Wyoming governor and U. S. Senator, John B. Kendrick, remarked in his memoirs that "I do not remember . . . seeing a wire fence between Fort Worth and the head of Running Water (*near Lusk*)." In 1976 there were wire fences in this lonesome prairie, but not much else. One Goshen county road, #159, connected Torrington on the south to Van Tassell. Drainage features were minor seasonal drains, the only river to be crossed was the North Platte at the south end of the project. We would need a lot more detail before the railroad could be built, but for now, we had a rough reconnaissance for a rail line that looked extremely favorable. If we only had some money.

▲ On March 24th I gave Ron Schardt of finance, background material on Project Yellow and some material on the South Omaha cutoff from last year. Brooker was sure taking his time doing anything with the coal project. R. S. (Bob) Sharp, strategic planning and once leader of this coal thing, finally made time to be briefed on the project (April 16th) and then got all excited by the prospects just as I was leaving for vacation on a flight to Paris. After my return May 11th, I found his enthusiasm had cooled because Larry Provo was telling people that we could not finance the coal line with the BN much less the 'secret' connector line. We had one year from May 17, 1976 to start the coal line project. Jim Brower reported that the Burlington Northern had commenced construction of the joint line on May 17th. Duerinck said on May 19th that the BN and C&NW had agreed to postpone the C&NW's payment of construction costs until December 1977, or until a new financial agreement could be negotiated. The coal project was not dead, but it was far from well. Meanwhile, Ray Gotshall was flirting with a daft idea to load coal in trucks and haul them east on TOFC cars.

While the coal project ebbed and flowed, there was a lot of other activity on the railroad and in our consulting group. I heard at the March 4th morning meeting that a railway electrician had been caught stealing the copper from the dismantling of the old direct current electrical system in the Chicago Passenger Terminal.

On March 17th I was saddled again with those dratted annual line density statistics. I had hoped that systems would have completed the PACT request I had submitted over a year ago to let the computer do the data collection and tabulation, but the legal deadline of May 9th would be here before the programmers had even parted their hair. Despite taking a short vacation to trip to California, we made the deadline. At that point I found that the chief engineering draftsman had made several large errors in the 1974 and 1975 maps. All through July our programmer, John Breider, struggled to get the program finished for the 1977 season. The 1975 statistics showed that a full third of our total traffic moved over the east-west main lines.

On April 8th Fred Yocum was sent to the Iowa Division as division manager; Schwarz was sacked because of an alcohol problem.

Jeff Koch, whom I had met at Kansas City, came to the Illinois division. The assistant division manager-transportation at Milwaukee got the boot. On April 9th I interviewed John Dudlak for Bob Fried's vacant job, but Bob Sharp charmed him into joining his marketing group. That same day I got the plans for a new coal-fired electrical generating station at Pleasant Prairie, Wisconsin. The plant was to be located between the C&NW and the Milwaukee Road main lines just north of the Illinois-Wisconsin state line which made an interesting exercise in designing trackage that came off the C&NW and yet blocked service from the Milwaukee. On September 3rd I worked out a TPC comparison of rail service to serve the plant which demonstrated clearly the C&NW's route was far better than the Milwaukee's.

▲ All through this commotion I used every spare moment to push the TPC coding through the railroads that could be involved in moving Powder River coal. I finished the ATSF-DRGW joint line south of Denver in May so that we had the BN's entire railroad into Houston, Texas, covered.

Dick Setchell, superintendent of stations, finally decided to get serious about cleaning out those obsolete stations and sent a bright woman over to learn about the maintenance of the consolidated file. At last someone was paying attention. That same day, May 17th, the Illinois division had a $1.7 million wreck at Glen Ellyn, Illinois. Liquid fertilizer (anhydrous ammonia) escaped from wrecked tank cars into the trackside ditches which led to the city's storm drains which in turn ran into Lake Ellyn behind the high school killing all the fish. The cost of the wreck was revised the next day to $2.8 million. On May 18th Jules handed me another task—sort out the pulpwood and chip car problems. That job had to wait until I got a new assistant on June 16th. Dave Nowicki was hired as summer help over Dave Leach's initial objections about hiring anyone because of expenses. Dave Nowicki eventually inherited the pulpwood and chip project and finished it August 13th. I had flown to Denver on May 25th to attend a Transportation Research Forum (TRF) meeting hosted by the truckers' tariff bureaus. It was a waste of time, however, and I came back early. By the end of May the party was about to open on Project Yellow.

▲ Zev Steiger, a rather new addition to our law department, was a clever young man who would develop into a real asset for the company. On May 27th in response to his inquiry, I gave him the presentation on Project Yellow, the secret plan to connect the C&NW and the Union Pacific in eastern Wyoming. A few days later on June 2nd, Larry Provo asked "to be refreshed" on the subject; but, first Dave Leach had to review the show before the 2 p.m. showing to the president and Richard M. (Dick) Freeman, the head of the law department. Provo and Freeman listened intently and seemed impressed. I got the distinct impression that the entire coal project somehow was floating in mid air. On June 7th Jim Brower came up with a cost estimate of $120 million for the rehabilitation of the 519 miles of old C&NW line from Shawnee, Wyoming, to Fremont, Nebraska, a six fold increase over his first estimate three years before. Jim completed his presentation to Provo and Freeman and then was told about Project Yellow which would eliminate the need to upgrade some 460 miles of the existing railroad. Jim, a bit stunned at the notion, quickly latched onto the concept and asked about details of the proposed connector. It was time to get some serious engineering underway. I was asked to get copies of the latest mapping for a detailed reconnaissance and location survey. I knew the USGS was working on new mapping of that area, but the work was far from ready for publication. The aerial work had been completed and the drafting was in progress. On June 19th I arranged for Don Groves, our Denver sales representative, to go out to the USGS office and secure copies of the maps in progress. The government people were most cooperative and made the maps available even though the lettering of the names and legends had not yet been applied. We did not care if the names were not on the maps, we just needed the topographic contour lines, meridians and parallels. By June 22nd the maps had not arrived—Emery Air Freight had misplaced them. They were found and delivered three days later. Since Brower was not in town, I had an opportunity to go over them and, to my relief, did not find anything sinister in this 56 mile stretch that would interfere with easy construction. This was a locating engineer's dream, but that was not my job, it was Jim Brower's.

▲ While waiting for the maps to arrive, I worked with Bob Sharp and Ron Boesen to develop a coal marketing plan (June 8th). The next day Ray Gotshall asked me to attend a meeting with Great Lakes Dredge and Dock Company who was interested in building a rail-to-water transfer facility. That afternoon two Union Pacific men showed up for a demonstration of the TPC program. The next day Walter E. (Bud) Braun, vice president-traffic, asked about Project Yellow—the secret was leaking. On June 14th Bob Sharp headed to Omaha to meet with the Union Pacific to acquaint them with the details of Project Yellow. On the 17th I heard that John Kenefick, president of the UP, and Provo had agreed to do a complete investigation of the proposed connector line and that our Bob Sharp and UP's Tom Graves would coordinate the program.

▲ Jim and his wife, Lee, worked at their kitchen table on the bicentennial Fourth of July to lay the line of railroad across those half-finished government maps. It went so easily that the line and profile were developed by late afternoon. Jim started the new line west of Van Tassell on the existing C&NW. The new line swept around a gentle righthand curve (there was lots of room), crossed U. S. 30 which ran parallel with the old railroad and headed south through the open, and empty, ranch land. The new line could be characterized as being a series of long tangents connected by easy curves. The most prominent land feature along the route was an isolated mesa called Spoon Butte about midway. Two thirds of the way south the new railroad crossed the Wyoming-Nebraska state line. Just before approaching the north bluff of the North Platte River, the line passed through a rippled area that looked, possibly, like ancient sand dunes. At the bluff the new railroad line crossed first U. S. highway 26 and then the Burlington Northern's single track railroad between Guernsey, Wyoming, and Scottsbluff, Nebraska. Fortunately, the proposed new railroad was on a natural elevation above the highway and BN so that overhead bridges were a practical solution with little earthwork required for the north approaches. The only major bridge structure, the North Platte River, required a substantial but not heroic structure. Just past the bridge the line curved around gently to the east to connect with the Union Pacific's North Platte subdivision near an old sugarbeet loading station called Joyce. The new railroad was indeed feasible and all downhill for the loads. Grading was minimal, culvert work simple, and construction of the several overhead roadway bridges presented no problems. Jim Brower finished the rough maps on July 7th and I translated the profile into the TPC program for test runs.

▲ Tom Graves of the Union Pacific came to my office June 28th to see my slide show on Project Yellow. He took some of the materials I offered and promised they would have a decision within 30 business days. He emphasized the extreme secrecy of this project (it was leaking out all over the C&NW). On July 16th Graves called Bob Sharp. A meeting was being held by the Union Pacific. They wanted to make a reconnaissance of the area. Dick Peterson of the Union Pacific came to see the Project Yellow material on August 5th. Suddenly, our coal line coordinator on the joint C&NW-UP project, Bob Sharp, announced July 16th that he was leaving the C&NW to become president of the Detroit, Toledo and Ironton Railroad; his last day was August 14th. What a time for a designated key character to leave the play—just as the curtain was going up. I would hear more from Robert A. Sharp in the future. I worked with Jim Brower on August 6th to get lines on maps for the Union Pacific. Three days later Larry Provo returned from a European vacation. Now that everyone was in place and primed with good information, fate stepped in. On August 17th Larry Provo was hospitalized for tests and had minor surgery on the 20th. Provo, a heavy smoker, had a biopsy taken and it came back positive—he had inoperable lung cancer. James R. Wolfe became president of the C&NW August 30, 1976. Provo was moved up to chief executive officer overseeing the money—accounting, finance, systems and personnel. The cancer had spread to his brain by October 1st; I last saw him on the 7th and he looked ghastly. Larry S. Provo died October 19, 1976 at age 49. His wake was held on the 21st. Attendees included twelve from the Union Pacific. Project Yellow had hit a bump in the road and seemed to be stalled.

Turning back to summer 1976, track changes in Elmhurst, Illinois, at HM tower were begun on June 15th to open up the area for construction of a long awaited highway underpass east of York Street. On June 22nd Ron Schardt became treasurer of the C&NW. On the 25th I ran a special TPC comparison for M. S. Reid on an alternate route south of Mason City via Eagle Grove and Marshalltown, Iowa, in connection with a branch line abandonment case. June 29th amidst some very bad summer storms, a big freight derailment occurred on the Illinois division at Flagg west of Rochelle. On June 30th Dave Nowicki and I visited the Cedar Lake shops at Minneapolis in connection with the covered hopper car study I had assigned to him. These were the old M&StL shops and were obviously not the best place for motivepower or car maintenance as they occupied a large level area below a bluff upon which a rather wealthy residential area was located. (The C&NW was to abandon Cedar Lake in 1983 and raze the buildings the next year). Dave and I drove south to Mankato, Minnesota, to look at the yard facilities there, stopped at Manly, Iowa, and then went on to Mason City. The hopper car problem involved proper cleaning to make them ready for the next load. Potash from Canada had to be washed out of the cars before grain could be loaded. I spent the morning of July 1st with Dallas B. Carlisle going over his plans for washing both potash and cement hoppers. We also discussed an alternate operating route south from Mason City via Eagle Grove. Dave and I drove over to Eagle Grove and looked at the condition of the railroad and tried to understand the problem of yard capacity there. On July 9th at East Chicago, Indiana, Dave and I watched GATX wash hopper cars with modern equipment that we could not afford to purchase. Back in Iowa on July 26th diesel locomotive #6912, an SD-40, dove into the 14 foot deep Wapsie River near Wheatland killing the engineer whose body was not recovered. There also was a head-on collision on the single track near Kennard, Nebraska that day.

Another little project which involved replacing the doubleslip switches at Nelson, Illinois, where the southern Illinois district began, was dumped onto our group. On July 20th I took Dave Nowicki to Nelson to survey the train operations at the interlocking plant. We developed a plan to identify the different movements and their frequency through the turnouts so that only crossovers that were really needed could be installed. Jules Eberhardt accepted my recommendations July 27th and I went over the plan with division engineer Dave Boger before making a general distribution of the report August 2nd. Dave Nowicki and I made a trip to see the BUC (Ballast Under Cutter) in action at Hahnaman, south of Nelson only to find it wasn't working because some part had broken. In April we also were faced with another track problem—what to do with old tie stubs? The new tie cutters chopped old ties in track into four pieces—two inside and two outside the running rails. The stubs were dangerous to leave lying about the right-of-way. The adjacent farmers didn't want the short pieces; they preferred the old full length ties for fence posts or crib walls. Burning or burying them was out because of environmental concern about releasing creosote, a suspected carcinogen, into the atmosphere. I never came to a conclusion about how to solve the disposition of this material. Today, old railroad ties cost more in garden centers than a new tie cost 50 years ago.

We got another new assignment July 29th. The Iowa division wanted CTC (but didn't get it). The next day our 1972 study on the Des Moines-Kansas City corridor was revived—again—and renamed the St. Paul-Kansas City project. Tom Juedes was now working with our group. I assigned him and Dave Nowicki to developing a plan for installation of centralized traffic control (CTC) from Fremont, Nebraska, to Missouri Valley, Iowa. That installation would have to wait another fifteen years, or so, when the Union Pacific would invest the necessary capital.

While my assistants were working on the CTC project, Kurt Ramlet from the tariff section again sought our help on generating a new mileage master tariff. This was a serious problem that could not be ignored as this tariff was a mandated legal requirement to provide the correct short line mileage for those freight shipments moving on mileage rates. The mergers of 1960 and 1968 had destroyed the validity of the old tariff by creating a bewildering number of alternatives through the tangle of branch lines. The mileage rule quite simply stated that the shortest aggregate of miles through a railroad's systems of lines was to be used in calculating mileage-based freight charges. The most obscure, weedy, seldom used intersection could be the key element in determining the short line mileage. With the aggressive branch line abandonment program, the short line mileage was in a constant state of flux. We had to have a network model for calculation of this tariff. The Segmented Railroad File (SRF) developed in 1969 became the base for the short line mileage tariff. It was quite interesting to see which little pieces of obscure lines were identified as key segments when determining the shortest distances between pairs of stations.

In finance Sue Frehleng, the misanthropic (yes, she seemed to

NETWORK EXAMPLE

Development for Short Line Mileage Tariff

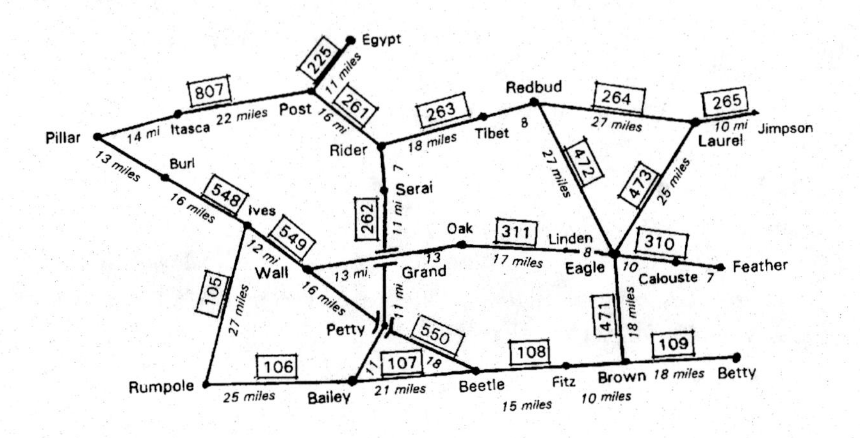

Segment Definitions

		Miles	Intermediate Stations
105	Rumpole - Ives	27	
106	Rumpole - Bailey	25	
107	Beetle - Bailey	21	
108	Beetle - Brown	25	Beetle - Fitz, 15: Fitz - Brown, 10
109	Brown - Betty	18	
225	Post - Egypt	11	
261	Post - Rider	16	
262	Rider - Bailey	40	Rider - Serai, 7: Serai - Grand, 11: Grand - Petty, 11: Petty - Bailey, 11
263	Rider - Redbud	26	Rider - Tibet, 18: Tibet - Redbud, 8
264	Redbud - Laurel	27	
265	Laurel - Jimpson	10	
310	Eagle - Feather	17	Eagle - Calouste, 10: Calouste - Feather, 7
311	Eagle - Wall	51	Eagle - Linden, 8: Linden - Oak, 17: Oak - Grand, 13: Grand - Wall, 13
471	Brown - Eagle	18	
472	Eagle - Redbud	27	
473	Eagle - Laurel	25	
548	Pillar - Ives	29	Pillar - Burl, 13: Burl - Ives, 16
549	Ives - Wall	12	
550	Wall - Beetle	34	Wall - Petty, 16: Petty - Beetle, 18
807	Pillar - Post	36	Pillar - Itasca, 14: Itasca - Post, 22

Advancing computer technology began to be helpful in designing network models. The Short Line Mileage Tariff was one of the first applications to address the legal requirement for updating mileage calculations in the railroad's rapidly changing tangle of lines.

NETWORK EXAMPLE

Development for Short Line Mileage Tariff

Segment Connections

Bailey	106, 107	106, 262	107, 262		
Beetle	107, 108	107, 550	108, 550		
Brown	108, 109	108, 471	109, 471		
Eagle	310, 311	310, 471	310, 472	310, 473	311, 471
	311, 472	311, 473	471, 472	471, 473	
Grand	0, 0 (does not connect)				
Ives	105, 548	105, 549	548, 549		
Laurel	264, 265	264, 473	265, 473		
Petty	0, 0 (does not connect)				
Pillar	548, 807				
Post	225, 261	225, 807	261, 807		
Redbud	263, 264	263, 472	264, 472		
Rider	261, 262	261, 263	262, 263		
Rumpole	105, 106				
Wall	311, 549	311, 550	549, 550		

Example

Given: shipment origin: Itasca; destination: Linden
Required: determination of short line mileage for rating
Known: Itasca is located in segment 807, 14 miles from Pillar and 22 miles from Post
Linden is located in segment 311, 8 miles from Eagle and 43 miles from Wall

Chain 1:

At Pillar	Itasca	14 miles	connections are: 548, 807		
Search 548					
At Ives	Itasca	43 miles	connections are 105, 548	548, 549	
Search 105					
Search 549					
At Rumpole	Itasca	70 miles	connections are 105, 106		
At Wall	Itasca	55 miles	connections are 311, 549	549, 550	
Search 106					
Match 311			Linden 43 miles from Wall		
At Bailey	Itasca	95 miles	discontinue, destination found		
Sum	Itasca	98 miles			

Chain 2:

At Post	Itasca	22 miles	connections are 225, 261	225, 807	261, 807
Search 225		no connections			
Search 261					
At Rider	Itasca	38 miles	connections are 261, 262	261, 263	262, 263
Search 262					
Search 263					
At Bailey	Itasca	78 miles	connections are 106, 107	106, 262	107, 262
At Redbud	Itasca	65 miles	connections are 263, 264	263, 472	264, 472
Search 106					
Search 107					
Search 264					
Search 472					
At Rumpole	Itasca	105 miles	discard, miles in excess of chain 1		
At Laurel	Itasca	92 miles	connections are 264, 265	264, 473	
At Eagle	Itasca	92 miles	connections are 311, 472	471, 472	
Search 265		no connections			
Search 473					
Match 311			Linden 8 miles from Eagle		
At Eagle	Itasca	117 miles	discard, miles in excess of Chain 1		
Sum		99 miles			
Discard Chain 2 account miles greater than found in chain 1					

Result of network search: short line miles Itasca to Linden = 98 miles.

The network model required accurate definition of segments and junctions for the logic of the program to function properly.

hate men) consultant left over from corporate development days, announced her departure August 9th—some faint cheers were heard from her colleagues. She was especially tart with Jerry Groner, her superior, claiming that he couldn't manage his way out of a paper bag. Jerry found out about her leaving by accident as she had not bothered to inform him. Despite Sue's caustic comments, Jerry stayed on in finance and played a key role in Project Yellow when it recovered from its swoon and in the other 511 loan programs that were soon to come.

On August 10th I started work on the Iowa corridor study in which I compared the TPC results of running a standard coal train over all the combinations of railroads between Council Bluffs and Chicago. We had not quite finished the Rock Island, but the track charts and timetables were in hand. By September 8th that line was in the computer

Al Myles had moved from labor relations to the engineering department as a quality inspector while Bill Clark of personnel was appointed omsbudman for the company. J. W. Alsop in operating was immobilized again. C&NW stock slipped to the $5\frac{1}{4}$–$4\frac{7}{8}$ range. We were in another traffic slump, too. A coal train wreck at Morrison, Illinois, was another half a million dollars lost. All track maintenance gangs were laid off as of September 1st; there was no money available for repair rail. In our office Joann Tsoodle, our super secretary and project helper, was interviewed by Bob Russell for a job in his department (August 20th). Bill McGovern and Jules Eberhardt had been at loggerheads since August; by November 17th it was clear that Bill was to be fired. I felt sorry for gentle Bill, but he was in the wrong place for his slow moving talents. Both Rich Howard and Jules were determined that he had to go.

On August 25th James A. Zito took three Union Pacific officials on an aerial tour of Proviso and Wood Street in a helicopter. Ron Boesen of coal marketing invited me to visit the Northern Indiana Power Company (NIPSCO) electrical generating station at Wheatfield, Indiana. Steve Dyslin and Glen Cordesh also went along to see the coal unloading and handling equipment. The operators of the rotary dumper said that unloading a unit train of hoppers worked fine unless an odd non-rotary car got into a consist or one of the rotary coupplered cars in the string was backwards. When a non-rotary coupler was encountered, the whole production line had to stop. Rotary coupler equipped hoppers always had that end painted a different color from the body of the car which made for some very colorful train sets. The orange ends of a black train consist were quite handsome. If two painted ends went into the rotary dump together, the chances were that the other ends of the cars were mated to solid couplers, which did not twist very well. Constant vigilance was required to catch those in-wrongs before wrecking something.

President Jim Wolfe now wanted more office space and took most of ours—he was an aggressive neighbor. Revenues continued to be poor all through September. By September 7th Ed Burkhardt had replaced departed Bob Sharp as assistant vice president-marketing and August (Gus) Malecha, became avp-transportation. Jerry Iwinski, formerly engineer of bridges, was now avp-track maintenance. I took Tom Juedes to Green Bay, Wisconsin, to learn about boxcar distribution and cleaning on September 8th. There I photographed the SLSF GP-7s which we had recently purchased still running about in their former home road paint. On the 9th I drove Warren Mott to Fond du Lac to see the car cleaning facility there—it was disappointing to see how little had been done to improve the operation. We then went to Butler on the 10th to interview car service people. None of this work and running about seemed important to me as I was focused on the coal project whose many demands needed, in my view, close attention.

Vice president of law, Dick Freemen, asked for our comments on a preliminary proposal by the Railway Service Planning Office of the Federal Railroad Administration (FRA) August 31st for loan guarantees to railroads. Our principal comment was centered on omission of fuel consumption as a criterium. Hearings were held September 13th in the Federal Building in Chicago amid complaints of the short amount of time allowed for responses to the proposal. That Iowa corridor study I had begun months ago now became a central feature of our application for FRA funds. On the 14th more hearings were held, but only three witnesses testified. I was working furiously to finish the study. On the 16th I shared the still incomplete study with James A. Zito, now vice president-operations, M. S. Reid and James A. Barnes, chief engineer. There was still a gap in the old CB&Q and a covert effort was made to secure the missing charts. On

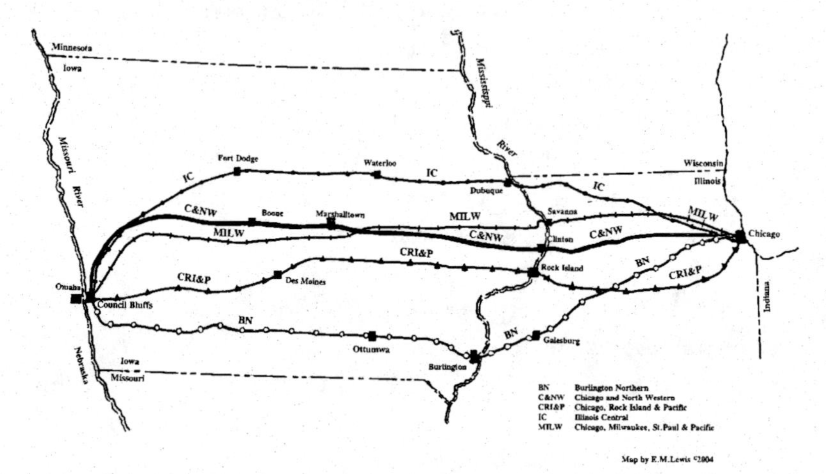

Five major rail lines across Iowa were rival candidates for Federal rehabilitation funds. One railroad would succeed in tapping that line of credit, one would fall into ruin and a third would just give up the route.

the 20th Lou Duerinck of law was briefed on the Chicago-Omaha study. From the completed study analysis the C&NW was clearly the most fuel efficient railroad from Omaha to Chicago by a substantial margin. Later in the month, on September 22nd, I expanded the corridor study to Chicago-Kansas City which meant that some new railroads would have to be drawn into the TPC database. The ATSF was the first to be finished on the 24th. That BN material I needed was still not in hand, but the grapevine was humming. I gathered in the N&W's Wabash line next and then the Rock Island and Soo Line. Some of the missing BN material appeared October 4th and was quickly absorbed into the TPC files. Christmas came early on October 15th when Doug Christensen showed up with the last Burlington track charts. There were so many parts of this treasure trove that I took most of the day just sorting them out. The grapevine had worked well and each participant deserved a "well done" even though many people must remain unidentified even at this remove.

Other events included an invitation September 15th from my old mentor M. C. Christensen to become engineer of bridges. That was a stretch; I didn't think I qualified but Chris insisted it was just an administrative job. Thanks, Chris, but not at this time. September 17th was Dave Nowicki's last day. He had done very well and I was sorry to lose him. The peripatetic Ed Garvey now announced he was leaving the C&NW—again—for a job with the railroad consulting firm of R. L. Banks in Washington, Ed and Kate moved to Annapolis, Maryland, and actually stayed there for several years. James J. Johnson, a fellow Elmhurst resident, was promoted to Gus Malecha's former job as head of quality control in operating. Another company wide scrap drive was launched in September. Tom Juedes was sent to the Lake Shore division, W. M. (Bill) McGovern did the Illinois, and I was excused. On the 22nd I met with Jim Barnes and Doug Oakleaf in engineering about a proposal to sell scrap ties and stubs to the Pyron Company who planned to manufacture reconstituted track ties using plastic resin filled with ground-up wooden tie fibers. We thought it was not a good idea because of difficulty in collecting the tie stubs spread all over the railroad. John Mruskovich of finance, one of Jerry Groner's people, asked for help with a joint facility study—we couldn't spare anyone, however. On October 6th I learned that Bill Zimmerman of systems had resigned to do what he always wanted to do—but no one seemed to know what that was. I attended his farewell dinner October 13th.

That Des Moines-Kansas City study of 1972 just would not stay dead. M. S. Reid advised me confidentially September 30th that Dick Freeman said he had never seen the study (which had been issued before he came to the C&NW) and that I was to expect further developments. I am still waiting.

Robert A. (Bob) Sharp, now president of DT&I, remembered what good results we had gotten from the TPC runs and now requested a special run for a piece of his new railroad and and parts of the old Erie and former Pennsylvania. He forwarded a heap of material that I had trouble linking together because of the widely different drafting conventions. I pleaded that I was extremely busy at the moment, but Bob, who was negotiating trackage rights, pleaded for a quick answer. Ken Schumacher, the former DT&I president, called on November 4th to inquire about the runs. Somehow I slipped them through the mill and sent the results to Sharp. Bob said he owed me a steak dinner—I'm still waiting.

On October 8th Jim Zito called Jeff Otto, Mike Arakelian, Ira Kulbersh, Tom Nixon and me together and created a new Track

Train Dynamics Committee, designating me as the chair. I could see that Dave Leach, the great purse watcher of our department, would have objections as this new group was going to need traveling expense money. The first trip came October 18th when I was invited to attend the grand opening of the AAR Test Track facility (FAST) at Pueblo, Colorado, on November 8–9th. On October 11th I began training Candace (Candy) Thomas from operating to run the TPC program. She was an Iowan from the small town of Traer and represented the "new look" in the company's effort to be a truly equal opportunity company. She was reluctant at first to get involved with the computer, but brightened-up considerably when she saw the ease with which Joann Tsoodle, our secretary, handled the operation. Before long she was adept as any of us, but the operating department remained a tough place for a woman.

▲ Ray Gotshall called me early in the morning October 4th with the information that Project Yellow was dead. Three hours later he called back to say it was very much alive. We had a meeting with Ed Burkhardt, Ray, Ron Schardt, and Jim Brower to either try and get organized on this project or quit kidding ourselves. There were too many outside events occurring to say the coal plan was alive or on life-support. We had a temporary new person working in our group, Dorothy Cranshaw. I put her to work coding Milwaukee Road trackage. Chris Sather of the Union Pacific came in from Omaha on October 20th to check our input to the TPC programs which had produced the Project Yellow runs. He was visibly impressed and came back the next day with questions about the fuel calculation module.

James R. Wolfe, now our president, brought with him a different style of management. At the morning meeting October 22nd a no smoking policy was instituted immediately which caused several attendees to stir uneasily. Revenues continued to be poor and a big interest payment was coming due. Jerry Conlon, vice president of public affairs. asked me about the Powder River By-Pass (he didn't quite have the concept right). I wondered where he had heard about Project Yellow which was supposed to be restricted information. I finished coding the mass of Burlington Northern material October 24th and did the Rock Island's main track from Inver Grove outside of St. Paul,

Minnesota, south to Albert Lea. Bob Kase at the Norfolk and Western was very helpful getting some old Wabash material that we needed (November 1st). On November 1st Lawrence F. (Larry) Kohn started with our group—another new boy to train in TPC coding as that is what we most needed to finish our corridor study.

▲ Wolfe did not seem to care about Project Yellow. When I asked him its status on October 28th; he replied, "That project was dead." "Why then," said I "are the Union Pacific people asking us so many questions?" He seemed surprised and said that he expected to see John Kenefick at an AAR board meeting the next day and would make inquiries. Chris Sather of the UP called on November 5th and said the project was definitely on with them and that its scope had been expanded. I passed this information on to Wolfe through Jules Eberhardt. Lou Duerinck opined that we were doing the right things. Ed Burkhardt and Tom Graves (UP) had a conversation about the C&NW's need for money. On November 11th Dick Peterson of the UP called—they were heavily into the coal project. I began to feel that my year's work was beginning to pay off. I pulled the corridor study together and wrote a summary report November 14th. The next day Jim Brower and I made a consolidated profile of the new connector line. Wolfe and Kenefick met on the 16th. John H. Ransom of the UP called wanting more materials; a day later he said they were ready to start our joint project. I was unaware there was a joint project as Wolfe hadn't said anything to us. On the 18th both John Ransom and Chris Sather called me to complain about the slow start; I was still mystified as there was no word from our front office. Ron Schardt, our treasurer, came in to ask about funding in 1977—how would I know? The next day our top financial man, John M. Butler, spent 40 minutes with me listening to my comparison of the Iowa corridors; John was looking for some answers to the riddle of the C&NW's poor operating performance. Dick Peterson of the UP called again on the 19th. This was getting comical; if there was to be a joint project, what were we supposed to be doing and who was in charge? The whole thing was still uncertain by the 22nd when I had two more calls from the UP. To fill in TPC coding gaps, I called Art Mennell of the Missouri Pacific to obtain more track charts for my growing library.

▲ After a short vacation visiting my daughter in California, I returned November 30th to find more urgent requests from the Union

Pacific for track decks as they were running their own version of TPC to check our results. Our data and theirs proved to be consistent. On December 2nd six business cars went west: three C&NW, three UP. The next day I learned from Maurie Reid that Provo had not told the directors everything about Project Yellow as well as some other activities which Maurie did not identify.

▲ Another coal user now popped-up—a coal gasification plant to be sited at Wichita, Kansas. That meant more segments of railroad to be jammed into our TPC library. Larry Kohn pushed track coding deep into the south over the Missouri Pacific which was slated to handle a great many coal trains. We added the results of the runs to our growing index file called UCTSIGMA. On December 15th the Union Pacific was reported to be completing their report to their management. That day their Bill Pietmeyer called asking about more details of the TPC program. We wished the UP would just acquire a TPC program like ours so that we could just send track decks to them. By December 27th we had over 2,000 TPC runs in our index file.

▲ I was starving for information about the coal project and wondered if I had been dropped out of the loop again. On a hunch I called Ed Burkhardt December 28th. He said the UP attorneys had been in Chicago all day on the 21st. The Union Pacific calculated that the capital investment for the new connector line, upgrading the old track from Shawnee to Van Tassell and the half interest in the new Burlington Northern line north to the coal fields would be $390 million. The return on investment (ROI) was favorable. Burkhardt was a little annoyed that Lou Duerinck had not included our group in the information flow—a typical C&NW habit of never involving those parties that knew anything about the subject under discussion. Attorney Zev Steiger cornered me a little later in the day and asked me what should be done. Of course I had some ideas; but, since I did not have the whole picture, I was reluctant to venture too strong an opinion. Ron Schardt, our treasurer, said the UP had ten people assigned full time to Project Yellow (they had adopted my code name). Here on the C&NW we were just muddling along with a ragged part time contingent struggling with poor communications and no direction from the top.

In November I had flown to Denver and changed planes for a short flight to Pueblo. Southern Colorado University hosted a big welcoming dinner November 8th for the inauguration of the Facility for Accelerated Service Testing (FAST). The crowd included railroad and government people, consultants and suppliers. The next day we were bussed 25 miles north to the FAST facility to see the new tracks, shops, test trains and meet the personnel. After the ceremonies and the group photograph, we all headed home. Our C&NW group met on December 16th and reported to Jim Zito on the 20th. Because I had not been in contact with Jim Zito for very long, I was pleasantly surprised to see how well he was handling his new duties as vice president of maintenance and operations. He was well acquainted with this track, train dynamics subject and gave our group some constructive ideas to work on. Zito's belief that good track meant good operations was right on the mark in my opinion. On the 18th Jeff Otto made his presentation to Wolfe on the crisis we were facing regarding the rapid wear of rail in our mainlines under the increasingly heavier car weights. We had another unexpected visitor on the 18th—Bill Gale, an ICC inspector, who insisted that this was just a social call. More importantly, the C&NW had decided to try and get some of that Federal funding provided under the so-called "four R" (RRRR) act. Gathering the huge volumes of material for this first application was not done very smoothly because the program was new to both the Government and to us. Sometimes the process seemed more like a Keystone Kops movie. We would get better at it next year.

Tom Nixon and I attended a Track Train Dynamics seminar at the Rosemont Holiday Inn December 1st. On the 6th I went to Washington for a Standard Point Location Code (SPLC) meeting at the ATA's building. The mood was fairly calm despite the controversies that had roiled the meetings over the past few years. The next day was the TDCC meeting where I saw Bill Allman, Al Wharton (DuPont), Elmer Matlox, Jack Neil (Hercules) and Nevil Black (Southern Railway). I came to Washington by way of New York after riding the *Metroliner* to the Big Apple. Jim Wolfe announced to us December 14th that a deal had been struck with the Regional Transportation Agency (RTA) for sale of our suburban fleet and service, but we were to keep it quiet until next week. Maurie Reid's branch line committee met on December 15th. He explained that the FRA rules regarding abandonments had changed with regard to filing applications and the kind of advance notification and mapping required. Alvera Hensler, Jim Wolfe's secretary, threw a Christmas party on the 17th and even included our group in the festivities. The best part of the year's end was an increase in pay to $26,100.

The year had been a wild ride of ups and downs, with many personnel changes, lousy financial returns, the death of a president and the breaking-in of a new one. We now faced the challenge of dealing with the powerful Union Pacific excited by prospects of a huge revenue from coal. The cash-poor underdog C&NW seemingly was unable to piece together their entry fee to the coalfields and still properly maintain its railroad and car fleet. In the CIE department we had increased our company's level of intelligence about our railroad competition and established a systematic approach for collecting and presenting that information. The next year would see us using all our carefully built-up intelligence to good advantage. The C&NW may not have survived without that valuable research into corridor economics which proved that its mainline from Chicago to Omaha was the superior route in terms of fuel efficiency and provided significant insights into the true operating costs and running capabilities of the railroad. The truism that you can't manage what you can't measure continued to be a guiding precept in our consulting work.

31

SMOKE AND MIRRORS—1977 . . . ▲ The new year began with no pause from the last. I concentrated on expanding the Train Performance Calculator database by coding Missouri Pacific tracks in Arkansas. The next day I delved into the intricacies of the new GP40 type diesel locomotive to determine if it would be more efficient in coal train service but soon found that the four wheel trucks did not provide enough traction at low speed. In the midst of our first fuel conservation meeting January 7th the Union Pacific called. They needed data on the BN-CNW-UP routes for a briefing they were giving their president, John Kenefick, the next morning. On the 10th they called again and wanted more data. I duplicated a Fort Worth and Denver track deck for them and mailed it to Omaha. On the 11th there were a growing number of complaints that there was no planning being done on the coal project—we were walking into a negotiating situation in our same old unprepared way. Ron Schardt, treasurer, reported that John Butler, vice president-finance, was very surprised at the high level of activity being displayed by the Union Pacific and had called Dick Freeman to join him to talk to Jim

---

▲ As noted before, this symbol will be used to denote material relating to the Western Coal Project.

Wolfe. It seemed strange to me to hear that our top officers were not communicating with each other on a project of this magnitude especially when we might be engaged in some intense sparring with another powerful railroad that just might become a partner in a giant enterprise. Marketing now came up with a new wrinkle; they needed a TPC analysis of Colorado to Chicago coal movements (January 12th). This time I called the Union Pacific's Dick Peterson and Dale Salzman to get some information. During that cold January I sat in my comfortable office and coded the ATSF west to Pueblo, Colorado—another potential coal route to be examined.

▲ On January 18th John Kenefick and Jim Wolfe agreed to proceed with Project Yellow—the new connector railroad in eastern Wyoming. Wolfe appointed a study group consisting of Ed Burkhardt (marketing), Zev Steiger (law), Jim Brower (engineering) and me (CIE). We suspected that there was a leak somewhere as C&NWT stock rose to $7½. On the 21st it rose another point and on the 25th, $9 was quoted.

At our January 6th Track Train Dynamics meeting Jim Zito expressed concern about the rash of derailments on other railroads involving the AMTRAK FP40 passenger locomotive. Since we expected to use the SD40 as our principal coal train engine and the FP40 used the same power trucks as the SD40, the committee was requested to dig into the causes for the derailments. Jim Wolfe came up with another hot project for us on January 10th—develop a reroute for the Kansas City line. Funny, how that old Des Moines-Kansas City study of 1972 kept reappearing after being officially declared dead five years ago. On January 12th I completed a draft of the Federal Railroad Administration (FRA) Corridor study. The C&NW route from Chicago to Omaha was by far the most fuel efficient railroad. Completed January 25th, Dave Leach approved the report and sent it to Jim Wolfe under a for-his-eyes-only cover. Another little task for us at this time was developing density statistics for the Regional Transportation Agency (RTA) to get funding for repairs to track in suburban territory. The finance department on January 19th asked for a TPC run on the Chicago and Illinois Midland Railway. Apparently, someone was looking at that property again as Commonwealth Edison had been trying to sell or give it away for several years. While I appreciated that early Christmas present last year of some important

Burlington Northern track charts, I received an early Valentine this year when Jim Chesner in operating came up with another vital link. The BN may have locked up their charts, but our grapevine crew knew where the keys were kept.

Mid January 1977 in Chicago was the coldest recorded period in 100 years. On the 16th the temperature dropped to −19° F which tied the record. By January 27th we had shivered through 33 days below zero, worse than the winter of 1917–18. Rivers and lakes froze to the bottom, killing all the fish; Ohio had a natural gas emergency; fuel and barges loaded with salt were frozen in place leaving road clearing crews short of that much needed commodity. The weather continued bad: by February 3rd there had been 38 days below freezing with 12 inches of snow still on the ground. Railroad revenues were actually up a bit because all the river barge traffic was stuck in the ice. The first thaw came February 9th after 43 days below frost—the old record had been 32 days. With the thaw came the danger of flooding in the Midwest while California was in the grip of a seven years' drought.

▲ In the late afternoon of January 20th I took a computer tape to O'Hare for fast delivery by United to John H. Ransom in Omaha. Jim Wolfe, who had not been in a hurry, suddenly on the 24th, gave us just 30 days to develop a complete planning timetable for Project Yellow. In addition to that project there were other sideshows—R. L. (Rich) Vasy of coal marketing needed help with a proposed coal fired power plant at Fort Atkinson, Wisconsin. (It was never built). On the 27th M. S. Reid, Ray Gotshall and I met with Steve Hill of Manalytics of San Francisco about providing management for the coal line project. We did not say anything about Project Yellow which completely removed the need for most of our old railroad east to Fremont, Nebraska. On February 2nd I went to Omaha and met with Dale Salzman, Dick Peterson, John Ransom and Bill Pietmeyer on Project Yellow. I was surprised they had not thought of some of the issues involved, but, after all, they were just bright young men who had never experienced a major construction job. Back in Chicago the TPC coding work reached a furious pace in early February with Denver-Julesburg on the Union Pacific and the Missouri Pacific's Pueblo line under study as possible Colorado coal routes. Larry Kohn finished the Missouri Pacific from Jefferson City (Missouri) to Kansas City on January 28th.

I got the bad news that I was again going to be stuck with development of the annual density statistics because the systems department had not yet finished that old PACT request. I asked Larry Kohn if he could handle it. With some guidance, he did a creditable job. We also had a question come up regarding the Burlington Northern's effort to merge the St. Louis-San Francisco Railway. By making a simple TPC run, the question (whatever it was) was adequately answered by February 3rd. On the 16th I made the mistake of going to work when I felt rotten; Dave Leach put a rush order on me for the annual density statistics because Wolfe had asked for it. I ran into Wolfe in the hallway; he said the Union Pacific had some secondhand heavy rail available and he needed the density map to decide where to have it installed. The next day (the 17th) I was asked to review a military rail network proposal. Surprisingly, the old CStPM&O Line from Sioux City, Iowa, to St. Paul, Minnesota, was a designated route for moving heavy armored equipment. My first thought on this information was perhaps we can get some government funds to keep the railroads in the military net in good condition.

▲ The pace on the coal line now speeded up especially in the analysis of the Missouri River crossings. We discovered we had overlooked coding the Missouri Pacific line from Omaha to Kansas City along the west bank of the Missouri River—a key link. The connection at Omaha between the Missouri Pacific and Union Pacific was an awkward reverse move for eastbound trains heading south on the MP. Some years later a major new connection joining the Union Pacific's main line to the former Missouri Pacific line south of downtown Omaha was constructed to facilitate coal train movements. Named the Wimmer Wye, it was constructed by Bill Wimmer, who had begun his career in the C&NW's engineering drafting room. Ed Burkhardt, Dick Freeman, Zev Steiger and I met on the 24th to discuss our strategy in the first meeting with the Union Pacific. Freeman said the first meeting should be all structure and no detail. On the 28th we flew from Palwaukee Airport to Omaha in the Aero Commander. In addition to Ed Burkhardt, Dick Freeman, Zev Steiger and me, Jim Brower, Ray Gotshall and John Mruskovich made up the delegation. Arriving at the UP offices on Dodge Street at 9:15 a.m. we had barely sat down only to hear the UP group flatly state that they wanted to take over the entire project and demanded that the C&NW

sell out. If we agreed, the UP would pay us a royalty based on the coal tonnage moved over the line. That was a short meeting. We flew back to Chicago wondering just what kind of a game the UP was playing. We met the next day to consider the proposal and concluded that their posture was just the Union Pacific's first position. Our resolve stiffened, we adopted a strategy of preparing *pro forma* scenarios based on different levels of coal tonnages and concentrated on the Missouri River crossings. John Mruskovich and his boss, Al Handwerker, worked all weekend developing the costs of the project.

▲ The C&NW coal committee prepared for the next Union Pacific meeting by meeting at 10 a.m. on March 9th to discuss strategy now that we had some costs assembled. The Union Pacific delegation arrived in time for lunch at the Tower Club which provided a pleasant atmosphere. Our pre-meeting conversations were good-natured but guarded. At 1:30 p.m. we reconvened in the boardroom with Dick Freeman acting as our chairman. The Union Pacific delegation then took off their party faces. Their position of February 28th was arrogantly restated, modified only to put a 2 cent per ton for 20 year number on the proposed royalty—if we would sell out. Freeman calmly asked if there was anyone present from the Union Pacific with the authority to discuss and negotiate the proposition. There was a pause; the UP delegation looked at each other and said no. Dick Freeman firmly closed his open portfolio, stood up and said, "this meeting is over and have a nice flight back to Omaha". The Union Pacific delegation sat there in stunned disbelief for a moment before they collected themselves and left. After they had gone, we caucused and adopted a resolution that we would go ahead on our own—somehow. Project Yellow now seemed to be a very sick, if not in fact, a dead horse. The concept had served one valuable function in that it had pulled the fractious C&NW departments together in a common cause. **We had reached a major turning point in the western coal project.** The failure of the C&NW and UP to reach a joint agreement meant we had to take another look at the entire concept and get back to basics. Western Coal took a breather.

Suddenly, I had a lot on my plate: helping Larry Kohn with his iron ore study, the western coal project, the Milwaukee corridor study, and a direct route for AMTRAK passenger trains operating on the C&NW's north line to the Chicago Union Station. This last item we disposed of quickly. On March 24th, a lovely Spring day, Larry and I hiked from Canal and Kinzie Streets where a track to Union Station crossed the C&NW's old North Pier line to the north through the former Grand Avenue yard now a parking lot. The old Pittsburgh, Cincinnati, Chicago and St. Louis (Panhandle), which became part of the Pennsylvania Railroad, once delivered cars to the C&NW at Grand Avenue over this route. We went on north to see where an incline to the C&NW's elevated Wisconsin Division mainline could be built. It could be done by knocking down a warehouse building or two, but the likelihood of that happening was remote. We just enjoyed the morning stroll, came back to the office and killed that project with a brief memorandum. The value of the TPC program was underscored when we were asked to make runs for iron ore trains by marketing in March. We were happy to oblige. On the 14th we were launched into a new project to study the capacity of the Milwaukee-Twin City corridor assuming the C&NW would have running rights on the Milwaukee Road. To start the affair properly we spent a few days collecting information and plotting train densities before George R. Hanson (operating), Jules Eberhardt, Larry Kohn and I went on an extensive inspection trip March 21st to Techny, Bain, South Milwaukee, Racine and Waukesha. We were especially interested in finding possible connections between the two railroads to provide by-passes around some of the tight spots we had identified. George, Larry and I made another trip on the 31st to Oconomowoc, Portage, Camp Douglas and LaCrosse. I left them at the latter place and flew back to Chicago. In April Larry learned the intricacies of plotting train movements using that old standby, the string diagram. I tried to slip into the background and become an advisor rather than the investigator because I was convinced that the western coal project needed my involvement more than this corridor study. Riding two horses at once proved to be a difficult feat. I went along with George April 26th on a hi-rail trip from the Milwaukee Road's Tower A20 to Bain where we drove back to Northbrook to hi-rail the Milwaukee mainline to Milwaukee and into their Grand Avenue Yard. On the southside of Milwaukee we stopped to extinguish a grass fire caused by hot oil spewing from a stuck injector on a laboring old F3 diesel (MILW #98A); the fire turned into a 2 alarm call. On the 27th George and I attended the C&NW's morning meeting at Butler before joining three Milwaukee Road men (Brodsky, Westmark and Frank Deutsch) for a hi-rail trip on the C&NW from BJ Tower to Wiscona and then on the Milwaukee's line south to Canco and Union Station through what the Milwaukee men called "Death Valley" where local poor minority residents regularly attacked slow moving freight trains with rocks and debris. That afternoon we took a look at the abandoned Chicago, North Shore and Milwaukee Railway's embankment adjacent to the Heil Company plant on Oklahoma Avenue as a possible connection from the Milwaukee mainline to the C&NW's Chase Yard. Possible, but expensive and hard to justify. We finished our rail tour with a fast ride from St. Francis on the C&NW to Butler and then went by highway to Elm Grove on the Milwaukee Road.

Gus Malecha, reaching for his Maalox bottle, still did not believe there were serious operating problems at Milwaukee (April 28th). Westmark, Schmidt and Anderson of the Milwaukee Road came in to see our string diagrams on May 2nd. We made it clear that the C&NW was not interested in using MILW trackage north from Proviso to Milwaukee; we had plenty of capacity and any coordination of operations would not produce identifiable benefits. Our position hardened May 6th when we learned that the Milwaukee Road might go under if the C&NW did not buy trackage rights. On May 11th I rode AMTRAK's *Turboliner* north with George Hanson and Larry Kohn. We were invited to ride the headend where we could observe the modern control system. There was a major defect observed with this type of shelf-like instrument board—the engineer's very large beer belly overlapped the control panel so far that it inadvertently hit the stop switch. The train slowed. The engineer finally noticed what was causing the problem, sucked in his gut to release the emergency stop handle and got the train back to speed. In Milwaukee we borrowed Jack Chester's car and drove over to the MILW's Muskego Yards for a walking inspection. The yard tracks were in poor shape—almost as bad as our yard at Adams. That afternoon in a meeting with the MILW folks, we all finally agreed that few C&NW trains could use the MILW to advantage. On May 12th a big party of ten toured the MILW mainline northwest on rail. One of the people was a Mr. Peacock, their division engineer, who was trying to hold on until his retirement in the very near future. It seemed the whole railroad was trying to hold on for something to save them. At Camp Douglas, Wisconsin, we split up. One group ran on to Wyeville and the other hi-railed to Necedah. On the 13th we attended the morning meeting at Butler. George Hanson did a traffic review with Keith Plasterer while I took Larry Kohn on a tour of the yard. This whole project began to be political rather than analytical. Larry ran a lot of TPC runs in May to verify our guesses about train operating characteristics. Finally, on June 3rd, the committee of George Hanson, Gus Malecha, Jimmie J. Johnson and I agreed that we would not recommend use of the Milwaukee Road mainlines in the Milwaukee-Twin Cities corridor or on any trackage west of Chicago. Carl Hussey, who was present at the meeting, said we were " a mean bunch of bastards". Our decision had to be kept quiet until Jim Wolfe had a chance to comment. George wrote a long report supporting our negative conclusions; John Butler approved it June 13th. I hoped that this investigation and report would end the perennial lure of the tottering Milwaukee Road which I viewed as a sideshow or diversion from more important matters. Long after I left the C&NW, the Milwaukee Road's tortuous death and revival under new ownership continued to be a mammoth and expensive struggle involving the SOO Line, the Canadian Pacific and the C&NW.

Our Track Train Dynamics committee so far had not contributed very much to the company's benefit, but now we really got into a worthwhile project—locomotive engineer training. On March 7th, M.X. (Mike) Arakelian, Jeff Otto, Tom Nixon, George Montgomery and I flew to Oklahoma City to see a locomotive simulator in action. We were met at the airport and driven 75 miles south to Duncan in a Halliburton Company car. When asked why we had not just flown to Duncan, the driver explained that the president of Halliburton had become so angry at the operator of the local airline that he swore that Halliburton would never again use their planes. That stubborn stance resulted in a lot of driving back and forth to Oklahoma City. We were here to see a Halliburton subsidiary company, Freight Master, which had developed a diesel locomotive simulation machine which they called TDA. After a day of testing which we filmed and taped, Mike Arakelian, formerly a road foremen of engines, declared that he liked the machine, found it realistic and easy to use. We were

Evaluating a locomotive simulator at Halliburton's in Duncan, Oklahoma, on March 8, 1977. Here part of the C&NW team (Tom Nixon, Jeff Otto, Gene Lewis and George Montgomery) consider the simulated in-train forces being displayed on a cathode ray tube monitor screen.

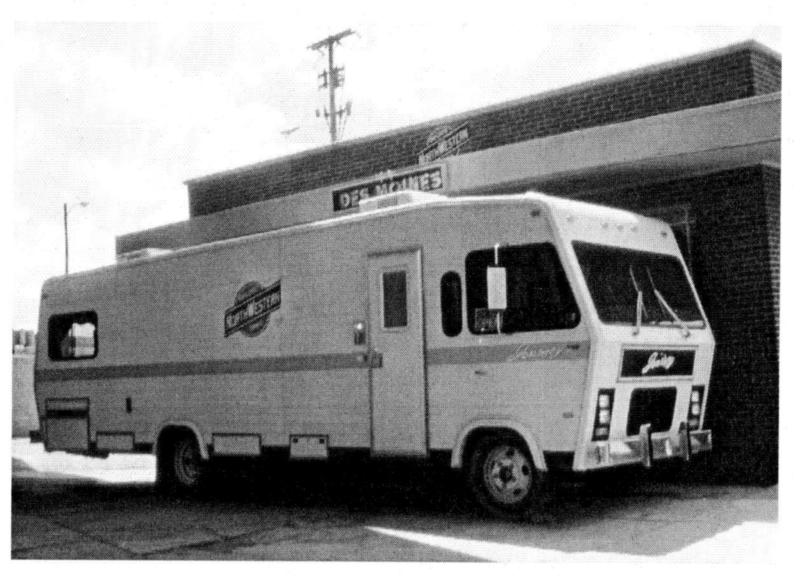

The Training Department's TDA train simulator was housed in a Winnebago van for easy movement around the divisions. Here it is parked at the former CGW Bell Avenue station in Des Moines, Iowa, on April 27, 1979 ready for the engine crews handling the Kansas City trains.

sold on the value of this equipment. On the way out of Duncan we spotted a retired Rock Island 4-6-2 #905 ensconced behind a sturdy fence looking for all the world like a caged beast or, perhaps, a cemetery monument. About a month later Jeff Otto, Frank Dickinson, who was in charge of locomotive engineer training at West Chicago, and I flew to Knoxville, Tennessee, on April 11th for a look at the Southern Railway's simulation machine. While similar to the Freight Master model, we felt that the Oklahoma machine was a better training device. On May 18th Mike and I attended an AAR seminar at my *alma mater,* the Illinois Institute of Technology, which demonstrated their locomotive simulation machine—we were not impressed as the device was poorly presented and seemed to lack features that Freight Master had available. On July 1st I drafted a contract for the purchase of the Halliburton machine and gave it to George Hollander, our contract attorney, to review and amend as necessary. On the 25th the purchase contract was ready. Now to get the funds to pay for it. I helped Mike prepare a request for an AFE (Authority For Expenditure) for the TDA and a Winnebago vehicle to house it. The AFE was turned down September 29th as the company's revenues had gone soft again. The TTD committee continued to function, however, and attended another seminar on September 27–28th at the O'Hare Hilton. It was so technical in nature that Arkelian and Dickinson were left in a fog. At this meeting I met John Gies of Morrison-Knudsen. This chance encounter would lead to some interesting side paths for me.

Returning to March 1977, Reid's branch line committee met on the 14th in response to presidential pressure to abandon a big bunch of low traffic lines. The abandonment scene had been relatively calm, but Wolfe now wanted a more aggressive program. The stockholders' meeting March 21st was enlivened by a very vocal owner who expressed his disgust at the high cost of operations at Clinton, Iowa. I learned April 1st that those pesky density statistics we had struggled with most of March were all wrong; systems had given us January through November instead of the whole year's data. There was no choice but to re-do the whole thing.

▲ March 23, 1977 was another pivotal day in the Western Coal Project. Now that the connector to the Union Pacific was dead—much to the UP's consternation—we were forced to face upgrading our existing railroad from Shawnee, Wyoming, to Fremont, Nebraska, all 520 miles. Because they were the obvious beneficiaries, we decided to ask the coal mining companies for help in the form of a loan of $187 million dollars up-front. The committee met with Ed Burkhardt in the early afternoon before sitting with Wolfe later in the day. I was appointed (again) to organize the planning for the upgrading effort with a deadline of December 1st. The first steps on my list included a review of the cost projections and to locate a construction manager skilled in handling such a large project. By March 28th some of people involved in this project began to understand that this was a HUGE endeavor. On March 29th I called on materials, personnel and communications to expand their involvement as there were just 245 days left before Wolfe's deadline. On March 30th Bob Russell, vice president-personnel, assigned himself to the project. Jim Brower, our construction engineer, now threw a giant monkey wrench into our planning by declaring that engineering would require three years, not two, for rehabbing the entire 520 miles. That seemed to stop our momentum. By April 4th I was very depressed as no one seemed to care about performing their parts of the planning. I could not find the start button. Day-to-day concerns gradually crowded Western Coal out of the spotlight. By April 15th the rebuilding cost of 520 miles of railroad had jumped another $33 million to $220 million. Zev Steiger, a C&NW attorney, said that Salomon Brothers, New York, were willing to put up the capital if the mining companies would guarantee the coal tonnages. Union Pacific, recovering from our rebuff and seeing our apparent inability to act, was now determined to crash the party. The potential was too great a lure to ignore. On April 20th Brower was directed to base his engineering planning for a two year rebuilding project. Brower hired an engineering firm to make an electronic survey of the line using an on-track vehicle equipped with equipment to record elevations and alignment. Jim was a little surprised at the results which differed somewhat from the old profiles and maps.

Tom Juedes, who was working under my direction, made his report and recommendation for boxcar cleaning. The director of car service, Tom Harvey, known as "the Silver Fox" because of his white hair and wily ways, was not pleased with our Tom's report because attention had been directed to some of the strange instructions he had issued to field forces. The boxcar cleaning problem would eventually go away over time mostly by itself. The 40 foot box was a doomed species especially in the grain trade. The sliding doors were not tight enough to keep kernels of grain from trickling out all along the route of car from elevator to destination despite the use of paper or wood grain doors. We usually had a pretty good "crop" growing right in the tracks especially on the weedy branch lines. The bird population liked the wastage, too. Economics killed the forty foot box car. Its capacity was not big enough in modern grain handling. Labor costs were higher for loading and unloading the boxcars. The older cars were all friction bearing cars subject to hot journals and the resulting troubles. The new giant grain elevators could not use the boxcars conveniently; the last users were the old country elevators that were fast disappearing. The C&NW initiated a program of scrapping these obsolete cars and replacing them with a fleet of modern covered hoppers. Many of the big grain companies were purchasing their own fleets of covered hoppers to ensure car supply for new 50 and 100 car unit train shipments.

On April 11th I received a major report prepared by Harza Engineering of Chicago regarding developments in the Yellowstone Basin which included the Powder River. They had completely neglected the coal line then under construction north of Douglas, Wyoming, and the extent of mining developments in that area. I was still active with the Standard Point Location Code (SPLC) policy council and attended a meeting in Washington April 18th. The truckers' new tariff manager, Jim Harkins, had replaced old Bob Hennell, but Jim adopted Bob's truculent refusal to cooperate in fixing the code's errors. Harkins was pretty upset when the council refused to accept his nine digit solution for logic problems. Adding a three digit suffix to a six digit code just added to the programming problems and accented the basic fact that the basic logic of the SPLC was flawed. I took the *Metroliner* to New York and came home from there. Lee Fox at the ICG sent a group of their engineering people to see our TPC machine in operation (April 29th).

Personnel changes in this period included the departure of the division manager at Milwaukee. His replacement was C. J. (Chris) Burger. Charlie Hellem went to Denver as regional sales manager while R. P. (Dick) McDonald went to the Twin Cities. On April 20th

I made an offer to Dave Nowicki for a permanent job in our CIE department. At this time Joann Tsoodle, our very capable secretary and project assistant, accepted a job at the AT&SF as a personnel officer at the annual rate of $36,000. She had achieved her goal to become an officer. The fact that she was getting a bigger salary than mine was a tribute not only to her ability but, probably, to her sex and Kiowa blood. Her last day was April 29th. Marilyn Jones from suburban Hinsdale took Joann's place. John Mruskovich of finance rejoined the C&NW after a brief sojourn with the Kansas City Southern—his wife wouldn't move to Kansas City from Chicago. On May 15th my good friend, Thomas Desnoyers, of the Transportation Data Coordinating Committee (TDCC) died in Virginia. To express our regrets to his wife Jean, Barbara and I went to Virginia to attend his service. Tom originally came from Glen Ellyn, Illinois, and had been a boyhood friend of my former colleague, Don MacBean.

▲ We were well aware that the Federal Railroad Administration (FRA) had money available for railroad reconstruction, but how could we tap this source for our coal line project? The terms for obtaining these long term loans were rigorous. In truth, the C&NW could have used the FRA's entire funding. We were dubious that we could qualify for this source of capital (May 4th). Two days later we ran a preview of the slide presentation we were to make to ARCO in Los Angeles on May 9th in a campaign to have the coal mining companies fund our reconstruction plan I heard the next day that the show was received rather coolly (Was the Union Pacific at work here?) ARCO took the position that we would have to first develop coal markets which seemed a bit odd given the extreme doubt we had even to rebuild our railroad. At this juncture, Ray Gotshall, my coal marketing stalwart, and Steve Dyslin, left the C&NW. Ray took a job with Old Ben Coal Company as director of sales. Our stock was hanging at 11½ with 24,000 shares traded on May 10th. Three days later 38,000 shares traded at 11⅞. My $2.25 average price per share still looked pretty good. On May 17th a new energy group was formed on the C&NW—just what mandate they had was still unclear two days later. The stock continued to creep up; it hit 12 on the 25th. An unsubstantiated rumor said that the Seaboard Air Line was buying a stake.

▲ On May 16th that proposal for a gasification plant at Wichita,

Kansas, supposedly ended. Marketing had never been very serious about this project, but it had encouraged us to extend our TPC coding into the area. At a meeting in Houston May 25th we learned that the Wichita proposal was faced with a lot of environmental issues that had to be settled before a plant of that nature could be constructed. On May 23rd I had decided that it might be a good idea to code the old Kansas Pacific line to Denver. John Sward of the ATSF kindly sent some Frisco track charts to me. Then on June 15th Ed Burkhardt advised that the Wichita proposal had new life, and we were to mount a sales campaign. The Union Pacific had agreed to work with us on the route beyond Fremont. We investigated the UP line south from Valley, Nebraska, to Wahoo and tested our own railroad south from Fremont toward Lincoln. To our surprise we found that our rickety branch line was an easier railroad to operate than the UP's line—at least beyond Wahoo where the two lines crossed. At the end of June we finished coding the Rock Island lines south of Lincoln, Nebraska, through Belleville and McFarland. At the end of June I met with Burkhardt and Brower to display what we had learned about the railroads to Wichita; Ed okayed my contacting the UP and Missouri Pacific to obtain operating details in that area. On July 6th George Hanson and I flew to Omaha to meet with Dick Peterson, Dale Salzman and John Ransom of the Union Pacific about the train service to Wichita, Kansas. After picking-up Fred Yocum, the division manager, our C&NW group visited South Omaha on a blazing hot day to look at the awkward Missouri Pacific interchange. Putting a coal train through there was going to be an adventure. The next day George, Fred and I hi-railed north from Omaha to Blair on the old Omaha mainline. We lost a wheel at Blair and tied up for repairs. I called my old friend the bank president, Howdie Hansen, who got mayor Al Sick of Blair over to see us at the stone and brick depot (built 1910; torn down October 19, 1987). The mayor had quite a lot to say to division manager Fred Yocum about the increase of trains through Blair. Fortunately, repairs to the hi-rail were quickly made, and we left town on the railroad west to Fremont where we switched over to the Lincoln line. We rolled across the Union Pacific mains, past Platte River where the Superior line had begun (abandoned in 1962) and went as far as Wahoo to look at the Union Pacific grade crossing there. On July 11th John Ransom called asking for track charts of our

Lincoln line. I horse-traded some of our track charts for a few Missouri Pacific items I needed. Two days later I suggested to John that they take a good look at the C&NW line south of Wahoo because it was better from a curve and grade point of view. Ed Burkhardt invited some Rock Island people over to see how they could participate in the Wichita movement. One them, Sangree, I had met before. They were quite taken with the depth of detail we had developed about their railroad—we knew more about it than they did. Sangree agreed to provide train service information July 18th, but by September 13th it was obvious that the Rock Island was not interested in planning joint coal train movements as they had more serious issues to face. (Frederick Yocum Jr. was named vice president-finance and administration and chief financial officer of the Montreal, Maine and Atlantic Railway in September, 2004).

In late May morale in our CIE group was not very good. Mark Johnson was quite annoyed that the promotion he had received earlier in the year was not accompanied by much of an increase in pay. Mark interviewed for a job with the Union Pacific in late May. Tom Juedes was not a happy employee of Rich Howard's industrial engineering group and interviewed with Allied Van Lines on May 27th. The rumblings of discontent must have reached our titular leader, David Leach, who attended our staff meeting on the 27th for only the third time in three years. On June 3rd I attended a showing in engineering of a new rail-laying train. Wolfe said we could have one if we could find the money. That was the type of equipment we would need for the reconstruction of the Shawnee-Fremont railroad. Meanwhile, in the real world, the Iowa division was struggling with a raft of slow orders imposed by the FRA because of bad track. In my own little personal world a landmark was reached June 1st. I now had established a continuous employment record of 360 months with the C&NW (one day in a month counted as a full month). This was an important benchmark in my private goal of retiring from the company. I now had the requisite time in the bank. However, the pension amount based on deductions for my age of 51 versus the plan's 65 would have slashed the monthly payment too drastically to make an immediate departure worth while. Besides, I wanted to see how this goofy coal project would finally play out.

▲ On June 7th (my 32nd wedding anniversary) I talked to Manalytics in San Francisco about their recent report developed for the

Electric Power Resources Board. They had made some poor assumptions about the capability of railroads to move quantities of coal. The TPC coding kept us all busy in our odd moments. I did the UP's Egbert branch in Wyoming on June 8th. This piece of railroad ran straight north from their mainline east of Cheyenne to Torrington near where the Project Yellow line was to have connected. It would have been a tough railroad to move a heavy coal train over, but it might have been a coal route—until I worked it through the program and saw the character of it. Actually, the UP did run some West Coast coal trains on this line in 1996. In the middle of June I cast my computer eye at the railroads operating along the Front Range of the Rockies as there was to be a lot of coal moving to Texas on these washboard profiles. Jim Barnes, vice president-engineering, called on June 16th to report that the UP was actively pursuing plans to build north from Lyman, Nebraska, along the Project Yellow route. No action ensued. About this time another coal customer prospect popped-up at Wausau, Wisconsin, which required our TPC analysis. Planning on the rehab of the Shawnee-Fremont's 520 miles, however, was almost non-existent. Tom Evans of communi-cations planned his part of it and Hank Hahn of engineering helped me with the public grade crossing issue, but everyone else was sound asleep.

▲ At the urging of Ed Burkhardt I began an assessment of the impact of moving 110 cars of loaded coal versus the 100 we had used as a standard for movement east over the Shawnee-Fremont line. After some comparative runs we learned the extra ten cars put operations into a marginal state when using four SD40-2 locomotives. This kind of work was typical of the marginal effort being put into the planning process that was supposed to be finished December 1st. By July 15th it was obvious that now only two people on the entire C&NW were doing any planning or devoting any thought to getting this coal line project into a coherent operation—Ed Burhardt and myself. Yet, the next week talks were again underway at the FRA in an attempt to pry loose some of the bag of money they had to spend. Both Jeff Otto and Jerry Groner were in the FRA talks (July 28th). Burkhardt reported that he had met with prospective receivers of western coal in Alexandria and New Orleans, Louisiana. Their interest level was high IF we could put the package together. Apparently, our activities stirred up some of the coal committee participants as more showed up at the July 31st meeting in the boardroom. Decisions were slowly being made; but no solid commitments were evident. I decided that we needed to turn up the heat and called the State of Nebraska and the Front Range people to dangle some bait about the economic impact of this project. Jim Wolfe attended the coal meeting August 15th which suggested that our top officers might be still interested. I was beginning to wonder about why there was no activity in some important aspects such as lining up a source for track ties, a long lead time item. Something was definitely screwy about this whole project.

On July 21st the question came up regarding use of the Burlington Northern's Missouri River bridge at Atchison, Kansas, instead of our existing swing span at Leavenworth for Kansas City traffic. The U. S. Corps of Engineers was pressing the C&NW to repair the old CGW span. The turning mechanism repeatedly failed causing blockage of river traffic—what there was of it. Train interference was not the problem as the span was usually lined for the railroad. For some reason I was asked to coordinate the investigation. On August 11th I met with Carl Hussey (operating), Jerry Iwinski (bridges), George Hanson (operating), Jules Eberhardt (CIE) and James J. Johnson (operating) to gather details and work up a strategy, if the route looked feasible. I took a look at making a track connection at the crossing with the BN at Beverly, Missouri, and looked at the span in the field. Built in 1890 it was well maintained, but a bit light for heavy trains crossing a major waterway. Extension of trackage rights on the Missouri Pacific to Atchison from Leavenworth would also be required and the connection in Atchison was almost as sharp as down river at Leavenworth. Jim Wolfe and Jim Zito added their comments August 25th by vetoing the idea. Instead Wolfe asked me to reactivate my old Des Moines-Kansas City study of five years ago and update it—quickly. That report just would not die. Updating the traffic data flowing through Kansas City, getting train and car counts and the other data from the 1971–72 report updated proved to be an annoyance to the division manager, Fred Yocum, who complained about demands on his people. I met with Robert W. (Bob) Schmiege of labor relations about the Kansas City labor pool. (Bob would become the C&NW's last president). I gave Dave Leach a briefing on the progress of the Kansas City investigation September 16th. Our relationship remained pretty starchy. To re-acquaint myself with the Kansas City situation, I flew there with George Hanson and Jimmy Johnson on September 20th. Ralph Eschom, trainmaster, took us through the complex starting at Airline Junction where southbound trains came off the Rock Island to go to the tiny CGW yard at Ohio Street. We visited the Union Pacific's Armstrong Yard, talked to Bill Apple of the Kansas City Terminal (KCT), toured their control center in the old Union Station, and went to Argentine on the ATSF for a look at their hump yard. Fred Yocum caught up with us for the hi-rail trip over the KCT from Big Blue Junction to the UP's Armstrong Yard. On the 21st amid rain and wind, Fred, Jim, George and I drove northeast to Polo, (Missouri) and Gallatin, Lineville, Allerton in Iowa, Melcher and Croydon sampling the condition of the Rock Island's track. While the line was laid with heavy rail (112 pound) on stone ballast, the track looked like it needed some attention. In Des Moines trainmaster Janovic showed us the place where crossovers could be installed in the parallel mains north of Short Line Tower to give C&NW trains a direct run onto the Rock Island's line. The crossovers would eliminate the current need to pull southbound C&NW trains into the Rock's yard and put the power on the other end before proceeding south with the train reversed. (The crossovers were installed April 1, 1980).

The project turned strange September 26th when the Rock Island suddenly became reticent about negotiating a trackage rights agreement south of Des Moines. Our negotiators grumbled that we had just better rebuild the Leavenworth bridge to the Corps of Engineer's specifications and stay on our own railroad. By the 29th the Rock Island was flatly refusing to negotiate further. Carl Hussey said that the Milwaukee Road would undoubtedly scotch any deal we made with the Rock Island through their joint ownership of the line from Polo to the Joint Agency Yard in Kansas City. When Jim Wolfe again buried the Des Moines—Kansas City study on November 14th that was the apparent end of the matter. Three years later C&NW trains were to be in full operation on that segment of the Rock Island.

Life on the railroad went on. There was a 58 car wreck at Oxford, Wisconsin. Ed Burkhardt told me on August 4th that he was holding on to his C&NW position just to see if the coal line would actually happen—we seemed to have the same idea. Our nominal leader, Dave Leach, and I had another run-in over an event that I did not want him to explore too carefully. In early August he asked for a preview of my presentation to a University of Wisconsin seminar September 14th. I had not written it yet. I was still considering the structure of the talk and which slides I would use to embellish the presentation. Using the excuse of a rush of urgent business, I fobbed him off. I threw the talk together on September 11th, wrote to Leach on the 12th, drove to Madison, Wisconsin, on the 13th, and delivered the talk on computerizing geography on the 14th to some 32 people attending Ed Marien's seminar. This was getting a little too close to my other life and the less said about computerized geography and standardization of codes to Dave Leach, the better. While I am sure he considered me to be an uncooperative nuisance, I did produce some pretty good work. Our Tom Juedes interviewed for a job at the Illinois Central Gulf on August 9th. This CIE group seemingly could not hold its staff. Tom announced November 11th that he was leaving for a job with FMC (formerly Food Machinery Corporation). Larry Kohn was promoted to Tom's job leaving me again short a man (November 28th). (It wasn't very long until Larry resigned from Rich Howard's staff and became an associate at A. T. Kearney, management consultants in Chicago.) The Burlington Northern Railroad was struck on August 29th, dumping a lot of commuters onto our west line trains. August was the wettest in 100 years. C&NWT stock on September 30th stood at 11¼. Fred Yocum, division manager of the Iowa, took a job with the United States Railroad Administration (USRA) in Washington in November. On December 12th John Clark of personnel left for the Oak Company in Crystal Lake, Illinois. There was good news in November—I finished updating that complicated mileage tariff on the 18th. Kurt Ramlet, the tariff publishing officer, was a pleased customer of CIE and Stu Gassner, a likeable but always cynical attorney, was also impressed.

Maurie Reid asked me to go to Nebraska with him and R. W. (Dick) Bailey to meet with the State of Nebraska's railroad commission staff. We flew to Lincoln August 23rd. With all this branch line abandonment in progress the state people were more than a little suspicious of our motives and seemed to be mildly hostile. We leveled with them about our abandonment plans and the rationale we followed. They seemed to appreciate our frank and open discussion along with some idea of the future of our rail lines in the state. In the discussions we learned that the Burlington Northern had left some deep scars in the way they handled abandonments.

▲ After Labor Day I started on the grade crossing data for the Shawnee-Fremont line. In order to properly site the passing sidings we would need the number, type and location of each crossing. Tom Evans was still actively planning communication needs and was deeply into the prospects of using microwave instead of copper line wire. The Union Pacific now told us they could not handle coal trains at their Kansas City Armstrong Yard. I did not worry about that item; the UP could solve their own problems. On September 19th the ATSF said they would work with us on handling coal trains through their Kansas City facilities.

▲ Jim Zito, vice president operations (and engineering), called me October 7th about taking a job in engineering. We agreed that the coal project was more important as he had just returned from a three day inspection trip with Jim Wolfe and the prime topic of discussion was the coal project. Jim wanted a full presentation on the state of the planning by November 1st. I was dismayed because to all appearances, there had not been very much planning. I called Dick Bailey in engineering; he delegated Jeff Otto to me to represent their part of the project. I also called Ed Burkhardt, the titular head of the project, and told him of Zito's request. Ed was ticked at that—why hadn't Zito asked him first? On October 10th my wife and I left Chicago on AMTRAK's #15, *Lone Star,* en route to Fort Worth via Oklahoma City. After attending an art exhibition in Fort Worth, we flew to Los Angeles for a few days before riding #14, *the Coast Starlight,* to Oakland. On the 17th I met John Gies of Morrison-Knudsen in San Francisco to discuss his company's possible interest in the engineering management of rebuilding the 520 miles of our railroad in Wyoming and Nebraska. M-K was interested in the project. Returning from California on October 31st to gather up the threads of the coal project, I found a curious lack of intensity, or even activity, in the project. Jim Wolfe was in Florida on vacation and would not be back until the 8th so that Burkhardt and I had a chance to round up the strays. I made exhibits on expected train frequency, annual tonnages on the 4th and 5th for the scheduled presentation on November 10th. That meeting drew 50 people including Jim Wolfe, Dick Freeman and Jim Zito. Ed Burkhardt introduced the program, I did my slide show and Jeff Otto filled in the details—nobody left the room. After lunch the meeting reconvened. Jim Johnson (operating), Del Swenumson, Glenn Kerbs, Dick Bailey, Jeff Otto, Ray Snyder, all from engineering, were present and paying attention. A computerized "Workbook" I had assembled listed all the types of items to be done including job numbers, locations, descriptions (line changes, bridge extensions) and proved to be very useful. To identify specific work sites I had included the SPLC as well as mileposts to provide keys for aggregating various types of tasks geographically. On November 14th Wolfe said he wanted the material presented at that meeting to be printed. While I was pleased at this, I should have been suspicious.

▲ Meanwhile, the Union Pacific publicly announced that they still wanted in and favored a delay to our project. At the morning meeting of November 15th I learned that Lou Menk of the Burlington Northern and John Kenefick of the Union Pacific had engaged in a shouting match over the UP's announcement. Perhaps the threat of the Union Pacific entering the mining area stimulated the Burlington Northern to agree on December 1st to extend the existing financial agreement covering the joint line expenses to November 30, 1979 while the C&NW was applying to the FRA for a $275 million loan guarantee to rebuild the 520 mile line to Fremont. On December 2nd a committee was formed to prepare the 511 application for the $275 million guarantee. (511 was the Federal Act number authorizing this activity). That same day George Hanson discussed the proposed operating plan at a staff meeting. On December 6th we had another new leader for the coal committee—Jules Eberhardt. Jerry Conlon, vice president of public affairs, opined that the political time was right for seeking the FRA guarantee. The magnitude of the enterprise was beginning to dawn on our official family along with a distinct probability that there would be legal snares along the way. During the first few days of December Jules and I worked on the cost estimates while a big winter storm raged outdoors. On December 9th the committee reviewed my slide show and recommended a couple of changes. As requested I called LORAM in Minneapolis December 12th and RAILCO in Portland, Oregon, to determine if they were interested in a management construction project. Two days later I called Steve Wight, Morrison-Knudsen's local representative, to arrange a presentation to our management.

On December 19th I heard the Milwaukee Road had filed for bankruptcy. Three days later CIE had a new project—select pieces of the Milwaukee that could be useful to the C&NW.

▲ The FRA application was completed. The request for a government guarantee of $275 million was an ambitious attempt to get the money needed to rebuild 520 miles of broken-down railroad from Shawnee, Wyoming, to Fremont, Nebraska. A large delegation was selected to carry the application to Washington and formally present it to the Federal Railroad Administration. The group included Dick Freeman (law), Don Joerger (finance), Doug Christensen (industrial development), Jeff Otto (engineering), Dave Weishaar (coal marketing), Kim Schlytter and me (both CIE) This application and the requisite copies were not simple sheets of paper—it was a sizeable package. While it could have been mailed, expressed or messengered to Washington, this presentation had to be done with a touch of class and enthusiasm. We collected at O'Hare and learned that a winter storm had closed all Washington's airports. We switched to a LaGuardia flight and landed in New York in the teeth of an ice storm. New York was then closed down—no hotel rooms were to be had and the railroad to Washington was shut down. AMTRAK would not even answer their telephone. We were stuck. I suggested that we rent cars and drive through the fog and rain to Washington. The group agreed and off we went in two cars. I drove as far as Delaware where we had a bite to eat; Jeff Otto finished the trip. We arrived in Washington very early in the morning of December 21st, checked into our hotel and tumbled into bed for two hours before going to the FRA meeting. The application was presented along with my slide show outlining the high points of the project contained in the application. With relief we made our way back to Chicago despite a work slowdown by baggage handlers which bolloxed plane schedules badly. The next day we were rewarded by learning that the FRA staff had been impressed by our professional presentation. What a way to end the year! Yes, I got a raise; all of 8% to the grand total of $28,200.

32

THE POKER GAME, 1978 . . . ▲ Encouraged by the prospects of actually securing the funding we so desperately needed to rebuild the Western Division for coal, activity level on the project picked up. The Federal Railroad Administration's initial sour look at our project for reconstruction of the railroad across Nebraska seemed to have moderated. On January 4th George Lindquist of Foley Brothers in St. Paul, Minnesota, called to inquire about the project; I asked him to send reference materials for the committee to review. George was the guy who had made that low level flight north from Douglas to the coal fields with Ray Gotshall in November 1972 and had said the terrain was level. Making eyeball surveys from airplanes can lead to big trouble. Fortunately, his assessment had been positive enough to encourage our further exploration. The public now knew about our new railroad. The *Chicago Tribune* even printed an article about the coal line. On the 6th I finalized arrange-ments with John Gies for Morrison-Knudsen's presentation to our management the next week. Again, concerned about alternative routings, I coded the old CB&Q branch line from Sioux City, Iowa, west to O'Neill, Nebraska, where it met our proposed coal route. Originally styled as

▲ As noted before, this symbol will be used to denote material relating to the Western Coal Project.

The Pacific Short Line, this piece of railroad possessed a colorful beginning as the Sioux City, O'Neill and Pacific with aspirations of continental grandeur before economic reality brought bank-ruptcy and ownership by the Great Northern. Transferred to the CB&Q, it was now just another light traffic agricultural branch serving a sparsely settled part of Nebraska. Rebuilt to heavier standards, it could have formed an important sub-route for western coal headed for the new George Neal power station on the east bank of the Missouri River south of Sergeants Bluff, Iowa—that is, of course, IF the C&NW's Shawnee-Fremont rehabilitation was done. In Illinois the winter weather was bitter with −45° F. wind chills. I was glad to be inside doing detail work rather than exploring on the western plains.

▲ Morrison-Knudsen presented their company's qualifications on January 10th in a bid for management of the coal line reconstruction. All of our vice presidents save two attended the meeting. John Gies arrived first to set the stage. After the affair, Jim Wolfe took four of the top M-K men to lunch at the Tower Club while I entertained the rest of the lower ranks at the nearby Regimental Grill. John Gies insisted on hosting a dinner at Binyon's that evening where Dick Bailey, Ray Snyder, Jeff Otto and I bombarded him with questions about managing the coal project. None of us had ever been involved with a project of this size and complexity while M-K made a good business of it. At the insistence of engineering I called Foley Brothers in St. Paul the next day to determine if they would be interested in the project. Foley brothers ran the Rock Springs, Wisconsin, quartzite ballast quarry for us and had been the operator of the company's old Algonquin, Illinois, gravel pit. I felt that they were too light weight for this size of job, but called them anyway. Franco of LORAM, a Canadian firm, came in on the 12th to set up a meeting for the 24th. News gets around quickly when a large job hits the market. Herzog, a track contractor, also came in to see me on the 12th about the reconstruction job. It was too big a project for them in those days; years later they had grown and prospered. Meanwhile, I "ran" ballast trains in TPC from Sioux Falls, South Dakota to the projected coal line. A couple of existing quartzite quarries at Sioux Falls were most inconveniently located for our project. The rail route for this rock involved a 62 mile eastward move over the old Omaha Road branch to the mainline at Worthington, Minnesota, where the train would turn south to Sioux City, another 91 miles. After another crew change, the ballast train would roll on to California Junction, Iowa, 69 miles, before turning west. And the train was still 31 miles from the starting point of the rehabilitation project at Fremont, Nebraska. This rock train movement was too circuitous and required too many crews. Coupled with massive repairs to track east of Sioux Falls, this ballast source was just too costly. As requested, I sent tele-copies of the FRA letter to Salomon Brothers in New York. Brad Huedepohl, formerly our company geologist, was now working as an independent consultant. He had located several quartzite sites in southwestern Minnesota and on January 16th was invited to develop a written presentation for development of these rock sources. Getting the proper ballast for a heavy traffic railroad in the needed quantities in the right place had the prospect of becoming a complicated and expensive undertaking.

The Burlington Northern now announced that they were taking over the St. Louis-San Francisco Railway, which they eventually merged in 1982. On January 11th Lou Duerinck and I had a chat about what the impact might be on the C&NW. While there were a few power plants on the SL-SF that would now be BN customers, the most serious impact would be the extension of the BN's system into Oklahoma and eastern Texas. Some good news in this frigid January—I got a new assistant, David Nixon, on the 16th. David was a very bright and laid-back young man who had been working at Ravenswood as a clerk. His one speed was forward—s l o w l y. I never found his accelerator button. A native of Boston where his father had been a signalman on the MTA, David never quite adjusted to the rough and tumble of railroad life. A gentle soul he eventually returned to Boston and opened a bicycle shop where he found much greater satisfaction in its more leisurely and personal pace. Now his calm demeanor was about to be tested by the bubbling cauldron of the coal project where he proved to be a great help in developing materials in connection with our increasingly hectic efforts to move that project along.

▲ The Federal Railroad Administration (FRA) sent a team to our office January 17th to monitor and guide our application for the loan guarantee. Loftus, Wofford, Vermut and Hanscome were the principal members of the group. It was clear from the beginning when they laid down their rigid set of conditions that they would be taking a hard look at alternatives, engineering estimates and marketing projections. It was their considered opinion that the Union Pacific and the Burlington Northern would file protests in this application. During this first meeting, Wolfe called me out twice to get Milwaukee Road material which I had. I wondered what that was all about; it was explained the next day when Wolfe attended a meeting called by Brock Adams, President Carter's Secretary of Transportation, to discuss the restructuring of Midwestern railroads. The collapse of the Milwaukee Road in December and the shaky condition of the Rock Island was triggering alarm bells in Washington.

▲ The alternative to rebuilding the old railroad was the first item to be addressed. They focused on any possible rail routes from the coal fields to the gateways of Omaha and Kansas City both single line and in combinations. Our TPC index file UCTSIGMA with its more than 2,000 test runs covering the Midwestern rail net proved to be invaluable in answering their questions. Marketing concentrated on their tonnage projections which were very hard to establish given the iffy character of our project. The figures they finally settled on, and which the FRA would accept, turned out, years later, to have been laughably far too low. Our chief engineer, James A. Barnes, called me to his office on January 23rd and said he wanted Ford, Bacon and Davis, a New York consulting firm, to be a candidate for the project manager contract. At the time I felt that Jim's request was a bit odd. The timing was strange and the way the request was made was not in character with the Jim Barnes I had known for so many years. He seemed uncertain or, perhaps, a bit furtive. I called the FBD contact in Media, Pennsylvania, and another of their principals, George Sargent, in Denver, Colorado. FBD was the company A. T. Peagan, a former Iowa division superintendent, joined after being urged to leave the C&NW by Mr. Heineman in 1957. I made an appointment for the next day to meet with FBD at their construction headquarters in Monroe, Louisiana, to introduce Jim Brower and the project to their management. Brower and I flew to Memphis that afternoon and made the switch to a smaller plane for the short hop to rainy Monroe. Despite the downpour outside, we gave the coal line project presentation to George Sargent and the FBD brass on the 24th. They were interested and agreed to make a bid on the job.

▲ As part of the alternate rail routes under study, it was necessary to quantify the population that might be impacted by the increased number of trains. Demographics were drawn using the index in the current Rand McNally road atlas by simply adding up the populations of the towns and cities traversed. Actually that simple approach yielded a huge overstatement of potential impact because not everyone in a large city would ever see much less be "impacted" by a freight train rolling through or near their municipality. FRA accepted the technique in lieu of anything better. LORAM added DeLeuw, Cather to their combination and made their joint presentation to our management January 30th despite a terrific snowstorm. Their pitch was good but their engineering and systems people went too far into detail. Jim Wolfe stayed until noon.

▲ Remembering that lesson I had learned from Carl Hussey in 1962, I made a personal inspection of the Burlington Northern's rail routes that would be handling coal trains. Flying in the western states in January could be interesting especially after the giant snowstorm we had on the 26th when O'Hare was closed down for only the third time in its history. My flight on the 30th to Denver was canceled. Looking at the weather forecasts I switched to a Lincoln, Nebraska, plane and, on arrival at that place caught a Frontier flight to Scottsbluff, Nebraska, where I found 4 inches of snow down. The next day was clear and bright as I drove northwest through Torrington, Wyoming, paralleling the BN's Northport-Guernsey main line. Nowhere were there any signs of upgrading this important railroad which was to soon carry a lot of heavy tonnage coal trains. Usually a railroad planning major repair work would have stockpiled ties and distributed rail and appliances along the way. The track was mostly 112 pound bolted rail with indifferent line and grade; it could carry coal safely for the time being, but would require rail and resurfacing within a reasonable time. Driving northwest I visited Guernsey, the BN's crew change and servicing point, found Wendover where the Colorado and Southern Railway (also BN) headed for Denver, Colorado, looked at Glendo and Orin Junction, Wyoming, before running into another snowstorm at Crawford, Nebraska. I managed to get to Alliance, Nebraska, the BN's major terminal in the area, bought some new "walking" boots and then returned to Scottsbluff. The next day was the first of February; I awoke to find three more inches of fresh snow, but Frontier's plane made light of it, flew to Cheyenne and on to Denver. This little expedition proved to be most instructive.

▲ The western coal committee, which had been renamed the 511 committee, met on the morning of February 3rd. Dan Swett convinced Jules Eberhardt that we should meet with the FRA before they visited the Burlington Northern or Union Pacific but Wolfe vetoed that move. Then, for some reason, he reversed himself and Jules was to fly to Washington. I had been concentrating on the empty train movements over the alternate routes and gave the results to Jules on the 6th for his trip. He had a terrible trip the next day because of another disruptive snowstorm but the reception at FRA was good. Back in Chicago the Ford, Bacon and Davis company made their presentation to the committee for the project management contract and did an able job. On February 8th our finance people announced they were satisfied with their project cost numbers so we had a base to work from.

Another serious meeting was held on the 8th about the Milwaukee Road's branch lines in South Dakota. Wolfe was to meet with the Burlington Northern, the Milwaukee Road, and Federal Railroad Administration the next week. At a management meeting on the 9th Jim Wolfe said there were two important things that must be done: The Milwaukee Road must go under and, secondly, the 511 western coal application must be completed. The South Dakota carve-up meeting on the 16th was attended by Wolfe and Zito.

▲ Tom Wofford of the FRA and his assistant, Postels, came into the office February 9th fresh from their visit with the Burlington Northern. They said little about their reception in St. Paul. The next day I prepared specifications for selecting a project manager which again proved to be a waste of time. In an afternoon meeting Wolfe, Zito, Dave Leach, Jules Eberhardt and I reviewed the three candidates without recourse to any written check-list. LORAM was the first to be discarded and Barnes had nothing good to say about Morrison-Knudsen which left Ford, Bacon and Davis—logic did not prevail. At another meeting on the 13th Jim Zito, Dick Bailey, Jim Brower, Ray Snyder, Jules and I decided that we should prepare specifications for application engineering and send them to both M-K and FBD. By the 16th it was obvious that our engineering officers had dumped their part of the western coal project planning on Jeff Otto before returning to their other near time needs of trimming the maintenance budget. Jeff was ill for several days so that the critical specifications were not finished until the 17th—four precious days were lost. I fired them off the same day I received them to M-K and FBD.

I went to the morning meeting on Washington's birthday. Jim Zito was running the show and acting rather badly. Chewing people out on a public broadcast is not considered to be the mark of a good manager. Jim just had a bad day and let his Italian temper get away. Jim Wolfe missed that performance—he was in Fort Dodge. The Iowa division called me after the meeting and requested some TPC runs to support their contention that train speeds had not deteriorated over the past two months as much as Jim Zito claimed. We could reproduce the conditions in each period by merely using the speed restrictions in effect at the time. On February 23rd Chicago's annual snowfall record was broken—77 inches had fallen. In retrospect it would appear that every winter was worse than the preceding given that records were broken in each succeeding season.

▲ Jules and Dick Freeman went to Washington again on the 23rd. The next day engineering received replies from M-K and FBD on the construction management job. Our chief engineer Jim Barnes was under considerable pressure to make a choice. I wondered if he was fearful of the costs involved to hire someone to work on a job that was so questionable that it might not even begin. Morrison-Knudsen pressed very hard by sending three of their top men to see Barnes. Jim was very uncomfortable being in this decision making position. M-K came back again on the 27th promoting their bid. Engineering was in a dither with their budget because the ICC had approved just 3% for the national freight rate increase. Jules Eberhardt reported at the 511 committee meeting on the 24th that the FRA was very interested in linking the Union Pacific to the coal line but, apparently, were not aware of our Project Yellow connector line. Meanwhile, I had coded the ATSF trackage all the way to Houston, Texas and the inestimable John Sward at the Santa Fe had come up with some more choice bits of the BN and some Missouri Pacific profiles. By February 28th I was running all the alternate routes using the Project Yellow connecting link. I had a strong feeling that the new railroad from Crandall to Joyce eventually would be built and that the broken-down

C&NW line east to Fremont, Nebraska, would never be fixed—it was just too much money. But I did not let on to the FRA that the Project Yellow link existed. On March 1st Wolfe shared a confidential Union Pacific report with Jules and me—Project Yellow was still very much alive—if not on the C&NW, it was a hot item on the UP. I kept my counsel.

▲ By February 28th the decision on a project manager still had not been made. I heard that the Ford, Bacon and Davis bid was $186,000 and Morrison-Knudsen came in with $85,000. Based on price I could not see what the problem was. After a boisterous meeting on March 1st, engineering still could not make a decision and sent the matter to Jim Wolfe who in turn sent the recommendations to our department to sort out. No one wanted to take a stand. Instead of a straight-forward business-like decision, the choice had turned political. I went back to calculating tonnage shifts as if Project Yellow were constructed and verified the Union Pacific's figures contained in that confidential report of March 3rd. Wolfe appeared to be wrestling with the choice of a project manager—trying to decide if FBD was worth the extra $100,000. Or was some other game in progress? If this were a foretaste of what we would face during actual construction, the prospects of success were rapidly diminishing. The selection of a project manager was still being kicked around on March 6th, which was the day the contractor was supposed to have begun work. Jules sent the recommendation back to Wolfe who then sent it to John Butler who was out of town—and so time leaked away. Finally on March 13, 1978 Ford, Bacon and Davis were selected to be the managers of the reconstruction of the railroad from Shawnee to Fremont—520 miles. I recommended to Wolfe that the 511 coal committee should meet with FBD as soon as possible. Jules Eberhardt was appointed as the coordinator with FBD who came to town on the 16th. I spent that day introducing them to the C&NW people with whom they would be working. On the 21st I put on the slide show as a preview to the following week's field trip over the line. We updated the "Work Book" which proved to be an essential tool in the weeks ahead. March 23rd was a day of meetings: Jules Eberhardt's coal committee and Maurie Reid's branch line group. I shuttled back and forth between them.

Our Mark A. Johnson announced March 2nd that he was leaving

on the 17th. This was a big surprise to his boss, Rich Howard, who was certain that he had made Mark into a good industrial engineer and loyal supporter. Mark, however, had gone to night school, secured his MBA and lined-up a decent job that had nothing to do with IE work. Good for Mark. On March 8th John Sward from the ATSF appeared with another armload of timetables, track charts and profiles. My library was overflowing with hard to get material. (I often wonder what happened to the collection).

▲ Early in March I pressed ahead with the tricky business of siting the passing sidings. We knew how many we needed from our simulation studies. I tested the locations by using TPC to judge the effect on train operations for each location. To scale the project I was also using marketing's projections for coal traffic for 1990—twelve years in the future. The size of the coal tonnage was unbelievable even in 1978. (By 1990 the actual coal tonnage moved was far higher than we had guessed in 1978). Jules reported from Washington on March 10th that the FRA had accepted our marketing projections. The next items for scrutiny were our financial and cost estimates. Jim Wolfe met with the Secretary of Transportation, Brock Adams, who assured him that "our" funds were not included in the CONRAIL bailout then under consideration.

▲ The big inspection tour with Ford, Bacon and Davis began March 27th when we all traveled to Denver and on to Casper, Wyoming. (My plane to Casper was a leased Aer Lingus aircraft named *St. Eugene*). That was a strange assemblage in Casper that night. The next morning a crowd of 19 people stuffed themselves into four hi-rail station wagons. We drove east over the highway to Douglas, Wyoming, where we transferred to rail. The caravan rolled east at 9:30 a.m. The trip was marked by frequent stops to inspect bridges, track and embankments. A lunch stop was made at Lusk, Wyoming, the only sizeable town on that part of the line. That first evening we tied up at Chadron, Nebraska, after traversing the scenic White River canyon. Some of the group took a side trip south of Lusk where geologist Toren Klinge found evidence of a rock formation that would prove to be our ballast source in 1983. The next day one of the cars developed electrical problems. Klinge split off to visit Rapid City and Sioux Falls to check the ballast sources we knew were in operation there leaving 18 of us to cram into the three remaining vehicles. We made a tour of the Chadron mechanical facilities—which did not take long—and headed east up Bordeaux Hill and past Hay Springs to Gordon, Nebraska, for lunch. Train #355 with units #844 and #839 eased past at Cody as we waited in the clear. The second night was spent at Valentine, Nebraska, where eastbound train #358 creaked past with diesels #4506 and #860. Before starting out on the third day we rearranged the party to get better coverage. Heading east across the high bridge of the Niobrara River, through the hilly sandhill grasslands, we stopped in Bassett for lunch. The afternoon run went through Stuart ("the hay capital of the world") to Oakdale. Getting off the rail we doubled back to O'Neill on the highway for the night. On March 31st to save time we took U. S. 275 to Neligh before getting on the rails so that we repeated only 6 miles from the day before. The weather was perfect for this trip. We got in the clear at Meadow Grove for westbound train #355 (units #4506 and #860 again). At Norfolk, we picked up R. M. (Bob) Milcik, division manager of the Iowa division on whose territory we were now traveling and arrived in Fremont trailing train #358. The Iowa division people with us were somewhat distracted as they were trying to deal with the wreck of #252 the night before at West Denison, Iowa. At Omaha on April 1st, Jim Brower, W. H. Huffman, Ray Snyder (all C&NW engineering), Hodges and George Sargent (FBD) and I flew back to Chicago. It had been a sobering trip as the track had not been improved very much from the last trip I had made, and recorded, four years earlier.

▲ On April 5th I took two of the FBD guys to the morning meeting which turned into a long affair as it was Wednesday chart day when each department explained its display. Later, FBD met with labor relations. I seem to have become the facilitator for FBD as I made the introductions and set up meetings on request. George Hanson and Gus Malecha got their chance on April 6th to explain how they intended to operate this railroad when it was rebuilt. FBD brought in a subcontractor, the Henry Company, to deal with communications and signals. On the 7th we met with our finance people. Then I hit a detour.

From April 10 through 14 I was stuck in a mandatory training class at West Chicago for the Kepner Tregoe problem solving technique. This came at a very inconvenient time for me, but there was no escape. The instructor was Jim McEleny. Our class included Vito Fillitti, Bob Kiley (personnel, Jim Hochstein (real estate) Bob Schmiege (labor relations and the future last president of the company), Jim Chesner (operating), Jack Harris (real estate) and Bob Hoffman. We were not allowed to go home at night and had to stay at a motel in St. Charles, Illinois, not far from the training center. We graduated on the 14th. Kepner Tregoe was one of Jim Wolfe's pet projects to make us all better managers. I was not that taken by the methodology but I learned the buzz words to use on appropriate occasions.

▲ Sprung at last from that Kepner Tregoe distraction, I spent most of April 17th with Jim Brower going over the siding location problems. Engineering also showed a film of one of the new rail laying machines in action which was encouraging as someone there was beginning to think about how this job might be accomplished. On the 18th we spent the day with labor relations and seemed to gather more participants as the day went along. On April 19th we resorted to that old method of the string diagrams to illustrate the proposed operation of trains. On a large cardboard the horizontal axis was time shown as a 24 hour period while the vertical axis was the scale miles of the line under study, in this case, all 520 miles from Shawnee to Fremont. Remembering my past experiences with string diagrams and the nuisance of repositioning the pins and strings every time an adjustment to a "train" was necessary, I made a bunch of cardboard templates for train operating characteristics which were very useful in quickly plotting the train diagrams. Some doubt was expressed at this innovation, but when the sceptics tried the templates they agreed they worked quite well. Sidings continued to be a major problem as their locations were never really established. By the 21st we had developed a staging plan and had a much better solution to the proper siding sites.

▲ I was increasingly uneasy with the progress of the planning—there was so much yet to be done. By April 24th I was very conscious of the fact that my position in this project was ambiguous at best because I had no specific directive to provide information or to lead work on any identified problems. I decided to go west and check out a couple of unanswered concerns that lingered in my mind. I flew to Sioux City, Iowa, and rented a car to drive along the Burlington

Northern branch to O'Neill, Nebraska. I sampled the branch, looked at bridges, track, station grounds as far as Randolph. The next day I went on to Osmond and chatted with the agent to learn if BN was planning any work on this railroad as some rumors had surfaced. The branch was pretty well maintained compared to some of our C&NW weed paths. My concern that this branch could become a coal train route was just my paranoia. (The BN's O'Neill branch was sold to the Nebraska Northeastern Railway, NENE, in July 1996). I came to our railroad at O'Neill and turned back east meeting eastbound #358 at Inman. Deciding I had some time I swung north through Verdel and Ponca Creek, Nebraska, crossed the Missouri River at Yankton, South Dakota, went on past Parker to Sioux Falls, the location of the giant quartzite quarry that might have to supply the eastern part of our project. Returning to Sioux City via Hawarden, I had a chance to look at the miserable collection of light iron which could never have carried a fleet of heavy ballast trains.

▲ In Chicago on the 27th FBD was still diddling with the siding solution without much obvious result The next day at the 10 a.m. 511 coal committee meeting the time had come to make some decisions. While some FBD reports were beginning to appear, there were three pressing questions on the table: 1) At least 400,000 track ties must be ordered—now; 2) a ballast quarry site must be found; and 3) a public relations campaign was urgently needed as wild rumors were shooting around the western area like demented jay birds. By May 2nd more FBD reports were being received and digested. I received a call from a worried Mr. Hamer of Ainsworth, Nebraska, who was concerned about the impact of so many trains running through their community. An estimated level of 25 million tons of coal per year translated to 14 trains per day—7 loaded and 7 returning empties. On May 4th DaleWorfel of FBD led an engineering overview session which got into the question of making real decisions. Inertia reigned. It was sort-of agreed that the ballast problem for the first year would draw on existing sources, like Sioux Falls, not withstanding the fact that the railroad serving the area was in poor condition. Jules Eberhardt was advised that the public relations aspect of the program must be faced as there were to be a substantial number of environmental concerns that must be addressed. On May 8th at a meeting of Wolfe and the vice presidents decisions were made to seek a rock ballast quarry at Lusk and to order track ties. Two days later FBD and the operating department hammered out an operating plan for the coal line. The next day the signal and communication costs began to arrive at a much higher cost level than expected. They decided to trim bridge improvements to reduce costs.

On May 12th we heard a rumor of a new office building to be constructed on the site of the Chicago Passenger Terminal at 500 West Madison Street. (The rumor was true and it was built in 1985). Other rumors had our real estate people looking at the former CB&Q office building at 547 West Jackson Boulevard as a possible site for C&NW offices. There was considerable consternation around the place May 24th when FRA track inspectors cited the railroad for 388 violations at Rapid City, South Dakota. The potential burden of a $1,000 per day fine for each was alarming. A decision was reached to embargo the line which put the FRA and the Interstate Commerce Commission into collision.

▲ We were expecting a letter from the FRA on May 15th accepting the material furnished to date. Later, we learned that the letter had not yet been drafted. Jules Eberhardt set up a work plan to complete the FRA 511 application by July 15th. On May 16th concerns were being expressed about the availability of electric power in northern Nebraska. I called a former acquaintance from my industrial development days, Bob Shively, who had once been the industrial representative for Norfolk, Nebraska. Bob felt that the Nebraska Public Power District distribution grid was adequate and that something could be worked out. He sent me a copy of the grid map which I forwarded to FBD's Media, Pennsylvania, headquarters. The operating plan continued to be reshaped and refined by Gus Malecha, George Hanson and Jimmy Johnson with some help from me. On May 23rd Ford Bacon Davis came up with a project cost of $288 million dollars just a bit more than the $275 million we had projected last December. That FRA letter still had not arrived by the end of May.

▲ On May 25th Worfel and Wagner of FBD came up with their final report which they presented to Jim Brower, Tom Tingleff, Kim Schlytter, Jules Eberhardt and me. Their charts showed that our decision making was way behind schedule. I read through the report on the 26th and found it full of typographical errors which, in my view, detracted from the report's impact and cast suspicion on the contents. Somehow, the FBD report was curiously unsatisfying. It mostly told us what we already knew. At this point I went to Greece with my wife and a group of friends for two weeks. When I returned June 13th, it was obvious that nothing to further the project had been done despite the consultant's report that said decision making was badly behind schedule. Something was radically the matter with this entire project.

▲ Maurie Reid, Jeff Otto and I made a hasty flight to Ainsworth, Nebraska, on the afternoon of June 15th in a small plane provided by Charlie Priester, owner of the flight service at Palwaukee airport where our company plane was based. Our regular plane was not available, and Charlie flew us out there quickly in his little jet bullet. We met with Mayor Burdick and the Mr. Hamer, who had complained to us earlier about potential impacts on the community from greatly increased train traffic over the grade crossings, increased noise, and air pollution from diesel fumes, roadway and coal dust. Dick Wozniak from the State of Nebraska was also present. We explained to the Council of Government meeting what we knew about the project and our plans to mitigate such impacts. It was late when we left and we arrived in Palwaukee at midnight. The next day the big problem was laid out for all to see—we had to spend $40 million this year if we were to meet the proposed opening date in 1980. Jim Wolfe could not authorize such a sum on his own; this was a major board decision. On the 19th in a boardroom meeting the decision was reached to advance funds only for engineering. Wolfe told Ray Snyder that the work must be telescoped and that Jim Zito and John Butler would see to getting the FRA application filed by August 15th. Lou Duerinck and I met on the 20th to go over the environmental requirements. Most of the vice presidents were away at a management retreat in Afton, Oklahoma.

At a group ceremony June 16, 1978 I received my gold 30 year service pin and was photographed with my "favorite" vice president, R. David Leach. At least that part of my retirement plan was cast in gold.

▲ Dale Worfel and George Sargent of FBD came back to Chicago with more details for the project. In an all day session the siding location question was settled and an operating plan was adopted. The daily coal meeting had gotten more popular so that our small conference room was now filling up. That long expected letter from FRA showed up on June 23rd containing no surprises. By June 26th decisions were reluctantly reached to commit for track ties, send out rate quotations for coal movements, and to obtain options on land at Lusk for the ballast rock quarry although we were not sure where, or if, suitable rock was to be found there. By June 27th it had become apparent that there was a shortage of C&NW personnel able to carry out all the tasks associated with the coal project. Meetings with FBD were now going continuously. They brought in more of their people to help do what should have been done two months ago (July 5th). David Nixon and I were extremely busy updating the project work book. We were happy to welcome our new hire, Mike Iczkowski, to our busy group to help assemble this growing pile of material (June 30th) which finally went to the printers on July 6th.

▲ I was moderately concerned when I hear that attorney Richard (Dick) Freeman was leaving the company. He had been a leading figure in the coal project, and we would miss his wise and clear vision. The U. S. Senate confirmed his appointment as the director of the Tennessee Valley Authority (TVA) October 13, 1978. The coal project itself was in a state of intense scrutiny by our Tommy Tingleff's auditors who questioned everything. FBD was supposed to produce that magic bottom line figure July 11th, but didn't. When the number was revealed, we were shocked. Our initial $275 million figure which had escalated to $288 million was now $440 million. Amazingly, the western coal project now had a life of its own and the higher cost estimate, while shocking, was not a deal breaker. We all realized (July 18th) that we had to look for alternatives that were less expensive. That library of TPC results was quite helpful in selecting combinations of routes possessing the best operating characteristics, least fuel consumption and running times. While I went through the combinations of TPC, the sharp-eyed auditors kept hacking away at our cost figures. On July 27th David Nixon completed exhibits for the alternative combination coal train routes, Mike developed reports on curvatures and I did the population charts. Jules Eberhardt went to FBD headquarters in Monroe, Louisiana, to learn about their computer cost control system which they were trying to sell to us. In Congress the effort to pass legislation to grant eminent domain to the coal slurry pipeline interests failed.

Work began July 20th to provide a receptionist station at the 7th floor elevator lobby. There had been numerous incidents of strangers wandering around the floor, wallets being stolen (mine included) and unannounced visitors at the president's office. On July 27th a film crew worked in Wolfe's office. Our stock stood at $16. Dave Leach was hospitalized by a mild heart attack on August 6th and his number two, Keith Waldron, was incapacitated by prostate trouble. On August 9th Dick Neu of marketing gave a line analysis on the subject of railroad line swapping with the SOO and the MILW in the iron ore country of upper Michigan.

▲ The Rock Island now attacked our FRA 511 loan guarantee application by filing a protest (August 4th). They claimed they had a superior route for moving coal trains. A quick review of our TPC files showed that they were not correct in that statement. I compared the Rock's against our line from Omaha to Chicago and sent it to Dick Freeman in Washington by telecopier. We had a lot of our people there fighting the Rock island's interference. On the 7th Jim Zito called from Washington triggering a hasty meeting with Lou Duerinck and Dick Freeman to assemble counter arguments. During our deliberations, Jimmy Zito called twice. The next day I attended a hearing in the Federal Building in Chicago at which the Rock Island was shocked to hear that the C&NW had obtained a $90 million loan guarantee from the FRA. The Rock tried desperately to stop the process, but the Illinois Central Gulf and the Burlington Northern refused to support them. It was an interesting sidelight that the loan agreement was signed 8 minutes before the Rock Island hearing commenced.

The Rock Island, however, would not go away and threatened to jeopardize a third of the loan guarantee claiming that their superior main line had as much claim on the funds as the C&NW's. Mike Iczkowski and I decided that we should take a look at this "mighty" Rock Island mainline. Unannounced and wearing hard hats, we inspected their yards at Blue Island, Riverdale and Joliet. The condition of the Rock Island's facilities merely underscored their terrible financial condition. The badly maintained track, shabby structures and equipment and even lack of housekeeping was plainly visible to anyone that cared to look. Rushing back to the office, I had to type the report on our inspection myself because our secretary had gone home because she was too tired to finish the day. Based on our findings Jim Zito requested the Rock Island to allow us to make a hi-rail inspection trip over their line. They would only agree to a business car tacked on the end of a freight train which wasn't good enough for us.

▲ The intense effort to complete the 511 application had picked up by August 10th just five days from the deadline set last June 19th. There was a rush to complete the exhibits accompanying the application. Jerry Groner and Jeff Otto scrutinized and edited our material before it went to press. I concentrated on speed restrictions, curves and bridges. On the 11th we began collating pages, but we were being held-up by John Butler's finance people. Then the Ravenswood computer went out again. On August 15th the ten inch high stack of paper that represented our 511 loan guarantee application for almost a half a billion dollars was completed—on time—and sent to Washington. We heard two days later that FRA was busy reading it. Just sending it in, however, did not finish our involvement—we had to go to Washington and make a formal presentation complete with a slide show. On August 22nd Jules Eberhardt, Jerry Groner, Jeff Otto, Doug Christensen and I met with the FRA at the L'Enfant Plaza hotel and gave our presentation show to some 30 people. Following the show a lot of questions and discussion ensued. It seemed important that we had to demonstrate determination and some passion if we were going to make this thing work. Little caucuses formed from time to time that afternoon as discussions on particular aspects of the application were held. For several days after that big effusion of energy, we all had post-meeting let-down, but the need to prepare additional copies of the application left little time to be depressed. Now we had to consider the problem of finding a construction manager. Certainly no one person on the C&NW was qualified to supervise and control a 520 mile long job. Morrison-Knudsen pushed hard for this job. Ray Snyder spent a lot of time trying decide what work could be handled through outside contractors and what work had to be done by union railroad labor.

The clerks' union struck the Norfolk and Western Railway but the stoppage was gradually being broken by a massive and determined management effort to operate their freight trains with non-union help. Our former manager from advanced systems, Bob Kase, was now an officer with the N&W and spent many days away from home working as a train crewman. The BRAC union threatened to widen the strike with new targets: C&NW, Chesapeake and Ohio and the Union Pacific. Our emergency teams were alerted but the threat dissolved in the late afternoon of August 25th. Four days later when the threat returned, Rich Howard and I were sent to Proviso at 9 p.m. I called locomotive crews and then stood around all night with a crowd at the Lake Street entrance to the yards used by truckers collecting and delivering trailers for the TOFC service. Our gang outnumbered the pickets sent by the N&W to block our yards. By 4 a.m. we were all pretty tired of the whole performance, but that is what J. W. Alsop had ordered. I arranged for someone else to play trains on the 30th; I had more important things to do. C&NW train operations were not in the least affected by this labor action. The strike re-erupted on September 5th. Howard Tell sent me to Proviso to act as chief clerk in the office. My job lasted three hours before the regular chief clerk decided that he had better come back to work. This fracas was the result of disciplinary actions taken against 100 clerks who had failed to protect their assignments the previous week. By noon Wolfe and the BRAC officials had reached an agreement which resulted in pulling the pickets. Commuter service was scrambled on September 6th, but was back to normal the next day. On the 21st there were more rumbles about a strike. George Maybee called me at 2:30 a.m. to help protect the railroad against a strike call. I had been down with influenza and begged off. The strike spread and the clerks refused to come back despite an injunction secured at noon on the 26th. Feeling marginally better, I called in and was assigned to Proviso the next day. At 5 a.m. I took over as chief clerk and served until 10 a.m. when the pickets were withdrawn During my tenure as chief clerk, the diligence of one female clerk caught my attention and I asked why she was not on strike. The reply was that this lady had one ambition in life—to own every twenty dollar bill ever printed. Injunctions and contempt orders filled the air as Rich Vasy and I drove into Chicago. C&NW trains ran on the 28th but the Burlington Northern was shut down. The clerks went back to work on the 29th—for ten days. Bill Alsop was in the midst of this turmoil and dealt with matters in his usual heavy-handed manner. Bill with his tobacco wrinkled stone face (I never once saw him smile) was under doctor's orders to quit smoking and lay off the booze especially after his dry-out in 1972. When business required seeing him, one had to be sure to knock on his door before entering. The smoke rising from his left hand top desk drawer was the telltale of his obeyance of the doctor's orders.

Things were looking up a bit. Revenues were holding at favorable levels with records being set in grain and iron ore movements. On September 15th CIE acquired a new project to which we quickly responded with information drawn from our TPC files. Information was wanted on operations on the Burlington Northern from Minneapolis to Duluth, Minnesota. Bob Fried called from CONRAIL on the 20th. He had been assigned to a project to justify a centralized traffic control (CTC) system and needed some ideas from us on how to approach the subject. Given our lack of success in similar work, I don't think we were of much help to him.

On the 19th I was drafted by the conductor of eastbound commuter train #20 at Elmhurst to act as fireman. Although I had been examined on the Rules of the Operating Department and possessed certificate #512 dated February 1, 1955 (permit to operate a track motor car), the basic signal aspects were still the same as contained in the January 1, 1953 book of rules making it an easy job to ride that left front seat at the end of the gallery car. It was the best view of the track ahead. Calling signal indications to the engineer was easy and brought a sense of really being involved in train operations. Staff jobs like mine were so often remote from actual railroading that I appreciated every chance I could get to be a part of it.

▲ As September began Rich Howard got involved in the coal project for the first time. We interviewed Ray Snyder and Jim Brower of engineering about their planning for the coal line and later talked to Vern Mitchell regarding the signals required especially at grade crossings. It was more and more obvious we had to have a strong construction manager. On the 6th we heard that the Union Pacific was prepared to negotiate again. I gave Mike Iczkowski a background lecture on Project Yellow on September 15th emphasizing the importance of doing good fact finding that would prepare us should the UP really enter negotiations again. On September 7th the east leg of the wye on the joint BN-CNW coal line at Shawnee was the subject of a meeting with Snyder, Brower and Jimmy J. Johnson. I wrote specifications for a project manager again on the 10th.

▲ On September 11th Jim Johnson, George Hanson and I, loaded down with a copy of the 511 application, flew to Norfolk, Nebraska, for a big company meeting held on the 12th. The meeting was packed with people who would be involved in the coming reconstruction. The presentation was closely followed by a crowd eager and wanting to believe that it was really going to happen. The next day W.R. (Bill) Otter and Mike V(?) the trainmaster took us out to see the Nucor plant north of town. It was an impressive steel manufacturing plant whose raw material was scrap steel. Zev Steiger asked me to look into the history of the Orin -Shawnee track construction (September 14th).

The Federal Railroad Administration (FRA) retained a consultant, Transportation and Distribution Associates, Inc. (TAD) of Media, Pennsylvania, to check details in our 511 application. They showed up at the office September 17th. One of the men was Charles L. Towle, well known in philatelic circles as an expert on early Arizona mail service and Railway Post Office history. Another was a Mr. Klaussing who was to check our TPC programs and contracts. A third was a Mr. Feiler whose speciality was risk management. By September 22nd I was privately convinced that the entire coal project was in a state of complete collapse. It was just not possible to rehabilitate 520 miles of railroad to coal train requirements within the estimated cost even with the Federal loan guarantee and still make our time deadline. No one on the C&NW would take any responsibility, action or make decisions. On September 25th president James Wolfe said in a coal meeting that he feared the Union Pacific was going to queer the coal project for us. He also warned against hiring a "big fish" project manager. Rigor mortis seemed to have seized the project. Brad Huedpohl, formerly our geologist, came in with quartzite samples he had obtained in southwestern Minnesota. Jim Brower grumped that Minnesota mining laws were very bad to work with—where he got that information was not mentioned. We continued to hold coal meetings on sheer inertia, but no progress was made on finding the urgently needed construction manager.

▲ Tired of the lack of any progress I fled to Omaha on a dawn flight. Renting a car I went out along the Union Pacific's old mainline at Lane, Nebraska, and followed it southeast to Gilmore Junction where the Missouri Pacific connected from the south. I crossed the Missouri River to Council Bluffs, Iowa, and went over the future coal train route which involved slow trains passing through many at-grade crossings on the C&NW line from its yard to the U. P. Pool Yard. Heavy tonnage train movement through this area was going to be awkward and have significant impacts on the motoring public. The next day I called John Ransom at the UP to ask about their grade crossings in Council Bluffs and learned that the FRA inspectors had been there the same day I was. On October 5th I went back to Omaha with George Hanson from operating. We met Jim Wilkinson, trainmaster, and talked to the Missouri Pacific about interchanging coal trains. J.R.(Jerry) Panning from the Iowa division's operating department caught up with us on the way. We hi-railed across the river on the Union Pacific bridge to the C&NW's yard over the route I had followed several days ago. By highway we went north to Blair and hi-railed south on the old Omaha line on the west side of the river to Florence with a stop at Fort Calhoun. (This piece of railroad had seen its last regular passenger service June 3, 1950 and the last passenger movement, a football special, in 1969. The line was formally abandoned in December 1980). Don York and I had a talk about environmental problems we were having at Council Bluffs (October 7th). There were many problems associated with moving coal trains through this congested area.

▲ The law department requested a description of the Train Performance Calculator (TPC) program. We were now getting a lot of requests for TPC runs from many directions—a tribute to the growing acceptance of this simulation tool. On October 11th the FRA requested permission to release some of our data to the Burlington Northern; Dick Freeman and John Butler took the matter under advisement. Early October brought more commotion and less progress toward getting the enormous coal route project started. The FDA planned their inspection of the Omaha-Council Bluffs area for the middle of October; I prepared a map for their guidance. The construction manager position was still open and would remain that way as Wolfe had instructed us not to hire anyone. Michael Baker Company from Pennsylvania sent a representative to inquire about bidding for the job on the 9th. The next day DeLeuw, Cather asked questions about the project. Three days later it was Peter Kiewit of Omaha. Even Bill Wilbur, our retired engineer of bridges, came in October 13th to seek information for his new employer, Envirodyne of Chicago. They were all given the same data but no decision was ever made to retain them. I read through the contractors' proposals November 2nd and was especially impressed with the Morrison-Knudsen presentation. But, the whole process was paralyzed.

Locomotive engineer training now got the big assist that it needed. We received authorization for purchase of the Freight Master TDA diesel locomotive simulator that we had looked at in 1977. At our track train dynamics committee meeting November 15th we approved ordering a Winnebago van to house the machine. There was the usual setback in our plans when the Freight Master people said that our TPC decks could not be used in the TDA machine's input which meant that we were faced with recoding thousands of miles of track in order to have a simulator that represented C&NW main lines. I was convinced that a software translation could be written—after all, we were only dealing with data files. I made a request on systems December 12th to develop a conversion software. Of course, it was given a low priority as systems was not in any position to determine what requested programs would be of the most benefit to the company and automatically dumped our humble work to the end of the line. Systems was like some sort of black hole where requests vanished and died as the needs for the requested help waxed and waned.

On Friday the thirteenth in October J. W. Alsop was appointed vice president-operations, one of several promotions announced that day. Jules Eberhardt was elevated to vice president-planning and my immediate superior, Rich Howard, was named assistant vice president of corporate industrial engineering. Like Oliver Twist, there was nothing in my pan. Despite a track record of never having produced a single worthwhile recommendation for improving the company, Rich Howard's promotion underscored management's continuing inability to put good managers into the right slots. While I did not covet the position—it would have interfered too much with my own activities—my seeds of separation began to sprout. By mid-October I was sure that my role in the western coal project was finished—the flag bearer had been shot. I determined that January 1, 1979 would be my target date for a "go" or "no go" decision. Because my private geographic business continued to do well, I could afford to hang around and see what happened until then. Shifting more toward my post railroad career, I lunched with Ken Bell of Kraftco and Thomas Hodgkins of Distribution Sciences on October 16th. Tom again said my job was ready whenever I was. The C&NW and the lure of the coal project were fading as my personal transition began. The coal project twisted and turned in the fitful winds of politics and intrigue, blew hot and cold, but never quite died. My role in the coal project was gradually turning into being just an information resource. I was never again to lead any coal project related tasks. My days of calling meetings for specific elements of the coal project were over.

I flew to Boston, Massachusetts, on October 19th to visit a concrete tie manufacturing facility at Littleton. They were making all the ties for the Northeast Corridor rehabilitation project. Dennis Sullivan of AMTRAK gave a very interesting talk and slide show after the reception and dinner. The subject was the CANRON TLS (Track Laying System) machine we were to see in action the next day. We went by bus on the 20th to East Greenwich, Rhode Island, where the TLS was steadily moving along the railroad ingesting the old jointed rail track with its wooden ties and dropping 700 pound concrete ties precisely into place with an authoritative clunk-clunk-clunk. This is exactly what the C&NW needed for rehabbing the 520 miles in Wyoming and Nebraska. The TLS machine, however, would remain a fantasy for the C&NW because it would never have the money to purchase or lease one. At this point, I was becoming just another consultant to the C&NW; I could proffer advice that I knew would never be heeded. I was becoming an outsider within in my company.

Rich Howard asked me to go to Richmond, Virginia, for him as one of a team of eight from the C&NW visiting the Chesapeake and Ohio's new CTC center on November 7th. Harmon Electronics had constructed the machine for this installation. I could see at a glance that the concept was good for that section of the C&O, but would never do for the coal line. The idea was growing in the industry that CTC control was the future and should be concentrated in a big central authority rather than scattered around the railroad piecemeal. The next day WABCO (the old Westinghouse Air Brake Company) came in to our Chicago office to describe their train control concepts which appeared better suited to our coal train operations. Preparation of another 511 application to the FRA for money to repair freight cars had Rich Howard's full attention which was just as well because it kept him out of my activities with the coal project. The FRA liked smaller projects such as fixing freight car and limited track repairs much better than that enormous elephant of the coal line rehabilitation. The C&NW was the most aggressive of the Nation's railroads in seeking government loans under section 511 of the so-called 4R Act.

▲ We were instructed to press ahead with planning for the coal line despite its lack of structure and crumbling facade. By November 9th the Union Pacific's proposal to build north to the coal fields had heated up again; they even were using our Project Yellow name for the proposed construction. The C&NW hired Washington consultants to assist Jerry Conlon through the political maze. I had a chance to meet them on November 10th—Bill and Alice Chambers and Max Parrish. Later that day I picked up the siding respacing problem again, but it was just an abstract exercise at this point. Something about the coal project was changing, but in my increasingly isolated position I could not quite make out what exactly was happening.

▲ The Union Pacific presented the C&NW with a proposed contract November 12, 1978. Our Project Yellow was back in play. We met on the 13th to review the UP proposal and to comment on the TAD report to the FRA. Brower reported on his recent visit to the UP. On the 16th we assembled in the boardroom to review TAD's proposal to rationalize the western coal tangle into a three railroad network. Their idea involved running C&NW and BN coal trains over the joint line to Shawnee, east on the rebuilt C&NW to Van Tassell, southeast on a new connector link to the BN's line near their station of Nonpareil, Nebraska, and thence to Alliance where some trains would turn east and others proceed to Northport where the Union Pacific could accept traffic. Nonpareil, originally the first town and county seat of Box Butte County, was no longer even a dot on the map. (The place was so tiny that the editor of its newspaper, *The Grip*, named the place for six-point nonpareil, a very small size of type. In 1978 it was the location of a passing siding on the old CB&Q line south of Crawford where the railroad climbs out of the White River valley). By building a new railroad from Van Tassell to Nonpareil, loaded trains would have avoided the drop down into the White River drainage to Crawford and the stiff climb back out of the valley that required helper service. The terrain to be crossed by this proposed railroad was rough, broken country that would present major challenges to railroad location. While the TAD idea might have made sense on paper, it was political poison to the three railroads involved. Involving the BN in a C&NW-UP show would have resulted in a transportation snarl with the BN holding control of the key section. Then, of course, the BN's important connection to their Colorado and Southern line south to Denver, Colorado, at Wendover would not be available. Even given multiple track construction and improved interchanges, the inherent grades across the intervening territory to the Platte River Valley worked against affording sufficient capacity for the immense tonnages that were to be hauled out of the Powder River Basin. The TAD proposal died at delivery. I had finished the siding thing—again—and fielded questions from Wolfe and Zito. TAD did not know about our Project Yellow plan and did not independently stumble on that critical connection. We now shifted our focus to the new connector line concept while still suggesting to the world that the huge 511 application to FRA was still our main effort. On November 21st at Jerry Groner's suggestion, we gave Peat Marwick, the accounting firm, answers to their questions regarding our capacity design. They seemed to agree with our response. By November 22nd the C&NW had resumed negotiations with the UP while we still clung to the 511 application as a fallback. By the 24th the marketing people from C&NW and UP had "nearly" agreed on a cooperative project. The whole project was scrutinized in every detail as the negotiations continued. On the 27th I was supposed to go to Washington, but that evening Jules Eberhardt called and asked me to stay in Chicago and help calculate the UP's capital expense requirements. On the 28th Dave Boger, Jim Brower, Ray Snyder and I worked through anticipated expenses the UP might face if we built Project Yellow's Van Tassell-Joyce connection. My treasure trove of track charts and timetables proved to be of great value. Zito disagreed with some of our assumptions, but a few adjustments eliminated his concerns on November 30th. I redid the siding spacing—again—using the Project Yellow connection. I hoped that a C&NW-UP deal was coming soon because there was now an 85% probability that we would be forced to re-do the FRA 511 application. The next day two men from ARCO (Atlantic Richfield) in Los Angeles came to see how things were going as construction of their mine in Wyoming was underway. We could only show them the 511 FRA application for the Shawnee-Fremont rebuild as the existence of Project Yellow was still kept a deep secret.

▲ It remained a secret until the *Wall Street Journal* broke the story December 5th. A week later, on December 11th, Wolfe revealed that he had put one over on John Kenefick and the Union Pacific. His poker hand now lay face up on the table. He knew the C&NW would never be able to reconstruct the 520 mile line across Nebraska within a reasonable time frame even with the Federal guarantees. Jim Wolfe had remained a firm believer in the Project Yellow concept and used the 511 application and Ford Bacon, Davis consultant study as smoke screens. He almost lost the game when we began to sniff out some of the inconsistencies of his position (delay in selecting a construction manager, delay in ordering track ties, seeming inability to make crucial decisions.) At this remove I believe that he selected Ford, Bacon Davis as the project consultants despite their higher bid price because they were likely to have had fewer personal connections to the UP at Omaha than either Morrison-Knudsen or Peter Kiewit which might have blown his bluff.

On the day of the big revelation, December 5th, I was in Washington at a TDCC meeting. The next day I flew to New York on the shuttle before returning to Chicago as there were always people to see in the Big Apple. On December 11th the Des Moines-Kansas City corridor problem was the operating department's main concern. I asked Mike Iczkowski to take the notes from that meeting. Earlier in November, Rich Howard, our new assistant vice president, had held his first staff meeting on the 17th. It was obvious that CIE was woefully understaffed to deal with the many large projects facing us. One of the assignments involved merger of the Lake Shore and Wisconsin divisions. My notes from the time speak of people unwilling to take the lead or any responsibility in many of these projects—the president had to decide everything. Wolfe may have liked that autocratic posture, but his unwillingness to delegate responsibility seriously undermined the company's ability to function well.

On December 12th Rich asked me if I was interested in the open director's job. That I was even asked given my marginalized position was surprising. I put on my game face and said certainly. He also had the temerity to say that some unnamed people had reservations about appointing me to that position. It was hard for me to conceal my utter contempt for his arrogant attitude. I had already decided that January 1st was a decision point for me and now the reluctance to promote me to the logical position in the department accelerated my determination to cut loose. If the coal project was going to happen, I wanted a chance to stay and see it through. As a director my position might be improved in becoming further involved. At the same time, the new title would not interfere with pursuit of my private affairs because being a director carried about the same level of responsibility as a manager, and it paid better. On the 19th he rather casually appointed me to the director's job. It was my last promotion on the railroad. Jules Eberhardt gave me a good appraisal on the 21st. I was now in a better position to monitor the coal project and participate where I could.

▲ In atrocious weather I flew to St. Paul, Minnesota, on December 20th to meet with Warren Stockton and Craig Kloer of the Burlington Northern about operating the joint coal line then nearing completion in Wyoming. They were deathly afraid of the Union Pacific and said the BN planned to oppose the recently revealed alliance with the C&NW. I flew back amid a press of Christmas travelers and wrote a memo to Jules about the BN's attitude. The memo was of great interest to our management and put them on guard for the coming legal battles in 1979 and 1980.

33

KEYSTONE KOPS, 1979 . . . ▲ Now that the pathway to the Powder River was clearly defined, the coal committee meeting was drawing record numbers of interested people. On January 2, 1979 despite deep snow and −13° F. cold, the committee meeting was crowded. Because Project Yellow, the coal connector line, was back in play, solving some nagging design problems was greatly simplified. I redid the siding positioning in one day. (That siding location problem was to bedevil the new railroad until complete double tracking many years later solved this intricate operating puzzle). On the 5th I worked with Gus Malecha and George Hanson of operating to set priorities for 1979 while Jules Eberhardt was being pressured to establish timing for the coal project.

Tom Hodgkins, president of Distribution Sciences and his colleague, Robert G. Ingersoll, encouraged me on January 8th to leave the C&NW and join their company as chief of geography. After a long and friendly conversation, I agreed that I would seriously consider their proposition, but I had just been promoted to director and there was a huge project afoot in which I had already invested seven years. While my future involvement in the coal project at this point was unknown, I wanted to stay on until it was clear

▲ As noted before, this symbol will be used to denote material relating to the Western Coal Project.

that I was no longer involved. Tom wanted my geographic business and me to manage it. I agreed to continue maintaining my geographic data base on DSI's computer because it had become a keystone of Distribution Sciences' software.

Dave Leach confirmed my appointment to director January 4th. At that time I obtained permission from him to continue my involvement in the national standards and coding efforts although that activity had tailed off to a great degree. My pay was now just over $31,000, triple what my salary in the traffic department had been in 1967. Of course there was inflation to consider when comparing dollars, but a threefold increase was hard to argue against. While I was pleased with the emergence of the coal line project after all the discouragements, there remained a faint air of regret that so many positive accomplishments could have been achieved more easily and sooner if only . . . Recognizing the capabilities and motivations of the personnel involved in this important work and understanding full well that my position in the process was of increasingly smaller influence, I determined to stand aside and let the anointed ones struggle with the project while I observed their progress, or lack thereof. From time to time, there would be a need to step into the project and, perhaps, contribute to easing it past tight spots, but only if asked. Being on the edge of the river sometimes has the advantage of perspective. The negative, of course, is that the river is flowing on without you. My separation from the company was underway. I consciously began to concentrate on the coal project and delegated other assignments to my aides as a good director should. Bob Russell called on the 14th to extend congratulations on my promotion. I tried to sound appreciative. My old manager's job was open; I talked to Lowderman of personnel about that on the 9th. He said there were not many candidates available.

Because of the extremely cold weather, January 12th's commuter service stood at a dismal 40% of on-time performance. On the 13th a big storm dumped 20.7 inches on snow on Chicago and northern Illinois. The center main from Elmhurst to West Chicago was out of service as it was impossible to plow it out without blocking the tracks on either side. On a Sunday evening January 14th at Park Ridge, Illinois, our highly regarded chief mechanical officer, Milton H. Crandall died in the cab of one his homemade creations, #502, from a sudden heart attack while trying to get the stalled engine running. Commuter service remained badly disrupted so that by the 17th the trains were running without schedules—just load and go. The entire week was a struggle against the elements. On the 24th we got another 12 to 18 inches of new snow. The C&NW spent $7 million just fighting snow. The Milwaukee Road and the Rock Island just about gave up.

Personnel changes in our small group came fast and furiously. Larry F. Kohn, my assistant announced January 15th that he was leaving on the 31st. I interviewed Kim Schlytter and Charlie C. Harmantas from finance for CIE work. Rich Howard agreed on Kim but doubted Charlie's abilities. Not having any other candidates, Rich finally approved Harmantas for the job April 10th. Charlie reported for my old project manager's job May 16th but didn't last very long; he resigned on July 31st to go into the real estate business. We had a new secretary, Sandra Swiela, a pleasant Polish girl from Chicago's northwest side. Roger Westburg joined our group in January to work for Rich Howard. Mike Iczkowski was drafted by the operating department to be acting yardmaster at Fortieth Street yard January 29th and 30th. Mike liked that experience a lot. He soon shifted to studying diesel locomotive characteristics which presaged his future career with the Union Pacific Railroad. I assigned Kim to pick up Larry's Kohn's work on the CTC investigation February 5th. That was not a project destined to go very far; it was declared dead on March 13th. We were now directed to consider a new yard at Ames—I gave it to Schyltter and Don Nelson (of operating) to chew on (March 30th). On April 12th the CTC program for the Iowa division revived briefly when the FRA turned down a 511 loan guarantee for yard construction at Ames. That denial effectively killed that proposal (May 2nd). So, it was back to considering the CTC again. This CTC project was like a Yo-Yo which goes up and down while making no forward progress. While the idea had suffered official death, here we were saddled with it again on May 4th. The original research had been done in 1978 by the now departed Larry Kohn. I put Kim Schlytter to work on it giving him guidance from time to time. By May 25th it was obvious that this study was not going well at all. By the 29th I was convinced that we knew too little about the kind of trains we were to control and the level of protection that would be required. As a first step I went through all the passing tracks that the Iowa division had scrapped from 1955 to 1978—there were a lot of them. Edward P. Evert was appointed a director of CIE April 17th. A former New York Central employee, he had worked at Ravenswood for several years before coming downtown.

▲ Mike Iczkowski developed the characteristics of the new GP35 diesels and tested them in the TPC program for use in coal trains from Shawnee to Fremont. They were not the best use for this kind of power—the good old SD40–2s did a much better job. At this point in development of the western coal project we were afflicted with a new mandate from management, the Interim Plan, which called for moving coal trains over the Shawnee-Fremont segment with the existing locomotive fleet in the 1980–1982 period before the new connector to the Union Pacific could be completed. The Interim Plan was a farce and consumed a year of staff time with absolutely no visible benefit unless the exercise could be said to have provided a training bed for the real event. The Shawnee-Fremont line was a junk pile, not a railroad. One result of Brower's reappraisal of the existing railroad was reduction of our original 520 miles to 519. That 519 miles was composed of 60.9 miles of 72 pound, 172.6 miles of 90 pound, 233.2 miles of 100 pound rail over which no loaded coal train could traverse with safety even if the tie condition was excellent which, of course, it wasn't. The balance of the rail was the 16.5 miles of 110 pound suspect heat rejects from the mainline laid on the grade down the White River canyon east of Harrison, Nebraska and 35.8 miles of fissure-prone 112 pound rail near Gordon. Theoretically, these sections were heavy enough for loaded coal trains, but the tendency of the defect rail to crystallize and break was too risky to operate very much heavy traffic over these sections. We were instructed to prepare a plan for beginning coal train operations on January 1, 1980 from Shawnee, Wyoming, to Fremont, Nebraska, identifying all the elements that required action and prepare a phased schedule for the necessary work. This was a task that had no foundation in reality. After recovering from our utter disbelief that such a plan could be proposed by our top management, George Hanson and I fell back on our hard-earned old C&NW experience to try and make something out of nothing. We listed the tasks to be done and figured out a plan to deal with this crazy concept. Since Project Yellow was not dead—

just dormant, I included a connecting track to the proposed new railroad at Crandall. Years later the remainder of the old railroad from Crandall to Van Tassell was used as a siding to store extra coal hoppers when the track east of Harrison was destroyed by the May 1991 White River flood. Had that flood occurred early in 1979 we would not have had to deal with Wolfe's Interim Plan. Behind the scenes Wolfe, apparently, was playing another poker game and needed this improbable plan to mask his moves. Even though we may have suspected something was going on, we had orders to pursue this impossible task. Gritting our teeth, we tried to make it work.

▲ Don York, our environmental engineer, asked us to justify our 19 train per day estimate on January 19th. That was a straight forward calculation based on the annual tonnages which marketing and the FRA had already agreed to. (In 1997 the actual average daily train loadings hit 24.7). While Jules Eberhardt, Jerry Groner and Jeff Otto took a revised FRA application to Washington on January 23rd, I stayed in Chicago to work on the environmental statement with Don York. On the 30th we were asked to collect any and all regulatory material that related to the coal project. Doug Christensen and David Weishaar of coal marketing invited me to accompany them to Kansas City to meet with Gulf States Utilities' consultants, Black and Veatch in Overland Park, Kansas. We presented the slide show and gave them the C&NW coal project, as now revised to show the Union Pacific connection at Joyce, Nebraska..

▲ On February 5th the bottom dropped out of the thermometer hitting –17°F. Snow covered the landscape because there had not been a thaw since December. Many freight trains were marooned on track 2 across Illinois. Mike Iczkowski and I investigated an alleged shortage of motive power. There was no shortage; disruption to normal locomotive distribution was the problem. On the 12th another storm broke the winter's record by running the season's total to 82.4 inches. During this bad weather the coal committee prepared the long missing public relations aspect of this project. Joe Marren of public affairs and I set up next week's visitation in Wyoming. We also had to prepare a logic statement that argued the use of Council Bluffs as the preferred interchange over Fremont (it gave the C&NW a longer haul). Meanwhile, the branch line committee was fussing about being required to handle freight other than coal on the proposed new connector line. This unwanted traffic would likely consist of bentonite off the Colony line. Moving general freight on the connector was a real threat to the coal line's track structure as bentonite clay had a tendency to leak out of cars and contaminate ballast. (From late in 1984 to May of 1991 some bentonite traffic destined for the west was handled by the Casper turn and set out at Barnes siding on the new coal line for further movement to South Morrill. When the White River took out the Harrison-Crawford railroad in May of 1991 that event coupled with the abandonment and sale of the railroad east of Chadron and the sale May 4, 1996 of the Dakota Junction-Colony lines to the Dakota, Minnesota and Eastern, shifted most of the clay traffic eastward from Rapid City). On February 9th David Nixon and I went to Ravenswood to dig out the density statistics we needed for the coal project report. We actually accomplished something thanks to David's prior experience in that chaotic operation.

▲ On Lincoln's birthday Jim Brower, Joe Marren and I flew to Denver despite the wretched weather. We transferred to a flight to Scottsbluff, Nebraska, where we met Larry Grapentine from the Chadron engineering office. Joe Marren and I visited with county officials in Scottsbluff to acquaint them with the coming changes to railroad operations in the area. We then drove west to Torrington, Wyoming and repeated the news to the Goshen County Association of Governments. The meeting was supplemented by press and radio interviews. We seemed to have brought better weather with us as the temperature rose to 66° on the 13th. Joe and I drove north to Lusk, Wyoming, for a noon meeting at a local restaurant. Twenty people were expected—ninety showed up. Our presentation was followed by a question and answer session that went on into the afternoon until we finally had to shut it down about 2:45 p.m. Based on our reception, Jim Brower was going to have a wasp's nest on his hands. Many ranchers feared their prized isolation was to be destroyed by great, growling mile-long coal trains. When the local newspaper featured a most unfortunate picture of me addressing the group, it typified the hostile attitude of the community and cast a shadow over the hoped-for easy progression for this project. After lunch we hi-railed east to Van Tassell. Joe and I then drove south over the highway to Orin and Wheatland to Little America near Cheyenne. On the 14th Joe

Marren and I called on the Wyoming state government at the capitol to share information regarding the proposed connector coal line. They expressed appreciation for our consideration. That was the beginning of the official public relations campaign and it was a flaming disaster. A storm of protest developed which, organized, caused great delay to the successful completion of the coal line. With the advantage of time and distance, a good case could be made that this introductory public affairs campaign should have been handled jointly with the more expert Union Pacific people who had a deeper relationship with Wyoming.

▲ Jim Brower, our construction engineer, became the focus of one rancher's ire on the 20th. In our early determination of the railroad's alignment, Jim had little knowledge of the ground details in that vast expanse of undeveloped country. Mr. Desenfants, a rancher east of Lusk, was very perturbed when he found the centerline of the proposed railroad came very near his ranch house. At this stage of the design it was relatively easy to shift the track some distance away from the dwelling. On the 20th we received ARCO's mine plan so that we could now plan railroad facilities to serve the tipple. Radio station KGOS at Torrington called me on the 21st to report that 55 ranchers had formed a Wyoming Landowners Association to oppose the railroad claiming adverse environmental impact. I reported this to public affairs for their response, but no one was willing to take any action to address the matter (typical of that crowd). I then turned to Jules Eberhardt as the project coordinator to formulate a response. By the 26th we still had not responded. Our studied inability to react exacerbated the whole situation. Two days later the adverse publicity in the Torrington paper landed right in Jules' lap. My role at this time was only that of an observer and reporter as I had no mandate to act. I elected to stand aside and let the deteriorating situation show who was able and who was not. My suggestion that a carefully prepared brochure describing the project and its public benefits would be useful in calming the uproar was not pursued. The locals were treated to a faceless Chicago railroad threatening their valued isolation. No one on the C&NW seemed to be in charge at this point. Jules Eberhardt stayed minimally involved despite his designated project leadership. I had lunch with Jim Pierce of Morrison-Knudsen and tried to figure out how our stumble-footed inability to handle this project in an orderly fashion could be papered-over for the public view. On March 1st Warren Wilson, a reporter for the *Casper Star Tribune* called to advise that anti-railroad forces were organizing in Wyoming. The Interstate Commerce Commission now entered the picture on March 2nd and expressed its displeasure at the public uproar. Jules finally bestirred himself and got through to the only person able to make a decision—the president, Jim Wolfe. Jim sent Jules Eberhardt out to face the wolves at a landowners' meeting March 15th which nearly turned into a lynch mob. Now he realized that we had a problem—a BIG problem. On his return he said we ought to position someone out there to deal with these local matters. Of course, that did not occur.

The horrible winter and its crippling effect on the railroad resulted in another thorough review to determine what had been done right and what was done badly. All our corporate industrial engineering (CIE) projects were put aside for concentration on this new presidential directive. I was supposed to address a class of trainees on February 16th about the work of our department. But, the commuter schedules were so badly disrupted that I was late to the affair and had to endure Sy Berman's description of the systems department. By February 20th the railroad was constipated with cars that connections could not take because of blockages on their lines. Our group began interviewing operating officers all over the railroad about the snow emergency. Kim Schlytter and Roger Westburg were sent to South Dakota; Mike Iczkowski and Rich Howard interviewed operating department personnel in Chicago. All through March our group asked questions and collected comments for the report.

Carl Hussey asked me on February 22nd to review the Burlington Northern line from St. Paul to Hinckley, Minnesota, in connection with the BN-SLSF merger case. If we could get operating rights, the C&NW would withdraw its opposition to the merger. Always happy to add to our store of knowledge, I coded the White Bear line into TPC. Three days later I handed Carl the results. Why was a project director doing simple TPC runs? Because all his help was out on the railroad working on the big snow job. On February 23rd I met with John O'Neil of motive power and some consultants who wished to set up a mechanics training course at the diesel shop M-19-A at Chicago Shops. That same day I got the 1978 density statistics map finished while David Nixon was working on a network model that might prove valuable to operating our railroad more efficiently. Too bad the model was so expensive. Jim Zito, however, was so impressed with our Network Model Report that he approved sending David Nixon to Stanford University for training (May 3rd).

▲ The Kansas City Southern was upset about the train operation material we had furnished to Black and Veatch on January 30th. Their railroad's Achilles Heel had been thoroughly exposed by our TPC runs. Those nasty grades through eastern Oklahoma were now public information, much to their annoyance (February 23rd). By March 5th the Burlington Northern was making claims that the C&NW-UP proposal outlined in Project Yellow would ruin the BN. Their complaint arrived on the 7th, a first reading of which gave us the distinct impression of hysteria. Chris Mills, a new C&NW attorney, was now handling the coal project (March 13th). Recognizing that the C&NW was going to have to learn some new techniques in coal train operations, I invited the Vapor Corporation to give us a presentation on locomotive creep control which was a real necessity when loading a coal train. A loading train must maintain an even pace through the flood tipple to provide equal loading per car. It is not easy to manually maintain a steady forward movement as each car when loaded adds more drag to the engine's power output requiring steady, but tiny, increases to the throttle settings. Diesel locomotives usually have only eight notches from idle to full power—far too crude for a flood loading operation. The lococontrol unit functioned in the lowest end of the throttle range performing the minute incremental adjustments that a human operator could not replicate.

Speaking of locomotives, our new training simulator from Freight Master arrived March 7th. Unfortunately, the conversion software to translate our TPC decks to the TDA input was not yet finished. Luckily, Halliburton now said they could convert the TPC data and I sent a load of tapes to them in Duncan, Oklahoma, on April 5th. Mike Arakelian could hardly wait to get his hands on the machine and show it off to Jim Wolfe. When Wolfe was invited to "run" a train on the machine, he quickly cranked a heavy coal train up to the full speed of 40 mph. Mike then quietly asked the beaming Wolfe what he intended to do about that slow order just ahead. Incautiously, Wolfe "wiped the clock", setting the brakes into emergency. The CRT monitor showed violent buff and tension forces rippling back and forth through the train. To his question of "what happened?" Mike said, "You have caused a million dollar wreck." Sobered by how quickly the "event" occurred, Wolfe became an enthusiastic supporter of the training program and the TDA machine. He said that he had not really appreciated the skills required to get a heavy train safely across the railroad. He also asked why we had not obtained the machine sooner. To our reply of "budget", Jim smiled because he already knew the answer to his question. Arakelian, a former locomotive engineer, loved his TDA simulator. Installed in a Winnebago van which was driven to crew change points all over the railroad, the TDA machine was introduced to the engine men by Mike in his big white Stetson hat and by instructor Frank Dickinson. Curiously, locomotive engineers were often anxious to re-run their trains in the simulator. Coming off duty, they would change clothes, go home for dinner and come back to use the machine. The simulator told them about train forces in braking and release and what happened through the dips and over the hills in their territories. Most enginemen had no concept of what happened in the waycar when slack ran in and out; their only information was an irate conductor or hostile rear brakie telling them off at the end of the run. With the advent of bigger diesels in freight service, the necessity of seat belts in the rear car of the train became apparent. The TDA machine was also very helpful in recreating actual wrecks when mishandling was suspected. This forerunner of a common computer game in later years was a huge success wherever trainmen gathered.

That C&NW Twin Cities-Kansas City corridor continued to plague the railroad. My solution of 1972 still held its premier spot but we did not own or have trackage rights on the Rock Island line south of St. Paul. On March 8th George Hanson, Don Nelson and I flew to Mason City to meet with Bob Milcik, Dallas B. Carlisle, Dick McDonald and Glenn J. Kerbs—all involved in moving trains north and south. The cost to patch-up the old CGW was estimated at $20 million. That afternoon an emergency was declared at Kansas City from Chicago. It wasn't the first time. On my return to Chicago I spoke to

Jules Eberhardt about the impasse; he said he had to talk to Wolfe— the only man on the railroad that could make a decision. In retrospect, I think Wolfe liked it that way; he relished his role as an autocrat. Over the years his presidency evolved into a regency as he ran the company like a royal potentate. Three days later on March 12th Jules reported that Wolfe had said the projected costs for repairs to the Kansas City line were out of sight and must be trimmed. This was another example of wasted staff time. Communications to the top man through intermediaries who filtered the information was not conducive to getting things done. Snyder and I tackled the cost problem April 2nd and found that engineering had given already given a great deal of practical thought to this rehabilitation project— the old CGW was in rotten shape. (In five years it would be abandoned).

On March 13th the tariff publication unit asked us to help re-run the C&NW mileage tariff which was now six years out of date. Because of the many branch line abandonments, the shortest mileage combinations had changed drastically affecting those freight rates based on miles. This was a difficult and complicated job even with the computer network in operation. I flew to Washington on March 14th for a TDCC meeting. The geocode committee was reluctantly chaired by Bill Rockwood of General Motors. There were about twenty in attendance, but no one from the trucking community was present which, at least, made a smoother affair of it. The next week I was summoned for jury duty at the DuPage County courthouse in Wheaton, Illinois. I knew that there would be a lot of sitting around waiting for a call to serve on a panel. I took my Duluth, Winnipeg and Pacific (DWP) profiles with me and coded track from Duluth north to Winnipeg, Manitoba. The county was satisfied that I had responded to their summons, and I was happy to produce some useful material while performing my civic duty. That TPC coding turned into a valuable resource when Ed Burkhardt negotiated with the Canadian National (DWP's owner) on a coordinated service from Winnipeg to South Itasca.

By March 23rd Rich Howard said that he was unhappy about the drift of our department away from industrial engineering applications. On the contrary, I was heartened by its gradual de-emphasis. Rich felt that our group's effectiveness was being diminished by the change to its new role as internal staff consultants. He was utterly mistaken—it was the industrial engineering that was at fault. Nothing he ever did in IE on the C&NW ever amounted to anything. The North Fond du Lac shop fiasco of 1975 was a black blot on his record. Frustrated at his growing loss of status, he systematically began editing and re-writing my letters and memos which irritated me beyond measure as his knowledge of the pertinent facts was often minimal. Rich's boss, Dave Leach, had a fit about some letter I had written only to find out his displeasure was misdirected. My letter was actually a rewrite by Mr. Howard who had not bothered to show the edited result to me before sending it on (March 29th). This arrangement was not working well and my relationship with Rich was becoming a good deal more prickly. Perhaps his impression that I was a bad director was true. Rumors of major layoffs were now current; I wished that one particular individual would be selected to be the first one to go.

▲ On March 29th I met with our lawyers working up a non-disclosure agreement with the U. S. Railroad Administration (USRA). John Prokopy of Peat Marwick and Mitchell, the accounting firm, showed up April 3rd. His firm had been retained by the FRA to review our latest revised application for a loan guarantee—another group to educate. On April 10th we were asked to examine the Colorado and Southern line south of Wendover, Wyoming, to Denver, Colorado. With the C&S profiles already in the computer, it was an easy task to identify and highlight the troublesome spots on the BN's coal train operations heading south to Texas. About this time someone suggested that we construct a railroad southeasterly parallel to the Burlington Northern's line to Northport, Nebraska instead of straight south to the Union Pacific. That was just an echo of the TAD suggestion of last year. Shifting the corridor east would have required more railroad construction, added to the existing trackage, involved more communities, crossed more public highways and required dealing with some pretty rough terrain. We documented these reasons and then put that proposal in the freezer. Jerry Groner worked on alternate routes across Iowa (April 11th) while I was writing a paper on capacity analysis.

On April 6th the operating department started a blitz campaign to get the cars-on-line count down. The resulting scene resembled what happens when the bottom drops out of a box of marbles. We had people running this way and that with no idea of what they were doing. After three days of intense concentration, there was no visible reduction in the count. By April 12th the car count had grown even larger because of a high proportion of uncontrollable industry owned cars. While that was going on, I was coding the old M&StL line from Peoria west to the Rapatee Mine because marketing was talking about a garbage train that would use the old strip mines as a dump. The infamous Winter Report (Snow Job) was still not completed by May 2nd. It was taking more time and effort than originally contemplated. By the 10th serious complaints were being made about the slow pace of completing the report. Of course, we were now in the middle of a heat wave; snow and ice were far from everyone's minds. Pushed to a completion near the end of May, Wolfe read it on May 31st. There were some real horror stories about bungled operations. Rich Howard asked our group to review the report on June 13th and comment on its content. The cars-on-line blitz was resumed June 15th. By September 4th the number was off the top of the chart. This blitz had no effect at all because no one ever understood the reasons for the high car count. The cause was, quite simply, poor and irregular train operations—a problem that was too big to be fixed overnight even if the correct decision was identified and implemented—just fix the track.

I heard a rumor on April 12th that the Milwaukee Road was about to collapse and passed it on to M. S. Reid and Jerry Conlon. They urged me to follow up but my contact had gone home for the Easter holiday. By the 19th the Milwaukee's demise seemed near. Unlike the Rock Island which liquidated its railroad in 1980, the Milwaukee decided to shear off pieces including their line to the Pacific Coast. The C&NW tried to acquire bits of the Milwaukee without much success. The Milwaukee in restructuring their system suggested on June 15th that they would like to swap some of their Iowa lines for some of our branches in Wisconsin. The ICC examiner's report June 20th caused the C&NW to junk its plans to acquire parts of the Milwaukee Road.

A new company film made by public affairs was released and CIE played a part. There was a segment in the film that had been shot by photographer George Gibson of our western coal simulation laid out in our conference room five years before. I was identified in the film as vice president-planning along with some other inaccuracies. The film was obviously a personnel department recruitment piece and was shown to new hires for many years after the content had ceased to have any meaning.

▲ Out in Wyoming entry to the ranchers' lands was planned for the next week to permit surveying. I dearly wished that I was with that location crew. There was trouble; the surveyors were physically barred from entry by the ranchers forcing the railroad to start legal action. (Keith Feurer's account of the long and complicated negotiations with the opposition WyoBraska ranchers and other non-association land owners was part of his May 1984 talk to the American Railway Development Association). Jules Eberhardt came back from an Arizona vacation on April 16th and called a staff meeting to get some momentum going on the coal project. Ed Evert, our new man, Murray Humphries and Bob Schmiege (labor relations), Rich Howard (CIE) and George Hanson from operating were there. As usual, trying to jump-start the project resulted in absolutely no progress at all. On May 7th hearings began in Wyoming on the access question; Jules attended the sessions. In Chicago I went over a yard plan for South Morrill, Nebraska, with George Hanson and Don Nelson from operating. This was to be our big interchange with the Union Pacific at the south end of the new connector. (The yard was constructed and became a key part of the coal line operation).

By April 20th freight revenues seemed to be improving, but the company still registered a $30 million loss in the first quarter largely due to the severe winter interuptions and heavy snow removal costs. I was again asked to write a synopsis of the Des Moines-Kansas City reroute. Our 1972 report was another idea that was just too good to go away. When I returned from a trip to Washington, I found that once again Rich Howard had rewritten my paper, completely destroying it in the process. I am glad that the concept was good enough to withstand his meddling. My next Washington trip was for a TDCC meeting to review the Federal Information Processing Standard Code (FIPS #55) as a geographic identifier. The SPLC, however, was better suited to transportation industry requirements. I included the FIPS codes in my private geographic database as just one more good access key. Rich Howard complained about my attending this meeting despite the agreement we had in January that I could continue my activities in national code programs. I let him know that he was becoming a pain to work for. Our already poor relationship deteriorated further.

The Rock Island was the next subject to be examined. Anticipating the day when it would collapse (in another year) I flew to Des Moines to see if our simulations accurately portrayed their mainline across Iowa. The weather turned bad in the west so I turned east and sampled track at West Liberty, drove south to Grandview and discovered a new track under construction to a power generating station. I cut back north through Muscatine to the mainline at Atalissa, Iowa, and then went to Des Moines. The bad weather swept past so I went west the next day through Urbandale and found a surprisingly good railroad. I also picked up a wad of discarded train orders which detailed all the slow orders in effect and talked with some of their field people I met along the way to Council Bluffs. Returning along the C&NW from Missouri Valley to Boone, I attended the morning meeting on the 27th much to Bob Milcik's surprise. Back to Des Moines via Ames was a quick trip. At Des Moines I saw the TDA locomotive simulator and its Winnebago in use under the tender care of Frank Dickinson. At least our Track Train Dynamics committee had done something useful for the railroad. I caught-up with David Nixon and Roger Westburg for the trip home. Updating our Rock Island TPC runs to reflect the slow orders I had picked up was easily and quickly done—the results replaced the older runs in our UCTSIGMA database. By May 10th we were expecting more trouble from the Rock Island who were opposing our FRA 511 coal line loan guarantee. On May 25th I put on a slide show of my Rock Island pictures for Doug Christensen, David Weishaar, Dave Boger and Ray Snyder which gave them a better appreciation of the condition of the Rock Island's track and facilities. The Rock could be characterized as a "used up" railroad with a lot of worn trackage and rundown buildings. On the 29th I completed a comparison of the C&NW and the Rock Island lines across Iowa anticipating that this track would become a bone of contention. I found that the C&NW's route was 18% more fuel efficient than the Rock Island's in our TPC runs between Council Bluffs and Chicago. I drove to Peoria May 18th to meet our trainmaster, Tim Lenzen, for an inspection of the Adams Street line (bad, bad, bad) and the Rock Island's branch running northwest to the Pioneer Industrial area. Later I went alone north to Bureau Junction inspecting the extensive flood damage along that line. (Gasoline was now 90 cents a gallon and rising- May 19th).

CIE was asked on May 7th to quickly prepare and submit an FRA 511 application for replacement of mainline track ties in Iowa. Mike Iczkowski and a man from the Union Pacific worked together to figure out the problems with the unbalanced locomotive hours account as the C&NW was in a huge deficit position (May4th). We were supposed to share locomotive power so that parity was maintained, but we were using more than we were loaning.

The restoration of the double track from Missouri Valley to West Denison, Iowa, came up again May 2nd. The congestion of trains in this single track bottleneck had been inflicted on the railroad by its former chief engineer B. R. Meyers who thought he would save on maintenance expenses. Meyers' decision had been conditioned upon the level of traffic at that time. He could not have forseen the huge jump in train movements occasioned by improving relations with the Union Pacific. I ran a sample on the train movements through this tight spot and wrote up my findings May 10th. There was nothing wrong with the C&NW that a big infusion of capital wouldn't cure. (After the Union Pacific merger in 1995, the restoration of the double track through this area became a high priority item). At a Saturday morning meeting June 2nd Jim Wolfe decreed that the Automatic Train Control (ATC) on lines west was to be dropped when CTC was installed. To Wolfe's frustration, the FRA refused to degrade train protection by allowing the ATC to be removed despite its age and high maintenance cost. The Iowa CTC project was the center of our attention in early June and by the 8th was as complete as far we could get it given the narrow scope we were allowed. Poor Kim—his writing was constantly being redone by Rich Howard who fancied himself a master craftsman in the use of the English language. Actually, his constant rewriting of subordinates' work was his way of exercising control and submerging his staff's creative work. The FRA said they would not entertain loan guarantees for any CTC installation program that called for elimination of the automatic train control system (July 26th). Earlier on June 12th I had prepared maps of the Illinois Central Gulf and the C&NW Iowa lines for Eberhardt. He said the next day that Wolfe was unable to negotiate for trackage rights on the parallel ICG line from just below Missouri Valley to

West Denison as he had no leverage. Wolfe said that we had better consider restoration of the second main track. (Crucial as it was, the restoration would have to wait another 16 years). Our application to the FRA for a loan guarantee to install CTC in Iowa was completed June 15th. Kim Schlytter took it to Washington on the 18th but somehow TWA lost it. We reprinted the entire package on the 19th. Kim was on his way back to Washington on the 25th trying to get this dead horse to run. A big inspection trip of the west end of the Iowa division by the C&NW, UP and FRA took place June 18th. Jim Wolfe threatened to cancel our application on August 7th unless we could kill the automatic train control requirement. The FRA did not bend and the application died.

Personnel were on the move again: Bob Milcik was relieved at Boone June 20th; Bill Otter was appointed avp-division manager at Chadron and George Maybee came to Boone. On June 19th General Motors made a presentation of their new GP50 diesel locomotive. During the presentation Jim Zito kept leaping up to take care of other matters which was disconcerting to the presenters to say the least.

▲ James Brower was promoted to project director for the coal line amid some reservations about his short fuse when dealing with people (June 11th). On the 14th I wrote a rejoinder to the coal slurry pipeline's claim that they would be so efficient in moving Wyoming coal to Texas. Where were they to get water in eastern Wyoming which averaged about 10 inches of rainfall each year? How much energy would be consumed in removing the suspended coal solids in the water when it got to Texas? And what did they intend to do with the residual black water—flood Texas streams with it? Peat Marwick and Mitchell's man Duffy was interested in exploring the impacts that 10 loaded coal trains per day in 1979, 18 in 1986 and 24 in 1991 would have on the Council Bluffs area. Factored into the equation was train delay, idle time, inspection, car repair and employment. From the advantage of distance we can smile at the gross underestimation of the number of coal trains that would pass through this area even though increasing use of the Fremont-Blair line took some of the traffic away from Council Bluffs. Jim and the FRA people went west in middle June. Marginalized again, I was not invited. The stories about that trip trickled in on June 21st. Brower had acted badly by arguing constantly with the FRA engineers. Our track engineering people were not at all happy with him either.

▲ That infamous Interim Coal project had not died of its own impractability—it was just dormant. On June 26th Jim Brower, George Hanson and I flew to Chadron in a new Beechcraft 125 from Palwaukee airport in just under two hours. Bill Otter picked us up for a meeting at the Chuck Wagon restaurant's private room where we discussed the Interim Plan and the so-called Big One which was the Project Yellow new railroad connector line to the Union Pacific. There were too many people for the hi-rail inspection party that headed west on the 27th so Jim and I stayed behind in Chadron to look at the mechanical facilities—which did not take long. What we did find was an open track joint in the mainline which may have saved a derailment. After the morning meeting we drove west on the highway to catch-up with the rest of the crowd at Shawnee. Switching folks around, we hi-railed into Douglas before hitting the highway to Casper. The company jet made it back to Palwaukee in two hours flat. By June 28th it looked like the Interim Plan was failing—patching up the 470 odd miles from Fremont to Van Tassell was too costly compared to anticipated revenues. On the section west of Van Tassell (which eventually was rebuilt) the passing sidings so carefully sited to best accomodate train operations were casually shifted hither and yon by Brower to accomodate some inconvenienced ranchers who had private crossings. I unsuccessfully tried to make the point that building a ranch road parallel to the railroad a short distance to avoid crossing two tracks by an unluckily located private crossing would be more cost effective than potentially impeding train operations. Brower would have none of that. Of course, the problem went away when the entire line was doubletracked some years later. We were told again on July 25th that the Interim Plan was to be in place by January 1, 1980 despite its high cost. I am firmly convinced that the Interim Plan silliness was a bone thrown to the coal mining companies in the Powder River to make them think we were really serious about getting C&NW unit coal train service to their properties over that ramshackle railroad before the connector could be constructed. On July 2nd Jim Brower requested TPC runs for the re-engineered line from Shawnee to Crandall. Project management, however, remained a problem. On July 6th I mentioned this to Jules Eberhardt; he disagreed. Three days later the lack of planning was beginning to show even more clearly. I decided to reintroduce myself into the process at least enough to help the group focus on what was needful. I made a detailed list of to-do items and began feeding them to Ed Evert a few at a time since he seemed to have better access to Eberhardt than I did. The whole month of July drifted away. I put it to use, however, by coding St. Louis Southwestern and Southern Pacific lines for our UCTSIGMA collection.

Lou Duerinck needed information about Northern Pacific and Great Northern Railway land grants; I found some information for him (June 29th). Jerry Groner needed main track mileages. I interviewed Tim Welles, a potential candidate for CIE who might be available next fall. Davalis of personnel came in July 27th to discuss minority hiring on the coal line—another wrinkle in this complex project we had not considered. On July 5th General Motors was back again with a presentation to Jim Wolfe on AC (alternating current) diesel locomotives. These machines had a future, but not on the C&NW—yet. On July 11th a rumor floated past that Jim Zito was going to be axed. Another rumor on the 16th contained a hint that the C&NW was going to seek 511 guarantees from the FRA fund to fix the Kansas City line—that old 1972 study still had legs. I got the old TPC decks out and made sure they were correct in case we needed them. The former CGW line still was cursed by slow speeds and few sidings. Bob Milcik was now avp-planning in Chicago, far from the tensions of running a division. He was a sensible fellow under that facade of being just a good ole country boy from Moberly, Missouri. On July 31st he was in the office talking about using the Rock Island to Kansas City. Join the club, Bob. I did another update for George Hanson on the same subject August 6th. When the April 1, 1980 takeover of the Rock Island occurred, the Kansas City reroute was the first action taken by the C&NW.

George Hanson now had that old proposal for a direct AMTRAK connection from Union Station to review again. On a hot July 16th morning an inspection party hiked from Kinzie and Canal Streets across the former Grand Avenue yard along the same route Larry Kohn and I had gone a year before. Some things never seem to die. If it were important enough to do, funding would have been found. Six of us (George Hanson, Bob Drengler, David Nixon, and a couple

I have forgotten) went all the way to the former Erie Street coach yard and up the incline to Clybourn station to catch a train back into the terminal. That July another rumor circulated that the company was interested in purchasing the old KRAFTCO headquarteres building on Peshtigo Court at the Outer Drive just west of the Navy Pier. Most of us did not think much of having to traipse across the Loop and North Michigan Avenue to our jobs. That rumor came up again December 27th. On July 27th Jim Feddick, grain marketing, left for a job at Archer Daniels Midland where he eventually became a vice president.

▲ Another attempt was made to bring the coal project before the public. The public relations disaster in February was fresh in our minds as Joe Marren (public affairs), Don York (environmental engineer) and I flew to Omaha July 30th. We picked up two Union Pacific men—Joe McCartney and Kurt Anderson—for the flight to Scottsbluff. Driving to Lyman just on the Nebraska state line, we called on the city officials to aprise them of the planned railroad construction east of their small town. The reception, we felt, was positive. That evening we met with the city council of Gering, Nebraska, which is located just south of Scottsbluff across the North Platte River on the Union Pacific. The meeting in Gering was held while a raging summer rainstorm swept through the area. Gering's attitude toward the promised significant jump in railroad traffic was one of caution, but the city fathers did not take a strong stand against the railroad. The Union Pacific stood ready to offer solutions to traffic blockages and other impacts so that serious public opposition did not develop.

I was invited to act as fireman on inbound #26 at Elmhurst on August 1st because the regular man had missed his call at West Chicago. Riding high up in the leading cab car, it was an easy job and a great way to see the railroad. I got a good look at the new signals installed for Provo Junction east of Mannheim Road at the Berkeley station.

As mentioned above, Charlie Harmantas did not last long in our CIE department having resigned July 30th. His suggested replacement was a woman, Jo DeRoche. I dimly remembered her as a slim, young blonde girl who had worked in finance in 1973. When she came in for her interview on August 17th, I was taken aback because she was no longer a lissome lass—she was now a self-assured, big lady.

Jo had decided that life in the finance department was not her future in the railroad and had applied for training as a trainmaster. Her timing was good because the company was in the throes of incorporating women and minorities into its corporate body. She got a place in the class and graduated becoming the first woman trainmaster in the United States. Assigned to the Council Bluffs terminal, she soon learned about the rough and tumble life facing a woman in what was a man's world. Every trick and excuse in the book was thrown at her. She perservered and imposed discipline on a very unruly bunch of tough trainmen. She paid the price, however, when her body responded to the irregular, hastily grabbed meals, the long hours and exposure to all weathers. Those middle-of-the-night safety train tests, the continual stress imposed by conniving train crews working the system for all it was worth, consequent disciplinary sessions, train emergencies and all the other unexpected and unforeseeable crises encountered in train service left her with a Junoesque body and teeth ground down in frustration. Having survived that field experience, she was now looking for a stress-free place to recuperate—a place to nest while she restored her physical being. If I had been smart enough to pick up on that, I would never have hired her. But I did. She reported for duty November 13th. She settled into her private office like a big pussy cat and slowly began to repair the ravages of her field experience. Assignments handed to her were treated diffidently. Her approach to a project was leisurely and circumspect. She did not interact often with other departments and really did not show much initiative. There was no sense of urgency in Jo's conduct of that job. She had her teeth repaired and dieted her way back to a more normal condition, all on our time. I never managed to find a way to motivate her from her own self-imposed pace. She eventually left the railroad, graduated from law school and became a labor attorney in Washington, DC. Somehow, I was just unable to motivate her in our work.

On August 6th Jim Wolfe said that coal was our greatest priority. Business in general was slowing down although grain and coal remained strong. Wolfe wanted to make up for the bad showing in the winter. It was obvious by August 14th that the railroad needed new revenue sources as the current base was not broad enough to support the company. Constant paring of expenses was slowly wrecking the railroad. Mike Iczkowski demonstrated to operating on August 20th how the locomotive hours deficit could be eliminated and our account balanced with the Union Pacific. On August 27th David Nixon and Don Nelson (operating) made a presentation of the network model. The concept was good; the motivation to use it was zero. This was another example of an investment that never yielded a return. At this point another new project landed in our laps—close the Fortieth Street yard in Chicago.

▲ George Hanson, Murray Humphries and I went to Casper August 21st. This was Murray's first trip and may also have been George's. Our mission was to update the mining situation and develop the Interim Plan to move coal over the Shawnee-Fremont line by January 1, 1980. Bill Otter took us to Shawnee the site of our proposed connection with the joint CNW-BN coal line; Orin, the BN interchange; and Bill, the place with its own ZIPcode. Bill was still mostly just a store, a sign advertising cigarettes and beer and two gas pumps, on Wyoming highway #59. The main attraction for us were the coal mines. As we passed an abandoned and tumbled-down ranch house along the highway south of Bill, George Hanson said that was the place he was going to build a saloon and call it the Golden Spike. That late afternoon we watched the loading of a Burlington Northern coal train from the mine operator's control cabin inside the loading silo at ARCO's Black Thunder mine. This was a new experience for us all—we had so much to learn about this business. After a night in Gillette, we looked at the BN's extensive terminal there before driving south to Reno Junction where the ARCO mine spur left the new joint coal line. ("Joint" in the sense that the C&NW would eventually pay for its half interest.) A new town, named Wright for a local rancher, had sprung up in the southwest corner of intersection of highways #387 and #59. Founded in 1976 to provide housing for coal miners, this town and Gillette were the only two incorporated cities in Campbell County. Wright attained a population of 1,347 by the year 2000. In 1979, however, the place had that raw, unfinished look all newly-founded western "towns" seemed to share. I talked to some real estate people about housing and commercial developments before we drove back to Casper by way of Midwest on U. S. highway #387. In six short years the demographics of the country had already

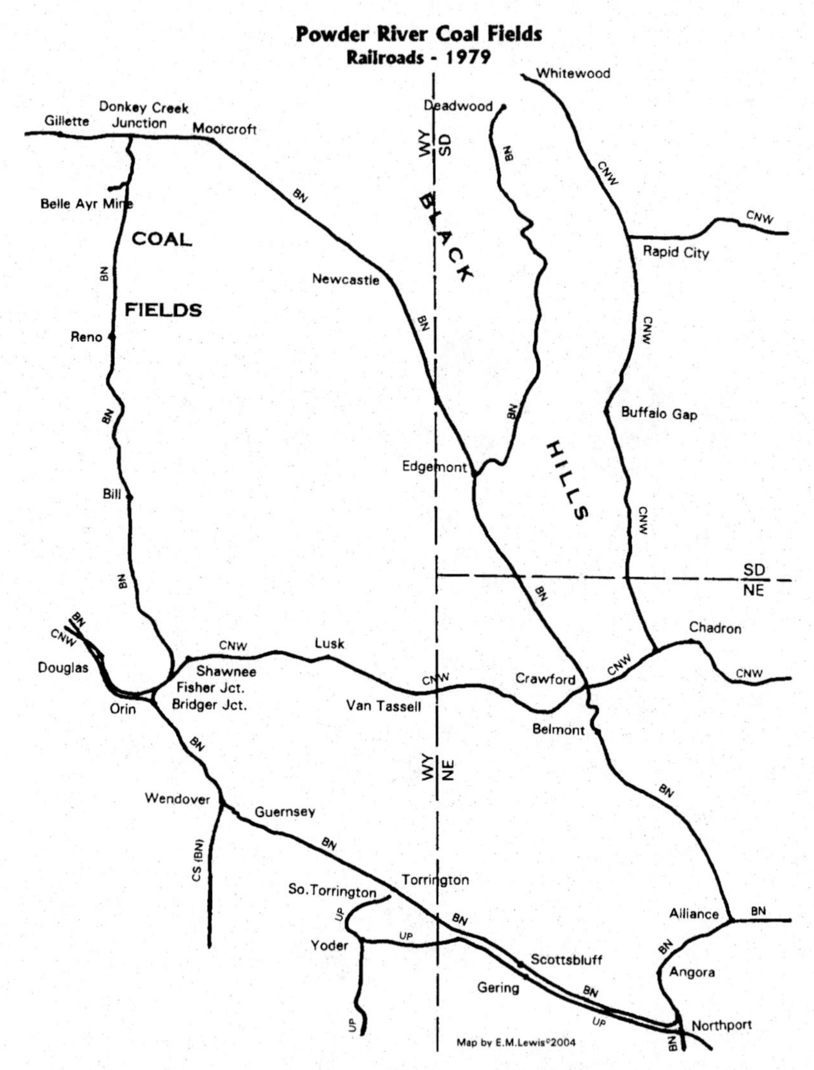

By 1979 the Burlington Northern had constructed a new railroad from a junction with the Belle Ayr mine spur south through Bill following the C&NW's projected alignment toward Shawnee. The C&NW had not yet found the money for its half share of the ICC mandated joint line. The new BN line swung south and west to join the C&NW's old main at Fisher Junction for a few miles to a new track connecting to the old CB&Q line at Bridger Junction. The C&NW's proposed new railroad south from Lusk to the U.P. east of Yoder remains unbuilt pending negotiation of financial agreements between the C&NW and U.P. and regulatory approvals.

changed. Back in Chicago Mike Iczkowski was string line diagraming the proposed operation of coal trains and empties (August 27th). This old technique was still an effective way to demonstrate complex train operations to an unsophisticated audience and point to where passing sidings had to be installed. George, Murray and I rewrote the Interim Plan based on our recent visit to the area. Wolfe must have sensed that we were very uneasy with the impossible task of stuffing coal trains over that single tracked abomination which was optimistically called a railroad. On August 30th Wolfe said we WOULD do it. The rewriting and editing continued. The Interim Plan was completely typed and ready for distribution when Bill Otter called from Chadron with some more changes. As part of the Interim Plan we had asked for departmental input to make the task as palatable as possible. We called this the George Hanson Request as he was the point man for this effort. The request contained a plea for all information and planning that would make the Interim Plan work. On September 10th Wolfe asked for copies of all responses to our requests. By September 12th about 65% of the responses were in. The breathtaking enormity of the project was laid out in detail. Although it was clearly impossible to start, much less finish, that work by January 1st., Wolfe turned a deaf ear to our protests and directed us to continue this doomed effort without spending any money. Since he controlled the money spigot, nothing further was done on the Interim Coal Plan that would entail expense. Not knowing how to make something out of nothing, we turned to other projects until Wolfe again tried to kick some life into the Interim Plan. Let us say that his plan to run coal trains across Nebraska by January 1, 1980 did not get much serious attention despite his demands.

On August 28th the Rock Island was struck by its unions. The end was in sight for that faltering railroad. On that same day the C&NW reorganized its management structure again. Jules Eberhardt was now to report to Jerry Conlon of public affairs instead of directly to Jim Wolfe, I. R. (Bob) Ballin retired and R. David Leach picked up the materials and purchasing function. Real estate went to the law department. Rich Howard was in a state of shock at the changes. On August 30th Jim Wolfe chortled that the C&NW had taken the Rock Island's premier West Coast auto parts train away from them. (A forlorn symbol of their pride, the train was simply known as number 57 on the Rock and as ARRO on the Union Pacific which took the train from Council Bluffs to Ogden). Some strange and intricate commuter train operations on the double track between Elmhurst and JN at Melrose Park were experienced on September 7th because of track work.

Mike Iczkowski also had problems working for the capricious Rich Howard and was offered a transfer to the operating department September 10th. However, he stayed with CIE and became deeply involved in developing the profile characteristics of the new GP38 diesels for TPC. The railroad eventually purchased 35 of these locomotives but not for coal service.

The Rock Island strike was still on September 10th complicated by a job action the same day by the clerks' union againist the Kansas City Terminal. Ed Evert, Don Fryer and I prepared maps of the Rock Island-Milwaukee-C&NW for presentation to the Governor of Iowa and were asked to secure information on all regulatory permits required if we should take over operations of the other railroads. Wolfe said September 12th that we either take over the Milwaukee and Rock Island or we go to the Gulf of Mexico. Two days later we heard that the Southern Railway had turned down the ICC request for directed service on the Rock Island and the tiny Kansas City Terminal had been selected as the designated operator. Lou Duerinck went to Washington on the 17th to promote the C&NW as the designated operator. Our unions were supportring that position. The decision went right up to the Carter White House. At home the C&NW was a congested mess looking a good deal like last winter's paralysis. On September 18th the car count showed 56,000 cars on line. I obtained the Rock Island's 1977 density statistics and recast them to match our figures to produce a consolidated map. The Rock Island was still struck September 25th. By September 26th the C&NW was choking on all the business offered to it. From all indications it looked as though we would be running the Rock's north-south railroad soon (we had to wait until April 1, 1980). I asked David Nixon to update the old TPC decks and, sure enough, requests began to arrive on October 1st. The Rock Island at this time was absolutely dead—nothing was running. On the 19th Jim Wolfe met with the FRA on the problem.

▲ In August Don York had reported that the environmental hearings on our application for the Project Yellow scenario had gone well and that the Burlington Northern's opposition did them no credit. My resource library was in great demand in early September as questions regarding grading costs, the new shop proposed for South Morrill, and environmental issues at siding locations were researched. Word came that the local people at Chadron were disgruntled because the coal trains would not be coming there under the Project Yellow scenario. They feared that Chadron would be doomed as a railroad center. They were correct in their concerns as it turned out. On September 10th Jim McDonald of public affairs was urged to get out of his comfy chair and go to Chadron to soothe local feelings. His public affairs activities were very close to being non-existent. I wished Frank Koval were still with us.

▲ George Hanson and I finished a commentary September 12th on the difference between 100 and 110 loaded cars in a unit train. The extra 10 cars could be handled but they pushed the safety margin too far making some grades marginal in bad weather. In the years ahead longer trains became the norm as lighter tare weight aluminum cars and immensely more powerful locomotives became available. Wolfe's Interim Plan was still alive even though everyone agreed, away from Jim Wolfe's ears, that running a coal train over that broken down excuse of a railroad was impossible especially by January 1, 1980 which was only three months away. Based on the returns to George's memorandum, we cast a second Interim Plan on September 14th. Five days later the thing had grown to 90 pages when it was sent to the printer. I was away at a meeting in Hot Springs, Arkansas, when George distributed the report. Of course Wolfe took exception to the horrendous cost and demanded that we slice more money out of it. Apparently, he needed to keep up the fiction as long as possible even though 1979's construction weather was fast vanishing. Wolfe was playing some sort of deep game, and we did not know either the rules or the players. On the real coal line project bids were due on the 14th for the construction management contract. Morrison-Knudsen did not submit a bid this time because they were disillusioned at the way the C&NW approached major projects. They were correct because, again, nothing was done on this subject. By October 29th Jim Zito was promoting Brower's choice of Peter Kiewit of Omaha despite advice against using this firm. Again, Wolfe was playing a delaying game.

In September I had gone to Hot Springs by way of Memphis on September 20th to attend a meeting of the Southwestern Car Accountants to give presentation on the Standard Point Location Code (SPLC). Because of bad weather I was able to stay over a night and visit my aged aunt before returning to Chicago. On September 25th I heard about the new 50 story office building to be erected on the Chicago Passenger Terminal at Canal and Madison Streets.

▲ That impossible Interim Plan to move coal through Fremont by January 1, 1980 continued to bumble along wasting everyone's time. On September 26th Roy Spafford (car department), George Hanson (operating), Murray Humphries (labor relations) and I flew to Omaha where we picked-up Glenn Kerbs (Iowa division engineer), George F. Maybee (Iowa division manager) and Dennis Waller (motive power) for a meeting with the Union Pacific to discuss interchange of coal trains. There was a lot of discussion about locomotives and timing before an agreement was reached that Council Bluffs was a better place to interchange as the UP had crewing problems if Fremont were to be used. Jim Zito and Ed Burkhardt agreed that our decision was correct—it didn't matter, of course. By October 2nd George Hanson wanted to write a third version of that idiotic operating plan. Nothing had been done to repair the railroad since this crackpot idea had been hatched last January. After finishing a comparison of Chicago-West Coast corrdors on October 3rd, I went on vacation to California.

On my return to Chicago I found the railroad management again changing. J. W. Alsop had really, and finally, retired. Roger Westburg and Mike Iczkowski were both being interviewed by AMTRAK. Mike wasn't interested but Roger went to Washington for more interviews and accepted a job at their Beech Grove, Indiana, car shops. Mike went to operations planning November 1st. A month later he was back in our group to finish a double tracking study he had been working on. (Mike later changed his name to Iden and became an officer in the Union Pacifics' motive power department). On October 22nd Doug Christensen was promoted to vice president-marketing, David Weishaar took over the coal marketing function and Edward Burkhardt became vice president-transportation on November 1st.

▲ October 19th was the day the Federal Railroad Adminstration was supposed to give us a GO or NO GO answer on our 511 application for a loan guarantee for the coal line. Jim Wolfe went on vacation and time slipped by. On the 25th we heard that President Jimmy Carter's close associate, Stuart Eizenstadt, had our application at the White House. Then, on November 5th, we got the bad news from the FRA—they could not act swiftly enough for the November 30th deadline set by the BN-C&NW agreement. An unfavorable article appeared in the *Wall Street Journal*. Our involvement in the joint line was really a question at this moment. Given the turmoil surrounding the Milwaukee Road-Rock Island collapses, the coal line's status was in serious jeopardy. The news that North American Car was constructing a car repair facility at Bill, Wyoming, just where we had planned our yard hardly mattered.

At the CIE departmental staff meeting October 26th our new superior, vice president Jerry Conlon, wanted research done on other railroads. Not just their physical location, track characteristics and motive power, but a complete dossier on their service territory, people, finances and prospects. This poetry quoting man understood the value of intelligence. I gave the task to Jo DeRoche when she came on board in November. Very little was accomplished beyond a feeble backhanded attempt on the MKT. On the 29th another meeting was held to address the restructuring of Iowa railroads. I started assembling statistics and maps for development of a grand scheme. Looking at the number of new leaders, we began to wonder if our old regime was finished. Jim Zito was testier than usual and Wolfe was still in Florida. On November 1st the Milwaukee Road shut down its lines in Iowa and in the west. Five thousand employees were furloughed and half of the railroad was closed. The next day Congress tried to revive the corpse, but the trustee did not want to reopen the closed lines. In our office I created a data base of Iowa material and a primitive digitized map showing the locations of the grain elevators. While computer technology was available for creating maps and drawings, we did not have access to that modern software. When creating my crude maps I did not give our systems group a thought and just did what I could in the short time available. By November 7th George Hanson had developed a presentation to Jim Zito on the grain lines of Iowa. The State of Iowa at this point wanted the Kansas City Southern to operate the grain lines rather than the C&NW. Our relations with the State of Iowa were never very good. Jerry Conlon suggested on November 23rd that I develop a plan to identify Milwaukee Road branch lines in Iowa that could be useful additions to the C&NW.

▲ Despite the negative response by FRA on November 5th, there really was no bad news about the coal project until the 14th when Jim Barnes, vice president-engineering, was told by Wolfe to halt any spending on the Western division in connection with the coal project. Again, I found that I was out of the coal project development stream. As the November 30th deadline for the C&NW to pay its half of the construction costs of the newly completed joint line into the coal fields neared, rumors of possible financing began to surface. On the 20th a rumor identified the Union Pacific and ARCO as our saviors. ARCO's mine was in operation and they needed C&NW service to avoid being a captive shipper. On November 29th Wolfe said all negotiations with the Union Pacific and ARCO had been suspended and that we would have to depend on the ICC to extend the deadline beyond November 30th. He added that the coal project still was a corporate objective and that we must all endure this difficulty until we can get affairs settled to our best advantage. On December 3rd the ICC said that the contract between the Burlington Northern and the C&NW was invalid and that the line must be operated as a joint facility. We had won a major point. On December 10th the *Wall Street Journal* published a favorable article about the C&NW. Wolfe said he had embellished it a bit. The Project Yellow coal project was far from dead. Financing arrangements were underway with the Union Pacific through a new corporate creature called Western Railroad Properties, Inc. (WRPI—pronounced "wor-pee"). The Interim Coal plan, flimsy and ridiculous as it was, had served its purpose as a smokescreen. Those of us forced to work diligently at it were merely unknowing actors in the charade Jim Wolfe was playing with John Kenefick who was led to believe we were really going to raise the Shawnee-Fremont line from the dead. I was somewhat annoyed at myself for not seeing through Wolfe's flim-flam, but realized that with my close association with Union Pacific people, I might have inadvertently given the show away had I known the real purpose of the Interim Plan. If Wolfe had set the target date later than January 1, 1980, I believe his deception would have been easier to carry-off. Any reasonable railroad man looking at our broken railroad on the ground would have seen that rebuilding of 470 miles was going to take a much longer period to accomplish. Wolfe was lucky that the UP folks did not inspect our ruin of a railroad; the deception would have been apparent. That old rule about visiting a site in person maintained its validity.

▲ John C. Kenefick, retired president of the Union Pacific, said in an interview with *Vintage Rails* (January-February 1998), " Provo asked if we would be interested in helping, and we turned some people loose on it. Somebody had a wonderful idea (he should have been rewarded but probably was not). The idea was to use the UP's North Platte branch and build a so called connector line to the North Western in extreme northwestern Nebraska, upgrade a little bit of the North Western, and buy into the BN joint line."

▲ For the record, the Shawnee-Norfolk line lasted as a through corridor until May of 1991 when the flooding White River east of Harrison severed the line forever. One source reported that in the 1980s the line was patched-up enough to permit operation of a few coal trains to an Iowa utility. The statement is incorrect; no unit coal trains from the South Powder River Basin ever operated over the former Fremont, Elkhorn and Missouri Valley line through the northern tier of Nebraska counties. The ICC approved abandonment of the Chadron-Norfolk portion March 21, 1992; the rails were removed from Merriman to Norfolk in 1995. Service on the east end from Norfolk to Fremont had ended in 1982 because of flood damage from the Elkhorn River. That is not to say, however, that no coal was ever handled in regular train service over parts of this line. Before 1984 weekly shipments of 45 to 60 cars of coal originating on the BN at the Fort Union mine north of Gillette were interchanged to the C&NW at Crawford and moved east to Norfolk where trackage rights south on the Union Pacific were used to regain the C&NW at Fremont. This coal was destined to Ames and Eagle Grove, Iowa. After completion of the coal line between South Morrill and Crandall in August 1984, this Fort Union coal traffic was merged with other blocks destined for Mankato, Des Moines and Clinton and handled as a normal unit coal train on the new coal line. In 1983 while construction of the new line was underway, a proposal was briefly considered by the C&NW to try another interim plan to run coal trains east on the old railroad for American Power and Light. Nothing came of this idea to many people's relief.

On November 13, 1979 the news at the morning meeting was about the ICG's train falling through their overhead bridge onto the C&NW's mainline at Woodbine, Iowa. Wolfe observed that maybe now was a good time to try and negotiate a joint track agreement with them for the use of their parallel line from Honey Creek to West Denison, Iowa, to substitute for our missing second main through that stretch. The ICG again rejected our overtures (January 24, 1980). On November 15th I heard that Jim Zito's special train had hit 100 mph on the Iowa division and that he wanted to do 110 mph on the Illinois until Wolfe heard about the stunt and told him to cut it out. Actually, Jim Zito had put his pudgy finger on exactly what the railroad needed—good railroad track which gave greater fluidity through higher train speeds. On December 4th the morning meeting was filled with appeals for more locomotive power and woeful tales about the increasing problems at Kansas City. Two days later an annoucement was made that only officers of assistant vice president and up would be allowed in the morning meeting. Also, CONRAIL was now to be linked in directly to the meeting. On December 18th the Chicago Transit Authority (CTA) was struck, throwing a heavy load of commuters on our Northwest Line trains. I gave David Nixon and Mike Iczkowski their annual performance reviews after Christmas. Jo DeRoche was mostly unavailable because of her extensive dental reconstruction. I could have used more Davids and Mikes and fewer Jos. Rich Howard was gloomy and not at all sure where the department was headed in 1980. Industrial engineering was obviously not well regarded after Rich's repeated series of project failures. The railroad needed more and better managers, not more half-baked studies.

The Iowa Restructuring plan was the hot item as the year came to an end. The database I was building was innovative and time consuming. All through the process I constructed logic trails to make the database as flexible as possible. As an exercise I calculated the probable center of corn production for 1985 in an effort to focus attention on the best lines to retain to maximize railroad business. By December 20th the map was circulating around for comments. The export business analysis showed that a lot of corn would be going to the SL-SF at Kansas City in 1980. On December 28th Bob Gallagher of the FRA in Washington, DC, called wanting density statistics on the Twin Cities-Kansas City line.

A severe blizzard in northern Nebraska provided a warning that Nature could interfere with the works of man. Bill Otter, the division manager, related that during a recent snowstorm one of his trains was caught in a cut and drifted over by the snow. Bill went up in light airplane trying to find the lost train and drop supplies to the marooned crew. Not having much success in even finding the railroad, he finally radioed the crew to goose the throttle on the idling diesel. A black geyser erupted from the snowy landscape giving Bill a target for his air drop. The crew was later rescued by snowmobile.

The year ended on a sad note when I learned that my stalwart supporter at the AAR, Peter Conway, had suffered a stroke on December 16th and died on the 20th. His diplomatic approach to knotty problems had bailed me out several times in the sometimes turbulent struggle with the truckers over the Standard Point Location Code (SPLC). His successor, Larry Lassman, was a former AAR auditor. I was destined never to meet him.

34

THE LAWYERS TAKE OVER, 1980 . . . The first business day of 1980 was January 2nd and the beginning of my gradual separation from the C&NW. The railroad itself was in pretty good shape that first morning meeting with only 47,500 cars on line and revenues up $200,000. Although I was only a director, I still was able to attend the morning meeting although restrictions on who might attend were to come later. The Peoples Agrarian Railroad was becoming less democratic under Jim Wolfe's rule. Our little CIE (Corporate Industrial Engineering) group was almost non-functional. Its leader, Rich Howard was drowning in a sea of numbers, his assistant Kim Schlytter floated by his side without any direction from his chief, Roger Westburg was absent a lot (soon to go to AMTRAK January 25th), and I was struggling with the Milwaukee Road-Rock Island collapse. My two assistants were out of action; David Nixon, was ill and Jo DeRoche was near paralysis. On January 4th repairs to a flood damaged fill at Blue Earth, Minnesota, were pegged at $60,000. Engineering estimated $80,000 a mile to fix the railroad from Mason City to Butterfield. That same day a new hiring freeze was announced despite which I managed to hire Bob Olson from systems with Jo's approval February 19th.

The news broke on January 8th of the planned merger of the Union Pacific and the Missouri Pacific. (The actual takeover occurred on December 20, 1982. A separate identity for the MP was maintained until January 1986; the corporations were not actually merged until January 1, 1997). After a friendly chat with Dale Salzman at UP and John Sward of the ATSF, I came to the conclusion that the merger would not have any bad effects on the C&NW. In fact the consolidation of the two would improve coal train movements headed south to Arkansas and Texas. At a strategy session on the 9th with Jerry Conlon, Jules Eberhardt, Ed Evert and Rich Howard my first impression seemed to be confirmed. On January 21st there was more merger news: the Union Pacific was acquiring the Western Pacific. The final 1979 revenue figures for the C&NW showed that there actually had been a profit through dint of creative accounting.

I now set my program to leave the railroad into motion. Rich Howard again delayed my annual appraisal on January 10th, an event I was not looking forward to in any case. I had lunch with my future employer, Tom Hodgkins of Distribution Sciences, and verified that he still wanted me and my geographic databases. He was ready for me to start the next day. While I would have liked to have done just that, I said the coal project was coming to a head, and I thought I still had a part to play so he would have to be patient. I am certainly thankful that he was willing to hold the door open so long. That delayed annual appraisal came the very next day on the 11th and it was ugly. Howard said I had submarginal ratings on motivating subordinates. Holding my rising temper, I mentioned his abysmal record for retaining people and that he had little reason to attack my "graduates' most of whom stayed with the company. Realizing that I was talking to a wooden post, I let him have his say and flatly refused to sign the appraisal. That appraisal was the final insult; there was nothing more the company could do to induce me to stay. I left him to explain to Eberhardt why I had refused to sign that bad review.

The Rock Island trustee was court appointed William M. Gibbons, a Chicago attorney. He was a tough negotiator and a feisty person by nature. He had presented a plan to the bankruptcy court to get the railroad running again. On January 14th our operating people developed their own plan to run the Rock. The trustee's plans were full of gaping holes and we poked a few more—the cost of fuel was enough to ruin his rosy projections. On January 24th the trustee's plan was ruled unworkable by the bankruptcy court, and he was ordered to liquidate the property. The Chicago, Rock Island and Pacific Railroad was dead. The circling buzzards moved in. On February 5th the Grand Trunk Western made a bid for the Rock's Chicago-Omaha mainline. Wolfe said, "Fight it. That main line to Omaha was better dead."

▲ On January 15, 1980 Jim Wolfe, Lou Duerinck, Jerry Conlon and Jules Eberhardt were in Omaha collecting $60 million from the Union Pacific for the Project Yellow coal line. That game had begun at last. On the 18th I helped Keith Feurer from real estate set up his database of landowners that he had to contact to secure right-of-way for the new connector line and the east leg of the wye at Shawnee. He now had authorization to proceed obtaining the options for the railroad and microwave tower sites. Keith would be aboard that first C&NW coal train in August 1984.

My heart may have been with the coal project but my body was working on the Rock Island problem. Making a map of Iowa with acetate overlays to show railroads, grain elevators, and different alternatives was not what I had rather be doing. The Coal Line Project now had a life of its own, and I was not a part of it. Anticipating my eventual departure, I gradually shifted more and more of my work onto David Nixon. If he wondered why, he never asked. I had heard from Hodgkins that the Distribution Sciences (DSI) board had approved acquisition of my geographic business and hiring me as its manager. Tom Hodgkins called the very day that the C&NW held a meeting to acquaint us about the company's supplementary pension plan. The existing plan called for 30 years of railroad service for a full pension. For every month short of age 65 the pension was reduced 2% which meant that if I retired immediately at age 54, my pension would be zero. What had not been anticipated by the creators of this plan was the case where a retiring individual with 30 years of service might forego any pension payments until reaching the age of 65. This looked like a good idea as I did not need any pension payments at the present time and could afford to wait out the 11 years without drawing any funds. My determination to take this course was strengthened when I learned January 23rd that I was not included in the executive bonus plan. On the 28th I flew to New Orleans, Louisiana, to use up some of my accumulated vacation time.

On February 5th the C&NW's bid on the KRAFTCO building on

Peshtigo Court was public news. Six days later we heard that the bid had failed to the great relief of nearly everyone on the railroad. Most employees rode the commuter trains which terminated west of the Chicago River at Madison Street. The extra trek across the Loop almost to the Navy Pier would have been an extra burden made worse by icy blasts off Lake Michigan in winter and brain-boiling heat in summer. Wolfe had a chauffeured automobile so that the office location made little difference to him. Train #26 pulled into Elmhurst without a fireman again on February 8th; the conductor drafted me to warm the seat and call signals. The railroad must be in good condition if you could believe the division reports at the morning meeting. The stock market thought so—C&NW stood at $20. My investment was now worth almost ten times my base cost. On April 23rd I heard that the company was planning a move again. This time it was to be to the former John Plain warehouse building on Canal Street just north of our present offices.

Wolfe had planned an inspection trip on the Missouri-Kansas-Texas in early January. Jo DeRoche had been given the assignment of pulling together a book of information about that railroad. Despite her languid manner, she actually finished the report in time for the trip scheduled about February 15th. She then was directed to do the same on the Kansas City Southern. Wolfe's ambition to get a rail line to the Gulf of Mexico was apparent. After a trip to San Francisco on a crew calling project, she started working on the KCS project (February 26th). Looking at the TPC ratings in our huge UCTSIGMA data base, the choice between the MKT and KCS was not a difficult one to make if operating characteristics were the only criterion; the MKT was a lot easier to operate. (The Union Pacific must have agreed as they acquired the MKT in 1988). Not wishing to let the Stanford network model drop completely out of sight, I asked David Nixon to put a manual together so that someday someone could figure out how the model worked.

Jules Eberhardt said on February 22nd that the FRA quickly wanted line capacity information of the C&NW's connections. The request was further refined on the 25th to include only those railroads carrying over 4 million tons *per annum*. With our valuable library we were able to respond to this request fairly easily despite all the other activities swirling around. The Milwaukee Road on February 11th had again reopened the prospect of coordinating freight service between Chicago and the Twin Cities, a subject we had rejected in 1977. It was a last gasp. On February 28th the Milwaukee embargoed non-core lines and began disposing of branches March 3rd. The C&NW was slated to operate the Milwaukee's Jefferson Junction to Marathon, Iowa, line—that is, until 300 broken rails were discovered in their track.

Our CIE department was asked on February 6th to assist Maurie Reid's branchline committee in abandoning 2,000 miles in 1980. That was shown to be impossible given regulatory requirements. Reid then drew up the list for 1,000 miles. Wolfe approved the list February 8th making it number 1 on his priority list. The annual management conference was held at the Midland Hotel in Chicago February 27th. Jim Wolfe told the gathering that our priorities were: 1) abandon 1,000 miles of branchlines, 2) the Western Coal project, and 3) midwestern railroad restructuring. At the time I guessed Union Pacific and C&NW would merge in 1982 (I was off by 13 years).

The FRA approved the C&NW takeover of many of the Milwaukee Road's and Rock Island's Iowa grain lines on March 12th. Also included in the order was the long desired north-south Rock Island mainline from the Twin Cities to Kansas City. I quickly added the new acquisitions into our segmented railroad model. The Rock Island trustee became a dog in the manger on March 17th by making some outrageous demands before the C&NW could begin operations. The FRA order was received on the 18th and a takeover meeting was held the next day. Lawyers sparring over the details postponed the takeover date to April 1st. On March 21st the C&NW also was directed to run the Rock Island's Chicago commuter service which operated out of the LaSalle Street station starting on March 24th and running through June 2, 1980. On the first day of C&NW operation the only announcement made at our morning meeting was a laconic statement that the Rock Island suburban district was on time. On May 9th the C&NW advised the Regional Transportation Agency (RTA) that they should begin running the Rock Island suburban lines after May 31st.

Our department was asked to participate in the Rock Island takeover. Ed Evert, Jo DeRoche and I went to Des Moines, Iowa, on March 28th. I went on north to Mason City in a rented car to help

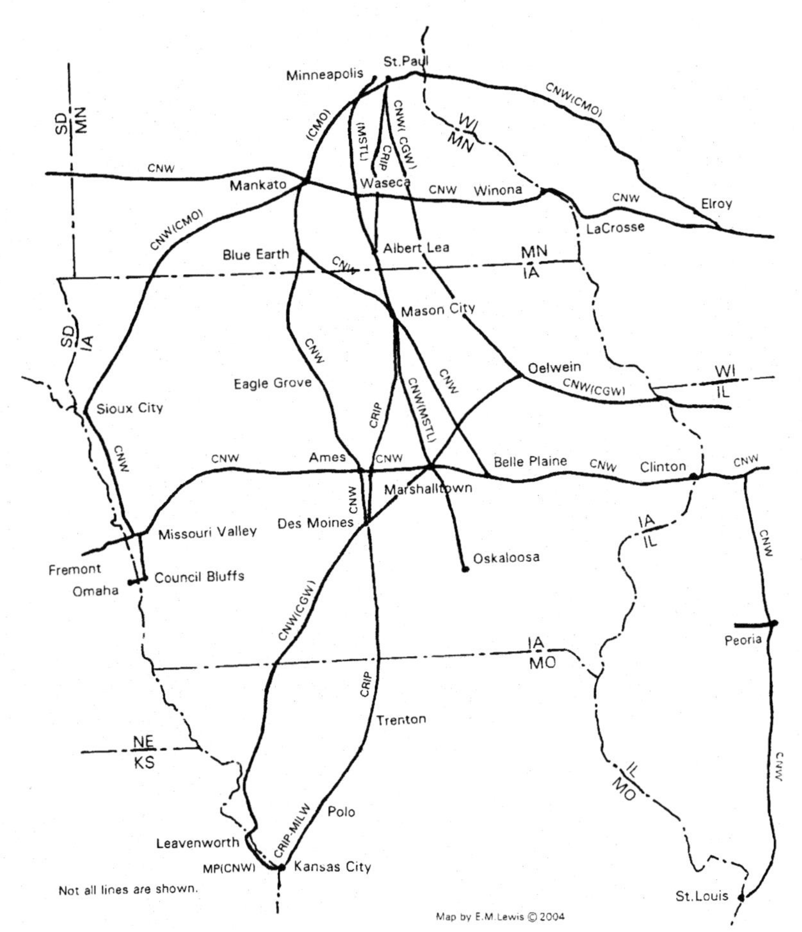

In 1980 the Rock Island finally collapsed and closed down. The C&NW took over operations of the Rock's main lines shown on this drawing April 1, 1980, solving its Kansas City problem. It did not take long to start pruning the redundant routes. The former CGW lines were the first to go.

the teams based there. We were issued ancient Polaroid cameras to photographic stock piles of materials so that the trustee could not later claim that we had stolen Rock Island property. My assigned territory was from Manly, Iowa, south to Des Moines. Operating as a one man team and knowledgeable about the railroad business, I quickly made my way down the line on the highways, stopping at every station. The tracks and property were now covered with a light snow cover. There was not much to account for at Iowa Falls, Nevada, or any other point on this line. The Rock Island just did not have extra supplies standing around. The silent railroad under its coating of new snow seemed like a deserted ruin.

On the last days of March, 1980, Jim Zito established a war room in a Des Moines hotel complete with a computer link and telephones. All of our top officers were present, ready to move in on the Rock Island. Ed Evert and I took a run down the Rock's Pella branch and also checked on the nearby Indianola line while a steady rain fell. Curiously, this was the strongest united team action I had ever seen on the C&NW. I could only speculate how much better the coal line development could have been if everyone had pitched in with the same enthusiasm they were demonstrating on this Rock Island campaign. I felt like an outsider. On March 31st we set up tables to begin signing up Rock Island train crews to work for the C&NW until June 2nd when a new order from the FRA could be expected while negotiations for purchasing the lines from the trustee were underway. Very few of the laid-off men would sign on that first day because they feared they would harm their Rock Island seniority rights should the Rock miraculously come back to life. I spent that day helping Zito organize the field notes of the many inspection teams and posted the results in a grid chart on a big blackboard—just like we had done on the Chrysler project in 1963. Jo DeRoche went to Kansas City. Mike Iczkowski was out in the "bow and arrow" country of Northern Iowa—so called because the train dispatcher had to use a bow and arrows to send messages to his train crews on those dark and remote lines. At 2 p.m. the media appeared for a news conference complete with television coverage. Many stayed up until midnight to see if any last minute injunctions would be filed. Most of us just went to bed.

At 00.01 CST April 1, 1980 the C&NW went in on 810 miles of Rock Island track in Iowa, Minnesota and Missouri. Dallas B. Carlisle, avp-engineering, was all prepared and immediately began installing a crossover at Easton Avenue north of Short Line tower in Des Moines to connect the Rock Island and C&NW main lines. Southbound C&NW trains could head directly south toward Kansas City eliminating that awkward pull-in and swap-ends move in the Rock Island's Short Line yard. The crowd watching the switch installation included Tom Judge from public affairs and me. That same morning our inspection teams fanned out by hi-rail on the now officially deceased railroad to inspect the Rock's tracks and inventory rolling stock. My section was Earlham to Newton, Iowa on the old east-west main line. Nick Couris and Rich Kennedy were my team. We collected car numbers and their locations and noted the many wrecks sprinkled along the right-of-way. Bill Gibbons, the Rock Island trustee, was exceedingly annoyed at our "high-handed" tactics. In reality, it was just a well organized and focused campaign to get the job done as quickly as possible. The next day, April 2nd, train crew hiring resumed. No trains were operated on the Rock Island. Jim Zito was threatened with jail by the trustee, but the crossover work was pushed to completion as quickly as possible. The first train operated April 3rd despite the trustee's continued recalcitrance to negotiate with Lou Duerinck and Jules Eberhardt. Our task in Iowa was finished; we all went back to work except that Jo was asked to help out in the hiring of Rock Island people. The C&NW purchased some of the newer Rock Island grain hoppers and boxcars in their bright blue paint scheme. Wolfe did not want the cars repainted. He said they were to be a reminder to the good people of Iowa and their state government that the Rock Island equipment now rudely stenciled "C&NW" was the result of their bad treatment of the railroad industry. Meanwhile, the Kansas City Southern was still hovering around the Iowa honeypot hoping to get something.

The Fortieth Street yard closure now became Rich Howard's tar baby. He found out first hand how hard it was to get decent train data from our records system. Exempted from the Rock Island takeover, Rich and Bob Olson spent a week in the field trying to find justifications for eliminating the facility. I could have given him one good reason—safety. The eleven foot track centers were dangerously tight for modern freight cars.

▲ On April 3rd I was asked to review recent testimony of Burlington Northern's Warren Stockton concerning line capacities. While I searched for flaws in his methodology, I realized that I seemed to

---

▲ As noted before, this symbol will be used to denote material relating to the Western Coal Project.

have once more become involved in Project Yellow. I spent most of April 8th with our Washington attorney setting up our rebuttal to the BN testimony. They really liked the color slides I produced from my archives. Four days later I received a Federal Express package from Washington full of testimony and got a long telephone call from attorney Larry Rudolph in Washington. After we had discussed the material at length, he made a few revisions and filed it with the ICC on April 14th. Jerry Conlon asked me April 21st to go with a group to Wyoming to call on the state's attorney general and discuss alternate routes which we had considered for the coal line. Chris Mills, our lead attorney in this action, briefed me the next day on the up-coming hearings. The involvement in Project Yellow I had so desired had arrived—too late for my railroad career. Realizing that my testimony might be attacked as being dated, I went to Wyoming to revisit the Burlington Northern's railroad and facilities, recalling that deeply ingrained lesson about the value of personal inspection. On the 24th I flew with Jerry Conlon, Ed Burkhardt, Jim Brower to Des Moines in the company jet. After Burkhardt got off, we went on to Omaha to pick up Union Pacific's Valerie Scott, an attorney, for the flight to Cheyenne. Valerie had been one of those astonished Union Pacific participants in the famous 1977 boardroom scene when Dick Freeman had announced there was nothing further to discuss. At Cheyenne Ed Sencabaugh, a jolly Union Pacific public relations front man, met the plane to escort us to the state capitol for our meeting with the attorney general. The meeting went well as we described the several alternatives we had considered. Later, I rented a car and headed north by myself to Glendo, a small station on the BN north of Wendover. I hiked down the BN's single track main past the junction switch at Wendover where their Colorado and Southern line headed for Denver and points south. I walked through tunnel #2 just west of Guernsey. This was a dank timber lined hole that burrowed under a state park which would make daylighting difficult. (The tunnel was opened and the line double tracked in 1999). Inside the tunnel the rail was all bolted 112 pound on a roadbed pockmarked with mud geysers that boiled up at the joints whenever a train passed. My poking around the railroad was not noticed by any Burlington personnel save a passing freight crew who waved as they passed. The next day I drove down the Union Pacific's line to Northport and back to

Gering, Nebraska. This Union Pacific branch's potential for easy handling of coal trains was obvious as the even valley terrain was downhill to the east. At lunchtime in Torrington I ran into Keith Feurer, Larry Grapentine and Mike McDonagh. They were just getting organized to contact landowners along the new railroad's route. That afternoon I swung south along the Union Pacific's Torrington line south to Cheyenne. I was ready to testify.

▲ On the 28th I was alerted for a fast trip to Washington, but that journey was canceled when the my submitted testimony went unchallenged by the Burlington. The hearings to be held at Cheyenne in mid-May were judged to be critical to the project. In a comical sidelight, the lead secretary in the law department, Sylvia Thomas, mixed up plane reservations for Louis T. Duerinck and E. M. Lewis resulting in Lou's trip being canceled (May 5th). Poor Sylvia had a hard time living that little gaffe down. On May 7th the meetings relating to the complicated Western Railroad Properties (WRPI)-CNW-UP application for construction were being pushed as a contract deadline was approaching. WRPI was the jointly owned C&NW-UP corporate shell for the coal railroad.

▲ Chris Mills, our energetic young attorney, two of our Washington attorneys, Tom Acey and Larry Rudolph, and I flew to Casper May 11th. Casper airport was weathered in so we diverted to Scottsbluff. Renting a car, Chris drove very fast indeed to Orin and then north to Reno where we visited the ARCO mine, got stuck in the gummy Wyoming mud along the highway, and finally tied-up at Lusk. These men had never seen this country before. The next morning, May 12th, there were three inches of snow on the ground. Later in the day, we caught up with Jim Brower and Larry Grapentine. While the party went south down the proposed connector line, I checked the grade crossings east from Shawnee on the existing railroad and made notes about the passing siding sites. We all met for lunch at Morrill, Nebraska, before returning south to Cheyenne to review our testimony for the next day's public hearing. ICC administrative law judge Richard H. Beddow Jr. opened the hearing on May 13th. The shipper witnesses were called to testify first, saying how much the new connector line would benefit their operations and how important transportation was in the coal business. Waiting in the rear of the room, I was unexpectedly trapped into an impromptu television interview by a Union Pacific public affairs man. The interview worried our attorneys but when it came out on the television that evening in a positive way, Chris was much relieved. The shipper witnesses finished up on May 14th. I was called next to discuss the Burlington Northern bottleneck between Orin Junction and Guernsey. My testimony was quoted at length in the newspaper. The public hearing concluded in Cheyenne and moved north to Torrington, Wyoming, for the next hearing. Reconvened on the 15th we went through our 47 witnesses. There must have been some high feelings outside the courtroom—someone slashed the tires on Larry Grapentine's truck. On May 16th landowner complaints were heard, in volume but not in substance. Then it was over, we went to Scottsbluff just down the road and all flew home in the company plane, N400T. That party included Chris Mills, Jack Osborne, Don York, Keith Feurer, Tom Acey and myself. Two days later Chris Mills, Don, Keith, Jim Brower, and I were on our way back to Wyoming. We held a review session at the motel in Cheyenne on arrival. These next hearings were the C&NW's rebuttals to Burlington Northern testimony. I was the third to testify on the subject of passing siding placement. There was little cross examination. While we were in Wyoming, the Environmental Impact Statement (EIS) was released in Washington with nothing of any significance to hinder construction. The rebuttal hearings finished up on May 20th without any major problems. We had been worried about Jim Brower as he had been bellicose and truculent during the previous evening's dry run. That episode and another strange outburst directed against me later in Torrington made me wonder about his mental stability. But he was the designated man in charge of construction, and I was not. Keith Feurer and I rode to Denver with Joe McCartney of the Union Pacific' public relations department. This was my last major contribution to the C&NW.

▲ A strange episode involving the ICC law judge, the Burlington's attorney and our legal staff centered around a low level flight in a chartered plane in mid May. Richard Beddow, the law judge, expressed an interest in seeing the country and the proposed construction sites. The Union Pacific quickly accommodated him despite fears that the eruption of Mount St. Helens in Washington State on the 18th might have put too much abrasive dust in the air. The

Burlington's attorney insisted that he also be included in the flight to insure that there would be nothing prejudicial to his case arise. Soon after take-off, he went soundly to sleep and did not awaken until the plane landed on its return. The judge apparently enjoyed the trip.

Back in Chicago I monitored the progress of the hearings to see if I would be further involved. As the calendar turned into June it became pretty clear that the coal project and I were at last finished. While waiting for something to happen, I made TPC runs for Cajun Electric of Louisiana and Iowa Electric at Sergeant Bluff, Iowa, dusted-off the Miles City, Montana, line extension of long ago for Jules Eberhardt, and got involved in a possible new connection between the Rock Island and the C&NW at Cedar Rapids for coal movements south. This last idea was patently ridiculous, but since I was killing time at this point while the western coal project settled down, I tried to appear busy without actually starting any new projects. Meanwhile, everything that I wanted to keep had been methodically removed from my files, piece by piece, over an extended period. There were no big boxes of stuff which might have attracted attention.

On June 10, 1980 with, I suspect, a self satisfied smirk on my face, I handed Rich Howard a letter stating my intention to retire from the C&NW effective June 30th. Visibly stunned, he hastened across the hall to see Jules Eberhardt who immediately conveyed it to his superior, Jerry Conlon. While Jules and Jerry may have been surprised, their poker faces didn't show any reaction. I think Howard was shocked because all my projects and involvements were now suddenly dumped onto him. The civil engineer had unloaded on the industrial engineer. As the remaining days passed and the word spread throughout the company, my friends exhibited one of three reactions: surprise, incredulity and envy ("how ja do that?"). I met with Bob Wilson of personnel to discuss retirement. He said there would be no pension for my age given the deductions for the eleven years I was short of age 65. I told him to keep the pension on ice for me; I would draw on it later. That option had never occurred to him and his face was a study. I had both escaped and managed to keep my pension entitlement. I also was quite careful not to reveal my new employment to anyone. When asked, I ducked the question saying "you can find me in the phone book in Elmhurst." Bob Russell, vice president of personnel and a long time colleague, seemed resigned to my departure. Howard while initially shocked now seemed pleased to have me out from underfoot. Eberhardt was on vacation, and I never knew how my retirement affected him, if at all. I lunched with Doug Christensen and David Weishaar of coal marketing on June 16th. At the branchline committee meeting which now was being run by Bob Milcik, I introduced David Nixon as my replacement. I fear that David was not much help there as he did not know the railroad very well. Milcik may not have run his first branch line meetings very smoothly, but he eventually became a main player in the extensive abandonment program. My few remaining days were largely social—everyone wanted an inside briefing on how I pulled off this coup. The curious included Jim Chesner, Bob Milcik, Bob Jahnke, Ed Evert, Dick Bailey and several others. My own gang arranged a farewell luncheon at the Apparel Mart restaurant and awarded me a model HO scale hopper car on an engraved plaque bearing the inscription of "Father of Western Coal". The little car was made by Mike Iczkowski and contained real western coal Jo DeRoche had obtained from a loaded car at Proviso. Chris Mills said on the 19th that I had saved them at least one year of litigation by my testimony at the hearings in May. Now, that made me feel that putting off Tom Hodgkins for six months had been worth it.

On June 24th Howard was off to Nebraska with an FRA delegation—his first trip west and his first involvement with the coal project. On the 25th the company organized a reception in our offices in my honor. It was nice to greet all my old friends even though they may have only wanted the cake and cookies. About 100 showed up including Bob Russell who casually told my wife that the company had never found the right job for me. That remark pretty well summed up my 33 year railroad career which I had found to be educational, entertaining, stimulating and, sometimes, frustrating. Just before my final departure I learned that I had been included in the executive bonus plan afterall, effective July 1st.

Yes, the new connector line to the Union Pacific was constructed, the old C&NW between Van Tassell (Crandall) and Shawnee was completely rebuilt, a new railroad from Shawnee to the jointly owned coal line at Shawnee Junction was established after three years of legal sparring. At first the C&NW's line was single track with long passing sidings. Soaring coal traffic soon stressed the capacity of the line. Double tracking eliminated the sidings which greatly improved operations. But, the coal tonnages continued to grow faster than expected requiring further expansion of capacity. On just the jointly operated Orin Line in a ten year span (1994–2004) the two railroads added 38.7 miles of double track and 46 miles of triple track and planned another 5.8 miles in 2005. Now, if we could just figure out how to make gasoline from the coal inexpensively. . . .

EPILOGUE
ONCE A RAIL, ALWAYS A RAIL

On Tuesday, July 1, 1980, I became a professional geographer for Distribution Sciences, Incorporated (DSI) in Schiller Park, Illinois. DSI was a computer software services company that provided a freight rating system for a wide spectrum of commercial clients. My new position was National Sales Manager for Geographic Systems. Later, the company moved to new quarters in Des Plaines, Illinois, and grew successfully until the deregulation of transportation under the Staggers Act changed the whole freight rate making industry.

Five of us at DSI decided in 1984 to form a separate organization to market a freight rate negotiation communication system to the railroad industry. Operating at first under the name DSI ★ RAIL this venture grew into a new corporation, Transportation Data Xchange (TDX), jointly owned by the Union Pacific, Conrail, Burlington Northern, CSX, Norfolk Southern, SOO Line, Chicago and North Western, Southern Pacific, Guilford and the American Short Line Railroads Association. I served as secretary—treasurer for four years until retiring in 1991 at age 65. The company was subsequently merged into RAILINC, the computer arm of the Association of American Railroads (AAR).

On June 16, 1983 the Union Pacific and the Goshen County Heritage Committee invited my wife and me to participate in a special train excursion July 3, 1983 from Cheyenne to Torrington. The trip was a commemoration of the commencement of construction on the "Wyoming Powder River Basin Coal Transportation Project" which was then the current name of my old Project Yellow. The train of the Union Pacific's fleet of passenger cars was headed by UP #8444. The

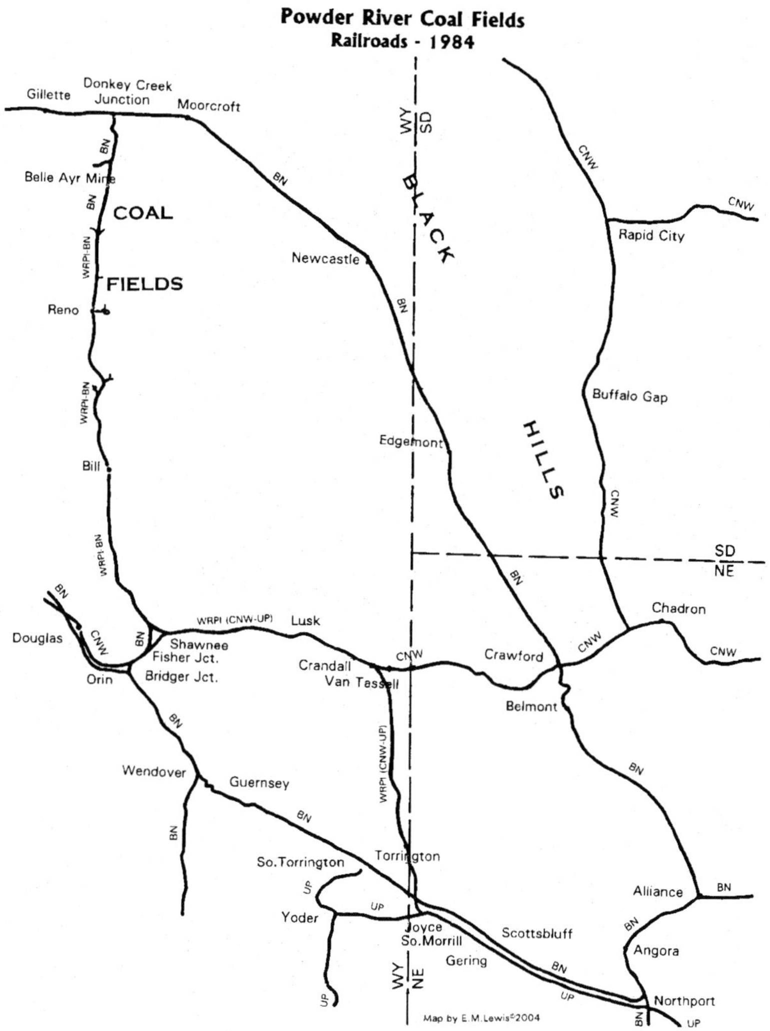

Powder River Coal Fields
Railroads - 1984

In 1984 the C&NW-UP combination began operations on their jointly owned coal line from the South Powder River coal fields to the U.P.'s North Platte subdivision at Joyce. The new construction known as "Project Yellow" extended from Crandall to South Morrill and was pressed into immediate heavy use as a major coal route.

4-8-4 Northern hauled nine cars of people, including Wyoming Governor Edgar J. Herschler, from Cheyenne to Egbert on the UP mainline and then north on the Yoder branch into South Torrington where ceremonies were held at the county fair grounds.

A year later, on August 16, 1984, the new connector railroad born

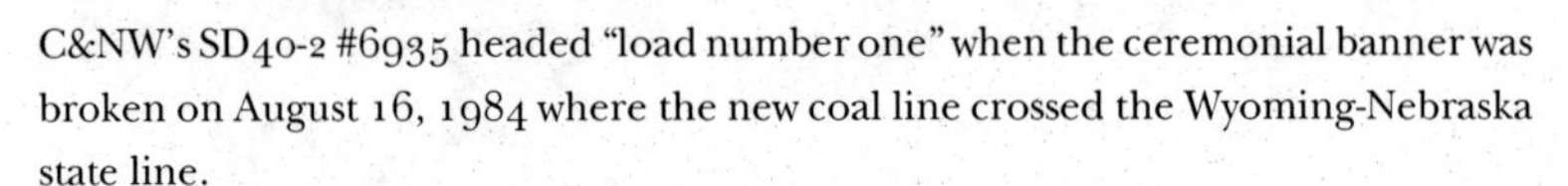

C&NW's SD40-2 #6935 headed "load number one" when the ceremonial banner was broken on August 16, 1984 where the new coal line crossed the Wyoming-Nebraska state line.

from the idea contained in Project Yellow was opened. At the kind invitation of Jim Ferrill of the Union Pacific, I rode the special passenger train from North Platte, Nebraska, west on the mainline to Gibbon, then up the North Platte subdivision to the new South Morrill connection and northwards to the Nebraska-Wyoming state line where the ceremonies were held. Robert Mickey of the C&NW's real estate department was also a guest. At the dedication site, under a blazing sun, we detrained and our special train backed down the new line to the junction at Joyce some miles south. After suitable speechifying, a cannon salute, and photographs, Load Number One headed by C&NW SD40-2 (#6935) and three Union Pacific units with 110 cars of Antelope Mine coal bound for Arkansas came through a banner stretched across the track. Keith E. Feurer, the C&NW real estate man who had acquired the necessary right-of-way in an epic acquisition campaign, was aboard that first locomotive as the honorary train pilot. Project Yellow was in business.

Gene Lewis with his railroad July 2003 (photo by Ray Barth)

C&NW first got interested in Powder River Coal in June, 1972; Project Yellow (the C&NW-UP connector) was conceived September 23, 1975; the construction effort began in June 1983 with groundbreaking February 2, 1984; and the first coal train ran August 16, 1984—a period of 12 years. Coal traffic soared in the last years of the decade, reaching the 25,000th train mark in 1992 and the 50,000th in December 1995. By the end of 2003 the C&NW-UP combination had moved some 1.7 billion tons of South Powder River Coal from the basin. Remembering the skepticism with which our estimates of future coal traffic made in 1973–74 were greeted, the enormous tonnage actually transported was a refreshing "we-told-you-so".

In November 1992 my wife of 47 years and I retired to California's North Coast. In April 1995 the Union Pacific took over the C&NW. In Wyoming the coal traffic continued to expand thanks in part to the Clean Air Act of 1970, amended in 1990 and 2000 which mandated reduction of sulphur emissions at coal-fired electric generating plants. Over 52% of the existing generating plants switched to low-sulphur coal with more to come as tighter air quality regulations came into effect after January 1, 2000.

The old adage that *you can have a life, or you can work on the rail-*

road; but, you cannot have a life and work on the railroad was not my experience. I had both a very interesting life and enjoyed working on the railroad. Perhaps my independent stubbornness adversely affected my career path, but, the 12,000 days invested with the C&NW is still regarded as a happy time as I look backwards from my elder years. My railroad life formed a solid foundation for my next two careers.

GLOSSARY

A2 Tower A2 at Western Avenue, Chicago

AAR Association of American Railroads

ACE Auditor of Capital Expenditures

ACL Atlantic Coast Line Railroad

AE CTC crossovers at Ashton East (Illinois)

AFE Authority for expenditure

ARCO Atlantic Richfield Co.

ARDA American Raiway Development Association

AREA American Railway Engineering Association

AS Alton and Southern Railway

ATA American Trucking Associations

ATLANTIC Type of steam locomotive 4-4-2 wheel configuration

ATSF Atchison, Topeka and Santa Fe Railway

AVP Assistant vice president

AW CTC crossovers at Ashton West (Illinois)

B&B Bridge and Building department

BLE Bessemer and Lake Erie Railroad

BN Burlington Northern Railroad

BO Bad Order

BO Baltimore and Ohio Railroad

BOCT Baltimore and Ohio Chicago Terminal Railroad

BRC Belt Railway of Chicago

BRT Brotherhood of Railroad Trainmen

C-1, 2, ETC. Car Department buildings at Chicago Shops

CAE Chicago, Aurora and Elgin Railroad

CBQ Chicago, Burlington and Quincy Railroad

CEI Chicago and Eastern Illinois Railroad

CGW Chicago Great Western Railway

CI Chihuahua, state of Mexico

CILCO Central Illinois Light Company

CIM Chicago and Illinois Midland Railway

CIW Chicago and Western Indiana Railroad

CJ Chicago Junction Railway

CLASP Consolidated Location & Station Project

CMO Chicago, St.Paul, Minneapolis and Omaha Railway

CMSTP&P Chicago, Milwaukee, St.Paul and Pacific Railroad

CNW Chicago and North Western Railway

CO Chesapeake and Ohio Railway

CO CTC crossovers at Cortland, Illinois, on CNW

CONSOLIDATION Type of steam locomotive 2-8-0 wheel configuration

CPT Chicago Passenger Terminal

CRIP Chicago, Rock Island and Pacific Railroad

CS Communications and Signals

CSTPMO Chicago, St.Paul, Minneapolis and Omaha Railway

CTC Centralized Traffic Control

CWI Chicago and Western Indiana Railroad

DF Distrito Federal, estado do Mexico

DF&PA District Freight and Passenger Agent

DMCI Des Moines & Central Iowa Railway

DNB Daily News Building (Chicago)

DSI Distribution Sciences, Inc.

EJE Elgin, Joliet and Eastern Railway

EM Estado Mexico, estado do Mexico

EMD Electro-Motive Division

FAST Facility for Accelerated Service Testing

FDDM Fort Dodge, Des Moines & Southern Railroad

FIPS Federal Information Processing Standard

FRISCO St.Louis-San Francisco Railway

FRA Federal Railroad Administration

FWD Fort Worth and Denver Railway

FX CTC crossovers west of Franklin Grove, Illinois, on CNW

GBW Green Bay and Western Railroad

GCU Galena and Chicago Union Rail Road

GE General Electric Company

GLSIDC Great Lake States Industrial Development Council

GM General Motors

GMO Gulf, Mobile and Ohio Railroad

GN Great Northern Railway

GP7 Type of general purpose GM diesel locomotive with 4 wheel trucks

GR Guerero, estado do Mexico

GX CTC crossovers west of Geneva, Illinois, on CNW

HF Heel of frog (back end of frog assembly)

HM Interlocking plant on CNW at Elmhurst, Illinois

HS Heel of switch (end of switch point rail where attached to heel block

HUDSON Type of steam locomotive 4-6-4 wheel configuration

HX CTC crossovers west of Creston, Illinois on CNW

IBM International Business Machines

IC Illinois Central Railroad

ICC Illinois Commerce Commission

ICC Interstate Commerce Commission

IHB Indiana Harbor Belt RR

IIT Illinois Institute of Technology

IPS Illinois Power System

ITC Illinois Terminal Company

JB Tower at EJ&E crossing in West Chicago, Illinois

JN Interlocking plant on C&NW at east end of Proviso Yards

KCS Kansas City Southern Railway

KEEL Marking crayon also called a lumber crayon. Came in blue, yellow, red and white colors.

LELAND Publisher of Leland's Open and Prepay Station List tariff in St.Louis, Missouri

LM Litchfield and Madison Railway

LN Louisville and Nashville Railroad

LSBC LaSalle and Bureau County Railroad

LX CTC crossovers west of LaFox, Illinois on the CNW

M-1 Chicago Shops machine shop

M-2 Chicago Shops Office and Test Lab

M-4 Chicago Shops powerplant

M-14 Chicago Shops heavy steam locomotive repair shop

M-16 Foundry at Chicago Shops on CNW

M-17 Enginehouse at Chicago Shops on CNW

M-19 Paint shop

M-19A Diesel shop designation at Chicago Shops on CNW

ME CTC installation at Meredith (east), Illinois on CNW

MH Michoacan, estado do Mexico

MIKADO Type of steam locomotive 2-8-2 wheel arrangement

MILW Chicago, Milwaukee, St.Paul and Pacific Railroad

MKT Missouri-Kansas-Texas Railroad

MOGUL Type of steam locomotive 2-6-0 wheel configuration

MP Missouri Pacific Railroad

MStL Minneapolis and St.Louis Railway

MW CTC installation at Meredith (west), Illinois on CNW

NI Interlocking plant at West Chicago, Illinois, on CNW

NKP Nickel Plate railroad (New York, Chicago and St.Louis Railroad)

NMFTA National Motor Freight Tariff Association

NORTHERN Type of steam locomotive 4-8-4 wheel configuration

NP Northern Pacific Railway

NQ CTC installation east of Nelson, Illinois, on CNW

NW Norfolk and Western Railway

NY West junction of Southern Illinois and Galena Divisions at Nelson, Illinois

NYC New York Central Railroad

OMAHA (THE) Chicago, St.Paul, Minneapolis and Omaha Railway

ORT Order of Railroad Telegraphers

PACIFIC Type of steam locomotive 4-6-2 wheel configuration

PC Point of curve (surveying term)

PCC Presidents' Conference Committee (a standardized modern street car design)

PF Point of frog (the dividing point in a railroad track fixture)

POC Point on curve (surveying term)

POT Point on tangent (surveying term)

PPU Peoria and Pekin Union Railway

PRAIRIE steam locomotive 2-6-0

PRR Pennsylvania Railroad

PS Point of switch (where switch rail begins)

PT Point of tangency (surveying term)

PTC Peoria Terminal Company (CRIP subsidiary)

RID Railroad Industrial Development, a social group

RLHS Railway Locomotive and Historical Society

S&C Signals and Communications

SD40 Type of GM 3,000 horsepower diesel with six wheel trucks

SI Southern Illinois subdivision of C&N W Galena Division

SIR Society of Industrial Realtors

SLBM St.Louis, Brownsville and Mexico Railroad

SLSF St.Louis-San Francisco Railway

SOO Minneapolis and Sault Ste.Marie Railroad

SOU Southern Railway

SP Southern Pacific Railroad

SPLC Standard Point Location Code

SRF Segmented Railroad File

SSW St. Louis Southwestern Ry.

TDCC Transportation Data Coordinating Committee

TDX Transportation Data Xchange

TEN WHEELER Type of steam locomotive 4-6-0 wheel configuration

TMERL The Milwaukee Electric Railway and Light Company

TOWER A-2 Interlocking plant at Western Avenue, Chicago, operated by the MILW

TP Texas and Pacific Railway

TPC Train Performance Calculator, a software program

TPW Toledo, Peoria and Western Railroad

TRRA Terminal Railroad Association of St.Louis

UP Union Pacific Railroad

USAAF United States Army Air Force

USGS United States Geological Survey

WAB Wabash Railroad

WX CTC installation west of West Chicago, Illinois, on CNW

YD CTC installation in DeKalb, Illinois, on the CNW

ZULU CNW class Z locomotive 2-8-0 wheel configuration

SOURCES

Information and materials for this account were drawn principally from personal daily diaries dating from January 17, 1950 to the present day. Earlier dated events were recaptured from daily work records that were preseerved from the first day of employment at the C&NW. Other important elements were found in a wide variety of printed works over the many years spanned by this work. Some of them included the following books, newspapers and magazines.

ADAMS, JAMES N. *Illinois Place Names* Publ.54, Ill. State Historical Soc, 1968

ALLEN, C. FRANK *Railroad Curves and Earthwork* McGraw-Hill, 1931

ANDREAS, A.T. *History of Chicago* A.T. Andreas, 1885

ANDREAS, A.T. *History of the State of* Nebraska A.T. Andreas, 1882

BARTELS, MIKE / BILL KRATVILLE / RICK MILLS / JERRY PENRY *C&NW Cowboy Line* South Platte Press, 1998

BERRY, BRIAN J.L. *Geography of Market Centers* Prentice Hall, 1967

BERRY, BRIAN J.L. / DUANE MARBLE *Spatial Analysis* Prentice-Hall, 1968

CASEY, R. / W. DOUGLAS *Pioneer Railroad* McGraw-Hill, 1948

CHORLEY, RICHARD J. / PETER HAGGETT *Models in Geography* Methuen, 1967

DERLETH, AUGUST *The Milwaukee Road* Creative Age Press. 1948

FOLLMAR, A. JOSEPH *Locomotive Facilities C&NW & CstPM&O* C&NW Historical Society, 1996

GALLAGHER, JOHN S. / ALAN H. PATERA *Wyoming Post Offices 1850–1980* The Depot, Burtonsville, MD, 1980

GANNETT, HENRY *Dictionary of Altitudes* Govt. Printing Office, 1906

SOURCES

GRANT, H. ROGER *The Northwestern* Northern Illinois University Press, 1996

HELBOCK, RICHARD W. *U.S.Post Offices, V. II* La Posta Pub., 1998

HELBOCK, RICHARD W. *U.S.Post Offices, V. III* La Posta Pub., 1999

HOLBROOK, S. *The Age of Moguls* Doubleday, 1953

KEYSER, LLOYD A. *C&NW in Color v.1 1941–1953* Morning Sun Books, 1997

KEYSER, LLOYD A. *C&NW in Color v.2 1954–1958* Morning Sun Books, 1999

AUGUST, LOSCH *Economics of Location* Yale Univ.Press, 1954

LOVETT, R.A. *Forty Years After* Newcomen Society, 1949

PIERSEN, JOSEPH / IRA KULBERSH *C&NW Final Freight Car Roster* C&NW Historical Society, 1999

QUIETT, G.C. *They Built the West* Appleton-Century, 1934

SEARLES, WILLIAM H. / HOWARD CHAPIN IVES *Field Engineering, a Handbook of the Theory and Practice of Railway Surveying, Location and Construction* John Wiley, 1946

STENNETT, W.H. *Yesterday and To-Day* C&NW Ry., 1910

STENNETT, W.H. *A History of the Origin of the Place Names connected with the Chicago and North Western and the Chicago, St. Paul, Minneapolis and Omaha Railways,* C&NW Ry., 1908

STRAUSS, JOHN F., JR. *The Burlington Northern, An Operational Chronology 1970–1995* Friends of the Burlington Northern, West Bend, Wi, 1998

TAYLOR, JEREMY *Powder River Coal Trains* Silver Brook Junction, 1997

TROTTMAN, NELSON A. *History of the Union Pacific* Ronald Press, 1923

ULLMAN, EDWARD L. *Geography as Spatial Interaction* Univ.of Washington Press, 1980

WELLER, PETER / FRED STARK *The Living Legacy of the Chicago, Aurora and Elgin* Forum Press, 1999

WILSON, N.C. / F.J. TAYLOR *Southern Pacific* McGraw-Hill, 1952

American Guide Series—Illinois A.C.McClurg, 1947

American Guide Series—Nebraska University of Nebraska, 1979

American Guide Series—Oregon Binsford and Mort, 1940

American Guide Series—Wyoming State Planning & Water Conserv. Board, 1941

C&NW Annual Reports—various

C&NW corporate records, minute books, legal files, *etc.*

SOURCES

C&NW Historical Society *Abandonments Book*, 1999

UP annual reports—various

Official Railway Guide—various

Moody's Railroads (1940–1948)

Poor's Manuals of Railroads (1871–1939)

Annals of Wyoming (Wyo.State Archives and Historical Dept.) (various)

C&NW Historical Society *Newsliner Magazine* (various)

Cheyenne Leader (various)

Chicago Daily News (various)

Chicago Sun Times (various)

Chicago Tribune (various)

Crain's Chicago Business (various)

Elmhurst Press (various)

Engineering News Atlas of Ry.Construction (1887–1889)

Fox River Lines, official publication of the Fox River Trolley Museum South Elgin, IL (various)

Industrial Marketing—August 1966

Info-News (Union Pacific Magazine) (various)

Journal of Commerce (various)

Laramie Boomerang (various)

Lincoln Journal (various)

Modern Railroads (various)

North Platte Telegraph (various)

Northwestern Lines—various C&NW Historical Society

Omaha World-Herald (various)

Progressive Railroading (various)

Railroad History (Railway Locomotive Historical Society) (various)

Railway Age (various)

Railway Track and Structures (various)

Santa Rosa Press Democrat (various)

SOURCES

Star Herald (Scottsbluff, NE) (various)

Trains (Kalmbach Publishing) (various)

Traffic World (various)

Vintage Rails (various)

Wall Street Journal (various)

Wyoming Tribune-Eagle (various)

Inside Story (International Stanley Corp., Sept. 1961)

UP *Coal Line Celebration*—publication for August 16, 1984 special train

BIOGRAPHIC INDEX

COLLEAGUES, CLIENTS, FRIENDS, ACQUAINTANCES AND OTHERS FROM THE RAILROAD CAREER OF EUGENE M. LEWIS

AASE, JAMES C&NW geologist, 390

ABBOT, RALPH C&NW finance officer & Northwest Industries, 431

ACEY, TOM Washington attorney, 873, 874

ADAMS IC Industrial Department, 377

ADAMS, "SPARKY" C&NW yardmaster, 223

ADAMS, BROCK Secretary of Transportation, 808, 813

ADAMSON, O. O. C&NW claims agent, 58, 174, 188, 192, 196

ADIK, STEVE C&NW mechanical officer, 675, 745

ALEXANDER king of ancient Macedonia, 648

ALEXANDER, WILLIAM C&NW locomotive engineer, 160, 161, 162, 163

ALLEN Chrysler engineer, 444

ALLEN, DILLIS C&NW rodman, Galena Div., 266, 268

ALLEN, WILLIAM Chrysler Fenton, MO. plant manager, 445

ALLISON, BENJAMIN D. C&NW electrical engineer, 287

ALLMAN, ALICE wife, Bill Allman, 551

ALLMAN, WILLIAM P. C&NW advanced systems director, 540, 541, 544, 545, 546, 549, 550, 551, 552, 554, 555, 556, 557, 558, 561, 562, 563, 564, 568, 569, 570, 571, 572, 574, 578, 579, 580, 581, 582, 584, 585, 586, 589, 590, 591, 593, 599, 600, 602, 603, 604, 605, 606, 608, 609, 610, 611, 612, 614, 615, 617, 618, 620, 621, 622, 623, 624, 625, 626, 627, 628, 635, 638, 639, 646, 648, 650, 651, 652, 654, 655, 657, 661, 666, 671, 689, 781

ALSOP, J. WILLIAM C&NW operating officer, 471, 497, 608, 613, 627, 651, 658, 725, 772, 822, 823, 826, 858

ALTORFER, JOHN Peoria industrialist and politician, 357, 358, 373

AMBURGEY, ROY AAR code clerk, 594, 596

ANDERSON president, Iowa Beef Co., 467, 469, 470, 471

ANDERSON MILW operating officer, 789

ANDERSON, CARL C&NW real estate agent, 353, 354, 363, 412, 439, 440, 441, 442, 443, 444, 445, 450, 451, 454, 455, 512, 513

ANDERSON, EUGENE C&NW attorney, 431, 458

ANDERSON, KURT UP public affairs, 851
ANDERSON, ROBERT C&NW sales agent, Peoria, 327
ANDERSON, WAYNE A. C&NW traffic agent, Sioux City, 307, 323, 332, 582
ANDREWS, JAMES A. C&NW rodman Galena Div., 38, 39, 46, 47, 96
ANHALT, LEN C&NW rodman Galena Div., 169, 170, 171, 176, 181, 185, 186, 189, 194
APPLE, BILL KCT, superintendent, Kansas City, 640, 800
ARADO, JOE EJE, industrial agent, 548
ARAKELIAN, MICHAEL X. C&NW locomotive training officer, 777, 790, 792, 840, 841
ARMSTRONG, GERRY A. C&NW traffic officer, New York City, 408, 529, 559
ARMSTRONG, WILLIAM F. C&NW engineer of buildings, 151, 181, 182, 194, 204, 207, 241, 246, 260, 262, 269, 280, 287, 291, 307, 327
ARNOLD owner, Arnold Cattle Co., 311
ARNTZEN, DON C&NW rail test engineer, 329
ARTHUR IMC officer, 494
ATKINSON, LEON N&W officer, 694
AUSTIN C&NW section foreman, 171
AUSTIN, HAROLD land developer, Peoria, IL, 351, 352, 354, 357, 368, 499
AXE, ED Susquehanna Corp, Riverton, WY, 476
AYLWARD, EDWARD President, Alyco Chemicals, Sullivan, IL, 365, 379
AZER, HOWARD C&NW material test engineer, 333

BACHENBURG Westcentral Grain Co., 428
BAEBLER Union Electric Company officer, 522
BAECHTEL, BOB Dun and Bradstreet, New York, 581
BAILEY, RICHARD W. C&NW engineering officer, 24, 26, 54, 69, 70, 80, 81, 88, 91, 97, 111, 112, 114, 128, 227, 238, 251, 720, 725, 801, 802, 803, 806, 810, 876
BAILEY, TOM C&NW sales agent, Marinette, WI, 300
BAIR, CLEM R. C&NW traffic officer, San Francisco, CA, 518
BAKER, NOEL citizen of DeKalb, IL, 227
BAKER, WILLIAM C&NW roadmaster, Galena Div., 227, 250, 251, 363
BALDAUF, FRED C&NW accounting officer, 284, 285, 550
BALL, EDWARD DuPont, 404
BALLIN, I. ROBERT C&NW real estate director, 333, 334, 349, 350, 364, 375, 383, 384, 401, 404, 411, 412, 421, 454, 476, 498, 508, 523, 530, 546, 700, 855
BAMMI, DEEPAK C&NW programmer, 569, 593
BANGS, EUGENE C&NW instrumentman, Wisconsin Div., 9, 579
BARNES, H. D. C&NW accounting officer, 141, 246
BARNES, JAMES A. C&NW engineering officer, 137, 141, 214, 246, 260, 270, 281, 727, 742, 775, 777, 798, 808, 810, 811, 860
BARR, HAROLD C&NW engineering officer, 268
BARRON, JOHN manager, Bi-State Agency, Granite City, IL, 364, 431, 489, 490, 522, 537
BARSEMA, MEL C&NW B&B foreman, 268

BATTEL, JACK C&NW marketing department, 668
BAUMANN, HARRY Chicago attorney, 513, 515
BAYER, GEORGE C&NW rodman Galena Div., 143
BEAN, LES C&NW operating officer, 281
BEAR, MAX C&NW engineering draftsman, 9
BEBJL, FRANK C&NW industrial engineer, 742
BECK, JOHN C&NW station agent, O'Neill, NE, 722
BECK, WALLY Armour, real estate, 395
BEDDOES, RALPH C&NW industrial agent, 41, 42, 294
BEDDOW, RICHARD H. JR. ICC law judge, 873, 874
BEEBE, TOM C&NW chief draftsman, 204, 246
BEHM, GEORGE C&NW real estate agent, Omaha, NE, 428
BELL, KEN Kraftco officer, 827
BELLMAR, TOM Northwestern Steel and Wire, traffic manager, 411
BENNETT grade crossing accident victim, 124
BENSON, ARTHUR E. C&NW roadmaster, Sterling, IL, 21, 37, 38, 39, 43, 227, 246, 277
BENSON, GORDON contractor State of Illinois, 155
BENSON, RAGNAR owner, Ragnar Benson & Sons, contractors, 243, 244
BERLIANT, KEN Fantus Company, 339
BERMAN, SY C&NW systems manager, 747, 839
BERNIER, DAVE C&NW materials dept., 713
BERNIKLAU, FRED land developer, Lincoln, NE, 347, 368, 377, 380, 387
BESINGER, FRED land developer, Carpentersville, IL, 310
BESTE, DON Peoria attorney, 333
BEXTEN, FRITZ industrial development, LaSalle, IL, 365, 379
BIEDERMANN Battelle Institute, 517
BIERMA, HARRY Real Estate Research consultant, 397, 411, 428, 430, 455, 526, 530
BIRD Chrysler officer, 491
BLACK, PERCY C&NW locomotive engineer, 166
BLACK, NEVIL Southern Railway systems manager, 619, 624, 625, 664, 781
BLAIR, JOHN I. Railroad builder, 308
BLAKEY, KEN construction supt. Overland Const. Co., 222
BLANDFORD, RHODES D. C&NW sales agent, Indianapolis, IN, 475
BLEIBTREY, GEORGE promotor, Motor Wheel Co., 427
BLUETT, KEN A. Bethlehem Steel, manager, 615, 624, 650, 728
BLUEWEISS International Paper Co. real estate, 352
BLUTT, DAVID J. C&NW rodman Galena Div., 277, 281, 283, 359
BOARDMAN, STANLEY B. C&NW traffic officer, 278, 408, 435, 518, 536, 559, 561
BOCKMAN, ARNOLD Funk Brothers, 403
BOE, NILS Governor, South Dakota, 537
BOEHM, BERNICE C&NW, chairman's assistant, 553, 699
BOERKE, ED real estate developer, 324
BOESIN, RON C&NW coal marketing, 692, 728, 767, 772
BOESLAGER C&NW dispatcher, Chadron, NE, 718

BOGER, DAVE C&NW engineering officer, 664, 770, 829, 846
BONE, FRED C&NW rodman, Galena Div., 128
BOOTH, ALBERT grade crossing victim, 120
BORNHAUSER, L. B. Chrysler vice president, 443, 444
BOWDITCH, NATHANIEL author of the Practical Navigator, 569
BOYD, ALAN W. IC president, 593, 594
BOYLES, JOHN D. C&NW sales agent, Casper, WY, 307, 330, 331, 405, 406, 476, 582
BRACHMAN, VERNON F. C&NW rodman, Galena Div., 135, 136, 137, 141, 143, 153, 154, 156
BRADEN, HOLMAN F. C&NW assistant engineer, Norfolk, NE, 307, 316, 380, 398, 399, 471
BRADY Chrysler officer, 444
BRALEY, RICHARD Canadian Freight Assn. manager, 668, 694, 732
BRANCH, OLIVE C&NW clerk, 61
BRANDT, TRUMAN C&NW vice president finance, 604
BRANNAN planning commission, Geneva, 512
BRAUN, WALTER E. C&NW traffic officer, 490, 517, 521, 531, 555, 556, 558, 561, 592, 612, 768
BREIDER, JOHN C&NW programmer, 765
BREMER bridge construction contractor for State of Illinois, 145
BRENNAN, J. ROBERT C&NW passenger traffic manager, 373, 388, 423, 537, 730
BRENT, ROBERT C. St. Joseph Paper Co. real estate, 404
BREZINSKI, RON C&NW systems manager, 618
BRIGHT Chrysler vice president, 444
BRODSKY MILW operating officer, 789
BROOKER, TOM C&NW director, 763, 765
BROOKES, OZZIE C&NW ore dock superintendent, 651
BROSNAN, D. WILLIAM president, Southern Railway, 625
BROWER, JAMES F. C&NW engineering officer, 702, 704, 710, 712, 714, 717, 720, 723, 725, 727, 732, 736, 741, 756, 765, 767, 768, 769, 778, 779, 784, 786, 793, 796, 808, 810, 814, 815, 817, 823, 824, 828, 829, 835, 837, 838, 848, 849, 857, 872, 873, 874
BROWER, LEE wife, Jim Brower, 702, 768
BROWN Dreyfuss Grain, 490, 504
BROWN, ADDISON freight car designer, 302
BROWN, WILLIAM F. C&NW rodman, Galena Div., 156, 158, 176, 185, 187, 194, 203
BRUCE farmer, Belvidere, IL, 442
BRUCE, HARRY Spector Motor Freight, 573
BRUCE WARREN C&NW personnel officer, 728
BRUGENHEMKE, JOHN C&NW traffic officer, 467, 494, 504, 506, 507, 520, 533
BUCHANAN, BUCK Wyoming booster, 349
BUCHHOLZ, HARVEY B. C&NW industrial development manager, 294, 297, 299, 300, 301, 324, 332, 333, 347, 373, 426, 435, 484, 491, 508, 548, 557, 593
BUCHHOLZ, RUTH wife, Harvey Buchholz, 426, 484

BUCKMAN, VERN D. C&NW sales agent, Des Moines, IA, 306
BULLINGTON, JOHN Northern Natural Gas Co., 457
BURDICK mayor, Ainsworth, NE, 818
BURGER, CHRIS J. C&NW operating officer, 794
BURKEE C&NW car distributor, Butler, WI, 616
BURKHARDT, EDWARD A. C&NW operating officer, 422, 599, 600, 601, 602, 604, 608, 617, 618, 619, 621, 626, 636, 638, 640, 641, 642, 644, 674, 678, 688, 697, 700, 707, 713, 715, 725, 732, 733, 745, 775, 778, 779, 780, 784, 786, 793, 796, 797, 798, 800, 802, 803, 842, 858, 872
BURKS, RAY C&NW roadmaster, 738
BUTLER, A. R. C&NW accounting officer, 137
BUTLER, JOHN M. C&NW vice president finance, 598, 622, 700, 725, 779, 783, 790, 812, 821, 825, *256*
BYHAM, CARL AAR manager, 539

CABAI, ARLENE C&NW secretary, 626, 695
CADY, KENDALL C&NW real estate manager, 308, 309, 316, 336, 353, 363, 364
CALLIHAN EDWARD Alaska RR accounting officer, 671, 672
CAMPBELL MILW engineering officer, 176
CAMPBELL, BUD C&NW sales agent, Omaha, NE, 313
CAMPBELL, O. D. C&NW (affiliation lost), 650
CAMPBELL, WILLIAM C&NW track inspector, Galena Div., 187
CANFIELD, KEN Atlantic Richfield officer, 695
CANNON, JAMES State of Illinois development officer, 310, 327, 333, 371
CANTWELL, JOHN R. C&NW operating officer, 196, 218, 223, 268, 276
CAPUTO, MIKE C&NW operating dept. clerk, 646
CARLAND, WILLIAM D. C&NW sales agent, Rapid City, SD, 494, 495
CARLEY, FLOYD president, Guardian Packing Co., 499
CARLISLE, DALLAS B. C&NW engineering officer, 495, 496, 642, 648, 738, 770, 841, 870
CARLTON, BRADFORD W. C&NW asst. to the president, 157
CARR, EDWARD J. C&NW traffic officer, 430
CARROLL C&NW conductor, 166
CARSON, JOHNNY TV personality, 307
CARTER, JACK General Foods traffic officer, 664, 711
CARTER, JIMMY President, United States, 808, 859,
CASE, R. E. C&NW industrial development manager, 294
CASEY chief State of Illinois Waterways Board, 310
CASSEN owner, Cassen Trucking, 457
CASSEN, RON C&NW systems designer, 549, 558, 561, 563, 574, 605
CASSIDY, JOHN Peoria attorney, 344, 345
CASTELVECHI, LEE C&NW materials manager, 674
CATHER, L. H. Principal, DeLeuw Cather & Co., 287
CATLETT, CHARLES IC industrial agent, 582, 583, 586
CERMAK, GENE F. C&NW director industrial development, 270, 271, 278, 287, 288, 291, 293, 294, 295, 296, 298, 301, 302, 304, 306, 308,

309, 311, 319, 322, 323, 324, 325, 326, 329, 330, 331, 332, 336, 337, 338, 340, 341, 342, 343, 346, 348, 349, 351, 352, 353, 355, 356, 357, 363, 364, 366, 368, 370, 372, 376, 377, 378, 380, 381, 383, 384, 387, 390, 391, 397, 400, 401, 404, 405, 406, 407, 410, 412, 415, 416, 419, 424, 425, 426, 427, 430, 431, 433, 435, 436, 437, 438, 439, 440, 441, 442, 444, 446, 450, 453, 454, 466, 470, 471, 472, 477, 478, 479, 482, 487, 490, 492, 494, 496, 497, 498, 502, 503, 505, 507, 508, 509, 511, 512, 513, 515, 516, 517, 518, 522, 523, 525, 526, 527, 528, 529, 531, 533, 536, 540, 542, 543, 544, 546, 555, 560, 561, 562, 564, 571, 572, 585, 586, 593, 610, 654, 742

CERMAK, PAT Gene's wife, 384, 446

CHABOT, GEORGE CNR systems manager, 581

CHADWELL Armour & Co. real estate, 371, 380

CHAMBERS owner, Chambers Bros. Manufacturing Co., 350

CHAMBERS, ALICE C&NW Washington lobbyist, 828

CHAMBERS, WILLIAM C&NW Washington lobbyist, 828

CHESNER, JAMES C&NW operating manager, 785, 815, 876

CHESTER, JACK C&NW trainmaster, 789

CHRISTENSEN, DOUGLAS A. C&NW vice president marketing, 580, 610, 707, 737, 776, 804, 821, 836, 846, 858, 876

CHRISTENSEN, MAGNUS C. C&NW engineering officer, 15, 18, 20, 27, 29, 30, 35, 36, 37, 38, 39, 40, 41, 42, 43, 46, 50, 53, 55, 56, 57, 60, 66, 68, 92, 96, 226, 246, 329, 359, 360, 407, 414, 416, 424, 443, 474, 522, 592, 720, 725, 776

CHRISTENSEN, PAUL Grain elevator owner, Fremont, NE, 428

CHRISTENSEN, RICHARD C. C&NW engineering officer, 494, 510, 738

CHRISTIE, ED C&NW labor relations officer, 758

CHRISTIE, ROBERT C. C&NW traffic officer, 388, 421, 496

CHURCH, DON Bureau of the Census manager, 539, 573

CLARK Illinois circuit court judge, 134

CLARK, JOHN C&NW personnel officer, 801

CLARK, WILLIAM C&NW personnel officer, 692, 761, 772

CLARKE, E. M. "PHONOGRAPH" C&NW locomotive engineer, 228

CLELLAND, LARRY C&NW sales agent, Detroit, MI, 582

CLEMMONS, LEE C&NW operating department clerk, 639

COAKLEY, ROLLINS W. C&NW public affairs, 758

COATES, H. C&NW real estate department, 376

COCKRELL, CLAYTON C&NW roadmaster, 748, 749

COE, BETTY JANE C&NW stenographer, 203

COFFMAN, HENRY W. NYC industrial manager, 348, 458

COFFMAN, MILT Wyoming rancher, 331, 349, 405

COLLINS, LESTER mayor, Springfield, IL, 328, 341

COLLOTON, JOHN C&NW rodman, Galena Div., 266, 267, 268, 283

COMSTOCK, MARJORIE wife, Bill Comstock, 265

COMSTOCK, WILLARD E. C&NW assistant engineer Galena Div., 128, 224, 226, 234, 243, 246, 254, 258, 262, 263, 265
CONARD, WALT M. C&NW rate department, 533
CONLEY, ROBERT C. C&NW operating officer, 291, 342, 352, 355, 356, 357, 358, 360, 443, 445, 635, 763
CONLON, JERRY C&NW vice president political affairs, 725, 778, 803, 828, 844, 855, 859, 860, 865, 866, 872, 875
CONWAY, PETER AAR officer, 584, 698, 863
COOK, BILL C&NW sales agent, Denver, CO & Omaha, NE, 483, 484, 500, 505, 520, 529, 547
COOK, CHARLES D. C&NW livestock agent, 142, 394, 406, 500, 505
COOK, WILLIAM C&NW trainmaster, 345
COOPER, DICK city official, Geneva, IL, 513
COOVER, EDWARD E. C&NW operating officer, 197
CORDESH, GLENN C&NW coal marketing, 772
COREY, (COL.) U.S. Corps of Engineers officer, 278
COSNAN EdApex Co., 434
COTTRELL, WILLIAM C&NW claims agent, 497
COULTER, J. RUSSEL TP&W president, 289, 303, 304, 339, 340, 341, 342, 343, 344, 379
COURIS, NICK C&NW engineering, 871
COX, GEORGE AT&SF industrial manager, 427, 526
COYNE, FRANK SP systems manager, 585
CRANDALL, MILTON H. C&NW mechanical officer, 594, 691, 712, 764, 833
CRANSHAW, DOROTHY C&NW industrial engineering, 778
CRAZY HORSE Cheyenne chief, 703
CREIGHTON, LARRY B. C&NW advanced systems chief clerk, 579, 605
CRIPPEN MILW president, 537
CROWN, HENRY chairman, Material Services Co., 172, 356
CRUMRINE, CARL C&NW finance officer, 381, 416
CUDAHY, RICHARD President, Patrick Cudahy Co., 436
CULBERTSON C&NW engineering officer, 214
CUMBEY, ROBERT E. C&NW rate department, 533, 560
CUNNINGHAM, F. EUGENE C&NW car department officer, 712
CUNNINGHAM, LESTER mayor, Belvidere, IL, 443, 444
CUSSEN, JOHN C&NW rodman, Galena Div., 196, 203, 204, 205, 206, 207, 215, 219

D'AMICO, JOHN Wyoming Highway Dept., 727
DAESCHLER, JOHN F. C&NW industrial agent, 270, 287, 293, 294, 302, 322, 325, 332, 335, 336, 337, 338, 346
DALEIDEN, ANTHONY C&NW, chief clerk, 8, 203
DALEY, CATHERINE C&NW division file clerk, 8, 15
DALEY, JAMES A. C&NW attorney, 342, 344, 345, 675
DALEY, RICHARD J. mayor, Chicago, IL, 260, 342
DANIELSON, JOHN C&NW attorney, 157, 172, 379, 435, 487, 585, 608, 642, 643, 663

DATESMAN, JACK P. C&NW engineering officer, 136, 281
DAVALIS C&NW personnel department, 850
DAVID IC, engineering officer, Waterloo, IA, 175
DAVID, AL C&NW programmer, 605
DAVIES, IAN International Paper Co., 338
DAVIS, AL Carnation Co. traffic manager, 521, 522
DAVIS, HERB president, Central Illinois Light Co., 289
DAVIS, HOWARD Kroger Company, Cincinnati, OH, 309
DAVIS, MAX C&NW marketing, 579
DAVIS, ROBERT Industrial representative, Springfield, IL, 327
DAVIS, WARREN W. C&NW programmer, 549, 555, 568, 583, 584, 594, 600, 602, 604, 605, 610, 612, 669
DAVISON C&NW medical director, 746
DEAN, GEORGE Gulf Chemicals, 525
DECAIRE, DICK C&NW systems development, 561, 574
DECAMARA, RICHARD P. IC systems director, 541, 544, 545, 557, 567, 568, 582, 586, 607, 631, 665, 735
DECOSTER, MAURY H. C&NW corporate consultant, 531, 533, 536, 540, 541, 542, 544, 550, 552, 554, 555, 558, 561, 562, 564, 567, 571, 593
DEEMER, RALPH Armour & Co. real estate, 395, 410, 411, 419
DEFIEL, GEORGE CB&Q industrial agent, 471
DEGNAN, RAYMOND E. C&NW traffic officer, 306, 311, 417, 467, 469, 533, 534, 559, 560
DEGRAZIA, VICTOR State of Illinois development officer, 372
DEMING, W. EDWARDS quality control consultant, 556
DENEEN, DON C&NW rodman, Galena Div, 197
DENO, LES C&NW engineering officer, 84, 207, 300
DEPIERRE, HOWARD C&NW operating officer, 742
DEROCHE, JO C&NW internal consultant, 851, 852, 859, 862, 864, 867, 868, 870, 876
DESENFANTS Wyoming rancher, 838
DESNOYERS, JEAN wife, Tom Desnoyers, 795
DESNOYERS, TOM N&W & TDCC systems manager, 568, 585, 599, 604, 620, 624, 660, 664, 795
DEUTSCH, FRANK MILW operating department, 789
DEVOL Industrial prospect, 493
DEWEY, JOE Kerr Magee traffic manager, 697, 698, 714, 720
DICICCO, ANN wife, John DiCicco, 454
DICICCO, JOHN Chrysler manager, 439, 442, 443, 444, 453, 454, 582
DICKINSON, FRANK C&NW training department, 792, 841, 846
DICKMAN C&NW attorney, 356, 424
DILLON, FRANK CB&Q systems manager, 570
DILLON, PAUL W. owner, Northwestern Steel and Wire, 252, 258, 259, 410, 411, 412, 413, 491
DIPALMA, RALPH automobile race driver, 196
DIRKSEN, EVERETT U.S. Senator from Illinois, 327, 358
DISHER, PAUL C&NW assistant trainmaster, 137
DIXON, ARTHUR Modine Manufacturing officer, 363

DIXON, JEREMIAH 18th century surveyor, 22
DONALDSON, BOB IBM Chicago representative, 540, 568
DONOHOE land surveyor, Geneva, IL, 514
DONOVAN, JACK Sumner Solitt Co. contractor, 288
DORSEY, BEN. L. Macoupin County entrepeneur, 44
DOTY, RAY H. C&NW B&B Supervisor, 66, 69, 91, 117, 132, 133, 137, 145, 169, 170, 171, 194, 277
DOUGHERTY, WILLIAM C&NW sales agent Duluth, MN, 582
DOUGLAS, STEPHEN A. U.S. Senator, 686
DOWNING, ROBERT Burlington Northern president, 700, 757
DOYLE W. R. Grace manager, 529
DRENGLER, ROBERT C&NW trainmaster, 297, 728, 850
DREWS, "BUTCH" C&NW roadmaster, 749, 750
DRINKARD, JAMES J. C&NW sales agent, St. Louis, MO, 522, 557
DUBOSE president, Testing Services Co., 444, 513, 514
DUCRET C&NW freight claims, 580
DUDLAK, JOHN C&NW marketing, 766
DUELL, GARTH Pacific Power and Light manager, 695
DUERINCK, LOUIS T. C&NW attorney, 673, 674, 693, 696, 698, 700, 702, 705, 706, 712, 714, 728, 733, 757, 761, 765, 776, 779, 780, 807, 818, 820, 850, 856, 866, 871, 873
DUFFY, PEAT Marwick and Mitchell accountant, 848
DUNN U.S. Steel manager, 712
DUNN, RALPH Reading (PC) systems manager, 649, 675
DUNTON, ALLEN H. C&NW staff services, 611
DYSLIN, ROYCE C&NW roadmaster, 748
DYSLIN, STEVE C&NW marketing, 774, 795

EARHARDT, WILLIAM B. C&NW locomotive fireman, 166
EBERHARDT, JULIAN S. C&NW industrial engineer, 598, 611, 612, 613, 616, 618, 628, 643, 649, 654, 655, 660, 661, 665, 669, 677, 689, 691, 697, 704, 708, 717, 719, 725, 731, 734, 735, 736, 737, 743, 754, 756, 757, 761, 770, 774, 779, 788, 799, 803, 810, 811, 812, 813, 816, 817, 819, 821, 826, 829, 831, 832, 836, 838, 839, 842, 845, 847, 849, 850, 855, 865, 866, 867, 871, 875, 876
EDEE, JEFF C&NW rodman, Wisconsin Div., 9, 246
EDMONDS, WIL Granite City Steel traffic manager, 490, 504, 509, 522
EDWARDS president Central Illinois Light Co., 356
EGBERT, RANDY TP&W chief engineer, 288, 289, 305, 339, 340, 341, 424
EHLERT, RON C&NW (affiliation lost), 675
EILERS Muirson Label Co. attorney, 356
EISLEY National Baroid officer, 496
EIZENSTADT, STUART Presidential advisor, 859
EK, TOM DeLeuw, Cather site engineer, 101, 130
ELLERBY Vipont officer, 476
ELLIOTT, JERRY W. C&NW sales agent, Seattle, WA, 695
ELLIS ED C&NW operating officer, 605

EMERY, JOHN (affiliation lost), 619
EMMERSON Canada Post official, 581
ENDERS, BILL C&NW B&B foreman, 268
ENDERS, NICK C&NW B&B foreman, 35
ENDERS, RAY C&NW accountant, 108, 189, 190
ERICKSON C&NW roadmaster, 748
ERICKSON, LLOYD C&NW freight accounting manager, 591
ERICKSON, ROY C&NW grain rates officer, 423, 490, 536, 560
ERNST, CARL C&EI industrial manager, 377
ERNST, ELLIS W. C&NW traffic manager, Pittsburgh, PA, 482, 510, 518, 742
ERNST, LEN Industrial Marketing, magazine, 540
ERSKINE Saginaw building supply land agent, 472
ESCHELMAN, LEE C&NW systems consultant, 716, 720, 730, 734
ESCHOM, RALPH C&NW trainmaster, 800
ESTES, SAM Commonwealth Edison industrial officer, 348
EVANS Atlantic Richfield officer, 714
EVANS, BILL National Alfalfa officer, 316
EVANS, TOM C&NW communications manager, 613, 615, 724, 796, 802
EVERT, EDWARD P. C&NW accounting manager, 591, 835, 845, 850, 856, 865, 868, 870, 876

FALCONER, GEORGE Gulf Chemicals manager, 525
FANNING Chrysler official, 442, 444, 453
FARRELL, GENE C&NW claims agent, 239
FAULKNER, LYNN International Paper Co. manager, 336
FEDDICK, JAMES R. C&NW traffic officer, 504, 559, 561, 565, 617, 851
FEENEY, JIM Standard Brands manager, 403, 408
FEILER TAD government consultant, 824
FERBER, LINCOLN NBC televison reporter, 445
FERGUSON, JOE C&NW sales agent, Peoria, IL, 373
FERRILL, JAMES P. UP energy development manager, 880
FEUCHT, PAUL E. C&NW president, 159, 201, 255, 257
FEURER, KEITH E. C&NW real estate agent, 526, 530, 673, 691, 705, 706, 845, 866, 873, 874, 880
FIEDLER passenger claiming injury, 176
FILLITI, VITO Head of Clinton Car Shops, 815
FISCHER Weyerhaeuser manager, 364
FISCHER, ROGER A. TP&W traffic officer, 305, 339
FISH, H. litigant, 124
FISHER professer of Harvard University graphics lab, 618
FITCH, LITTLETON H. Mobil Oil Company systems manager, 615, 624, 719
FITZGERALD, ARTHUR zoning consultant, 300
FITZGIBBON, CHARLES T. C&NW sales agent Sioux Falls, SD, 493
FITZPATRICK, CLYDE J. C&NW president, 255, 257, 260, 269, 274, 276, 288, 291, 303, 304, 309, 316, 341, 342, 343, 344, 352, 353, 355, 356, 357, 359, 366, 378, 411, 412, 414, 438, 442, 454, 467, 470, 471, 472,

477, 478, 479, 490, 491, 511, 520, 529, 604, 635
FLISS, DAN C&NW computer operations manager, 502, 655
FLYNN, JAMES C&NW freight claims manager, 579, 580
FORBES, SID PPG planning engineer, 510
FORD, GERALD K. President of the United States, 741, 756
FORD, NANCY Modern Railroads, reporter, 562, 564, 661
FOUTTS, WALT L. C&NW perishables agent, 280
FOX Union Pacific officer, 709
FOX, LEE J. C&NW engineering officer, 642, 667, 669, 670, 671, 675, 687, 688, 691, 696, 697, 704, 711, 725, 728, 735, 737, 738, 743, 745, 751, 794
FOXEN, JOHN J. C&NW industrial agent, 292, 293, 301, 303, 322, 332, 347, 435, 494, 498, 548, 742
FRANCO LORAM consultant, 806
FRANCOIS, WALLY Real Estate Research consultant, 384
FRANK, WARNER C&NW claims agent, 42, 141, 154, 250
FREEMAN, ORVILLE L. Governor of Minnesota, 298
FREEMAN, RICHARD M. C&NW attorney, 757, 767, 775, 777, 783, 786, 787, 803, 804, 811, 819, 820, 825, 872
FREHLENG, SUSAN C&NW finance consultant, 580, 581, 733, 771, 774
FREIDLE, LUIS Rand McNally librarian, 502
FRIED, ROBERT S. C&NW internal consultant, 711, 714, 717, 718, 719, 720, 729, 734, 736, 737, 744, 745, 747, 753, 758, 761, 766, 823
FROMKNECHT, DON R. C&NW traffic, Des Moines, IA, 306
FROST, JACK S. IC industrial director, 348, 368, 377, 426, 427, 582
FRYER, DON C&NW operating officer, 558, 601, 720, 856
FUHS C&NW division mechanical officer, Chadron, NE, 718
FULTON, MAURICE Fantus Company president, 338, 339, 340, 341, 364, 382, 387, 393, 400, 435, 487, 492, 503, 507, 520, 528, 529, 565
FUNK Elgin Paper Co. president, 537
FYFE, RUSS C&NW station agent, Rochelle, IL, 336, 416, 526

GAGE, TOM IC industrial agent, 377, 437, 438, 586
GALBREATH Sutherland Lumber Co., 479
GALE, BILL ICC inspector, 781
GALL, JOHN author, 590
GALLAGHER, BOB FRA officer, 862
GALLANO farmer, Belvidere, IL, 441
GALLAWAY, PHIL P&PU Ry. President, 345
GALLUP, JOE C&NW trainmaster, 342, 616
GARD, MICHAEL C&NW local attorney, Peoria, IL, 342, 344, 345, 359
GARDELLA, HAROLD P. C&NW sales agent, Rockford, IL, 516
GARIS, LEO G. C&NW bridge inspector, 266, 267
GARVEY, EDWARD A. C&NW systems planner, 549, 551, 571, 597, 605, 611, 612, 613, 618, 626, 649, 777
GARVEY, KATE Ed Garvey's wife, 777
GASSNER, STUART F. C&NW attorney, 801
GASTLER, HAROLD L. C&NW operating officer, 572, 582, 603, 608, 621,

639, 642, 644, 651, 663, 665, 671, 674, 691, 693, 701, 704, 761
GAUEN, RALPH E. C&NW rates officer (x L&M), 499, 533
GAULOCK, JOSEPH C&NW accountant, 109, 145, 280
GENERELLA, JOE labor contractor, 52
GEPHART, WALTER Northern Natural Gas industrial manager, 526, 538
GERBER, FRED Atlantic-Richfield manager, 758
GERRARD, RALPH Northern Illinois Gas industrial manager, 527
GIBBONS Oklahoma Gas and Electric manager, 714
GIBBONS, WILLIAM M. CRI&P bankruptcy trustee, 865, 871
GIBSON, F. W. Flying Tigers systems manager, 615, 624
GIBSON, GEORGE C&NW photographer, 717, 844
GIEGOLD, VERNON C. C&NW industrial agent, 287, 288, 289, 292, 293, 302, 303, 304, 305, 306, 309, 311, 312, 313, 316, 317, 319, 323, 324, 325, 326, 327, 329, 332, 339, 346, 347, 350, 351, 364, 370, 373, 391, 419, 425, 435, 436, 462, 497, 508, 518, 524, 548, 549
GIES, JOHN Morrison-Knudsen engineer, 792, 802, 805, 806
GIFFORD C&NW dispatcher, 632
GIFFORD, BRUCE TP&W assistant to president, 288
GILCHRIST, JAMES developer, Peoria, IL, 339, 373, 402
GILLESPIE, EUGENE C&NW sales agent, Denver, CO, 697
GILMAN, ALEX C&NW instrumentman, Wisconsin Div., 9
GIVEN, JOHN Standard Brands manager, 404, 492
GLEIM, PAUL Western Railroad Association manager, 753
GLEISSNER chemical plant manager, 463
GLOVER, G. R. CB&Q vice president, 471
GOBEL, JOHN C&NW attorney, 224
GOERS, MARTIN C&NW rodman, Wisconsin Div., 256
GOINZ, JAMES B. C&NW sales agent, Detroit, MI, 327, 453, 454, 533
GOKEY, J. G. Structo Mfg. Co., Freeport, IL, president, 157, 178, 179
GOLDAMMER, BILL C&NW finance department, 590
GOLDBERG, ARTHUR U.S. Secretary of Labor, 414
GOLTERMAN, RALPH W. C&NW rodman Galena Div., 120, 128, 157, 158, 219, 224, 227, 246, 254, 421
GOLTERMAN, RICHARD State of Illinois highway engineer, 120, 219, 421, 422
GOODWIN, JACK E. C&NW vice president, 82, 84, 159, 197, 214, 258
GORDON, JIM Sackley & Co. construction engineer, 211
GORDON, KATHY C&NW secretary, 590
GOTSHALL, RAYMOND E. C&NW coal marketing manager, 649, 673, 674, 678, 681, 682, 687, 693, 695, 697, 700, 704, 712, 714, 715, 728, 731, 733, 744, 752, 760, 762, 764, 765, 768, 778, 785, 786, 795, 805
GRAN, IRVIN Green Giant officer, 429
GRAPENTINE, LARRY C&NW engineering, 837, 873, 874
GRAVES, TOM UP attorney, 769, 779
GREENBERG, WILLIAM B&LE systems manager, 694, 732

GREENE International Minerals and Chemicals officer, 496
GREY, ALLEN Terra Chemicals president, 477
GRIMES, ED IC traffic agent, Springfield, IL, 328
GRIMM, CHAUNCEY C&NW dispatcher, Galena Div., 191
GRONER, JERALD B. C&NW finance manager, 598, 599, 605, 606, 613, 659, 671, 689, 725, 774, 777, 798, 821, 829, 836, 843, 850
GROVES, DON S. C&NW traffic officer, Denver, CO, 462, 697, 767
GROVES, KAREN C&NW station agent, Lusk, WY, 703
GRUMHAUS, DAVID DEAN Distribution Sciences, vice president, 692
GUENTHER, DONALD E. C&NW industrial agent, 293, 327, 347, 370, 379, 410, 435, 462, 470, 476, 508, 527, 536, 540, 559, 572, 582, 608, 654
GUERIN (affiliation lost), 624
GUILBERT, EDWARD A. Transportation Data Coordinating Committee director, 664
GUNVALSON, DONALD L. C&NW traffic Rockford, IL, 305, 348, 350, 373, 437, 440, 442, 443, 444, 447, 450, 452, 472, 506, 516, 572
GURNON, JOE C&NW sales agent Omaha, NE, 311
GUSTAFSON farmer, Belvidere, IL, 441

HABERBECK, LOUIS C&NW systems manager, 570
HAHN, HENRY C&NW engineering officer, 798
HAHN, JOHN Northern Natural Gas Co. industrial agent, 457, 475, 484
HAKALA, HUGO mayor, DeKalb, IL, 319, 385, 530
HALL Kerr Magee officer, 697, 714
HALL, BOB Monsanto manager, 489
HAM International Minerals and Chemical manager, 496
HAMER citizen of Ainsworth, NE, 816, 818
HAMMERSTROM public relations flack, 499
HANDWERKER, AL C&NW finance manager, 474, 571, 582, 787
HANSCOME FRA consultant, 807
HANSEN, CLIFFORD P. Governor, Wyoming, 531
HANSON, GEORGE R. C&NW operating officer, 632, 633, 634, 637, 639, 641, 642, 732, 751, 753, 788, 789, 790, 796, 799, 800, 803, 814, 817, 824, 825, 832, 835, 841, 845, 849, 850, 853, 855, 857, 858, 859
HANSON, HOWARD, JR. Blair Bank president, 518, 526, 547, 548, 796
HANSON, HOWARD American musican and composer, 307
HANSON, MARY MAUD wife, Howard Hanson, 526
HARDCASTLE Armour and Co. engineer, 392
HARDING, AL state senator, Wyoming, 405, 406, 407
HARDWIG, E. B. Alton and Southern Railway superintendent, 522
HARKINS, JAMES National Motor Freight Tariff Bureau manager, 663, 794
HARMANTAS, CHARLES C. C&NW finance dept., 834, 851
HARNEY, EDWARD E. C&NW traffic officer, 303, 495
HARRINGTON, JOHN J. realtor, Chicago, 510, 511, 512, 513, 515
HARRIS, ARTHUR E. C&NW engineer of bridges, 54, 55, 92, 93, 211, 359, 571

HARRIS, B. R. CGW industrial manager, 499
HARRIS, JACK C&NW real estate agent, 815
HARRISON, F. E. C&NW operating officer, 278, 279
HARTENBERGER Alton and Southern Ry., 522
HARTMAN Leath Furniture Co. manager, 519
HARVEY SPLC chairman, 1974, 719
HARVEY, BILL C&NW Galena Division conductor, 22
HARVEY, THOMAS W. C&NW operating officer, 559, 763, 793
HASLAM Hawkeye Chemical Co., 390
HASSELMAN Northwestern Steel and Wire, 506
HASTINGS, LOWELL C&NW vice president- law, 157
HAUFF, FRANK C&NW operating officer, 344
HAUPT Strobel Construction Co. president, 217
HAWKEYE farmer, Belvidere, IL, 442
HAY, WILLIAM W. professor of railroad engineering, Univ. IL, 328
HAZELWOOD, ROBERT International Paper Company real estate manager, 313, 336, 338, 421
HEIDLE International Minerals and Chemicals, 495, 496, 530
HEIDKAMP, JAMES C&NW real estate agent, 493, 500, 525, 547
HEINEMAN, BEN WALTER C&NW chairman, 255, 257, 260, 269, 287, 294, 303, 309, 319, 324, 335, 340, 356, 366, 368, 376, 378, 383, 384, 391, 400, 403, 412, 413, 414, 420, 423, 431, 435, 442, 462, 470, 471, 477, 478, 492, 507, 512, 513, 516, 522, 523, 529, 530, 532, 536, 537, 540, 541, 543, 552, 553, 572, 584, 585, 587, 593, 606, 610, 612, 632, 659, 666, 690, 699, 808
HEITZ, EDWARD C&NW real estate manager, 364
HELLEM, CHARLES C&NW operating officer, 738, 794
HELLIKER, JAMES C&NW sales agent Tulsa, OK, 390
HENN, DONN City of DeKalb official, 319
HENNELL, ROBERT G. National Motor Freight Tariff Assn. manager, 594, 615, 625, 663, 677, 794
HENSLER, ALVERA C&NW presidential secretary, 782
HEPBURN, JOSEPH National Can Company officer, 505
HERMANN, OTTO SP marketing manager, 599, 675, 732
HERON, ROBERT C&NW operating officer, 196, 442
HERRON realtor, Geneva, IL, 515, 516
HERSCHLER, EDGAR J. Governor, Wyoming, 879
HESSER FRANK friend of Sam Lezak, 430
HESSLER, KEN IMC manager, 529
HICKS, JIMMIE C&NW welding supervisor, 62, 64, 256
HILL, LEO Northwestern Metals, president, 378, 380
HILL, RICHARD C&NW resources agent and corporate secretary, 390
HILL, STEVE Manalytics consultant, 785
HILLEGASS, NORMAN W. C&NW rodman, Galena Div., 234, 236, 240, 241, 242, 243, 244, 246, 248, 249,

264, 266, 267, 268, 270, 280, 281, 597, 662

HILLMAN, JORDAN JAY C&NW attorney, 478, 479

HINCHCLIFF, RICHARD National Motor Freight Tariff Assn. manager, 712, 731

HINTZ, PAUL C&NW sales agent Green Bay, 559

HISE, CHARLES E. C&NW engineering officer, 128, 132, 133, 142, 169, 197, 198, 209, 217, 247, 261

HITZMAN, WALTER J. C&NW sales agent Rochester, MN, 298, 299

HOCH, CLINTON Fantus Company, New York, 381, 421

HOCHSTEIN, JAMES C&NW real estate agent, 815

HODGES, FORD Bacon and Davis consultant, 814

HODGKINS, THOMAS Distribution Sciences president, 635, 636, 692, 762, 827, 832, 865, 866, 876

HOFFMAN Judge Circuit court, Chicago, 461

HOFFMAN, LARRY C&NW rodman Galena Div., 256, 257, 265

HOFFMAN, RICHARD P. C&NW operating officer, 664, 666, 667

HOFFMAN, ROBERT C&NW (affiliation lost), 815

HOFFMAN, WALTER J. C&NW traffic officer, 261, 333, 355

HOLDEN, ANNIE C&NW traffic clerk, 549

HOLDEN, CHARLES S. CNR systems manager, 581, 698

HOLLAND, VINCE V. C&NW sales agent, New York, 535

HOLLANDER, GEORGE C&NW attorney, 792

HOLMAN owner, Hiland Potato Chip Co, Des Moines, IA, 304

HOLMSTROM, JOHN C&NW local attorney, Rockford, IL, 439, 445

HOOVER, REED K. C&NW traffic manager Portland, OR, 335, 510

HOPPER Morton Salt Company officer, 372, 373

HOUX, FRED L. SR. C&NW operating officer, 172

HOWARD, BILL State of Illinois construction engineer, 256

HOWARD, RICHARD J. C&NW industrial engineer, 678, 691, 697, 704, 716, 730, 734, 735, 742, 742, 754, 774, 797, 801, 813, 822, 823, 826, 827, 830, 831, 834, 839, 842, 843, 844, 845, 847, 862, 864, 865, 871, 875, 876

HUBBARD, FREEMAN railroad author, 193

HUDSON, THOMAS H. C&NW traffic manager, New York, 408, 559, 622

HUEDEPOHL, E. BRADLEY C&NW geologist, 317, 333, 349, 371, 390, 674, 728, 807, 824

HUFFMAN, WILLIAM H. C&NW engineering officer, 147, 279, 291, 282, 412, 474, 520, 563, 651, 654, 659, 665, 671, 687, 693, 814

HUGHES Grace Chemicals manager, 464

HULTGREN, BOB land developer, Rochelle, IL, 336

HUMBURG, ARTHUR C&NW instrumentman, Galena Div., 15, 18, 20, 27, 226, 246, 254, 256, 257, 264, 265, 266, 267

HUMPHRIES, MURRAY C&NW labor relations, 845, 853, 855, 858

HUNGERFORD, CLARK JR. (affiliation lost), 581
HUNT Judge circuit court, Peoria, IL, 344, 345
HUNT, DOROTHY wife of E. Howard Hunt, 675
HUNT, E. HOWARD Watergate scandal figure, 675
HUNT, JERRY C&NW sales agent Pittsburgh, PA, 560
HUNT, TOM C&NW sales agent Pittsburgh, PA, 482
HUSS U.S. Post Office manager, 538
HUSSEY, CARL R. C&NW operating officer, 278, 279, 345, 355, 587, 635, 640, 714, 790, 799, 800, 809, 839

ICZKOWSKI, MICHAEL C&NW internal consultant, 819, 820, 823, 830, 834, 833, 836, 839, 847, 853, 855, 856, 858, 862, 870, 876
IDEN, MICHAEL UP locomotive manager, 858
IGOE, MICHAEL L. Judge, circuit coutrt, Chicago, 177
IND, CHARLES State of Illinois highway engineer, 256
INGERMANSEN, BOB CRI&P systems manager, 732
INGERSOLL, ROBERT G. Distribution Sciences manager, 832
INGRAM, GEORGE E. C&NW internal consultant, 613, 626, 637, 655, 661
IWINSKI, JERRY C&NW engineering officer, 760, 775, 799

JACOBY C&NW locomotive engineer, 173
JACOBSEN, HERMAN C. C&NW rates officer, 364, 417, 424, 431, 517, 527, 533
JACOBSON, LARRY C&NW forester, Rapid City, SD, 510
JACQUES, BILL C&NW systems manager, 675
JAHNKE, ROBERT A. C&NW operating officer, 876
JAMES, WILLIAM philosopher and author, 714
JANER, RAY Penn Dixie Cement Co. manager, 504
JANOVIC C&NW trainmaster, 800
JAUCH, MAX C&NW car accountant, 500
JAVACHEFF, CHRISTO artist, 705
JEFFRIES, EDWARD H. C&NW sales agent Fort Dodge, IA, 477, 504, 559
JENKO, FRANK C&NW traffic officer, 536
JENSEN, HAROLD W. C&NW engineering officer, 16, 30, 137, 142, 158, 191, 195, 257, 279, 368, 415, 735
JEVNE, CHARLES P. C&NW real estate technician, 241, 283, 398, 412, 439, 440, 449, 450, 451, 476
JODAT Patrick Cudahy & Co. manager, 465
JOERGER, DON C&NW insurance manager, 804
JOHNSON industrial prospect, 493
JOHNSON, A. G. C&NW operating officer, 325, 376, 483
JOHNSON, AL C&NW roadmaster Sterling, IL, 277, 586
JOHNSON, ART B&O industrial manager, 565
JOHNSON, ERNEST J. C&NW rate officer, 463
JOHNSON, GUS C. C&NW B&B supervisor, 172, 197
JOHNSON, HELEN C&NW stenographer, 293
JOHNSON, JAMES J. C&NW operating officer, 651, 777, 790, 799, 800, 803, 817, 824

JOHNSON, LYNDON B. President, United States, 523
JOHNSON, MARK A. C&NW internal consultant, 678, 681, 682, 687, 689, 691, 698, 701, 704, 712, 714, 734, 742, 797, 812, 813
JOHNSON, OWEN attorney Belvidere, IL, 440, 441
JOHNSON, RALPH C&NW operating officer, 620, 663, 689
JOHNSON, WALTER F. C&NW operating officer, 209, 218, 223, 250
JOHNSTON, STANLEY H. C&NW rodman & industrial agent, 245, 247, 249, 250, 254, 255, 256, 348, 421, 435, 436, 493, 520, 529, 537, 546, 547, 549, 597, 651
JOHNSTON, WAYNE IC president, 197, 582
JONES, FRED C. C&NW trafic manager Peoria, IL, 249, 250, 339, 342, 348, 355, 417, 464, 520, 537
JONES, JACK Southern Railway systems manager, 663
JONES, MARILYN C&NW secretary, 795
JONES, S. CHASEY C&NW vice president, 276, 398, 446
JUDGE, THOMAS J. C&NW public affairs, 871
JUEDES, THOMAS J. C&NW internal consultant, 771, 775, 777, 793, 797, 801
JULIAN, K. Commonwealth Edison industrial manager, 491

KAHN, SID C&NW rodman Galena Div., 169, 170, 171, 172, 174, 175, 185, 188, 194
KANE, JOHN J. C&NW vice president, 563, 568, 572, 582, 592, 593
KAPLAN, BURT M. S. Kaplan Co. manager, 323
KARCHER, DR. American Humates Co. president, 405
KARRAS, TONY C&NW section foreman, 19, 20
KASDORF, GLENN State of Illinois Highway Dept., 261
KASE, ROBERT H. C&NW systems manager, 568, 569, 581, 599, 605, 671, 694, 779, 822
KASSEL, RUDY C&NW rodman, Wisconsin Div., 248
KATZ, BOB Revlon Co. manager, 381, 408, 421
KAUTH, BETTY C&NW stenographer, 433
KEATHLEY, SYLVESTER A. C&NW traffic manager, St. Louis, MO, 365, 402, 432, 489, 490
KEELER, HAROLD C&NW engineering officer, 592
KELBERG, HAROLD C&NW sales agent Omaha, NE, 313
KELLY, W. H. C&NW claims attorney, 134
KEMP Kerr Magee manager, 714
KENDRICK, JOHN B. Former Wyoming governor, 764
KENEFICK, JOHN C. UP president, 690, 757, 758, 768, 779, 783, 784, 803, 830, 861
KENNARD Fel-Tex Fertilizer manager, 473
KENNEDY, JOHN F. President United States, 414, 415, 434, 444, 445
KENNEDY, RICH C&NW track worker, 871
KENNEDY, ROBERT N. C&NW sales agent Sioux City, IA, 559
KENNY (affiiliation unknown), 423
KENT, DAN DeLeuw, Cather site engineer, 72

KENZIL, LOU PennCentral systems manager, 585, 595
KERBS, GLENN J. C&NW engineering officer, 738, 739, 803, 841, 858
KERNER, OTTO J. Governor of Illinois, 343, 373, 454, 467, 522
KERR St. Louis Chamber of Commerce, 363
KHAITAN, VISWANATH C&NW programmer, 549, 579, 604
KIERNAN Montgomery Ward manager, 524
KILEY, ROBERT C&NW personnel manager, 815
KILLINGER, TIM C&NW industrial engineer, 675
KIMES, RAY C&NW fireman, 160, 161, 163
KING ADM officer, 340
KIONKA, ROBERT C&NW rodman, Galena Div., 508, 672
KISS Superior Coal Co., 251
KLASSEN, CLARENCE State of Illinois water engineer, 310
KLAUSSING TAD consultant, 824
KLEMME, DR. Northern Natural Gas Co. systems director, 325, 397, 418
KLINGBERG C&NW locomotive fireman, 173
KLINGE, TOREN geologist consultant, 813
KLOER, CRAIG BN systems manager, 831
KLOVER, ERVIN G. C&NW sales manager Sioux City, IA, 467, 469, 532, 559
KLUENDER, WILLIAM A. C&NW resource manager, 317, 371, 372, 390, 500, 509, 510, 693
KNECHT, GENE C&NW traveling engineer, 344, 345
KNIGHT, ARTHUR E. C&NW sales agent Des Moines, IA, 304, 319, 476, 532
KNIGHT, WILLIAM C&NW systems, 598
KNORST International Minerals and Chemicals, 416, 417
KNOWLES, WARREN P. Governor, Wisconsin, 491
KOCH, JEFF C&NW trainmaster, 638, 766
KOEHLER U.S. Post Office manager, 538
KOENIG, MILES C&NW engineering officer, 299
KOEPKE C&NW freight conductor, 234
KOHLER, ELMER C&NW trainmaster, 174
KOHN, LAWRENCE F. C&NW internal consultant, 779, 780, 785, 786, 787, 788, 789, 790, 801, 834, 850
KOOP, HOWARD C&NW real estate consultant, 428, 430, 512, 530
KOVAL, FRANCIS V. C&NW director public affairs, 86, 157, 250, 261, 278, 562, 675, 700, 857, *200*
KOWALSKI International Paper Co. manager, 408
KRAEGEL, NORB E. C&NW accounting officer, 549, 550
KRAUSE, LARRY D. C&NW industrial agent, 425, 438, 497, 518, 547, 548, 549, 610
KREIDLER C&NW programmer, 598
KREILING, ED C&NW finance department, 582
KRISH, FRANK C&NW yardmaster, 161
KRUGER Chamber of Commerce Fremont, NE, 380
KRUSE, FRED C&NW instrumentman, Galena Div., 267, 275, 277, 278, 281, 283, 289, 290, 291, 310, 319, 474
KUHN, ALFRED J. C&NW rodman, Galena Div., 247, 250, 266, 267

KULBERSH IRA A. C&NW mechanical department, 746, 747, 777
KUNKEL, J. ROBERT C&NW coal rates officer, 339, 342, 344, 423, 491, 497, 509, 523, 527, 559, 571, 593
KVITINKAS, DAN See: Kent, Dan, 72

LABAGH, PETE California Packing Co. traffic manager, 368
LAIDLAW Peoria investor, 499
LAMPORT, LEONARD R. C&NW engineering officer, 6, 114, 246, 257, 265, 267, 278, 279
LANDGREN, ED Armour and Co. manager, 410
LANDOW, HERBERT T. IC systems designer, 552, 553, 565, 581, 604, 615, 621, 625, 653, 691
LANE farmer, Belvidere, IL, 440, 441
LARMON General Foods officer, 322
LARSON, JOHN C&NW yardmaster, 209
LASSMAN, LARRY AAR manager, 863
LATTA, BOB City of Peoria representative, 348
LAUGHLIN Northwestern Steel & Wire Co. manager, 411
LAURIE, JOHN promoter, Sioux City, IA, 497
LAUTENSLAGER, DALE ADM plant manager, 345
LAWHORN, ROBERT Bethlehem Steel systems designer, 615
LAWRENZ, IRVING O. C&NW rate officer, 339, 417, 465, 517
LAWSON, ROY E. C&NW sales agent, Chicago, 446, 742
LAWTON, ROBERT RALPH C&NW engineering officer, 96, 121, 124, 130, 131, 132, 133, 136, 138, 139, 147, 151, 153, 156, 158, 159, 160, 163, 164, 166, 167, 182, 186, 187, 188, 196, 198, 199, 203, 204, 206, 226, 227, 234, 236, 240, 241, 242, 254, 257, 260, 266, 267, 282, 284, 285, 286, 329, 474, 664, 672, 688, 737, 738
LEACH, R. DAVID C&NW vice president, 378, 550, 552, 557, 558, 562, 569, 570, 571, 574, 585, 595, 606, 608, 610, 611, 612, 615, 667, 669, 704, 714, 731, 734, 735, 742, 744, 754, 755, 756, 757, 766, 767, 778, 784, 786, 797, 799, 801, 810, 818, 820, 833, 843, 855
LEHNERTZ, ROBERT N. C&NW systems designer, 570, 581, 604, 671, 751, 752
LELAND, PETER publisher of station tariff, 576
LEILICH, GEORGE M. Western Maryland vice president, 732
LENSKE, HAROLD A. C&NW public affairs, 261, 359
LENZEN, TIM C&NW trainmaster, 846
LESEUR, GENE photographer, Chicago, 196
LEWIS, BARBARA K. wife of author, 5, 53, 316, 326, 358, 388, 404, 408, 414, 426, 442, 465, 524, 534, 604, 612, 613, 616, 620, 622, 659, 661, 712, 729, 742, 747, 755, 795, 802, 818, 871, 876, 881
LEWIS, BARRY C&NW psychologist, 655, 675
LEWIS, BARRY (MRS.) wife of Barry Lewis, 676
LEWIS, DENVER B. author's father, 4, 284
LEWIS, MERIWETHER explorer, 22
LEWIS, RICHARD C&NW systems designer, 561, 598
LEWIS, SARA author's daughter, 53, 408, 545, 695, 756, 779
LEWIS, WILMA M. author's mother, 4

LEZAK, SAM attorney and land developer, 429, 430, 492, 523
LIBBER CB&Q enginehouse foreman, 181
LIDDY, G. GORDON Watergate burglar, 568
LILLIG, ED C&NW systems manager, 565, 599
LIND, DENNIS Phyllis' Nyboer's husband, 617
LINDEN, JOHN EDWARD DeLeuw Cather engineer, 287, 305
LINDHOLM mayor, Geneva, IL, 512, 514, 515 *215, 216*
LINDHOLM, ROBER C&NW photographer, 360, 506, 540
LINDQUIST, GEORGE principal of Foley Brothers of St. Paul, MN, 681, 684, 805
LINN, GERRY C&NW engineering officer, 325, 483
LIPPOLD, ROBERT Commonwealth Edison industrial agent, 418
LIPSITZ, WILLIAM N. C&NW rodman, Galena Div., 97, 98, 135, 182
LIST, WILLIAM N. See: Lipsitz, William N., 98, 182
LITHGOW, MARION BOYD C&NW engineer and industrial agent, 15, 31, 40, 49, 53, 54, 100, 114, 198, 270, 293, 294, 297, 298, 299, 304, 305, 332, 346, 347, 383, 384, 435, 469, 477, 484, 492, 498, 544
LITHGOW, MARY wife, Marion B. Lithgow, 298, 544
LOFTUS FRA consultant, 807
LOHR, LENOX director, Museum of Science and Industry, 86
LONGDON, RUSS C&NW traffic office clerk Rockford, IL, 175
LONGINI Mellon Bank officer, 481, 482
LONGMAN, CHARLES C&NW operating officer, 147, 185
LOONEY, DICK Dun and Bradstreet manager, 581
LOUCKS International Minerals and Chemicals manager, 496
LOVELESS, HERSCHEL C. Governor of Iowa, 325
LOWDERMAN C&NW personnel department, 833
LUND, GEORGE W. C&NW sales manager Kansas City, MO, 582
LYDON, THOMAS C&NW real estate manager, 455, 729

MAACK, DON farmer, West Point, NE, 398, 399
MAAS, NICK C&NW roadmaster, 94, 196, 209, 243
MACBEAN, DONALD A. C&NW industrial department, 433, 435, 438, 457, 480, 481, 482, 484, 498, 502, 503, 505, 511, 517, 527, 529, 533, 536, 537, 538, 539, 540, 542, 545, 549, 551, 552, 556, 562, 564, 568, 569, 570, 572, 581, 595, 598, 599, 750, 795
MACCARTHY, RAY J. C&NW sales manager, Green Bay, WI, 300, 517, 559, 742
MACEN, RUDY Warren Petroleum Co., 328
MACK ROBERT T. C&NW traffic department, 333
MACKAY city engineer Wheaton, IL, 203
MACLEAN, BARRY MacLean-Fogg Locknut Co. officer, 418
MACOMBER, BOB C&NW freight claims manager, 559, 579
MAGRITTE, RENE Belgian surrealist artist, 746

MAIR, PAT State of Illinois construction engineer, 211, 222, 239, 243
MALECHA, AUGUST M. C&NW operating officer, 557, 724, 725, 741, 744, 753, 763, 775, 777, 789, 790, 814, 817, 832
MALO, EDWARD J. C&NW rodman, Galena Div., 10, 11, 13, 14, 15, 24, 27, 41, 56, 57, 67
MANCHESTER, MILES U.S. Department of Commerce systems designer, 698, 731
MANFIELD scrap dealer, Sterling, IL, 281
MANGOLD landowner, Peoria, IL, 354
MANN, JAMES C&NW personnel dept., 741
MARIEN, EDWARD J. Univ. Wisconsin professor, 731, 801
MARQUARDT, EDWARD C&NW chief clerk, 8, 267, 609
MARQUARDT, EDWARD W. family doctor, 53
MARREN, JOE C&NW public affairs, 751, 836, 837, 838, 851
MARSHALL, LEW C&NW personnel, 593
MARTIN AGRICO manager, 500
MARTIN, TOM C&NW finance, 590
MARTIN, WILLIAM C&NW operating officer, 276
MASON, ART C&NW real estate clerk, 376
MASON, CHARLES 18th century surveyor, 22
MATHEWS, HARRY Armour & Co. traffic officer, 410
MATLOX, ELMER Hercules manager, 781
MAY, PHIL land developer, Rochelle, IL, 336
MAYBEE, GEORGE F. C&NW operating officer, 559, 579, 666, 822, 848, 858
MCCARTNEY, JOE UP public relations, 851, 874
MCCLINTIC, JOHN C&NW sales agent Lincoln, NE, 316, 331, 559
MCCORD, JACK C&NW roadmaster, 207, 219, 228, 229, 247, 251, 281, 443
MCCRACKEN, DONALD D. C&NW freight conductor, 729, 730
MCCULLAGH, KEITH IC industrial agent, 377, 378, 417, 452, 548
MCCULLOUGH General Foods system manager, 664
MCDERMOTT, BERNARD F. C&NW operating officer, 398
MCDONAGH, MICHAEL C&NW real estate agent, 873
MCDONALD, JAMES C&NW public affairs, 700, 712, 857
MCDONALD, RICHARD P. C&NW operating officer, 738, 794, 841
MCDONOUGH, ROLLIN P. C&NW operating officer, 360, 392, 497, 509, 522, 616
MCELENY, JIM C&NW training department, 815
MCGEHEE, CHARLES D. C&NW traffic officer, 423, 465, 500, 559
MCGILL Elgin Softener, 431
MCGOVERN, WILLIAM C&NW industrial engineer, 735, 737, 758, 774, 777
MCGOWAN, CARL C&NW attorney, 356
MCHARGUE, LURAY AAR code manager, 625
MCINTYRE, J. C. (PETE) C&NW operating officer, 704, 717, 718,
MCKECHNIE, ART IC systems manager, 567
MCKENNA, WILLIAM R. C&NW industrial engineer, 611, 613, 618, 621, 627

MCKNIGHT, BOB Railway Age reporter, 649
MCNEAR, GEORGE TP&W president, 343
MCPHERREN Allis-Chalmers officer, 419
MCQUEEN, THOMAS contractor, 211, 215, 309
MEINDERS, BRUCE C&NW rodman, 203
MEIER, BRUCE Armour & Co. manager, 410
MENK, LOUIS WILSON Burlington Northern president, 698, 757, 803
MENNELL, ART MP engineering, 779
MENZE International Minerals and Chemicals officer, 495, 496, 530
MERCER Judge, circuit court, Peoria, IL, 342, 343, 344, 356, 357
MERRILL, FRANK Barber-Greene manager, 492, 530
MERRITT, BILL C&NW materials officer, 674
MERRYMAN, W. R. Grace manager, 529
MESA, JOSE C&NW industrial engineer, 691, 730
MEYERS, BERNARD R. C&NW vice president, 129, 130, 131, 159, 180, 193, 194, 196, 197, 207, 209, 211, 217, 242, 244, 256, 266, 267, 271, 275, 278, 279, 285, 287, 288, 290, 291, 324, 352, 353, 383, 398, 407, 412, 427, 443, 444, 445, 458, 495, 509, 847
MICHALEK, BONNIE C&NW stenographer, 613, 624
MICKEY, ROBERT W. C&NW Real Estate manager, 880
MICHUDA City of Chicago bridge engineer, 278
MILCIK, ROBERT M. C&NW operating officer, 617, 619, 623, 664, 707, 735, 742, 758, 761, 814, 841, 846, 848, 850, 876
MILES, CHARLES C&NW chief special agent, 675, 676, 725
MILLER, C. A. C&NW traffic manager, New York, 317
MILLER, JUNE C&NW stenographer, 433, 508, 549
MILLER, "RED" Black Hills Bentonite principal, 406
MILLER, ROY C&NW Land Department manager, 665
MILLION, ROBERT C&NW engineering manager, 758
MILLS, CHRIS C&NW attorney, 840, 872, 873, 874, 876
MINARD, TOM C&NW clerk, 560, 628, 697
MITCHELL, VERN C&NW signals engineer, 279, 356, 474, 724, 823
MOLLICA General Motors Real Estate, 454
MONTGOMERY, GEORGE C&NW training department, 790
MOODIE, PLINY attorney, West Point, NE, 398, 399
MOORE, "DINTY" C&NW yardmaster, 186
MOORE, SARA LEWIS author's daughter, 756
MOORE, SARAH attempted Presidential assassin, 756
MOORE, WILLIAM H. PennCentral president, 618
MORLOCK, JULIUS C&NW steel erection crew foreman, 53
MORLOCK, WILLIAM C&NW steel erection crew foreman, 53, 114
MORRISON, FRANK Governor of Nebraska, 365, 366, 370, 374, 425, 452
MOSES Goodyear manager, 698

MOTT, WARREN C&NW operating clerk, 775
MOUDRY Litchfield and Madison Railway officer, 304
MRUSKOVICH, JOHN C&NW finance officer, 777, 786, 787, 795
MUELLER, ERNEST J. C&NW rate officer, 555
MULLEN, JOE C&NW sales agent Minneapolis, MN, 582
MUNKRES, DEAN mayor, Bill, WY, 686
MURPHY, HARRY CB&Q president, 252
MURPHY, JOE C&NW truck driver, 138, 158
MYLES, AL C&NW labor relations officer, 642, 670, 671, 729, 774

NEBEN, JERRY DeLeuw Cather of Canada railroad consultant, 736
NEIL, JACK Hercules officer, 781
NELSON, C. P. C&NW car department officer, 195
NELSON, DON C&NW operating department, 834, 841, 845, 853
NELSON, JOHN C&NW sales agent, 547
NELSON, ROBERT D. C&NW engineering officer, 58, 59, 61, 62, 84, 114, 151, 279
NERI mayor, Northlake, IL, 365
NETZEL, ART C&NW roadmaster, 124, 125, 205
NEU, RICHARD C&NW marketing officer, 820
NEWGENT, WILLIAM R. C&NW sales agent St. Louis, MO (x L&M), 522, 527
NEWHOFER Plasser International manager, 525
NEWMAN, ROY Curtiss Candies officer, 492
NEWSOME, PAUL Nebraska Consolidated Blenders manager, 547
NEWTON, LOUIE N&W officer, 694
NICE, BANKER Sterling, IL, 412
NICK, GEORGE C&NW industrial agent, 582
NIXON, DAVID C&NW internal consultant, 807, 819, 837, 840, 846, 850, 853, 856, 862, 864, 866, 867, 876
NIXON, RICHARD M. President, United States, 606, 655, 709, 734
NIXON, TOM C&NW safety department, 777, 781, 790
NORDAN, EDWARD L. C&NW internal consultant, 568, 569, 585, 596, 597, 600, 601, 602, 603, 604, 605, 610, 611, 613, 616, 617
NOWICKI, DAVE C&NW internal consultant, 766, 770, 771, 777, 795
NYBOER, PHYLLIS L. C&NW secretary, 552, 561, 590, 613, 617

O'DONNELL, THOMAS A. C&NW traffic officer, 197
O'GORMAN Superior Coal Company officer, 251
O'HEARN, C. H. C&NW auditor capital expenditures, 103, 257
O'NEIL, JOHN C&NW motive power officer, 839
OAKLEAF, DOUGLAS E. C&NW engineering officer, 89, 92, 109, 213, 248, 727, 736, 777
OGDEN, WILLIAM BUTLER C&NW president, 1
OGLE, RICHARD C&NW systems designer, 571, 573, 605, 622
OLESON, CARL K. C&NW rodman, Galena Div., 204, 207, 219, 225, 246, 267, 283
OLLMANN farmer, Belvidere, IL, 439

OLSEN, IVER S. C&NW sales manager, 304, 342, 349, 364, 442, 463, 469, 487, 490, 517, 532, 586
OLSON, A. O. C&NW industrial development director, 294
OLSON, ARTHUR P. C&NW instrumentman Galena Div., 15, 22, 23, 30, 34, 35, 40, 54, 62, 64, 96, 109, 110
OLSON, EDWARD A. C&NW vice president, 290, 311, 322, 342, 349, 356, 357, 383, 393, 405, 412, 416, 420, 433, 437, 442, 446, 454, 457, 477, 478, 479, 482, 484, 490, 550
OLSON, GEBHARD C&NW B&B supervisor, 290
OLSON, JOHN C&NW assistant engineer Galena Div., 352, 353, 359, 360
OLSON, ORVILLE C&NW B&B supervisor, 268
OLSON, ROBERT C&NW internal consultant, 864, 871
ORR, JAY C&NW sales agent Green Bay, WI, 559
ORRICO CRI&P industrial manager, 526
ORSI, LEO Armour & Co. engineer, 397
OSBORNE, JACK (affiliation lost), 874
OSDENDORF, GEORGE Litchfield and Madison Ry. locomotive engineer, 168
OSENBERG, WARREN associate of Sam Lezak, 492
OSTBY Wisconsin lobbyist, 491
OSTERHOUT, DAVID State of Nebraska industrial director, 347, 370, 374, 397
OTTER, WILLIAM R. C&NW operating officer, 824, 848, 849, 853, 855, 863
OTTO, JEFFREY S. C&NW internal consultant, 641, 642, 643, 644, 649, 659, 661, 662, 663, 664, 667, 669, 675, 677, 682, 691, 704, 710, 735, 752, 762, 777, 781, 790, 792, 798, 802, 803, 804, 806, 810, 818, 821, 836
OWENS, STEVE C&NW engineering officer, 354, 355, 356, 407, 443, 444, 445, 455, 459, 466, 469, 471, 674, 678, 682, 687, 693, 695, 696, 697, 698, 700, 704

PACEY, DON City of Elgin, IL, industrial director, 537
PACKARD, BRUCE G. C&NW engineering officer, 142, 153, 155, 156, 157, 158, 169, 170, 171, 177, 178, 179, 188, 190, 203, 209, 211, 214, 257, 424
PADGETT, RUSS C&NW rodman Galena Div., 267, 282
PALLAS, MICKEY commercial photographer, 376
PALMER, ROBERT city manager Elmhurst, IL, 619
PANNING, JERRY C&NW trainmaster, 825
PARKER, DOROTHY authoress, 581, 699
PARKS, "LIGHTNING" C&NW brakeman, 163
PARRETT General Motors manager, 454
PARRISH, MAX C&NW lobbyist, 828
PARSONS U.S. district judge, 359
PATTON, CHARLES landowner, Peoria, IL, 354
PATTON, MADGE wife, Charles Patton, 354
PAUL, GEORGE C&NW vice president, 572
PEABODY, R. V. Smith-Douglass Co. manager, 418
PEACOCK MILW engineering department, 789

PEAGAN, A. T. C&NW operating officer, 269, 270, 286, 808
PECK, TOM farmer, Geneva, IL, 511, 512, 515, 516
PERINO, RALPH C&NW track supervisor, 289
PERKALITIS, VARIS C&NW rodman, Terminal Div., 286, 321
PERLMAN, ALFRED J. president, New York Central, 535
PERRIER, JACK L. C&NW engineering officer, 256, 258, 265, 277, 309, 407, 497, 509, 510, 689
PERRIN, DEL C&NW operating officer, 276, 277, 286, 604
PERRY, JACK newspaperman, Casper, WY, 331
PERRY, JOHN C&NW instrumentman Galena Div., 15, 24
PERRY, SAM Judge, Federal district court, Chicago, 245
PETERS, JACK M. C&NW traffic manager Omaha, NE, 313, 316, 428, 463, 497, 529
PETERSON C&NW engineering department Chadron, NE, 495
PETERSON, KEITH C&NW chief clerk, 536, 562,
PETERSON, RICHARD B. UP service planning manager, 769, 779, 784, 785, 796
PETRASH, ROBERT A. AAR Data Systems director, 596, 599, 675, 698, 732
PEW, ARTHUR E. C&NW rodman, Galena Div., 545, 573, 574,
PICKETT, TOM oil well supplier, Riverton, WY, 331, 407
PIERCE, CAREY C&NW real estate manager, 669, 670
PIERCE, JIM Morrison-Knudsen officer, 838
PIERSON, EARL K. C&NW roadmaster, 135, 189, 207, 227, 246, 281
PIETMEYER, BILL UP systems manager, 780, 785
PIKE, HAROLD Iowa Beef manager, 470, 471
PIQUET, NORMAN Grace Chemicals manager, Memphis, TN, 419, 423
PLASTERER, KEITH C&NW sales agent, Milwaukee, WI, 790
PLOETZ Harnischfeger Corp. manager, 459
PLUMMER, WILLIAM K. C&NW sales agent, Rockford, IL, 257
POLCHOW, WILLIAM F. C&NW instrumentman, Galena Div., 40, 58, 97, 100, 102, 133, 135, 145, 203, 240
PONGRACE, OTTO NYC industrial manager, 426, 427
POPE, FRANK Dun and Bradstreet of Canada, 581
PORTER, BILL C&EI industrial agent, 458
POSTELS FRA consultant, 810
POWELL, HOWARD S. IC vice president, 582
POWELL, RUDY Ford Motor Company, 454
PREISSER, VICTOR L. C&NW corporate consultant, 531, 541, 572, 580, 586, 593, 613
PRIESTER, CHARLIE flight service owner, 818
PRITCHARD, OTIS C&NW engineering officer, 653
PROKOPY, JOHN Peat Marwick and Mitchell, 843
PROVO, LARRY S. C&NW president, 378, 431, 435, 503, 508, 510, 530, 550, 556, 558, 563, 582, 584, 588, 590, 592, 606, 609, 611, 613, 614,

619, 624, 639, 642, 643, 644, 648, 650, 657, 659, 661, 662, 664, 665, 667, 669, 670, 674, 687, 688, 689, 690, 693, 695, 697, 698, 699, 700, 706, 711, 712, 713, 714, 715, 725, 729, 732, 733, 736, 739, 740, 743, 744, 745, 752, 757, 758, 760, 761, 765, 767, 768, 769, 780, 861
PTASHKIN, BARRY C&NW industrial engineer, 704, 730

QUINN, WILLIAM J. CB&Q and MILW president, 529, 537, 609

RADIVICH, JOE attorney, Geneva, IL, 512, 513, 514, 515
RAMLET, KURT C&NW tariff officer, 533, 578, 771, 801
RANSOM, JOHN H. UP service planning manager, 779, 785, 796, 825
RASMUSSEN, PETE C&NW stations department, 577, 580
RATHBUN, HOWARD F. C&NW brakeman, 157
RATHENAU, TONY C&NW systems manager, 573, 598
REED Nebraska State geologist, 316
REED, JIM Southwest Port Development, 522
REED, SMITH MP industrial manager, 522
REED, WILLIAM WP engineering, 241, 260, 334
REESE, JACK C&NW chief linesman, 207, 416
REID, MAURICE S. C&NW engineering officer, 71, 84, 159, 214, 217, 218, 222, 224, 227, 239, 240, 243, 244, 252, 258, 266, 268, 269, 279, 443, 474, 616, 619, 692, 713, 745, 769, 775, 777, 780, 781, 785, 792, 801, 812, 818, 844, 868
REIFFLER, RON Fantus Co., 493, 523, 631
REIMERS, JOHN F. C&NW sales agent Kansas City, MO, 582
RENO, ARTHUR W. C&NW traffic manager, 533, 534
REYNOLDS farmer, Belvidere, IL, 440
REZEK, JACK C&NW programmer, 549, 568, 577, 579
RICE, P. TP&W assistant to president, 338, 340
RICHARDS, CHARLES State of Illinois engineer, 256
RIPPLE, JACK C&NW materials engineer, 662, 663, 688 713
RISING, LEE industrial representative, Lincoln, NE, 370, 380
RISSMAN, HAROLD attorney, DeKalb, IL, 282
ROBB C&NW car distributor, Milwaukee, WI, 616
ROBERTS, W. H. C&NW safety officer, 140
ROBINSON professor, American University, 573
ROCK, LLOYD vice president, Guardian Packing Co., 499
ROCKWOOD, BILL General Motors manager, 842
ROEPKE, HOWARD professor Univ. Illinois, 427, 542, 544, 631
ROGERS, CARL MILW systems manager, 649
ROGERS, PAT Pettibone Mulliken Co. engineer, 597
ROMAN, JOHN C&NW locomotive engineer, 163, 164
RONAYNE, JAMES M. C&NW traffic manager, 487, 533, 534, 559, 618, 623
ROOT Guardian Packing Co. officer, 499

ROOT, EARL C&NW roadmaster, 702, 703, 728
ROSE, BOB attorney, Casper, Wyoming, 705
ROSENBAUM C&NW trainmaster, 559
ROSENTHAL, BOB owner, Rosenthal Lumber, Crystal Lake, IL, 348
ROSER, JOHN C&NW systems designer, 555, 561, 571, 579, 732
ROSS, THOMAS A. C&NW corporate secretary, 477
ROSSMAN, GUY K. C&NW sales agent St. Paul, MN, 297, 298
ROSTENSTOY, DON IMC manager, 529
ROTH, HACK Northern Illinos Gas Co. manager, 348, 434, 523
ROTH, RAY M. C&NW sales manager Chicago, 494, 495
ROWAND National Baroid manager, 496
RUCHE, ED C&NW sales agent Kansas City, MO, 582
RUDOLPH, LARRY attorney, Washington, DC, 872, 873
RUDOLPH, RON C&NW internal consultant, 626, 670, 691
RUNDELL, WILLIAM C&NW land surveyor, 172
RUPP, ADOLPH coach, Univ. of Kentucky, 326
RUSSELL, GEORGE Container Corp. of America manager, 310
RUSSELL, ROBERT W. C&NW attorney and vice president-personnel, 177, 250, 287, 341, 342, 343, 359, 403, 424, 564, 573, 593, 598, 605, 609, 611, 617, 627, 654, 655, 667, 675, 699, 725, 728, 729, 742, 774 793, 833, 875, 876
RUTH, BYRON E. C&NW rodman, Galena Div., 156, 157, 169, 175, 178, 179, 180, 188, 189, 190, 198, 199, 203, 222, 225, 246
RUTH, HAL C&NW assistant engineer, Wisconsin Div., 156, 182, 256, 258, 265
RUTKIN, NORM Revlon Co. manager, 381
RYAN, RICHARD C&NW traffic manager Lincoln, NE, 347, 368, 370, 374, 380, 425, 533, 653
RYE International Paper Co. manager, 408
RYE, VIC AAR Data Systems manager, 699
RYMAROWICZ, VIC See: Rye, Vic, 570

ST. MARTIN, R. C&NW yard conductor, 148
SALZMAN, DALE W. UP transportation research manager, 784, 785, 796, 865
SANGREE CRI&P officer, 797
SANTORE C&NW locomotive engineer, 148, 149
SANTUCCI Contractor, 187
SARGENT, BOB owner, Sargent Grain Co, Des Moines, IA, 304
SARGENT, GEORGE Ford, Bacon and David consultant, 808, 814, 818
SATHER, CHRIS UP systems manager, 778, 779
SAWICKI, PHIL State of Illinois bridge engineer, 210, 219
SAYRE, CHUCK State of Nebraska public power agent, 380, 387, 461
SBARBARO, LENA C&NW stenographer, 433
SCHACTNER, TONY C&NW insurance manager, 244
SCHAFFER, RAY C&NW public affairs, 192, 431, 580

SCHALLER C&NW sales agent St. Louis, MO, 431
SCHARDT, RON C&NW finance officer, 510, 560, 700, 712, 733, 765, 769, 778, 779, 780, 783
SCHARENBURG, NAOMI C&NW secretary, 293, 508
SCHARENBURG, VERA C&NW clerk, 8
SCHENK Susquehanna Western officer, 529
SCHENK Kroger Co. Officer, 309
SCHIRRA, WALLY astronaut, 705
SCHLENKER, WALTER General Electric systems manager, 660
SCHLYTTER, KIM C&NW industrial engineer, 804, 817, 834, 839, 847, 848, 864
SCHMIDT MILW operating officer, 789
SCHMIDT, JOE C&O systems manager, 698
SCHMIDT, PHIL CRI&P industrial manager, 325, 427, 458, 510, 511
SCHMIEGE, ROBERT C&NW labor relations and last president, 799, 845
SCHONNING, HARROLD C&NW sales agent Minneapolis, MN, 582
SCHRADER Grace Chemicals technician, 464
SCHROEDER, HENRY J. C&NW rates officer, 342, 492
SCHULLER citizen South Sioux City, NE, 532
SCHULZ, LARRY C&NW manager of blueprint shop, 127
SCHUMACHER, KEN DT&I president, 777
SCHWARTZ, VAN C&NW (affiliation not remembered), 649
SCHWARZ, DON C&NW operating officer, 651, 653, 765
SCHWEIN, OSCAR T. C&NW sales agent Green Bay, WI, 300
SCOTT Kroger Co. Officer, 309
SCOTT, R. friend of Chas. Towle, 493
SCOTT, VALERIE UP attorney, 872
SEEGERS, BOB owner, Seegers Grain Co., Crystal Lake, IL, 382, 429
SEIBERT, ROBERT T C&NW sales agent Little Rock, AR, 548
SELMAN, WILLIAM B. C&NW freight claims manager, 559, 670
SENCABAUGH, G. EDWARD UP special representative to the president, 872
SETCHELL, RICHARD C&NW superintendent of stations, 766
SHAFFER, JACK C&NW director industrial development, 536, 546, 547, 548, 549, 551, 552, 556, 562, 564, 565, 568, 570, 572, 583, 585, 586, 596, 606, 661
SHANKLIN, JOHN C&NW instrumentman, Galena Div., 15, 36
SHANNON, CHARLES C. C&NW operating officer, 255
SHAPPERT, FRED contractor, Belvidere, IL, 244, 256, 440, 442, 443, 454
SHARP, ROBERT A. C&NW marketing director, 673, 683, 693, 698, 700, 704, 705, 706, 707, 714, 725, 733, 736, 745, 763, 765, 766, 767, 768, 769, 775, 777
SHATTUCK farmer, Belvidere, IL, 439, 442
SHEAREN General Foods, president, 322
SHEARER Atlantic Richfield officer, 697
SHEEDS, JACK C&NW sales agent Detroit, MI, 582
SHELGREN, RICHARD Trailer Train manager, 666
SHEPARD, ALAN astronaut, 375
SHEPARD, EUGENE P. C&NW switchman, 133

SHEPHEARD, JIM Chrysler manager, 454
SHERWOOD, BILL Wyoming Highway Dept., 727
SHREFFLER, KEITH C&NW trainmaster, 464
SHINROCK, HOWARD Fremont, NE, industrial representative, 331, 347, 375, 380
SHIVELY, BOB Nebraska State Power Authority manager, 817
SHUSTER owner, meat processing plant, Gordon. NE, 483
SICK, AL mayor, Blair, NE, 796
SIDOR, BARBARA C&NW secretary, 619, 624
SIEBERT W. R. Grace Chemicals, Memphis TN, 423
SIERZGA, JOHN C&NW rodman Galena Div., 239
SILVERMAN C&NW personnel manager, 596
SILVERTHORN Chrysler manager, 444
SIMANDL, AL C&NW chief clerk, 268, 742
SIMMONDS MILW attorney, 176
SIMONS, JAMES L. C&NW rodman Galena Div., 281, 283, 289, 303, 359, 411, 474, 616, 718, 720, 721, 727, 728
SINGER, NORMAN A. C&NW sales agent, Milwaukee, WI, 465
SJOSTROM Contractor, Rockford, IL, 454
SKOGLUND, DORVAN C&NW industrial agent, 365, 397, 425, 433
SKOGLUND, HOWELL P. E. North American Life & Casualty president, 425
SLATER, DRENNAN J. C&NW attorney, 172, 249
SLONECKER, DON Nebraska State Power Authority manager, 387, 461
SMITH, AL Dun and Bradstreet manager, 527
SMITH, C. T. International Minerals and Chemicals manager, 403
SMITH, CECILIA E. C&NW secretary, 605
SMITH, ELBERT T. CRI&P industrial agent, 327, 374, 464, 472
SMITH, H. B. C&NW operating officer, 197, 252, 260
SMITH, H. F. (BOOMER) C&NW sales agent, Omaha, NE, 500, 504, 505
SMITH, HAROLD attorney, Geneva, IL, 512
SMITH, MORT C&NW assistant engineer, Wisconsin Div., 9
SMITH, OTTMAR W. C&NW engineering officer, 100, 124, 128, 157, 185, 188, 190, 260, 450
SMITH, WORTHINGTON MILW president, 689
SNYDER, RAYMOND E. C&NW rodman Galena Div., 266, 267, 303, 616, 632, 637, 743, 748, 750, 803, 806, 810, 814, 818, 821, 823, 824, 829, 842, 846
SODAY, DR. Hawkeye Chemical Co. director, 390
SOMMER, JOHN L. C&NW industrial agent, 347, 373, 392, 435, 548, 557
SORENSON Statex officer, 428
SOUTH, DAVE C&NW marketing, 611
SPAFFORD, ROY C&NW car department, 858
SPENCER farmer, Belvidere, IL, 441
SPITZENBERGER, WILLIAM CB&Q industrial agent, 424, 526
SPORRY, ERNEST Thomas McQueen construction superintendent, 210, 212, 217, 218, 219, 229, 232, 233
STALEY, JOHN Quaker Oats Co. officer, 374

STAPLETON, ROBERT T. Industrial agent, Clinton, IA, 351, 390, 586
STEADRY, FRED O. C&NW attorney, 344, 350, 383, 512, 513, 742
STEELE, BOB Atlantic Richfield manager, 695, 712, 714, 720
STEIGER, ZEV C&NW attorney, 767, 780, 784, 786, 793, 824
STEIN, JOSEPH J. C&NW operating officer, 147, 197, 229
STEINER, RAY J. C&NW traffic manager St. Paul, MN, 297, 299
STEINER, SETH CA&E engineering, 174
STEINHOFF, VIRGIL C&NW industrial agent, 433, 536
STEMMERMAN, RICHARD C&NW rodman Galena Div., 135, 153
STERN, FRANK I. C&NW advanced systems manager, 549, 550, 552, 553, 554, 555, 557, 559, 560, 563, 567, 571, 572, 573, 577, 578, 580, 582, 586, 589, 590, 595, 599, 604, 605, 606, 613, 615, 650, 654, 655, 657, 659, 661, 664, 688, 691, 694, 697, 704, 741
STEUTZ, HERB C&NW instrumentman, Galena Div., 508
STEVENS, BILL C&NW marketing director, 567, 580
STEWART W. R. Grace Chemicals, 423
STEWART, BRUCE C&NW personnel officer, 608
STEWART, CHARLES F. C&NW traffic manager, 373, 742
STIFT, LEROY C&NW operating officer, 154
STOCKDALE, WARREN B&O systems manager, 573
STOCKTON, WARREN BN systems manager, 831, 871
STOLL, EDWARD J. MILW industrial manager, 369, 423, 438, 440
STONE ARNOLD THOMAS C&NW tapeman, 224
STOVO Chrysler manager, 453
STRATTON, WILLIAM G. Governor of Illinois, 237
STRAUSS, AL Lehigh Cement manager, 421
STRIEBEL C&NW locomotive engineer, 207
STRUVE, TED A. C&NW corporate consultant, 531, 541, 580
STUBBS, ROBERT C. C&NW traffic officer, 322, 332, 342, 349, 357, 383, 393, 402, 409, 412, 418, 420, 433, 436, 442, 446, 471, 479, 490, 491, 496, 506, 507, 508, 509, 516, 523, 524, 528, 529, 533, 536, 546, 549, 550, 556
STUSEK C&NW systems manager, 598
SULLIVAN Judge, 224
SULLIVAN, DENNIS AMTRAK track engineer, 827
SULLIVAN, JAMES C&NW traffic manager Detroit, MI, 453, 582
SUNDERMAN Union Carbide manager, 408
SURMA, "RED" C&NW track gang foreman, 81
SWAIN, TIM Peoria attorney, 342
SWARD, JOHN AT&SF systems, 751, 796, 811, 813, 865
SWATEK, JEAN daughter of Wilton Swatek, killed by train, 172
SWATEK, MARILLA duaghter of Wilton Swatek, killed by train, 172
SWATEK, WILTON F. Outsider, killed by train, 172
SWENUMSON, DEL C&NW engineering officer, 632, 634, 639, 803

SWETT, DAN C&NW (affiliation lost), 810
SWIELA, SANDRA C&NW secretary, 834

TAMURA, SUSUMU. R. C&NW programmer, 561, 594, 602, 604, 605, 609, 618, 622
TARGALL, KAMI Chrysler engineer, 444
TATE, ROBERT R. C&NW traffic manager, 419, 560
TAYLOR Oklahoma Gas and Electric, 714
TAYLOR, RICHARD C&NW real estate manager, 476
TELL, HAROLD O. C&NW operating officer, 620, 621, 822
THELANDER, PETER VICTOR C&NW engineering officer, 240, 245, 278
THELANER, DAVID C&NW systems manager, 562
THOMAS, ARCHIE C&NW chief clerk, 49
THOMAS, CANDACE C&NW operating department, 778
THOMAS, SYLVIA C&NW law department secretary, 873
THOMPSON Blair Cattle Co. principal, 505
THOMPSON, BOB TRW consultant, 724
THOMPSON, HOWARD Blair, NE, industrial representative, 496, 519, 526, 547, 548
THOMPSON, JAMES W. C&NW sales agent St. Louis, MO, 365, 445, 489, 490, 509, 763
THOMPSON, ROBERT Fantus Company, 339, 487
THOMPSON, RUTH wife of Howard Thompson, 526
THOMPSON, WILLIAM H. IC vice president operations, 619
THORMAHLEN, JOHN real estate agent, 429, 430, 492
THORPE, JIM famous Native American athelete, 224
THORSON, HARRY Black Hills Bentonite Co. principal, 405, 406
TIEMAN, LEO G. C&NW engineering officer, 260, 268, 651
TIETJEN Jones and Laughlin Steel Co. officer, 482
TIGHE Union Carbide manager, 408
TINGLEFF, THOMAS C&NW auditor, 817, 819
TOBIN, JIM C&NW sales agent, Chicago, 519
TONDRYCK, VINCE C&NW claims agent, 134
TOREN, JACK C&NW marketing, 642
TORGERSON C&NW sales agent, 140
TORME, OMER C&NW sales agent Green Bay, WI, 559
TOWLE, CHARLES L. TAD consultant, 824
TOWLE, CHARLES M. C&NW industrial agent, 328, 335, 346, 347, 384, 405, 406, 415, 419, 421, 490, 493, 506, 515, 536, 546, 548, 567, 582, 586
TOWNSEND, LYNN president, Chrysler Corp., 442
TOWNSEND, MARK Industrial agent Central Illinois Light Co., 237, 338, 339, 341, 343, 344, 348, 365, 373, 379, 386, 387, 461, 586
TREPTOW, DARRELL C&NW secretary, 695, 730
TRIBBEY, FRANK H. C&NW traffic officer, 297, 457, 559, 572
TRIMBERGER, FRANK C&NW rodman, Galena Div., 182, 185, 189, 209
TRIMBLE, LARRY Commonwealth Edison industrial manager, 654

TROTTMAN, NELSON C&NW attorney, 479, 480
TROYKE, JOHN C&NW traffic officer, 468, 474, 533
TRUSHEIM, BOB Olin Mathiesen manager, 548
TSOODLE, JO ANN C&NW secretary, 730, 751, 774, 778, 795
TUCKER SL&SF chief architect, 124
TUMABARELLO, VINCENT A. C&NW sales agent, 383
TWIST, OLIVER character in Dickens novel, 826

UPLAND, EDWARD C&NW grain traffic manager, 316, 516, 565, 748
URKHARDT, AL C&NW auditor, 137

VAN HORNE, JOHN C&NW rodman Galena Div., 22, 23
VAN NOSTRAND, ROBERT American Cyanamid manager, 343, 461
VANDENBURGH, EDWARD C. C&NW chief engineer, 127, 130, 133, 158
VANNEMAN, EDGAR C&NW attorney, 355, 356, 357, 358, 359, 379
VASY, RICHARD L. C&NW coal marketing, 785, 822
VAVRA, SHIRLEY MAY C&NW stenographer, 8
VERDIN, LOUIS C&NW sales agent Houston, TX (x L&M), 496
VERMUT FRA consultant, 807
VERONA, JOSEPH C&NW dining car service manager, 593
VETTER, RUBE landowner, Peoria, IL, 354
VIHSTADT Brunswick Corp. manager, 425
VIOLANTE, DOMINICK D. C&NW systems planner, 549, 561, 584, 691
VOGENTHALER (Unknown) chemical plant prospect, 463
VOLKERT, RICHARD C. C&NW industrial agent (ex-M&StL), 365, 420
VON ROSEN, JOHN Chrysler manager, 437, 453, 454
VOREL Griffin Wheel Co. manager, 242
VOSE citizen, South Sioux City, NE, 532

WAGGENER, O. O. CB&Q industrial manager, 369, 406, 418, 433, 438, 440, 470, 471, 473
WAGNER, FORD Bacon and Davis, 817
WAGNER C&NW division engineer, 207
WAGNER, MAL B&O industrial agent, 192
WAHL, JIM General Mills manager, 574
WALDO, JOSEPH Milwaukee Gas Company manager, 333
WALDRON, KEITH C&NW systems director, 580, 820
WALKER interested in Blair, NE property, 308
WALKER Allis-Chalmers manager, 419
WALKER, ROBERT GTW systems manager, 582, 586
WALLER, DENNIS C&NW motive power officer, 858
WALSH farmer, Geneva, IL, 512
WALTERS, G. E. C&NW locomotive fireman, 163, 164
WALTHERS, RALPH C&NW draftsman, 196
WARD, CARL Grace Chemical manager, 464
WARD, WILLIAM C&NW B&B supervisor, 290

WARDEN, EDWARD L. C&NW attorney, 127, 128 133, 172, 496, 512, 515, 539
WARE, TONY Kane County (IL) planner, 430
WASHINGTON, GEORGE President, United States, 22
WATERMAN, NATE L. C&NW operating officer, 59, 81, 82, 197, 218, 250
WATSON, CHARLES Patrick Cudahy & Co. officer, 436, 465
WATSON, T. J. C&NW conductor, 646
WAUTERLEK, JACK NYC industrial agent, 552, 596
WEBB Blair Cattle Co. principal, 504, 505
WEILAND, JEAN CB&Q industrial agent, 471
WEIS, C. O. State of Illinois highway engineer, 154
WEISHAAR, DAVID G. C&NW coal marketing, 804, 836, 846, 858, 876
WELLES, TIM candidate for C&NW employment, 850
WELLINGTON president CILCO, 461
WELTMAN, LES C&NW conductor, 228
WERNER, JULIUS Litchfield and Madison Ry. locomotive fireman, 168
WESTBURG, ROGER C&NW internal consultant, 834, 839, 846, 858, 864
WESTMARK MILW operating officer, 789
WETHERALL, WILLIAM C&NW mechanical dept. officer, 645, 662, 693, 750
WHARTON, AL W. DuPont systems manager, 781
WHEELAND, ROBERT State of Illinois Highway Dept., 261
WHEELER, CARL AGRICO officer, 408
WHITE, GEORGE C&NW traffic manager Kansas City, MO, 525
WHITE, JOHN Costain Concrete Tie Company salesman, 713
WHITE, JOHN C. C&NW sales agent, 560, 570
WHITE, W. L. C&NW, corporate secretary, 84
WHITEHOUSE, BERNIE C&NW fire inspector, 246
WHITTEN, HERB economist consultant, Washington, DC, 565, 573, 574, 619
WHITTLE, JACK developer, Chicago, 526
WIESE, WALTER developer of a paper warehouse at Madison, WI, 548
WIGHT, STEVE Morrison-Knudsen officer, 804
WIGNOTT, JAMES H. C&NW sales agent New York City, 559
WILBENS International Minerals and Chemicals manager, 495, 496
WILBUR, WILLIAM C&NW engineering officer, 8, 9, 47, 266, 269, 270, 271, 274, 275, 277, 281, 288, 289, 290, 291, 352, 353, 356, 359, 443, 612, 691, 826
WILDER, NELSON BUDD C&NW rodman, Wisconsin Div., 258
WILEY Patrick Cudahy & Co. manager, 465
WILKINS, CARL C&EI industrial manager, 458
WILKINSON, JAMES C&NW trainmaster, 825
WILKINSON, JOHN A. C&NW roadmaster, 115, 116, 188, 197
WILKINSON, JOHN E. C&NW roadmaster, 115
WILLIAMS C&NW agent Rockford, IL, 157, 257
WILLIAMS, ROWLAND L. C&NW president, 3, 84, 150, 193, 201
WILLIAMSON, JAMES F. C&NW sales manager Washington DC, 538

WILLIS farmer, Belvidere, IL, 439, 446
WILLSON, CHARLES City of Elgin, IL industrial manager, 348, 425, 512, 526, 537
WILSON Montgomery Ward manager, 524
WILSON, BOB C&NW personnel officer, 746, 875
WILSON, HINKLE MP cashier, Jackson, MO, 4
WILSON, ROBERT C&NW superintendent of stations, 471, 498, 559
WILSON, WARREN reporter Casper Star Tribune, 839
WIMMER, GEORGE Sioux City, IA, industrial manager, 323, 332, 364, 470
WIMMER, WILLIAM C&NW engineering officer, 353, 454, 521, 551, 552, 786
WINKELHOUSE, L. C. C&NW architect, 151, 204, 246
WINKRANTZ, W. F. C&NW sales agent Des Moines, IA, 226
WINN attorney, International Paper Co., 357
WITT, ROBERT C&NW instrumentman, Wisconsin Div., 9, 204
WOFFORD, TOM FRA consultant, 807, 810
WOJAN, CARL C&NW rodman Galena Div., 203, 206, 207
WOLF, SHARON C&NW systems manager, 731
WOLFE, JAMES R. C&NW president, 639, 642, 648, 650, 655, 660, 661, 662, 665, 665, 667, 669, 670, 671, 672, 674, 687, 693, 695, 697, 698, 699, 704, 705, 706, 724, 744, 769, 775, 778, 779, 781, 782, 783, 784, 785, 786, 790, 792, 793, 797, 799, 800, 802, 803, 806, 808, 809, 810, 811, 812, 813, 815, 816, 818, 820, 822, 825, 829, 830, 836, 839, 840, 841, 842, 844, 847, 848, 850, 852, 855, 856, 857, 859, 860, 861, 862, 864, 866, 867, 868, 871
WOODRUFF, JAMES G. C&NW corporate consultant, 556, 584
WOODS, JAMES AT&SF systems manager, 599, 675
WOODS, O. H. C&NW pass bureau manager, 660
WOOLSEY farmer, Belvidere, IL, 442
WORFEL, DALE Ford, Bacon & Davis consultant, 816, 817, 818
WOZNIAK, ED CRI&P industrial agent, 520, 548
WOZNIAK, RICHARD Nebraska state official, 818
WRIGHT Wyoming rancher, 853
WRIGHT, G. WILLIAM GATX traffic officer, 598

YASEEN, LEONARD Fantus Co. founder, 338
YEAGER, DON C&NW traffic department, 641
YOCUM, FRED C&NW operating officer, 735, 762, 765, 796, 797, 799, 800, 801
YORK, DON C&NW environmental engineer, 825, 836, 851, 856, 874
YOUNG, WILLIAM Chrysler real estate manager, 436, 438, 439, 440, 441, 442, 453, 454, 457
YULE, CLINTON TV weatherman, 241

ZABOROWSKI, ANTHONY B. C&NW assistant engineer Galena Div., 143, 144, 175, 177, 185, 194, 198, 199, 203, 214, 218, 238, 239
ZACK, JOE photographer, Chicago Daily News, 195

ZAMBA Atlantic Richfield manager, 714
ZANUCK, DARYL F. movie director, 307
ZEIGLER Chicago realtor, 597
ZELENDA, JOSEPH JR. C&NW sales agent, 333, 365, 407, 507
ZIMA, JOHN J. C&NW rodman Galena Div., 156, 157, 158
ZIMMERMAN, HENRY American Distillers, Pekin, IL, officer, 417, 418
ZIMMERMAN, WILLIAM C&NW systems director, 574, 593, 598, 606, 609, 724, 754, 763, 777
ZITO, JAMES A. C&NW vice president, 664, 688, 742, 745, 764, 774, 777, 781, 784, 799, 802, 803, 810, 811, 818, 820, 821, 829, 836, 840, 848, 850, 857, 859, 862, 870, 871
ZITTER Kerr McGee manager, 697
ZOGG, RICHARD C&NW systems manager, 615, 616, 755

INDEX

ACKNOWLEDGMENTS

Numberless individuals and organizations influenced and encouraged the creation of this highly personal account of life on a Midwestern railroad in the latter half of the twentieth century. Detailing all of their contributions would double the size of the work so that I can only highlight a few of the many.

Firstly, Barbara, my wife of 60 years, has not only read this typescript several times, but has provided valued guidance in assembly of the account. She lived through this period with me and knew many of the people who are mentioned along the way. An artist, I only trust that dedication to this work did not seriously interfere with her painting.

The key to getting this work into the public consciousness had to be Joseph Piersen, archivist for the Chicago and North Western Historical Society. His energy and determination made the publication possible. Others in the C&NWHS were supportive and provided many details that fleshed-out the bare recital of events.

When Adrian Ettlinger, webmaster of the Railway & Locomotive Society, heard that I had written this account, he traveled across the continent to make my acquaintance. After reading a rough draft, he confirmed his earlier opinion that it should be published as a historical account.

The advent of the Internet proved to be a major factor in confirming dates, names and places that appear in the text. The generous responses of all the railroad interest groups that were contacted is gratefully appreciated.

The thousands of individuals I knew that made up the Chicago and North Western Railway from 1947 to 1980 formed a flowing river through time. They included swimmers and sinkers, leaders and followers, dreamers and cynics—all caught up in the vast enterprise of railroad transportation through a period of changing times. Our community of railroad people varied from day to day as some were lost and others joined. I was privileged to record my impressions of some of the events along the way albeit 15 years short of the final juncture with the Union Pacific RR in 1995. That period, so far, remains an untold story in any detail.

My recollections enhanced by personal notes may be at variance with the memories of those still with us who were also involved in this glorious enterprise. If I have bruised any feelings, I apologize because no hurt was intended in an honest attempt to report the past. As Sgt. Friday used to say on television, "Just the facts, sir."

All in all, I had a marvelous railroad career. Thank you, one and all.

Eugene M. Lewis
Fort Bragg, California
December 2005

ERRATA

as of August 31, 2007

1. Page 351 at chapter break, “Perora” should be Peoria

2. Page 408 at chapter break. “I had sent my wife and fourteen year old...”

3. Page 421 last paragraph: Delete “to” in second line.

4. Page 488 wrong caption for map. The correct caption is on page 514.

5. Page 494, 16th line: R. J. Christenson.

6. Page 510, 1st line: R. J. Christenson.

7. Page 514 wrong caption for map. The correct caption is on page 488.

8 Page 632, ninth line of first paragraph. Eliminate one period after “operations”

9. Page 738, 3rd line: R. J. Christenson.

10. Page 833, last line, add “of” ...”in the cab of one of his...

11. Page 871, 5th line: Nick Louris

12. Page 880 second line should be “Jim “Farrell”

13. Page 898, 2nd col. R. J. Christenson.

14. Page 899 change Couris to Louris and move to correct alpha sort.

15. Page 902 Ferrill, James P. should be “Farrell, James P.” - move to correct alpha sort position.

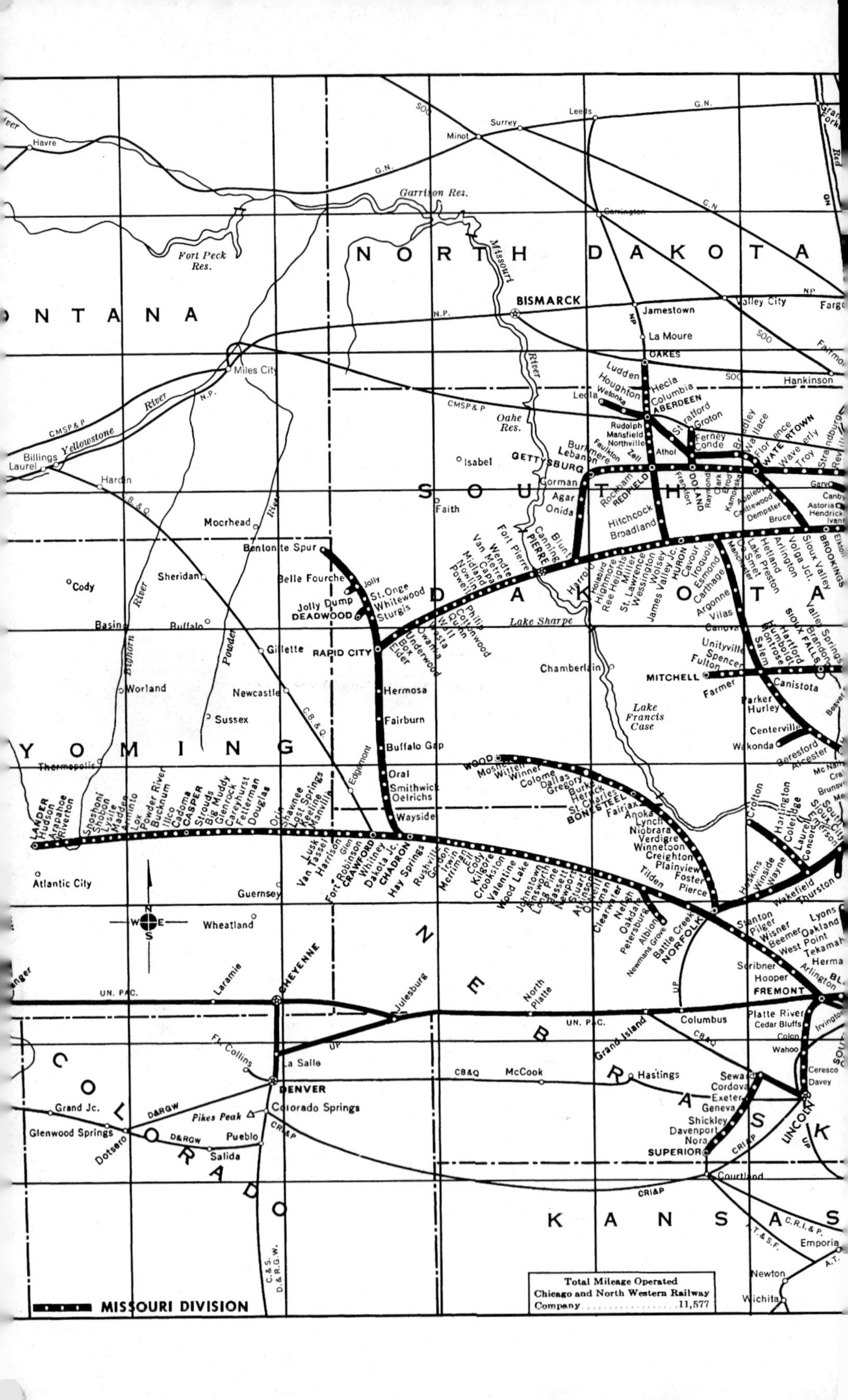

NORTH DAKOTA
MONTANA
SOUTH DAKOTA
WYOMING
NEBRASKA
COLORADO
KANSAS
Havre
Minot
Surrey
Leeds
Grand Forks
Garrison Res.
Fort Peck Res.
Carrington
BISMARCK
Jamestown
Valley City
Fargo
La Moure
OAKES
Hankinson
Miles City
Billings
Laurel
Hardin
Yellowstone River
Missouri River
Oahe Res.
Isabel
Faith
GETTYSBURG
ABERDEEN
WATERTOWN
BROOKINGS
HURON
PIERRE
REDFIELD
DOLAND
Mocrhead
Bentonite Spur
Belle Fourche
St. Onge
Whitewood
Sturgis
Jolly Dump
DEADWOOD
RAPID CITY
Hermosa
Fairburn
Buffalo Gap
Oral
Smithwick
Oelrichs
Wayside
Cody
Sheridan
Basin
Buffalo
Gillette
Worland
Newcastle
Sussex
Thermopolis
Lake Sharpe
Chamberlain
Lake Francis Case
MITCHELL
SIOUX FALLS
Canistota
Centerville
Wakonda
WOOD
Winner
Dallas
BONESTEEL
LANDER
Riverton
CASPER
Douglas
Lusk
CRAWFORD
CHADRON
Hay Springs
Rushville
Valentine
Atlantic City
Guernsey
Wheatland
NORFOLK
Battle Creek
Fremont
FREMONT
CHEYENNE
Laramie
Julesburg
North Platte
Grand Island
Columbus
Ft. Collins
La Salle
DENVER
McCook
Hastings
LINCOLN
SUPERIOR
Courtland
Grand Jc.
Glenwood Springs
Dotsero
Pikes Peak
Colorado Springs
Pueblo
Salida
Emporia
Newton
Wichita
UN. PAC.
CB&Q
D&RGW
CRI&P
C.&S. D.&R.G.W.
N.P.
G.N.
SOO
CMSP&P
Total Mileage Operated Chicago and North Western Railway Company 11,577
MISSOURI DIVISION